Edited by
Rüdiger Wagner, Jürgen Marxsen, Peter Zwick, and Eileen J. Cox

Central European Stream Ecosystems

Related Titles

Moss, B.

Ecology of Fresh Waters

A View for the Twenty-First Century

2010

ISBN: 978-1-4051-1332-8

Maltby, E., Barker, T. (eds.)

The Wetlands Handbook

2009

ISBN: 978-0-632-05255-4

Rice, S., Roy, A., Rhoads, B. (eds.)

River Confluences, Tributaries and the Fluvial Network

2008

ISBN: 978-0-470-02672-4

Edited by Rüdiger Wagner, Jürgen Marxsen, Peter Zwick, and Eileen J. Cox

Central European Stream Ecosystems

The Long Term Study of the Breitenbach

With contributions from Georg Becker, Heino Christl, Thomas G. Horvath, Reimo Lieske, Michael Obach, Joachim Reidelbach, and Hans-Heinrich Schmidt

WILEY-BLACKWELL

The Editors

Prof. Dr. Rüdiger Wagner
University of Kassel
FB 10 Natural Sciences - Biology
Heinrich-Plett-Str. 40
34132 Kassel
Germany

Dr. Jürgen Marxsen
Justus Liebig University
Department of Animal Ecology
Heinrich-Buff-Ring 26-32
35392 Giessen
Germany

Prof. Dr. Peter Zwick
Schwarzer Stock 9
36110 Schlitz
Germany

Dr. Eileen J. Cox
Natural History Museum
Department of Botany
Cromwell Road
London SW7 5BD
United Kingdom

Cover
The cover picture shows a view (looking upstream direction) into the Breitenbach valley during spring, with the first greenhouse constructed over the stream for collecting emerging insects in 1969 in the background (photo: Jürgen Marxsen). The simplified scheme of carbon flow was designed after Figure 6.2 from this book. The photo of the Trichopteran species *Chaetopteryx villosa* was kindly provided by Dipl.-Biol. Brigitta Eiseler.

Library of Congress Card No.: applied for

British Library Cataloguing-in-Publication Data
A catalogue record for this book is available from the British Library.

Bibliographic information published by the Deutsche Nationalbibliothek
The Deutsche Nationalbibliothek lists this publication in the Deutsche Nationalbibliografie; detailed bibliographic data are available on the Internet at <http://dnb.d-nb.de>.

Wiley-Blackwell is an imprint of John Wiley & Sons, formed by the merger of Wiley's global Scientific, Technical, and Medical business with Blackwell Publishing.

Cover Design Adam-Design, Weinheim
Typesetting Toppan Best-set Premedia Limited, Hong Kong
Printing and Binding Fabulous Printers Pte Ltd, Singapore

Printed in Singapore
Printed on acid-free paper

Print ISBN: 978-3-527-32952-6
ePDF ISBN: 978-3-527-63467-5
ePub ISBN: 978-3-527-63466-8
Mobi ISBN: 978-3-527-63468-2
oBook ISBN: 978-3-527-63465-1

Foreword

This is an astonishing book and a great achievement by the small band of scientists who have written it. The Max Planck Society's "Flußstation" (River Station) worked from 1951 to 2006 at Schlitz, in Hessen in central Germany, very largely on the ecology of a single small stream, The Breitenbach (in English "Wide Brook" – which it really is not!) This concentration of effort was partly because the larger River Fulda, of which the Breitenbach is a first order tributary, had become very polluted, and it offered a more "pristine" ecosystem for study. The program was initiated by two giants of early stream ecology, whose names may not now be well known to modern students, certainly non-German ones. Joachim Illies was the founder of the "river zonation" school of lotic ecologists and Karl Müller was the originator of the eponymous "Müller's colonization cycle". Their pioneering work was carried on by a small team of scientists (five at any one time) plus assistants, mainly under the leadership of Peter Zwick, and this book is a testament to their dedication, persistence, and skill. This book is a goldmine, a unique compendium of detailed information on a stream (crucially including the underlying data that will be made publicly accessible). I urge students of the next generation to exploit and pore over this material because quality and reliability (of taxonomy and methodology) shine through the pages; these were all very highly skilled aquatic ecologists working with the best equipment, and the results are the best.

Let me mention a few highlights from the more than 600 pages. The list of animals from this tiny stream contains more than 2000 species! There are 820 species of insects alone! This is an order of magnitude greater than the species list from "my own" Broadstone Stream (Hildrew, 2009), another well-known stream community and a system of a similar size. The Breitenbach is in a highly diverse area of central Europe (not a biotically impoverished island), is not acidified or excessively enriched, and almost every animal group has been studied by an expert at some time over the last 50 years. Emerging adult insects were also trapped and specifically identified continuously from 1969 to 2006, thus accumulating rare species. Indeed, the high quality of the taxonomy applied to this system is a real highlight of the research program. This also applies to the algal work, in which live diatoms were identified, greatly increasing one's confidence that the many species identified from the different microhabitats were active in the stream, and not merely allochthonous inputs. The lesson is that the less exhaustive studies that we routinely see must greatly underestimate total diversity because the sample size is small and because "difficult" groups are ignored. But what is the meaning of this diversity? Ecosystem processes in less diverse systems seem to proceed perfectly adequately. Does this indeed point to a high redundancy in stream ecosystems?

For those concerned with climate change, mean water temperature in the Breitenbach has increased only by about 0.85°C in 37 years (0.023°C $year^{-1}$), whereas air temperature has increased by 1.8°C in only 20 years (0.08°C $year^{-1}$). Warming in this spring-fed system has thus been rather slight and, perhaps not surprisingly, evidence for biological responses is muted. The date of emergence has not changed significantly for most species, though many show some indication of earlier (and a few of later) emergence. Exhaustive chemical analyses of the water have been undertaken. Not surprisingly, nutrient concentrations are low, but alkalinity and acid-neutralizing capacity showed some sign of decline over the whole record (in an area where acid depositions have been quite serious though decreasing over the last 20 years or so). The stream is circumneutral, however, and is not acidified (pH 6.5–7.8).

The macrofaunal community has also been remarkably persistent, the dominant species having remained almost unchanged over 37 years, though numerical fluctuations have been large. Year to year fluctuations in stream discharge, through a variety of direct and indirect effects but mainly via determining habitat availability for the guilds of scrapers and filterers, seem proximally associated with these shifts in density. Very large populations of trichopteran grazers, such as *Agapetus fuscipes* and *Tinodes rostocki*, dominate the epilithic community when clean stone surfaces are available, and the former appears to regulate algal biomass via grazing and to compete intraspecifically for food.

The microbiology (particularly bacteriology, there is less information on the fungi) of the system has been exceptionally well described, as have organic budgets. There have been no particular surprises here, but lotic systems for which there are reliable and complete measurements of flows in the microbial loop are few, and this is an absolutely prime example. Bacteria are responsible for 78% of heterotrophic respiration and 59% of total system respiration, accounting for 36% of heterotrophic biomass and most (71%) of heterotrophic production. About 50% of bacterial production enters the higher food web, 35% via protozoa, and 15% by direct consumption via the macrofauna. About 10% of carbon assimilated by the macrofauna is bacterial, the latter getting most carbon direct from dead organic matter. Bacterial production is similar to that of phototrophs, even in this well-lit, open stream, at about 0.2 kg m^{-2} $year^{-1}$. The system is dominated by allochthonous inputs, with about 1.02 kg m^{-2} $year^{-1}$ of dissolved organic matter and 0.42 kg m^{-2} $year^{-1}$ of particulate organic matter. The chemical nature of dissolved organic matter has been unusually well characterized, as has the nature of flowing and interstitial water. Work on the microbiology and organic carbon budget of the Breitenbach has been of an international standard and deserves much more recognition. Overall, the carbon budget of the Breitenbach for the period 1973–1998 balances, with total inputs of 1.88kg m^{-2} $year^{-1}$ and outputs of 1.84 kg m^{-2} $year^{-1}$, with allochthonous inputs about three times that from autochthonous sources. None of this is particularly surprising, a small, headwater stream dominated by allochthonous inputs and bacteria being major metabolic drivers of the system, but the quality of the measurements underlying these figures is first rate.

As it was at its completion, the "Breitenbach project" was in some senses science from an earlier age, frankly largely descriptive and driven by an intimate knowledge of natural history and conventional taxonomy. This is at the same time its strength, and it has hitherto been very much underestimated as a "model ecosystem", and its weakness. Along with the solidity, reliability, and sheer quantity of the data, there have been disappointments and blemishes, which are partly scientific and partly, I suspect, due to the dynamics of the team. It was a frustration that the measurement of insect emergence, motivated initially by the notion that it would lead to a "short-hand" measurement of secondary production, did not in the end do so. The book spends some time detailing the problems of measuring emergence quantitatively. Surprisingly also, the measurements of emergence were not accompanied by a sufficient campaign to measure benthic density. I feel that the authors are too pessimistic in their dismissal of more direct ways of measuring secondary production. There is almost nothing about vertebrates in this book, and they are more or less dismissed as having "no role". This is surely an overstatement; there is a population of brown trout in the lower reaches of the Breitenbach, yet it has not been characterized quantitatively. Work on the fish would surely have led to an appreciation of links between the aquatic food web and the riparian system, via subsidies to fish production from terrestrial resources. The team also had unrivalled measures of potential reverse subsidies, of the stream to riparian predators via insect emergence, long before they were appreciated elsewhere.

Finally, of course, we have to mention the problem of replication. This is one system, though some attempt is made to compare it with one or two other streams in the area. This is always a problem when producing a model ecosystem (though the book itself never describes the Breitenbach as a model). These are essential in ecology but only of use when they can be compared with others that differ in well-characterized ways. With hindsight, it would surely have been wise to set up a network of systems where comparisons with the "jewel in the Crown", the Breitenbach itself, would have been possible, thus increasing the generality of the inferences drawn.

I want to end my foreword by looking forward – since I really hope this is not a dead end for research on this system. As I said at the outset, the data are there and offer wonderful opportunities for the imaginative and curious. But if I were standing at the threshold of a new program, with all the battery of techniques in ecology now available, the Breitenbach and its heritage would be a wonderful place to start. I have no remit to develop a new program, but an obvious start would be to characterize the diversity using molecular methods (this was begun with the bacteria) and to place this stream within a meta-community of its neighbors. I would look further at groundwater–stream linkages – for groundwater is the new frontier for freshwater ecologists – and particularly look at the possibility of subsidies to the food web from chemoautotrophic production (via methane, for instance).

My third line would be to develop a food web for the Breitenbach, embedded within the terrestrial system, using stable isotopes, measurements of fatty acids markers, and the rest. This book should be a new beginning, not an epitaph.

Alan Hildrew

Reference

Hildrew, A.G. (2009) Sustained research on stream communities: a model system and the comparative approach. *Advances in Ecological Research*, **41**, 175–312.

Contents

List of Contributors

Georg Becker
Universität of Cologne
Cologne Biocenter
Department of General Ecology
Zülpicher Str. 47 b
50674 Köln
Germany

Heino Christl
2 Poplar Way
Harrogate HG1 5PR
United Kingdom

Eileen J. Cox
The Natural History Museum
Department of Botany
Cromwell Road
London SW7 5BD
United Kingdom

Thomas G. Horvath
Director of Environmental Sciences Program
State University New York College at Oneonta
Oneonta, NY 13820
USA

Reimo Lieske
Im Alpenblick 10
8400 Winterthur
Switzerland

Jürgen Marxsen
Justus Liebig University
Department of Animal Ecology
Heinrich-Buff-Ring 26-32
35392 Gießen
Germany

Michael Obach
San Telmo 11-3° D.
E-20750 Zumaia
Spain

Joachim Reidelbach
Negelerstrase 53
72764 Reutlingen
Germany

Hans-Heinrich Schmidt
Schlesische Str. 22
36110 Schlitz
Germany

Rüdiger Wagner
University of Kassel
FB 10 Natural Sciences – Biology
Heinrich-Plett-Straße 40
34132 Kassel
Germany

Peter Zwick
Schwarzer Stock 9
36110 Schlitz
Germany

Acknowledgments

The Max-Planck-Gesellschaft zur Förderung der Wissenschaften (MPG), its gremia and officers, and the directors and staff at the head office of the former Max-Planck-Institute of Limnology (Plön) are sincerely thanked for long-lasting funding and support of the Limnologische Fluss-Station Schlitz. We continued and developed a study initiated by the late Prof. J. Illies, who is gratefully remembered. After his sudden death, Prof. J. Overbeck (Plön) arranged for the continued existence of the Schlitz laboratory for which we are very grateful.

Visiting colleagues and students completing their theses at the Fluss-Station funded by the MPG, the Deutsche Forschungsgemeinschaft (DFG), or the Deutscher Akademischer Austauschdienst (DAAD) contributed to our study. They are too numerous to be listed here but their names appear in the text and the references of this book. We thank them all!

We are most grateful for the busy dedicated work of the team at the Fluss-Station Schlitz, everybody contributing in his specific capacity. We particularly thank our highly specialized and experienced technicians who participated directly in our everyday scientific work, some for decades:

Ingrid Aszmutat, Evelyn Etling, Dr. Beate Knöfel, Carla Kothe, Birgit Landvogt-Piesche, Carmen Möller, Agnes Palotay-Ries, Hannelore Quast-Fiebig, Gisela Stüber, Irene Tade, Elke Turba.

We are pleased to acknowledge taxonomic and ecological expertise, advice, information, and help from the following:

Dr. W. Barkemeyer (Flensburg), Dr. C. Becker (Aachen), R. Bellstedt (Gotha), Dr. R. Brinkmann (Schlesen), Dr. R. Gerecke (Tübingen), Dr. T. Gregor (Schlitz), Dr. P. Havelka (Karlsruhe), Dipl. Biol. M. Hecht (Herborn), Dr. R. Heiß (Frankfurt/Oder), H. Hergersberg (Hürtgenwald), Dr. H.-J. Krambeck (Plön), P.-W. Löhr (Mücke), Dr. P. Martin (Kiel), Dr. A. Piechocki, Łodz, Dr. A. Pont (Oxford), Dr. H. Reusch (Suhlendorf), W. Schacht (Schöngeißing), Dr. M. Spiess (München), Dr. U. Werneke (Kleve), Dr. K.-P. Witzel (Plön), Prof. F. Wojtas † (Łodz), Prof. Dr. T. Zatwarnicki (Wrocław), Dr. H. Zwick (Schlitz).

Schlitz, November 2010 *The Editors*

1
Introduction

Peter Zwick

1.1
History of the Limnologische Flussstation Schlitz

After World War II, Germany was divided into four occupation zones and free travel to neighboring countries was not possible. At that time, the Rivers Weser and Fulda formed the only major German river continuum that was accessible over its entire length. However, most of the second constituent tributary of the Weser, the River Werra in the Soviet Zone, was inaccessible. Therefore, the Fulda and Weser were the natural choice as study objects for a group of five biology students at the University of Göttingen who hoped to found an institute dedicated to stream limnology.

Martin Scheele, Joachim Illies, Wolfgang Schmitz, Karl Müller, and Ernst-Joseph Fittkau received local support from Prof. Demeter Beling and Dr. Adelaide Beling, German ichthyologists and microbiologists who had previously worked on the Dnjepr in Russia. Prof. August Thienemann, head of the famous Hydrobiologische Anstalt der Max-Planck-Gesellschaft (MPG) at Plön, soon became mentor and supporter of the enthusiastic group.

In 1949, the Belings and the five students sampled the River Fulda during what became a real expedition, under the adventurous conditions of post-war Germany. The group made contact with sport fishermen at Schlitz who expressed interest in, and eventually funded, an exhibition of freshwater fauna and flora entitled "*Das Leben unserer Heimatgewässer*" which was shown in the sportshall at Schlitz, in the autumn of 1949. The illustrious Otto Hartmann Graf von Schlitz, genannt von Görtz, visited and decided to provide the young students with a building to serve as a base for further studies of the River Fulda. He had his sculptor grandfather's former studio (first built in 1876) completely rebuilt and donated this plus some land and fishing rights to the MPG (Figure 1.1).

The opening ceremony of the Schlitz institute was held on 4 June 1951, in the presence of Count and Countess v. Görtz, Otto Hahn, President of the MPG, A. Thienemann from Plön, D. v. Denffer of the Justus-Liebig-Universität at Giessen, and many other guests. The choice of name, "*Limnologische Flussstation*"

Central European Stream Ecosystems: The Long Term Study of the Breitenbach, First Edition. Edited by Rüdiger Wagner, Jürgen Marxsen, Peter Zwick, Eileen J. Cox.

Figure 1.1 The original building of the Limnologische Flussstation in 1951, and the name plate on its front wall.

(Figure 1.2)[1], anticipated a change in scientific emphasis, which manifested itself years later when the long-established Hydrobiologische Anstalt at Plön became the Max-Planck-Institut (MPI) fuer Limnologie.

J. Illies held the single scientist's position at the Limnologische Flussstation Schlitz, but the salary was shared between the five founders until the other four found themselves different positions. Later, a second scientist's position was installed by the MPG. Since 1982, the payroll included 15 positions, of which five were scientists. The original building soon became too small. In 1959 the MPG added a large extension to the original building, and in 1969 Graf Otto Hartmann donated a former mill opposite the Flussstation (Figure 1.2). The MPG had the Hallenmühle transformed into a laboratory and office building. Great efforts were made to turn the millrace running through the building into a living stream laboratory. However, at that time the poor water quality of the River Schlitz precluded the maintenance of the stream fauna or any undisturbed experiments. Operating artificial streams with recirculating river water from a large reservoir was not a long-term viable alternative. Littoral filtrate of the river water was used for several years, mainly to run biofilm experiments. Eventually the room was dedicated to other technical equipment. In the main building, laboratory space was at a premium, until the MPG added a dedicated laboratory section in 1995.

The scientific activities of the Limnologische Flussstation are evidenced in the publication list, with contributions from staff members, visiting scientists, and, not least, graduate students working on master and doctoral theses. Research focused on a variety of subjects, with a change of emphasis over time.

In the first years, under Joachim Illies, the focus was on methodological studies, regional limnology and the regular sequence of characteristic biocoenoses along rivers. Much of this appeared in the *Jahresberichte* (later: *Berichte*) *der Limnologischen Flussstation Freudenthal,* the station's own periodical. The required taxonomic expert knowledge of stream fauna was largely developed by members of

1) See Fittkau (1992, 2001) for the history of the epithet "Freudenthal", further donors and additional offices operating on the Weser and so on for some limited time.

Figure 1.2 The first extension building (1959; top left) is plastered and stands at a right angle to the original half-timbered building. The laboratory section added in 1995 extends the old building longitudinally and copies its half-timbered style (bottom). The Hallenmühle (top right) stands across the road opposite the main building.

the Schlitz group themselves. Taxonomic expertise, an indispensible precondition for ecological studies, always remained a stronghold of the Flussstation.

Based on intimate knowledge of the Mölle stream in North-Rhine-Westfalia and of the River Fulda, J. Illies developed a concept of the biocenotic structure of streams (Illies, 1955), which he later extended as "*Versuch einer Allgemeinen Biozönotischen Gliederung der Fließgewässer*" (Illies, 1961). Only after organisms have been identified can their functions and roles in the ecosystem be analyzed. Illies' (1961) concept of river zonation therefore logically preceded the *River Continuum Concept* (Vannote *et al.*, 1980). The first describes the discontinuous distribution of biocoenoses along streams, the second the continuous change of functions along river continua. Although at first glance the concepts may seem contradictory, they are actually two sides of the same coin.

From 1957 onwards, J. Illies worked in the main institute at Plön while K. Müller led the Flussstation. Studies on organismal drift and fish biology then predominated. In 1965, the Flussstation became an outlier of the new Department of Microbial Ecology of J. Overbeck at Plön. J. Illies returned to Schlitz, as Prof. Overbeck's local representative, but because of these changes, several studies performed at Schlitz by K. Müller and collaborators were published elsewhere and are missing from our publication list (http://edoc.mpg.de/ins/22/col/399).

For some years J. Illies and his students resumed their studies on the River Fulda before work at the Flussstation concentrated on two, first-order streams near Schlitz. Meanwhile the River Fulda had become heavily polluted while the small Breitenbach and Rohrwiesenbach were hardly disturbed and, because of their small size, more amenable to quantitative ecological studies. In both streams, Chordata play no role and invertebrates, especially insects and amphipods, dominate. J. Illies attempted to quantify the secondary production of stream insects by using emergence traps, initiating a series of emergence trap studies. Differences down the Breitenbach required several traps along its length, at the expense of work on the Rohrwiesenbach. A general survey of the Breitenbach fauna was performed and, for some time, amphipod ecology also received special attention (M.P.D. Meijering and students, compare the publication list of the Flussstation [http://edoc.mpg.de/ins/22/col/399]).

In June 1982, J. Illies suddenly died. As part of J. Overbeck's department the Limnologische Flussstation Schlitz was not closed and, in 1983, P. Zwick became head of the station and chose to continue work on the Breitenbach, to fully exploit previous work done there. When J. Overbeck retired, the MPG decided to continue the Schlitz station as an independent working group of the MPI of Limnology at Plön.

The various scientific activities of the LFS attracted visitors from all continents. A few spent sabbaticals in Schlitz, but most guests were funded by the MPG, the Deutsche Forschungsgemeinschaft (DFG) or the Deutscher Akademischer Austauschdienst (DAAD), staying between one month and two years. In a few cases, external funding from the DFG was available for longer periods. The Flussstation hosted several German and international limnological congresses, until limnological associations became too large to be accomodated within the small township of Schlitz. Among other congresses held at the LFS, were the First International Congress on Groundwater Ecology organized in conjunction with the Third International Colloquium on *Gammarus* and *Niphargus* (1975), and the Sixth International Symposium on Plecoptera (1977). The Deutsche Diatomologen Treffen was initiated in Schlitz in 1987, meeting annually since then and going on to become the Central European diatomists meeting. The Rhithron Ecology Group was also founded at Schlitz (1988).

Staff of the LFS were always actively involved in academic teaching. Graduate students from all parts of Germany came to work at Schlitz for their Diploma or Doctorate. Students of the LFS were treated as in-faculty students by the universities at Giessen, Kassel, Kiel, and Marburg. Cooperation with other universities was no exception.

The choice of a successor after the retirement of Prof. Overbeck indicated that the MPG was redirecting the institute at Plön, and it has now become the *Institute of Evolutionary Biology*. When the heads of the *Department of General Limnology* (Prof. W. Lampert), the *Working Group on Tropical Ecology* (Prof. W. Junk), and the LFS retired in short succession, from autumn 2006 onwards, limnology was discontinued in the main institute and the LFS was closed, after 56 years.

The present book summarizes some of the work done on the Breitenbach by the Schlitz River Station.

2 The Breitenbach and Its Catchment

Jürgen Marxsen, Rüdiger Wagner, and Hans-Heinrich Schmidt

2.1 Study Area

The Breitenbach is a small first-order stream, situated in eastern Hesse (Germany) between the Vogelsberg and Rhön mountains, approximately 4 km east of the town of Schlitz (10 000 inhabitants) and 100 km northeast of Frankfurt am Main (Table 2.1, Figures 2.1 and 2.2). It was selected for a detailed ecosystem study not only because of its close vicinity to the Limnological River Station, but mainly because it is a typical Central European stream containing typical communities of organisms (Zwick, 1998a).

The catchment area is part of the "Fulda-Haune-Tafelland" (Fulda-Haune-plateau), which belongs to the larger "Osthessisches Bergland" (Eastern Hesse upland; Klausing, 1974). The Fulda-Haune-Tafelland is a large plateau (up to about 500 m a.s.l.) dominated by Middle Bunter Sandstone but occasionally perforated or superposed by basaltic rock. Pleistocene and Holocene debris, as well as eolian und fluvial sediments, are the predominant uppermost geologic layers (Kupfahl, 1964, 1965; Motzka, 1968a, b). Several streams and rivers carve valleys into the sandstone layers, down to about 200 m a.s.l. These valleys are mostly used for agriculture, sometimes including the gentler hillsides and their summits. About half of the region is covered by forests (Seibert, 1954; Bohn, 1996).

The Fulda-Haune-Tafelland experiences the typical temperate Central European climate. The winter temperatures are somewhat lower than in north-west, south-west and south Germany, but summer temperatures are similar to north-west Germany. However, the precipitation is lower and the typical westerly winds are less dominating (Schönhals, 1954). The annual average temperature of the region is between 7 and 8 °C, the mean values for January and July are between −1 to −2° and 16 to 17 °C, respectively (Deutscher Wetterdienst in der US-Zone, 1950). The maximum air temperatures remain below 0 °C for 20–40 days $year^{-1}$, the minimum temperatures fall below 0 °C for 80–120 days $year^{-1}$. Ten to 30 days $year^{-1}$ have temperatures above 25 °C. Annual precipitation in the region fluctuates between 550 mm and 750 mm. Ten to 15% of the precipitation falls as snow.

Central European Stream Ecosystems: The Long Term Study of the Breitenbach, First Edition. Edited by Rüdiger Wagner, Jürgen Marxsen, Peter Zwick, Eileen J. Cox.

Table 2.1 Geographical and physical characteristics of the Breitenbach.

Variable	Value
Latitude	50°39′N
Longitude	9°38′E
Catchment area	8.3 km^2
Stream length	6300 m
Stream length, main channel	4200 m
Stream bed area, main channel during base flow	3225 m
Average gradient	0.033 $m\,m^{-1}$
Mean annual water temperature	7.3 °C
Mean annual discharge above mouth at GT6	26 $l\,s^{-1}$
Mean annual precipitation	63 cm
Mean annual air temperature	8.0 °C

Figure 2.1 Network of major rivers in Germany and the position of the study area (star) in the upper reach of the River Fulda.

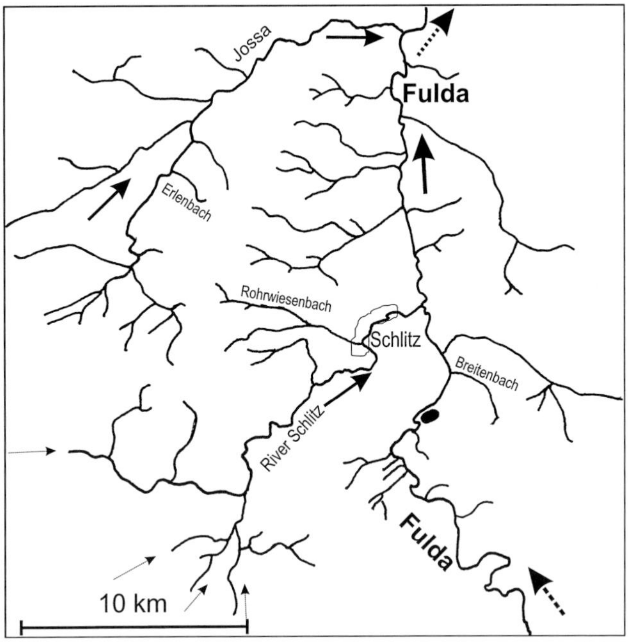

Figure 2.2 Map of the River Fulda catchment close to the town of Schlitz, showing streams studied by the staff of the Limnologische Fluss-Station and mentioned in this book. Arrows indicate flow direction.

2.2 The Stream and the Catchment

The Breitenbach is part of the Fulda-Werra-Weser drainage system (Figures 2.1 and 2.2). The stream enters the River Fulda, about 66 km below the latter's source, which is at about 850 m elevation in the Rhön Mountains about 30 km southeast of the Breitenbach. The River Fulda is one of the headstreams of the Weser, merging after about 220 km with the second Weser headstream, the River Werra at the town of Hannoversch-Münden, to form the River Weser (Hessischer Minister für Landwirtschaft und Forsten, 1964; Pusch *et al.*, 2009). The water travels for a further 432 km from here to the Weser River estuary, where it enters the North Sea close to the town of Bremerhaven.

The Breitenbach valley is carved into a slightly undulating sandstone plateau (Figure 2.3). The plateau east of the River Fulda slopes down somewhat to the north. Thus the plateau south of the stream lies between 400 and 420 m a.s.l., whereas the hills north of the stream are only about 360 m a.s.l.

The stream system originates on the plateau from a number of rheo- and helocrene springs at about 430 m a.s.l. (Figure 2.3), and enters the River Fulda 6.3 km downstream at 220 m a.s.l., exhibiting a mean gradient of 3.3% (Figure 2.4). The

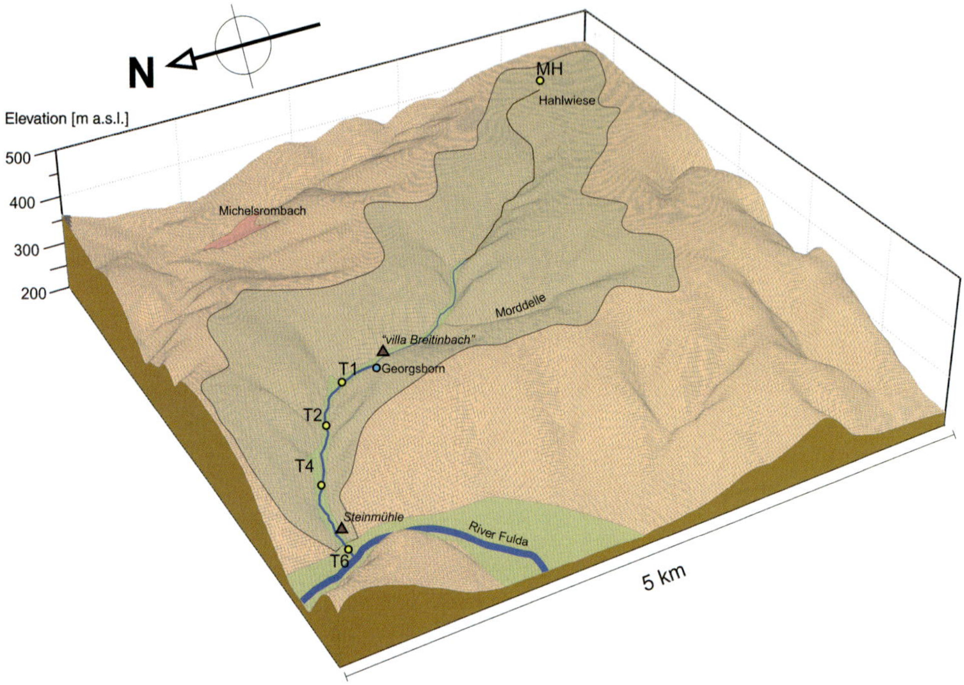

Figure 2.3 Three-dimensional image of the Breitenbach catchment with relevant locations. The village of Michelsrombach is the location of a measuring station of Deutscher Wetterdienst. Details of the other mentioned sites are given in the text. For abbreviations see Table 2.2. Source: Dig. Geländemodell © Hessisches Landesamt für Bodenmanagement und Geoinformation, 2008.

springs in the uppermost course are unable to feed the stream with sufficient water for most time of the year. Thus the first 2.1 km are mostly dry (intermittent flow). The main channel of the Breitenbach begins 4.2 km above the confluence with the River Fulda (Table 2.2, Figure 2.5). However, the next 2.1 km of the upper course (Figure 2.6) also exhibit intermittent flow. Here the streambed usually falls dry in late summer and early autumn in most years. With increasing frequency the period of desiccation has become more extended over recent decades, culminating in 2004, when discharge at the lower end of the upper course was observed for a few weeks in May only (Marxsen *et al.*, 2010). Annual average discharge from 1990 to 2005 at this point was 165 000 m^3 (Figure 2.4).

The middle course of the stream (Figures 2.7–2.9) with perennial flow begins about 2050 m above its mouth, at the conjunction of the upper course with several permanent springs (Table 2.2). The largest of them, named the Georgsborn (Figure 2.10), enters the stream from southwest at 270 m a.s.l. The city of Hünfeld extracts drinking water about 400 m upstream of the Georgsborn, pumping up to 352 m^3 groundwater daily from 120 m below the surface (~170 m a.s.l.; data pro-

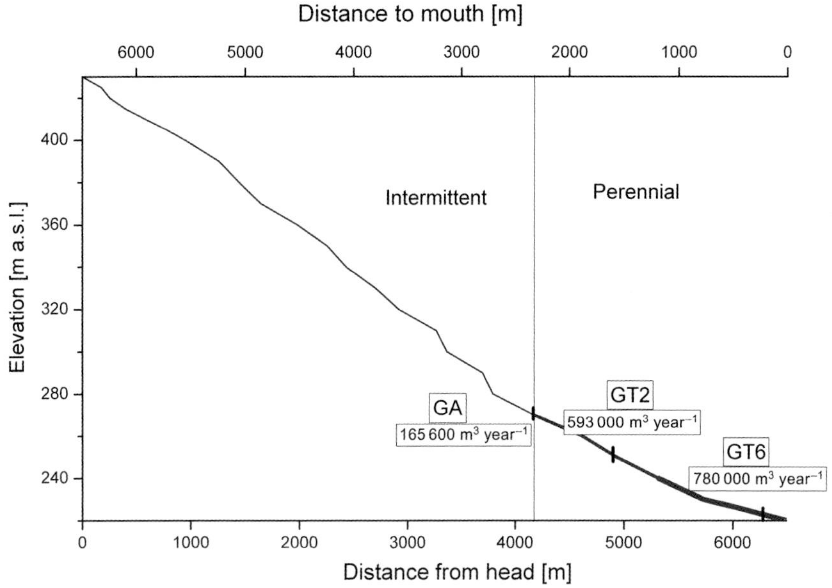

Figure 2.4 Upper (intermittent flow), middle, and lower courses (perennial flow) of the Breitenbach with indications of elevation and distances from head and mouth. Average annual discharge data from 1990 to 2005 are given (for details on sites GA, GT2, and GT6, see Table 2.2).

vided by the City of Hünfeld), a maximum potential water extraction of about 128 000 m^3 year^{-1}. Although it has repeatedly been assumed that this water abstraction has reduced the supply to the Breitenbach, this has never been proved.

Beginning with the entrance of the Georgsborn spring, the Breitenbach is fed for approximately 500 m by water from an extended spring horizon, contributing 356 300 m^3 year^{-1} to the stream (Figure 2.5). The water provided by these springs is sufficient for permanent flow downstream of the confluence of the upper course and the Georgsborn, even in very dry summers, together with groundwater from the adjacent slopes, which enters by diffuse perfusion through the streambed along most of the downstream course (Fiebig, 1995).

Another spring horizon is located about 930 m above the confluence with the River Fulda (Figure 2.5). Here two more rheocrene springs provide permanent groundwater discharge points. The larger spring enters the Breitenbach from the south. On the northern stream bank, close to the smaller spring, another intermittent rheocrene spring delivers water, mainly in winter and spring (Fiebig 1995). The annual amount of water supplied by these springs is about 205 000 m^3. This entrance of springs marks the interface between the middle and lower courses of the stream. About 150 m above its mouth (at the lowest measuring point), average annual discharge of the Breitenbach was 780 000 m^3 (Figure 2.4), which is probably a good estimate of the volume of water entering the River Fulda.

Table 2.2 Study sites in the catchment area and along the Breitenbach. Distances given as meters of stream length, based on measurements by Schwank (1983) for the middle and lower reaches, and by Marxsen (1980a) for the upper reach. T1–T6 = insect emergence traps (greenhouses) constructed across the stream.

Site	Code	Altitude (m a.s.l.)	Latitude	Longitude	Distance from mouth (m)	Distance from Georgsborn spring entrance (m)
Plateau measuring station	MH	419	50°37′99″	09°39′77″		
Spring of main channel		373	50°38′31″	09°38′41″	4172	2125
Gauge at end of upper course	GA	272	50°39′52″	09°37′78″	2061	14
Georgsborn spring	BS	272	50°39′52″	09°37′77″	2072–2075	25–28
Gauge below the Georgsborn spring	GS	271	50°39′52″	09°37′78″	2057	10
Outlet of the Georgsborn spring	U/A	270	50°39′52″	09°37′78″	2047	0
T1	T1	267	50°39′56″	09°37′79″	1968–1974	73–79
Gauge at T2	GT2	255	50°39′70″	09°37′47″	1470	577
Measuring station at T2	MT2	253	50°39′71″	09°37′44″	1422	625
T2 first trap	T2	253	50°39′71″	09°37′44″	1416–1422	625–631
T3 sand trap	T3	248	50°39′76″	09°37′34″	1261–1267	780–786
T3 stone trap	T3	246	50°39′78″	09°37′26″	1138–1144	903–909
T4 trap at lower fish ponds	T4	242	50°39′82″	09°37′16″	972–978	1069–1075
Spring entrance between middle and lower course		240	50°39′84″	09°37′13″	933	1114
T5 trap below alder trees	T5	233	50°39′87″	09°36′90″	596–602	1445–1451
Measuring station at T6	MT6	222	50°39′88″	09°36′59″	202	1845
T6 trap above mouth	T6	222	50°39′88″	09°36′59″	196–202	1845–1851
Gauge below T6	GT6	221	50°39′88″	09°36′55″	149	1898
Mouth into River Fulda		220	50°39′87″	09°36′42″	0	2047

At 645 m below the confluence of the upper course and the Georgsborn spring, water was extracted from the Breitenbach to feed three small fishponds, which were used with very different intensities throughout the period of investigation on the Breitenbach. The water from these ponds re-entered the stream about 430 m downstream of the withdrawal point, about 1200 m above the stream's mouth. In 1993, when the area was designated as a nature reserve, fishpond use ended. Water

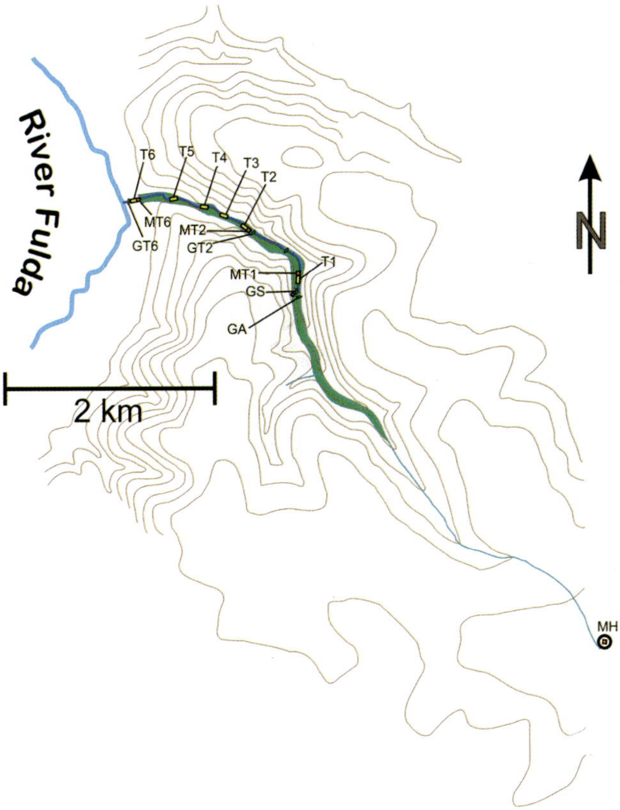

Figure 2.5 Map of the Breitenbach with locations of collecting sites (traps T1–T6), gauges (GA, GS, GT2, GT6), and measuring stations (MT1, MT2, MT6 at traps T1, T2, T6; MH on the plateau).

flow to the ponds was stopped and the ponds were allowed to fall dry, receiving stream water only during floods and otherwise receiving only rain water and leachate. Another four, larger, fishponds are situated south of the stream, at the interface between the middle and lower courses. They are fed by cold spring water, which leaves the ponds via the spring channel, entering the stream 930 m above its mouth.

The stream sediments are derived from Bunter sandstone, which dominates the catchment geology. Sediments range in size from large stones, some dm in width, to fine grained sand or even silt. However, two types dominate (Marxsen, 2001). One is sand with a mean grain size of about 0.5 mm. The other is gravel and pebble, approximately 0.5–6.0 cm in diameter, often mixed with cobbles, up to about 20 cm or more, also containing considerable fractions of sand. At the lower fishpond site small dolomitic limestones fragments were deposited on the streambed surface, presumably dropped by the fishpond owners.

Figure 2.6 View showing the location of the earlier settlement "villa Breitinbach" in the upper course of the Breitenbach (summer, view looking downstream). The village was mentioned in 1339, but documented as deserted in 1478. The Georgsborn spring (indicated by an arrow) is situated on the opposite (southern) side of the meadows on the valley floor, where the deciduous trees begin at the edge of the forest. The stream flows close to the forests on the northern slope, along a straight, probably artificial, bed. In the foreground are extensive areas covered by periwinkle (*Vinca minor* L.). This is typical of sites that were previously populated.

Figure 2.7 Middle course of the Breitenbach valley in spring (looking downstream). Two greenhouses (traps T2 and T3) are visible, with measuring station, MT2, in front of T2.

Figure 2.8 Middle course of the Breitenbach (looking upstream direction) in spring with the first greenhouse constructed in 1969 (trap T2).

Figure 2.9 Middle course of the Breitenbach (looking upstream) in spring with the first greenhouse constructed in 1969 (trap T2).

Figure 2.10 The Georgsborn spring pool (summer) showing the presence of allochthonous inputs of particulate organic matter.

The Breitenbach has a catchment area of $8.3\,km^2$. It is almost completely forested, chiefly by beech (*Fagus sylvatica* L.) and pine (*Pinus sylvestris* L.). Between 1990 and 2005, the mean annual precipitation in the catchment was 630 mm. Annual evapotranspiration in this region is 450–500 mm, leaving about 130–180 mm for discharge, which broadly agrees with the discharge leaving the valley (Marxsen *et al.*, 1997). The mean annual discharge for the Breitenbach (1990–2005) of about $26\,l\,s^{-1}$ is equivalent to about 100 mm of precipitation. Kupfahl (1965) estimated that between 63 and 92 mm of precipitation leaves the catchment via subsurface discharge.

The Breitenbach is a typical Central European stream, which means that its environment has probably suffered drastic human impacts for 1000 years or more (Brehm and Meijering, 1996). The forests, which dominate the vegetation of the catchment, are not natural but developed under strict human control. Pristine forests in this area were dominated by beech (*Fagus sylvatica*), mainly as *Luzulo-Fagetum*. In the higher reaches of the catchment small areas of stagnosols occur. Along with *Fagus sylvatica*, *Quercus robur* L. and *Qu. petraea* Liebl. formed the tree layer and *Molinia caerulea* L. was a typical plant of the ground layer (Bohn, 1996). This natural situation had been changed over the centuries of human use to forests that are still dominated by beech, but mixed mainly with hornbeam (*Carpinus betulus* L.) and oak (*Quercus robur* and *Qu. petraea*; Seibert, 1954; Bohn, 1996). Large areas were also planted with coniferous trees, mainly pine (*Pinus sylvestris*) and spruce (*Picea abies* [L.] H. Karst), but also larch (*Larix decidua* Mill.).

In 1993, the major part of the catchment became a nature reserve of the state of Hesse, to allow scientific study of a typical stream ecosystem, close to its natural situation (Obere Naturschutzbehörde, 2004). At that time an initial overview on

Table 2.3 Numbers of plant and animal species as listed in the nature conservation maintenance plan for the Breitenbach catchment (Wagner, 1993).

Taxon	No. of species	Taxon	No. of species
Plants		Invertebrates	
Mosses/liverwort	16	Mollusca	12
Higher plants	167		
		Insecta	
Sum plants	183	Odonata	19
		Hymenoptera	17
		Saltatoria	11
Vertebrata		Coleoptera	501
Mammalia	14	Heteroptera	60
Aves	47 (+16)	Lepidoptera	46
Reptilia	4	Diptera	
Amphibia	5	Mycetophilidae	171
Pisces	3	Syrphidae	115
Sum vertebrates	73 (+16)	Sum invertebrates	952

the flora and fauna of the area was provided as the basis for a maintenance plan for the nature reserve (Wagner, 1993). Over a short period a very rough survey of plants and animals in the area recorded more than 1500 species, excluding the stream flora and fauna of the Breitenbach (Table 2.3). The maintenance plan also contained suggestions for changing the forest communities in the area in general, and particularly adjacent to the stream. The main goal was to reduce, or even completely remove the coniferous trees, and convert the community to mixed or deciduous forests, similar to the inferred natural situation. Recent storms (2005–2009; possible harbingers of global change) seem to have dramatically accelerated this process.

Parts of the valley directly adjacent to the stream were clear-cut and used as farmland during medieval periods. There was a small settlement, probably three or four farmsteads, in the valley close to the Georgsborn spring, which provided drinking water for the residents (Figures 2.6 and 2.10). It was mentioned as "villa Breitinbach" in 1335, but documented as a deserted village in 1478 and 1498, and was never recolonized (Fischer, 1990). However, restructuring of the valley surface at that time, including removal of the stream channel to the north to irrigate meadows, is still visible.

Another settlement directly south to the Breitenbach, just before the stream enters the River Fulda valley, was occupied for a longer period (Fischer, 1990; Figure 2.11). A water mill, "Steinmühle" (stone mill), was first mentioned as being in use here in about 1150, and also several times later until 1548. But by about 1650 the mill had been abandoned. However, stone ruins remained, until at least 1725 when a map documented "Rudera von der Stein mühl" (ruins of the stone mill). A weir for irrigating the meadows on the valley's southern slopes was

Figure 2.11 Lower course valley with the "Steinmühle" site in summer (looking upstream from the country road crossing the Breitenbach). The stream is flanked by alder trees revealing the short stretch where it flows across the valley, from the southern to the northern slope, between two almost right-angled bends. The medieval water mill was located just below this stream section, between the stream and the forest.

maintained until about 1870, but probably not much later. No subsequent documents mention such a construction. But the straight track of the stream in this area, with two almost right-angled bends reflects the earlier existence of the "Steinmühle" in the lower course of the Breitenbach.

Since its reclamation, the main agricultural use of the clear-cut part of the valley was as unfertilized grassland, although this state was never documented. The limnological investigation of the area roughly coincided with its intensive agricultural use between about 1950 and 1990. In this period artificial fertilizers were generally applied to the grassland. The former meadow vegetation of the valley can only be deduced from casual notes of local teachers interested in botany (Gregor, 1992) and from comparisons with similar areas. The lower part of the valley with rather good soil was presumably covered by a *Trisetion* community, with *Phyteuma nigrum* F. W. Schmidt, *Trisetum flavescens* (L.) P. Beauv., and *Alchemilla monticola* Opiz as typical species. This grassland was later used as highly productive grassland dominated by *Bromus hordeaceus* L. and *Holcus lanatus*. In the upper part of the valley, where the soil is less productive, poor meadow vegetation dominated by *Festuca rubra* and *Agrostis tenuis* L. can be inferred for the preindustrial period. This meadow type is still present in some parts of the valley. An unknown extent of low productive grassland dominated by *Nardus stricta* L. occurred, with species like *Arnica montana* L., *Calluna vulgaris* (L.) Hull, and *Carex pilulifera* L. As such grassland is very unproductive, grazing was probably the common form of management, rather than mowing. Parts of the upper valley, planted with spruce (*Picea abies*) in the nineteenth century, probably carried this

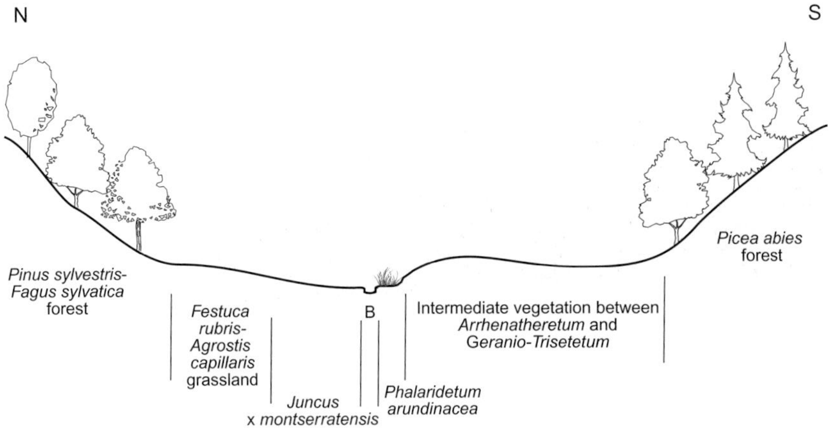

Figure 2.12 Typical vegetation profile across the Breitenbach valley above trap T2 (1990, cf. Figure 2.7). B = Breitenbach.

type of vegetation. Except close to the stream, damp meadow grassland is not found in the area, although it certainly existed. From casual notes we know that *Dactylorhiza majalis* (Rchb.) P. F. Hunt and Summerh. and *Eriophorum angustifolium* Honck. occurred in the area, species which grow in wet meadows of the *Calthion* formation. This type of vegetation very probably existed in the lower fish pond area.

During the time when limnological research occurred, the grassland in the valley was mainly used for making hay. It was composed of typical plants for such sites in Central Europe, mainly *Arrhenatherum elatius* (L.) P. Beauv. ex J. Presl and K. Presl and *Trisetum flavescens*, forming an intermediate vegetation between *Arrhenatheretum* Braun and *Geranio-Trisetetum* Oberd. (Wagner, 1993; Figures 2.7–2.9, 2.12). Close to the stream the vegetation today is often dominated by *Juncus* × *montserratensis*, a hybrid of *Juncus acutiflorus* Ehrh. ex Hoffm. and *J. articulatus* L. It is unknown whether its occurrence is related to the intensive agricultural use of the valley, or if this hybrid, common in eastern Hesse today, occurred before the onset of modern agriculture (Nowak, 1990; Gregor, 1992).

The Breitenbach is lined by *Phalaris arundinacea*, sometimes interspersed with *Sparganium erectum*. The moss flora is rather species-poor, most species occurring sporadically on disturbed soil (Gregor and Wolf, 2001).

The upper course of the Breitenbach is flanked by forests on at least one side (Figure 2.6). In this region, from 0 to 600 m above the stream's confluence with the Georgsborn spring, stream width ranges during base flow from 0.11 to 1.20 m, and is on average 0.60 m wide (Marxsen, 1980a). The middle and lower courses flow predominantly through grassland. The forests along these courses are mainly about 50 m away from the stream (Figures 2.7–2.9), but there are also stretches where the stream is directly flanked by alder, *Alnus glutinosa* (L.) Gaertner (Figures 2.11 and 2.13), or where it runs very close to the forest. Measurements of stream

Figure 2.13 River Fulda valley showing the entry point of the Breitenbach (summer, looking south). The stream is revealed by the alder trees along its banks. The Breitenbach enters the River Fulda (defined by a few trees, mainly willow, along its banks) close to the edge of the forest where the valley floor begins to rise (left side of the picture).

width in this area are available for the upper middle course only, from the entrance of Georgsborn spring to just below trap T3 (about 900 m). Base flow width was between 0.35 and 1.64 m, on average 0.82 m (Marxsen, 1980a). Stream width gradually increases further down, but no detailed measurements are available.

About 250 m above its mouth, the Breitenbach valley enters the much broader River Fulda valley (Figure 2.3). For about the last 150 m the northern Breitenbach slopes are no longer forested, but are used as farmland, whereas the southern slopes are forested to the end of the valley. The transition from the Breitenbach to the Fulda valley is marked by a country road which crosses the Breitenbach over a small bridge. The Fulda valley, which is about 250–300 m in width in this area, is intensely used as grassland (Figure 2.13), but the Breitenbach is flanked by trees, mainly alder, and shrubs for the whole stretch from the country road to its mouth (Figure 2.13).

2.3 Sampling Sites

Although the total stream length is about 6.3 km, the investigations performed by members of the Limnologische Fluss-Station were focused on the lower 2.1 km, beginning with the Georgsborn spring. Thus mainly the middle and lower courses were considered (cf. Table 2.2, Figure 2.5). The greenhouse study sites and the measuring stations are indicated in Figure 2.5. These are listed, together with other important sites, in Table 2.2. The first greenhouse (T2) installed for insect emergence studies in 1969 was sampled continuously until 2006. More greenhouses were constructed later and sampled over different time periods (for details, see Section 7.1.2). The main location for continuous monitoring and/or regular analysis of the climatic, hydrologic, physical and chemical characteristics of the Breitenbach was site T2, too, although less detailed measurements were performed at other sites (covered in detail in Chapter 3). However, many more sites were selected for investigations on different aspects of the stream's ecology over the years. These are described in the relevant chapters of this synopsis of almost 50 years of Breitenbach research.

3
Environmental Characteristics

Hans-Heinrich Schmidt

3.1
Climate and Weather

There is no universally applicable definition of the term climate. Climate comprises a combination of all meteorological causes of possible weather conditions, including their typical seasonal changes and daily fluctuations at a site. Describing climate requires many years' data. In contrast, weather describes a momentary condition of the atmosphere as the result of atmospheric conditions over a particular time period. Atmospheric conditions are described via a number of variables, which affect the physical conditions in the air.

Climatic effects and weather conditions in the region and in the Breitenbach catchment are the highest level of a structure that has a decisive influence on the biocoenoses. Irradiance, precipitation, wind and air temperature contribute.

Alongside the geological characteristics of the catchment area, these four factors determine the prevailing conditions and the future development of the ecosystem. Each individual factor has a local component; nevertheless it is above all the extent of the external influences that work on the region and its living communities (Figure 3.1).

3.1.1
Global Irradiance

Irradiance data from station MH are available for the upper Breitenbach section for 1995–2005. The average total annual irradiance is 888 KWh m^{-2}; minimum 785 KWh m^{-2} (2005); maximum 1029 KWh m^{-2} (1999) (Figure 3.2).

Comparative regional values are available from weather stations of the Hessischen Landesanstalt für Umwelt und Geologie (HLUG, http://www.hlug.de). The weather stations Wasserkuppe and Grebenau lie within an approximately 30 km radius of the Breitenbach valley. Direct comparisons are not possible because of the altitudinal differences, measuring equipment type and the different number of measurements; nevertheless the table gives an impression of the order of magnitude of the irradiance exposure of the individual localities (Table 3.1).

Central European Stream Ecosystems: The Long Term Study of the Breitenbach, First Edition. Edited by Rüdiger Wagner, Jürgen Marxsen, Peter Zwick, Eileen J. Cox.

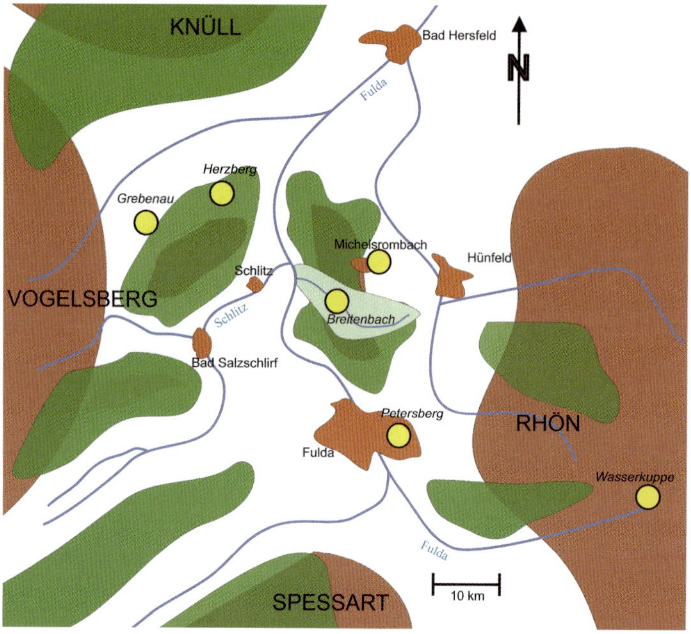

Figure 3.1 Map showing the Breitenbach catchment in relation to measuring stations of Michelsrombach, Grebenau, Fulda, and Wasserkuppe.

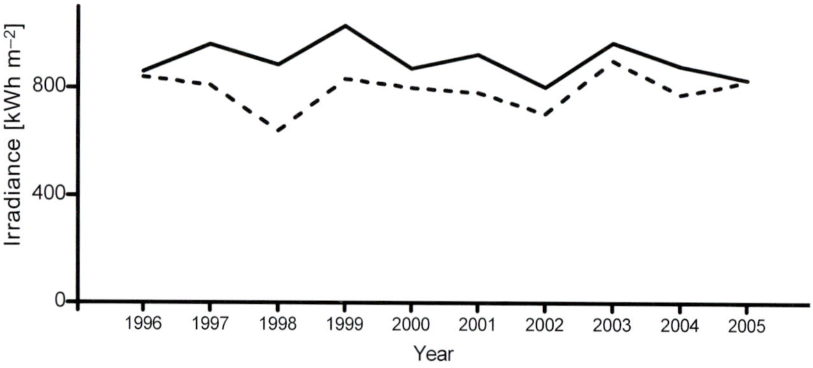

Figure 3.2 Global irradiance in the Breitenbach catchment 1996–2005 (MH full line; T2 dashed line).

Table 3.1 Annual irradiance at two sites in the Breitenbach compared to the official weather stations Wasserkuppe and Grebenau (compare Figure 3.2) and data obtained from HelioClim. (1) slight shading in east–west direction; increasing tree shading; (2) valley oriented south-east/north-west; strong shading from south; (3) irradiance calculated for BTB using the Heliosat method, resolution about 20 km (www.helioclim.net).

Location/site	Altitude m a.s.l.	Position of instruments (m above ground or stream at T2)	kWh m^{-2}
Wasserkuppe	950	3.5	1031
Grebenau	373	3.5	1013
MH (1)	420	1.0	888
T2 (2)	271	1.0	805
Helioclim (3)		Meteosat images	804

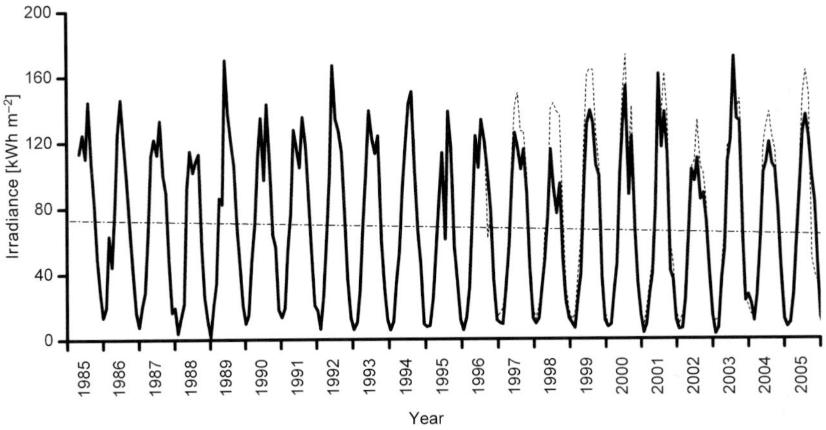

Figure 3.3 Global irradiance in the Breitenbach valley over 21 years (mean monthly). Full line: valley of Breitenbach (T2). Dashed line: measuring station Hahlwiese (MH). Dashed-dotted line: long-term trend.

Helioclim data comprises data collected from the weather satellite METEOSAT, processed using Heliosat. The daily values demonstrated such amazing agreement with our data that they can always be used to compensate for gaps in our measurements.

When considering long-term trends, 10 years is a very short time. However, a series of hourly measurements over a 21-year period are available for measuring station T2 (Figure 3.3). From a mathematical perspective the seasonal components of these data show a slight negative trend. However, based on observations from 12 European irradiance monitoring stations over more than 50 years, Ohmura (2006) suggested a slight increase in irradiance. This is a good example of the problem with calculating trends over time series of less than 40 years.

3.1.1.1 Local Effects of Irradiance in the Catchment

Irradiance at T2 in the Breitenbach is strongly affected by the valley topography and orientation. In the main part of the investigated stretch the Breitenbach flows in a south-easterly (130°) direction to the River Fulda. The bordering mountain ridges rise about 100 m above the valley bottom to the south and north. This means that from mid-November direct sunlight no longer reaches the stream. As a result of the prevailing low light intensity this has only a slight effect on the annual irradiance balance. Considering the diurnal variation on a bright summer's day, there is a slope difference between the MH and T2 curves, on both the rising and falling slopes. The reason for this is that at T2 the morning sun shows a rather gradual increase in irradiance, such that, until the sunlight strikes the valley directly, the increase remains slight (Figure 3.4). Towards 0900 CET the shading ends and the curve rises steeply. The reverse occurs in the afternoon with the decreasing slope. This effect decreases as the sun's path changes during the annual cycle. It produces the annual pattern seen in Figure 3.5. While the irradiance curve on the high plateau (MH) nears the expected time curve, in the summer months near T2 in the valley the curve is distinctly flattened. Thus the reduction in irradiance on the two sides of the diurnal curve can be explained.

3.1.2 Precipitation

Precipitation records are available from the Breitenbach valley since 1990 at measuring station T1 while the measuring equipment at MH was installed in 1996. Records were taken from both stations until 2005. The Joss–Tognini precipitation sensors work with a precision of ±2%, but based on our experience and also due

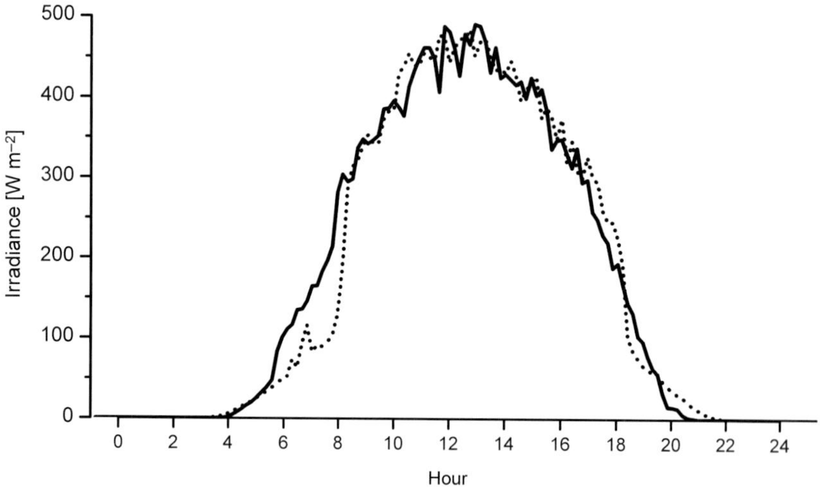

Figure 3.4 Irradiance during the course of a day in July 2003 to 2005 (10 min means) at measuring station MH (plateau, full line) and T2 (in the valley, dotted line).

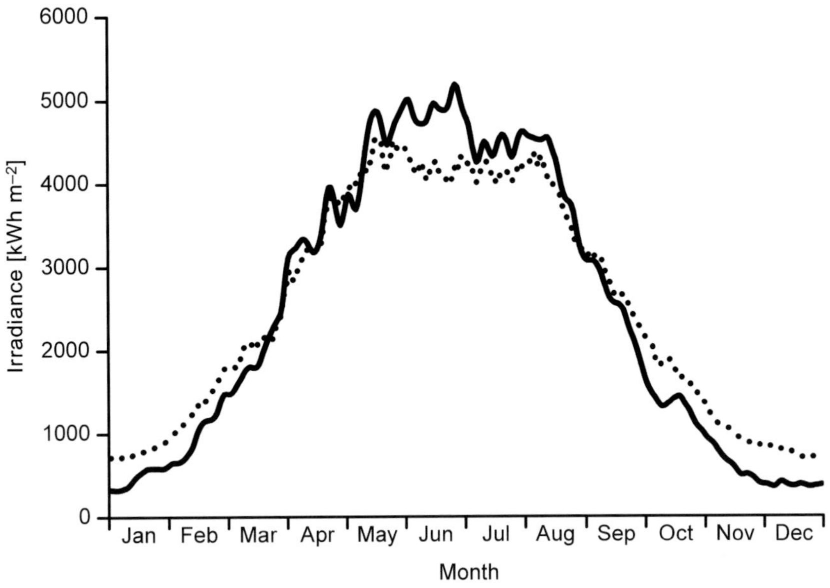

Figure 3.5 Irradiance in the course of one year at two sites at the Breitebach: MH (plateau, full line) and T2 (in the valley, dotted line).

Table 3.2 Average annual precipitation at sites MH and T1 at the Breitenbach compared to other localities (1996–2005) in the area. For both MH and T1 the minimum is in 2003 and maximum in 2002.

	Mean	Minimum	Maximum
T1	617	472	807
MH	657	453	793
Michelsrombach	662		
Schlitz	632		
Fulda	749		

to local conditions, for field installations, differences between two comparably installed instruments can be expected to be up to 10%. Values from the different local measuring stations in Michelsrombach, Schlitz, Petersberg, as well as super-regionally from the city of Meinigen (Thuringia), are available for comparison (all stations of the German Weather service, DWD; Table 3.2).

Given the measuring uncertainties, the values from stations T1, MH, Michelsrombach and Schlitz are not significantly different. This is not the case for values for Fulda/Petersberg, with a mean of 749 mm that be explained by other regional precipitation patterns.

The relatively low precipitation values in the study area are partly explained by the fact that, because of the north-westerly direction of rain-bearing winds, the Breitenbach lies in the rain-shadow of the Westerwald and Vogelsberg mountains.

The annual total precipitation varies very strongly from year to year. The minimum of 472 mm in 2003 and a maximum of 832 mm in 1984 mark the extremes. Only 16 annual cycles are available for the Breitenbach, which is too little to draw any long-term conclusions. Data from the DWD are available for Schlitz and Michelsrombach from 1969. Figure 3.6 clearly shows the large fluctuations and no clear trend can be determined.

Over the year there is a minor increase in precipitation events towards summer (Figure 3.7). However the random variation is so high that no significant relationship can be determined.

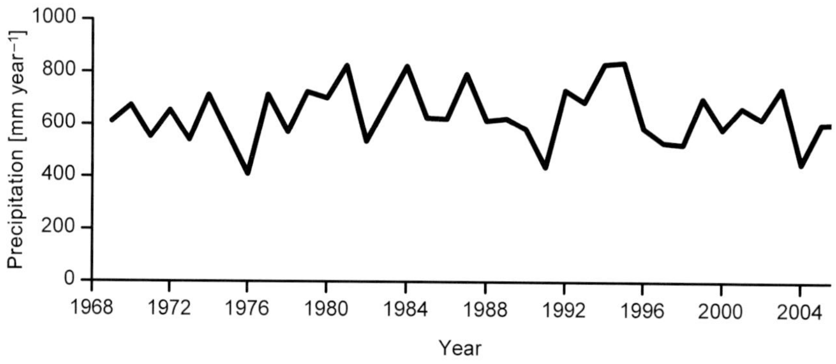

Figure 3.6 Yearly precipitation 1969–2005 for area of Schlitz and Michelsrombach.

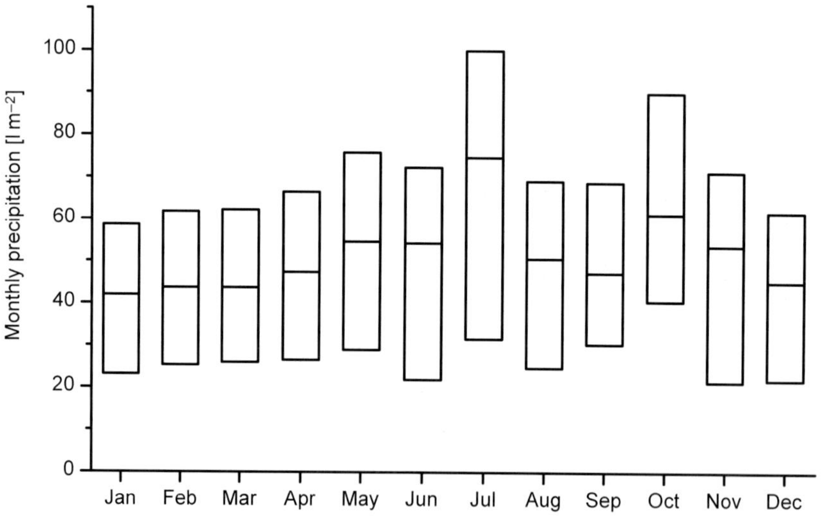

Figure 3.7 Monthly precipitation sum at the Breitenbach (mean, min, max).

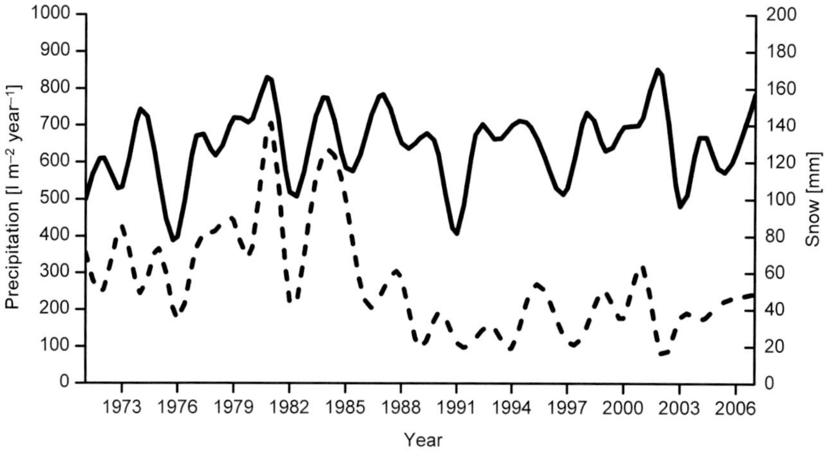

Figure 3.8 Total yearly precipitation (full line) and proportion of snow (dotted line) at the Breitenbach for the period 1970–2006.

A comparison of precipitation values (Figure 3.8) shows that, until the end of the 1980s on average about 11% (81.7 mm) precipitation fell as snow. Since 1990 this has fallen to 24.6 mm.

3.1.3
Wind

It is impossible to evaluate the data from the wind records from MH and the Breitenbach valley (T2) using means of the individual values. The reason for this is the very unstable directional signal in the valley at low wind speeds that would usually be interpreted as wind-still from the mean value. Thus pairs of values, wind direction and wind speed were recorded every 10 min, but also in 5-min intervals for short periods. Such a short recording interval provided important information for interpreting the wind system in the Breitenbach valley.

As expected, wind directions from the north-west (Figure 3.9) are most often observed on the high plateau at MH. The influence of the Fulda and Breitenbach valleys are slight there, 420 m a.s.l.. The wind speed ranges from wind-still (about 75% all observations) to a maximum of 26 m s^{-1}.

Figure 3.10 shows the seasonal variation in wind direction and speed for the year 2000. Directional events are shown on the *x*-axis by open circles; the circle diameter is proportional to wind speed. Strong winds from the sector around 250° dominate wind direction in the winter months. Noteworthy frequent occasions with high winds, from the 80° to 160° sector, were recorded in the early part of the year.

Evidence of the dependence of frequency of precipitation on wind direction is readily seen from the clear dominance of north-westerly winds (Figure 3.11).

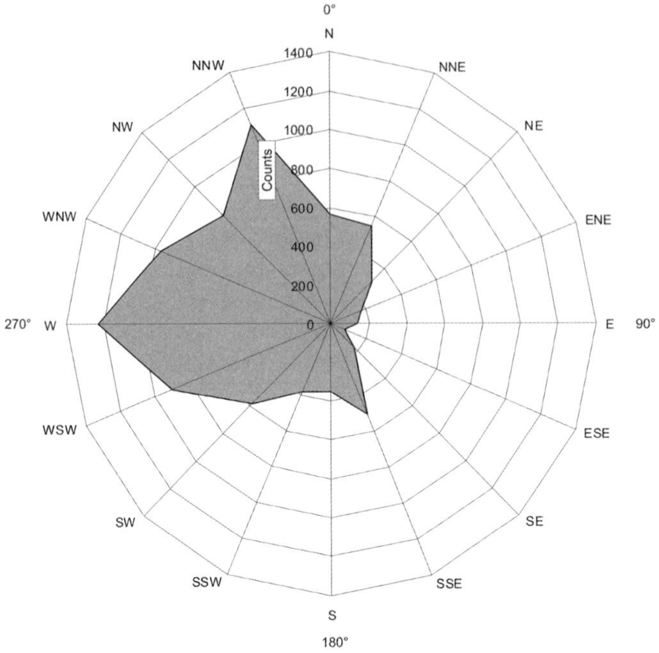

Figure 3.9 Counted frequencies of wind directions at measuring station MH in the year 2000.

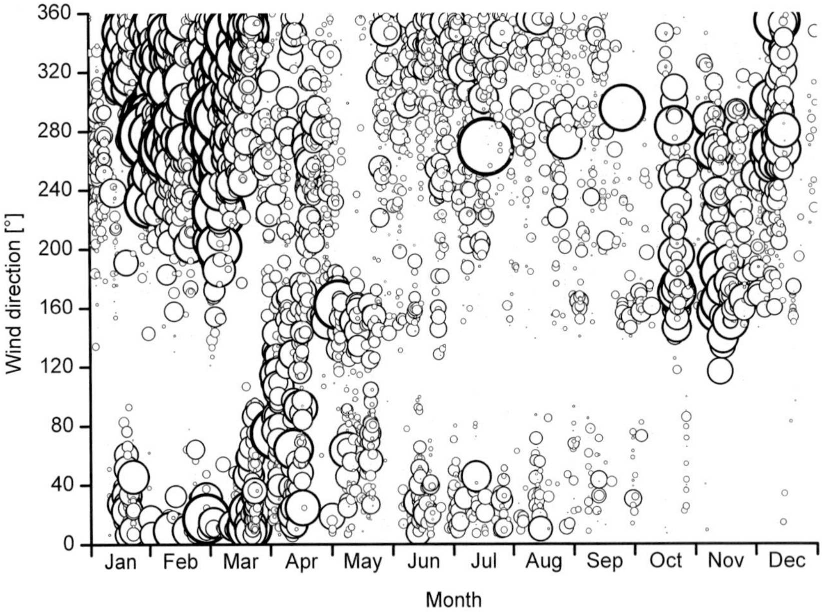

Figure 3.10 Seasonal distribution of wind speed (circle size) and wind direction (compare Figure 3.9) at measuring station MH in the year 2000.

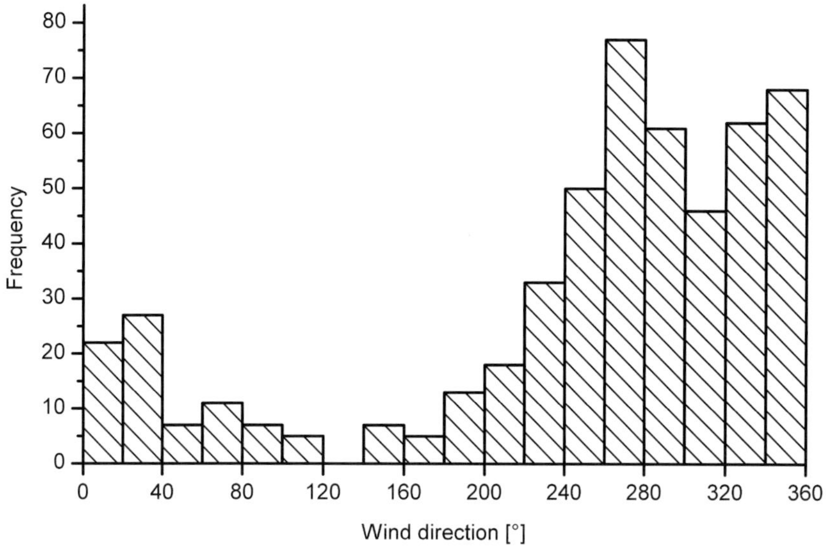

Figure 3.11 Frequency of precipitation events related to wind direction at measuring station MH in the year 2000.

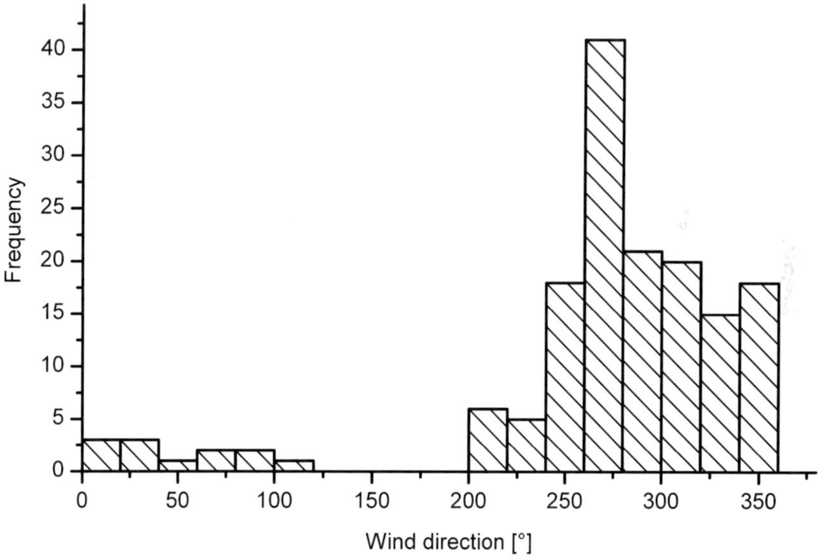

Figure 3.12 Frequency of precipitation events above $10\,l\,m^{-2}h^{-1}$ related to wind deirection at measuring station MH in the year 2000.

When precipitation above $19\,l\,m^{-2}h^{-1}$ is considered, the events come from the sector, that is, a westerly wind direction (Figure 3.12).

The situation in the Breitenbach 200 m below T2 is quite different. Here the main wind direction is from the south east (125 °, Figures 3.13 and 3.14). This wind blows regularly over the whole year, from the evening through the entire

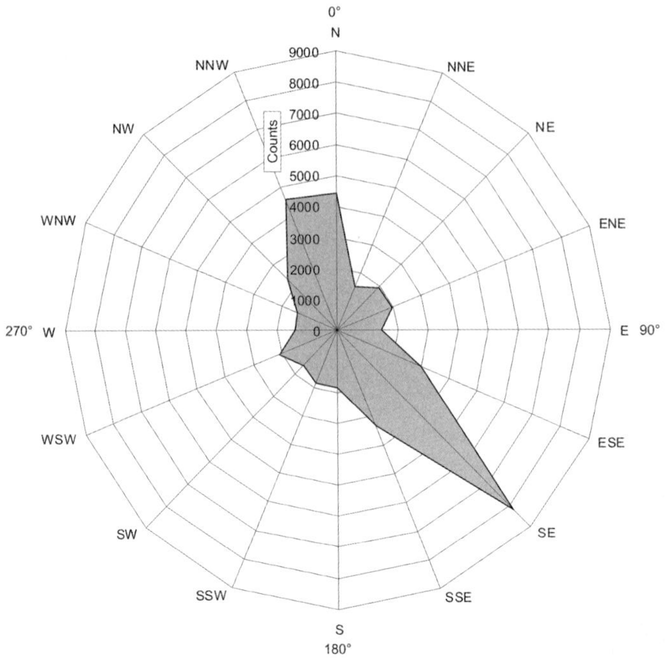

Figure 3.13 Frequencies of wind directions at measuring station T2 in the year 2000.

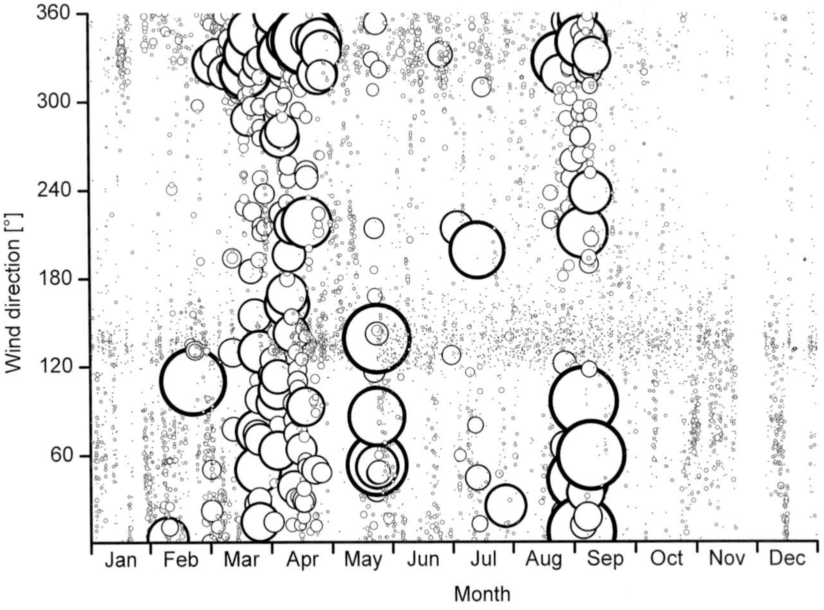

Figure 3.14 Seasonal distribution of wind speed (circle size) and wind direction (compare Figure 3.12) at measuring station T2 in the year 2000 (please notice accumulation of events at the 130° line).

night, creating a downcurrent (Figure 3.15, Bendix 2004). The wind speed is 2–3 m s^{-1}. Higher wind speeds have also been recorded, but these cannot be separated from the effects of large-scale wind systems.

Under sufficient light intensity (>50 W m^{-2}) another wind system develops during the morning, which crosses the valley diagonally and establishes an unstable air current from the north-north-west. As the sun starts to shine directly into the valley, the night-time down-valley current ends. In the summer this occurs at about 0800, in the winter at about 1100 (Figure 3.16). This wind can continue to dominate until the afternoon, but is then rapidly overtaken by other air currents. Apparently this is a local expression of a thermal wind system that is always observed in valley systems (Bendix, 2004).

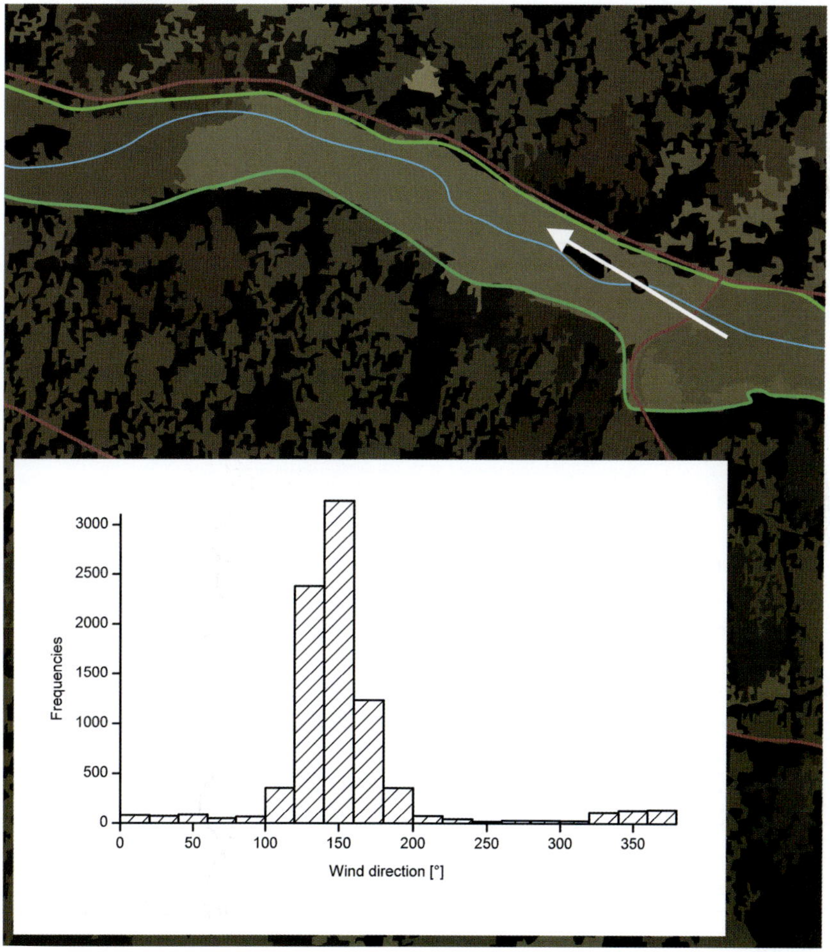

Figure 3.15 Nightly wind blowing down the valley (arrow) and frequencies of nightly (2100 to 0400) wind directions in July inserted.

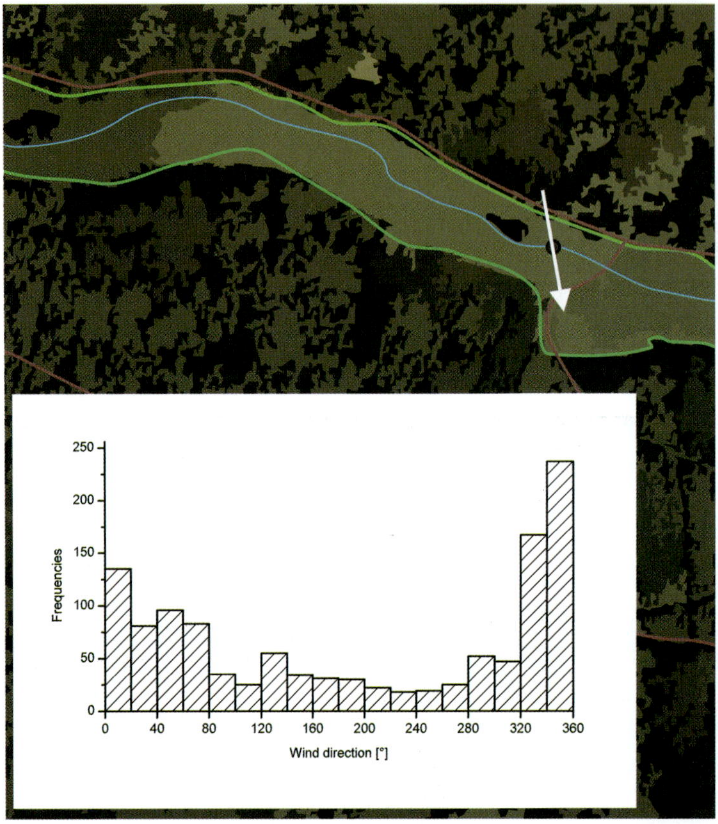

Figure 3.16 Wind system crossing the valley before noon (0800 to 0900) in July.

3.1.4
Air Temperature

There have been separate air temperature measurements for T2 since 1986. The long-term statistics from this series show the following range: mean 8.01 °C, minimum −22.3 °C, maximum 40.6 °C. The minimum temperatures occurred in the years 1986–1987, while the maximum of 40.6 °C was recorded in 2003 (Figure 3.17).

An increase in the annual mean of 1.8 °C over 20 years can be calculated from the available data (Figure 3.18), an average increase of 0.08 °C year^{-1}. An investigation of the seasonal effects on the development shows that the summer maximum (May to September) shows the strongest trend with a 2.6 °C rise. The 1.3 °C increases in the winter minimal and maximal temperatures remain below the average.

Comparative results from the three air temperature measuring sites were only possible from 1996 to 2004 (Table 3.3).

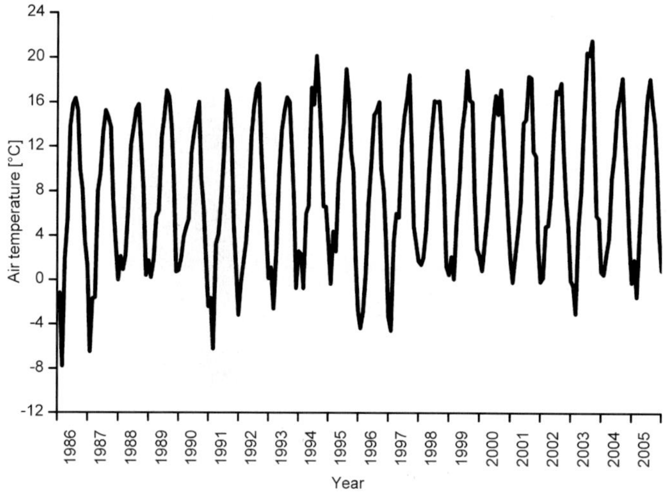

Figure 3.17 Monthly mean air temperature 1986–2005 at measuring station MT2.

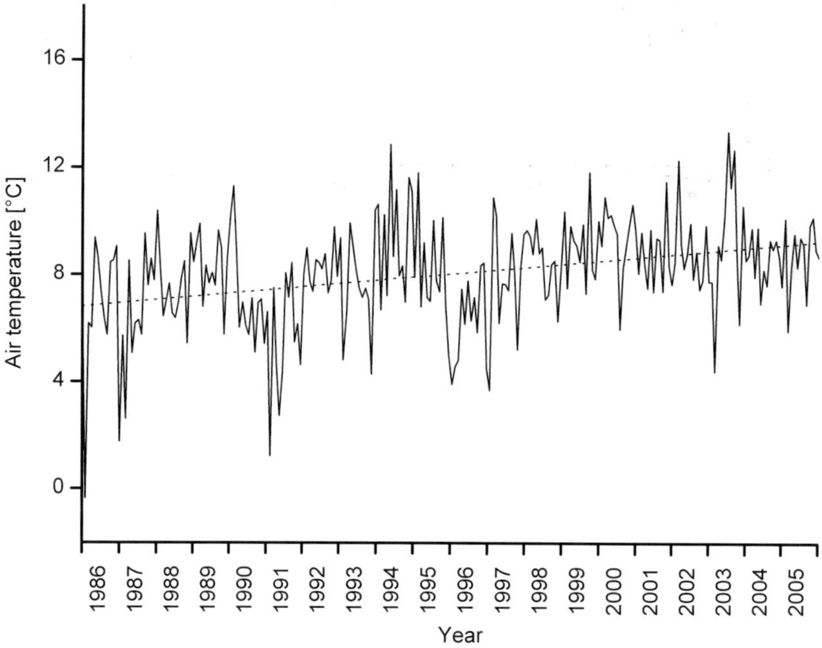

Figure 3.18 Air temperature 1986–2005 without periodic seasonal components with trend $y = 0.0102{*}x + 6.81$.

Table 3.3 Ranges of air temperature (°C) at three sites at the Breitenbach.

	T1	T2	MH
Mean	7.2	8.6	7.4
Min	–20.0	–21.7	–20.2
Max	33.3	40.7	35.9

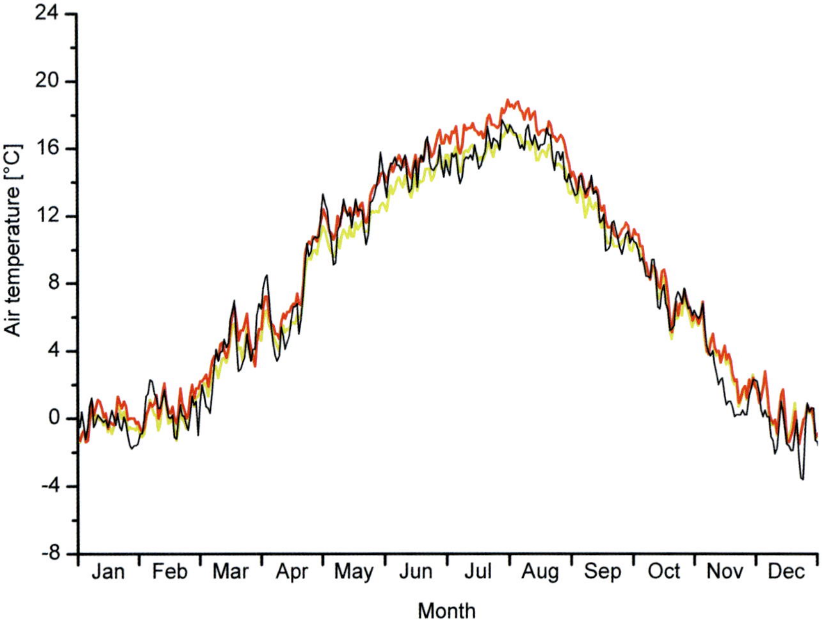

Figure 3.19 Mean air temperature at measuring sites T1 (yellow), T2 (red), MH (black).

T2 is the site with the highest average air temperature, the other two stations differ only slightly, despite an approximately 200 m altitudinal difference (Figure 3.19). The highest air temperatures occur in July/August, one month after the irradiance maximum. Analogously, the daily temperature maximum also occurs about one hour after the irradiance maximum (Figure 3.20).

3.1.5 Discharge

The flowing wave, with all its characteristics, plays a central role in the combination of factors affecting lotic ecosystems. In this chapter, the focus of the discussion will be the hydrological interrelationships. How does the catchment area, and thereby the stream, react to the climatic and weather-dependent events? The

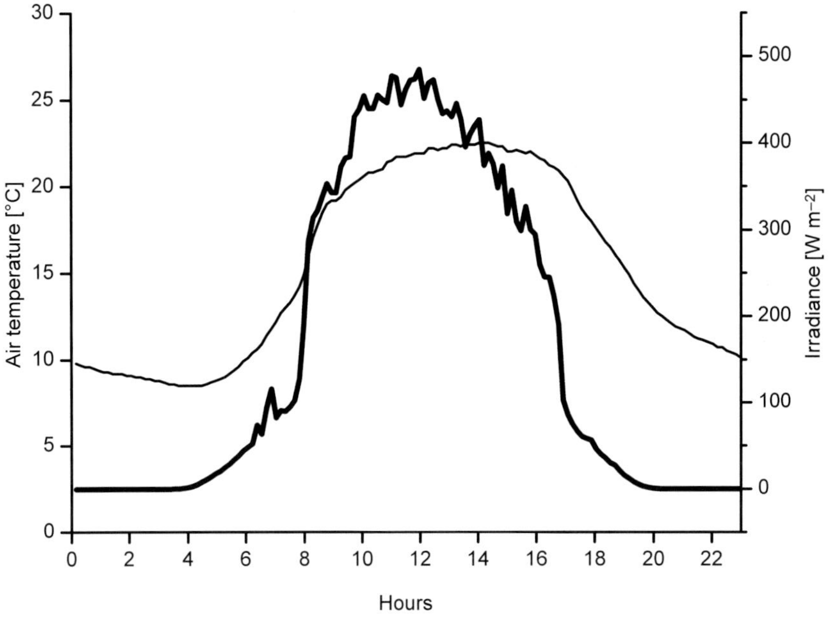

Figure 3.20 Daily cyclic fluctuation of irradiance (thick line) and air temperature (thin line) at measuring station MT2 in July.

analysis of the continuous discharge recorder builds a very detailed picture. Figure 3.21 gives an overview of the areas where spring water is contributed to the Breitenbach.

Discharge data since 1972 are available. The original measuring installations were read sporadically and already flooded by flows exceeding $60\,l\,s^{-1}$. From 1986 values obtained by the above method at T2 and T6 were continuously recorded. In 1990 measuring weirs were introduced at the upper reach (GH) and the spring (GS). As a small flowing water on Bunter sandstone, with a catchment area less than $10\,km^2$, the Breitenbach only flows continuously along its lower stretch, approximately 2 km.

The total water volume that reaches the Fulda below T6 was measured at the GT6 weir. On average, $780\,000\,m^3\,year^{-1}$ were measured there. However, this could be as much as $1.8 \times 10^6\,m^3\,year^{-1}$ (1987/1988) in wet years. The lowest annual runoff was $304\,000\,m^3\,year^{-1}$ (1997/1998).

Given the framework of the ecological questions, investigations were predominantly undertaken in the middle section of the Breitenbach. The inflow regime into this section was represented by the measurements at T2. Water was collected from an area of $6.5\,km^2$, reaching a long-term annual average of $593\,000\,m^3$, minimum $160\,900\,m^3$ (1995/1996), maximum $1\,440\,000\,m^3$ (1987/1988) (Table 3.4).

Figure 3.22 shows the dynamics of annual flow over the weir at T2. Higher water flow phases clearly occur predominantly in the months of February to May. Periods

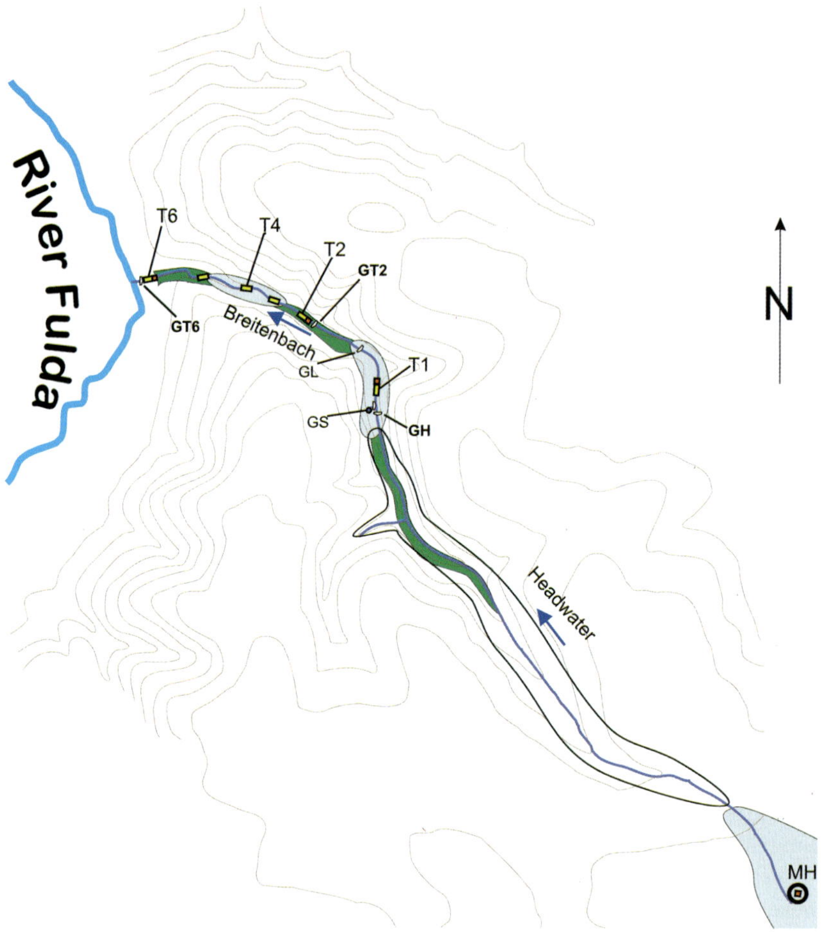

Figure 3.21 Spring areas of the Breitenbach (light blue shaded).

of lower water flows occur with the greatest likelihood in September. Given this seasonal regularity, when considering the timing of the inflow regime it logical to use not the calendar year, but the hydrological year with a start in October.

In the annual precipitation to runoff balance, annual runoff represents a long-term annual average of 15.2% of the annual precipitation. The largest share of the precipitation is returned to the atmosphere as evapo-transpiration (Figure 3.23).

To characterize flow regimes, hydrology uses a series of calculations, a few of which will be discussed here in relation to the use of routine data.

3.1.5.1 **Base Flow and Base Flow Index (BFI)**

According to Gordon *et al.* (2004) the outflow curve of flowing water includes both "direct runoff" and "base flow". The relationship of base flow to direct flow is described as the base flow index (BFI). Two different calculations were used to

Table 3.4 Ranges of hydrological measures in the Breitenbach.

	Annual runoff rate (m^3)			Discharge ($l\,s^{-1}$)		
	Min	Mean	Max	Min	Mean	Max
Headwater	0.0	165.6	410.0	0.0	5.3	179.3
Spring	52.6	188.7	352.4	0.1	6.1	175.4
GT2	160.9	593.8	1445.8	0.2	18.8	479.1
GT6	304.2	780.1	1844.0	2.6	25.4	521.1

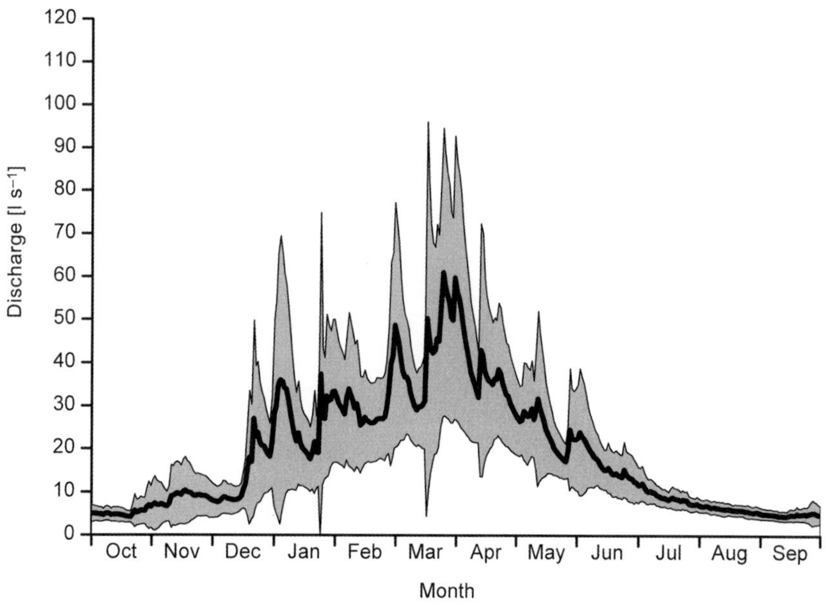

Figure 3.22 Mean (thick line), upper and lower limits of discharge at T2.

determine the BFI for the Breitenbach data. Eckhardt compared different algorithms for base flow calculation from outflow time series. To what extent the calculated indices are really linked to base flow remains unclear. More often the method allows the outflow characteristics of different flowing waters to be compared. The program 2prdf.exe (Eckhardt, 2008) and program BFI (Hisdal *et al.*, 2004) deliver the values shown in Table 3.5. The theory states that the index is a measure of the type of overall outflow. Low values indicate a porous bed and unstable inflow regime, high values indicate an outflow regime with steady water flow and unrestricted supply. On the basis of these, the two main sections of the Breitenbach are clearly distinct. The upper reach has a BFI from 0.42 to 0.45; the middle and lower reaches are stable, with a value around 0.7.

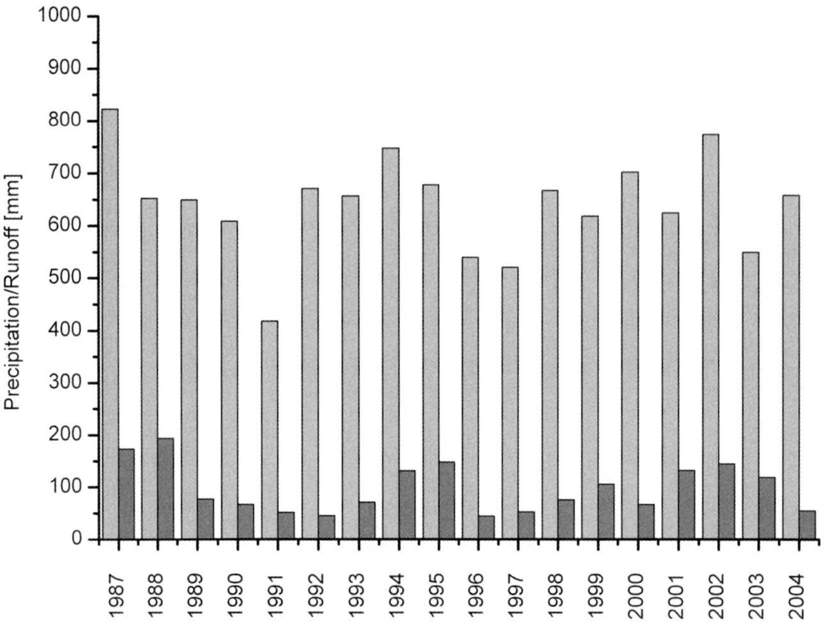

Figure 3.23 Precipitation (light gray) and runoff (gray) at the Breitenbach in the period 1987–2004.

Table 3.5 Base flow indices for some sections of the Breitenbach (Hisdal *et al.*, 2004; Eckhardt, 2008).

	Hisdal	Eckhardt
Upper reach	0.45	0.42
Source region T1	0.73	0.62
T2	0.74	0.64
T6	0.76	0.72

3.1.5.2 **Falling Limb**

The recession curve is an additional characteristic for the hydrological evaluation of a catchment area. The relationship between time and discharge is calculated using the formula from Linsley *et al.* (1975):

$$q_t = q_0 * K_r^t$$

where q_t is the discharge, t the time interval from q_0 and K_r the regression constant.

According to the calculation model and the collected data there are few differences in the calculation of the constant. An average value of 0.88 was obtained for

Table 3.6 Slope of the rising limb of eight flood events at the Breitenbach.

Rising limb		Time	Mean	Max
	$\Delta\, l s^{-1}$	(h)	$(\Delta\, l s^{-1})\, h^{-1}$	$(\Delta\, l s^{-1})\, h^{-1}$
1986 March	170	6	28.3	76.0
1988 March	143	10	14.3	37.5
1994 April	162	10	16.2	136.6
1995 January	199	3	66.3	216.4
1997 February	214	14	15.3	35.0
2001 February	130	12	10.8	20.7
2002 February	174	35	5.0	32.2
2002 December	193	9	21.4	54.9

the middle section of the Breitenbach, and a slightly higher value of 0.9 for the entire catchment area. This means that, with a recession constant of 0.88, water input of $200\,l s^{-1}$ over 7 days is reduced to half that volume of outflow. With *K* of 0.9, the same reduction in outflow requires one more day. The same relationship in the upper section reliably produced a constant of 0.85. Here the half-life was only 5 days.

3.1.5.3 Rising Limb

The rising edge of a spate is a very different. Table 3.6 shows individual values for selected flood events.

There can be considerable differences in the time for a spate to reach its peak. Even when the increase in inflow is comparable (mean of $170\,l s^{-1}$) the increase in outflow in January 1995 was $199\,l s^{-1}$ in 3 h, whereas in February 2002 it was $174\,l s^{-1}$ in 35 h. This represents about a 10-fold difference in the duration of build up of the spate. The reason for this is that each time different combinations of determining factors occur, that is, precipitation, temperature, water level and absorbance capacity of the soil in the catchment.

Figure 3.24 shows the typical input curve for a spate. As a rule, after a period of low water flow, often favored by frost, the water level rises several-fold in a few hours. In 1994 the flood water reached a rise of $136.6\,l s^{-1} h^{-1}$. One year later, heavy rain on frozen ground led to the maximum measured increase of $216\,l s^{-1} h^{-1}$. These were however extreme events; the average rise was lower, about $40\,l s^{-1} h^{-1}$.

3.1.5.4 Flow Duration Curve

The Flow duration curves (FDC, Figure 3.25) clearly show the irregularity in the occurrence of the various outflow values. In more than 50% of cases the outflow does not reach the $10\,l s^{-1}$ level. This curve is helpful in that it gives a good overview of the distribution of spates, which are ecologically relevant.

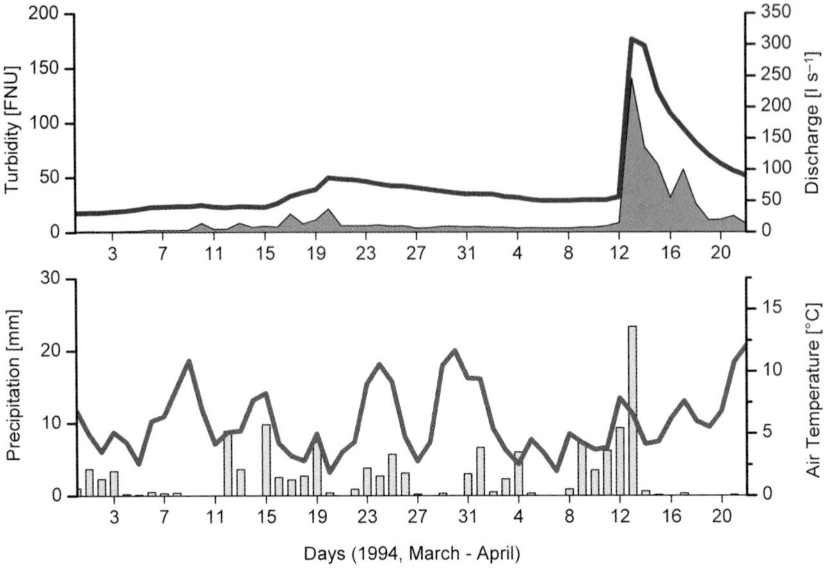

Figure 3.24 Typical course of a flood event at the Breitenbach. Example for March–April 1994. Lower figure shows daily precipitation sums (slim columns) and air temperature (line). Upper figure shows discharge (line) and turbidity (area).

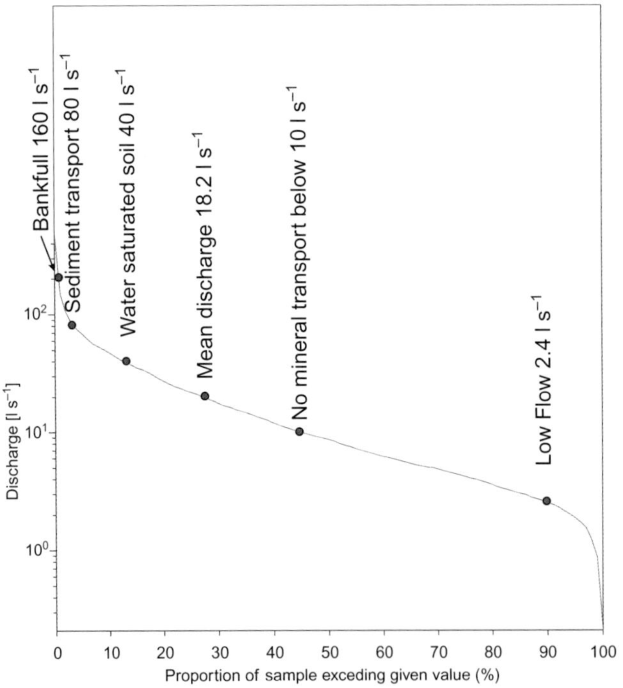

Figure 3.25 Flow duration curve for GT2.

Threshold value. The range 1 to about $500\,l\,s^{-1}$ outflow is divided into six classes on the basis of morphologically controlled characteristics:

$160\,l\,s^{-1}$: *Bank-full*: At about $160\,l\,s^{-1}$ or more the Breitenbach floods its banks in the main observation area. Additional increases in input result in no significant increases in the water column. Rather the stream spreads over the adjacent meadow.

$80\,l\,s^{-1}$: *High water*: On the basis of detailed observations along a transect beside T2 it was found that no significant sedimentation occurs from the draining water with input of $80\,l\,s^{-1}$. This corresponds with Schäffer's (1999) statement, based on his calculations of the risk of sediment transport for the Breitenbach. His results indicate that from about $80\,l\,s^{-1}$ the risk of sediment particle transportation increases up to 100%. On the basis of the uniform morphological effects outflow of more than $80\,l\,s^{-1}$ is described as high water.

$40\,l\,s^{-1}$: *Saturating inflow*: With input of 40–$50\,l\,s^{-1}$ and above everything indicates that the ground of the catchment has reached high water saturation and in the case of further precipitation direct in put into the water is likely.

$20\,l\,s^{-1}$: *Input average*: A mean value of $18.8\,l\,s^{-1}$ was obtained from long-term investigations. Statistically an input range of 15–$30\,l\,s^{-1}$ occurs most often in the frequency spectrum.

$10\,l\,s^{-1}$: *Lower limit of sediment transport*: On the basis of detailed sediment investigations, there is no significant transport of mineral material below $10\,l\,s^{-1}$.

$2.5\,l\,s^{-1}$: *Dry flow*: This is a hydrological term that describes a boundary below which the stream shows drought characteristics. After the FDC exceeds 90% measured values a low flow value of $2.4\,l\,s^{-1}$ is obtained, and from the mean annual minimum (MAM) in the area of T2 this value is $2.5\,l\,s^{-1}$ (Hisdal *et al.*, 2004).

3.1.5.5 **High Water**

Based on the above definition high water events occurred 34 times during the 20 years of investigation. Figure 3.26b shows clearly that high water events tend to occur in winter and spring. Exceptionally, the ecosystem was once surprised by a flood in June. To this irregularity belongs also the persistence of these events for almost two years (1992, 1996, 1997, 1998, 2000, 2004, 2005). Using the program AQUAPAK (Gordon *et al.*, 2004) the frequency of high water events from 1986 to 2005 were collated and the likelihood of their repetition calculated. An average likelihood of repeat high water occurring in the middle section of the year of seven months and six days was obtained. Thus, water input of about $80\,l\,s^{-1}$ or more is likely twice a year. However, based on our observations it is not simply the height of the flood but the ecological significance of the event that is important. The duration of the event often has more effect, in terms of long-lasting disturbance of the stream bed. The stability of the sediment is also dependent on how much time the vegetation has to penetrate the substratum with above- and belowground shoots (Table 3.7).

Observations from 1988 show that on 10 April the stream bed was cleared of all sediment of 2 cm diameter. It is also clear from the table how variable the size of the high water events can be. The $200\,000\,m^3$ spring high water of 1988 accounted

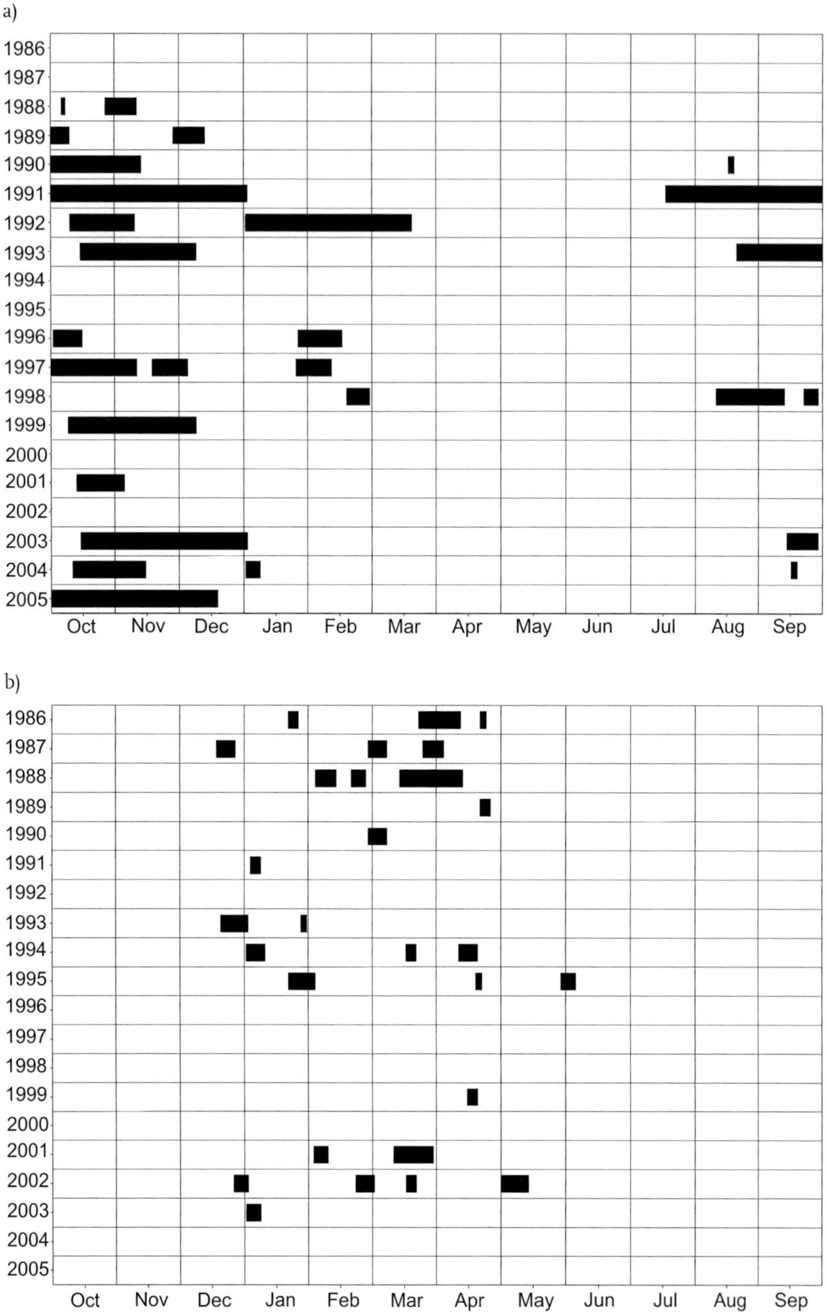

Figure 3.26 (a) Distribution of spells below $2.8\,l\,s^{-1}$ (low flow). (b) Distribution of spells above $80\,l\,s^{-1}$ (flood).

Table 3.7 Date and duration of floods, interval to the next flood and amount of discharge flow-through of the respective flood.

Date	Duration (days)	Interval (days)	m^3 (×1000)
15 March 1988	29	17	200
19 December 1993	22	323	152
11 March 2001	20	31	138
24 March 1986	19	58	131
1 May 2002	14	39	97
26 December2002	13	225	90
21 January 1995	12	275	83
25 March 1987	10	17	69
3 February 1988	10	39	69
11 April 1994	10	20	69
27 February 1990	9	66	62
28 February 1987	8	17	55
18 December 1987	8	126	55
22 February 2002	8	328	55
20 February 1988	7	7	48
30 May 1995	7	37	48
1 February 2001	7	652	48
8 February 1989	6	301	41
21 April 1989	6	66	41
15 April 1999	6	68	41
20 January 1986	5		35
18 March 2002	5	16	35
3 January 1991	4	301	28
18 March 1994	4	67	28
22 April 1986	3	10	21
27 January 1993	3	16	21
20 April 1995	3	77	21
23 February 1997	2	628	14
2 January 1987	1	252	7
10 February 1987	1	38	7
13 August 1987	1	131	7
22 December 1989	1	239	7
10 January 1993	1	734	7
5 February 1999	1	710	7

for a good one-third of the long-term average outflow and more than the lowest outflow volume of a whole year, while the winter high water which reached the 801 level with 7000 m^3 only matched one-thirtieth of the 1988 event.

3.1.5.6 **Bank-Full**

Periods during which the water input exceed 160 $l\,s^{-1}$ fall into the high water category. Based on the above definition, bank-full occurs more rarely, 12 times in 20 years. Statistically flooding of the Breitenbach meadows can be expected every 1.6 years.

3.1.5.7 Water Saturation of the Valley

When water input falls to 40–60 l s^{-1} the water-holding capacity of the adjacent meadows falls to a minimum and following precipitation events immediate changes in the water level must be anticipated due to direct runoff. Exceeding this threshold of about 40 l s^{-1} would be expected on 10% of all days over the 20 years.

3.1.5.8 Input Average

The frequency of exceeding the threshold on about 30% of all daily individual observations reveals the strong shift in the distribution of the individual values around the mean.

3.1.5.9 Low Water

Less spectacular, but still regular and, in contrast to the above described extreme events, described by absent dynamics are the periods of low flow. From the large choice of methods to identify a threshold value, exceeding which defines the outflow as low water outflow, the Q_{90} value (Hisdal *et al.*, 2004) is used here. While the calculation of this index frequently gives completely equivalent results, on the other hand water flow of 2.5 l s^{-1} hardly provides solid evidence of water movement. Significant current is only measurable at the measuring weir. The result of visual observation, and the calculations of the threshold value are congruent. As with high water, there are phases during which no dry period occurs (Figure 3.26a). Spring and early summer show this particularly. There are also years in which no dry period occurred: 1986, 1987, 1994, 1995, 2000 and 2002.

3.2
Attempt to Define Substratum Dynamics

Water flow dynamics are evident from the succession of various long phases of falling and rising waters. This succession of different flow regimes is the cause of morphological changes in the stream bed over time. These include not only the spectacular events, such as high or low water, but also the unremarkable, gentle structural changes that are the result of continual variation. If one was to observe the bed structure more precisely it is clear not only that the current and input relationships shape the bed, but also that the sediment itself has its own “memory”. The fact is that the remobilization of a stabile sediment aggregate requires higher hydrological force than is necessary for its transport. The following section describes an attempt to describe sediment dynamics in the studied area with the help of a “memory” function. The basis of the model is on the one hand the fact that changes in input lead to changes in substratum strength but always with the longer or shorter long-term properties must first be brought into consideration. For the calculations, the daily input mean must be checked, whether the next day’s value matches it, whether it is lower or higher, or whether a significant change in the outflow regime has occurred. The number of days with the same development would be separated by trend direction and added in two groups and recorded for the study period, usually a hydrological year (from October to September). The result is the total number of days per trend and counts for the number of similar episodes for both groups. The product of both the total day quotients (positive or negative) and the number of episodes (positive or negative) gives a measure of the outflow dynamic of a year. An extreme case would be an outflow year without a single increase in water input giving a value: $Vari = \frac{0}{1} * \frac{0}{365} = 0$. More frequently the falling water phase would be interrupted by raised inputs, for example, precipitation, thereby increasing the divergence of the value from zero. A value of 1 is mathematically possible, but hydrologically almost impossible. Comparison with the coefficient of variation (Lake *et al.*, 1985) is similar, but for example would not reveal the obvious differences in outflow regime between years such as 1998 and 1999 (Figure 3.27). This variability is an important factor that affects the habitat characteristics. In contrast to the coefficient of variation this estimate indicates not only a statistical instance, but also the above intimated detailed indirect measurement of the substratum modification dynamic. It is clear that the two measures of variability show parallel paths, but also contrasting sections. The usefulness of these measures will be further discussed in relation to ecological considerations.

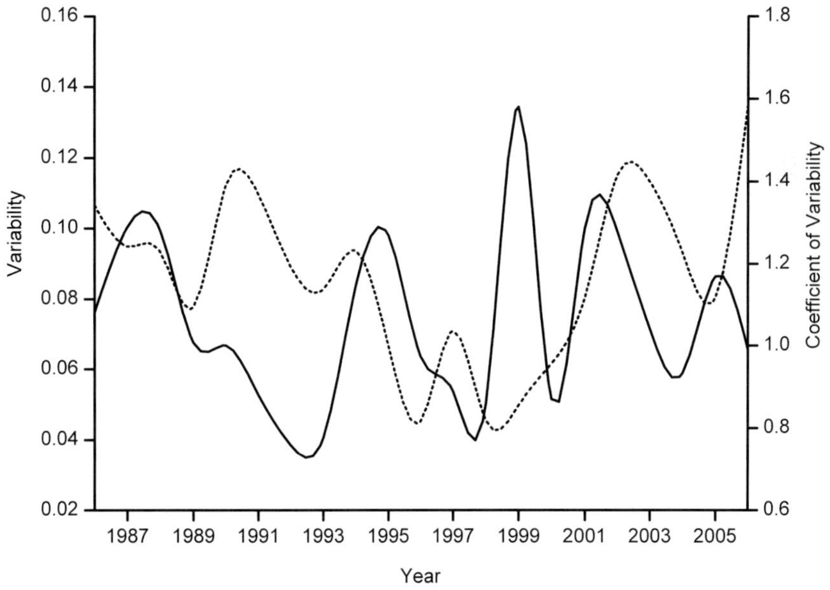

Figure 3.27 Attempt to characterize stream bed sediment structure from discharge variability (left y-axis, full line) versus coefficient of variability (dotted line).

Table 3.8 Ranges of morphological and hydrological measures.

	Upper	Middle	Lower	All
Cross-sectional area of flow (cm^2)	385	748	1175	765
Average width (cm)	43	60	105	69
Average depth (cm)	8	12	11	10
Wet sediment area (m^2)	529	1637	1265	3273
Current speed ($cm\,s^{-1}$)	20.7	18.5	19.7	19.5
Froude number	0.253	0.188	0.201	0.213

3.2.1
Cross-Section Measurements

From 1997 to 2004 regular monthly stream cross-sections analyses were undertaken at sampling sites SU1 to SL2. Three cross-section profiles were made at 1 m intervals at each sampling site (Table 3.8).

The following results were obtained. The Breitenbach has an average depth of 10 cm. The shallowest value was 1.3 cm, the deepest more than 28.7 cm. The latter value was under normal water levels whereas during high water periods depth was estimated at 60–70 cm, although this was not encountered during the sampling run. The average width of the stream is 69 cm, only 43 cm in the upper reach, 60 cm in the middle and 105 cm in the lower reach. That gives an average through-flow channel of 765 cm^2 (upper reach 385 cm^2, middle 748 cm^2 and below T5 1175 cm^2). Based on these data it is apparent that the stream profile is shallower below T5. The breadth:depth ratio above T5 is 5.2, whereas downstream it rises to almost 10 (9.6). The current speed does not change markedly and is on average 20 $cm\,s^{-1}$. An average Froude value of 0.213 indicates a basically laminar current flow. A maximum value of 1.5 was measured once, against a minimum of 0.01.

3.2.2
Wet Stream Area

Multiplying the wet area by the stream length estimates the order of magnitude of the colonizable stream bed (Figure 3.28). The value also provides an estimate of the area of wet stream bed relative to water input. For a total length of 3711 m, the maximum colonizable stream area is 6137 m^3, with a mean of 3420 m^3, and for low water 2291 m^3. It is interesting that when water input changes by two orders of magnitude, the colonisable stream areas changes only by a factor of 2. The reason for this relationship is the typical shape of the stream profile. With increasing water input, it is primarily water depth that changes. The stream cannot expand laterally. The wet area does not therefore increase to the same extent.

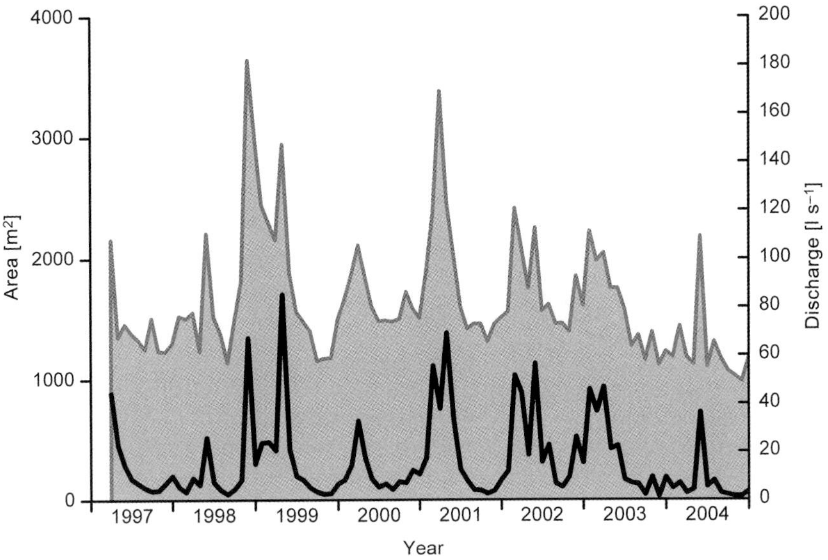

Figure 3.28 Variation of stream discharge (line) and colonizable area (area) on the stream bottom 1997–2004.

3.2.3 Water Temperature

3.2.3.1 Water Temperature in the Sediment

Water temperature was measured 30 cm deep in the sediment at T2. The average temperature of 8 °C was slightly under the surface water temperature. In February 2006 the minimum temperature was 0.22 °C, and a maximum temperature of 14.3 °C was measured in August 1990. During the course of a year, sediment temperature reaches its maximum in the middle of August, somewhat later than the water temperature. The minimum temperature behaves similarly (Figure 3.29). However the diurnal sediment temperature curve is quite different. With respect to the typical variation in global irradiance in July, the temperature at 30 cm deep follows with a delay of about 11 h. While, over the course of a year, the amplitudes of water and sediment temperatures are almost the same, in July water temperature shows an average daily range of 3 °C, while sediment temperature hardly shows a long-term daily variation profile, only about 0.4 °C.

3.2.3.2 Water Temperature in the Stream

Water temperature is influenced by a complicated combination of factors. In the region of the spring, to which the larger part of the study area belongs, the influence of the generally stable groundwater (and sediment) temperature is particularly strong and evident. As variables that have a significant effect on the fluctuation of water temperature, air temperature, global irradiance and discharge

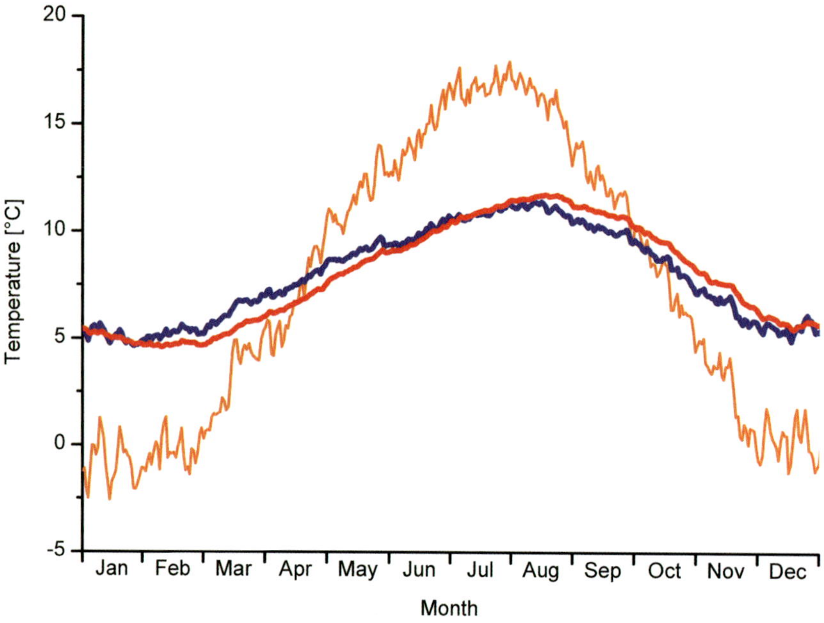

Figure 3.29 Seasonality of air (orange) and water temperature in the stream (blue) and in the bottom (red).

should be considered. Over the course of a year, a pair of extreme situations strikes the observer: in spring the stream bottom is exposed and usually, when unprotected by high water, is open to irradiance and air temperature. While irradiance and air temperature affect water temperature in a similar way with a diurnal rhythm, the water body has a buffering effect. This buffering capacity is reduced in late summer by the steadily decreasing discharge, but is partly offset by the macrophyte cover of the stream course. Neither irradiance nor air temperature have direct contact to the water body. In extremely cold winters it is possible that, with low water, the surface of the water has already frozen by T2. This is a situation that seldom occurred, but that underlines the outstanding role of discharge on the course of the water temperature. The influence of precipitation on water temperature will be discussed more below in conjunction with the minimum temperature at T1.

Water temperature has been recorded along the length of the stream since 1969: from 1969 to 1985 with mechanical drum recorder from Firma Lambrecht, Göttingen; from 1985 with an automatic measuring station using PT 100 sensors at T2 and T6, additionally at T1 from 1990. Digital daily values are available for T2 for the time from 1969 to 1985, for individual years for the other study sites, T1, T4, T5 and T6. Thus for T2, water temperature measurements are available for a period of 37 years. For 1969 to 2006 the long-term average water temperature was 7.99 °C. The lowest water temperature, –0.70 °C, was measured in December

Table 3.9 Mean, minimum and maximum water temperature (°C; 1990 to 2004) at sites T1, T2, and T6.

Site	Mean	Minimum	Maximum
T1	8.0	2.6	14.3
T2	8.3	−0.7	17.6
T6	8.7	−2.0	20.1

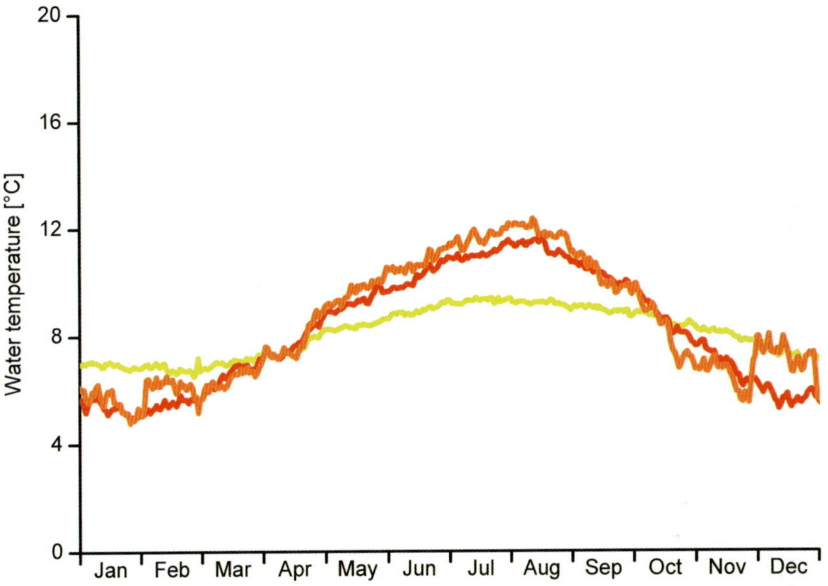

Figure 3.30 Yearly course of water temperature at sites T1 (yellow), T2 (red), and T6 (orange).

1991, while the water temperature in the middle stretch of the Breitenbach reached a maximum of 17.6 °C. Both extremes were the result of low water. Over the total period of 37 years there has been a measurable temperature rise of 0.851 °C, which represents about 0.023 °C per year. The measuring uncertainties, particularly for the period 1969 to 1985, do not permit any further statistical guarantee of the results. A reproducible standardization of the mechanical recorder better than ±0.3 °C was hardly possible (Table 3.9).

Figure 3.30 shows the average course of water temperature at T1, T2 and T6. As expected, the annual curve for T1 has the flattest path. The leveling influence of the spring water and the narrowness of the valley limit the maximum measured value to 13.9 °C. The long-term minimum is 4.1 °C. The absolute minimum of

2.6 °C was obtained in December 1993, after the upper reach experienced an increase in discharge of $87\,l\,s^{-1}$ within 3 h, as a result of heavy precipitation (32 mm in 24 h). The cold precipitation mixed with the similarly strongly increased volume of spring water. It was concluded that the sudden increase in spring water discharge was also primarily due to the contribution of cold surface water. The average annual value is 8.0 °C, close to the spring water temperature of about 7.8 °C. About 500 m downstream at T2, the average temperature rises to 8.3 °C and at T6 reaches a value of 8.7 °C. Here the annual amplitude rises to 20.1 °C. At T2 and T6 the water temperature reaches its maximum at the beginning of August, whereas the lowest temperature was always experienced in February.

Over the course of a day the water temperature follows the photoperiod. Figure 3.31 shows an example of average July water temperature values together with the variation in air temperature. The phase shift between the irradiance maximum and the highest daily temperature is 1–4 h. The maximum air temperature follows the height of the sun with a one hour delay. Water temperature at T2 reaches its daily maximum 2 h after the maximum global irradiance, the maximum is delayed by 3 h at T1, and the highest water temperature at T6 follows irradiance with a 4-h delay (Figure 3.32). The actual water temperature depends upon temporal and spatial factors: time of day and season, as well as proximity to the spring or mouth of the stream. The important relevant factors are air temperature, global irradiance, valley orientation and discharge, but also, as described above, precipitation. Air temperature and global irradiance fluctuate with daily and sea-

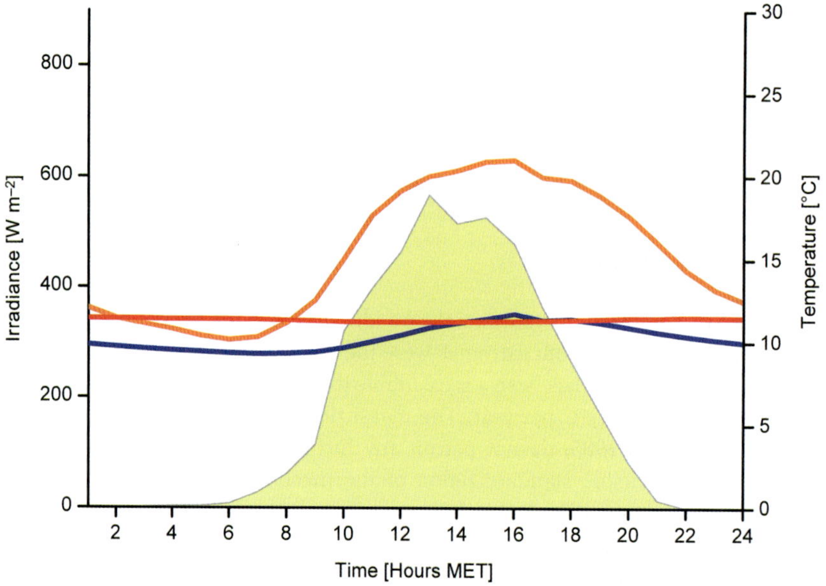

Figure 3.31 Daily course of irradiance and temperatures of air (orange), water (blue) and water in the stream bottom (red) in July at the Breitenbach.

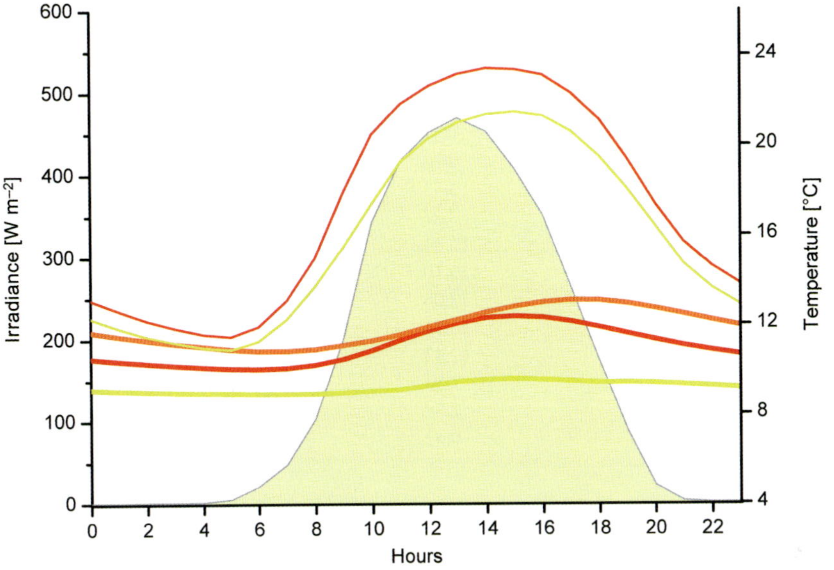

Figure 3.32 Phasing of selected temperatures to irradiance. Water temperature at T1: thick yellow line. Water temperature at T2: thick red line. Water temperature at T6: thick orange line. Air temperature at T1: thin yellow line. Air temperature at T2: thin red line.

sonal rhythms, the effect of valley orientation is topographically determined, varying with changing solar altitude, variation in discharge responds to known seasonal patterns, but by no means to the same extent as global irradiance and air temperature.

In order to determine the influence of individual factors from different combinations, correlation tables of air temperature, water temperature, global irradiance and discharge were compiled for different seasons. Three scenarios were tested: low, intermediate and high flow over the course of the year. The boundaries of the particular categories under low, intermediate and high water were established according to the significance of the correlation coefficient (Spearman), optimized for the best correlation between water temperature and discharge. The following distinctions were used: low flow (LF) – less than $6\,\mathrm{l\,s^{-1}}$ ($N = 2637$), high flow (HF) – more than $20\,\mathrm{l\,s^{-1}}$ ($N = 1960$), intermediate (IF) – 6 to $20\,\mathrm{l\,s^{-1}}$ ($N = 2472$).

In order to have a sufficiently large number of cases for the correlation analysis, months were paired on the basis of their corresponding average monthly temperature (Table 3.10).

This gave the following pairs: January/December, March/November, April/October, May/September and June/August.

It was tested whether the trio, air temperature, global irradiance and discharge, on the one hand influenced the daily average, or on the other hand influenced the daily amplitude of water temperature. The correlations for low and high flow were

Table 3.10 Mean monthly air temperatures (t) at T2.

	t
January	–0.42
February	0.22
March	3.53
April	7.27
May	12.34
June	15.10
July	17.05
August	16.54
September	12.43
October	7.84
November	2.88
December	0.42

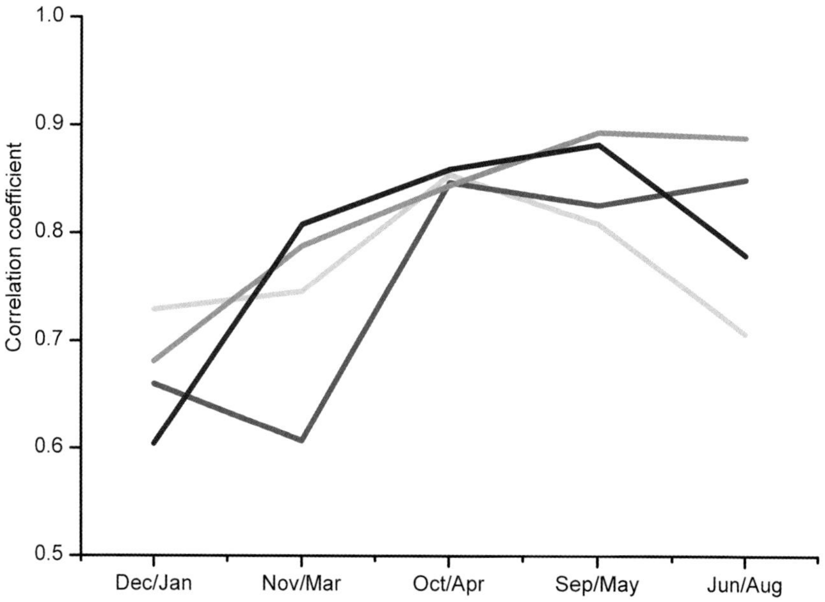

Figure 3.33 Correlation coefficients (Pearson) for daily mean and range at low and high flow between air and water temperature in pairs of months.

particularly considered for the analysis. With only a few exceptions, statistical significance was shown to be high ($P = 0.01$). The results of the intermediate flow conditions were rarely such clear values obtained and thus, as transitional scenarios, were not considered further.

As expected, throughout the year air temperature controlled the daily average and amplitude of the water temperature curve. Figure 3.33 shows the results. The

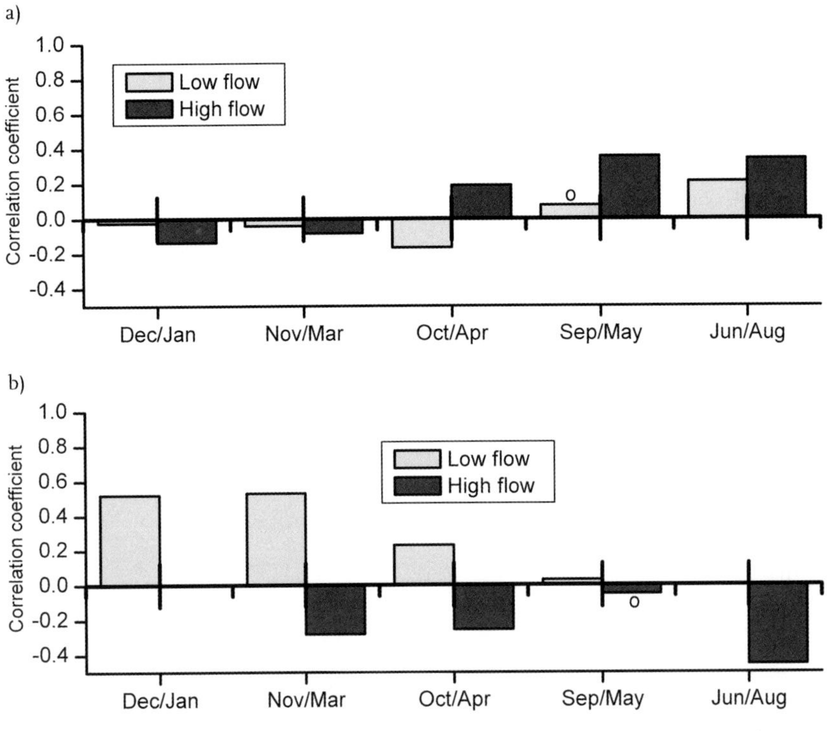

Figure 3.34 Correlation coefficients between irradiation and water temperature (a) and between discharge and water temperature (b) under conditions of high and low flow in particular pairs of months ($P = 0.01$, *o* = not significant).

correlation coefficients, close to 0.8, are in general highly significant ($P = 0.01$). Slight differences in the daily average values emerge in winter.

The results of tests of the relationship between global irradiance and water temperature show two clearly different characteristics (Figure 3.34a). The daily averages are negatively correlated –0.16 (HF) in winter. This relationship changes with increase in daily irradiance and reaches a maximum of 0.34 in summer under moderate and high water (HF). The point at which the correlation changes from negative to positive occurs in the months April/October for HF, and May/September for LF. In contrast, there is a clear positive correlation between the daily amplitude of water temperature and global irradiance (Figure 3.35a). The effect appears somewhat stronger in spring and autumn than in summer and winter.

The relationships between water temperature and discharge are more complicated. Throughout almost the entire year, there is a negative relationship between moderate to high flow with average water temperature (Figure 3.34b). There is no clear finding for the winter months December to February. The relationship reverses with low flow. In winter months there is a positive correlation, which

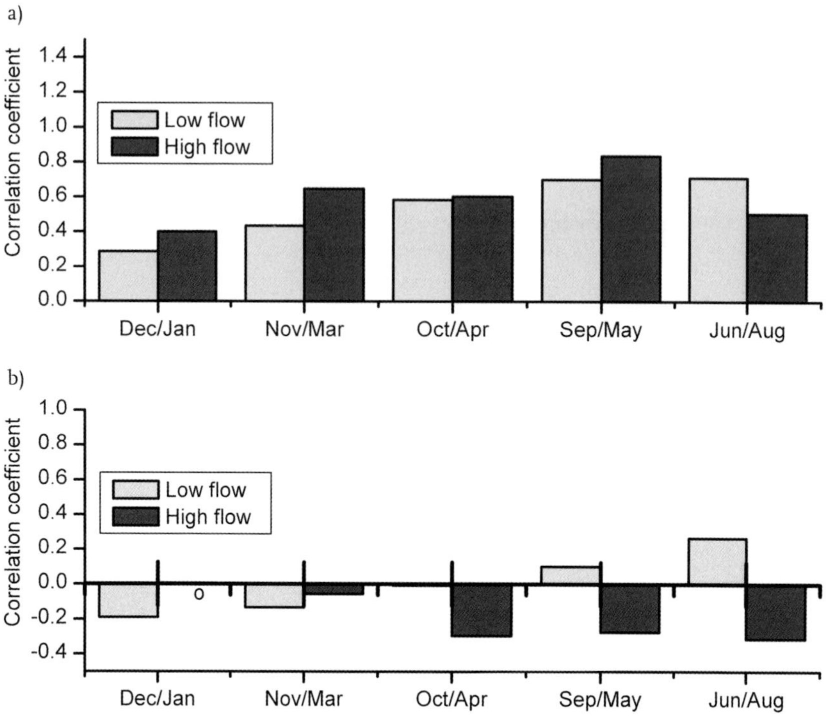

Figure 3.35 Correllation coefficients between irradiation and daily water temperature amplitudes (a) and between discharge and daily water temperature amplitudes (b) under condition of high and low flow in particular pairs of months ($P = 0.01$, o = not significant).

diminishes in summer to a neutral value. The highly negative correlation is a result of the high buffering capacity of the water body under moderate and high water. With low water the flowing current loses its leveling roles and cannot counteract the effect of air temperature on water temperature. As with the moderate flow, high discharge similarly shows a negative correlation with the daily amplitude of water temperature (Figure 3.35b). A clear effect is lacking in winter. In contrast, the relationship between water temperature and low water is different. There is a clearly negative value in winter, April/October is transitional, and there is a clear positive relationship in summer. The buffer capacity of the water body also plays an important role in the daily amplitude of water temperature.

Influenced by the sinus-shaped annual temperature curve (Ward 1985), all measurements can be expected to show a double peak within any year. This phenomenon is more or less pronounced (Figure 3.36) according to the differing statistical spread of measurements with medium (i.e., water/air) or topology. In our studies a clearly symmetrical double peak is seen at T2. The interesting aspect of this distribution is the transition region of the peak temperatures of 6° and 11°, as well as values from 7° to 8°. On an hourly basis this temperature regime is

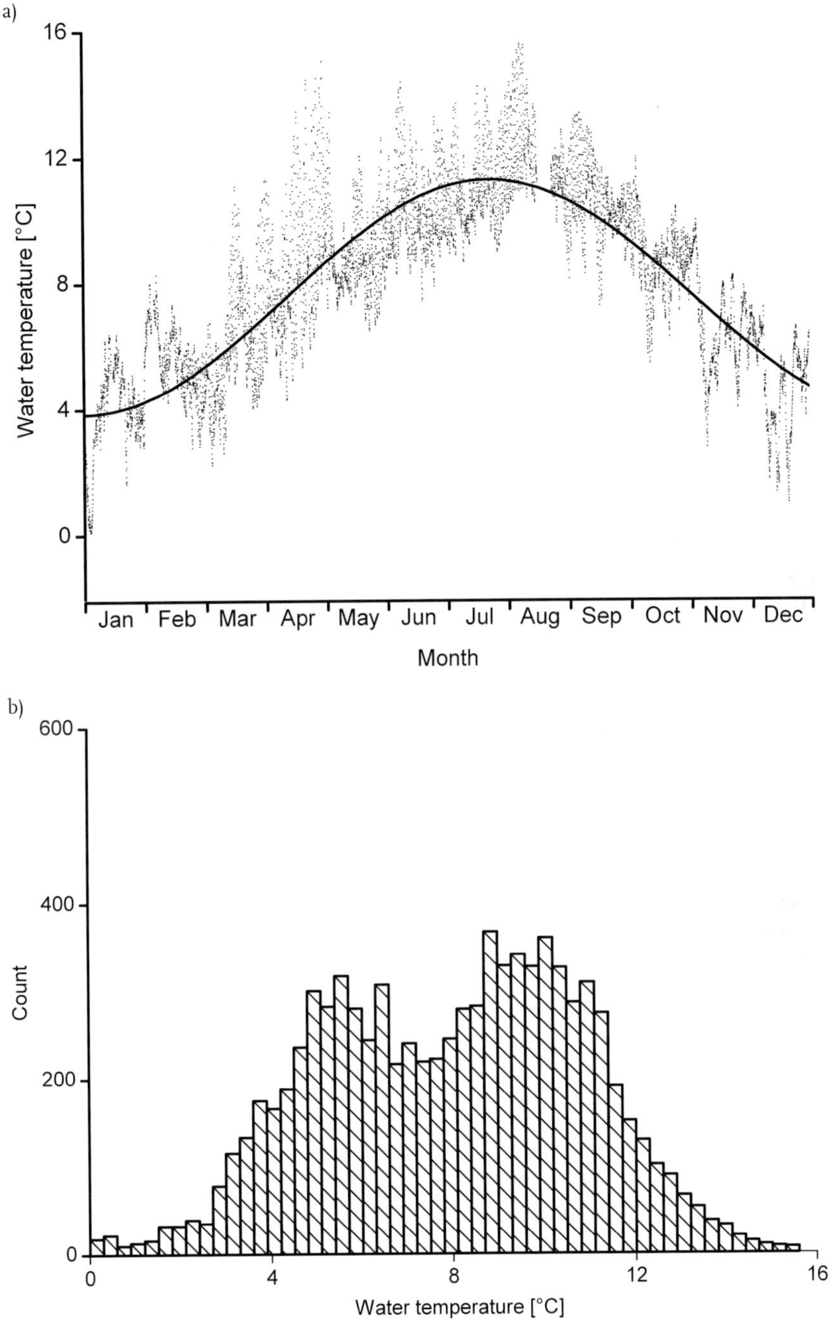

Figure 3.36 (a) Sinoidal fitting of water temperature course at T2. (b) Distribution plot of water temperature readings at T2.

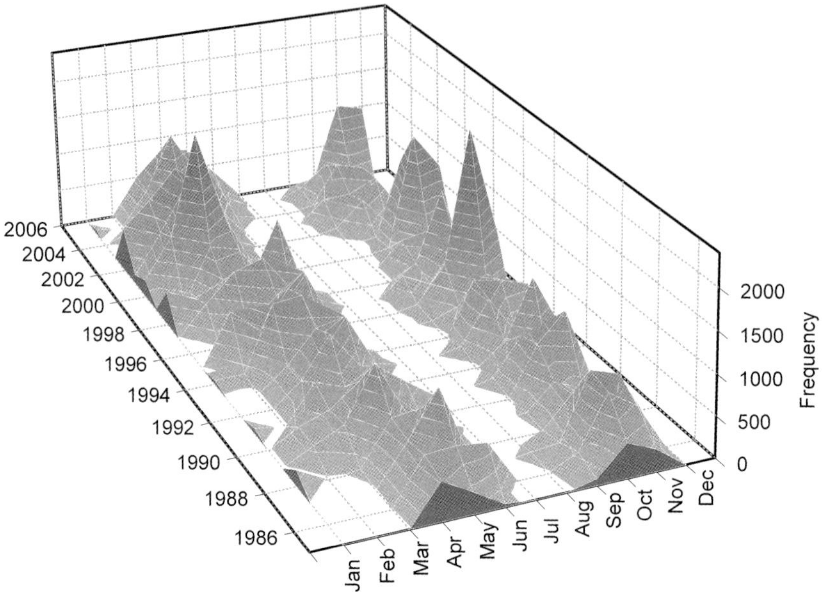

Figure 3.37 Montly frequencies of water temperature measures between 7 °C and 8 °C, 1986–2006.

much more rarely recorded both in summer and winter. In the month of July, out of 16 268 hourly measurements over 20 years, this interval has apparently only been recorded twice. Figure 3.37 provides an overview. A similar finding occurs for January, although only 1588 values were recorded for this period. That a temperature of 7.8 °C matches the constant spring temperature, it is assumed that, apart from the transitional seasons, spring and autumn, this base temperature is either rapidly exceeded in summer, or quickly decreased in winter.

3.3 Chemistry

Since 1986 the measuring program included a series of important chemical and physico-chemical parameters. These include pH, buffering capacity (ABC), oxygen content and electrical conductivity. In addition, mostly at monthly intervals, nutrients such as phosphorus as soluble reactive phosphorus (phosphate), nitrogen as ammonium and nitrate, and silicate were analyzed. Since 1991 the cations sodium, potassium, magnesium and calcium and the anions, bicarbonate, chloride and sulfate have been included. Investigations of the diurnal changes in these dissolved components were also performed.

3.3.1
Buffering Capacity (ABC – Acid Binding Capacity)

During the course of the year, buffering capacity is particularly influenced by flow regime and photoperiod. In the Breitenbach values from 0.15 to 0.51 mmol(eq) l^{-1} were obtained, and discharge has a significant effect on the fluctuations. The lowest values were obtained during high water, in drought periods the highest values reached 0.5 mmol(eq) l^{-1} and above (Figure 3.38). This is analogous to the strong, similarly discharge-dependent, fluctuation in pH. The dependence of the ABC value on discharge regime is however more conspicuous.

For further consideration of the long-term development of binding capacity, ABC values were transformed against a constant reference value of 20 l s^{-1} (long-term discharge average) using the compensation function, $abc = 0.1863 + 0.2364\, e^{(-0.08\, q)}$, where q = discharge. The resulting curve is shown in Figure 3.39. After excluding the residual seasonal periodicity, there is a negative trend whose statistical strength must be tested. However, because the data set is not homogeneous, the use of statistical methods is not worthwhile. Until 1993 a series of methodological changes were implemented, which cannot be fully compensated by comparing the methods. However, two time periods (1987–1993, 1994–2007) can be tested.

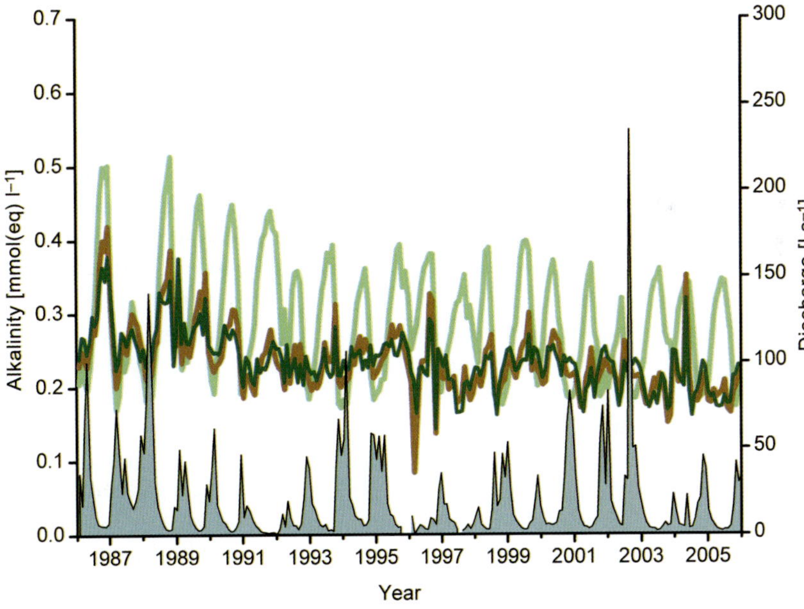

Figure 3.38 Long-term (1986–2006) development of alkalinity in relation to discharge at the Breitenbach. Discharge: light blue area. Alkalinity: light green line. Alkalinity compensated by discharge: brown line. Alkalinity compensated by discharge and seasonal components: dark green line.

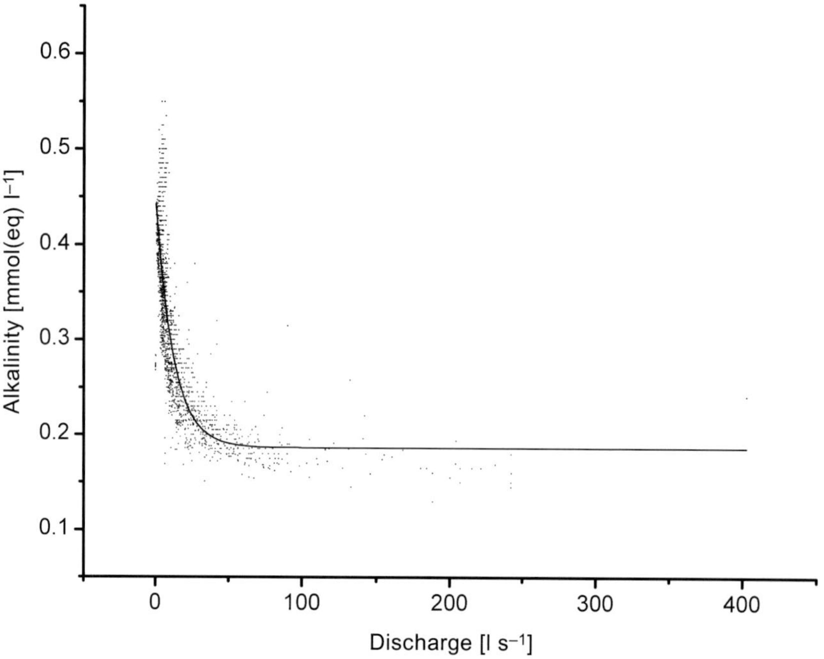

Figure 3.39 Dependence of alkalinity on discharge in the Breitenbach with compensation function abc = $0.1863 + 0.2364\ e^{(-0.08\ q)}$, where q = discharge.

For the period 1986–1993, a significant negative development ($P = 0.034$) is shown by the Mann–Kendall test. This trend cannot be statistically proven for the following period, 1994–2006. As a result of these investigations it can be concluded that, visually, there is a slight trend towards lowered buffering capacity, although this cannot at present be verified.

In the light of the above-described methodological uncertainties, no clear statement on whether there is increasing acidification (loss of buffer capacity in the catchment) can be made. The increase in bicarbonate values from 1993 to 2005 indicates the same trend.

The question of the mechanism, which is the basis of the observed fluctuations remains to be proved. Investigations of the physico-chemical relationships of the pore water in the Breitenbach sediments compared with the stream water over a year have shown that, with increase in sediment strength, the interstitial water samples from the deeper levels with lower oxygen concentrations always show higher ABC values. Values of 2 mmol (eq) l^{-1} were sometimes obtained. It seems that, according to the penetration of the moving water body at the junction between sediment and flowing water, there is a balance in the exchange of buffer capacity. The link between interstitial and flowing water is clear from abrupt discharge events, e.g., in late autumn 1988 (Figure 3.40). Short-term discharge increases

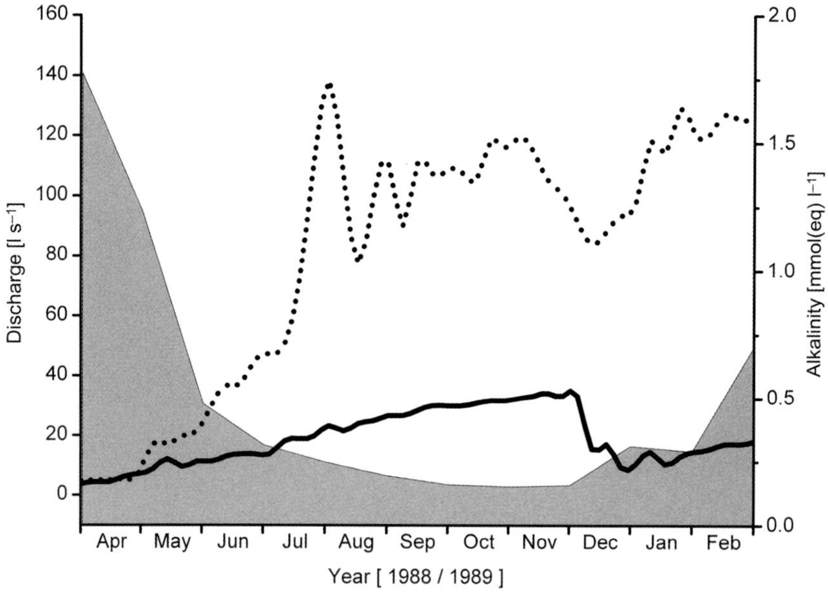

Figure 3.40 Development of alkalinity 1988/1989 dependent on discharge (area). Alkalinity of stream water–full line. Alkalinity of pore water–dotted line.

disturb the uppermost sediment layers, and the alkalinity is briefly reduced and then re-established as a result of the exchange processes.

The carbonate system is considered an important component of the buffering system. Analyzed values confirm a close link between ABC values and the bicarbonate concentration, although there are considerable differences in the contribution of bicarbonate to the ABC value. The range varies along the stream from 55% in the upper reach to almost 95% from the spring tributaries (Figure 3.41). The upper reach (SU1, SU4; Figure 3.48) carries water with very low buffer proportion. The increasing difference between ABC and bicarbonate concentration reveals that, with distance from the spring surface, the flowing water is increasingly influenced by the interface of surface water and interstitial water. In this section, in addition to the carbonate system, a series of organic and inorganic reactions come into play for the buffering capacity (Hildrew, 1995).

The diurnal fluctuations revealed while monitoring daily events can clearly be linked to observations on photosynthetic activity and will be later discussed in this context.

3.3.2
pH

The pH of the flowing current in the area of T2 varies between 6.5 and 7.8. Near the spring tributaries it is somewhat lower. pH increases slightly towards the lower

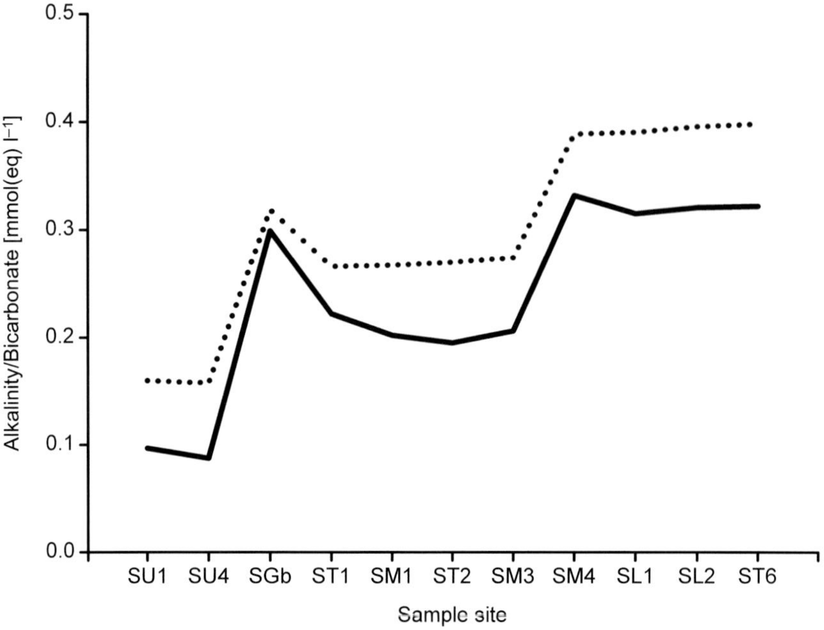

Figure 3.41 Bicarbonat concentration and total alkalinity at different sites (compare Figure 3.47) along the Breitenbach.

reaches of the stream. Over the course of the year important relationships between pH, alkalinity and discharge are apparent (Figure 3.42a,b). Spate clears out the buffering streambed sediment and lowers the pH of the flowing water. Increasing sedimentation during the summer leads to the restitution of the buffering capacity and via the exchange process on the sediment surface to a rise in pH. The close relationship between ABC and pH produces a discharge-influenced pH curve (Figure 3.43). Beside the buffering-controlled annual rhythm, pH is subject to irradiance-dependent diurnal variation. The fluctuation in pH as an indirect measure of CO_2 also reflects the carbon uptake mechanisms of photosynthetically active plant cells. Thus photosynthetic events can be very precisely followed in the flowing water. Above all, the parameters that are involved in this are pH, conductivity and oxygen content. These interrelationships will be discussed alongside the explanation of fluctuation in oxygen content.

3.3.3
Oxygen

Continuous measurements are available since 1986. The recording interval of 1 h provides a good picture of the diurnal fluctuations. Breitenbach water contains 85–100% dissolved oxygen, and over-saturation can occur briefly on summer

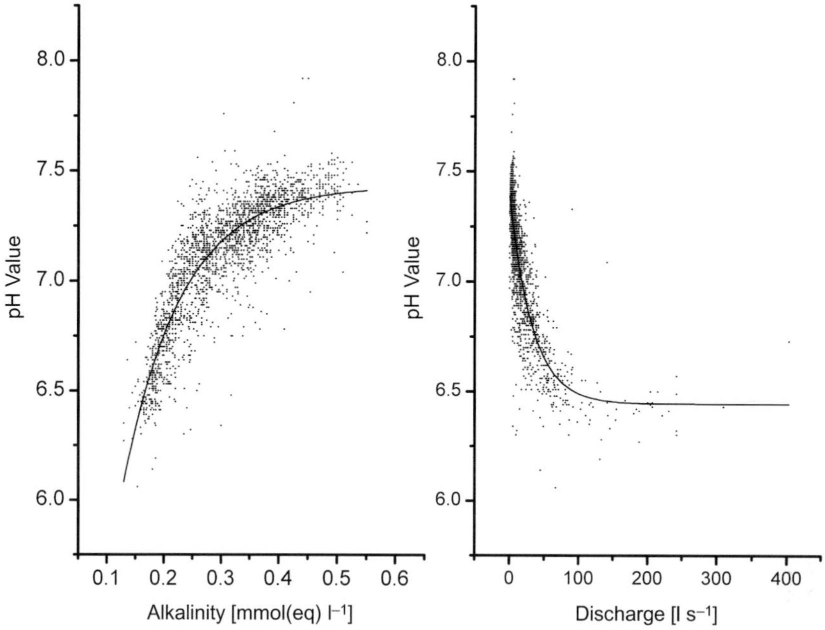

Figure 3.42 pH-value against alkalinity (left) and discharge (right) at the Breitenbach.

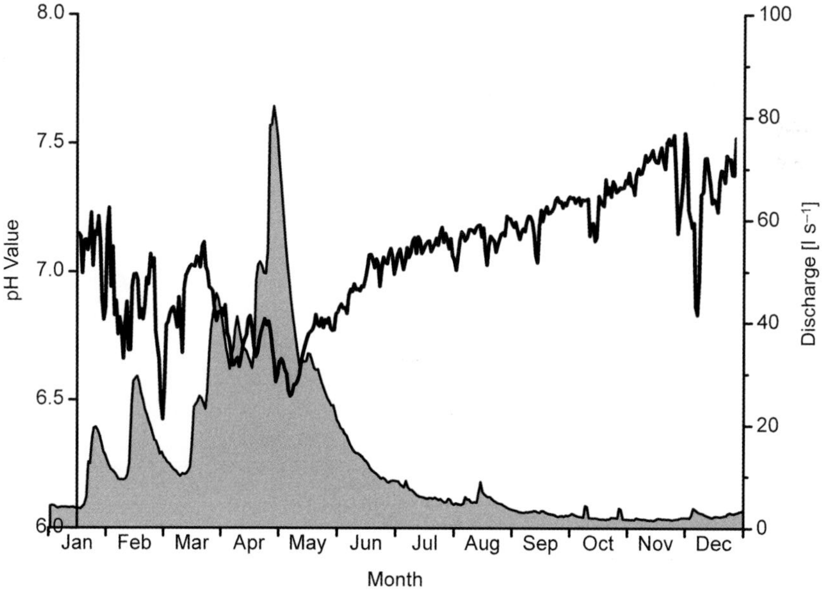

Figure 3.43 Courses of discharge (area) and pH (full line) in a year.

afternoons. On average diurnal fluctuations span 1 $mg\,l^{-1}$ oxygen, but occasionally in summer months the amplitude is up to about 2 $mg\,l^{-1}$.

The close relationship between water temperature and oxygen solubility is shown in standard tables (ISO/DIS 5814). Concentrations in the Breitenbach are between 8 and 14 $mg\,l^{-1}$ O_2. Figure 3.44 shows the typical course of irradiance, water temperature and oxygen. A summer day (24 July 2004) was chosen for this graph. With the increase in irradiance in the morning the oxygen content of the water increases, but the increase reaches a maximum at about 10 $mg\,l^{-1}$ in the morning and then, despite increased irradiance, falls. The daily minimum in the oxygen curve is reached in the evening. This curve closely resembles the water temperature versus 100% saturation curve (Figure 3.44). Against that, the saturation curve follows the irradiance intensity. This curve is typical for the Breitenbach and also observed, if somewhat weaker, in the gloomier times of year. Odum (1956a) described synchronous change in both oxygen curves for flowing waters. However that does not fully agree with the situation in the Breitenbach. The reason for this rests with lower primary production. In 1990 and 1991 preliminary investigations were performed to estimate primary production. The results reveal the

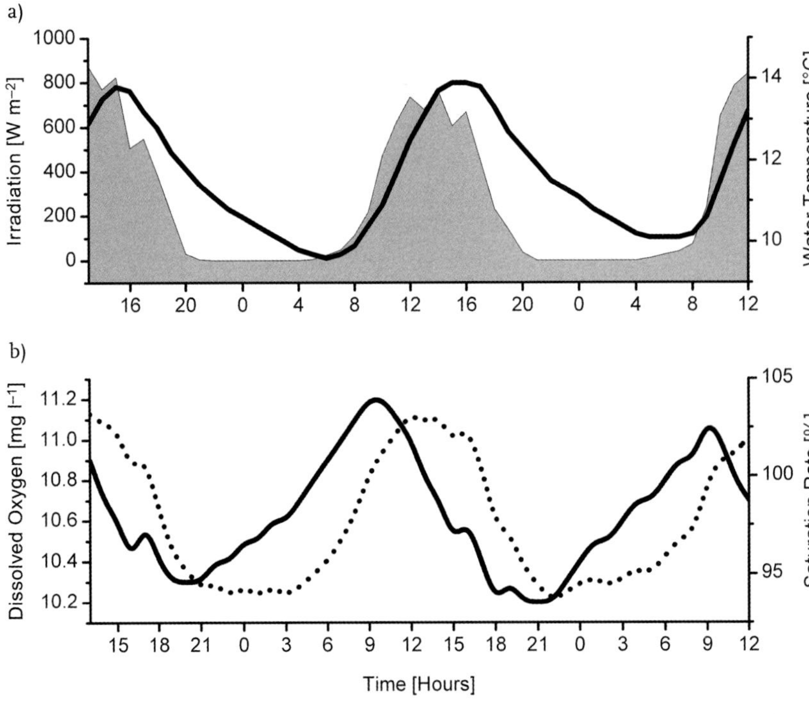

Figure 3.44 Course of the day for: (a) irradiation (area) and water temperature (line) and (b) dissolved oxygen (full line) and oxygen saturation rate (dotted line) on a summer day at the Breitenbach.

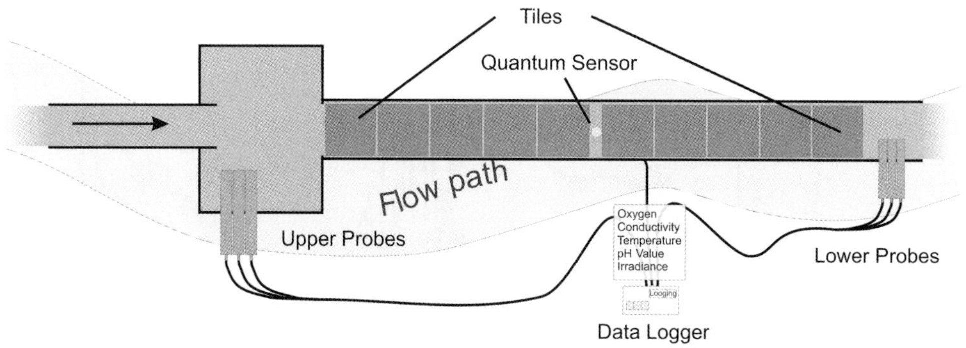

Figure 3.45 Construction of a channel system to estimate primary production in the Breitenbach (arrow indicates flow direction).

response of the oxygen-water-system to higher production. As an example a bright July day in 1991 was chosen.

Twenty stone tiles were conditioned in the described channel construction for three weeks. After this the tiles were heavily overgrown by an algal turf. Macrozoobenthos were largely excluded. Instruments to measure oxygen, pH and water temperature monitored the water conditions in the supply tank and below the stone tiles at the outflow. A datalogger recorded irradiance, oxygen concentration, pH and water temperature at short intervals (45 s; Figure 3.45).

Both oxygen curves followed the saturation curve of the morning and fell slightly with falling temperature overnight (Figure 3.46). As irradiance increased during the morning the curves diverged. The input curve was flatter and due to the temperature rise had already reached a maximum of $7\,mg\,l^{-1}$. Gaseous exchange processes modified the oxygen concentration. The outflow curve rose further with increasing irradiance and reached its peak towards $10\,mg\,l^{-1}$. There was no further increase in oxygen concentration here either, despite the increased irradiance. During the morning and afternoon the channel was occasionally darkened, once for one hour, similarly for half an hour. Changes in the measurements occurred immediately, that is, the reaction occurred within the first measurement interval of 45 s. Clouds are a natural cause of such events. Water temperature hardly changed during this shading. During the darkened phase the oxygen curve dropped to precisely the level of the input curve. There was a slight difference in the speed of equilibration between pH and oxygen concentration. Unfortunatetly there were no further measurements of the characteristics of the test substrata. In this context it should only be noted that, in the same water body, the additional oxygen production from the channel substratum had shifted the peak by 2 h. The variation in oxygen concentration shown by this investigation allows us to conclude that the running water in which Odum made his measurements had high production and respiration rates, which does not match the Breitenbach (Figure 3.47).

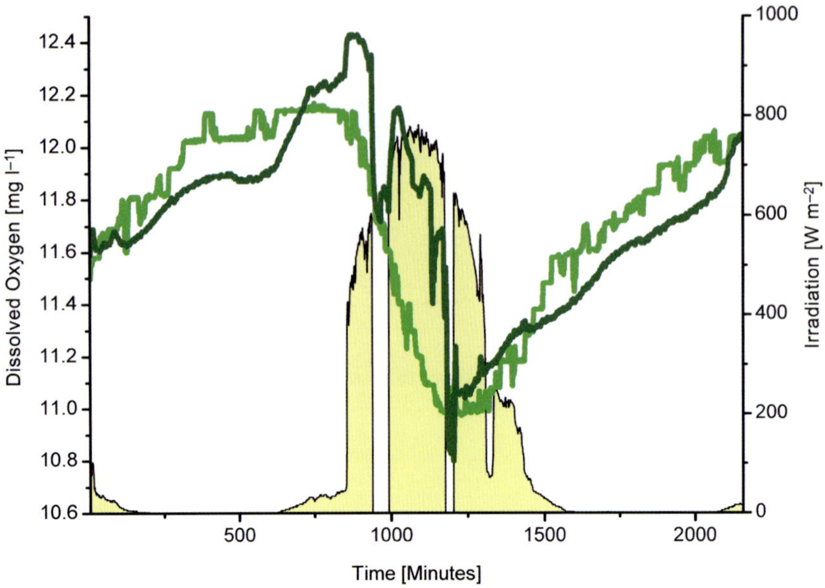

Figure 3.46 Effects of irradiation on primary production on amounts of dissolved oxygen. Irradiation: coloured area. Upper oxygen probe: light green line. Lower oxygen probe: dark green line.

3.3.4 Soluble Inorganic Elements

3.3.4.1 Methods

The analysis of nitrate, ammonium, orthophosphate and silicate were performed using a continuous flow method following the Deutsches Einheitsverfahren. Samples for chloride, bicarbonate and sulfate were analyzed by ion chromatography (Waters), as were dissolved sodium, potassium, magnesium and calcium.

The following discussion relates to a series of measurements from 1997 to 2005 from 11 sampling stations distributed along the stream. The samples were taken monthly. Figure 3.47 shows the position of the sampling sites. The nomenclature follows the above used scheme with a prefixed S. The letters U, M, and L stand for upper, median and lower reach. Table 3.11 provides an overview of minimum, mean and maximum concentrations dissolved nutrients determined in the upper, middle and lower stream sections.

3.3.4.2 Nitrogen

Combined nitrogen was routinely measured as ammonium and nitrate, the dataset comprising monthly samples. In addition investigations into diurnal variation were performed. Results from the flowing water were supplemented by sediment investigations in 1988/1989.

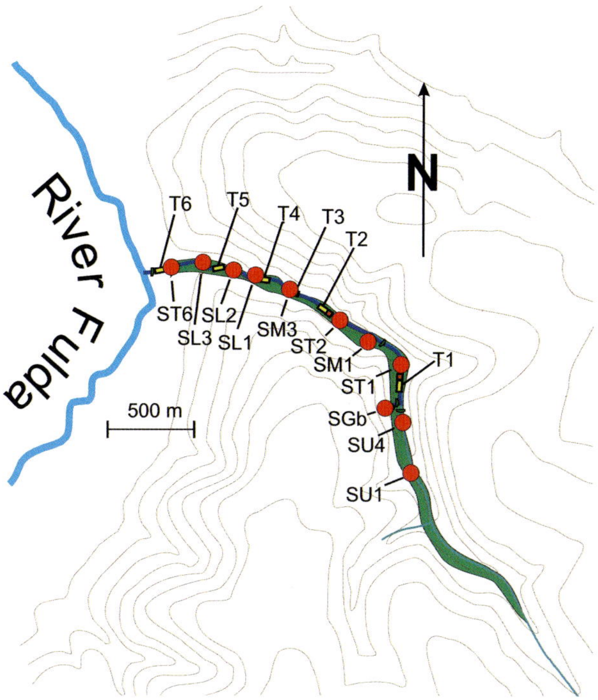

Figure 3.47 Plan with positions of sampling sites (with short codes) for dissolved nutrients along the Breitenbach. For comparison positions of emergence traps (T1–T6) are also given.

Table 3.11 Mean, Maximum and minimum concentrations (mg l^{-1}) of dissolved nutrients in the upper, middle and lower parts of the Breitenbach.

Reach		Ammonium	N nitrate	N SRP	P chloride	Cl sulfate	Silica Si	Ca	Mg	Na	K
	Max	0.055	1.6	0.194	8.9	16.2	6.9	17.9	4.6	7.6	5.7
Upper	Mean	0.008	0.8	0.041	7.1	12.9	3.4	14.4	3.6	5.0	2.4
	Min	0.001	0.1	0.009	5.2	9.0	0.7	10.7	2.6	3.5	2.0
	Max	0.020	1.2	0.103	10.4	6.2	6.8	23.2	5.5	8.2	5.3
Middel	Mean	0.007	0.8	0.042	7.5	13.1	4.3	15.1	3.7	5.3	2.5
	Min	0.001	0.4	0.010	6.2	11.0	0.8	13.0	3.1	4.7	2.1
	Max	0.057	1.2	0.069	10.3	15.9	6.7	21.3	4.9	7.0	3.2
Lower	Mean	0.015	0.8	0.040	7.5	12.5	4.5	15.8	4.1	5.4	2.5
	Min	0.002	0.3	0.026	6.4	10.1	0.7	13.7	3.3	4.9	2.1

3.3.4.2.1 **Nitrate**

The nitrate content of the surface water of the Breitenbach averaged 0.9 mg l^{-1} N, with a fluctuation of 0.5 mg l^{-1} N. Values of 1.0 mg l^{-1} were regularly measured in the spring while low values were obtained in late summer (Figure 3.48b). This seasonality is very slight and cannot be clearly linked to any particular cause. Particularly in spring, the increasing likelihood of high water is a possible factor, in summer, the increased nutrient uptake from the vegetation (Figure 3.49). Since

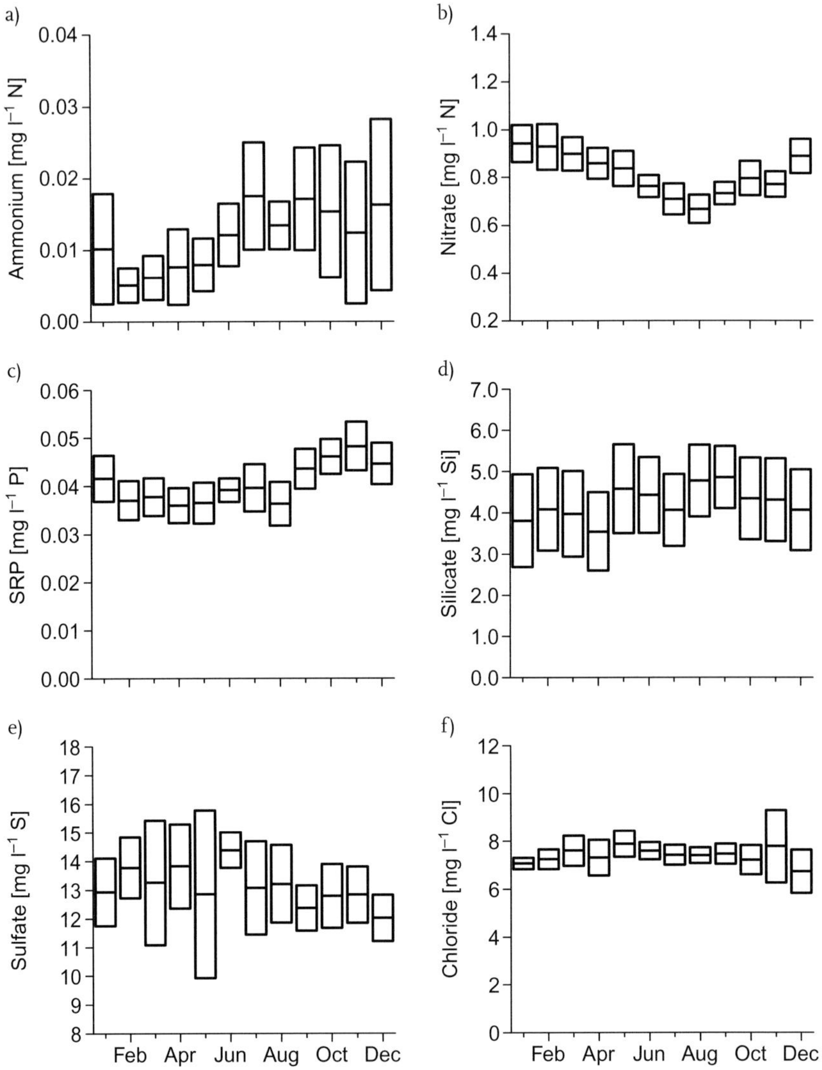

Figure 3.48 Minimum, mean and maximum concentrations of ammonium (a), nitrate (b), soluble reactive phosphate (c), silicate (d), sulfate (e) and chloride (f) from monthly samples during one year at site T2.

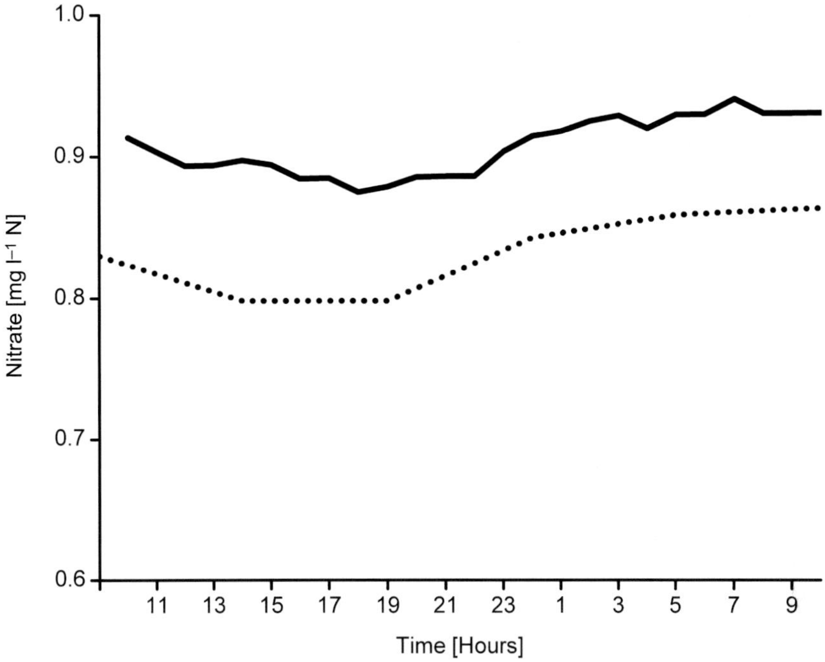

Figure 3.49 Diurnal nitrate concentrations at T2 in April 1987 (dotted line) and May 1988 (full line).

at present there are no measurements available for high water events, the sort of observations made by Elwood and Turner (1989) on the Walker Branch catchment, cannot be confirmed. These authors described a clear, marked increase in the concentration of almost all dissolved compounds during a spate.

The situation during the diurnal cycle in the Breitenbach is quite different. Here a clear synchronisation with the photoperiod can be observed. Data on diurnal cycles are available from two years (1987 and 1988). The responses shown in Figure 3.50 demonstrate a reduction in nitrate concentration in the afternoon. The nitrate content rises again overnight. The relationship between irradiance and nitrate concentration is particularly clear here, because only on those days when there was sufficient irradiance was this pattern shown.

Along the length of the stream, there is an important nitrate source in the spring region near T1 (Figure 3.51b). The upper reach provides the middle reach with a minimum of $0.6\,mg\,l^{-1}$ N. Concentrations of more than $1.0\,mg\,l^{-1}$ come in from the spring tributaries, which are however steadily reduced, and even with the incoming water from the spring region near T4, not significantly affected.

3.3.4.2.2 **Ammonium**

The ammonium concentration in the Breitenbach water is about $0.015\,mg\,l^{-1}$ N. During low water periods this increases to more than $0.02\,mg\,l^{-1}$ N. Minimal values

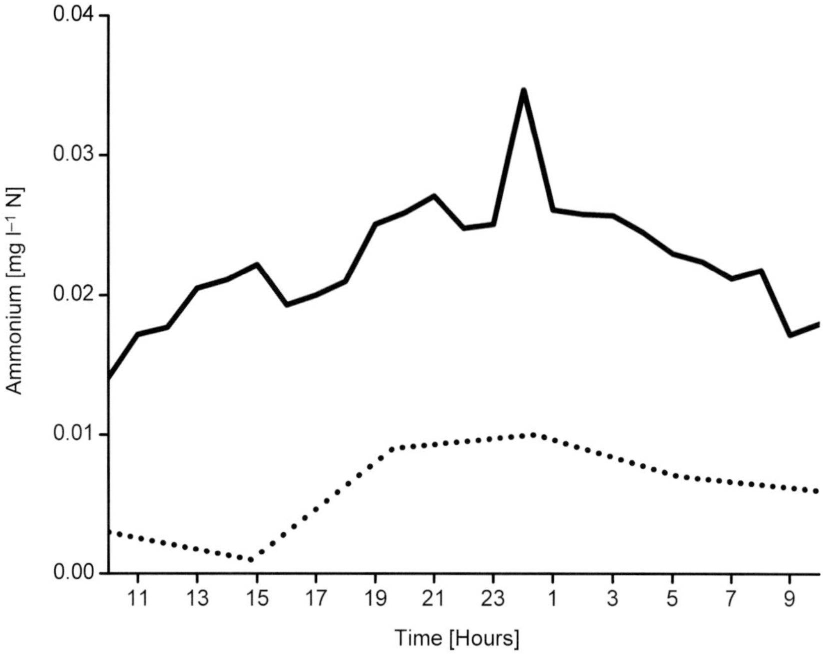

Figure 3.50 Diurnal cycle of ammonium concentration in Breitenbach stream water at T2 in April 1987 (dotted line) and May 1988 (full line).

of $0.005\,mg\,l^{-1}\,N$ are encountered in the spring months, and there is a strong seasonality over the course of a year. Figure 3.48a shows the typical fluctuation in ammonium concentration. Due to the dependence on discharge, low ammonium values are highly likely in the months of February and March. Concentrations increase during the course of the year, and then from October/November, with the start of higher winter outflows, fall again. During the course of the day a concentration increase is observed at night (Figure 3.51a).

From a longitudinal aspect, the upper reach carries a relatively high ammonium concentration. The spring horizon at T1 carries a very low ammonium content. It increases strongly as far as sample site SL1 and is then again strongly enriched by water from the spring region near T4. This ammonium input is presumably linked to the fishponds that discharge into the stream in this area (Figure 3.51a).

Additional measurements of ammonium and nitrate concentrations in the interstitial are also available for the years 1988/1989. Figure 3.52 shows the relevant concentration curves. A spate in the middle of March 1988 cleared the entire streambed down to the clay bottom. With the start of sedimentation at the end of April/start of May, probes were installed at different sediment depths. The extracted interstitial water was analyzed for oxygen, ammonium and nitrate. With increasing sediment thickness the sediment oxygen content steadily decreased. The nitrate content decreased similarly, while there was a complementary increase

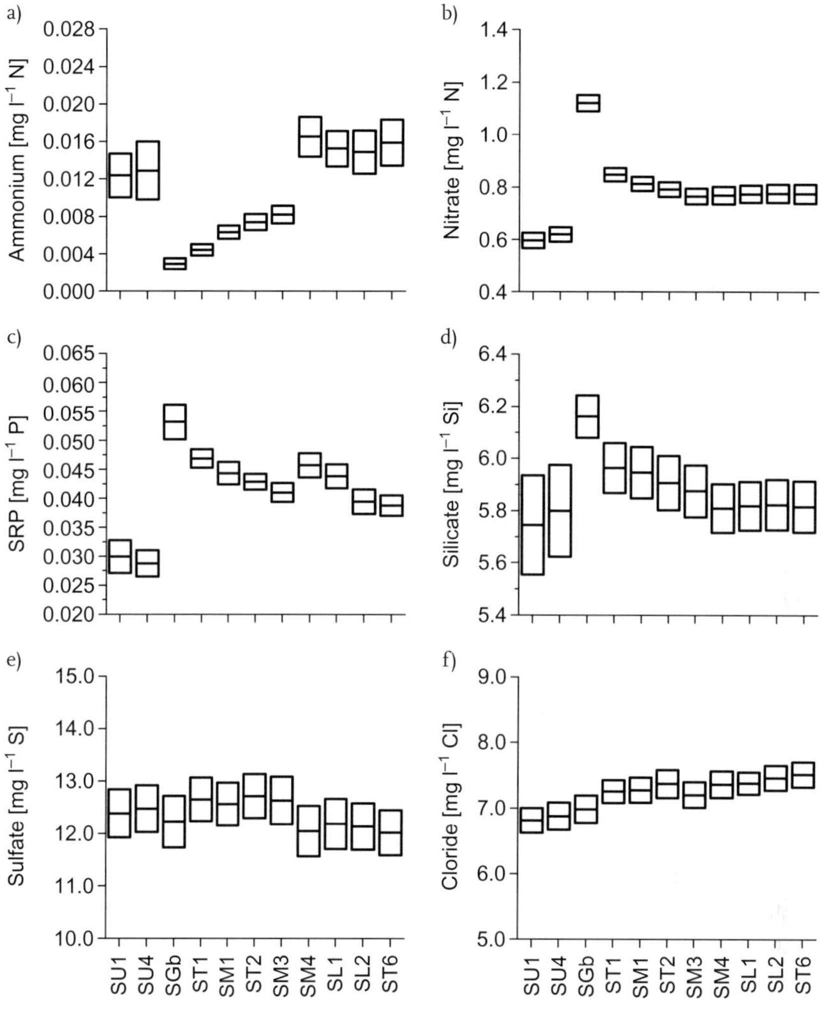

Figure 3.51 Minimum, mean and maximum concentrations of ammonium (a), nitrate (b), soluble reactive phosphate (c), silicate (d), sulfate (e) and chloride (f) sampled at sites along the Breitenbach in a year.

in ammonium concentration. Only at the end of the year, with the onset of precipitation and a rise in discharge is a reaction in ammonium and also nitrate noticeable in the flowing water. During summer stagnation, sediment oxygen content, only a few cm below the surface falls to less than $2\,mg\,l^{-1}$.

3.3.4.3 **Phosphate**

The long-term average phosphate content of the Breitenbach water is $0.041\,mg\,l^{-1}$ P. A minimum of 0.036 and a maximum of $0.052\,mg\,l^{-1}$ P were measured. There is

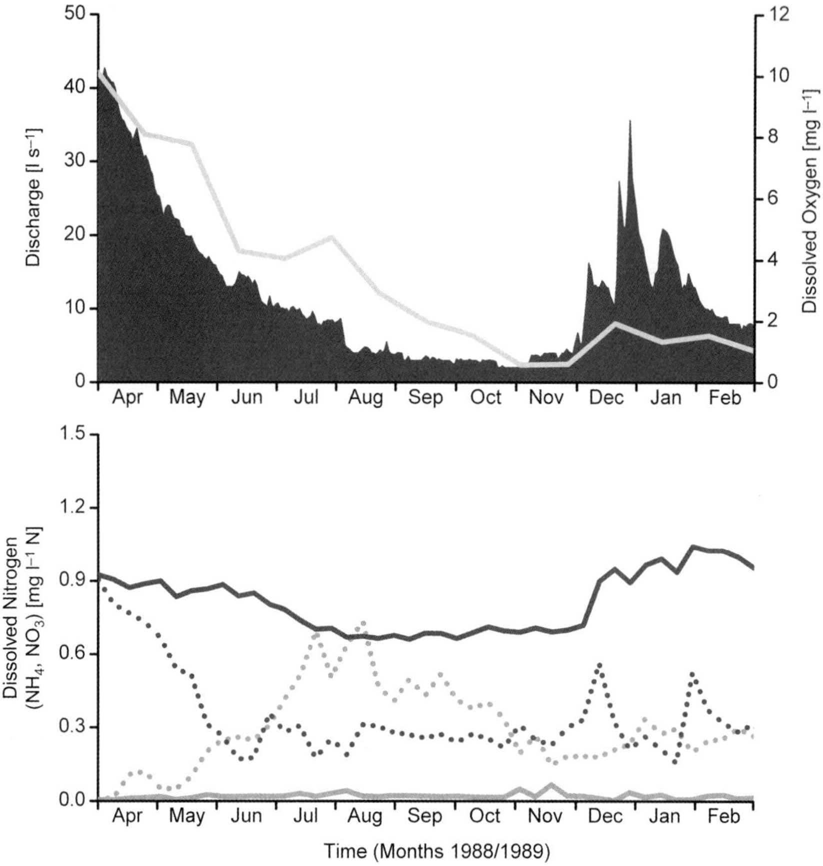

Figure 3.52 Oxygen, ammonium, nitrate in stream and pore water 1988–1989. Upper figure shows discharge (dark area) and oxygen (line). Lower figure shows nitrate and ammonium in stream water (full lines) and pore water (dotted lines). Ammonium level starts at 0. Nitrate level starts near $0.9\,mg\,l^{-1}$ N.

a slight trend towards raised concentrations in winter (Figure 3.48c). Over a 24-h period, on sunny days, a slight decrease is observed in the evening and at night.

Figure 3.51c,d shows the longitudinal distribution of phosphorus concentrations. The lowest values of only $0.28\,mg\,l^{-1}$ are found in the region of T1 in the Oberlauf. The spring horizon provides relatively phosphorus-rich water. Towards station SM3 the concentration decreases to $0.04\,mg\,l^{-1}$ but is then again increased by higher concentrations from the spring horizon T4. The phosphorus load decreases steadily to T6 (Figure 3.51c).

3.3.4.4 **Silicate**

There is some uncertainty about the quantitative evaluation of silicate because of a change in method. Only the post-2000 values provide a homogeneous data set.

Previously obtained values are hardly up to the definitive standard. The silicate concentration of Breitenbach water averages 5.7 mg l^{-1} Si, fluctuating between a maximum of 7.8 mg l^{-1} and a minimum of 4.2 mg l^{-1} Si. Regular seasonal variation could not be substantiated from the data (Figure 3.48d). Similarly no definite pattern in a diurnal rhythm could be demonstrated. Based on a series of preliminary investigations in 1987, over the course of the day a reduction in the silicate concentration was observed in the early morning, corresponding to the timing of diatom cell division, although this result could not be confirmed.

Silicate shows a similar pattern to nitrate and phosphate along the stream; lower values in the Oberlauf, higher load from the spring inputs and decreasing concentrations along the long stretch (Figure 3.51d).

3.3.4.5 **Sulfate**

A sulfate concentration of 37.1 mg l^{-1} was calculated for the total stream, with little variation, 2.1 mg l^{-1} (Figure 3.48e). The concentration increases down the length of the stream to the long stretch. Waters from the spring regions at T1 and T4 have minimal sulfate concentrations (Figure 3.51e). There is a slight increase in sulfate concentration in the stream section between the springs.

3.3.4.6 **Chloride**

The average chloride content of the stream water is 7.4 mg l^{-1} Cl. Extreme values of 12.1 mg l^{-1} (maximum) and 4.7 mg l^{-1} Cl (minimum) were recorded (Figure 3.48f). Hardly any seasonality could be detected. A steady increase in chloride concentration is seen along the length of the stream but any influence of the spring inputs is slight (Figure 3.51f).

4
Detrital Energy Sources

4.1
Organic Matter in Streams

Jürgen Marxsen

Detritus consists of nonliving organic matter, containing particulate and dissolved materials (Wetzel, 1983). Originally defined as particulate organic matter (POM) only, detritus is now considered also to include dissolved organic matter (DOM). There is a large amount of detritus available in most aquatic ecosystems, especially running water ecosystems, which provides an important source of matter and energy to the heterotrophic members of their communities. This includes not only the organic matter from primary producers within the ecosystem, which is not utilized by herbivores and enters the pools of dead particulate and dissolved organic matter (Wetzel *et al.*, 1972), but also organic matter of allochthonous origin, produced outside the particular system. Especially for fluvial systems the much larger fraction of detritus is usually provided by allochthonous sources (Webster and Meyer, 1997a), resulting in the dominance of heterotrophic processes in most streams and rivers, and in the great importance of detritus-based food chains (cf. Chapter 6 and Section 8.7). However, most of the dead organic matter in stream systems consists of substances that are difficult to degrade (Figure 4.1). The occurrence of these resistant materials results in special adaptations of stream communities with typical detrital food chains.

The separation of particulate from dissolved organic matter is not defined chemically, but is purely pragmatic. Material that does not pass through a 0.45 μm membrane filter is defined as particulate, but that which passes through this filter is regarded as dissolved (Lock and Hynes, 1975). Thus viruses, small bacteria and colloidal organic matter may be included in the dissolved fraction.

Central European Stream Ecosystems: The Long Term Study of the Breitenbach, First Edition. Edited by Rüdiger Wagner, Jürgen Marxsen, Peter Zwick, Eileen J. Cox.

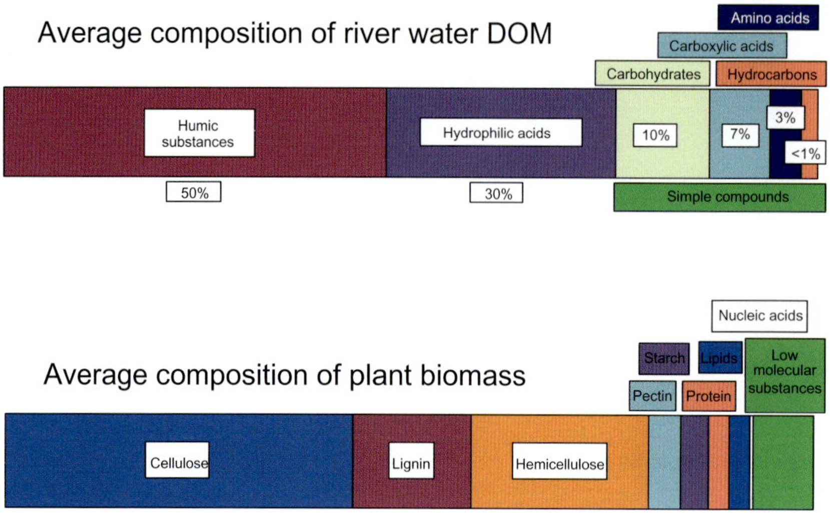

Figure 4.1 Average composition of river water dissolved organic matter (after Thurman, 1985) and average composition of plant biomass (after Fritsche, 1985). Most of the compounds are difficult to utilize (e.g., humic substances and hydrophilic acids, or cellulose, lignin and hemicellulose) and thus undergo special degradation processes (see Section 6.3.2).

4.2
Particulate Organic Matter

Jürgen Marxsen and Rüdiger Wagner

4.2.1
Particulate Organic Matter in Streams

Particulate organic matter (POM) is one of the most important sources of carbon and energy in streams, not only in woodland streams, but also in those crossing open grassland or other landscapes, such as deserts, that allow unrestricted insolation to streams (Webster and Meyer, 1997b, c). However, autochthonous macrophytes in particular may produce a significant amount of POM to the detrital food chain in the latter. POM is the nutritional basis for many benthic invertebrates, especially shredders and filter feeders (Hynes, 1970a, b; Petersen and Cummins, 1974). Even the scraping and epilithic larvae of the caddisflies *Agapetus fuscipes* and *Silo pallipes* obtain large amounts of POM in their food (Becker, 1990). But the microbial, fungal and bacterial, communities, as well as the protozoa, are important consumers of POM, too (Petersen and Cummins, 1974; cf. Section 8.7).

Particulate organic matter inputs, as well as material produced within the stream, are usually divided into two categories. These are coarse particulate organic matter (CPOM), which is larger than 1 mm, and fine particulate organic matter (FPOM), which passes a 1 mm mesh size net. Both categories can be further subdivided (e.g., Cummins, 1974), although such dividing lines are arbitrary

(Allan, 1995). For the purposes of the Breitenbach Ecosystem Study no further division was usually made. Thus for this contribution, particulate organic matter is only divided into FPOM and CPOM.

POM is mostly measured as mass (ash free dry mass = AFDM), after drying to constant weight at 105 °C and then ashing at 500 °C. This is the usual procedure followed in the studies performed on the Breitenbach. Particulate organic matter can be converted to particulate organic carbon (POC) by assuming that organic carbon amounts to 40–50% of organic matter. The direct approach to determine particulate organic carbon is by combustion in an oxygen stream, measuring the resultant CO_2 (e.g., Bretschko and Leichtfried, 1987), but this was not usually done for the Breitenbach studies.

Following Webster and Meyer (1997b), for this study we generally assumed that 2 g of organic matter (OM) correspond to 1 g of organic carbon (OC). This is in good agreement with results from Neumann (1995), who determined that the relationship for FPOM from the Breitenbach sediment followed the equation: OC $[\text{mg g}^{-1}] = 0.52 \times$ AFDM $[\text{mg g}^{-1}] - 1.88$ ($r^2 = 0.99$).

4.2.2 POM Inputs to the Breitenbach

4.2.2.1 Measuring Periods and Methods

The input of coarse particulate organic material into the Breitenbach has been repeatedly determined. Marxsen's (1980a) data are from the period 1973–1974 and cover the entire stream (Figure 4.2). The traps, 0.09 or 0.16 m^2 in size (Figure 4.3), were placed in various parts of the stream, allowing lateral blow-in and direct inputs to be determined. A second set of data is available for October 1982 to December 1985, published by Koch (1990) for 1985 and Wantzen and Wagner (2006) for 1982 to 1984. During this period two or three pairs (left and right bank) of traps (cages, 1 mm mesh size, 20 × 50 × 60 cm [wide, long, high], open on top) were placed along the stream at three sites with representative riparian vegetation (managed pasture, alder rows, and fallow land).

Cages were placed at the bank with the upper side at the elevation of the terrestrial environment. Thus, not only leaves falling directly into the stream were trapped but also leaves that fell to the ground at some distance from the stream and were transported into the stream by the wind. For both studies traps were emptied every 5–14 days, depending on the season. The material was separated into different categories, dried (105 °C) and weighed. Lateral inputs are the main sources of allochthonous CPOM at the Breitenbach, since it largely lacks a closed canopy, and CPOM traveling distance is comparatively short due to high in-stream retention (Horvath, 2004).

The vegetation in the various sections along the stream, including aspects of the riparian macrophytes, were described in detail for algae by Cox (1990) and for macrophytes, which are the most important instream sources of POM in the Breitenbach, by Ringe (1974) and Horvath (2004; cf. Chapter 5).

The organic matter budget for the Breitenbach was summarized by Marxsen *et al.* (1997) and Marxsen (2006), including gross primary production

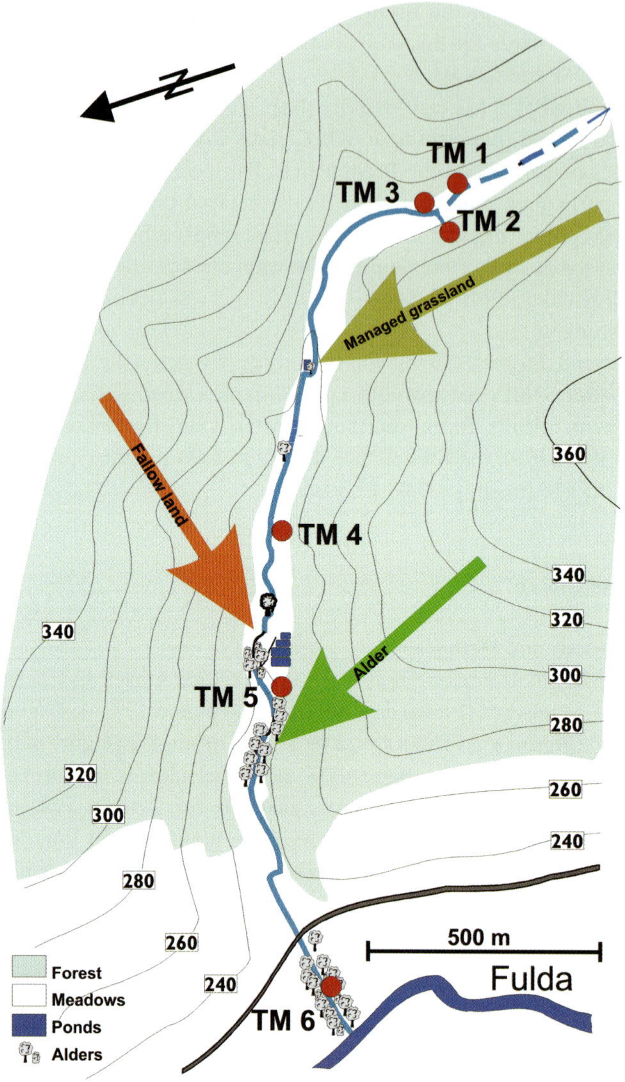

Figure 4.2 Sites where POM input has been determined. Dots: investigation from Marxsen (1976, 1980a) during 1973/74, arrows: investigations by Wagner (published by Wantzen and Wagner, 2006) and Koch (1990) during 1982–1985.

(800 g OM m^{-2} $year^{-1}$), inputs of POM (835 g OM m^{-2} $year^{-1}$) and DOM (2040 g m^{-2} $year^{-1}$) by groundwater (cf. Section 8.7).

4.2.2.2 Seasonal Distribution of CPOM Inputs

Composition, amount and seasonal distribution of CPOM input are characteristic for individual sites (Figures 4.4–4.7, Table 4.1). The upper reaches of the Breiten-

Figure 4.3 Trap for determining coarse particulate organic matter input during the 1973/74 studies (Marxsen, 1980a). Photo from Marxsen (1976).

bach in particular are strongly exposed to the surrounding woodland, receiving higher quantities of CPOM than most downstream parts. However, the middle and lower areas, particularly those with alder trees at the stream bank, also receive large amounts of CPOM. Although alder is the dominant tree species along the stream, occasional winter storms blow oak (*Quercus robur*) and beech (*Fagus sylvatica*) leaves into the stream from distances of up to about 50 m. These input pulses can contribute more than 50% of the total annual CPOM input (Wantzen and Wagner, 2006).

Leaf fall is mainly seasonal and occurs predominantly in autumn and early winter (September to December; Figure 4.4). However, summer rainstorms may provoke a moderate input of green alder (*Alnus glutinosa*) leaves, and long dry periods in the summer may cause earlier leaf fall. Usually up to 70% of the total annual input enters the stream between September and the end of November (Figure 4.4), with up to 300 g OM m^{-2} $month^{-1}$ at some sites. Although the main POM input occurs in autumn, there is a small but continuous input of POM of 15–30 g OM m^{-2} $month^{-1}$ (whole stream average) into the stream throughout the year.

The seasonal distribution of POM input was similar in the alder and fallow land reaches in the 1980s. However, at the managed grassland sites there were large grass inputs in July which contributed more than 30% of a year's total input. This is a clear effect of the type of management in the near-stream environment and demonstrates how stream POM budgets may be altered in environments subject to different agricultural intensity. Clear cutting of streams affects the amount and composition of POM and probably affects parts of the stream community.

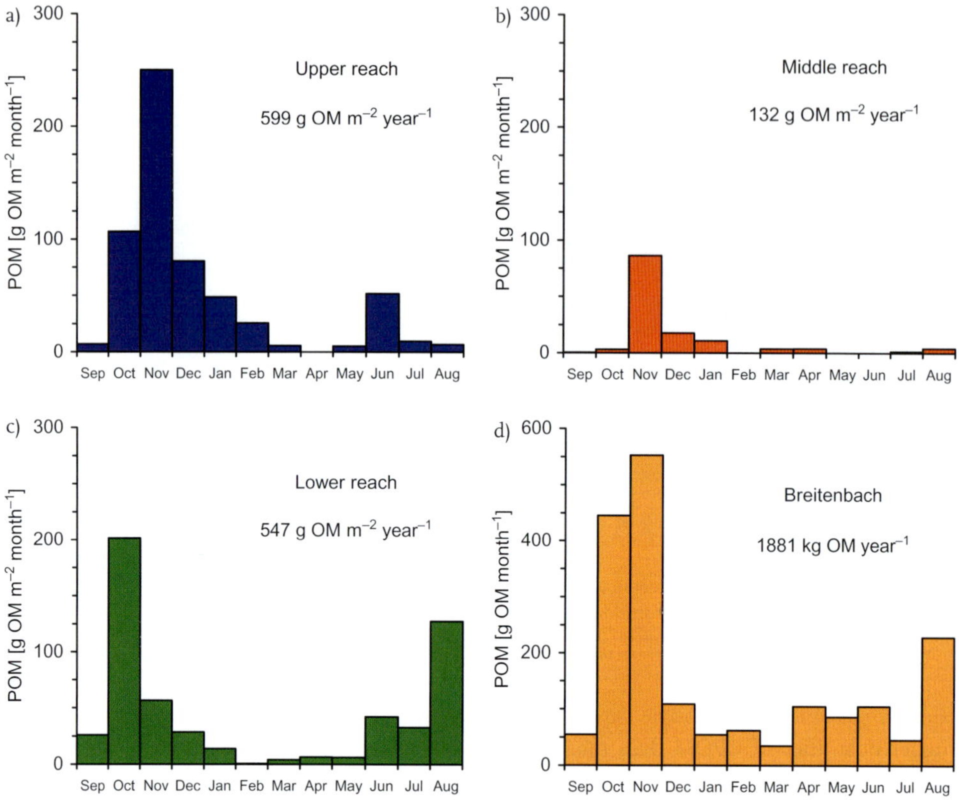

Figure 4.4 Seasonal dynamics of coarse particulate organic matter input to the Breitenbach, at: (a) a woodland site in the upper reach (TM1), (b) a grassland site in the middle reach (TM4), (c) a site with alder trees besides the stream in the lower reach (TM6), (d) for the Breitenbach as a whole, from September 1973 to August 1974. For details on the location of the sites compare Table 4.1 and Figure 4.2.

4.2.2.3 **Amounts of CPOM Inputs in Various Years and Stream Sections**

In 1973/1974 CPOM input at different sites was between 73 and 1137 g OM m^{-2} year^{-1} (Table 4.1). It was highest in forested reaches (TM1 and TM5), with approximately 1100 g OM m^{-2} year^{-1}, and lowest at the sites outside the forests (TM 3 and TM 4), around 100 g OM m^{-2} year^{-1}. From 1982 to 1985 CPOM input was highest in an alder reach (1605–2682 g OM m^{-2} year^{-1}), followed by managed grassland (287–767 g OM m^{-2} year^{-1}) and fallow land areas (about 233–421 g OM m^{-2} year^{-1}; Figure 4.7). But forested reaches were not included in this study.

Between year variation may be high at individual sites, up to more than 100%. Variation from 1982 to 1985 was not similar at the different sites (Figure 4.7). Whereas left and right bank traps in the same area usually received similar amounts of POM, due to mowing activities in the managed grassland sites different amounts of grass often entered the left or right bank traps.

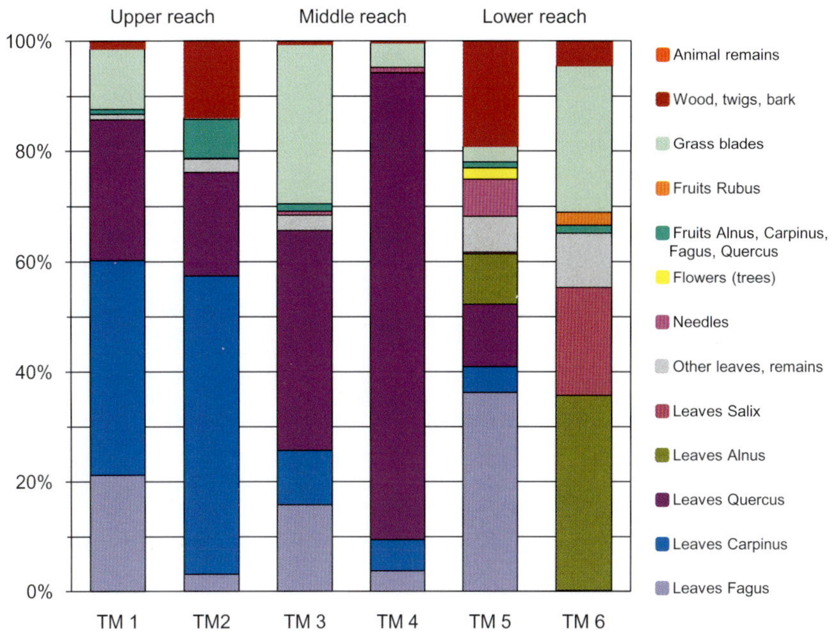

Figure 4.5 Contribution of individual components to total CPOM input along the Breitenbach 1973/74. For details on sites compare Table 4.1 and Figure 4.2.

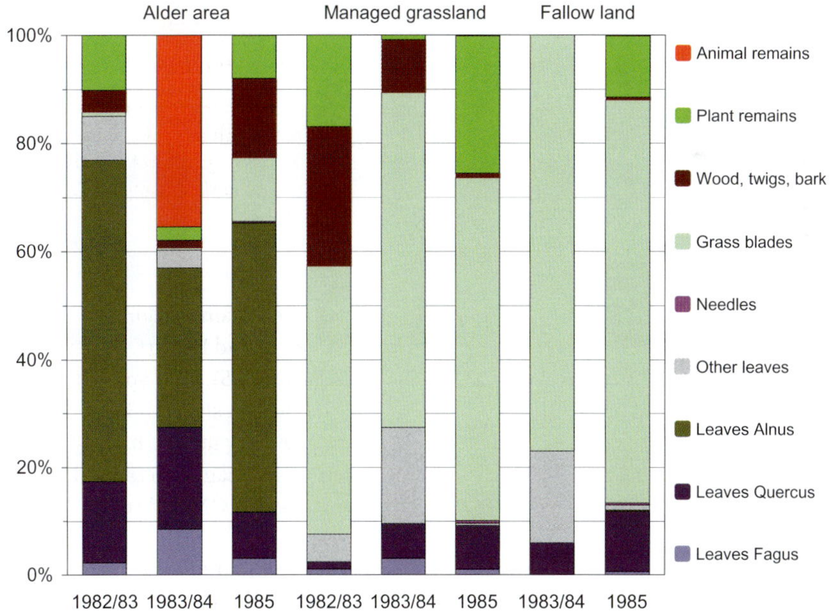

Figure 4.6 Composition of the CPOM input into the Breitenbach 1982–1985 in regions with different bank vegetation (1982–1984 data by Wagner; 1985 data by Koch). For details on sites compare Figure 4.2.

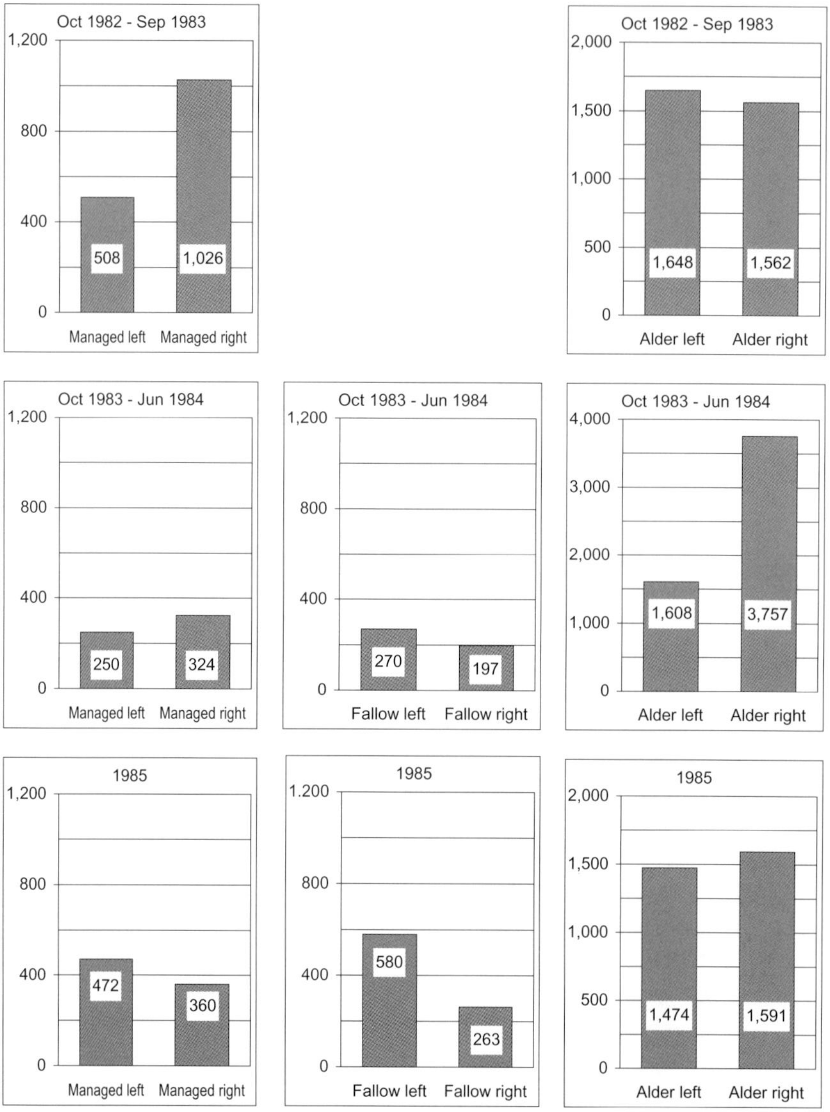

Figure 4.7 Total CPOM input into paired traps on the left and right banks of the Breitenbach for an alder area, managed grassland and fallow land (October 1982 to December 1985; *y*-axes: $g\,OM\,m^{-2}\,year^{-1}$). Note: different scales of *y*-axes for alder areas. For details on sites compare Figure 4.2.

Further, exceptional interannual variation in litter input at individual sites may also depend on climatic variation. In February 1984, during a winter period without snow cover, a heavy storm blew an extraordinarily large amount of *Fagus* leaves from the forest, from distances greater than 50 m, into the stream. This increased the usually less than $300\,g\,OM\,m^{-2}\,year^{-1}$ CPOM input to $580\,g\,OM\,m^{-2}$

Table 4.1 Annual inputs of particulate organic matter at different sites of the Breitenbach from August 1973 to August 1974. For location of traps compare Figure 4.2.

Site	Input (g OM m^{-2} $year^{-1}$)
TM 1: Upper reach 40 m above gage GA	599
TM 2: Georgsborn spring pool	1081
TM 3: Confluence of upper reach and Georgsborn spring outlet	73
TM 4: Middle reach at T3[a)]	132
TM 5: Lower reach about 100 m above trap T5	1137
TM 6: Lower reach near gauge GT6	547

a) The trap was located where trap T3 was later constructed.

$year^{-1}$ in the wind exposed trap, but not in the trap on the other side of the stream (Figure 4.7).

4.2.2.4 Composition of CPOM Inputs

The composition of the CPOM collected from the various traps along the Breitenbach mirrors the vegetation in the respective surroundings; between site differences depended on the differing plant communities of the bank vegetation (Figures 4.5 and 4.6). In the alder area in the middle and lower reaches, leaves accounted for about 50–75% of the CPOM inputs. They were predominantly alder (*Alnus glutinosa*, 30–60%), followed mostly by oak (*Quercus robur*) and beech (*Fagus sylvatica*; Figures 4.5 and 4.6). In the upper reach inputs by leaves dominated even more, accounting for at least 70% of total CPOM input (Figure 4.5). But alder leaves were missing here; mainly leaves from hornbeam (*Carpinus betulus*), oak and beech entered this part of the stream. Surprisingly a high amount of animal remains was observed falling into the Breitenbach in the alder area below trap T5 in winter 1984 (Figure 4.6). These consisted mainly of oligochaetes and Diptera larvae which were blown into the stream together with a lot of alder leaves during storm events in February and March.

Leaves were rare in the managed grassland area traps (Figure 4.6). Here, blades of grass and bank vegetation dominated the CPOM input; however, in some years, wood, "other leaves" or "plant remains" of different origin were abundant. Furthermore, the seasonal management of the grassland affected the timing and composition of the input. The period of grass cutting, and whether or not parts of the bank vegetation were cut, clearly affected the amount and composition of the input.

Bank vegetation was the dominant fraction of CPOM input in the fallow land, followed by fruits, wood and "other organic remains", but with year to year differences (Figure 4.6). At the managed grassland and fallow land sites the intact bank vegetation prevented most of the regular autumnal input of leaves that were usually blown into the stream from the forests at some distance from the banks.

4.2.2.5 **Total Inputs of POM**

As average total input of CPOM 583 g OM m^{-2} year^{-1} (corresponding to 0.29 kg C m^{-2} year^{-1}; Table 4.2) was obtained for the whole stream during 1973/74 (Marxsen, 1980a and unpublished data). Koch (1990) determined input to the middle and lower reaches at 558 g OM m^{-2} year^{-1} in 1985. This is about 200 g OM m^{-2} year^{-1} more than the 1973/74 CPOM inputs for this stream area, which were 361 g OM m^{-2} year^{-1}. This is a credible increase because a considerable number of trees had developed near the stream between these studies. Alder (*Alnus glutinosa*) appeared directly on the river banks along the middle and lower reaches, but other species also grew up in the middle reach, between the stream and the forests on the hills. Changes were less distinct in the upper reach. Also, due to the changed grassland management in the middle and lower reaches, because of increased mowing activities more grass entered the stream during summer. Thus, as a conservative estimate of total CPOM input to the Breitenbach for 1985, the sum of the input to the upper reach from 1973/74, and to the middle and lower reach from 1985, was calculated (697 g OM m^{-2} year^{-1}, corresponding to 0.35 kg C m^{-2} year^{-1}; Table 4.2).

The trap techniques used for measuring POM inputs did not allow quantification of fine POM (FPOM), and no other FPOM input data are available for the Breitenbach. In a forested upland stream in Austria (Oberer Seebach, Lunz), Bretschko and Moser (1993) determined that FPOM contributed between 9 and 24% of the total POM input. Assuming that FPOM constitutes 16.5% of POM inputs to the Breitenbach, total POM input can be conservatively estimated to 698 and 835 g OM m^{-2} year^{-1} (corresponding to 0.35 and 0.42 kg C m^{-2} year^{-1}) for 1973/1974 and 1985, respectively (Table 4.3).

Table 4.2 Coarse particulate organic matter inputs to the Breitenbach. Koch (1990) performed input measurements for the middle and lower reaches only. To achieve total CPOM input data to the whole stream for 1985, the input measurements to the upper reach from 1973/74 (Marxsen, 1980a, unpublished data) were combined with the data from 1985. No substantial changes occurred with respect to the vegetation in the catchment surrounding the upper parts of the stream.

Year	Stream area considered	Input		References
		(g OM m^{-2} year^{-1})	(kg OM year^{-1})	
1973/74	Entire stream (3225 m^2)	583	1881	Marxsen (1980a, unpublished data)
	Upper reach (1360 m^2)	888	1207	
	Middle and lower reach (1865 m^2)	361	674	
1985	Middle and lower reach	558	1041	Koch (1990)
1973/74 + 1985	Combined data: upper reach 1973/74 and middle/lower reach 1985	697	2248	Marxsen (2006)

Table 4.3 Inputs and outputs of particulate organic matter to the Breitenbach, including coarse as well as fine particulate matter.

	Value ($g\,OM\,m^{-2}\,year^{-1}$)	References
Input	835	Marxsen (1980a, 2006), Koch (1990)
Standing stock annual average	608[a)]	Koch (1990), Marxsen (2006)
Output	680	Marxsen *et al.* (1997), Marxsen (2006)

a) ($g\,OM\,m^{-2}$).

4.2.3
Standing Stocks of POM in the Breitenbach

The standing stocks of particulate organic material in the Breitenbach are highly variable, not only spatially but also seasonally. The total particulate benthic organic matter was intensely studied in the stream's middle reach in 1984/85 (Koch, 1990). Temporal changes in streambed structure were recorded and used for selecting typical sampling locations. These locations were sampled using a surber sampler. The macroscopic benthic fauna was excluded before the samples were dried and AFDM was determined. 608 $g\,m^{-2}$ total particulate organic matter was obtained as annual average, corresponding to 0.30 $kg\,C\,m^{-2}$. This result is in good agreement with measurements in specific habitats during particular studies on the Breitenbach (Pieper, 1978; Knöfel, 1983; Wolf, 1991; Neumann, 1995).

Microfauna, microflora and bacteria were included when calculating POM. From determination of phospholipid content, Neumann (1995) estimated that the total biomass of these organisms contributes on average 6–7% of the total amount of FPOM in the sandy, more lentic sediments of the Breitenbach, and 9–10% in the lotic areas with coarse deposits. Subtracting microbial biomass, which is 14 $g\,OM\,m^{-2}$ (0.007 $kg\,OC\,m^{-2}$), results in average total particulate organic matter for the whole stream of 608 $g\,OM\,m^{-2}$ (0.30 $kg\,C\,m^{-2}$; Table 4.3).

4.2.4
Outputs of POM from the Breitenbach

The FPOM export from the Breitenbach was determined monthly from September 1985 to August 1986 at the stream's confluence with the River Fulda (Marxsen *et al.*, 1997; Marxsen, 2006, unpublished data). AFDM was determined after filtering the stream water through glass-fiber filters (Marxsen *et al.*, 1997), giving an annual transported loss of FPOM for the whole Breitenbach of 1770 kg. Gerhardt (1986) estimated CPOM export (>1 mm) at 410 kg from September 1985 to May 1986. Thus, there is a lack of data for the summer months from June to August. But this is a period of low water flow in the Breitenbach and hence low POM, especially low CPOM export. Summing the FPOM and CPOM losses gives 2180 kg total POM export, that is, 680 g OM m^{-2} $year^{-1}$ (0.34 kg C m^{-2} $year^{-1}$) based on streambed area (Table 4.3). This is probably an underestimate, especially because elevated outputs during storms were not included.

4.3
Dissolved Organic Matter

Jürgen Marxsen

4.3.1
Dissolved Organic Matter in Streams

Dissolved organic matter (DOM) is a natural biological product in streams. It not only originates from in-stream processes, such as leaching from POM (e.g., leaves), extracellular enzymatic degradation or extracellular release (mainly from plants), but predominantly from external sources, such as soil and groundwater around a stream. It is of great importance as a carbon and energy source, mainly for bacteria. However, a large proportion is highly refractory (Figure 4.1) and may therefore be of limited biological significance (Thurman, 1985).

The earliest measurements on dissolved organic matter in limnological science were by chemical oxidation, thus results were usually given as DOM. Later studies usually relied on automated carbon analysis by combustion of water samples and values were reported as dissolved organic carbon (DOC). Data collected for the Breitenbach were determined by high temperature catalytic combustion after filtration through 0.45 μm filter membranes, acidification and purging (Fiebig, 1995; Marxsen *et al.*, 1997; Katzner, 2011). The resultant CO_2 was measured by infrared detection. A Beckman 915B Total Carbon Analyzer was used for the samples from 1985 to 1990, later a Dohrman DC-190 model (until 1997) and then a Shimadzu TOC-5000A analyzer. For comparison with data from other studies, which gave results as dissolved organic matter, these are converted to DOC by multiplying by 0.5 (Webster and Meyer, 1997b).

The dissolved organic matter of average stream and river water (Thurman, 1985) comprises about 50% fulvic and humic acids and of 30% hydrophilic acids (Figure 4.1). Of the remaining 20%, usually about half (10% of total DOM) are carbohydrates. Carboxylic acids contribute 7%, amino acids 3%, and hydrocarbons less than 1%. These values are only a guide, for example, the percentage of carbohydrates in different environments has been estimated to represent 1–30% of the dissolved organic carbon pool.

4.3.2
Concentrations and Dynamics of Dissolved Organic Matter in the Breitenbach

Concentrations of dissolved organic carbon in Breitenbach stream water were within the typical range for such unpolluted small streams (Thurman, 1985).

Table 4.4 Concentrations of dissolved organic carbon in Breitenbach water at different sites along the stream. All values (Mean = annual mean, Min = annual minimum, Max = annual maximum, s.d. = standard deviation) are given in $mg\,C\,l^{-1}$. GS = gauge at the end of the Georgsborn spring brook, at the beginning of the middle reach, GA = gauge at the end of the upper reach, T1 = trap T1 about 200 m below the entrance of the Georgsborn spring brook, T2 = trap T2 in the middle reach, T4 = trap T4 at the end of the middle reach, GT6 = gauge below trap T6, about 200 m above the mouth to the River Fulda.

		GS	GA	T1	T2	T4	GT6
1985	Mean	1.18	2.08	1.61	1.65	1.68	1.77
	Min	0.40	1.60	1.00	0.90	0.80	0.90
	Max	1.80	2.80	2.30	2.10	2.40	2.40
	s.d.	0.32	0.32	0.35	0.28	0.32	0.34
	N	33	36	23	36	36	35
1986	Mean	1.45	2.89	2.08	2.17	2.27	2.35
	Min	0.66	1.60	1.00	1.27	1.29	1.40
	Max	3.30	12.0	12.4	8.80	9.60	8.80
	s.d.	0.56	1.83	1.75	1.19	1.31	1.15
	N	45	37	45	47	47	47
1987	Mean	1.67	2.60	2.20	2.37	2.40	2.51
	Min	1.04	1.81	1.70	1.80	1.82	1.80
	Max	2.73	3.20	3.00	3.10	3.55	3.55
	s.d.	0.59	0.49	0.49	0.40	0.54	0.53
	N	11	11	11	11	11	11

Values were lowest in the Georgsborn spring outlet, where concentrations were usually around 1.0 $mg\,C\,l^{-1}$ (Table 4.4, Figure 4.8). Higher values, up to 3.3 $mg\,C\,l^{-1}$, were observed during spates, when stream water from the upper reach entered the spring pools. Highest annual averages were determined in the upper reach (measurements at gauge GA; Table 4.4 and Figure 4.8), intermediate values after the confluence of spring and stream water (sampling site at trap T1). Downstream DOC increased slightly and continuously, reaching the highest values of the middle and lower sections just above the entrance to the River Fulda (sampling site GT6 at the gauge below trap T6).

Although distinct seasonal fluctuations of DOC were observed in many years, there was no clear annual trend (Figures 4.8 and 4.9). No correlation could be found between temperature and DOC concentrations (data not shown), but there were distinct correlations between discharge and DOC concentrations in the middle and lower reaches, as well as in the spring pool outlet (Table 4.5, Figure 4.10). However no clear trend was identifiable in the upper reach, although the calculated r^2 was slightly significant ($P < 0.05$). Positive correlations between

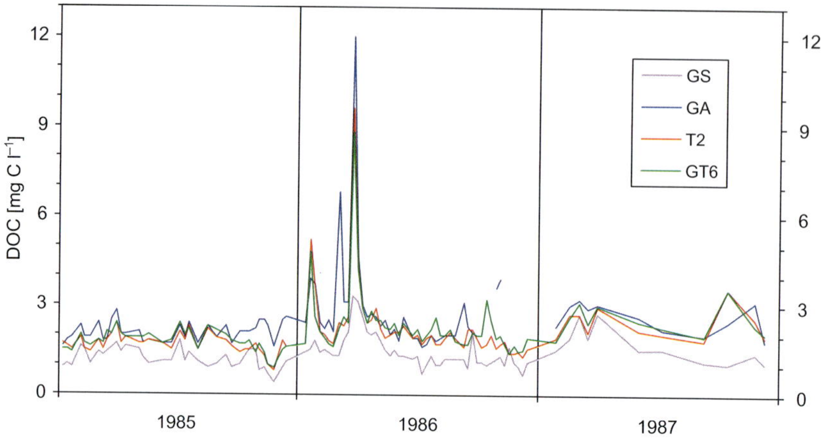

Figure 4.8 Concentrations of dissolved organic carbon (DOC) in Breitenbach water at different sites along the stream during the years 1985 to 1987. GS = gauge at the end of the Georgsborn spring brook, at the beginning of the middle reach, GA = gauge at the end of the upper reach (no discharge during several weeks at the end of 1986 and at the beginning of 1987), T2 = trap T2 in the middle reach, GT6 = gauge below trap T6, about 200 m above the mouth into the River Fulda.

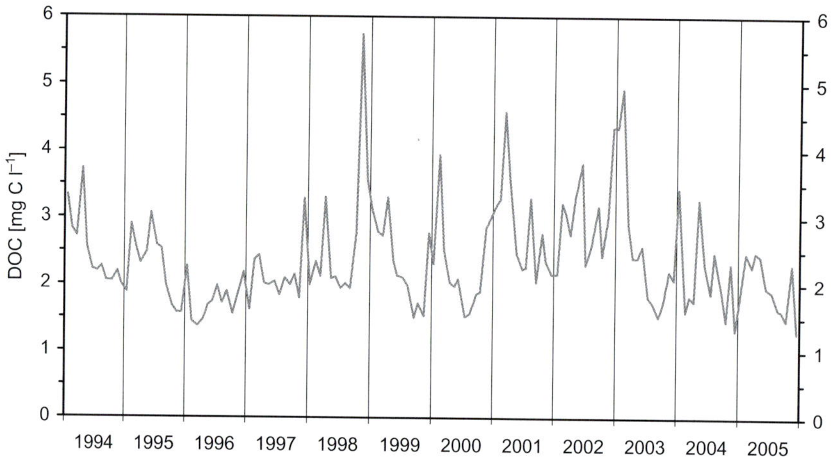

Figure 4.9 Concentrations of dissolved organic carbon (DOC) in Breitenbach water (middle reach at trap T4) from 1994 to 2005.

discharge and DOC concentrations have been reported from many streams (e.g., Meyer *et al.*, 1988; Fiebig *et al.*, 1990; Sedell and Dahm, 1990), but this trend is not universal (McDowell and Wood, 1984).

Over a two year period (1989–1990) Fiebig (1995) compared the DOC concentrations in groundwater adjacent to the stream with those in stream water. Ground-

Table 4.5 Correlation between discharge and dissolved organic carbon in Breitenbach stream water at different sites along the stream. Regression was calculated by using the equation $y = A\,x^B$. GS = gauge at the end of the Georgsborn spring brook, at the beginning of the middle reach, GA = gauge at the end of the upper reach, T2 = trap T2 in the middle reach, GT6 = gauge below trap T6, about 200 m above the mouth to the River Fulda, * = $P \leq 0.05$, *** = $P \leq 0.001$.

	Site	A	B	r^2	N
1985–1987	GS	0.7128	0.3061	0.49***	89
	GA	2.134	0.0544	0.06*	84
	T2	0.9896	0.2280	0.53***	84
	GT6	0.9876	0.2275	0.38***	83
1994–2005	T2	1.526	0.1838	0.47***	142

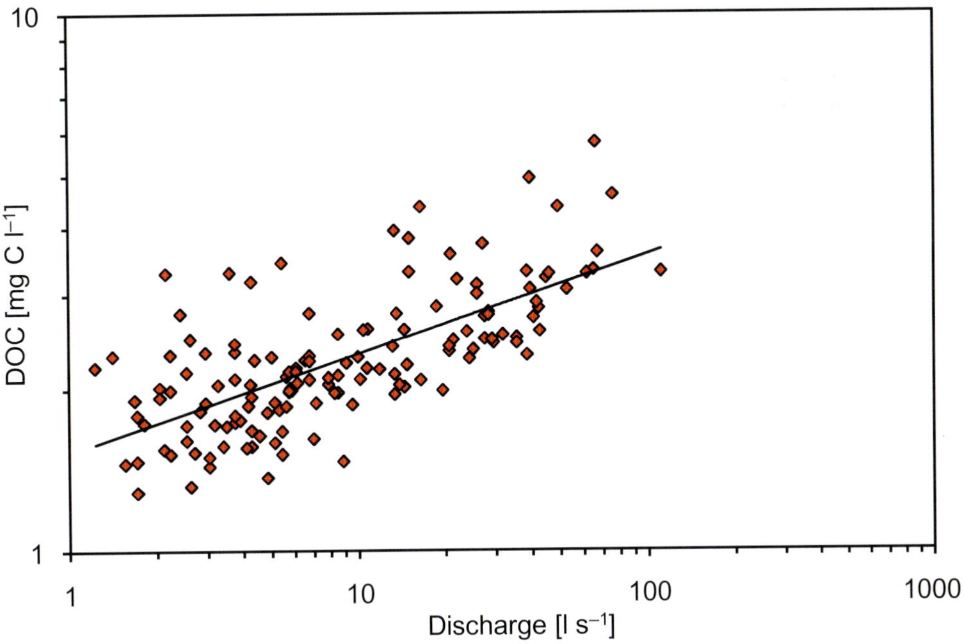

Figure 4.10 Relationship between discharge and dissolved organic carbon in water from the Breitenbach middle reach at trap T2. Data from 1994 to 2005 ($y = 1.526\,x^{0.1838}$, $r^2 = 0.47$, $P < 0.001$, $n = 142$).

water was obtained from ten sampling wells (at about 60 cm depth) installed about 0.5 m beside each side of the stream, about 50 m above trap T4 in the stream's middle reach. DOC concentrations in the individual groundwater wells varied from 0.7 to 27 mg C l^{-1}, with an annual mean of 5.5 mg C l^{-1} for all wells. DOC in the groundwater consistently exceeded that in the stream water which averaged

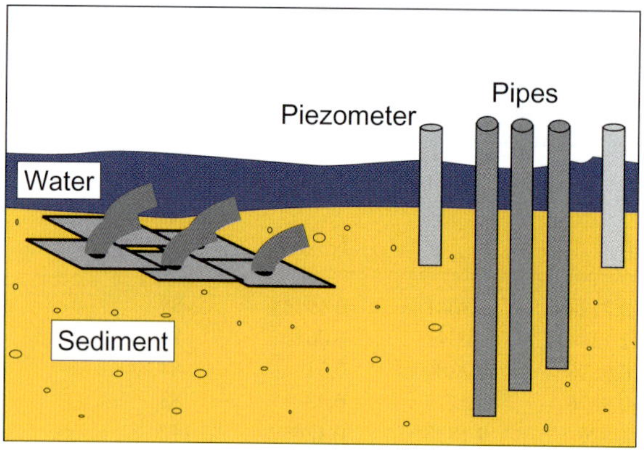

Figure 4.11 Sampling site scheme for DOC measurements in different habitats of the Breitenbach, performed from May 1999 to May 2000 (Katzner, 2011). Samples were taken from stream water and from interstitial water of layers of upper sediment (0–15 cm depth) and deeper sediment (40–120 cm depth) about 3 m above gauge GT2 and about 50 m above trap T4. For sampling groundwater the wells installed for Fiebig's (1995) study at the site above trap T4 were used.

1.4 mg C l^{-1} over the two-year period. There was neither a significant relationship between DOC concentration in the wells and time of year, nor between the rate of groundwater discharge to the Breitenbach and the DOC value.

From May 1999 to May 2000, dissolved organic carbon concentrations were compared weekly for different habitats in the middle reach of the Breitenbach, 3 m above gauge GT2 and about 50 m above trap T4 (Katzner, 2011). Samples were always taken from stream water and from interstitial water of up to five different depths in the upper sediment layers, down to about 15 cm. For sampling interstitial water special types of disks were used which had been installed in the sediment before the investigations began and which remained there for the whole period (developed by H.-H. Schmidt; for details, see Chapter 3). Three steel pipes had been installed for sampling deeper interstitial water, at depths of 40, 80 and 120 cm (Figures 4.11 and 4.12). At the site above trap T4 the wells installed for Fiebig's (1995) study were used for sampling groundwater.

DOC concentrations in the stream water ranged from 1.6 to 8.6 mg C l^{-1} (average 3.2 mg C l^{-1}) at GT2, whereas in the interstitial water close to the surface (0–15 cm depth) higher values, from 1.7 to 16.7 mg C l^{-1} were measured (average 4.5 mg C l^{-1}; Table 4.6). DOC concentrations in the deeper layers (40–120 cm) ranged from 0.3 to 18.0 mg C l^{-1} (with averages in different depths between 4.7 and 5.5 mg C l^{-1}). Above trap T4, DOC concentrations ranged from 1.6 to 8.2 mg C l^{-1} in stream water, from 1.2 to 10.6 in upper interstitial water (0–15 cm) and from 1.0 to 22.0 in deeper interstitial water (40–120 cm) (averages 3.3, 3.7, 3.6 mg C l^{-1}, respectively). At this site, groundwater from 60 cm depth contained the highest DOC concentrations measured in this study, ranging from 2.2 to 35.9 mg C l^{-1}, with 9.7 mg C l^{-1} on average.

Figure 4.12 Site 3 m above gauge GT2 for DOC sampling (photo by Katzner). Three steel pipes for sampling deeper interstitial water (80–120 cm) are visible on the right hand side of the picture, together with a PVC tube serving as a piezometer (Fiebig, 1995). Tubes from the sampling disks (installed for investigating interstitial water between 0 and 15 cm depth) are visible to the left of the piezometer (cf. Chapter 3).

Compared to the amount of POM (annual average: 0.30 kg C m^{-2}; Table 4.4), the quantity (m^{-2}) of dissolved organic matter in the stream is far less. If both stream and interstitial water down to 7 cm are taken into account, mean annual DOC is about 0.8 g C m^{-2}. However, because of the high through-flow rates, this relationship does not say much about the importance of the two fractions. Considering the average DOC input of 1.02 kg C m^{-2} $year^{-1}$ for the Breitenbach (Table 4.7) compared with the particulate organic matter input of 0.42 kg C m^{-2} $year^{-1}$ (Table 4.4), the availability of dissolved material to the microbial community is much higher.

4.3.3
Inputs and Outputs of Dissolved Organic Matter to and from the Breitenbach

Dissolved organic matter inputs into the Breitenbach were determined as DOC by Fiebig (1995). Measurements were performed in the stream and adjacent groundwater at 6-week intervals over two years (1989/90). Mass balance calculations including storm flows indicated that an average of 337 g C m^{-2} $year^{-1}$ (or 1087 kg C $year^{-1}$ for the whole stream) entered the stream via groundwater discharge. The DOC input from the springs in the middle reach was 657 kg C $year^{-1}$ (Marxsen *et al.*, 1997), corresponding to 204 g C m^{-2} $year^{-1}$. Together the fractions total 541 g C m^{-2} $year^{-1}$. The input rates are highly dependent on groundwater discharge. Because stream discharge (reflecting groundwater discharge) was especially low

Table 4.6 Dissolved organic carbon (DOC) in different habitats at two sites in the middle reach of the Breitenbach, about 3 m above gauge GT2, and about 50 m above trap T4. For details, see Figure 4.10. Weekly sampling was from May 1999 to May 2000 (Katzner, 2011).

Site, habitat, depth	DOC (mg C l^{-1})		
	Mean	s.d.	Range
GT2 stream water	3.17	1.34	1.62–8.60
GT2 interstitial water 0–15 cm	4.88	2.48	2.10–16.71
	4.36	2.42	1.71–16.42
	4.35	1.76	1.85–8.98
	4.23	1.71	2.18–8.81
	4.74	2.06	2.33–9.74
GT2 interstitial water 40 cm	5.53	2.58	2.78–14.80
GT2 interstitial water 80 cm	4.65	3.04	0.28–18.02
GT2 interstitial water 120 cm	5.03	2.44	2.11–15.32
T4 stream water	3.28	1.40	1.64–8.21
T4 interstitial water 0–15 cm	3.60	1.78	1.35–10.61
	3.89	1.61	1.58–8.63
	3.55	1.51	1.27–7.77
	3.76	1.55	1.15–8.27
	3.67	1.46	1.63–9.09
T4 interstitial water 40 cm	4.34	3.79	1.91–21.99
T4 interstitial water 80 cm	3.37	2.63	1.30–16.72
T4 interstitial water 120 cm	3.20	2.50	0.96–16.11
T4 groundwater right bank	6.17	3.09	2.18–15.76
T4 groundwater left bank	13.13	5.34	6.15–35.90

Table 4.7 Inputs and outputs of dissolved organic matter to the Breitenbach (from Marxsen, 2006). Data are for the years 1985–1987 and 1989–1990. Import by inter- and overland flow, and release from algae and macrophytes, was not determined.

Source	Value (kg C m^{-2} year^{-1})	Value (kg C year^{-1})	References
Import via groundwater and springs	1.02	3290	Fiebig (1995), Marxsen *et al.* (1997)
Import by inter- and overland flow	?	?	
Release from algae and macrophytes	?	?	
Export by transport to the River Fulda	0.74	2390	Fiebig (1995), Marxsen *et al.* (1997), Marxsen (unpublished data)

during the investigated period, the measured inputs were probably much lower than the long-term average. Thus, when all the years with DOC output measurements are taken into account, the input of DOC via groundwater is markedly higher (Marxsen, 2006): about 3290 kg C $year^{-1}$ (total stream) or 1020 g C m^{-2} $year^{-1}$ (Table 4.7).

The amount of DOC provided by inter- and overland flow has not been measured but is probably much less. The amount of DOC liberated from algae and macrophytes is also unknown, but excreted DOC can represent a considerable proportion of primary production, especially in oligotrophic waters (Cotner and Biddanda, 2002), and thus can contribute significant amounts of organic matter to streams (Kaplan and Bott, 1989).

Three different studies of DOC export were performed. For each study, stream water was sampled close to the confluence of the Breitenbach with the River Fulda. Between September 1985 and August 1986 monthly samples, taken simultaneously with FPOM measurements, resulted in an annual export value of 1650 kg DOC. Weekly samples from 1985 to 1987 revealed annual DOC exports of 1573, 3751, and 4109 kg (Marxsen *et al.*, 1997; Marxsen, 2006, unpublished data). For 1989 and 1990, DOC concentrations were modeled on a daily basis and Fiebig (1995) estimated that 1333 and 1195 kg DOC, respectively, were exported. Export of DOC is hydrologically variable, which is demonstrated by the distinct contrast between 1987 (4109 kg, with annual stream discharge of $1.17 \times 10^6\,m^3$ in the middle reach) and 1990 (1195 kg, annual stream discharge of $0.48 \times 10^6\,m^3$). By combining the data for 1985–1987 and 1989/90, the average DOC export from the Breitenbach equals 2390 kg C $year^{-1}$, corresponding to 740 g C m^{-2} $year^{-1}$ (Marxsen, 2006).

4.3.4
Composition of Dissolved Organic Matter in the Breitenbach: Carbohydrates

Dissolved carbohydrates are important components in freshwaters because they provide a significant source of carbon and energy, especially to heterotrophic bacteria. The technique of HPLC with pulsed amperometric detection (PAD), which combines high sensitivity and accuracy (Mopper *et al.*, 1992; Jørgensen and Jensen, 1994; Gremm and Kaplan, 1997), was used to determine their composition in the Breitenbach. Thus, even for water from an unpolluted stream like the Breitenbach, it was possible to quantify at least the most abundant free saccharides without pre-concentration, thus reducing the risk of contamination (Katzner, 2011). From May 1999 to May 2000 the composition of the pool of dissolved mono- and disaccharides, and their contribution to dissolved organic carbon (DOC), was analyzed in water from different habitats by weekly sampling: in stream water, interstitial water of the upper sediment layers (down to about 15 cm), interstitial water of deeper sediment (40–120 cm depth), and groundwater adjacent to the stream (Figures 4.11 and 4.12). The HPLC system used (Dionex DX 500) consisted of a GP40 gradient pump, an AS40 automated sampler, a LC10 chromatography organizer with rheodyne injection valve and an ED40 electrochemical detector, which was connected to KromaSystem 2000 (Kontron) for data acquisition and evaluation. It was equipped with a CarboPac PA-100 separation column (4 × 250 mm), packed with microporous polymeric resin (90 μeq anion exchange capacity), and a CarboPac PA-100 guard column (4 × 50 mm). The detection limits of the system were between 5 and 10 nmol l^{-1}, depending on sugar.

In addition to the weekly determination of individual monosaccharides by PAD-HPLC, a spectrophotometric approach was applied bi-weekly to analyze the sum of dissolved mono- and polysaccharides (Avigad, 1968; Myklestad *et al.*, 1997). Total monosaccharide concentrations were measured after oxidation at alkaline pH. The sum of mono- and polysaccharides was determined using the same procedure, after hydrolysis of the polysaccharides with hydrochloric acid. The concentration of polysaccharides was calculated from the difference between total saccharides and monosaccharides. For analytical reasons (disturbance of the reactions by iron from the steel pipes) no measurements were available for deeper interstitial water (40–120 cm) using this method.

Table 4.8 gives the annual average concentrations of DOC and carbohydrates for stream and interstitial water (0–15 cm) at the two sampling sites from the middle reach. Average total dissolved carbohydrates contributed 0.38 and 0.41 mg C l^{-1} in stream water and at 0.48 and 0.35 mg C l^{-1} in interstitial water (0–

Table 4.8 Dissolved carbohydrates (annual means) in stream and interstitial water at a sampling sites 3 m above gauge GT2 (GT2) and 50 m above trap T4 (T4). Data from bi-weekly measurements from May 1999 to May 2000 (Katzner, 2011). All data in mg C l^{-1} except percentages.

	Site	Stream water		Interstitial water (0–15 cm)	
		Mean	s.d.	Mean	s.d.
Total dissolved organic carbon (DOC)	GT2	3.17	1.34	4.51	2.06
	T4	3.28	1.40	3.69	1.55
Total dissolved carbohydrates (TCHO)	GT2	0.38	0.14	0.48	0.15
	T4	0.41	0.14	0.35	0.13
Total dissolved free carbohydrates (TFCHO)	GT2	0.20	0.07	0.28	0.11
	T4	0.22	0.08	0.17	0.07
Total dissolved combined carbohydrates (TCCHO)	GT2	0.18	0.13	0.20	0.14
	T4	0.19	0.13	0.18	0.13
TCHO relative to DOC	GT2	12.8%	6.5%	11.5%	0.3%
	T4	13.2%	5.6%	10.0%	0.3%
TFCHO relative to DOC	GT2	6.6%	2.7%	6.5%	0.3%
	T4	6.8%	2.8%	4.8%	0.4%
Sum of individual free dissolved carbohydrates (FCHO)	GT2	0.0067	0.0058	0.060	0.052
	T4	0.0079	0.0061	0.061	0.042
Sum of FCHO relative to DOC	GT2	0.24%	0.27%	1.56%	1.60%
	T4	0.29%	0.31%	1.78%	1.40%

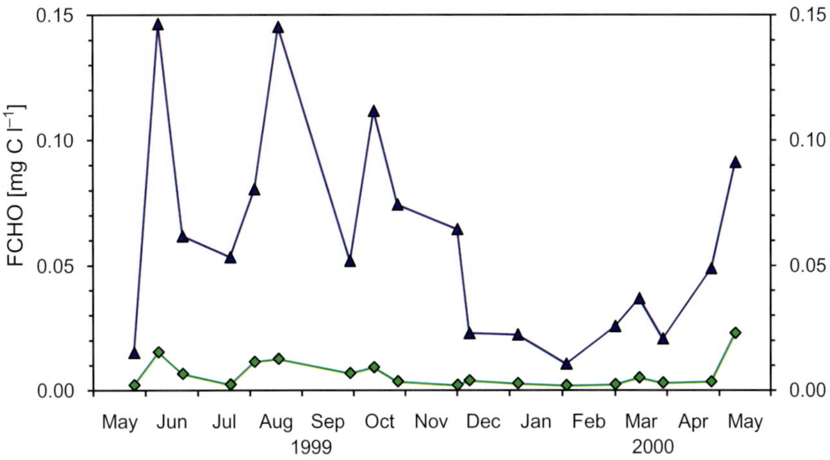

Figure 4.13 Concentrations of the sum of free dissolved carbohydrates (FCHO) in stream and interstitial (0–15 cm depth) water at gauge GT2 from May 1999 to May 2000 (Katzner, 2011).

15 cm). This represents 10.0–13.2% of total dissolved organic carbon. At least half of the total carbohydrates (sometimes more), are free dissolved.

The sum of the individual free saccharides determined by HPAE-PAD was much lower than the concentrations determined with the colorimetric assay for total carbohydrates (Table 4.8): Only about 7–8 μg C l^{-1} (annual average) were measured in stream water and about 60 μg C l^{-1} (annual average) in interstitial water (Figure 4.13, Table 4.8). This is a well documented effect (e.g., Jørgensen and Jensen, 1994). One reason might be that not all carbohydrates are identified using HPLC, but it has also been suggested that total carbohydrate assays overestimate the concentrations because there is interference from other compounds (noncarbohydrates).

Seasonal fluctuations in the saccharides are shown as the sum of the concentrations of the individual compounds, measured by HPAE-PAD against time for site GT2 in Figure 4.13. There is a trend to higher concentrations in the summer and also during autumn, and to lower concentrations during winter. The observation of temperature dependent occurrence of free dissolved saccharides is an indication that most of the compounds originate from products of biological degradation from high molecular weight organic matter.

Fifteen different saccharides could be identified and determined quantitatively by HPAE-PAD: one disaccharide (saccharose) and 14 monosaccharides (Table 4.9). Although distinct deviations between individual sugars occurred, concentrations measured were highly variable for all compounds at all sites and habitats. Standard deviations achieved typically a magnitude similar to the respective average concentrations or were even higher.

Only six saccharides were found regularly or at least mostly in stream water, as can be seen from the annual distribution and the annual averages of individual

Table 4.9 Saccharides identified by HPAE-PAD after separation on a CarboPac PA-100 column, with their abbreviations.

Sugar	Abbreviation	Formula
Fucose	Fuc	$C_6H_{12}O_5$
Rhamnose	Rha	$C_5H_{12}O_5$
2-Deoxy-glucose	dGlc	$C_5H_{10}O_4$
Galactosamine	GalN	$C_6H_{13}NO_5$
Arabinose	Ara	$C_5H_{10}O_5$
Glucosamine	GlcN	$C_6H_{13}NO_5$
Galactose	Gal	$C_6H_{12}O_6$
Glucose	Glc	$C_6H_{12}O_6$
Mannose	Man	$C_6H_{12}O_6$
N-Acetyl-glucosamine	GlcNAc	$C_8H_{15}NO_6$
Xylose	Xyl	$C_5H_{10}O_5$
N-Acetyl-galactosamine	GalNAc	$C_8H_{15}NO_6$
Fructose	Fru	$C_6H_{12}O_6$
Saccharose (sucrose)	Sacch	$C_{12}H_{22}O_{11}$
Ribose	Rib	$C_5H_{10}O_5$

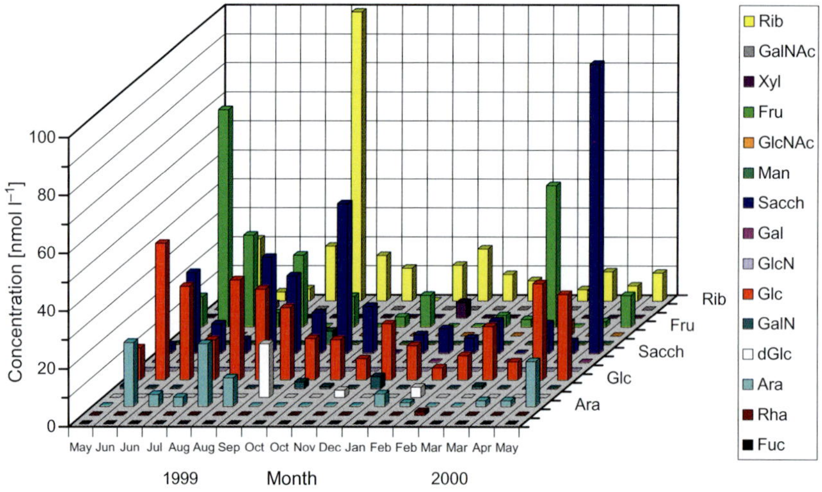

Figure 4.14 Concentrations of individual free dissolved carbohydrates in stream water at gauge GT, from May 1999 to May 2000 (Katzner, 2011). For abbreviations, see Table 4.9.

free dissolved carbohydrates (Figures 4.14 and 4.15). These saccharides were: saccharose, glucose, ribose, fructose, 2-deoxy-glucose, and arabinose, reaching concentrations up to about 100 nmol l^{-1}.

In the interstitial water (0–15 cm) at site GT2 the concentrations of the individual carbohydrates were up to ten times higher, and additional compounds occurred

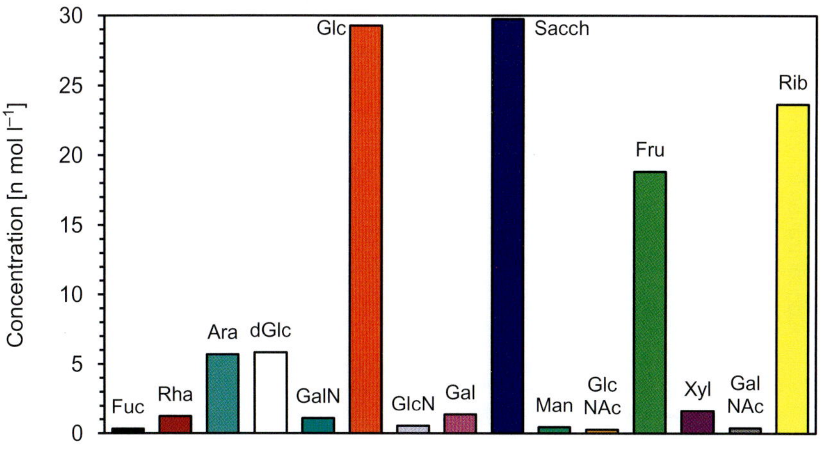

Figure 4.15 Annual averages of individual free dissolved carbohydrates concentrations in Breitenbach stream water (six sampling sites along the whole stream) from May 1999 to May 2000 (Katzner, 2011). For abbreviations, see Table 4.9.

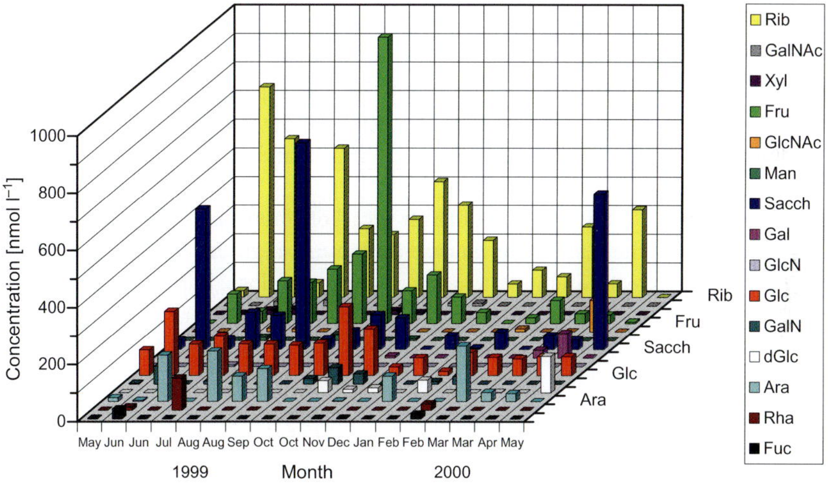

Figure 4.16 Concentrations of individual free dissolved carbohydrates in interstitial water (averages of measurements from 0 to 15 cm depth) at gauge T2 from May 1999 to May 2000 (Katzner, 2011). For abbreviations, see Table 4.9.

above the detection limit (Figures 4.16 and 4.17). But there were also a lot of saccharides whose concentrations remained below the detection limits. Besides the six most important saccharides, which are the same in both habitats, three more carbohydrates, with annual mean concentrations between 8 and 19 nmol l^{-1}, were regularly detected (Figure 4.17). These were: galactose, galactosamine, and

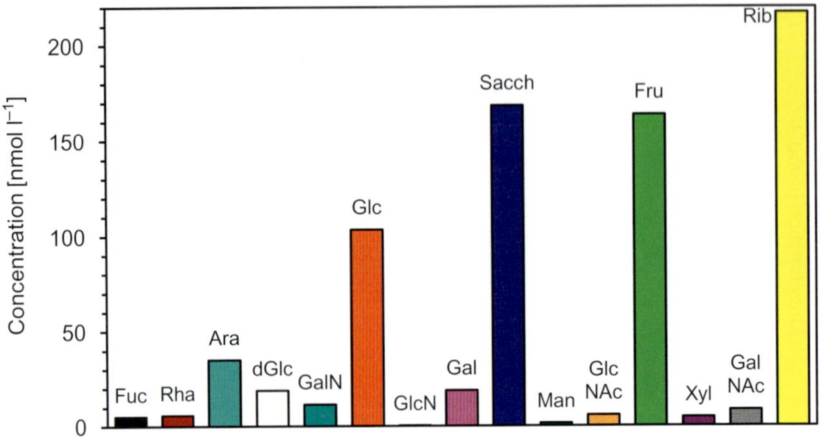

Figure 4.17 Annual averages of individual free dissolved carbohydrates concentrations in interstitial water (averages of measurements from 0 to 15 cm depth) at gauge GT2 from May 1999 to May 2000 (Katzner, 2011). For abbreviations, see Table 4.9. Similar concentrations were measured at the site about 50 m above trap T4 (data not shown).

N-acetyl-galactosamine. Similar results were obtained at the site about 50 m above trap T4 (data not shown).

Some general trends are apparent from Katzner's pioneer study with respect to the factors that determine the distribution of individual free dissolved carbohydrates over time and space, their origins and fates. However the final picture is not yet clear.

Glucose is one of the most abundant free carbohydrates in the Breitenbach, but considering that there are many glucose-containing, high molecular weight, compounds in the stream and its catchment, mainly cellulose, but also starch or glycogen (Neumann, 1995; Marxsen, 2006), and also considering the very high activities of extracellular enzymes involved in the degradation of these macromolecules (Marxsen and Fiebig, 1993), much higher concentrations of glucose (than other saccharides) should occur. However, the concentrations of compounds in the pool of organic matter are not only determined by inputs to the pool, but also by the outputs, especially their utilization by heterotrophic microorganisms (Marxsen, 1980c).

The main difference in the occurrence of single saccharides between the two habitats (stream and interstitial water, 0–15 cm) were the much higher concentrations of ribose and fructose in pore water, which increase to about ten-fold that in stream water. Increases for arabinose, saccharose, 2-deoxy-glucose, and glucose were only three- to sevenfold. Ribose is a component of RNA and many nucleotides. These molecules occur in greater amounts in small, metabolically active organisms than in higher plant tissues where they are overshadowed by structural polysaccharides (Cowie and Hedges, 1984). Thus the remarkably high concentrations of ribose in interstitial waters, which have not been reported before, were probably the result of highly active sediment communities which contain much RNA. High degradation activities, especially high extracellular enzyme activity (e.g., Marxsen and Fiebig, 1993; Marxsen and Schmidt, 1993) and high bacterial production (e.g., Marxsen, 1996, 2001, 2006) have been consistently determined in the Breitenbach sediment habitats.

Whereas ribose occurred at considerably high levels, its DNA counterpart, deoxyribose, could never be detected. But deoxyribose is a very labile sugar that is oxidized very quickly (Mopper, 1977). It is probably included in the high quantities of ribose, although its contribution is unknown.

Elevated concentrations of galactose and arabinose were observed in stream water and mainly in surface interstitial water. These compounds are indicative of a nonwoody angiosperm origin of organic matter (Cowie and Hedges, 1984), because nonwoody tissues are usually richer in pectin (a polysaccharide in which arabinose and galactose are the predominant neutral sugars) than woods. Thus the presence of elevated levels of both compounds reflects the dominance of angiosperms in- and outside the stream. Macrophytes within the stream are almost entirely angiosperms, while particulate organic matter inputs to the Breitenbach are composed primarily of angiosperm tissue (cf. Chapter 5 and Section 4.2).

A higher percentage of mannose is produced by gymnosperms than angiosperms, due to the greater abundance of glucomannan in their tissues. Thus mannose is indicative of a gymnospermous origin of organic matter (Cowie and Hedges, 1984). That mannose was usually below the detection limit correlates with the low importance of gymnosperm tissue within the stream's macrophyte flora and among the allochthonous inputs of particulate organic matter to the stream.

5
Primary Producers

Eileen J. Cox, Jürgen Marxsen, and Thomas G. Horvath

5.1
Primary Producers in Streams

From an ecological point of view, the autotrophic organisms that acquire their materials from inorganic sources are referred to as primary producers. In the Breitenbach these are mainly photoautotrophic organisms that receive their energy from sunlight. However, chemoautotrophic production, which is based on energy from redox-reactions, has never been investigated. However it is known that, in this type of stream, photoautotrophic production dominates and chemoautotrophic production is quantitatively unimportant (Wuhrmann, 1974).

The major primary producers of the Breitenbach are large plants (macrophytes) and various small organisms, referred to as phytoplankton when occurring in suspension and as periphyton ("aufwuchs") when associated with surfaces. Among the macrophytes, aquatic angiosperms dominate, but other vascular and nonvascular plants, such as pteridophytes and bryophytes (mosses and liverworts) also occur. Some lotic algae may also become sufficiently large to be treated as macrophytes, but these do not usually occur in the Breitenbach. The algae composing the photoautotrophic periphyton and plankton comprise a diverse group of eukaryotes, especially diatoms and green algae, but contain also the prokaryotic group of cyanobacteria.

During the decades of research on the Breitenbach this group of organisms was not investigated continuously. There were a few early, highly specialized studies, in the 1960s, especially on algal drift (Müller-Haeckel, 1966), but the main research on algae was performed between 1985–1989 by Cox (1987, 1988, 1990a–d, 1991). These investigations focussed on the composition of the algal communities and their abundance, including the impacts of environmental factors that potentially influence the communities, such as substratum, current or the scouring effects of floods. Macrophytes were considered even less. There were only a few occasional studies with respect to community composition, abundance and production, for example, from Koch (1990) or Horvath (2004, unpublished data). A floristic survey of bryophytes is available from Gregor and Wolf (2001).

Central European Stream Ecosystems: The Long Term Study of the Breitenbach, First Edition. Edited by Rüdiger Wagner, Jürgen Marxsen, Peter Zwick, Eileen J. Cox.

No systematic approach was followed to measure primary production. A very early investigation of periphytic production is available (Marxsen, 1980a), and two studies of macrophyte production were performed later (Koch, 1990; Horvath, 2004, unpublished data). These were the basis for the tentative estimates of total stream primary production by Marxsen *et al.* (1997) and Marxsen (2006). Finally an estimate of primary production was made based on diurnal oxygen curves from 1997 to 2004 (Schmidt, unpublished data).

Thus the focus of this chapter on primary producers and primary production are the detailed studies of algae in the Breitenbach performed from 1985 to 1989 by Cox (1987, 1988, 1990a–d, 1991).

5.2 The Communities

5.2.1 Algae

5.2.1.1 Algal Assemblages in Streams

Benthic algae are usually the dominant primary producers in unshaded streams (Lamberti, 1996), and are therefore crucial to the development of higher trophic levels. Such algal assemblages develop primarily on solid substrata, such as rocks and large stones, and on sediments that accumulate under lower flow velocities. Where macrophytes (angiosperms or bryophytes) occur microalgae may also grow epiphytically, or be loosely associated in and around submerged stems, branches and leaves. True plankton does not develop as flow rates are too great for active growth *in situ*. Cells found in suspension will usually have been entrained from the benthos, or washed in from other aquatic habitats.

Over the past 80 years there have been many studies of algae in lotic systems, with varying emphases on species assemblages (Budde, 1928; Butcher, 1940; Douglas, 1958; Moore, 1974, 1976; Patrick, 1949, 1961; Cox, 1990a), zonation (Johnson *et al.*, 1975; Aykulu, 1982; Kawecka, 1986; Ward, 1986) and temporal variations (Antoine and Benson-Evans, 1986; Burkholder and Sheath, 1984; Moore, 1976, 1977, 1981; Sheath and Burkholder, 1985), effects of discharge and flooding (Cox, 1988; Biggs and Close, 1989; Biggs and Smith, 2002), nutrient regimes (Deniseger *et al.*, 1986; Descy and Mouvet, 1984; Lowe *et al.*, 1986) and grazers (Cox, 1990b; Lamberti *et al.*, 1987a, b) on both standing crop and species occurrences. Direct field observations have been complemented by *in situ* manipulations, particularly dosing experiments and the use of artificial substrata (McIntire, 1968; Marker and Casey, 1982; Steinman and McIntire, 1982; Pringle and Bowers, 1984; Lowe *et al.*, 1986), and occasionally by laboratory experiments (Cox, 1993, 1994). With the development of water quality monitoring programmes (e.g., Kelly and Whitton, 1995; Stevenson and Pan, 1999) diatom assemblages have been used as a proxy for nutrients, particularly phosphorus and pollution (e.g., Prygiel, 1991; Round, 1991; Sabater *et al.*, 1991; Schiefele and Schreiner, 1991), and diatom analyses have also been incorporated into large-scale freshwater surveys in particular geographical regions, for example, the Himalayas (Jüttner *et al.*, 1996, 2003; Rothfritz *et al.* 1997). Yet, perhaps because of the varied climatic, geological and hydrological factors which determine the immediate environment of any one stream, simple comparisons can be difficult, even within particular stream types. Entirely field-based observations are also inherently limited; they allow inferences to be made but cannot confirm causal relationships (Cox 1991).

Understanding the factors controlling the presence and abundance of benthic algae in streams is compounded by the intimate interactions between the algae and their immediate, microenvironment. Growing at the interface of a substratum and overlying water, they are potentially influenced by both, and also by any grazers in the system. Thus the standing crop of algae at any site on any occasion is the net outcome of primary production versus grazing (and dislodgement). Reductions in grazing pressure will be reflected in increases in algal biomass, while flood events will drastically reduce it. Meanwhile the composition of the assemblage will be a result of individual species' responses to chemical and physical variables expressed through their differing growth rates. Stochastic effects, such as first colonization events, may also be significant.

Algal communities are discriminated not only temporally and spatially, but also according to their mode of growth, whether attached, and if so to what type of substratum (epilithon, epiphyton, epipsammon), or free-living (epipelon, plankton). The study of different habitats requires different sampling strategies, and also presents challenges to the quantification of algal populations. Approaches to studying different groups of algae have resulted in a split between diatoms and other groups. Diatoms are most frequently studied only after samples have been oxidized with acid, thereby destroying all other algae. As a result, most studies of benthic diatom communities have expressed results as percentage abundances of cleaned samples. The study on the Breitenbach was unusual in that diatoms were studied from fresh material and the presence of other algal taxa could be noted simultaneously (see algal list, Table 5.1). However, diatoms were the dominant group on most occasions and algal work on the Breitenbach focussed on them.

5.2.1.2 Spatial and Temporal Patterns of Diatom Occurrences in the Breitenbach

From March 1986 to December 1988 epipelic samples were taken at 2- or 4-week intervals from 12 sites along the Breitenbach, downstream from the Georgsborn (including just above its junction with the upper reach), and investigated using Round's (1953) 1-day count method as modified by Happey-Wood and Priddle (1984). Taxa were counted as fresh specimens but identifications were checked using cleaned material from the same sample. Assemblage composition was assessed based on counts of at least 300 live cells whenever possible, and species with >2% occurrence on any occasion in any habitat are given in the Appendix[1)].

1) Supplementary material is available at www.wiley-vch.de/home/breitenbach and http://pubman.mpdl.mpg.de/pubman/item/escidoc:969687.

Table 5.1 Percentage similarity between epilithic diatom samples at different sites and substrata (for location of sites, see Chapter 2; $T4_s$ = site T4, sandstone, $T4_l$ = site T4, limestone).

28 January 1988

T1	74.4			
T2	65.2	67.1		
$T4_s$	48.7	56.1	66.4	
$T4_l$	39.8	49.4	37.5	39.0
	G	T2	$T4_s$	$T4_l$

10 March 1988

T1	71.7			
T2	43.4	57.7		
$T4_s$	53.1	64.2	71.8	
$T4_l$	15.8	25.5	57.1	37.9
	G	T2	$T4_s$	$T4_l$

5 May 1988

T1	72.9			
T2	48.0	43.7		
$T4_s$	59.5	54.2	77.2	
$T4_l$	41.4	39.1	89.6	71.4
	G	T2	E_s	E_l

7 July 1988

T1	64.8			
T2	46.0	61.1		
$T4_s$	45.2	75.9	73.4	
$T4_l$	37.8	57.9	45.3	63.7
	G	T2	$T4_s$	$T4_l$

6 September 1988

T1	29.4			
T2	19.4	51.9		
$T4_s$	24.7	56.5	65.2	
$T4_l$	14.7	55.5	57.0	72.0
	G	T2	$T4_s$	$T4_l$

3 November 1988

T1	28.0			
T2	53.1	26.8		
$T4_s$	68.8	24.8	59.2	
$T4_l$	56.9	24.4	58.6	76.2
	G	T2	$T4_s$	$T4_l$

From April 1987 to December 1988 epipelic samples were taken from defined areas with a short surface corer, allowing cell densities (numbers cm^{-2}) to be estimated (Cox, 1990a).

Three sections, the upper reach, Georgsborn and the middle reach had been recognized within the stream system and this separation was reflected in the algal assemblages (Cox, 1990a). The crenic habitat of the Georgsborn represented the most stable environment, in which water temperature remained at about 8 °C throughout the year. The upper course is cool and shaded and may freeze in winter, whereas the middle and lower sections are winter-cool and summer-warm, unshaded for much of its length. There is greater similarity (about 60%) between the diatom floras (>2% level) in the upper and middle reaches than between either of these and the Georgsborn (about 30% similarity), with some diatoms found only in the latter habitat, while others species are invariably absent from the spring. Comparisons between the middle and lower reaches revealed that while the floras may be similar, species occur at contrasting abundances in the different sections.

The relative stability of the Georgsborn environment was reflected in the almost consistently abundant presence of several diatom species on the sediments, particularly *Planothidium lanceolatum, Amphora copulata, Neidium amphirhynchus, Placoneis pseudanglica, Staurosira elliptica* and *Sellaphora pupula* (see Figure 4 in Cox, 1990a). However, there were greater seasonal fluctuations in species in the upper reach (see Figure 6 in Cox, 1990a) and in the middle reach (see Figures 7,

8 in Cox, 1990a). In the upper reach, *Surirella brebissonii* and *Diatoma mesodon* were abundant in the spring, followed by a number of *Navicula* species in the summer, *A. copulata* in the autumn and *Placoneis clementis* in the winter. The number of taxa per site increased downstream in the middle reach, with *Navicula* and *Nitzschia* species becoming commoner. Some of the seasonal patterns seen in the upper reach were repeated in the lower reach, for example, spring abundances of *S. brebissonii* and *D. mesodon*, while peaks of the *Navicula* spp were often stronger in the middle than the upper reach (Figures 5.1–5.9). For most of the period May 1987–December 1988, cell densities at all sites fluctuated around 10^4 cells cm^{-2} sediment (Figure 5.10), although numbers were lowest in January–February and again after spring flood in the upper and middle reach, whereas in the Georgsborn cell densities were lowest in July–August 1988. Although cell numbers did not drop markedly at spring flood in the Georgsborn, there was a subsequent sharp increase in numbers (Figure 5.10).

Studies of epilithic diatoms revealed remarkable uniformity in the diatom assemblages in the stream, which were usually dominated by small naviculoid species that were not easily identified. The three most abundant taxa were *Eolimna minima*, *E. subminuscula* and *Fistulifera pelliculosa*, and at most sites these species consistently contributed at least 60%, often more, of the assemblage. Only at one site, T4 at the end of the middle reach, was the assemblage noticeably modified by the presence of *Amphora pediculus* and *Reimeria sinuata* (see Figure 11 in Cox, 1990a). There was relatively little effect of spatial location on the epilithic flora, but differences in substratum, for example, between sandstone and limestone (present

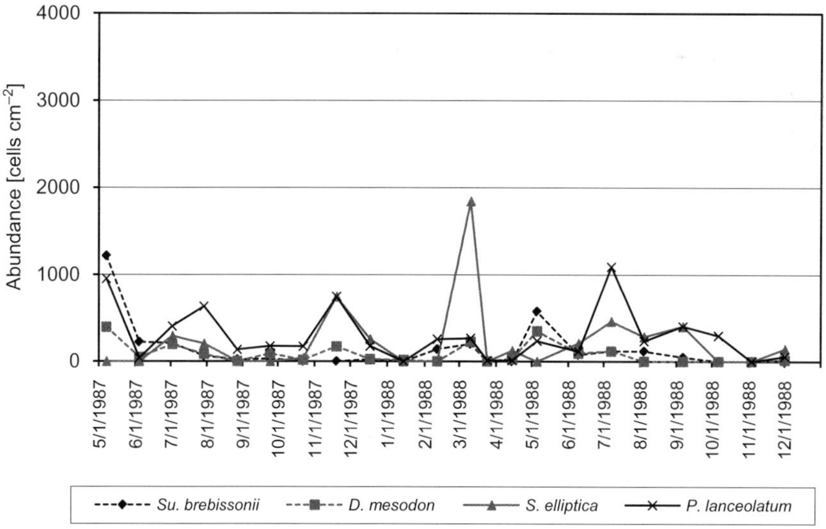

Figure 5.1 Estimated cell densities for four species in the upper reach over a 20-month period.

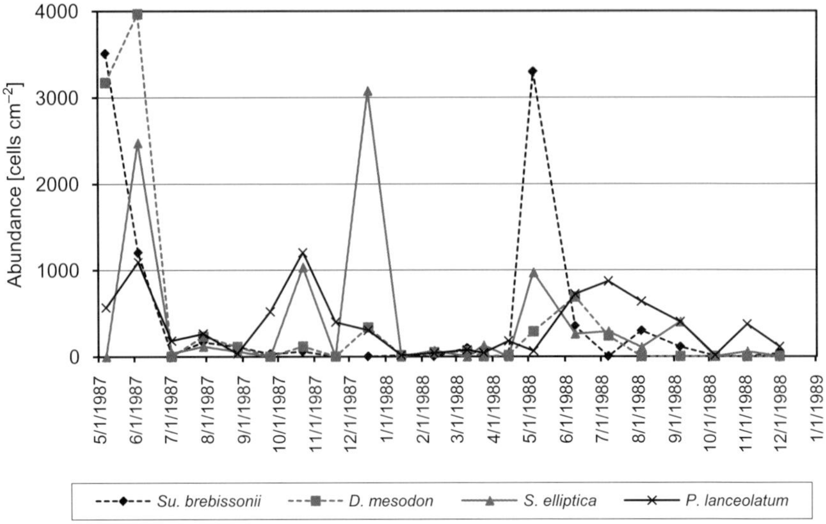

Figure 5.2 Estimated cell densities for four species in the middle reach (T2) over a 20-month period.

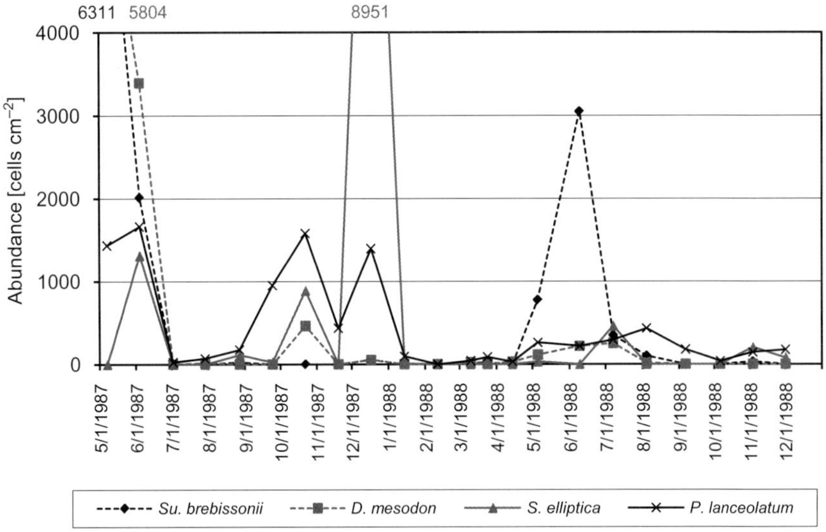

Figure 5.3 Estimated cell densities for four species in the middle reach (T3) over a 20-month period.

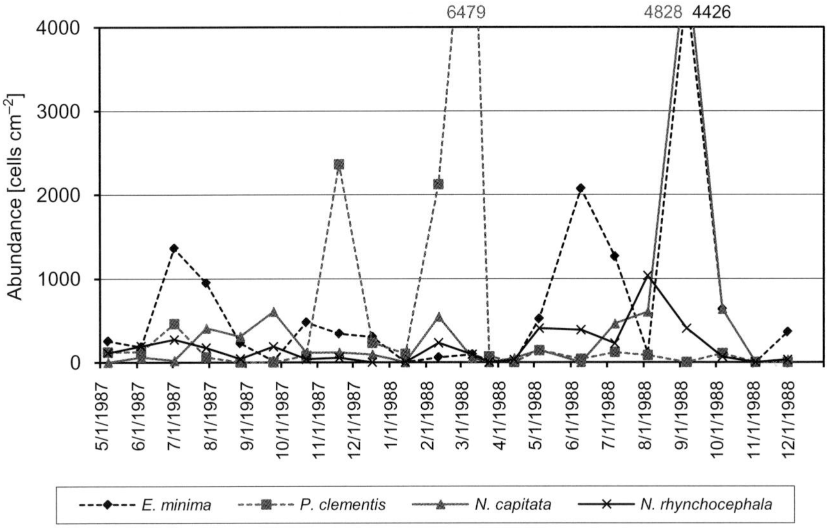

Figure 5.4 Estimated cell densities for four species in the upper reach over a 20-month period.

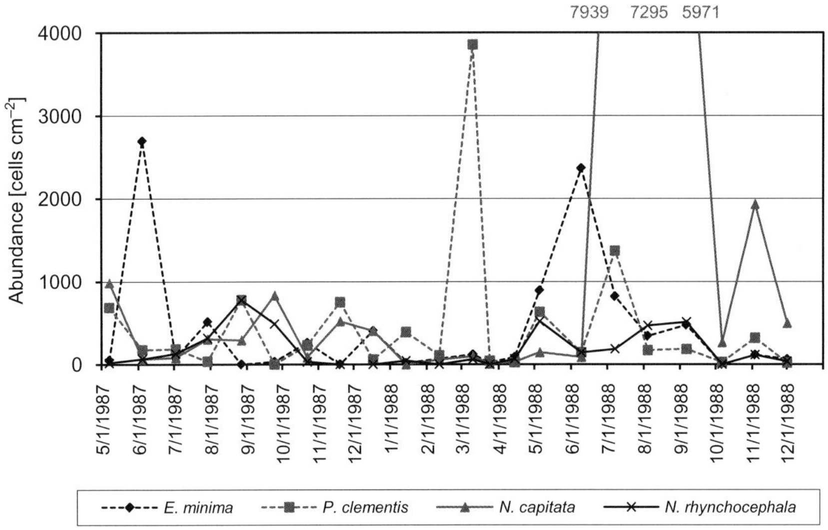

Figure 5.5 Estimated cell densities for four species in the middle reach (T2) over a 20-month period.

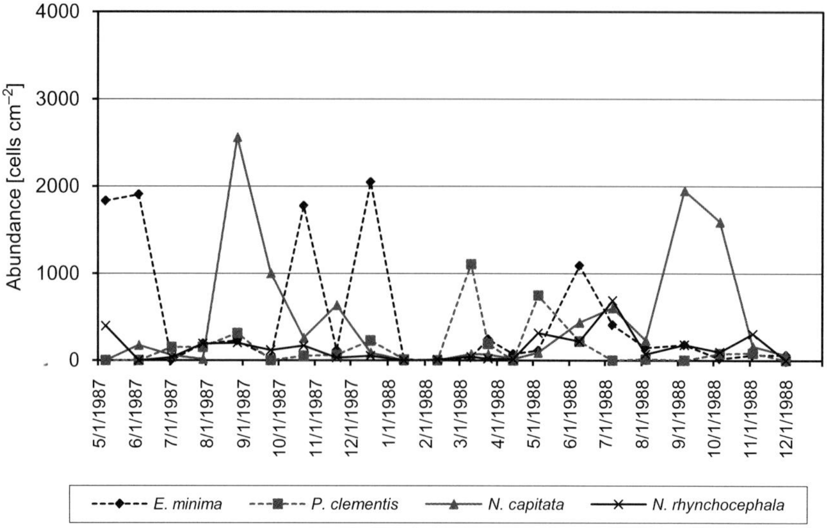

Figure 5.6 Estimated cell densities for four species in the middle reach (T3) over a 20-month period.

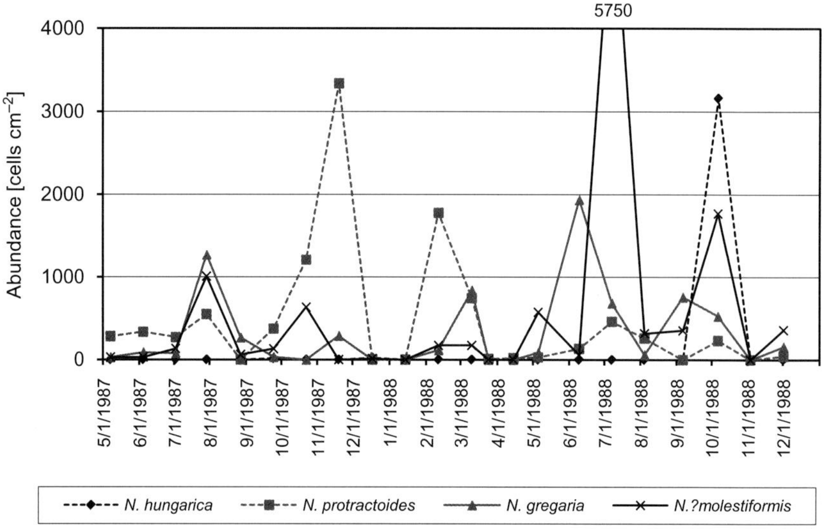

Figure 5.7 Estimated cell densities of four species in the upper reach over a 20-month period.

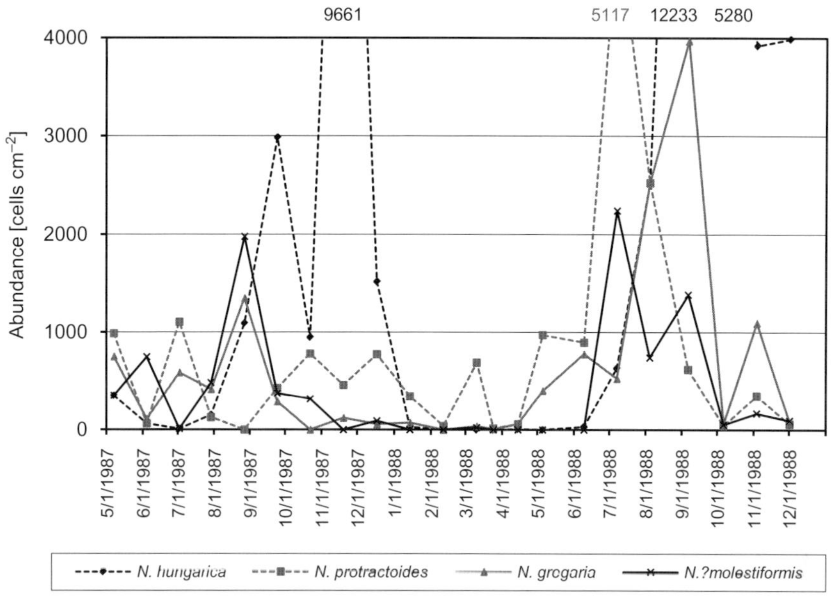

Figure 5.8 Estimated cell densities for four species in the middle reach (T2) over a 20-month period.

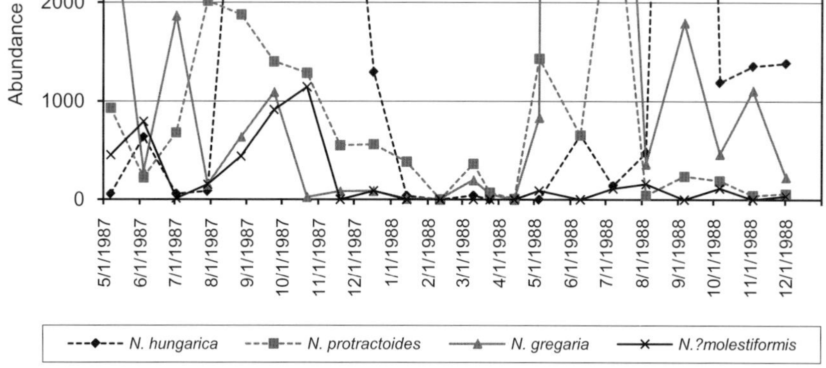

Figure 5.9 Estimated cell densities for four species in the middle reach (T3) over a 20-month period.

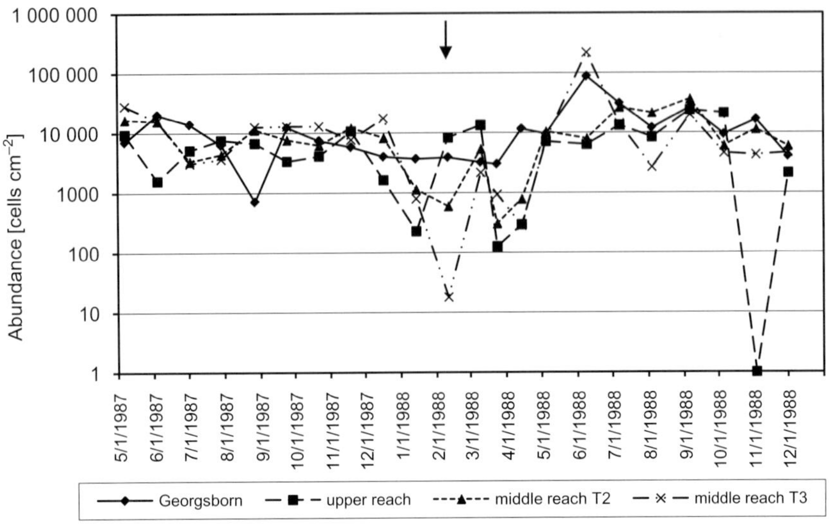

Figure 5.10 Graph showing fluctuation in total diatom cell numbers at four sites: Georgsborn, upper reach, middle reach T2 and T3. Arrow shows time of spate flood.

at the most downstream site T4), markedly reduced the between-site percent similarity (Table 5.1). Substratum type can therefore have a greater influence on the flora than location.

The epiphytic flora in the Breitenbach was relatively species-poor, usually dominated by *P. lanceolatum* and *Eunotia minor* (Table 5.2). Diversity often increased in the autumn, with the growth of *Synedra* and *Encyonema* spp. but despite the number of macrophytes in the stream, dense epiphytic growth was rare, particularly with respect to the monocotyledons. Surface examination of macrophytes frequently revealed very patchy distributions over plant surfaces, reflecting the limited motility of many epiphytic diatoms. Submerged bryophytes supported a contrasting diatom flora with higher abundances of *Pinnularia* spp. (Cox, 1990d), but these could also be contaminated by epipelic species when fine sediment had accumulated around the mosses.

5.2.1.3 Variation in Algal Biomass

Algal biomass was estimated over a 16 month period at four sites (Georgsborn, upper reach, two sites in the middle reach at T2 and T3), from February 1987 to May 1988, using chlorophyll-a (chl-a) extractions from sediments and stones (Cox, 1990b). For sediments the percentage of organic matter was also calculated by loss on ignition. Other shorter term studies investigated short-term fluctuations in chl-a in sediments and accumulation on sandstone tiles.

Regardless of the site, maximum chl-a concentrations were usually markedly higher on sediments (from 15 to >40 μg cm^{-2}) than on stones (<7 μg cm^{-2}), although

Table 5.2 Diatom species found as epiphyton or metaphyton, grouped according to their "preferred" habitat.

Epiphyton	Metaphyton
Achnanthidium minutissimum	*Diatoma mesodon*
Eolimna minima	*Encyonema silesiacum*
Eunotia minor	*Fragilariforma virescens*
Fistulifera pelliculosa	*Martyana martyi*
Navicula minuscula	*Melosira varians*
Planothidium lanceolatum	*Navicula cryptocephala*
Rossithidium linearis	*N. gregaria*
Synedra rumpens	*N. lanceolata*
	Nitzschia palea
	Staurosira elliptica
	Surirella brebissonii
	Synedra ulna
Species which are not consistently more abundant in one or other habitat	
Gomphonema angustatum	*Gomphonema truncatum*
G. parvulum	*Synedra vaucheriae*

the seasonal patterns differed between substrata and sites. In the Georgsborn and the upper reach there were peaks in sediment standing crop only in the spring (April/May), with lower, relatively stable chl-a concentrations (<5 μg cm^{-2}) between August and March (Figure 5.20). However, in the middle reach of the Breitenbach, there were two distinct peaks in standing crop, one in the spring (May) and a second in early autumn (September), although the relative heights of the peaks differed between the two sites (Figure 5.11). At T2 the peaks were comparable, whereas at T3 the spring peak was higher and the autumn peak lower than at T2. Chl-a was very low at both sites (~1 μg cm^{-2}) immediately after spring floods, indicating the scouring capacity of the floods.

The seasonal pattern of chl-a concentration on stones was similar between sites (Figures 5.12–5.13), but contrasted with that on sediments. At the start of the monitoring period (June 1987) chl-a on all stones was between 3–7 μg cm^{-2}, decreasing to the end of September after which there was a variable increase until November or December, depending on site. In the Georgsborn, chl-a concentration was decreasing before the spring flood, although more sharply with the flood, whereas it was increasing in the upper reach before flood. The middle reach sites showed an increase in chl-a on stones from October to November, decreasing to a low in January, after which chl-a was increasing until the spring flood occurred. Following a marked increase in chl-a at all sites after the spring flood (April–June), concentrations fell to a fairly constant low level until the end of the year (December 1988).

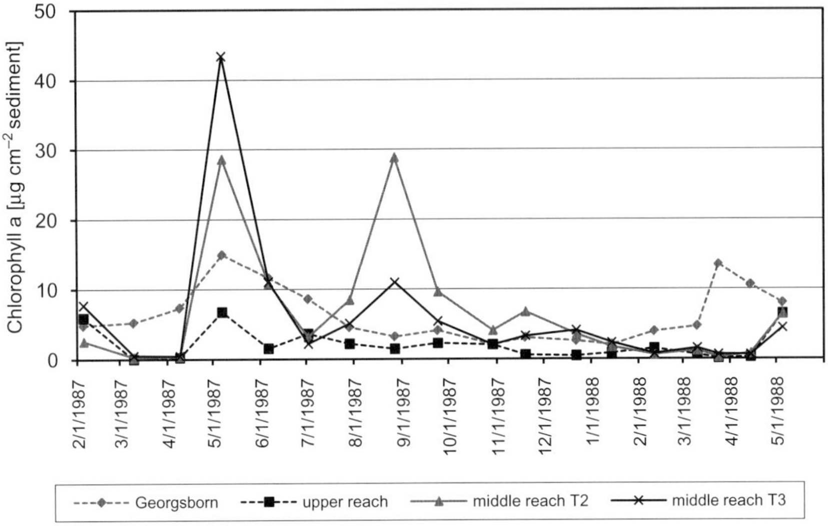

Figure 5.11 Graph showing variation in chlorophyll-a in upper 1 cm of sediment at four sites.

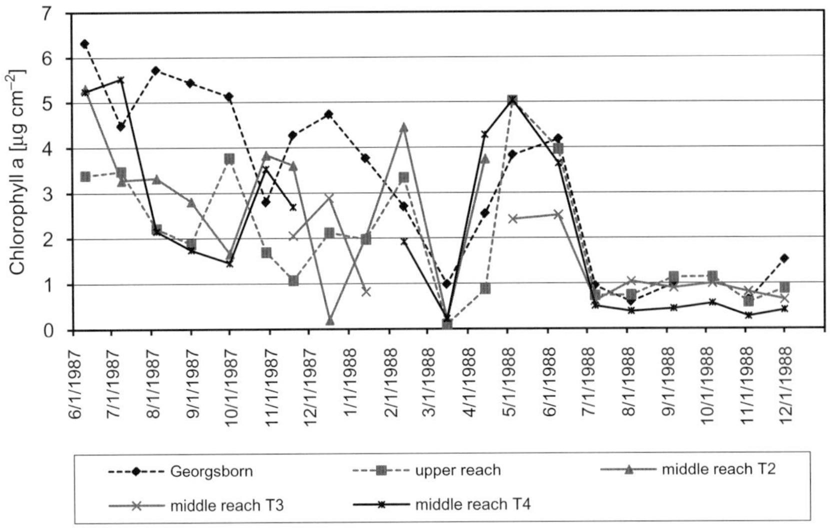

Figure 5.12 Graph showing changes in chlorophyll-a cm^{-2} stone at different sites in the Breitenbach.

On sediments the percentage of organic content largely followed the chlorophyll concentrations although there was an autumnal increase in organic content in the upper reach that was unrelated to any chl-a increase, probably due to leaf fall. A larger autumnal increase in organic matter at one middle reach site may also have been enhanced by allochthonous inputs (Figure 5.14). Organic content was not

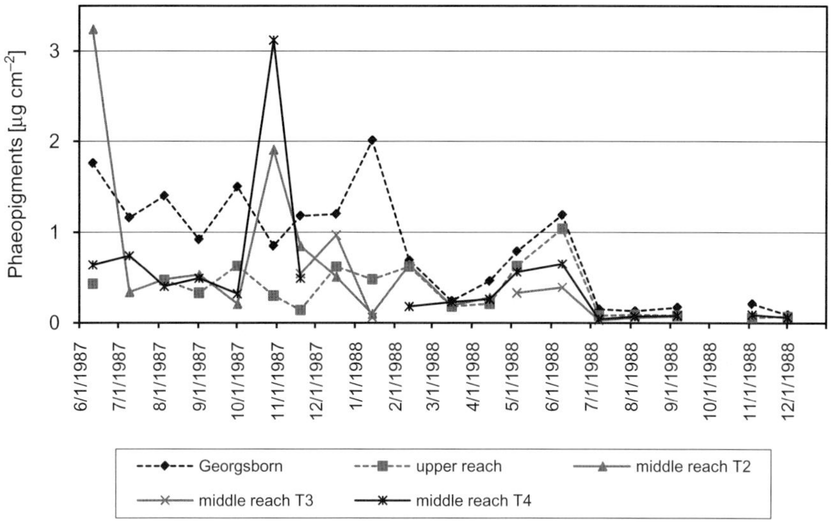

Figure 5.13 Graph showing changes in phaeopigments cm^{-2} stone at different sites in the Breitenbach.

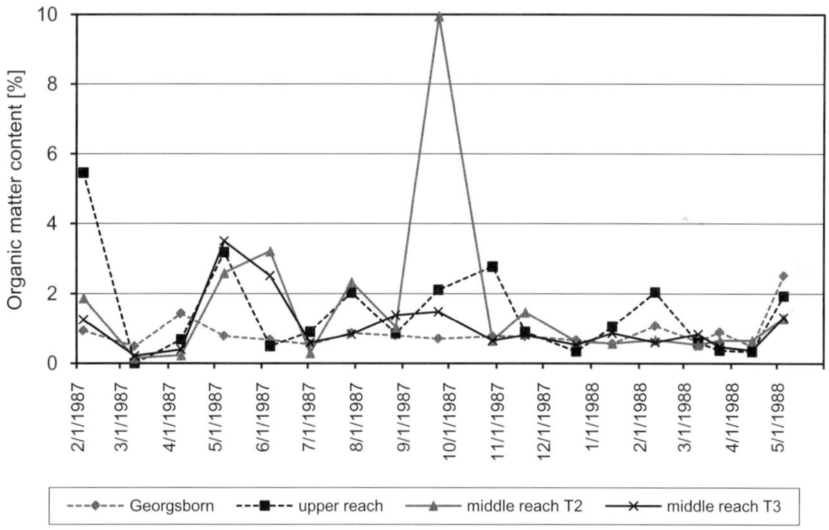

Figure 5.14 Graph showing variation in organic matter (%) in upper 1 cm of sediment at four sites.

estimated for stone standing crop, but variation in phaeopigments compared to chl-a indicated that there was an increase in dying or dead algae concomitant with, or towards the ends of, the chl-a peaks (Figures 5.12–5.13).

Short term studies on standing crops showed that chl-a concentrations in sediments could fluctuate strongly over a matter of days, for example, increasing from about 5–75 $\mu g\,cm^{-2}$ in less than a week, and also halving in a few days (Cox, 1990b). Accumulation rates on stones were not as striking, but varied with season and site (Cox, 1990b), the fastest accumulation in the middle reach occurring in the spring (April–June), that in the Georgsborn in summer (June–August) and autumn (October–November).

5.2.1.4 **Effects of Discharge on Benthic Diatom Assemblages**

The effects of discharge on benthic diatoms are most frequently recognized when substratum scouring removes diatom assemblages during spates, with the speed of recovery and the resulting assemblage composition being dependent on time of year (diatom growth is usually faster in the spring and autumn than summer and winter) and the nature of the substratum, although grazing pressure may also be important. Whether there is invariably a particular sequence of colonization is less clear. Some studies have suggested that opportunistic taxa such as *Achnanthidium minutissimum* are the first colonizers (Stevenson, 1983), or that there is a progression from low diversity, low-growing adnate assemblages to more diverse assemblages including apically attached and/or stalked diatoms (Stevenson, 1984; Oemke and Burton, 1986; Steinman and McIntire, 1986). Most stream and river studies have focussed on epilithic habitats (Stevenson, 1984; Peterson and Stevenson, 1989) and the epipelon has effectively been ignored, perhaps because most workers have assumed that these are restricted to relatively quiet waters (Round, 1981). Yet extensive epipelic assemblages that can play a significant role in sediment stabilization (Grant *et al.*, 1986) develop in high-energy estuarine ecosystems (Smith and Underwood, 1998; Consalvey *et al.*, 2004) and even in moderately fast-flowing systems or at lake margins, characteristic epipelic and epipsammic assemblages occur (Cox, 1984, 1988, 1990a).

In lotic systems, water movement sorts sediments by particle size, the stronger the current, the larger the particles. Thus, according to water velocity, particles of different sizes will settle and create varying microenvironments for aquatic organisms. Where there is a regular change in discharge with season, sediment accumulation will follow a predictable pattern and different assemblages will develop according to the prevailing microhabitat. Changes in assemblage therefore reflect modification of the substratum rather than the more obvious seasonal factors such as day length, insolation and temperature. In addition to variation in sediment mobility linked to grain size, adsorptive properties of the sediment modify the chemical environment, finer silts having higher nutrient absorption and therefore providing a more nutrient-rich environment.

As might be expected, grain size analysis showed less fluctuation in the Georgsborn, changes in the finest and coarsest fractions accounting for much of the variation (Figure 5.15). This site also had a consistently higher proportion of coarser particles (>0.6 mm) and rarely more than 20% of the finest fraction (<0.12 mm). The other sites showed stronger fluctuations in grain size with a higher proportion of finer particles; coarser fractions rarely exceeded 50% and the finest fraction usually varied between 20–40%, sometimes exceeding 60% after periods of low flow (Figures 5.16–5.18). Grain size variation at T2, where discharge was moni-

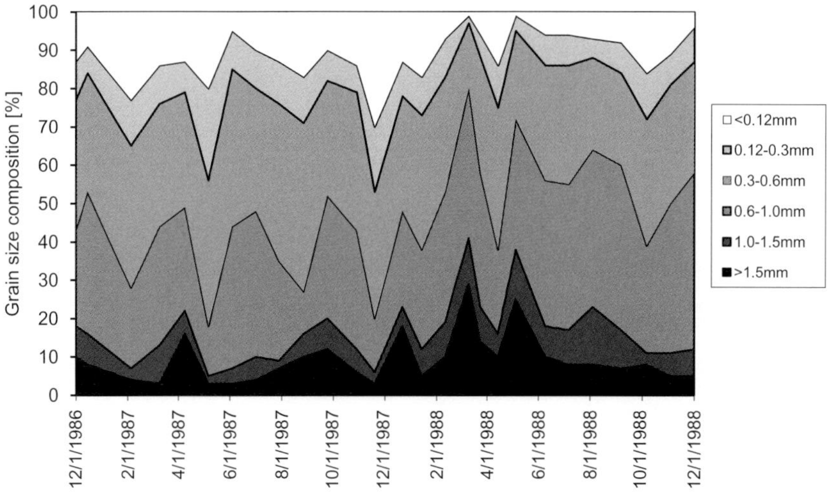

Figure 5.15 Graph showing grain size analysis in the Georgsborn from December 1986 to December 1988.

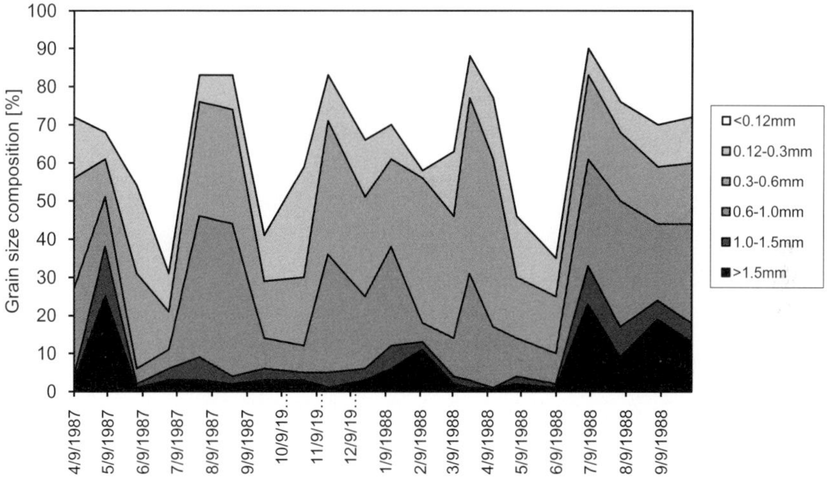

Figure 5.16 Graph showing grain size analysis in the upper reach from April 1987 to October 1988.

tored, clearly showed the effect of spring flood in 1988, with a short sharp increase in coarser particles, followed by a gradual increase in the percentage of the finest fraction (Figure 5.17).

No clear seasonal pattern of variation, or correlation of diatom assemblage with grain size could be discerned in the Georgsborn, and there was a marked

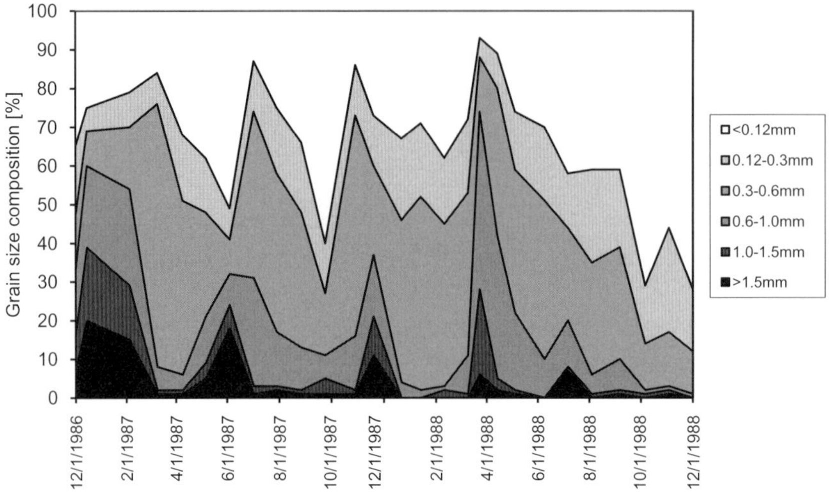

Figure 5.17 Graph showing grain size analysis in the middle reach at T2 from December 1986 to December 1988.

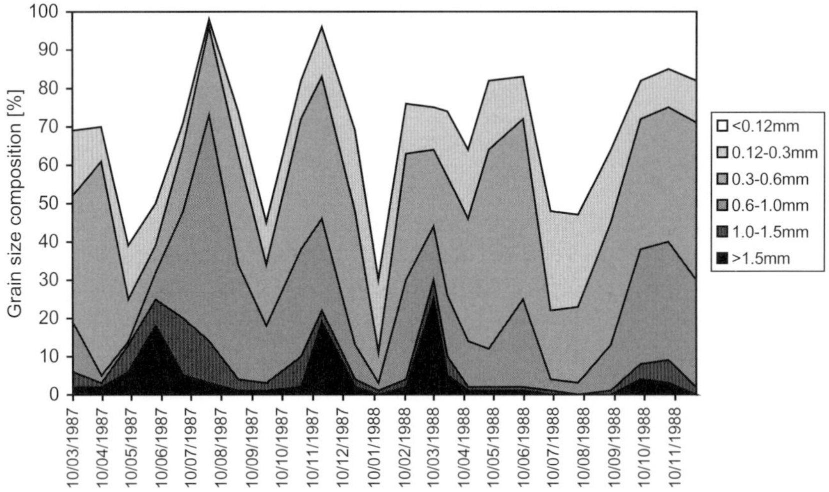

Figure 5.18 Graph showing grain size analysis in the middle reach at T3 from March 1987 to December 1988.

difference between the two years, with much lower abundances of *Amphora* in 1988 than 1987 (Figure 5.19). Compared to the in-stream sites, there was markedly greater abundance of *Planothidium* and *Eunotia*, taxa that were often attached to larger sand grains, and far less abundance of *Navicula* spp. The in-stream sites were usually dominated by *Navicula* spp., although there were strong vernal peaks

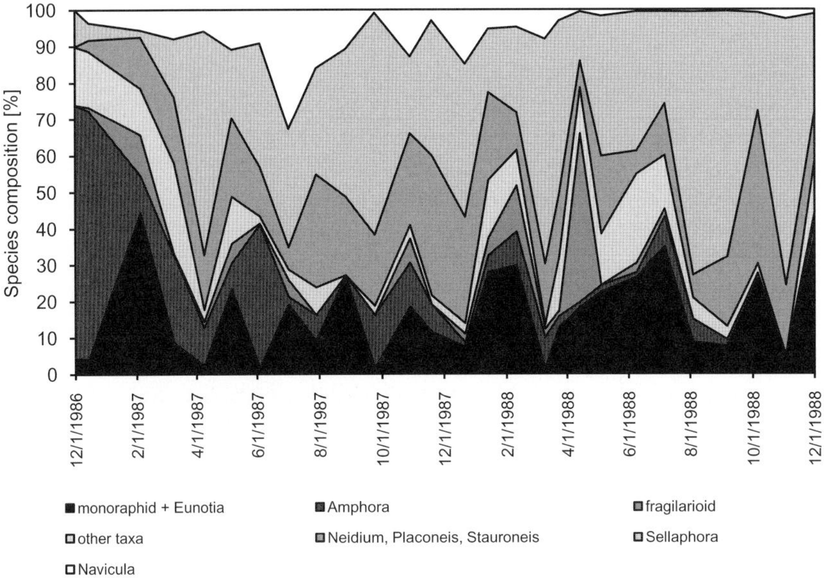

Figure 5.19 Graph showing variation in species composition (%) in the Georgsborn from December 1986 to December 1988.

of *Diatoma* and *Surirella*, late winter–early spring increases in *Amphora*, and increases in *Nitzschia* spp. generally later in the summer (August–September). Increases in the percentage of *Navicula* spp. tended to follow increases in finer sediments, but this relationship is a composite of the differential responses of individual species (Cox, 1990c). For example, *Navicula* cf. *hungarica* is far more abundant on the finest sediment particles, whereas *N. capitata* is often found on intermediate sized particles and *N. protractoides* on coarser sediments. The species-specific responses to particle size are reflected in their changing relative abundance at a site over time after flood; *N. protractoides* is initially most abundant, followed by *N. gregaria* and then *N.* cf. *hungarica* as the finest particles accumulate (see Figure 16 in Cox, 1988). Differences in distributional patterns along the stream also reveal that some species are most successful at particular sites, for example, *N. gregaria* reached its highest densities (>10^5 cells cm^{-2}) at site T3 (Figure 5.9), more than an order of magnitude higher than anywhere else (<10^4 cells cm^{-2}) in the Breitenbach (Figures 5.7 and 5.8).

5.2.1.5 **Microhabitats**

Most studies of lotic habitats relate diatom assemblages to water quality, assuming that this is the primary determinant affecting species occurrences and abundance. However, well illuminated stream systems are dominated by benthic algae, living at the interface of the substratum and the overlying water. Motile taxa of the epipelon are also capable of migrating within the uppermost layers of the sediments (Round and Palmer, 1966; Consalvey *et al.*, 2004), potentially experiencing a different chemical environment of the interstitial water. Comparisons of live floras with cleaned material from the same site reveal that frustules of many taxa are found in the sediments, but the taxa are not represented in the live population at that site (Cox, 1990a). Some specimens will have remained *in situ* from earlier assemblages, but others will have been washed in, either from similar or dissimilar habitats upstream, and accruing over time between periods of high water (Cox, 1988). Only by examining live samples from individual microhabitats is it possible to begin to understand the potential ecological specificity of many benthic diatoms.

In defining microhabitats it is important to consider the size relationships of diatoms and their surroundings. Thus, variation in sediment grain size will not only affect the mobility of a sediment and its potential to adsorb nutrients (Cox, 1988, 1990c), but also the surface area available for attachment. Similarly, epiphytic environments may be subtly different from epilithic ones, and both differ from the water flowing over them. Scanning electron microscopy shows clearly the variation in sediment and macrophyte surfaces that are potentially colonized by diatoms (see Figures 2–11 in Cox, 1988), illustrating the variability in topography between different sediments and plant surfaces and the relative sizes of diatom colonizing them.

In addition to the regular sampling of soft sediments throughout the stream, diatom floras on different substrata, in and beside the Breitenbach, were examined intermittently between September 1985 and July 1988 (Cox, 1990d). Epilithon (sandstone and dolomitic limestone), epiphyton, macroalgae, submerged, aerial and subaerial bryophytes and soil beside the stream were sampled, coverslips being used to capture the motile taxa as before. Again, live cells only were counted in fresh samples, and where cleaned material was counted, only whole frustules were included (Cox, 1990d). The latter samples were prepared by gently oxidizing diatoms on the coverslips using hydrogen peroxide (Cox, 1990d).

As mentioned above, the epilithic flora showed greater uniformity than the epipelon, and this was also the case for epiphytic samples on angiosperms throughout the stream, with small naviculoid diatoms (*E. minima, F. pelliculosa*), *P. lanceolatum* and *E. minor* most consistently present. *Gomphonema, Synedra* and fragilarioid (*Fragilaria, Fragilariforma, Staurosira, Staurosirella*) spp. became more abundant in the autumn. However, the non-angiosperm floras were more variable, with *Navicula cryptocephala* and *Pinnularia* spp. occurring more frequently and abundantly in submerged mosses than elsewhere. Growth on other algae varied with the host taxon and probably depended on its surface characteristics. Thus, the mucilage surrounding *Batrachospermum moniliforme* enclosed *Pinnularia subcapitata, N. cryptocephala* and *Nitzschia palea*, the latter also occurring abundantly within *Tetraspora* colonies, again a very mucilaginous alga. The diatom flora on other, non-mucilaginous, algae more closely resembled the angiosperm flora (Cox, 1990d).

Bryophytes, whether submerged or at the stream margin, characteristically supported the growth of *Pinnularia* and *Eunotia* spp. which were usually absent from the epipelon, although *E. minor* occurred as part of the epilithon. This probably reflects differences in the microhabitat within moss clumps. Because of their tightly packed leaves, mosses probably increase the retention time of water around them, increasing the likelihood that they can modify the water chemistry within the clump. *Sphagnum* is known to acidify its immediate environment, and the fact that the published pH preferences of many of the species found on bryophytes are lower than that commonly encountered in the stream suggests that other bryophytes may also lower the pH in their immediate vicinity.

5.2.1.6 **Species Occurrences and Ecological Inferences**

Most ecological inferences using diatoms assume that there is a direct relationship between species occurrences and water chemistry. While this may be true for planktonic diatoms which live suspended in water, it is more problematical for benthic diatoms. As described above, depending upon the type of substratum, site within the stream and time of year, contrasting diatom assemblages occur, yet the water chemistry is relatively stable and uniform throughout the stream. However, elution experiments on sediments in the stream showed marked differences in available silica and total P, as well as variation in nitrate and orthophosphate (Cox, 1988), therefore epipelic assemblages probably reflect the characteristics of the sediment (grain size, absorptive properties) and the interstitial water rather than the overlying water. The relative constancy of the epilithic diatom flora in the Breitenbach (except on dolomite, which is not a natural feature of the area) supports the use of epilithic assemblages as a water quality monitoring tool, although the accumulation of cells that were growing under different conditions remains a potential source of error.

Water quality monitoring also generally relies on cleaned samples, and assumes that in lotic habitats water flow prevents the accumulation of dead cells (Patrick *et al.*, 1966), and that the diatom flora at a particular site reflects the species that grew there. Yet it is apparent from comparisons with the sediments that diatoms are transported into and within the stream system as dead cells, for example, both *Tabellaria flocculosa* and *Pinnularia borealis* were found in the main part of the Breitenbach although live cells of the former were only found in the upper part of the upper reach, and the latter was confined to terrestrial or sub-aerial habitats adjacent to the stream (Cox, 1990d). Furthermore the usually common epiphytic diatom *Cocconeis placentula* was only found as dead cells, never in the living assemblages (Cox, 1990a). Thus, if only cleaned samples are evaluated, the subsequent interpretation of water quality may be skewed, while allochthonous diatom inputs have a similar potential to modify the assemblage used for any water quality analysis.

However, the habitat specificity or seasonal patterns of particular taxa in the stream (Cox, 1990a) indicate that diatoms can respond in a species-specific manner to their immediate environment. As mentioned above, field data can only generate inferences; confirming ecological preferences requires testing under controlled conditions (Cox, 1993, 1994). For example, *Meridion circulare* is usually described as a vernal species, and work by Lowe *et al.* (1986) on two stream systems suggested that it was intolerant of high light intensity, whereas *Synedra ulna* was limited by low light. However, they recognized that light and temperature tend to be correlated and their effects could not be separated in field studies. Observations in the Breitenbach revealed that *M. circulare* could grow under high light intensity later in the year at a site just downstream from the Georgsborn (~8°), while *S. ulna* was most abundant in the middle course of the stream in the autumn. Subsequent experimental studies revealed that *M. circulare* was unable to grow under high temperature and high light, but could tolerate higher light intensity if water temperature was low, whereas *S. ulna* grew faster at higher temperature and light intensities (Cox, 1993). Thus, low water temperature had allowed *M. circulare* to grow under higher light intensity, and the autumnal abundance of *S. ulna* was a function of raised water temperature. Unfortunately, the growth responses of very few diatoms have been determined experimentally and thus most ecological "preferences" remain inferences.

Some attempts to test diatom responses more experimentally have involved *in situ* nutrient enrichments (Pringle and Bowers, 1984) or translocation experiments (Iserentant and Blancke, 1986) to study diatom responses to changing conditions. A diatom translocation experiment carried out in the Breitenbach (Wendker, 1992) showed contrasting responses of different species to the new conditions (from upstream near the Georgsborn to downstream near its entry into the River Fulda), the assemblage on translocated glass slides showing high adaptation to the new conditions after one month. Thus, *Diatoma mesodon* decreased in abundance after translocation compared to slides at its original locality while *E. minor* increased in abundance, although *P. lanceolatum* was the most abundant diatom at both sites throughout the investigation. Wendker (1992) discussed the observed changes in relation to the effects of differential accumulation of detritus (more upstream than downstream) and potential grazers (more downstream than upstream), but did not discuss whether the decrease in *D. mesodon* could have been a seasonal effect. This taxon also decreased in abundance at the upstream site over the course of the investigation: the experiment was performed in May–June as the relative abundance of *D. mesodon* usually declines (cf. Figures 5.20–5.22).

With any *in situ* experiment, conditions are only partially controlled: changes in assemblage composition may reveal the inability of taxa to survive in new conditions (Iserentant and Blancke, 1986), but cannot discriminate between offspring of a taxon in the original assemblage that are growing faster in the new environment from representatives of the same taxon at the new site that have colonised the translocated substratum and thereby increased its numbers. Similarly, it is difficult to separate the effects of translocation from concomitant seasonal responses unless the same experiment is conducted on a number of occasions at different times of the year.

However there is a potentially more fundamental problem in basing inferred preferences on percentage abundance data that has rarely been addressed (but see Descy, 1984; Cox, 1991): percentage abundance is a relative measure and provides no information on cell numbers per unit area or volume. The percentage abundance of any particular taxon will be affected by the total number of taxa in the assemblage. If a species is present at a particular cell density in a sample, but is the only taxon present it will have a percentage abundance of 100, but if another

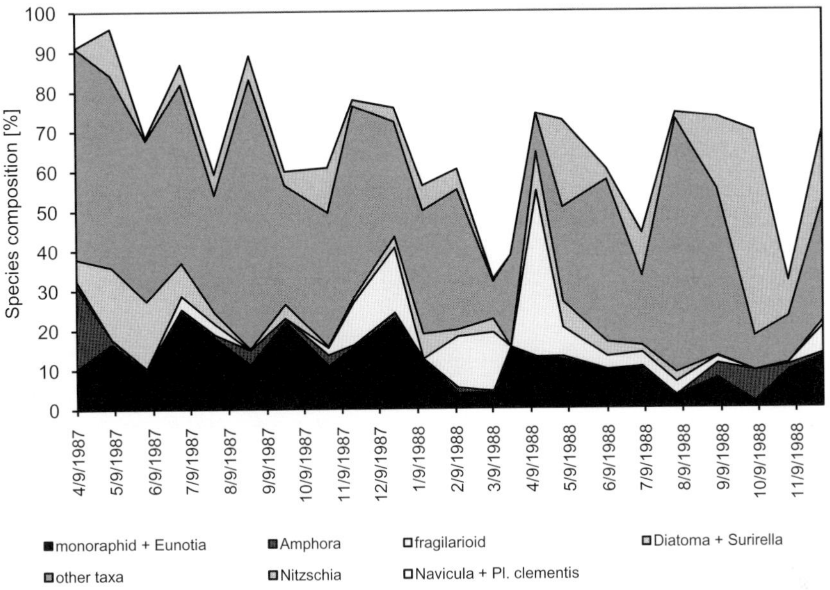

Figure 5.20 Graph showing variation in species composition (%) in the upper reach from April 1987 to December 1988.

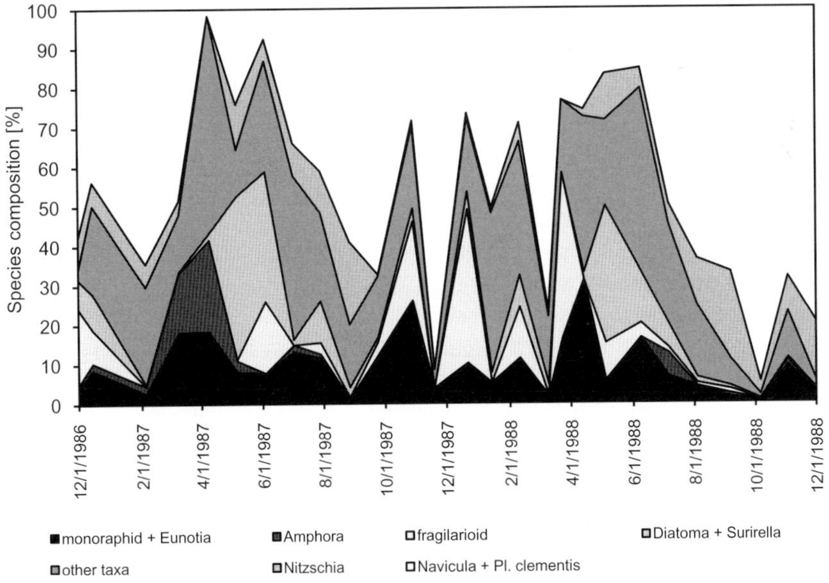

Figure 5.21 Graph showing variation in species composition (%) in the middle reach at T2 from December 1986 to December 1988.

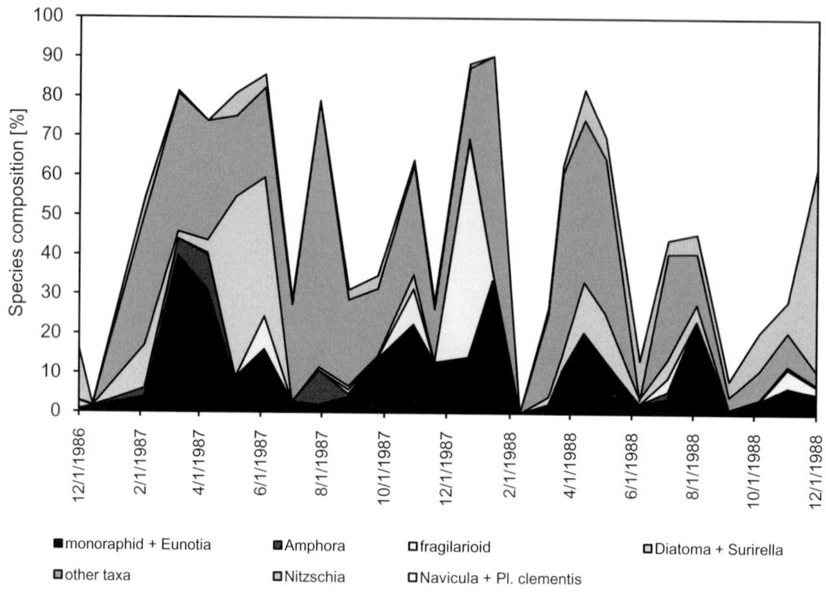

Figure 5.22 Graph showing variation in species composition (%) in the middle reach at T3 from March 1987 to December 1988.

taxon is present at the same density, its abundance will drop to 50%. With three taxa at the same density, each contributes 33.3% to the assemblage, and percentage abundances will be similarly modified as more taxa are included. Yet the performance of the taxon in terms of density is exactly the same. While changing percentage abundance data may provide a reasonable picture of changes in assemblages at a single site over time, it does not allow the performance of a single taxon to be evaluated against environmental factors across a range of sites, unless all members of the assemblage are competing in the same way for particular environmental variables. Within the plankton, competition for limiting nutrients may drive successional changes, but there is little evidence that this situation applies in the benthos, except perhaps when assemblages become extremely dense.

Comparisons between cell numbers cm^{-2} and percentage abundances of some *Navicula* spp in the Breitenbach (Figures 4, 6–8 in Cox, 1990a; Figures 5.1–5.9 and 5.19–5.22) reveal that cell density and percentage abundance do not correspond. The temporal changes in several *Navicula* species at a single site is shown as percentage abundance and cell numbers in Figures 5.23 and 5.24. Thus, because *N. gregaria* was almost the only diatom recorded in June, its percentage abundance was very high (86%), but that corresponded to an estimated density of only 420 cells cm^{-2}. At the same site a month later, its percentage abundance had fallen (44%), but its population was much denser (6200 cells cm^{-2}). Similarly, comparing percentage abundances and cell numbers for two species in September and October shows how these measures are not necessarily correlated. *Navicula*

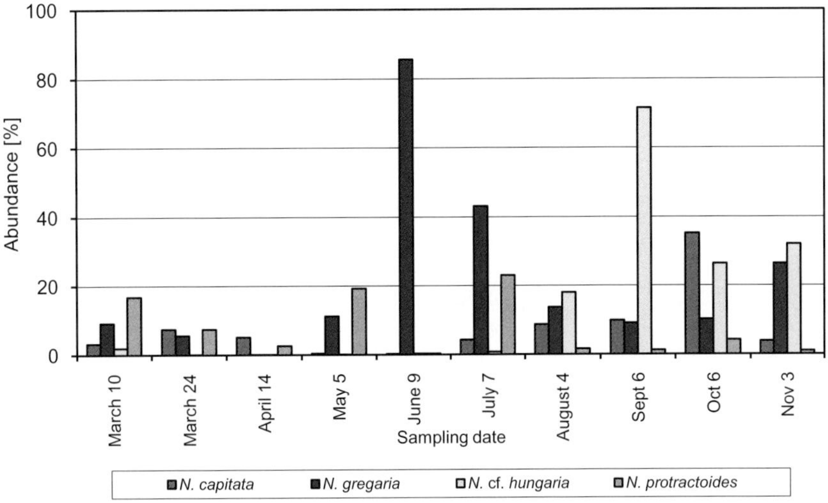

Figure 5.23 Graph showing the abundance (%) of selected *Navicula* species at Breitenbach T3.

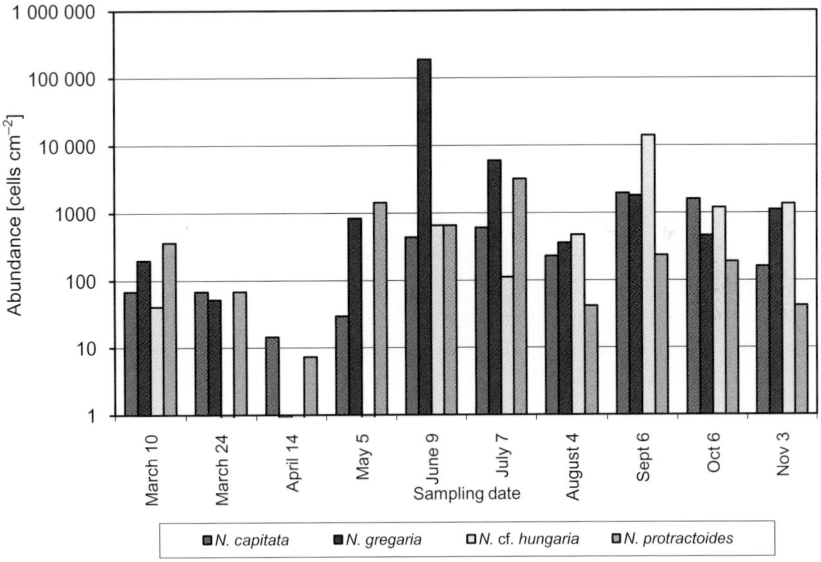

Figure 5.24 Graph showing the cell density of selected *Navicula* species at Breitenbach T3.

capitata contributed 10% of the assemblage and *N. gregaria* 9%, with 2000 and 1800 cells cm^{-2} respectively in September, but in October, while *N. capitata* had increased to 36% of the assemblage its cell density had decreased to 1600 cells cm^{-2}, and *N. gregaria* with 10% had fallen to 460 cells cm^{-2}. Results based on percentage abundance must therefore be interpreted with care. They are helpful for within site comparisons of entire assemblages, but do not invariably reveal individual species responses to changing conditions. They cannot be used to evaluate individual species responses to particular environmental variables because of the confounding effect of other species in the assemblage.

5.2.1.7 Impact of Grazers on Algal Assemblage and Biomass

As mentioned above (cf. Section 5.2.1.1) the standing crop of algae at any site on any occasion is the net outcome of primary production versus grazing (and dislodgement), and grazing activity has been recognized as a strong determinant of algal assemblages (Lamberti and Resh, 1983; Steinman *et al.*, 1989; Steinman, 1996). But determining the effect of grazers on assemblage composition and biomass usually requires experimental manipulations to exclude grazers, or comparisons between areas under different types of grazing pressure. In general, small adnate diatoms are usually considered more resistant to grazing pressure because of their prostrate growth form, whereas more erect or loosely attached algae will be more susceptible because they are easier to remove. The predominance of adnate *Planothidium* and small naviculoid diatoms in the Breitenbach epilithon can be explained in this way, as can the growth of *E. minor* under higher grazing pressure in the lower reaches of the stream (Wendker, 1992). To what extent epipelic diatom associations are modified by grazing pressure is unclear, and it is also possible that, rather than being grazed by insect larvae or snails, they are eaten by larger protozoa, such as ciliates, whose populations were observed to fluctuate in the sediments (Cox, unpublished data).

The interaction between grazers and algal biomass was also evident in an experiment in which sandstone tiles were exposed for varying lengths of time at sites in the upper and middle courses of the Breitenbach (Cox, 1990a; Wagner and Cox, unpublished data). At higher grazing densities, algal biomass (as chl-a) was markedly lower (Wagner and Cox, unpublished data). That invertebrate larvae utilize microalgae was also shown by the longer survival rates of caddisfly larvae whose cases contained microalgae than those with newly formed "clean" cases (Cox and Wagner, 1989). However, the most striking example of the potential grazing pressure of the invertebrates in the Breitenbach was seen after the accidental introduction of insecticide into the stream system. Following the death of much of the invertebrate fauna (Zwick, 1992), huge blooms of *Microspora amoena* and other filamentous green algae occurred in large stretches of the stream, dead *Agapetus* cases were rapidly overgrown by algae and chl-a accumulation on exposed sandstone tiles increased (Cox, 1990a). As the invertebrate fauna recovered, the algal blooms disappeared and algal epilithic biomass dropped back down.

5.2.1.8 General Conclusions from a Particular Ecosystem

As in many other lotic ecosystems, diatoms are usually the most abundant algal group in the Breitenbach, growing on a variety of substrata and showing marked fluctuations in species composition and biomass during the course of a year. Unlike most stream studies, because diatoms were not subjected to oxidation before evaluation, the data from the Breitenbach reflect the live population at the time of collection, uncomplicated by the retention of dead cells from preceding populations, or inwash of allochthonous representatives. Counts on cleaned sediments invariably revealed taxa that were absent from the live counts, the diversity of taxa in the sediments increasing between consecutive flood events. It was also clear that some common, widespread taxa appear to be absent from the Breitenbach, most notably *Cocconeis placentula*. Care is therefore required when only cleaned material is evaluated, particularly when this is being used to make inferences about the prevailing environment, or about the ecological "preferences" of the taxa found.

The use of percentage abundances is also shown to be problematical for benthic populations, especially if comparisons are being made about individual taxa across sites. Percentage abundance is not invariably correlated with cell density, the latter being the better gauge of successful growth. Cell count comparisons between markedly different-sized cells should also be interpreted carefully; large cells often divide less frequently than small ones, and therefore may attain lower population densities, but total cell volume may nevertheless be comparable or greater. Although quantitative sampling in natural benthic habitats has been considered problematical, it provides greater insight into the growth performance of individual species across sites and should therefore be attempted. Even if between replicate variability is higher than for planktonic samples, nevertheless, it will allow species-specific between-site comparisons (at least to order of magnitude) without the implicit assumption of percentage abundances, that is, that the highest percentage necessarily equates to optimum growth.

Year to year comparisons revealed that there can be marked differences in species assemblages and standing crop because of variation in weather patterns and discharge regimes. Discharge patterns during the first and third of the three years of intensive study of the diatom flora of the Breitenbach were more similar to each other than to the intervening year. Had the study only taken place for two consecutive years, it would have been very difficult to draw any conclusions about typical seasonal cycles, but based on the three years, and assuming that the middle year was anomalous, it was possible to draw some general conclusions on the seasonal fluctuations in the diatom assemblages. This emphasizes the need for long-term investigations if general patterns of species responses are required. Short-term studies preclude any such deductions.

Comparisons between epipelic, epilithic, epiphytic and moss floras revealed the potential microhabitat specificity of many diatom species, while many, more recent, taxonomic revisions have also led to the subdivision of a number of species and the description of many more new ones. Thus, although it was recognized that *S. pupula* in the Breitenbach comprised at least two morphologically distinct entities occupying contrasting habitats (sandy and finer sediments), these were not separated in the published analyses, with an inevitable loss of ecological information. Taxonomic studies on *Navicula* spp. in the stream also revealed the presence of hitherto undescribed species (Cox, 1987). The latter were illustrated and therefore it is possible for future workers to compare their species with these. Where illustrations are not included in ecological studies (as is very often the case) it becomes more difficult to know whether different authors are referring to precisely the same taxon or not. It would therefore be advisable to initiate some kind of reference collection or voucher system to minimize the likelihood that the same name is applied in different senses, with the resulting loss of ecological precision.

The results of algal work on the Breitenbach may not be directly comparable to that on other stream systems, and in fact it may never be possible to extrapolate from a single to all comparable ecosystems, but detailed, long-term studies on a single system will reveal more of the biological interactions in that system which would otherwise probably go unnoticed. If many ecological processes are essentially stochastic, more detailed data at a finer scale are almost certainly necessary in order to understand all the interactions that can come into play and affect the fluctuations in composition and biomass of algal populations and their dependent higher trophic level consumers.

5.2.2
Macrophytes

The macrophyte biomass in the Breitenbach was determined for six different locations along the middle and lower reach of the stream in July and October 1997 and in January and April 1998 (Horvath, unpublished data). The annual average was 239 g AFDM m^{-2}. Because the upper course of the Breitenbach contains hardly any macrophytes, this value corresponds to a whole stream average of 138 g AFDM m^{-2}. These data agree very well with earlier measurements by Koch (1990), who determined the whole stream average macrophyte standing crop as about 153 g OM m^{-2} in autumn 1984.

A tentative estimation of total plant biomass was presented by Marxsen *et al.* (1997) and Marxsen (2006). It was assumed that, although algae probably contribute at least a similar amount of primary production as macrophytes, because of the much higher turnover rates their biomass is much lower (P/B for annual macrophytes 1.0–2.0, for planktonic algae 9–353, mean 113; Wetzel, 2001). Thus a good conservative estimate of the total average plant biomass in the Breitenbach is probably not more than 150 g OM m^{-2} (75 g C m^{-2}). This is in agreement with data from Cox (1990b; cf. Section 5.2.1.3) who found between 1 µg chl-a cm^{-2} immediately after spring floods, occasionally up to about 40 µg chl-a cm^{-2} on sediments. Assuming that chlorophyll-a constitutes about 1–2% of algal organic matter (e.g., Desertová, 1981; Riemann *et al.*, 1989), her values correspond to a span of 1–80 g algal organic matter m^{-2}.

The biomass from macrophyte production is probably less available to consumers within the Breitenbach than algal biomass. Unlike periphyton, macrophyte biomass must undergo conditioning, such as microbial activity and shredding by macroinvertebrates (Kaushik and Hynes, 1971) before it is available to consumers. In this respect, macrophytes are similar to allochthonous inputs such as deciduous leaves and wood. However, based on pilot study data (Horvath, unpublished data) involving litter bag techniques (Boulton and Boon, 1991), many of the species

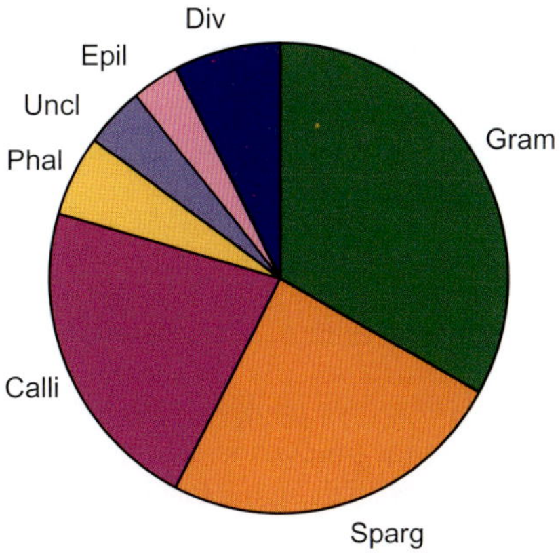

Figure 5.25 Composition of detrital POM inputs to the Breitenbach from autochthonous macrophytes in 1984/85 (after Koch, 1990). Gram = Gramineae, excluding *Phalaris arundinacea*, *Glyceria fluitans* and *Agrostis stolonifera*, Sparg = *Sparganium erectum*, Calli = *Callitriche platycarpa*, Phal = *Phalaris arundinacea*, Uncl = unclassified autochthonous macrophytes, Epil = *Epilobium*, mainly *E. obscurum*, but also *E. adenocaulon*, *E. roseum* and *E. nutans*, Div = contains *Juncus effusus*, *Sium erectum*, *Myosotis palustris*, *Cardamine amara*, *Glyceria fluitans*, *Filipendula ulmaria*, *Agrostis stolonifera*.

found in and along the Breitenbach are slow decomposers, much slower than the common deciduous tree species (*Fagus sylvatica*). Thus, macrophyte biomass must remain in the Breitenbach for extended periods of time for it to contribute to food web dynamics. Based on particulate matter transport in the Breitenbach (Horvath, 2004) and unpublished observations (Horvath, unpublished data), senescing macrophytes often fragment and are poorly retained. It is unclear what contribution they make to secondary production.

The composition of the macrophyte vegetation was incidentally described during other research studies, for example, by Benedetto Castro (1975) or Koch (1990). The latter author found that about 80% of the annual macrophyte input in 1984/1985 was from *Callitriche platycarpa*, *Sparganium erectum* and several, undetermined gramineous species (Figure 5.25). However, there was large variation in site specificity. A detailed macrophyte species list became available during 1997 and 1998 from Horvath's investigations (Horvath, unpublished data; cf. Appendix).

During a floristic survey in 1996 and 1997, Gregor and Wolf (2001) found 56 moss species (cf. Appendix). The aquatic moss *Brachythecium rivulare* was most dominant, accompanied by other aquatic mosses, *Amblystegium fluviatile*, *A. tenax*, *Calliergon cordifolium*, *Chiloscyphus polyanthos*, *Fontinalis antipyretica*, *Pellia epiphylla*, *Platyhypnidium riparioides* and *Scapania undulata*. A large number of short lived bryophyte inhabited open areas on the banks, mainly produced by floods. Mosses can be important habitats for invertebrate species (Allan, 1995). This was noted for the Breitenbach by Nolte (1990, 1991), who observed chironomids living in populations of *Brachythecium rivulare* and *Amblystegium tenax*, and by Becker (1990) who observed the caddisfly, *Micrasema undulata*, using *Philonotis fontana*, *Chiloscyphus polyanthos* and *Scapania undulata* as habitats. However, as food resources, mosses are unimportant for these organisms (Allan, 1995).

5.3 Primary Production

Primary production in the Breitenbach mainly comprises photoautotrophic production by algae and macrophytes. Chemoautotrophic production can be considered as quantitatively unimportant (Wuhrmann, 1974), although it has never been measured in the Breitenbach.

From Koch's (1990) measurements of 153 g OM m^{-2} macrophyte standing crop in autumn 1984, the annual net primary production (NPP) of macrophytes for the Breitenbach can be estimated to be at least 150–200 g OM m^{-2}, using the approach of Westlake (1963, 1969). Several years later, in July and October 1997, and January and April 1998, the macrophyte standing crop was again measured (Horvath, unpublished data; cf. Section 5.2.2). These data allow (after Westlake, 1963, 1969) the calculation of annual net macrophyte production for the middle and lower reach of 266 g AFDM m^{-2}. This corresponds to 154 g AFDM m^{-2} $year^{-1}$ for the whole stream, because hardly any macrophytes grow in the upper reach. Thus, although performed with a gap of 15 years, both measurements are in fairly good agreement (Marxsen, 2006).

Periphyton primary production was determined monthly during 1973 and 1974 (Marxsen, 1980a). Parts of polyethylene films, which had been exposed on stones in the stream for six weeks, were exposed to ^{14}C-carbonate in bottles placed in the stream for 24 h. From these ^{14}C-incorporation data, annual NPP was calculated as 11 g C m^{-2} (Table 5.3). But this value represented only epilithic and some epipsammic production, and did not include epipelic or epiphytic production. Net primary production of suspended algae in the stream water was much lower than epiphytic production, <1 g C m^{-2} $year^{-1}$ (Marxsen, 1980a).

Table 5.3 Net primary production in small streams in eastern Hesse. Annual averages with ranges in brackets (after Marxsen, 1976, 1980a).

Site	Periphyton (mg C m^{-2} $year^{-1}$)	Water (mg C m^{-3} day^{-1})
Breitenbach middle course	31 (4.8–103)	3.8 (0.32–11.3)
Rohrwiesenbach upper course	61 (0.23–314)	2.2 (0.20–5.1)
Rohrwiesenbach middle course	55 (4.0–159)	64 (4.9–297)[a]

a) The site where the measurements were performed was just below the outlet from fishponds. There were a lot of typical planktonic pond algae in the water samples from this site. This explains the exceptionally high values of primary production in the water (Marxsen, 1976, 1980a).

There was a clear seasonal trend in algal primary production, much more distinct for planktonic than for periphytic production, with lowest production rates in winter and highest during summertime. Significant relationships between temperature (t) and production (PP) were calculated for both periphytic and planktonic primary production (Marxsen, 1980a):

$$\text{periphyton primary production PP} = 3.06 \times e^{0.240t} \; (r = 0.63,\ P < 0.05,\ Q_{10} = 11),$$

$$\text{planktonic primary production PP} = 0.119 \times e^{0.347t} \; (r = 0.71,\ P < 0.01,\ Q_{10} = 32).$$

The much clearer fluctuation of planktonic production with temperature is also visible from its higher Q_{10} value, which gives the relative increase of production rates when temperature increases by 10 °C. The increase in planktonic production rates was ascribed to the higher periphytic biomass during the warmer season, not only because of the greater abundance of algae cm^{-2} (cf. Section 5.2.1) but also because of the increased area for algal colonization on macrophytes and possibly also an effect of invertebrate grazing.

Comparisons of macrophyte production to epiphytic or whole stream primary production were performed by Mann (1975), Wetzel (1975), Fisher and Carpenter (1976), and Naiman (1983). These few studies demonstrated that macrophyte production was always less than 50% of the total primary production. Thus, considering all the studies from 1973 to 1998, a tentative conservative estimate of 400 g OM m^{-2} $year^{-1}$ NPP in the Breitenbach, including macrophytes as well as suspended and attached algae, was considered a reasonable value (Marxsen, 2006), equivalent to the earlier estimate (Marxsen *et al.*, 1997). As a reasonable first approach for aquatic environments, it is generally assumed that net primary production equals respiration (Waters, 1977; Webster and Meyer, 1997b). Thus gross primary production (GPP) in the Breitenbach can be estimated to be about 800 g OM m^{-2} $year^{-1}$ or 400 g C m^{-2} $year^{-1}$.

In a more recent study, Schmidt (unpublished data) calculated the community metabolism from daily oxygen curves in the middle reach of the Breitenbach at T2 for the years 1997 to 2004. Calculations were based on the single station method described by Bott (1996) and used hourly measurements of oxygen and temperature, as well as data for cross-sections, average depth and flow velocity on a monthly basis. The data obtained as oxygen concentrations were converted to carbon using the equation:

$$\text{g C} = \text{g } O_2 \times 1/\text{PQ} \times 12/32$$

where PQ is the photosynthetic quotient (moles O_2 released per mole CO_2 incorporated during photosynthesis), and assumed to be 1.2, and where 12 and 32 are the atomic/molecular weights of carbon and oxygen, respectively. With few exceptions, results for gross primary production and community respiration (CR) were always available for one week per month during the investigation period.

GPP exhibited a significant annual trend, with the highest rates in spring and summer (April–July; Figures 5.26 and 5.27, Table 5.4), and lowest rates in winter. Average annual production was 216 g C m^{-2}, with minimum and maximum values

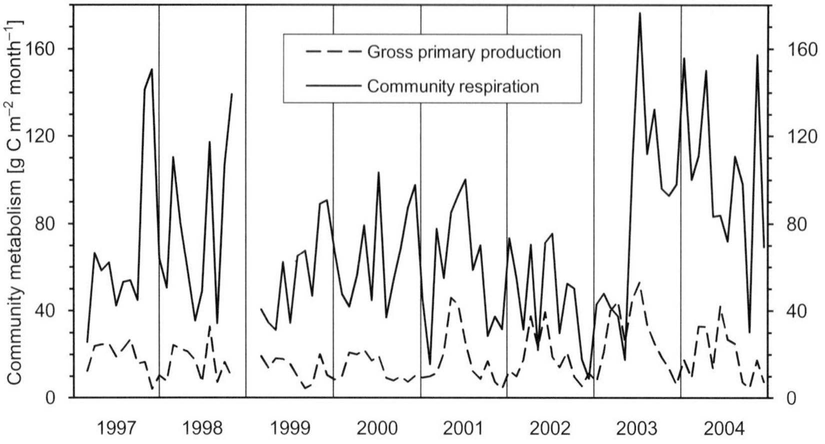

Figure 5.26 Gross primary production and community respiration in the middle reach of the Breitenbach at T2. Metabolic rates per month are given (Schmidt, unpublished data). These are based on one-week evaluations of continuous oxygen concentration measurements using the method of Bott (1996).

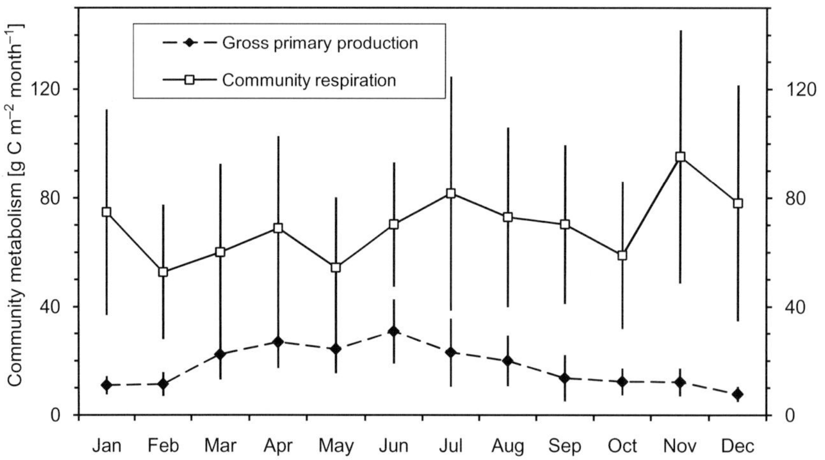

Figure 5.27 Average annual variation (including standard deviation; s.d.) of gross primary production and community respiration in the middle reach of the Breitenbach at T2, from 1997 to 2004 (Schmidt, unpublished data). For more details compare Figure 5.26.

of 159 and 335 g C m^{-2} in 1999 and 2003, respectively. 2003 was a year with exceptionally high temperatures in Central Europe.

The average value for gross primary production is somewhat below the estimated average GPP of 400 g C m^{-2} $year^{-1}$ for 1973–1998 (Marxsen, 2006). However, a large proportion of emergent macrophytes among the Breitenbach plant community photosynthesize above the water. Thus a large amount of macrophyte biomass production is connected to direct gas exchange with the atmosphere (CO_2

Table 5.4 Community metabolism: gross primary production and community respiration with monthly and annual averages (Ave), standard deviations (s.d.), minimum (Min), and maximum (Max) values (Schmidt, unpublished data). Annual P/R ratios were on average 0.27 with s.d. = 0.07, minimum = 0.19 and maximum = 0.40.

	Gross primary production				Community respiration			
	Ave	s.d.	Min	Max	Ave	s.d.	Min	Max
	($g C m^{-2} month^{-1}$)							
Jan	11	3	7	18	75	38	43	156
Feb	11	4	8	21	53	25	15	100
Mar	22	9	11	40	60	33	26	111
Apr	27	10	14	44	69	34	34	150
May	24	9	13	46	54	26	18	85
Jun	31	12	17	46	70	23	35	109
Jul	23	13	7	53	82	43	34	176
Aug	20	9	9	34	73	33	30	117
Sep	14	9	4	27	70	29	34	132
Oct	12	5	5	19	59	27	28	107
Nov	12	5	5	20	95	47	18	157
Dec	8	3	4	11	78	43	9	151
	($g C m^{-2} year^{-1}$)							
Year	216	52	159	335	838	195	558	1222

uptake, O_2 release). Such emergent biomass is, for the most part, not included in the primary production calculations based on oxygen measurements in stream water, although it is included in the community respiration calculations because it decays within the stream (Owens and Edwards, 1961; Edwards and Owens, 1962). Thus the earlier estimate of $400 g C m^{-2} year^{-1}$ for GPP is probably not too far from reality, although it should probably be adjusted downwards.

Community respiration was distinctly higher than gross primary production, as shown by Schmidt's measurements from 1997 to 2004 (Schmidt, unpublished data). The annual average was $838 g C m^{-2}$, varying from $558 g C m^{-2} year^{-1}$ (2002) to $1222 g C m^{-2} year^{-1}$ (2004). There was no statistically significant seasonal difference between months, although highest respiration rates often occurred in autumn and winter (November–January; Figures 5.26 and 5.27, Table 5.4). This trend of high respiration rates in autumn and early winter, despite decreasing temperature, is in good accordance with the main input of particulate organic matter in autumn, delivering a lot of detrital material for mainly microbial utilization (cf. Section 4.2.2). The average annual community respiration rate calculated from daily oxygen curves is close to the estimate of $800 g C m^{-2} year^{-1}$ (Marxsen *et al.*, 1997; Marxsen 2006; cf. Section 8.7) and confirms this long-term (1973–1998) average community respiration rate.

6
Bacteria and Fungi

Jürgen Marxsen

6.1
The Role of Bacteria and Fungi in Streams

After Juday's (1940) and Lindeman's (1942) seminal papers on the energy budgets and trophic dynamics of ecosystems, intensive research based on their concepts began on trophic–dynamic interactions in aquatic environments. Whereas the first attempts concentrated on measurements of primary production (Odum 1956a, 1957a, b), the importance of allochthonous organic matter as a basis for running water food chains soon became obvious (Teal, 1957; Nelson and Scott, 1962; Hynes, 1963; Fittkau, 1964). At this time grazer and detritus food chains were assumed to dominate trophic interactions in streams and rivers, as in other ecosystems (Figure 6.1; Odum, 1962; Hynes, 1963; Cummins *et al.*, 1966). The grazer food chain connects the primary producers, mainly algae and macrophytes, to the primary consumers, and within the detritus food chain, the detritivores utilize detritus of both autochthonous and allochthonous origin. The detritus food chain is especially important in almost all running waters (Teal, 1957; Cummins *et al.*, 1966; Hynes 1970b) in contrast to other aquatic systems, such as lakes and oceans. But bacteria and fungi were not included as parts of any food chains. Bacteria in particular were only mentioned in their role as decomposers of organic matter (Thienemann, 1955) and as organisms of high metabolic activity (Zeuthen, 1953), but as occurring only in very low biomass in aquatic systems (Odum 1956b, 1957a).

But from the 1960s onward, investigations showed that bacterial biomass was of great importance among aquatic food webs. For example, Kusnetsov and Romanenko (1966) determined that bacterial production exceeded primary production in the Rybinsk Reservoir (River Volga, Russia), and Overbeck (1972, 1974) showed the importance of bacteria, not only as decomposers, but also as producers of biomass in lakes. For marine environments, Pomeroy (1974) suggested the

Central European Stream Ecosystems: The Long Term Study of the Breitenbach, First Edition. Edited by Rüdiger Wagner, Jürgen Marxsen, Peter Zwick, Eileen J. Cox.

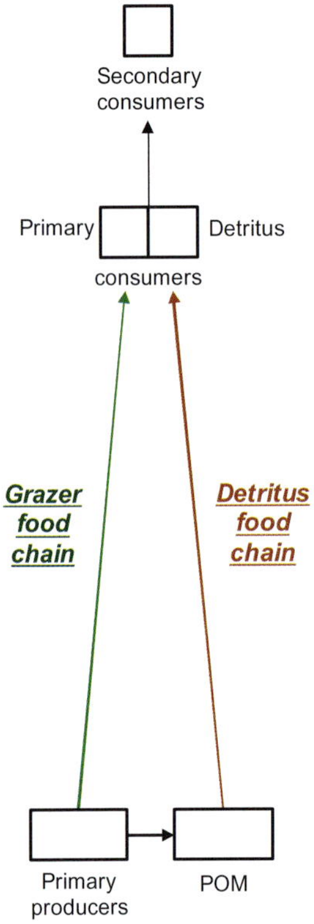

Figure 6.1 Traditional concept of food chains in aquatic ecosystems.

existence of a parallel food chain to the grazer food chain, from algae via zooplankton to fish. This pathway connects dissolved organic matter to bacterial biomass, which in turn is connected to flagellates, ciliates, and metazoa. The concept was designated the "microbial loop" (Azam *et al.*, 1983), but this pathway can also be regarded as the "microbial food chain" (Pomeroy, 1994; Cotner and Biddanda, 2002). Thus trophic–dynamic interactions can be shown in a simplified diagram (Figure 6.2; Marxsen, 2006), with three food chains: the grazer food chain, the detritus food chain, and the dissolved organic matter based microbial food chain.

Whereas the first investigations on the microbial food chain were performed in pelagic systems of lakes and oceans, there were only a few attempts to apply this concept to running waters (summarized by Meyer, 1994). In lotic environments, the microbial food chain is mainly dependent on carbon sources from surrounding terrestrial habitats. Hynes (1970b) summarized that allochthonous organic matter is of major importance as a food resource for higher trophic levels in streams and rivers, compared to autochthonous primary production. Cummins (1974) published a diagram which included bacteria and fungi, not only as consumers of organic matter, but also as a nutrient resource for insects in the food web of a headwater stream. Since then research on this pathway in streams and rivers has

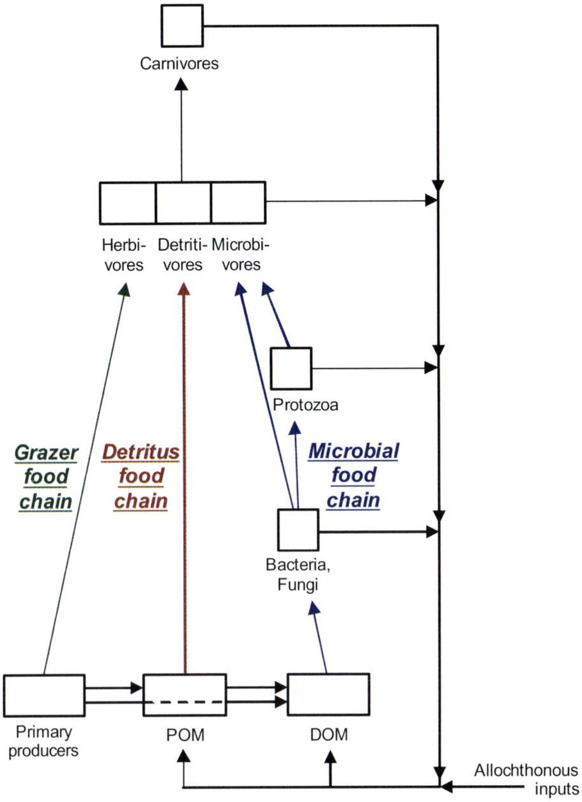

Figure 6.2 Simplified scheme of carbon flow in a stream. Sources and major pathways of organic matter are shown, with emphasis on the role of microorganisms in the microbial food chain. Viruses, which are not explicitly mentioned, are regarded as catalysers which should be included in the paths that connect organisms (at all trophic levels) to particulate and dissolved organic matter (Wilhelm and Suttle, 1999). POM = particulate organic matter, DOM = dissolved organic matter. After Marxsen (2006).

become much more frequent. However, many open questions on the role of bacteria and fungi in streams remain. In particular the question of whether bacteria and fungi are an important food resource for higher trophic levels is still a matter of controversy (Findlay and Sobczak, 2000; Hall *et al.*, 2000).

This chapter integrates the many investigations performed on bacteria at the Limnologische Fluss-Station over the past decades, mainly on the Breitenbach, but also on similar streams in Eastern Hesse. The results from these typical Central European streams serve as an example of the general importance of bacteria to stream ecosystems. They demonstrate how bacteria are interconnected with the carbon flow of a typical stream ecosystem. Unfortunately, the probably important role of fungi in stream systems was remarkably understudied in the Breitenbach.

6.2
The Communities

6.2.1
Bacterial Abundance and Biomass

6.2.1.1 Abundance in Different Habitats

The methodology for determining bacterial abundance in aquatic environments has undergone different stages of development over the past decades. The first technique to determine bacterial numbers was by cultivation of diluted samples on agar plates. The number of colonies observed on the surface of the agar (spread plate technique) or within the agar (pour plate procedure) was taken as a measure of bacterial abundance in the sample. However, the numbers of colony forming units (CFU) were found to represent only a very small proportion of the total number of bacteria (Overbeck, 1973; cf. Table 6.1), depending particularly on the type of agar used and the trophic status of the environment (Daubner, 1969). Thus the membrane filter technique for quantification of bacterial cells by transmitted light microscopy was an important step towards obtaining more reliable data on the size of aquatic bacterial communities (Razumov 1932, 1947). This technique was used for the first investigations on bacterial abundance in the waters of the Breitenbach and Rohrwiesenbach (Marxsen 1976, 1980b). Improvement of the technique by the application of fluorochromes, such as acridine orange (Zimmermann and Meyer-Reil, 1974), 4′,6-diamidino-2-phenylindole (DAPI; Porter and Feig, 1980) or SYBR green (Noble and Fuhrman, 1998) in combination with epifluorescence microscopy, made counting bacteria much easier, but did not change the number of cells detected. However, fluorescence microscopy techniques enabled the estimation of bacterial biovolumes (Marxsen 1982a, b) and the subsequent calculation of bacterial biomass (Norland, 1993; Bratbak, 1993; Loferer-Krößbacher *et al.*, 1998).

The different habitats of the Breitenbach are colonized by bacteria at different intensities (Table 6.1). The densest bacterial communities were observed on the surfaces of particles, where communities comprising not only bacteria but also

Table 6.1 Bacterial abundance in different Breitenbach habitats.

	AODC (cells ml^{-1})	Colony forming units (CFU ml^{-1})
Spring water	$0.023–0.13 \times 10^6$	$0.29–1.3 \times 10^3$
Stream water	$0.29–5.4 \times 10^6$	$5.3–120 \times 10^3$
Sandy sediment	$0.9–11 \times 10^9$	$5.0–30 \times 10^6$

AODC = acridine orange direct counts (epifluorescence microscopic counts after staining with acridine orange).
After Marxsen (1980b, 1982a, b, 1988a), Marxsen and Moaledj (1988), Höppner (1999), Beier *et al.* (2008).

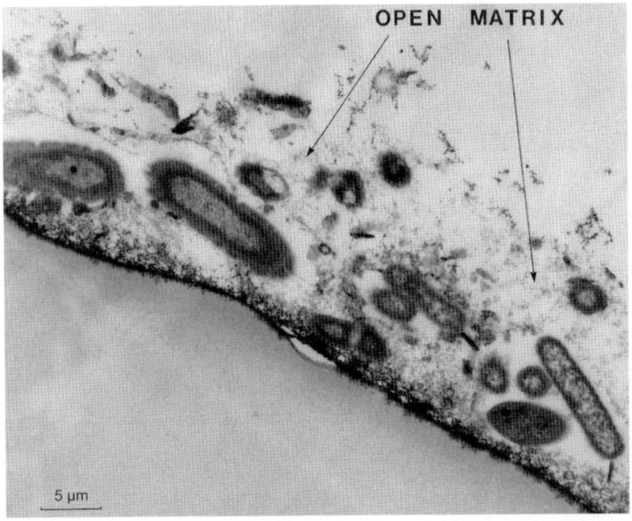

Figure 6.3 Electron micrograph of a thin section through a biofilm from the Breitenbach. A biofilm, about 15 µm thick, is growing on a solid substratum (lower left corner), composed of single microorganisms surrounded by a diffuse polysaccharide slime matrix. Photo by S. Ditner and M. A. Lock (University of Bangor, Wales).

protozoa and fungi develop, and are organized as biofilms (Figure 6.3). Where light reaches particle surfaces, these communities also contain photoautotrophic organisms, such as cyanobacteria and algae.

The total abundance of bacteria on coarse particles (stones, gravel) in the Breitenbach was found to be between about 5 and 80 × 10^6 cells cm^{-2} (Freeman *et al.*, 1990; Romaní and Marxsen, 2002; Marxsen, unpublished data). Highest numbers were observed in autotrophic biofilms on stones (Figure 6.4a) in autumn, at about 60–80 × 10^6 cells cm^{-2} (Romaní and Marxsen, 2002). The data from the Breitenbach are in the same range as for other running water environments (Geesey *et al.*, 1978; Goulder, 1988; Ainsworth and Goulder 2000b; Romaní *et al.*, 2004; Mahmoud *et al.*, 2005; Sabater *et al.*, 2005; Oliveira and Goulder, 2006; Pohlon *et al.*, 2010). Somewhat higher abundances are only occasionally detected, especially in polluted environments (Goulder, 1988; Ainsworth and Goulder 2000a; Oliveira and Goulder, 2006). Because microscopy of natural biofilms on stones is difficult (see Figure 6.4), most of these results were obtained from artificial substrata, which had been exposed in streams for several weeks to achieve biofilms that were similar to natural ones.

The surfaces of sandy particles are colonized by fragmentary biofilm structures (Marxsen 1982b, Figure 6.4b), with dispersed patches of bacteria (and other microorganisms), with much space in between, as described for marine (Weise and Rheinheimer, 1978; Rheinheimer, 1985) and groundwater (Marxsen 1982a, 1988b) sandy sediments. These habitats are thus, on average less well colonized by

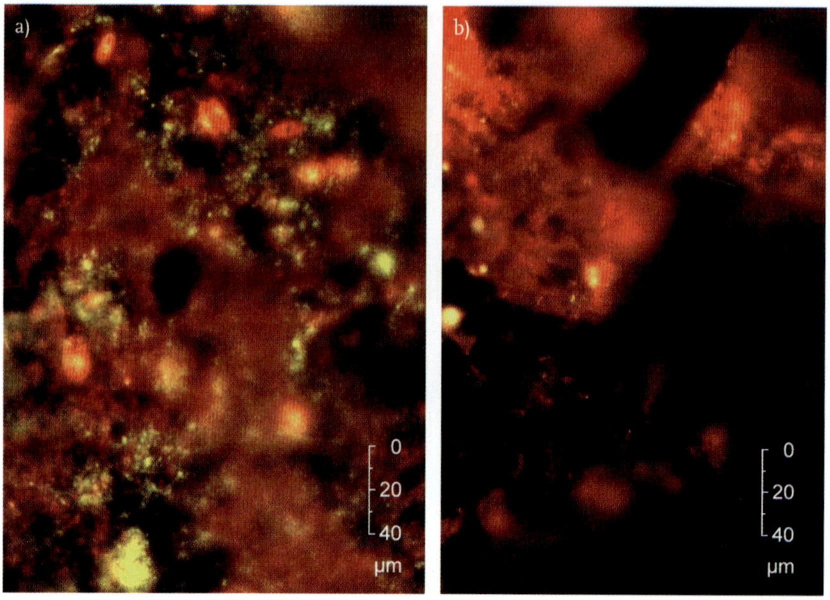

Figure 6.4 Bacterial colonization on stones and sand particles from the Breitenbach. Microscopy of such habitats is extremely difficult because of the very rough surfaces of particles. Even more difficult is the photographic documentation which allows the visualization of only very narrow sections of the objects under investigation. (a) Bacterial colonization of the surface of a stone from the Breitenbach. Using a chisel, small pieces were removed from a stone, stained with acridine orange and observed under the epifluorescence microscope during excitation with blue light (Zeiss filter set no 9: excitation filter BP 450–490 nm, beamsplitter FT 510 nm, emission filter LP 520 nm). Because of the fundamental technical problems such photos are always unfocused. However, the intense colonization by yellow to green fluorescing cells is visible. (b) Bacterial colonization on the surface of a sand grain from the Breitenbach (staining and microscopy as described in Figure 6.4a). The colonization of sand grain surfaces by bacteria is usually recognizably lower then on stones and gravel.

bacteria relative to particle surface area, by about $1.2–16.0 \times 10^6$ cells cm^{-2}, although the abundance of cells within the patches can be similar to the highest densities in biofilms on stones.

Within both habitats (coarse gravel and sandy deposits) bacteria are the dominant heterotrophic microorganisms. Almost no fungi occur. This differs from the situation on decomposing leaves where, besides a large number of bacterial cells, a significant amount of fungi is also growing (Figure 6.5; cf. Section 6.2.3). Decomposing leaves contain about $0.5–5.0 \times 10^{10}$ bacterial cells g^{-1} AFDM, corresponding to about 0.1–1.0 mg C g^{-1} AFDM, whereas fungal biomass is mostly much higher (Findlay *et al.*, 2002).

No data on the number or biomass of epiphytic bacteria are available from the Breitenbach. The few investigations from other running water environments show

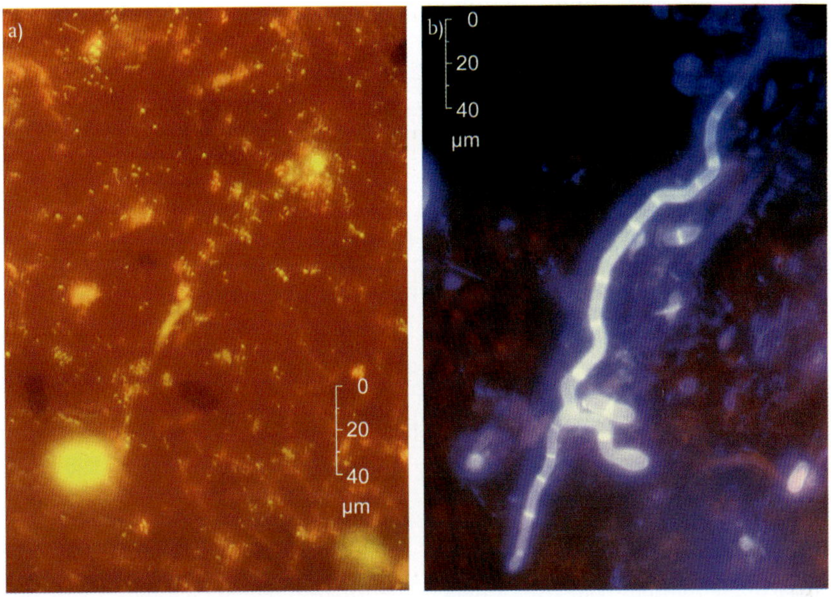

Figure 6.5 Bacteria and fungi on a degrading alder leaf after four months in the Breitenbach. (a) High intensity of bacterial colonization. The yellow fluorescing bacteria were stained with acridine orange and observed by epifluorescence microscopy during excitation with blue light (for methodological details, see Figure 6.4). (b) Leaves are also densely colonized by fungi, as is demonstrated by this photograph of the section shown in Figure 6.5a. Fungi were stained with Calcofluor white M2R (Sigma) and observed by epifluorescence microscopy during excitation with UV light (Zeiss filter set no 1: excitation filter BP 365/12 nm, beam splitter FT 395 nm, emission filter LP 397 nm).

that epiphytic bacterial abundance is dependent on the age of macrophytic stems or leaves (besides seasonal variation) and that it can attain values equivalent to those for epipelic bacteria on gravel and stones (Hossell and Baker, 1976a, b, c, Rimes and Goulder 1986a). In a study on bacterial production in different habitats of the upper Spree River, where macrophytes and their surfaces are more important than in the Breitenbach, Fischer and Pusch (2001) observed that bacterial production on epiphytes only made up between 0 and 2.7% of total bacterial production. Thus it can be concluded that the epiphytic biofilms are not of major importance for the Breitenbach. It should be noted however that, in a calcareous English stream with high macrophyte productivity, where epiphytic bacteria are more important, such bacteria are assumed to deliver a substantial contribution to the abundance of suspended bacteria (Rimes and Goulder 1986b).

The abundance of suspended bacteria in the stream water of the Breitenbach is similar to other headwaters, for example, the nearby Rohrwiesenbach (Table 6.2). This is also true for the bacterial numbers in the Breitenbach spring water (Georgsborn). Compared to rivers the Breitenbach stream water contains high amounts of bacteria. It is not surprising that bacterial numbers in the Breitenbach exceed

Table 6.2 Bacterial abundance in the water of different limnic environments.

Environment	No. of bacteria (cells ml^{-1})	Source
Breitenbach upper reach	290 000–5 400 000	Marxsen (1980b)
Breitenbach middle reach	490 000–2 700 000	Marxsen (1980b)
Rohrwiesenbach upper reach	200 000–9 200 000	Marxsen (1980b)
Breg spring	19 000–514 000	Deufel (1972)
Danube at Bratislava	110 000–2 200 000	Daubner (1969), Daubner and Trzilova (1973)
Rio Negro	200 000–300 000	Schmidt (1970)
Rio Solimões	10 000–50 000	Schmidt (1970)
River Elbe above Hamburg	3 000 000–8 000 000	Rheinheimer (1977)
River Trave	2 000 000–6 500 000	Rheinheimer (1977)
Lake Plußsee (eutrophic)	1 100 000	Overbeck (1965)
Lake Schöhsee (mesotrophic)	500 000	Overbeck (1965)
Eutrophic Russian lakes	2 000 000–8 000 000	Rodina (1961))
Oligotrophic Russian lakes	90 000–1 000 000	Rodina (1961)

those in the nutrient-poor river waters of the Amazon region (Schmidt, 1970), but they also surpass bacterial abundance in the River Danube (Mucha, 1967; Daubner, 1969; Daubner and Trzilova, 1973). Higher bacterial numbers in rivers usually occur in polluted environments (e.g., Rheinheimer, 1977 from the River Elbe or Mucha, 1967 and Deufel, 1972 from the Danube). Compared to lentic environments, bacterial numbers in the Breitenbach are close to those in eutrophic lakes (e.g., Overbeck, 1965; Rodina, 1961), although the concentration of organic matter in the Breitenbach is clearly below the concentrations known from meso- to eutrophic lakes (Ohle, 1972). It is difficult to explain these high abundances of almost only heterotrophic bacteria suspended in stream water, especially when the short bacterial residence time in the Breitenbach water before entry into the River Fulda is considered. Residence time from the end of the upper course to the stream mouth is about 4–10 h (Marxsen 1976, 1980a; Beier *et al.*, 2008; unpublished data). Furthermore, the temperature is low in the middle and lower reaches of the stream even in summer (daily fluctuations in summer 2002 at trap T2 in the middle reach 8.2–14.4 °C, at trap T6 close to the mouth 6.5–13.6 °C; cf. Chapter 3). But the assumption that the suspended bacteria originate from other habitats, such as streambed sediments and/or the terrestrial environment, is contradicted by the composition of suspended bacterial communities which differs from that in these other habitats (Beier *et al.*, 2008, for a detailed discussion see below Section 6.2.2.6).

Most streamwater bacteria occur in association with particles, mainly organic matter (detritus flocs; Marxsen 1980b; Figure 6.6), only a few occur as single cells. Most cells, often more than 90%, are cocci, which is indicative of unpolluted environments (Daubner, 1972). According to Jannasch (1955) aquatic environments are secondary habitats for bacteria, which to a greater extent occur at such sites as starved cells. These small coccoid cells (less than 1 μm) may develop into

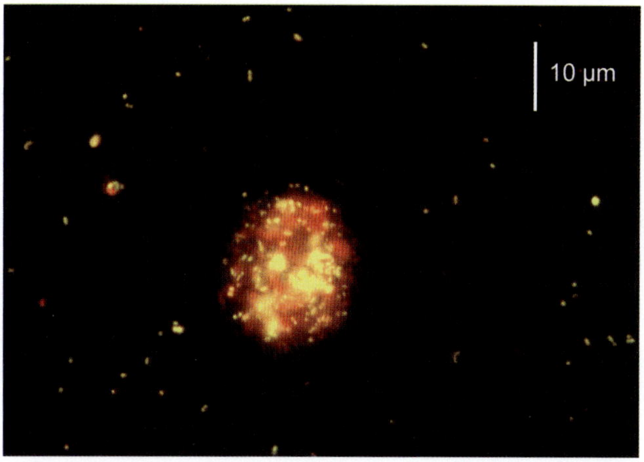

Figure 6.6 A small detritus particle from Breitenbach stream water colonized by bacteria. Breitenbach water was concentrated on a polycarbonate filter (0.2 μm pore size), stained with acridine orange and observed by epifluorescence microscopy during excitation with blue light (for methodological details, see Figure 6.4). Many single cells were retained around the densely colonized particle on the filter surface.

larger "normal" bacteria under suitable nutrient conditions, not only into larger cocci, but also into cells of other shapes, such as rods and spirals (Jannasch, 1955; Gray and Williams, 1971). Meanwhile very small bacteria (ultramicrobacteria with a cell size <0.1 μm^3) have been observed in large numbers in freshwater (Hahn, 2003; Hahn, 2006; Hahn *et al.*, 2003). Sequences belonging to these *Polynucleobacter*-like organisms were detected in the Breitenbach, too. Thus the typically small-sized bacterial community in the Breitenbach and other streams might perhaps not be composed mainly of starved cells, but have evolved from special small-sized organisms.

6.2.1.2 **Effects of Temperature and Discharge on Suspended Bacterial Abundance**

There is a distinct seasonal dynamic of bacterial abundance in the stream water of the Breitenbach upper reach, but not in its downstream sections (Figure 6.7). This correlates with the small seasonal temperature variations in the middle and lower courses, against the much larger fluctuations upstream of the entrance of the Georgsborn springs. There is also a much more distinct, temperature-dependent, variation in bacterial abundance in the upper reach waters compared to the middle and lower sections (Q_{10} = 3.8 vs 1.8; Figure 6.8, Table 6.3). The bacterial dynamics in the Breitenbach upper reach are similar to those found in the small woodland stream, the Rohrwiesenbach, situated some kilometres to the west of Schlitz (Figure 2.2), which exhibits similar large temperature fluctuations to the Breitenbach headwater (Figures 6.7 and 6.8).

Unlike the Rohrwiesenbach, a distinct relationship between discharge and bacterial abundance in stream water was not observed for the Breitenbach (Marxsen

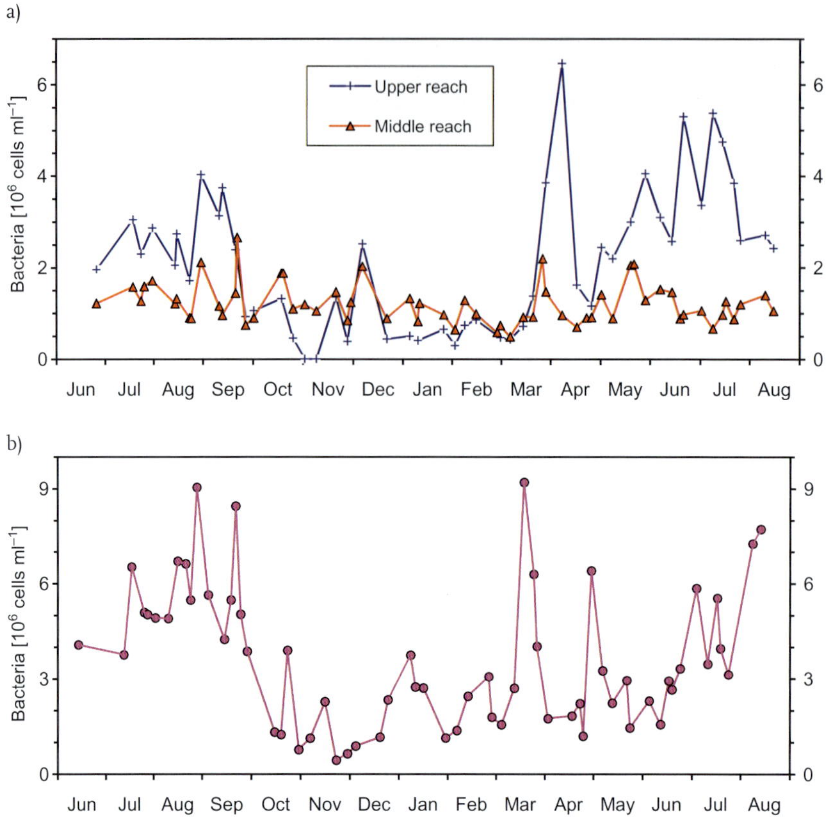

Figure 6.7 Seasonal variation (June 1973 to August 1974) of bacterial abundance in stream water. (a) Breitenbach upper reach above gage GA and middle reach at trap T2. (b) Rohrwiesenbach upper reach (woodland site). Data from Marxsen (1980b).

1980b). At sampling sites in the Rohrwiesenbach upper reach, discharge and bacterial numbers were most significantly correlated (Figure 6.9, Table 6.4), but only weak correlations were found for the Breitenbach. Opposite effects were observed between the upper and the lower parts (Figure 6.9, Table 6.4). In the upper reach an increase in bacterial abundance occurred with increasing discharge, whereas in the middle reach the number of bacteria decreased slightly with increasing discharge. However, only results during low and middle water flow were available for the Breitenbach middle reach. Thus it was suggested (Marxsen 1980b) that the observed relationships were only two branches of a u-shaped curve: bacterial abundance decreases with increasing discharge at low water flow, but increases again when discharge increases further. Such curves have been found for fine particulate organic matter in streams (Fisher and Likens, 1973; Bormann *et al.*, 1969) and for fecal bacteria in land-drainage water (Evans

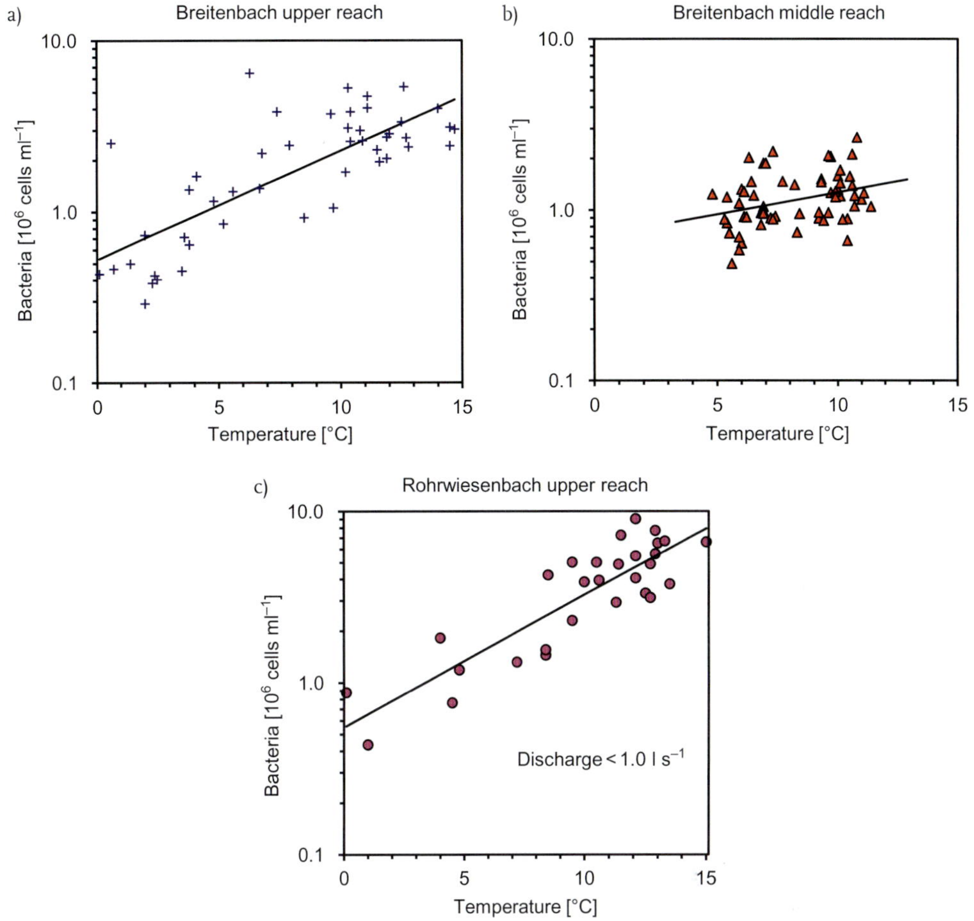

Figure 6.8 Relationship between temperature and abundance of suspended bacteria. (a) Breitenbach upper reach above gauge GA ($r = 0.77$, $P < 0.001$). (b) Breitenbach middle reach at trap T2 ($r = 0.33$, $P < 0.05$). (c) Rohrwiesenbach upper reach (woodland site, $r = 0.89$, $P < 0.001$). Data from Marxsen (1980b).

and Owens, 1973). This would support the assumption that the suspended bacteria in the Breitenbach originate mainly from stream sediment and the terrestrial environment around the stream.

6.2.1.3 **Bacterial Biomass**

Bacterial biomass in the Breitenbach was determined using a volumetric microscopic approach after staining the cells with acridine orange (Marxsen 1982a, b). The data on bacterial biovolumes (Marxsen 1988a; Marxsen and Moaledj, 1988; unpublished data) were converted to bacterial carbon using a conversion factor of

Table 6.3 Relationships between temperature and bacterial abundance (microscopic counts) at Breitenbach and Rohrwiesenbach stream water sites calculated using the equation $B = b \times e^{at}$, where B = number of bacteria ml^{-1}; t = temperature in °C; a and b are constants.

Site	a	b	r^2	Q_{10}
Breitenbach upper reach (above gauge GA)[a]	0.133	461 200	0.59 ($P < 0.001$)	3.8
Breitenbach middle reach (at trap T2)	0.0593	701 800	0.11 ($P < 0.05$)	1.8
Rohrwiesenbach upper reach[b]	0.175	571 800	0.79 ($P < 0.001$)	5.8

a) at discharge values <0.75 l s^{-1}.
b) at discharge values <1 l s^{-1}.
Q_{10} = factor giving the increase per 10 °C.
From Marxsen (1980b).

100 fg C μm^{-3} (Findlay *et al.*, 2002). Thus an annual average of 4 g bacterial C m^{-2} was obtained for the Breitenbach (Marxsen, 2006). This is within the range Findlay *et al.* (2002) found in a cross-system comparison of microbial biomass in streams, and is not far from the data for Walker Branch (Tenn., USA). The latter is the most similar stream to the Breitenbach in that study with respect to the standing stock values of organic matter.

Summation of all long-term, standing crop, annual average data gives about 390 g C m^{-2} total standing crop of dead and living organic matter in the Breitenbach (Table 6.5). Bacteria make up 36% of the total heterotrophic biomass in the Breitenbach: 4 g bacterial C m^{-2} of 11 g total heterotrophic C m^{-2}. The importance of bacteria to the heterotrophic biomass accords well with studies from other benthic habitats. For example, Kemp (1990) listed several marine sediments that show similar dominance of bacterial biomass.

6.2.2
Bacterial Community Composition

6.2.2.1 Methodological Approaches

The bacterial communities in small streams like the Breitenbach, and most other natural streams worldwide, are dominated by heterotrophic organisms. Only in polluted environments or in mine drainage streams does a clear shift to, or dominance of, chemoautotrophic bacteria occur (Wuhrmann, 1974). However, the considerable number of photoautotrophic cyanobacteria are not included here (see Chapter 5). Until recently there were few attempts to identify the composition of bacterial communities in running waters. This was probably due to the fact that definitive identification of bacteria was among the most difficult ecological procedures. In contrast to higher animals and plants, bacterial morphology is very simple. The direct microscopical identification of individual cells is possible only in very rare cases, and it is usually necessary to isolate pure cultures of these organisms. After successful isolation many biochemical and physiological tests

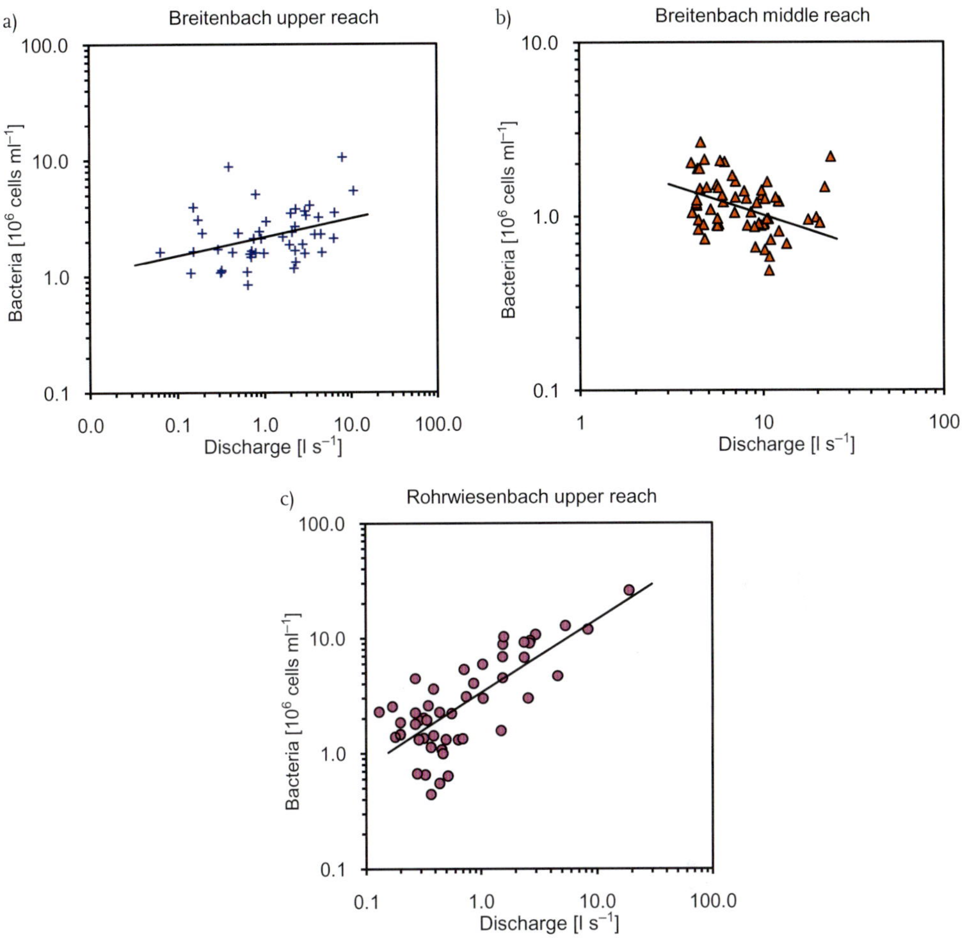

Figure 6.9 Relationship between discharge and abundance of suspended bacteria. (a) Breitenbach upper reach above gage GA ($r = 0.36$, $P < 0.05$). (b) Breitenbach middle reach at trap T2 ($r = -0.42$, $P < 0.001$). (c) Rohrwiesenbach upper reach (woodland site, $r = 0.76$, $P < 0.001$). Data from Marxsen (1980b).

must then be performed, results of which are often ambiguous. All these processes are very time-consuming, but one never knew whether the isolated organisms were really relevant for the environment from which they were isolated. At best only a very few percent of the organisms in the samples could be cultivated (Overbeck, 1974; Amann *et al.*, 1995). And because the conditions for growing them were so different from natural conditions, these were probably not the bacteria that were responsible for the biological processes in the natural systems. Thus, for the Breitenbach, only one attempt has been made with low nutrient isolation

Table 6.4 Relationships between discharge and bacterial abundance (microscopic counts) at Breitenbach and Rohrwiesenbach stream water sites calculated using the equation $B = b \times D^a$, where B = number of bacteria ml^{-1}; D = discharge in $l\,s^{-1}$; a and b are constants.

Site	a	b	r^2
Breitenbach upper reach (above gauge GA)	0.160	2 180 000	0.13 ($P < 0.05$)
Breitenbach middle reach (at trap T2)	−0.341	2 230 000	0.18 ($P < 0.001$)
Rohrwiesenbach upper reach	0.431	4 190 000	0.49 ($P < 0.001$)

From Marxsen (1980b).

Table 6.5 Standing crop annual averages for the Breitenbach.

Source	Value ($kg\,C\,m^{-2}$)
Particulate organic carbon	0.30
Dissolved organic carbon	0.001
Plants (algae and macrophytes)	0.075
Bacteria	0.004
Fungi	0.001
Protozoa	<0.001
Metazoa (invertebrates only)	0.005
Total standing crop	0.39

From Marxsen (2006).

techniques (Moaledj 1984, 1986) followed by rapid microcultivation and numerical taxonomy (Marxsen and Moaledj, 1988).

Meanwhile molecular biology techniques have been developed and have to a large extent overcome the old limitations of bacteriological investigations. These include approaches using denaturing gradient gel electrophoresis (DGGE) and temperature gradient gel electrophoresis (TGGE) of DNA fragments, which enable the investigation of genetic diversity in a single habitat and comparisons between different habitats (Muyzer and Smalla, 1998), the microscopical visualization of bacterial communities via fluorescence *in situ* hybridization (FISH) or catalyzed reporter deposition FISH (CARD-FISH) of larger taxonomic groups (Amann *et al.*, 1995; Pernthaler *et al.*, 2004), or the direct sequence analysis of DNA fragments and its affiliation to known sequences (Ludwig *et al.*, 2004). All these approaches have been applied to at least some samples from the Breitenbach, to achieve a rough insight into its bacterial community composition using these newly developed techniques.

6.2.2.2 **Cultivation**

In 1984 bacterial strains were isolated from different habitats along the Breitenbach and, for comparison, from the surrounding meadow soil, to obtain an impression of the bacterial community composition on the basis of classical cultivation approaches (Marxsen and Moaledj, 1988). In all, 137 strains were examined for 72 colony and cell morphological features by conventional techniques and for 145 physiological–biochemical features by rapid microtechniques established by Moaledj (1984, 1986). These data, together with colony and cell morphological features, were used for numerical taxonomical analysis. Although the application of numerical taxonomic methods produces a classification and grouping of strains, not an identification (Witzel *et al.*, 1981, 1982), nevertheless the data can be used for a rough taxonomic assignment. The data from that study were then re-evaluated taxonomically. The strains were from four sampling sites: two samples from water (outlet of Georgsborn, GS, the spring at beginning of the middle reach, and at trap T2 in the middle reach), one from surface sediment (upper 2 cm) at trap T2 and one from the upper 2 cm of meadow soil, about 5 m from the stream close to trap T2.

Most of the strains (70%) isolated from stream and spring water belonged to the *Bacteroidetes*, mainly to the genera *Cytophaga* and *Flavobacterium* (Figure 6.10). Similar results were obtained from the water of streams south of Stuttgart, where Kracht (1983) found about 40% *Cytophaga*, while Cousin *et al.* (2008) determined that 59–74% of isolates from the Westerhöfer Bach in the Harz mountains (Germany) belonged to the *Bacteroidetes*. Strains from this phylum were only isolated in low numbers (11%) from Breitenbach sediment and were not isolated from the soil sample. Of the water isolates, 18% belonged to the *Betaproteobacteria* which were the dominant strains from sediment (71%) and to a lesser extent from soil (52%). *Gammaproteobacteria* occurred in all habitats at a similar proportion (7–10%). *Actinobacteria* were the second most important group among the soil strains (41%), but achieved only a low percentage of the sediment isolates (9%) and were negligible among the water strains (1%).

Thus distinct differences were revealed between the investigated locations, although only one medium was used to isolate all strains. The high proportion of aquatic strains, which were arranged in a cluster without any isolates from sediment and soil habitats, was particularly striking. However, Holder-Franklin (1981) found similar differences between sediment and water populations of the Dunbar, a small Canadian woodland stream, where she identified 29% of the strains from stream water communities as *Cytophaga*, but only 1% from sediment.

6.2.2.3 **Fluorescence *in situ* Hybridization**

Through the molecular biology developments, promising tools have been made available that overcome the severe limitations on the analysis of aquatic bacterial community composition. For example, FISH techniques (Amann *et al.*, 1995) have been applied to obtain an insight into the bacterial community composition of different aquatic environments (Zwart *et al.*, 2002). The most common type of probe, fluorescently labeled, rRNA-targeted oligonucleotides, was used successfully to identify prokaryotic cells in eutrophic aquatic environments such as sewage treatment plants (Wagner *et al.*, 1993). But in samples from oligotrophic environments the numbers of bacterial cells which could be visualized by this technique

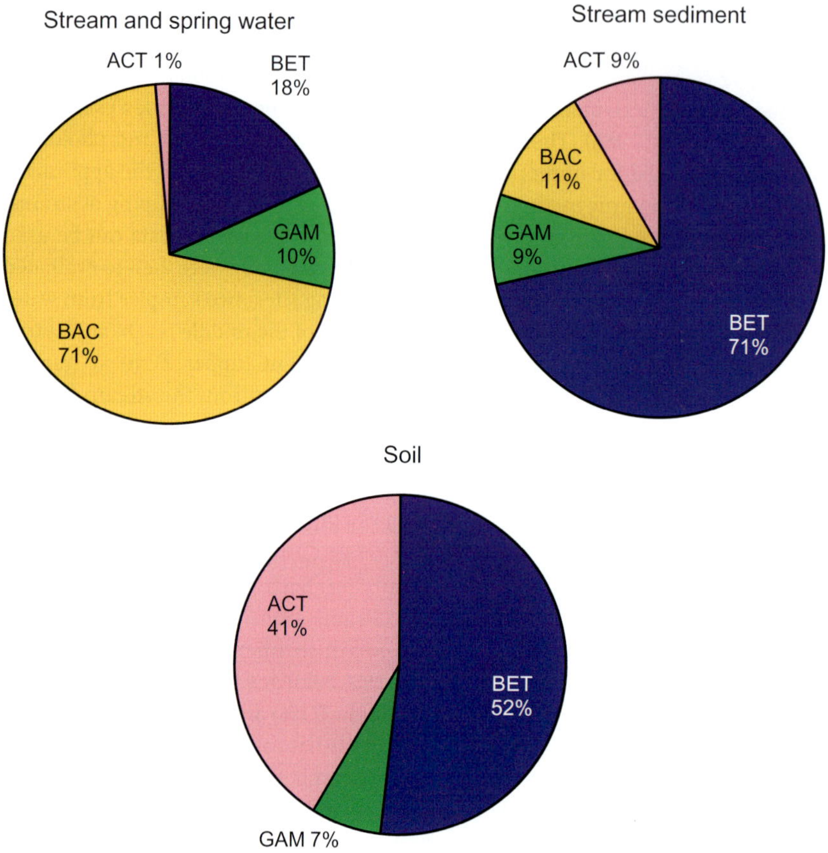

Figure 6.10 Taxonomic distribution of strains isolated from stream and spring water, stream sediment and soil adjacent to the stream. BET = Betaproteobacteria, GAM = Gammaproteobacteria, BAC = Bacteroidetes, ACT = Actinobacteria. Data from Marxsen and Moaledj (1988).

were often very low, probably due to the occurrence of small bacteria with low rRNA content (Glöckner *et al.*, 1996). Also the occurrence of high amounts of humic matter and clay, which interfere with molecular biological reactions, poses many difficulties for this approach. However, with the sensitive carbocyanine dye, CY3, and special fluorescence filters (Glöckner *et al.*, 1996), the FISH technique could successfully be applied to evaluate the bacterial community composition of most of the Breitenbach habitats (Figure 6.11).

With a specific probe for the domain, *Bacteria* (EUB338, *Eubacteria*; Table 6.6), up to 82% of bacterial cells visualized by DAPI staining were detected (Figure 6.12). The highest percentages were found in biofilms, for example, on average 71% in biofilms on microscopic slides exposed in a laboratory stream fed with Breitenbach water, or 82% in biofilms on slides that were exposed for at least six

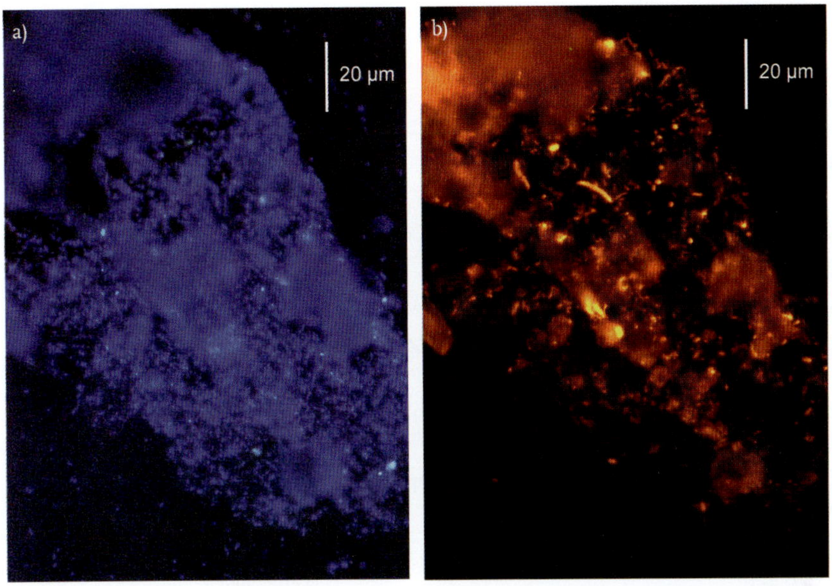

Figure 6.11 Sediment particle from the Breitenbach colonized by bacteria. Identical microscopic fields are shown. Bacteria were hybridized with the CY3-labeled oligonucleotide EUB338 (Table 6.6) and stained with DAPI. (a) Epifluorescence microscopic image of DAPI-stained bacterial cells (Zeiss filter set no 1: excitation filter BP 365/12 nm, beam splitter FT 395 nm, emission filter LP 397 nm). (b) Image of cells hybridized with probe EUB338 specific for Bacteria (CY3 filter set, Chroma HQ 41007: HQ 545/30, Q 565 LP, HQ 610/75).

weeks in the Breitenbach itself (Packroff, 1994) to achieve biofilms comparable to natural autotrophic biofilms. Similar high values of hybridized cells (about 50–65%) were found in stream water, spring water and sandy sediments. Fewer cells (<20%) were detected in decomposing *Fagus* leaves collected from the Breitenbach, due to the interference of fluorescence from chlorophyll and its degradation products remaining in leaves, and perhaps due to the high amounts of humic matter in leaves, which may disturb the hybridization process. With a probe for *Archaea* (Table 6.6) hybridized cells were not detected indicating the very low abundance, or even the absence, of this group in the investigated Breitenbach habitats. This agrees with results from Gao *et al.* (2005) who observed only around 1% of benthic bacterial cells to be *Archaea* in North American streams using the FISH approach.

By applying specific probes that can be designed for all phylogenetic levels (Amann *et al.*, 1995; Amann and Fuchs, 2008) a more detailed view of the bacterial community composition can be achieved. This was done for different Breitenbach habitats with probes for *Alpha-*, *Beta-*, and *Gammaproteobacteria* and for *Bacteroidetes* (*Cytophaga–Flavobacteria* group) (Table 6.6). Cells hybridized with these

Table 6.6 Oligonucleotide probes applied to investigations on bacterial community composition in different Breitenbach habitats by FISH and CARD-FISH.

Probe	Specificity	Sequence	References	FISH	CARD-FISH
NON338	Negative control	ACTCCTACGGGAGGCAGC	Wallner *et al.* (1993)	+	+
EUB338	*Bacteria*	GGTAAGGTTCTGCGCGTT	Amann *et al.* (1990)	+	+
ALF968	*Alphaproteobacteria*	TTCCGTCTAACCGCGGG	Neef (1997)	+	
ALF1b	*Alphaproteobacteria*	CGTTCGYTCTGAGCCAG	Amann *et al.* (1995)		+
BET42a	*Betaproteobacteria*	GCCTTCCCACTTCGTTT	Manz *et al.* (1992)	+	+
GAM42a	*Gammaproteobacteria*	GCCTTCCCACATCGTTT	Manz *et al.* (1992)	+	+
CF319a	*Bacteroidetes* (*Cytophaga-Flavobacteria* group)	TGGTCCGTGTCTCAGTAC	Manz *et al.* (1996)	+	+
HGC69a	*Actinobacteria*	TATAGTTACCACCGCCGT	Amann *et al.* (1995)		+
LGC354b	*Firmicutes*	CGGAAGATTCCCTACTGC	Meier *et al.* (1999)		+
ARCH915	*Archaea*	GTGCTCCCCCGCCAATTCCT	Amann *et al.* (1995)	+	+

The probes were 5′-aminolinked oligonucleotides labeled with CY3 (FISH) or horseradish peroxidase (CARD-FISH).

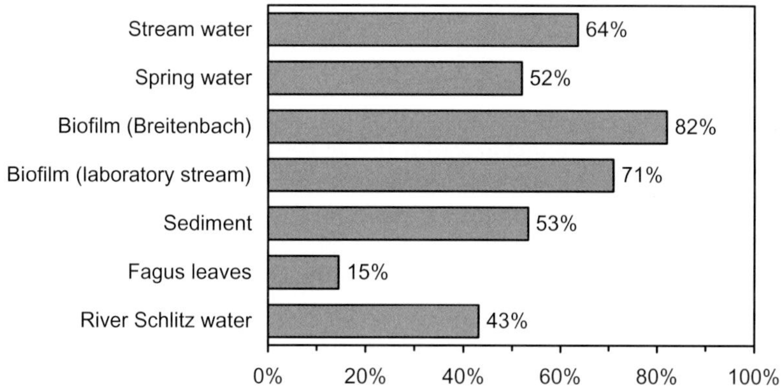

Figure 6.12 Percentage of bacterial cells hybridized with probe EUB338 (specific for domain Bacteria, Table 6.6) relative to DAPI cell counts in different Breitenbach habitats. The experimental laboratory stream was fed with Breitenbach water. Data from the River Schlitz are given for comparison.

probes contributed between 38 and 96% of the cells hybridized with the EUB338 probe for *Bacteria* (Figure 6.13).

As is usual for freshwater (Wagner *et al.*, 1993; Glöckner *et al.*, 1996), *Betaproteobacteria* were most frequent in spring and stream water, as well as in the biofilm of the laboratory stream (27–35%), but they were not usually the most dominant group (Figure 6.13). Cells from the *Bacteroidetes* phylum were dominant on degrading *Fagus* leaves. This agrees with the well-known ability of bacteria of the genus *Cytophaga* to decompose macromolecules such as cellulose, which occur at high concentrations in leaves (Reichenbach, 1999). Bacteria from the *Bacteroidetes* were also detected at high levels in water samples, but only at very low percentages in sediment. This agrees with results from earlier investigations using strain isolation and cultivation (Marxsen and Moaledj, 1988; see Section 6.2.2.2). High amounts of *Alphaproteobacteria* (12–24%) were detected in most samples, which is not typical for such environments, but these bacteria were observed also in other headwaters of Europe and North America (Battin *et al.*, 2001; Fazi *et al.*, 2005; Gao *et al.*, 2005). Cells belonging to *Gammaproteobacteria* occurred less frequently (except in the artificial laboratory stream), which is usual for these freshwater environments (Fazi *et al.*, 2005; Gao *et al.*, 2005).

In sediment samples cells hybridized with the four specific probes made up only 38% of the hybridized cells with the EUB338 probe for the *Bacteria* domain (Figure 6.13). Spring water samples contained 69% from those groups, whereas for all other habitats this percentage is between 83 (stream water) and 96% (*Fagus* leaves). Sediment is the most heterogeneous stream habitat with the highest diversity of physical and chemical conditions. In general aquatic sediments are considered to contain the most diverse bacterial communities on earth (Lozupone and Knight, 2007). Thus it is appropriate to assume that the microbial community in the Breitenbach sediment is also the most diverse, and that more bacterial groups may develop than in other stream habitats. This was confirmed by results from DNA sequence analysis. Beier *et al.* (2008) determined that only 36% of bacterial sequences belonging to *Alpha-*, *Beta-*, *Gammaproteobacteria* and *Bacteroidetes* occurred in sediments from the Breitenbach and similar streams in eastern Hesse, whereas they found 77% of sequences of these groups in water samples.

The more elaborate CARD-FISH technique was applied to the upper Breitenbach sediment once, during an experiment on rewetting desiccated streambed sediments, which showed only very moderate changes in bacterial community composition over the first four days of rewetting (Marxsen *et al.*, 2010). With this approach *Archaea* were found to contribute only about 1% of total bacterial abundance. *Betaproteobacteria* were found to be dominant (45%) among the *Bacteria* (Figure 6.14), followed by the *Alphaproteobacteria* (27%). *Bacteroidetes* reached 12%, *Actinobacteria* 8%, *Gammaproteobacteria* 7%, and *Firmicutes* only 1%. Together, these groups made up 53% of the total number of cells hybridized with a *Bacteria* probe. Thus the results on bacterial community composition are similar to that found with the initial FISH technique, except for the high abundance of *Betaproteobacteria*. However, it is questionable whether the missing percentage is really accounted for by the more or less abundant groups of *Bacteria* that had been

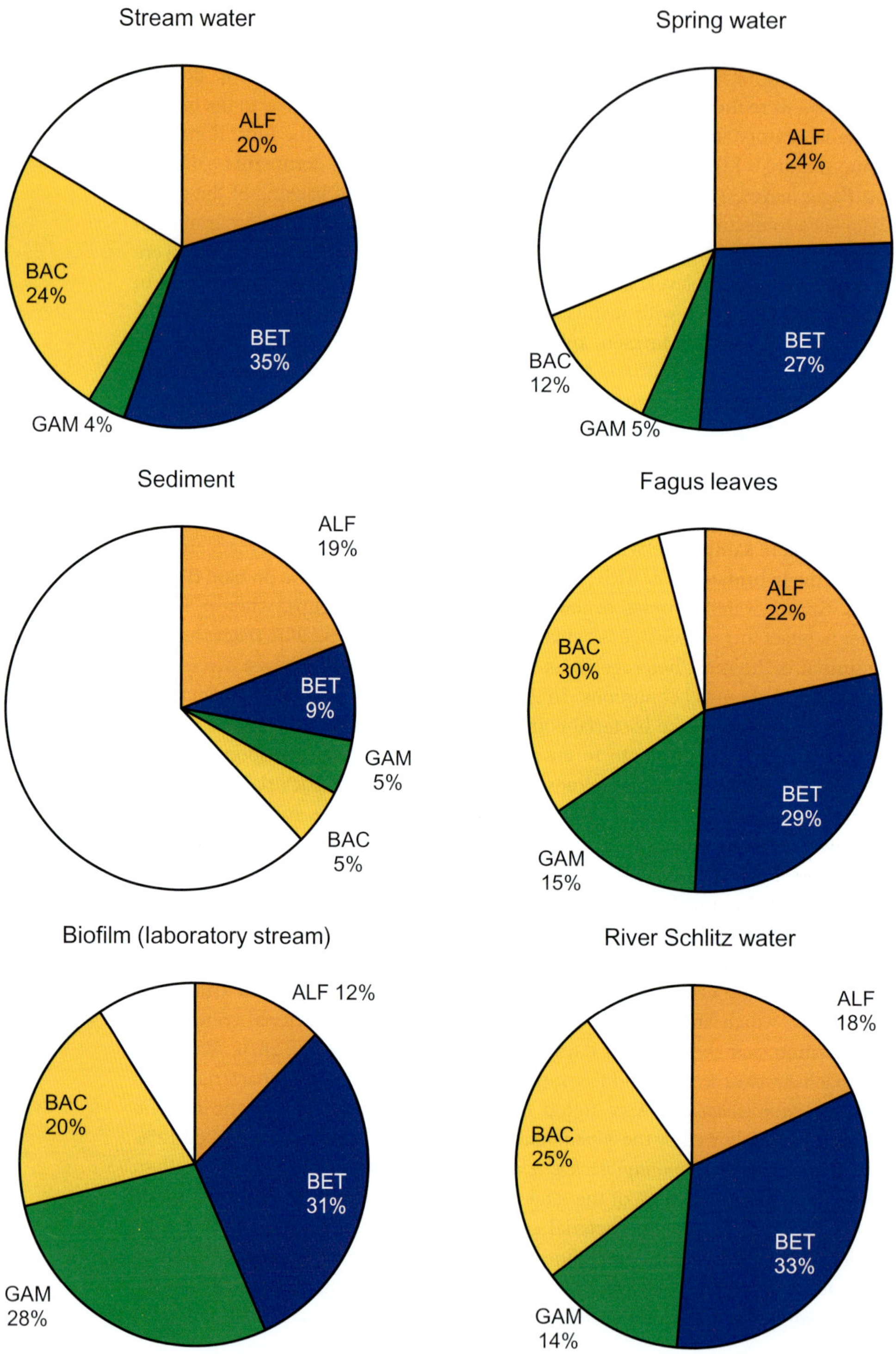
Stream water
ALF 20%
BET 35%
GAM 4%
BAC 24%
Spring water
ALF 24%
BET 27%
GAM 5%
BAC 12%
Sediment
ALF 19%
BET 9%
GAM 5%
BAC 5%
Fagus leaves
ALF 22%
BET 29%
GAM 15%
BAC 30%
Biofilm (laboratory stream)
ALF 12%
BET 31%
GAM 28%
BAC 20%
River Schlitz water
ALF 18%
BET 33%
GAM 14%
BAC 25%

Figure 6.13 Percentages of cells hybridized with specific probes relative to the abundance of cells hybridized with probe EUB338 (specific for domain Bacteria), ALF = Alphaproteobacteria, BET = Betaproteobacteria, GAM = Gammaproteobacteria, BAC = Bacteroidetes (for probes used, see Table 6.6).

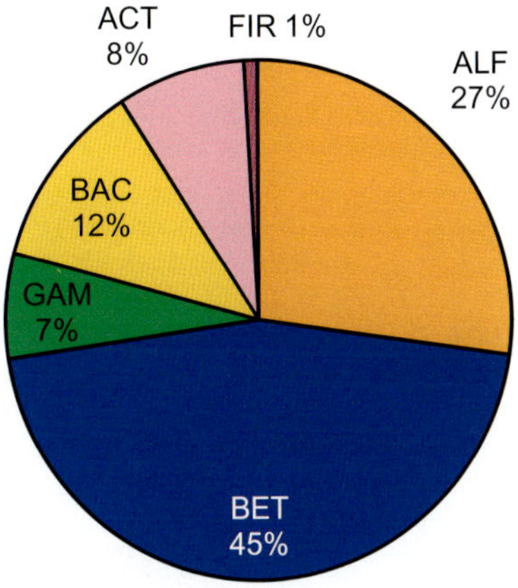

Figure 6.14 Average taxonomic composition of the bacterial community in sediment from the Breitenbach upper course during rewetting (November 2004) analyzed by CARD-FISH (Marxsen *et al.*, 2010). Percentages of affiliated cells are given. ALF = Alphaproteobacteria, BET = Betaproteobacteria, GAM = Gammaproteobacteria, BAC = Bacteroidetes, ACT = Actinobacteria, FIR = Firmicutes (for probes used, see Table 6.6).

detected by sequence analysis (especially *Acidobacteria, Deltaproteobacteria, Verrucomicrobia*; Beier *et al.*, 2008; for details see Section 6.2.2.5). Methodological constraints may also explain this. For example, the CF319a probe (Manz *et al.*, 1996) covers only 44% of the target *Bacteroidetes* (Amann and Fuchs, 2008).

6.2.2.4 **Temperature Gradient Gel Electrophoresis**

The structure and diversity of bacterial communities in different habitats of the Breitenbach, in the headwaters of the Rohrwiesenbach and Jossa, and in the River Fulda (all in Eastern Hesse; cf. Figure 2.2), were compared by TGGE (Beier, 2003; Beier *et al.*, 2008). Like DGGE, this genetic fingerprinting technique is highly suitable, for comparing temporal and spatial population dynamics (Muyzer *et al.*, 1993; Muyzer and Smalla, 1998). By treating the TGGE bands as operational taxonomic units (OTU) and surrogates of bacterial species, the approach allows the comparison of samples and an estimation of the complexity of their population structure. But it does not generate information on community composition, phylogenetic relationships or functional abilities of organisms.

Complex patterns of up to 40 bands per sample were observed in the TGGE gels, suggesting the occurrence of a large number of bacterial species at most of

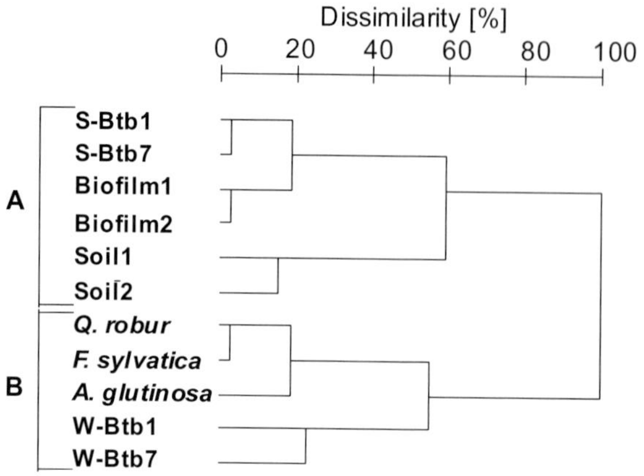

Figure 6.15 Bacterial community composition in different habitats of the Breitenbach compared by cluster analysis of TGGE profiles prepared with 16S rDNA fragments. W: water samples; S: sediment samples; Btb1: Breitenbach upper reach about 500 m above gauge GA; Btb7: Breitenbach lower reach at trap MT6. The dissimilarity matrix for each sample was calculated using squared Euclidean distances, and clustering was performed by the Ward method. The highest dissimilarity value between all samples was set to 100%. After Beier *et al.* (2008).

the investigated sites, although the number of bands does not necessarily represent the precise number of phylotypes (Clayton *et al.*, 1995; Ferris and Ward, 1997; Wang and Wang, 1997). It confirmed the results from the numerical taxonomy (Marxsen and Moaledj, 1988) and sequence analysis (Beier *et al.*, 2008), which suggested high bacterial diversity in the Breitenbach.

Cluster analysis for the different Breitenbach habitats, that is, stream water, sediment, autotrophic biofilms on large stones, degrading in-stream leaves of alder (*Alnus glutinosa*), beech (*Fagus sylvatica*), and oak (*Quercus robur*), and soil 2 m from the stream, showed closely related TGGE patterns among identical habitats, but clear differences in bacterial community composition between habitats (Figure 6.15). The most significant differences occurred between the communities on leaves and in stream water compared to stream sediment, biofilm, and soil communities. Leaf communities differed distinctly from water ones, as did those in soil from both sediment and biofilm. Similarity between biofilm and sediment habitats was in the same range as between unrelated samples within the other habitats. These findings agree with the results from the numerical taxonomy and the sequence analysis studies, which demonstrated particularly marked differences between water and sediment.

Cluster analysis of water and sediment samples at nine locations along the Breitenbach, from source to mouth, together with samples from the other environments under investigation, revealed two main groups of bacterial communities (Figure 6.16). One (cluster A) contained only water samples from the Breitenbach and the River Fulda, whereas the other (cluster B) contained all sediment samples

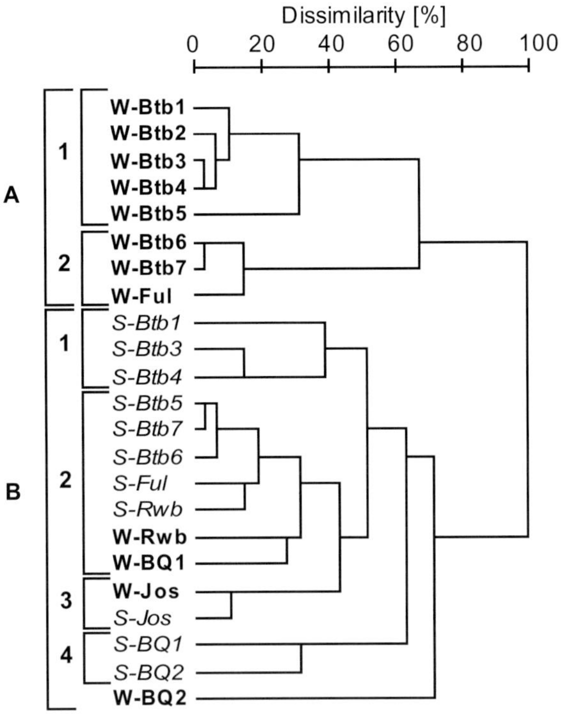

Figure 6.16 Bacterial community composition compared by cluster analysis of TGGE profiles prepared with 16S rDNA fragments in water (W-) and sediment (S-) samples from Breitenbach (Btb1–Btb7: from the upper reach about 500 m above gauge GA to the lower reach at trap MT6 about 150 m distance from the mouth), Breitenbach spring (Georgsborn, BS1, BS2), Rohrwiesenbach (Rwb), Jossa (Jos) and Fulda (Ful). Dendrogram was prepared as in Figure 6.15. After Beier *et al.* (2008).

and the water samples from the Breitenbach springs, from the Rohrwiesenbach and the Jossa. These TGGE data suggested that a distinctive suspended bacterial community occurred in the upper reach of the Breitenbach and that this changed only slightly, albeit continuously, downstream. Similarly for sandy sediment communities, the distribution of samples from the uppermost site to that closest to the mouth suggested continuous downstream variation of bacterial communities. Similar continuous variation of bacterial communities had been observed in a study determining the pH dependence of CFU occurrence on agar plates from sediment samples along the Breitenbach. This showed a slight but continuous downstream shift (Marxsen, 1989).

The entrance of spring water (Georgsborn, sites BS1 and BS2) at the start of the stream's middle reach did not noticeably change the TGGE pattern. This was not surprising when the much lower bacterial numbers of spring water were considered (Table 6.1). The samples taken from the Georgsborn spring (BS) itself were distinct from those at all other sites, except one water sample. The springs are fed by deep groundwater and the spring pool sediments are well perfused, with a short residence time (minutes) and low, but constant, temperature (annual variation: 7.7–8.4 °C). Thus it would be expected that this community would differ from those in the main stream channel, although this was not shown by numerical taxonomy and FISH analysis. However, the latter approaches only attributed bacteria to higher taxa; it does not exclude the possibility of distinct differences at the genus or species level.

6.2.2.5 **Sequence Analysis**

The genetic diversity of bacterial communities in the Breitenbach, in the nearby Rohrwiesenbach and Jossa streams, and in the River Fulda (with a focus on the Breitenbach; cf. Figure 2.2) was estimated by sequence analysis of 16S rDNA fragments by Beier (2003) and Beier *et al.* (2008). DNA was isolated from different stream habitats and amplified by PCR, followed by the establishment of a clone library, sequencing, and the construction of a phylogenetic tree (Beier *et al.*, 2008). This allowed the taxonomic classification of the sequences isolated (Tables 6.7 and 6.8).

Stream water communities mainly comprised *Betaproteobacteria* (33%), whereas sediment communities were dominated by *Betaproteobacteria* and *Acidobacteria* (about 25% each; Table 6.7, Figure 6.17). Sequences related to *Alpha-* and *Deltaproteobacteria*, to *Verrucomicrobia*, *Actinobacteria*, *Spirochaetes*, *Planctomycetes*, and *Chloroflexi* occurred in sediments, and sequences from *Acidobacteria*, *Alpha-* and *Gammaproteobacteria*, *Actinobacteria*, and *Bacteroidetes* in water samples at lower percentages. The representation of different bacterial taxa resembled widespread freshwater clusters, indicating the presence of typical freshwater organisms in small first-order streams (Glöckner *et al.*, 2000; Zwart *et al.*, 2002; Warnecke *et al.*, 2004). However, a considerable number of sequences from water and sediment were related to sequences known from soil bacteria, also as observed by Glöckner *et al.* (2000), Warnecke *et al.* (2004), and Hullar *et al.* (2006). The close connection between soil and stream sediment habitats is typical, especially for small streams, which probably contain species that are able to proliferate in streams as well as in soil (Beier *et al.*, 2008).

Composition varied when specific functional groups of bacteria were considered. The majority of genes encoding for chitinases (glycosyl hydrolases family 18A), which were isolated from different Breitenbach habitats, were predominantly affiliated to the *Actinobacteria* (nearly 70%), but also to the *Betaproteobacteria*, *Gammaproteobacteria*, *Firmicutes*, and *Bacteroidetes* (lower percentages; Table 6.16; for details, see Section 6.3.2). The chemolithoautotrophic ammonia-oxidizing bacteria (AOB) on the other hand are known to belong almost entirely to the *Betaproteobacteria* (Woese *et al.*, 1984; Head *et al.*, 1993; Teske *et al.*, 1994), except for *Nitrosococcus* species, which are *Gammaproteobacteria* (Woese *et al.*, 1985) and have so far not been observed in freshwater. In order to detect the occurrence of AOB in the Breitenbach, Rohrwiesenbach and Jossa, Höppner (1999) used the NitA and NitB primers to amplify 16S rDNA from ammonia-oxidizing bacteria (Ward *et al.*, 1997). Two cloned fragments from the Breitenbach showed close similarity (>97%)

Table 6.7 Phylogenetic affiliation of bacterial 16S rRNA partial gene sequences amplified from running water environments in eastern Hesse contained in Beier *et al.* (2008).

Accession no.[a)]	Origin[b)]	Taxonomic assignment: phylum class order family genus[c)]	Nearest neighbor BLAST matches[d)]		
			Identity/strain[e)]	Origin[f)]	%[g)]
DQ017910	Rwb sed	*Acidobacteria Acidobacteria Acidobacteriales Acidobacteriaceae*	Uncultured clones DQ532357/ EU861884	Air, Kennedy Space Center, FL, USA/soil, CO, USA	95
DQ017911	Rwb sed	*Chloroflexi Anaerolineae Caldilineales Caldilineacea*	Uncultured clones EU134153/ EU917659	Soil, OK, USA/ marine stromatolite, Bahamas	98/90
<u>DQ017912</u>	Btb sed	*Proteobacteria Betaproteobacteria Burkholderiales Incertae sedis Leptothrix*	*Leptothrix* sp. DQ241397/ uncultured clone DQ833473	Mine drainage water/reservoir sediment, China	98
DQ017913	Btb sed	*Verrucomicrobia Verrucomicrobiae Verrucomicrobiales*	Uncultured clones AJ401114/AY509518	River biofilm, Elbe, Germany/lake plankton, Sweden	94/89
<u>DQ017914</u>	Btb sed	*Proteobacteria Alphaproteobacteria Sphingomonadales Sphingomonadaceae Novosphingobium*	*Alphaproteobacterium* AB452969/ uncultured clone EU937972	Freshwater plankton, Lake Biwa, Japan/ riparian iron oxidizing biofilm	97
DQ017915	Btb sed	*Acidobacteria Acidobacteria Acidobacteriales Acidobacteriaceae*	Uncultured clones EU861837/ EU150232	Soil, CO, USA/ soil, CO, USA	98
<u>DQ017916</u>	Btb wat	*Proteobacteria Alphaproteobacteria Rhizobiales Hyphomicrobiaceae Devosia*	Uncultured clones GQ010485/FJ535114	Human skin/ washing water of carrot washing plant, Germany	99/98
<u>DQ017917</u>	Btb wat	*Bacteroidetes Flavobacteria Flavobacteriales Flavobacteriaceae Flavobacterium*	*Flavobacterium* sp. AM167560/ *Flavobacterium succinicans* AM230492	Westerhöfer Bach, stream, Germany/ type strain: DSM 4002	98
DQ017918	Btb wat	*Acidobacteria Acidobacteria Acidobacteriales Acidobacteriaceae*	Uncultured clones EF019471/ DQ451491	Aspen rhizosphere, WI, USA/soil, Taiwan	99

Continued

Table 6.7 *Continued*

Accession no.[a]	Origin[b]	Taxonomic assignment: phylum class order family genus[c]	Nearest neighbor BLAST matches[d]		
			Identity/strain[e]	Origin[f]	%[g]
DQ017919	Btb wat	*Proteobacteria Betaproteobacteria Burkholderiales Comamonadaceae Curvibacter*	Betaproteobacterium AF236004/ Uncultured clone EF667920	Lake Hibara, Japan/Lake Pohlsee, Germany	100/99
DQ017920	Btb wat	*Proteobacteria Gammaproteobacteria Legionellales Coxiellaceae Aquicella*	Uncultured clones EU834786/FJ516869	Aerobic phosphorus-removing batch reactor/lower sediment, Spain	98/97
DQ017921	Btb wat	*Proteobacteria Betaproteobacteria Burkholderiales Burkholderiaceae Polynucleobacter*	*Polynucleobacter necessarius* subsp. *asymbioticus* AB269809/AJ550657	Sakura River, Japan/Yangtze River, China	100
DQ017922	Btb wat	*Proteobacteria Gammaproteobacteria Xanthomonadales Xanthomonadaceae*	Uncultured clones AB240508/ EU662365	Sosei River, Japan/ cave microbial mat, VI, USA	99
DQ017923	Rwb sed	*Proteobacteria Betaproteobacteria Burkholderiales (Incertae sedis Thiobacter)*	Uncultured clones EF020061/AY326600	Aspen rhizosphere, WI, USA/soil, Western Amazon, Brazil	98/97
DQ017924	Rwb sed	*Acidobacteria Acidobacteria Acidobacteriales Acidobacteriaceae*	Uncultured clones DQ017934/ AJ532721	Jossa stream sediment, Germany/uranium mining waste pile, Germany	99
DQ017925	Rwb sed	*Acidobacteria Acidobacteria Acidobacteriales Acidobacteriaceae*	Uncultured clones FJ624936/DQ499285	Soil/Frasassi cave wall biofilm, Italy	96
DQ017926	Rwb sed	*Proteobacteria Betaproteobacteria Burkholderiales Comamonadaceae Curvibacter*	Uncultured clones DQ017929/ EF632936	Rohrwiesenbach stream sediment, Germany/Lake sediment, Laguna de Piacota, Chile	97
DQ017927	Rwb sed	*Acidobacteria Acidobacteria Acidobacteriales Acidobacteriaceae*	Uncultured clones DQ648911/FJ152864	Soil, Spain/alkaline saline soil of former lake Texcoco, Mexico, following flooding	98

Table 6.7 *Continued*

Accession no.[a]	Origin[b]	Taxonomic assignment: phylum class order family genus[c]	Nearest neighbor BLAST matches[d]		
			Identity/strain[e]	Origin[f]	%[g]
DQ017928	Rwb sed	*Proteobacteria Betaproteobacteria Burkholderiales Comamonadaceae Rhodoferax*	Uncultured clones AY212612/FJ517721	Stream water, OH, USA/Hydra epithelium, freshwater, Germany	99
DQ017929	Rwb sed	*Proteobacteria Betaproteobacteria Burkholderiales Comamonadaceae Curvibacter*	Uncultured clones DQ017926/ EF632936	Rohrwiesenbach stream sediment, Germany/Lake sediment, Laguna de Piacota, Chile	97
DQ017930	Jos sed	*Verrucomicrobia Verrucomicrobiae Verrucomicrobiales*	Uncultured clones AY792311/FJ592707	Humic lake, WI, USA/wetland soil (elev. 3661 m), at Socompa Volcano, Andes, Argentina	98/96
DQ017931	Jos sed	*Proteobacteria Betaproteobacteria Burkholderiales Oxalobacteraceae*	Uncultured clone FJ624877/AJ575695	Boreal soil/lake water Große Fuchskuhle, Germany	96
DQ017932	Jos sed	*Proteobacteria Deltaproteobacteria Myxococcales Polyangiaceae Byssovorax*	Uncultured clones EF516177/EF516117	Soil, CA, USA/ soil, CA, USA	97
DQ017933	Jos sed	*Proteobacteria Alphaproteobacteria Rhizobiales*	Uncultured clones DQ083106/FJ152767	Soil/alkaline saline soil of former lake Texcoco, Mexico, following flooding	97
DQ017934	Jos sed	*Acidobacteria Acidobacteria Acidobacteriales Acidobacteriaceae*	Uncultured clones DQ017924/Z95718	Rohrwiesenbach stream sediment, Germany/sediment	99/98
DQ017935	Btb spr sed	*Verrucomicrobia Verrucomicrobiae Verrucomicrobiales Verrucomicrobiaceae*	Uncultured clones EU335362/ DQ450785	Soil, TN, USA/ saturated alpine soil, CO, USA	95
DQ017936	Jos sed	*Proteobacteria Betaproteobacteria Burkholderiales Comamonadaceae*	Uncultured clones AF289156/ AY328727	Parker River, MA, USA/drinking water, OH, USA	97/95

Continued

Table 6.7 *Continued*

Accession no.[a]	Origin[b]	Taxonomic assignment: phylum class order family genus[c]	Nearest neighbor BLAST matches[d]		
			Identity/strain[e]	Origin[f]	%[g]
DQ017937	Jos sed	*Actinobacteria Actinobacteria Actinomycetales Propionibacteriaceae Propionibacterium*	*Propionibacterium acnes* AE017283/ AB097215	Human skin/ hospital, Japan	99
DQ017938	Jos sed	*Proteobacteria Alphaproteobacteria Rhizobiales*	Uncultured clones EU801228/AJ582029	Chesapeake Bay, MD, USA/soil, Germany	99/98
DQ017939	Jos sed	*Acidobacteria Acidobacteria Acidobacteriales Acidobacteriaceae*	Uncultured clones FJ004679/EF515895	Rhizosphere, Netherlands/soil, CA, USA	97/96
DQ017940	Rwb sed	*Acidobacteria Acidobacteria Acidobacteriales Acidobacteriaceae*	Uncultured clones FJ437944/FJ517007	Green Lake sediments, NY, USA/water, Spain	98
DQ017941	Rwb sed	*(Planctomycetes)*	Uncultured clones GQ094006/ EU135040	Human skin/soil, OK, USA	93/91
DQ017942	Rwb sed	*Proteobacteria Betaproteobacteria Burkholderiales Incertae sedis Leptothrix*	Uncultured clones EU015089/ AB294325	Bioreactor/stream, Japan	98
DQ017943	Btb spr sed	*Spirochaetes Spirochaetes Spirochaetales Leptospiraceae Turneriella*	*Turneriella parva* AY398688/ uncultured clone FM201055	Porcine uterus/ bioreactor	96/95
DQ017944	Btb spr sed	*Proteobacteria Alphaproteobacteria Rhizobiales*	Uncultured clones EU937836/ EU335286	Riparian iron oxidizing biofilm/ soil, TN, USA	98/99
DQ017945	Btb sed	*Acidobacteria Acidobacteria Acidobacteriales Acidobacteriaceae*	Uncultured clones EU937897/FJ004730	Riparian iron oxidizing biofilm/ rhizosphere soil, Netherlands	99
DQ017946	Btb sed	*Acidobacteria Acidobacteria Acidobacteriales Acidobacteriaceae*	Uncultured clones DQ017945/ EU937897	Breitenbach stream sediment (Germany)/ riparian iron oxidizing biofilm	99

Table 6.7 *Continued*

Accession no.[a)]	Origin[b)]	Taxonomic assignment: phylum class order family genus[c)]	Nearest neighbor BLAST matches[d)]		
			Identity/strain[e)]	Origin[f)]	%[g)]
DQ017947	Btb sed	*Actinobacteria Actinobacteria Actinomycetales*	Uncultured clones AM176853/ EF590048	Mangrove sediment, China/ Songhua River sediment, China	99
DQ017948	Btb sed	*Proteobacteria (Deltaproteobacteria)*	Uncultured clones FJ592579/EU134544	Wetland soil (elevation 3661 m), at Socompa Volcano, Andes, Argentina/soil, OK, USA	96/95
DQ017949	Btb sed	*Proteobacteria Deltaproteobacteria Myxococcales*	Uncultured clones DQ017950/ EU234271	Breitenbach stream sediment, Germany/Wang Yang River, China	99/97
DQ017950	Btb sed	*Proteobacteria Deltaproteobacteria Myxococcales*	Uncultured clones DQ017949/ EU234271	Breitenbach stream sediment, Germany/Wang Yang River, China	99/98
DQ017951	Btb wat	*Proteobacteria Betaproteobacteria Burkholderiales Burkholderiaceae Polynucleobacter*	*Polynucleobacter cosmopolitanus* AJ550667/AJ550651	Canal Roanne plankton, France/ Lake Victoria, plankton, Uganda	99
DQ017952	Btb wat	*Actinobacteria Actinobacteria Actinomycetales Microbacteriaceae*	Uncultured clones EU800315/AJ575528	Delaware Bay, NJ, USA/Lake Niegocin, Poland	99

a) Accession number in NCBI GenBank (http://www.ncbi.nlm.nih.gov/Genbank/). Numbers in bold indicate sequences with identities of at least 97% to sequences of named species (for details see Table 6.8).
b) Btb = Breitenbach, Rwb = Rohrwiesenbach, Jos = Jossa, spr = spring, wat = water, sed = sediment.
c) Taxonomic assignment of sequences was performed at the highest level of resolution possible (from phylum to species level) via Ribosomal Database Project (Release 10.15) website (Cole *et al.*, 2007, 2009) with Naive Bayesian rRNA Classifier Version 2.0, July 2007 at http://rdp.cme.msu.edu/classifier/ (Wang *et al.*, 2007) in August 2009. Taxa are given at a confidence threshold of 80%, but of 60% only for taxa in brackets.
d) The two nearest neighbor matches by NCBI BLASTN (2.2.21) nucleic acid search in Nucleotide collection (nr/nt) using Megablast (Zhang *et al.*, 2000) are given (performed August 2009).
e) Including GenBank sequence accession number.
f) If available.
g) Maximum identity.

DNA had been amplified with primers EUBI and EUBII (5′-GAGTTTGATCCTGGCTCAG-3′/5′-AGAAAGGAGGTGATCCAGCC-3′, *E. coli* positions 9 through 27 and 1525 through 1542, respectively; Liesack *et al.*, 1991).

Table 6.8 Bacterial species detected in the Breitenbach and other lotic environments in eastern Hesse based on sequence similarity analysis of 16S rDNA (for details see Table 6.7).

Phylum	Class	Order	Family	Genus and species	Accession no/site[a)]
Actinobacteria	Actinobacteria	Actinomycetales	Microbacteriaceae	*Candidatus rhodoluna* limnophila	DQ017952 Btb wat
			Propionibacteriaceae	*Propionibacterium acnes*	DQ017937 Jos sed
Bacteroidetes	Flavobacteria	Flavobacteriales	Flavobacteriaceae	*Flavobacterium succinicans*	DQ017917 Btb wat
Proteobacteria	Alphaproteobacteria	Rhizobiales	Hyphomicrobiaceae	*Devosia riboflavina*	DQ017916 Btb wat
		Sphingomonadales	Sphingomonadaceae	*Novosphingobium* sp.	DQ017914 Btb sed
	Betaproteobacteria	Burkholderiales	Burkholderiaceae	*Polynucleobacter cosmopolitanus*	DQ017951 Btb wat
					AJ508171 Jos wat
				Polynucleobacter necessarius subsp. *asymbioticus*	DQ017921 Btb sed
					AJ508179 Btb wat
			Comamonadaceae	*Curvibacter delicatus*	DQ017919 Btb wat
					DQ017926 Rwb sed
					DQ017929 Rwb sed
				Rhodoferax ferrireducens	DQ017928 Rwb sed
			Incertae sedis	*Leptothrix* sp.	DQ017912 Btb sed
		Nitrosomonadales	Nitrosomonadaceae	*Nitrosomonas europaea*	AJ508181 Btb wat
					AJ508178 Btb wat

a) Accession no in NCBI GenBank (http://www.ncbi.nlm.nih.gov/Genbank/). Btb = Breitenbach, Rwb = Rohrwiesenbach, Jos = Jossa, wat = water, sed = sediment. Sequences with accession numbers beginning with DQ are from Beier *et al.* (2008), those beginning with AJ from Höppner (1999).

The same species was assumed when the sequence identity was at least 97% (Rosselló-Mora and Amann, 2001). Comparison of sequences was performed via NCBI BLASTN (2.2.21) nucleic acid search in nucleotide collection (nr/nt) using Megablast (Zhang *et al.*, 2000) in August 2009.

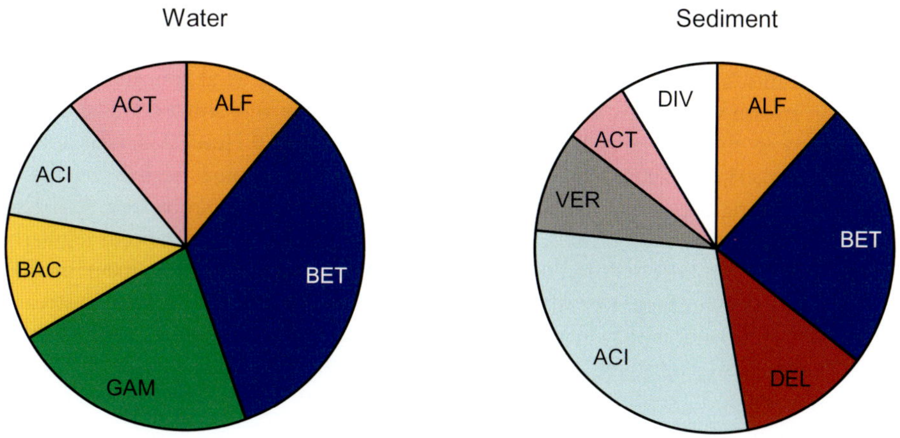

Figure 6.17 Taxonomic distribution of clonal 16S rDNA sequences from water and sediment of streams in eastern Hesse (mainly Breitenbach). ALF = Alphaproteobacteria, BET = Betaproteobacteria, GAM = Gammaproteobacteria, DEL = Deltaproteobacteria, BAC = Bacteroidetes, ACI = Acidobacteria, VER = Verrucomicrobia, ACT = Actinobacteria, DIV contains Spirochaetes, Planctomycetes and Chloroflexi. Data from Beier *et al.* (2008).

to sequences of *Nitrosomonas europaea* and another *Nitrosomonas* species (Table 6.8), whereas a fragment from the Jossa was related to *Nitrosospira multiformis*. A total of 33 sequences was retrieved from the streams using this approach; all but two were *Betaproteobacteria*, the two were from the *Gammaproteobacteria*. In addition to the four sequences belonging to the AOB group *sensu strictu*, five sequences showed highest similarity to genera phylogenetically related to this group (*Herbaspirillum, Ultramicrobacterium, Sterolibacterium*). All other clones showed greatest similarity to different groups of clonal sequences, often from hitherto unidentified bacteria, that have been retrieved from various freshwater and soil habitats.

6.2.2.6 **Water versus Sediment Bacterial Communities**

TGGE patterns, data from sequence analysis, FISH and numerical taxonomy, all suggest that streams like the Breitenbach harbor typical freshwater bacteria. But, especially in the sediment habitat, bacteria that are considered typical soil bacteria can also proliferate. There are only slight differences within the same habitats along the stream, but clear differences between different habitats, such as stream water, sandy sediment, biofilms on coarse substrata, and decaying leaves. However, the occurrence of distinctive water communities, which are markedly distinct from those in other habitats, remains an unexplained paradox. The water residence time of 4–10 h from the upper reach (water gauge GA) to the mouth (Beier *et al.*, 2008), together with the low nutrient concentrations and low temperatures (below 15 °C at almost all sites during the day, even in summer) probably makes the proliferation of introduced cells over several generations impossible.

Marxsen and Moaledj (1988) had suggested that bacterial cells are continuously selectively replenished from the surrounding habitats such as sediments and biofilms, and even soil, thus establishing the typical bacterial community found in stream water. This assumption is supported by the very small cell size of bacteria in the Breitenbach, which is on average about 0.1 μm^3 (Marxsen 1980b, 1982a). Hahn (2003) discovered that organisms from a pure freshwater cluster containing bacteria of the genus *Polynucleobacter* are ultramicrobacteria (cell size about 0.1 μm^3). *Polynucleobacter*-like cells were meanwhile found to occur regularly in large numbers in different freshwaters (Hahn *et al.*, 2003; Hahn, 2006), including the Breitenbach. Here their 16S rDNA-sequences have been detected in stream water and sediment by Beier *et al.* (2008), but also earlier by Höppner (1999), when investigating nitrifying bacteria (Table 6.8). Their small cell volume facilitates selective cell seeding and can cause enrichment of the stream water with ultramicrobacteria.

6.2.2.7 Community Composition: Conclusions

Some of the fundamental results of the earlier studies with classical approaches still appear valid. These are, for example, the discovery of distinctly different bacterial communities between different habitats, or of the slight, but continuous variation of communities downstream. But the application of molecular biology techniques to the Breitenbach demonstrated the value of these newer approaches. The taxonomic affiliation of sequences obtained by PCR amplification of 16S rDNA was different from that derived from numerical taxonomy. TGGE and sequence analysis in particular have allowed much finer resolution. The picture obtained with the new approaches is much more exact, detailed and reliable, because the techniques are less selective and less time-consuming. From the application of molecular biology methods to determine bacterial community composition in streams new insights into the importance of bacteria in running water environments can be expected.

6.2.3 The Fungal Community

Conidia numbers of aquatic hyphomycetes and community assemblages were monitored over one year (1998) in the Breitenbach (Callies, unpublished data), filtering monthly water samples (5.0 μm pore size membrane filters). For the longitudinal analyses of the Breitenbach fungi, water was collected at four sites along the stream. Seasonality in conidia numbers increased with increasing distance from the source and with increasing nutrient concentrations in the water. In the effluent water of the Georgsborn spring, only 14–168 conidia l^{-1} (minimum to maximum) were found during the year. The other sites showed the lowest amounts of conidia per liter in the months of January to August (below trap T2: 90–1800 conidia l^{-1}; at former trap T5 below alder trees: 405–2127 conidia l^{-1}; below trap T6: 464–1803 conidia l^{-1}). In October/November sporulation reached its maximum (T2: 2556 conidia l^{-1}; T5: 8681 conidia l^{-1}; T6: 13 213 conidia l^{-1}). Twenty-three species of aquatic hyphomycetes were identified. Dominant taxa were: *Flagellospora curvula, Clavatospora longibrachiata, Tetrachaetum elegans* and *Tumularia aquatica.*

For comparison, samples were also taken from three other small streams (Figure 2.2), which differ in pH: Rohrwiesenbach (pH = 8), Erlenbach (pH = 7) and Jossa headwater (pH = 4). No stream exhibited any seasonal trend in its fungus community composition. The Erlenbach community was similar to that of the Breitenbach, but the maximum conidia numbers were only 2430 l^{-1}, and 20 species were identified. Higher maximum numbers of conidia l^{-1} were found in the Rohrwiesenbach (14470 l^{-1}), but only 14 species were identified. The species found in the acidic Jossa headwater differ from all other streams. Only seven species (most not identified) and 32–2470 spores l^{-1} were detected. The lack of seasonality can be explained by the very low water column in summer and in the Rohrwiesenbach by the continuous input of riparian leaves from *Alnus glutinosa* while in the Erlenbach the continuous input of material from the adjacent meadow plays an important role. The low sporulation activity in the Jossa headwater results probably from a combination of low pH with a low input of leaves.

A rough estimation can be given for fungal biomass in the Breitenbach on the basis of data from Hendel (1999) and Hendel and Marxsen (2000). The ergosterol content of leaves and wood in the Breitenbach was determined to be about 30 $\mu g\, g^{-1}$ OM. The mean POM standing crop in the Breitenbach is 621 $g\, m^{-2}$. By using the conversion factor of Gessner and Chauvet (1993) (biomass/ergosterol = 182), the average total biomass of fungi in the Breitenbach can be calculated to 1.7 $g\, C\, m^{-2}$. But the Breitenbach contains large amounts of FPOM, which is populated much less by fungi (Findlay *et al.*, 2002). Thus this value is probably an overestimation. Therefore as a very rough, first estimate, 1 $g\, C\, m^{-2}$ of fungal biomass in the Breitenbach (Table 6.5) was assumed by Marxsen (2006). Compared to a cross-system comparison performed by Findlay *et al.* (2002), this estimate seems comparatively low and detailed investigations on the fungal biomass in the Breitenbach might show a higher level. However, it is within the span of fungal biomasses of 0.1–6.5 $g\, C\, m^{-2}$ determined by Suberkropp (1997) in Walker Branch (Tenn., USA) and 0.7–1.2 $g\, C\, m^{-2}$ determined by Methvin and Suberkropp (2003) in two woodland streams in Alabama, USA.

6.3 Microbial Metabolic Activity

6.3.1 Approaches to Microbial Metabolic Activity Measurements in the Breitenbach

The difficulty of separately investigating specific microorganism species in natural environments is a well known problem of microbial ecology. It is often not only impossible to identify the complete composition of microbial communities, but often even impossible to identify the most important ones (cf. Section 6.2.2). Thus microbial ecology has focussed on quantifying microbial processes over long periods. This was also the preferred approach during the decades when the Breitenbach Ecosystem Study was performed.

Two particularly important processes were investigated in detail at the Breitenbach. One was extracellular enzyme activity. Extracellular enzymes are involved in the main processes of inorganic nutrient regeneration (Marxsen and Schmidt, 1993) and utilization of macromolecular organic matter to make it available as carbon and energy resources for heterotrophic organisms (Marxsen and Fiebig, 1993). Following this approach enzymes from different groups of organisms were usually measured simultaneously (Marxsen *et al.*, 1998). The second important process investigated in detail was bacterial carbon production (BCP). Bacteria are not only involved in the degradation of organic matter, but they have also been shown to play an important role in the food chains of streams like the Breitenbach (Marxsen, 2006). However, we also performed some specialized investigations, for example, on the uptake of dissolved organic substances in stream water (Marxsen 1980c), on the utilization of amino acids in sediments (Fiebig 1992, 1997; Fiebig and Marxsen, 1992), on heat production of biofilms (Freeman *et al.* 1990, 1993) and on nitrification (Höppner, 1999), which cannot be covered in this synoptic contribution.

6.3.2
Extracellular Enzyme Activity

6.3.2.1 The Function of Extracellular Enzymes in Streams

In streams large quantities of dissolved and particulate organic matter are available to heterotrophic organisms (cf. Chapter 4), but only low molecular weight compounds can pass through the cell membranes of such organisms by active or passive transport. In natural undisturbed streams these compounds generally constitute less then 20% of the standing stock of dissolved organic matter (Thurman, 1985). If particulate material is included (Figure 4.1, Table 6.5), then the quantitative significance of low molecular weight material to the total available organic carbon in streams is extremely low (Marxsen and Fiebig, 1993). Thus only a very small percentage of organic matter can be utilized directly by microorganisms. But these organisms are able to increase their accessibility to the macromolecular pool of organic matter by producing extracellular enzymes, mainly hydrolases, which depolymerize dissolved and particulate macromolecules, including polysaccharides, proteins, nucleic acids, and humic matter (Münster, 1984; Figure 6.18).

The cleavage of macromolecular organic compounds by extracellular enzymes is often a crucial initial step in the microbially-mediated degradation of organic matter (Overbeck and Chróst, 1990; Marxsen and Witzel, 1990) and its subsequent transformation to carbon dioxide and microbial biomass (Marxsen and Fiebig, 1993), not only in streams but generally in aquatic environments (Hoppe 1983, 1993). Such cleavage is also highly important for the regeneration of inorganic nutrients (Chróst and Overbeck, 1987; Münster *et al.*, 1989; Boon 1989, 1991a, b,

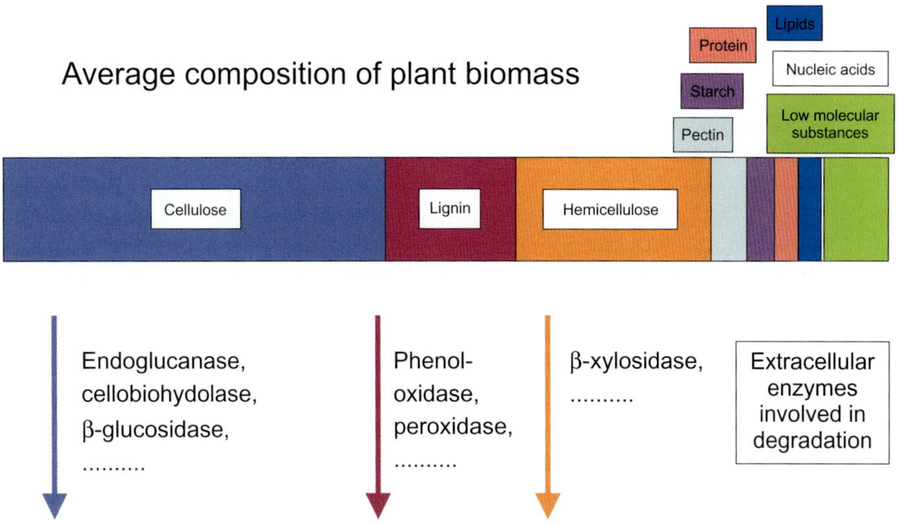

Figure 6.18 Average composition of plant biomass and extracellular enzymes involved in the degradation of polymeric compounds from plant biomass (after Fritsche, 1985).

Chróst, 1991; Marxsen and Schmidt, 1993). Thus (cf. Section 6.1 and Section 8.7), the activity of extracellular enzymes at the basis of the microbial food chain is a most important, often limiting, step for the carbon and energy supply of organisms within the chain.

"Extracellular enzymes" are defined as enzymes whose function is external to the cell and which may be free or cell-bound (Wetzel, 1991). In addition to the extracellular enzymes which are produced directly by microorganisms such as bacteria, fungi, and algae, organisms of the micro- and macrofauna are also able to release such enzymes (Marxsen *et al.*, 1998). Enzymes which are liberated by the lysis of any organism can also be important for the occurrence of extracellular enzymes (Jansson *et al.*, 1988).

As in other ecosystems, two main groups of extracellular enzymes with different ecological roles occur in running water environments. Enzymes of the first group, for example, phosphatases and sulfatases, are responsible for the remineralisation of nutrients. The activity levels of such enzymes can be used as indices of the availability of nutrients. High activities indicate depletion of nutrients, whereas low activities occur when sufficient nutrients are available (Jansson *et al.*, 1988; Whitton, 1991; Romaní and Sabater, 1998).

The other group contains enzymes which cleave macromolecular organic matter, such as polysaccharides, proteins, lipids, and humic substances, making them available as a carbon resource for heterotrophic microorganisms (Table 6.9). Such enzymes are mainly hydrolases, like α- and β-glucosidase or peptidases (Hendel and Marxsen 2005a, b), but other reaction types also occur, for example, oxidizing

Table 6.9 Main model substrates used in the Breitenbach studies with 4-methylumbelliferone (MUF) or 7-amido-4-methyl-coumarin (MCA) as fluorigenic compound, the corresponding natural substrates, and the enzymes which cleave these molecules.

Model substrate	Natural substrates	Enzymes
MUF-α-glucoside	Maltose, starch	α-Glucosidases
MUF-β-glucoside	Cellobiose, cellulose	β-Glucosidases
MUF-β-xyloside	Hemicellulose, xylanes	β-Xylosidases
MUF-β-N-acetyl-glucosaminide	Chitobiose, chitin	β-N-Acetylglucosaminidase
Leucine-MCA	Peptides, proteins	Peptidases
MUF-laurate	Lipids	Lipases
MUF-phosphate	Polyphosphates	Phosphatases
MUF-sulfate	Sulfate esters	Sulfatase

enzymes like phenoloxidase or peroxidase systems (Hendel *et al.*, 2005) which are involved in the degradation of humic compounds. The activity of these enzymes is commonly regarded as a crude index of the relative importance of the suite of available macromolecular materials (Chappell and Goulder, 1994; Findlay *et al.*, 1997; Boschker and Cappenberg, 1998; Burns and Ryder, 2001).

6.3.2.2 **Methodological Approach**

The extracellular enzyme activity in aquatic environments is mostly measured using model substrates, which are available for a suite of different natural compounds (Table 6.9). The usual model substrates (Hoppe, 1983) consist of a molecule of 4-methylumbelliferone (MUF) or 7-amido-4-methyl-coumarin (MCA) and another compound (e.g., glucose, phosphate, amino acids; Figure 6.19). These substrates, which are added to the samples to be investigated, are cleaved like the natural oligomeric or polymeric substances, but the liberated compounds, MUF or MCA, are fluorescent. They can be quantified fluorimetrically and indicate the level of extracellular enzyme activity in the sample.

This technique was used to determine the activity of different extracellular enzymes in the Breitenbach. It has been described in detail by Marxsen and Witzel (1990), Marxsen and Fiebig (1993), Marxsen and Schmidt (1993), Hendel and Marxsen (1997), Marxsen *et al.* (1998), Romaní and Marxsen (2002), Hendel and Marxsen (2005a, b), and Hendel *et al.* (2005). Two approaches were used:

1) The model substrates were added to several replicates at increasing concentrations in order to determine the dependence of the cleavage velocity on the substrate concentration. At low concentrations a rapid increase in splitting activity is usually observed with increasing concentration. This increase slows down with further increasing concentration, achieving its maximum value at the maximum efficiency of the enzyme. Using this kinetic approach, values for the maximum hydrolytic velocity (saturation velocity) V_{max} and the Michaelis constant, K_m, a measure of the affinity of the enzyme for its substrate, are achieved.

2) Measurements were often performed at a single concentration. This concentration was always in the saturation range for the particular enzyme–substrate combination, which had to be determined first. Thus a hydrolytic velocity, value v^*, was determined, close to the maximum hydrolytic velocity, V_{max}.

There are some general problems that make the comparison of data from different studies on extracellular enzyme activity difficult (Marxsen *et al.*, 1998). One major problem is the use of different model substrates. Fluorigenic compounds (especially MUF or MCA compounds) are usually employed, but often, particularly in older studies, *p*-nitrophenyl (pNP) and *p*-nitroanilide compounds were used (e.g., Boon, 1990; Sinsabaugh and Findlay, 1995). With respect to measurement sensitivity, the fluorigenic compounds are preferable because fluorimetric techniques are much more sensitive (by about three orders of magnitude) than chromogenic methods (Schwedt, 1981; Tank *et al.*, 1998). Furthermore, it is possible that different substrates are variously susceptible to the enzyme for which they are designed. This was only tested by Boon (1990), who obtained higher extracellular enzyme activity rates with *p*-nitrophenyl phosphate than with methylumbelliferyl phosphate, but with a nonlinear relationship. The differences appeared to be proportionally greater at higher activities. In general it must be noted that the application of nonnatural substrates, like MUF and pNP compounds, means that the obtained rates are not necessarily synonymous with activities on natural substrates (Chróst, 1990).

Other important problems with respect to the comparability of different studies rest with the detailed experimental conditions. Incubation temperature and duration, as well as pH, are the most important factors and have to be chosen with great care (Marxsen *et al.*, 1998). The possibility of using much shorter incubation times with fluorigenic substances, because fluorimetry is much more sensitive, is the main advantage of this technique. The long incubation times necessary for chromogenic measurements (up to 24 h or even more) carry a high risk of drastic qualitative and quantitative changes in the microbial communities. Thus the obtained results might not reflect natural situations.

For comparison only studies that ensured that the experimental conditions, such as temperature and pH, were well controlled and close to natural situations were considered. Furthermore only studies that present extracellular enzyme activity rates as V_{max} values (maximum activity rates) from kinetic measurements or at least v^* values obtained at or near saturation concentrations were considered.

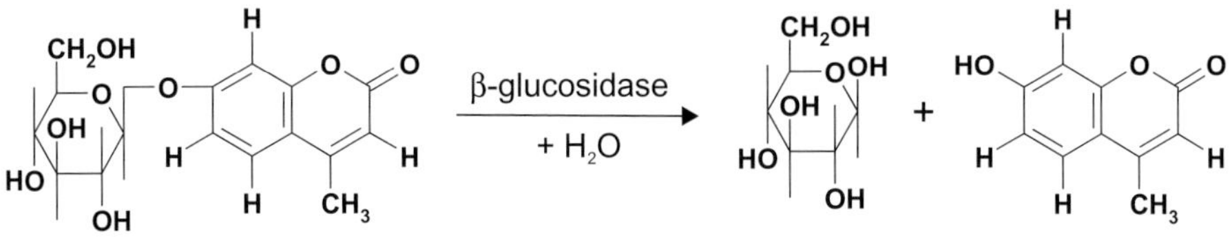

Figure 6.19 MUF-β-glucoside, as an example of a MUF model substrate, and the hydrolytic cleavage of the non-fluorescent molecule into an oligo- or monomeric (like glucose) compound and the fluorescent MUF (from Hendel, 1999).

Table 6.10 Stream habitats to be considered for microbial ecology investigations.

Habitat		Type of metabolism
Water column (suspended)		Mixed auto-/heterotrophic
Biofilms	On stones (epipelon)	Mixed auto-/heterotrophic or purely heterotrophic
	On sand surfaces (epipsammon)	Mixed auto-/heterotrophic
	On macrophytes (epiphyton)	Mixed auto-/heterotrophic
	On dead wood (epixylon) and leaves	Mixed auto-/heterotrophic or purely heterotrophic
Within sandy sediments		Heterotrophic
Within degrading leaves and wood		Heterotrophic
Gut contents		Heterotrophic

In this table, autotrophic metabolism only refers to photoautotrophic metabolism. Chemoautotrophy is quantitatively of very low importance in streams like the Breitenbach (Wuhrmann, 1974; Marxsen, 2006).

6.3.2.3 Spatial Distribution

Natural streams usually contain many extremely different habitats, especially in low-order systems (Table 6.10). Habitats vary over short distances and contain different communities. Most detailed investigations on extracellular enzyme activity have been performed in the "planktonic" habitat, better considered as *suspended*. The communities in the water column environment consist of photoautotrophic and heterotrophic organisms (Table 6.10). But there are also several different *biofilm communities*: (i) on the surfaces of stones and gravel (which may be purely heterotrophic on the bottom sides of stones, or within stony and gravelly deposits, or mixed auto- and heterotrophic on the upper sides where light is available), (ii) on the upper surfaces of sandy deposits, (iii) on macrophytes, (iv) on the surfaces of dead wood and leaves (the last three all mixed auto- and heterotrophic). There are also entirely heterotrophic communities *within the sandy deposits* of streambeds, which are often regarded as biofilms, but are not typical biofilms. They are mostly hot-spots of microorganismal assemblages, whereas most of the sand grain surfaces remain uncolonized (cf. Section 6.2.1). There are also heterotrophic communities *within degrading leaves and wood,* as well as microbial communities *within gut contents,* which are hot spots of extracellular enzyme activity that should not be neglected.

These habitats and communities can be of very different relative importance in different running water systems. Their importance changes drastically over the course of a river, from its source to the mouth. With respect to the role of biological activity, there is a shift from the dominance of streambed communities in headwaters (biofilms, sediments, accumulations of leaves and wood) to an increasing importance of suspended water communities in large, slowly moving rivers. However, there are few publications that present detailed data on the different habitats of one stream (or river) system to prove these generally accepted assumptions. Thus we must rely on the Breitenbach data alone for our comparative account.

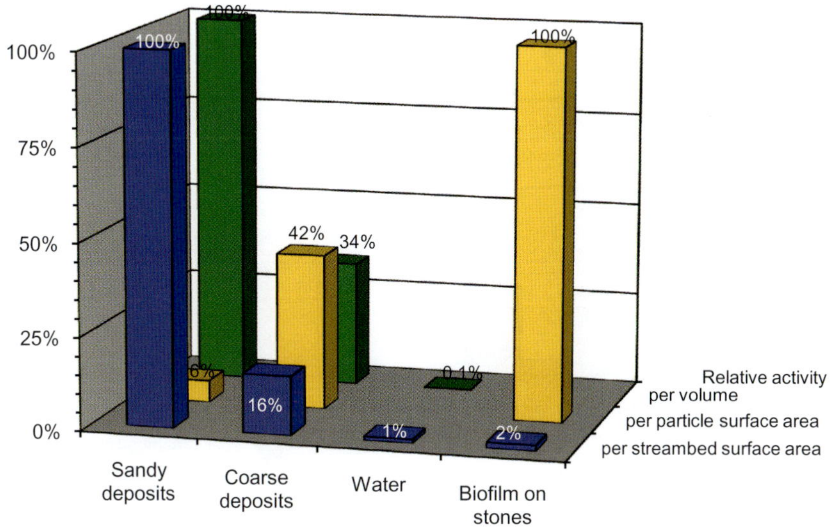

Figure 6.20 β-Glucosidase activity in four different Breitenbach habitats: sandy deposits, coarse gravelly deposits, water, and biofilms on stones. The data were calculated for the habitats (if appropriate) on a per volume basis (nmol h^{-1} ml^{-1}), per particle surface area (nmol h^{-1} cm^{-2}), and per streambed surface area (nmol h^{-1} m^{-2}), but are given here as percentages relative to the highest activity value in the habitat. Data combined from Freeman *et al.* (1990, 1993), Marxsen and Witzel (1990, 1991), Thomas (1990), Romaní and Marxsen (2002), Marxsen (unpublished data).

Comparison of β-glucosidase activity (Figure 6.20; data summarized from Freeman *et al.* 1990, 1993; Marxsen and Witzel 1990, 1991; Thomas, 1990; Marxsen, unpublished data), calculated on a per volume basis, shows the highest extracellular enzyme activity in the fine grained sandy deposits of the Breitenbach, followed by coarse gravelly deposits. By far the lowest activity is measured in the water column. These data are correlated with microbial biomass.

Recalculation of the data on the basis of particle (grain) surface area gives the lowest activity on sand particles, but much higher activity on the surfaces of coarse streambed deposits (the aquatic habitat is not considered in this case). Even higher activities were measured in another habitat: the autotrophic biofilm on the upper surfaces of stones in the streambed (Figure 6.20; Romaní and Marxsen, 2002). Extracellular enzyme activity increases in different Breitenbach biofilms with a decreasing ratio of bacterial to algal biomass: from sandy deposits, through coarse gravelly sediments, to large stones. This is a general trend found in river biofilms (Romaní and Sabater, 2000).

To estimate the relative importance of the different habitats for the Breitenbach with respect to extracellular enzyme activity, the spatial occurrence of these habitats in the stream must be taken into account. Thus, sandy deposits are by far the most important, followed by deposits of coarse gravel and stones. The relative

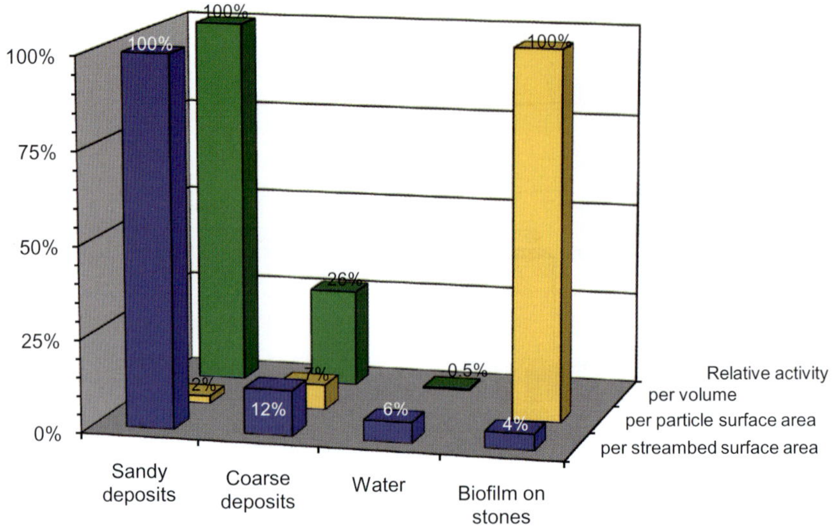

Figure 6.21 Phosphatase activity in four different Breitenbach habitats: sandy deposits, coarse gravelly deposits, water, and biofilms on stones. For further details see Figure 6.20. Data combined from Freeman *et al.* (1990, 1993), Marxsen and Witzel (1990, 1991), Scholz and Marxsen (1996), Romaní and Marxsen (2002), Marxsen (unpublished data).

activities in the water column are quantitatively unimportant, as are the autotrophic biofilms on stones. This trend, which was shown for β-glucosidase, is more or less similar for other enzymes, including phosphatase (Figure 6.21).

6.3.2.4 **Activity of Different Enzymes**

Many extracellular enzymes act in streams and other aquatic environments, but only a few are regularly found exhibiting high activities. Of the enzymes involved in nutrient remineralization, phosphatases and sulfatases, and those acting on polymeric matter, peptidases (which are also important for N remineralisation), esterases, α- and β-glucosidases, β-galactosidases, β-xylosidases, β-N-acetylglucosaminidases, and cellobiosidases are dominant. Different studies often use very different approaches, thus it is difficult to compare all these data directly. However, it is possible to compare the relative importance of the enzymes in a specific stream or river. For the comparison performed here, only data from investigations which used hydrolytic reactions on fluorigenic or chromogenic model substrates were included. Attempts using other methods, like viscosimetry (Hendel and Marxsen 2005a), or oxidation–reduction techniques (Hendel *et al.*, 2005), were not considered because of the difficulty of obtaining absolute activity values from such approaches, or even directly comparable data to those achieved with hydrolytic techniques.

Table 6.11 presents a list of extracellular enzyme activities in stream and river waters, which are arranged roughly (as well as possible) from small headwaters to

Table 6.11 Relative activity of different extracellular enzymes in stream and river waters (arranged from small headwaters to large rivers).

Environment, location, reference	Characteristics	Order of enzyme activity
Breitenbach, Ger, 1	First order	LPep > Pho > βGlu
Birk Gill, UK, 2	Upland	Pho > βGlu > βXyl > βGal > Sul
Long Gill, UK, 2	Upland	Pho > βGlu > βGal > βXyl > Sul
Schlitz, Ger, 3	Fourth order	LPep > Pho > βGlu
Ter River, Spain, 4	Fourth order	Pho > βXyl > βGlu
Horloff River, Ger, 5	5 m width	APep ≈ βGlu > Pho ≈ αGlu > Est
Driffield Beck, UK, 2	Lowland	Pho > βGlu > βXyl > βGal > Sul
Derwent River, UK, 6		LPep > Pho > βGlu > βXyl > βGal
Ouse River, UK, 6		LPep > Pho > βGlu > βXyl > βGal
Kiewa River, Australia, 7		LPep > Pho > Est
Elbe River, Ger, 8	20 m width	LPep > Pho > βGlu > αGlu > Cell
Ottawa River (and others), Ohio, USA, 9		Est > LPep > Pho > βGlu ≈ αGlu
Hudson River, N.Y., USA, 9	Estuary	Est > Pho > LPep > βGlu ≈ αGlu
Hudson River, N.Y., USA, 10	Estuary	Est >≈ LPep > Pho > βGlu ≈ αGlu
Lake Plußsee, Ger, 11	Eutrophic	LPep > Pho > βGlu
Lake Mekkojärvi, Fin, 12	Polyhumic	Pho > βGlu >≈ LPep > αGlu
Marine, 13		LPep > Pho > βGlu > βAcg > Est > αGlu

Abbreviations: αGlu, α-glucosidase; APep, alanine-aminopeptidase; βGal, β-galactosidase; βGlu, β-glucosidase; βAcg, β-N-acetylglucosaminidase; βXyl, β-xylosidase; Cell, cellobiosidase; Est, esterase; LPep, leucine-aminopeptidase; Pho, phosphatase; Sul, sulfatase.

References: 1 Marxsen and Witzel (1990, 1991), 2 Chappell and Goulder (1994), 3 Hendel and Marxsen (1997), 4 Romaní and Sabater (1999), 5 Kuhbier *et al.* (1999, 2002), 6 Chappell and Goulder (1995), 7 Boon (1990), 8 Wilczek *et al.* (2005), 9 Sinsabaugh *et al.* (1997), 10 Findlay *et al.* (1998), 11 Chróst (1990), 12 Münster (1991), 13 Hoppe *et al.* (2002).

Typical data from lake and marine environments are also given. The enzymes are arranged along rows in order of activity, beginning with the enzyme of highest activity.

large rivers. The enzymes were sorted in order of decreasing activity. There is a general trend in more or less all these studies with the highest activities for peptidases, followed by phosphatases, then by β-glucosidases, and then by other enzymes involved in hydrolyzing polymeric carbohydrates, such as β-xylosidases, β-galactosidases, α-glucosidases. This trend is very similar to that observed in lake and marine waters, with the exception of polyhumic lakes. A limitation on this comparison is that different model compounds were sometimes used to determine the activity of the same enzyme. This is particularly relevant for esterases, where the results were often very different, dependent on the particular model compound used. The variable ranking of esterase in Table 6.11 is probably due to this: Sinsabaugh *et al.* (1997) used MUF acetate for the North American rivers Ottawa and Hudson, but Boon (1990) used pNP palmitate for the Australian Kiewa River.

If enzyme activity measurements from mixed auto- and heterotrophic biofilms are listed the enzymes are quite similarly ranked (Table 6.12), but deviations occur if sediments are considered (Table 6.13). Two main changes to the order of activi-

Table 6.12 Relative activity of different extracellular enzymes in mixed auto- and heterotrophic biofilms of streams and rivers (arranged from small headwaters to large rivers).

Environment, location, references	Characteristics	Order of enzyme activity												
Breitenbach, Ger, 1+14	First order			LPep	>	Pho	>	βGlu	>	βXyl				
Naent Waen, UK, 15	First order	Est	>					βGlu	>	βXyl	>	EPep		
Seven calcareous streams, UK, 16	First/second order					Pho	>	βGlu	>	βXyl	>	βGal	>	Sul
Eight milestone-grit streams, 16	First/second order					Pho	>	βGlu	>	βXyl	>	βGal	>	Sul
La Solana, Spain, 17	Second order					Pho	>	βGlu	>	βXyl				
Riera Major, Spain, 4	Second order					Pho	>	βGlu	>	βXyl				
Ter River, Spain, 18	Fourth order					Pho	>	βGlu	>	βXyl				
St. Regis River, N.Y., USA, 19	Fourth order			LPep	>	Pho	>	βGlu βAcg αGlu	>	βXyl	>	Cell	>	Sul

Abbreviations: αGlu, α-glucosidase; βGal, β-galactosidase; βGlu, β-glucosidase; βAcg, β-N-acetylglucosaminidase; βXyl, β-xylosidase; Cell, cellobiosidase; EPep, endo-peptidase; Est, esterase; LPep, leucine-aminopeptidase; Pho, phosphatase; Sul, sulfatase.

References: 1–13 see Table 6.11, 14 Freeman *et al.* (1990), Romaní and Marxsen (2002), 15 Jones and Lock (1993), 16 Chappell and Goulder (1992), 17 Romaní and Sabater (1998), 18 Romaní and Sabater (2000), 19 Sinsabaugh and Linkins (1988).

No data are available from lake or marine environments. The enzymes are arranged along rows in order of activity, beginning with the enzyme of highest activity.

Table 6.13 Relative activity of different extracellular enzymes in stream and river sediments (arranged from small headwaters to large rivers).

Environment, location, reference	Characteristics	Order of enzyme activity												
Breitenbach, Ger, 1	First order			Pho	>	βGlu	>≈	LPep	>	βAcg	>	αGlu βGal Cell	>	βXyl
Three streams, New Zealand, 20	Upland					βGlu	>			βAcg	>	βXyl	>	Cell
La Solana, Spain, 18	Second order			Pho	>	βGlu	>					βXyl		
Riera Major, Spain, 17	Second order			Pho	>	βGlu	>					βXyl		
Horloff River, Ger, 5	5 m width	APep	>	Pho	>	βGlu	>					αGlu	>	Est
Elbe River, Ger, 8	20 m width	LPep	>	Pho	>	βGlu	>					αGlu	>	Cell
Hudson River, N.Y., USA, 21	Estuary					βGlu	≈			βAcg	>	βXyl	>	Cell
Lake sediments, 22	Eutrophic	Est	>			βGlu	>	βXyl	>	EPep	>	Cell	>	αGlu
Marine sediments,13		LPep	>	Est	>	βAcg	>	βGlu	>			αGlu		

Abbreviations: αGlu, α-glucosidase, APep, alanine-aminopeptidase, βGal, β-galactosidase, βGlu, β-glucosidase, βAcg, β-N-acetylglucosaminidase, βXyl, β-xylosidase, Cell, cellobiosidase, Epep, endo-peptidase, Est, esterase, Lpep, leucine-aminopeptidase, Pho, phosphatase, Sul, sulfatase.
References: 1–19 see Tables 6.11 and 6.12, 20 Findlay *et al.* (1997), 21 Sinsabaugh and Findlay (1995), 22 Boschker and Cappenberg (1998).
Typical data from lake and marine environments are also given. The enzymes are arranged along rows in order of activity, beginning with the enzyme of highest activity.

ties were found for sediments of different aquatic environments. One is between β-glucosidases and β-N-acetylglucosaminidases. The first are enzymes involved in the degradation of cellulose, the others in chitin degradation (thus often referred to as chitinases). As mentioned before, the activity levels of such enzymes can be regarded as crude indices for the availability of their target compounds (e.g., Findlay *et al.*, 1997; Burns and Ryder, 2001). Thus, the usually high activity levels of β-N-acetylglucosaminidases relative to β-glucosidase in marine sediments demonstrate that chitin is much more important in marine environments than in streams and rivers, where the dominance of cellulose as an organic carbon resource is well known (e.g., Allan, 1995; Marxsen, 2006).

Another important shift is that of aminopeptidase. This enzyme exhibits highest activity (much higher than β-glucosidase) in the sediments of large rivers and of

Table 6.14 Relative activity of different extracellular enzymes in typical stream and river habitats (arranged from small headwaters to large rivers), from lakes and the sea.

Environment, location, reference	Characteristics	Order of enzyme activity												
Breitenbach, water, 1	First order	LPep	>	Pho	>	βGlu								
Breitenbach, biofilm, 1+14	First order	LPep	>	Pho	>	βGlu	>	βXyl						
Breitenbach, sediment, 1	First order			Pho	>	βGlu	>≈	LPep	>	βAcg	>	αGlu Cell	>	βXyl
Elbe River, water, 8	20 m width	LPep	>	Pho	>	βGlu	>			αGlu	>	Cell		
Elbe River, sediment, 8	20 m width	LPep	>	Pho	>	βGlu	>			αGlu	>	Cell		
Ouse River, water, 6		LPep	>	Pho	>	βGlu	>	βXyl	>			βGal		
Hudson River, water, 10	Estuary	LPep	>	Pho	>	βGlu	≈	αGlu						
Lake Plußsee, water, 11	Eutrophic	LPep	>	Pho	>	βGlu								
Lake Mekkojärvi, water, 12	Polyhumic			Pho	>	βGlu	>≈	LPep	>	αGlu				
Marine, water, 13		LPep	>	Pho	>	βGlu	>	βAcg	>	αGlu				
Marine, sediment, 13		LPep	>			βAcg	>	βGlu	>	αGlu				

Abbreviations: αGlu, α-glucosidase, βGal, β-galactosidase, βGlu, β-glucosidase, βAcg, β-N-acetylglucosaminidase, βXyl, β-xylosidase, Cell, cellobiosidase, Lpep, leucine-aminopeptidase, Pho, phosphatase.
References: see Tables 6.11–6.13.
The enzymes are arranged along rows in order of activity, beginning with the enzyme of highest activity.

marine environments (Table 6.14), but only similar, or even lower, values of aminopeptidase compared to β-glucosidase activity occur in the Breitenbach streambed. In Breitenbach stream water, β-glucosidase activity achieves only about 2% of peptidase activity (Figure 6.22), but in sandy sediments, the activities of both enzymes were similar. β-Glucosidase usually showed somewhat higher activity values (Marxsen and Witzel 1990, 1991, Marxsen unpublished). In water of the River Elbe similar relationships exist between the three enzymes (Wilczek *et al.*, 2005), but β-glucosidase activity remains far below peptidase activity in the sediment, although β-glucosidase and phosphatase are more important than in the water column (Figure 6.22). This suggests that proteinaceous material in the sedi-

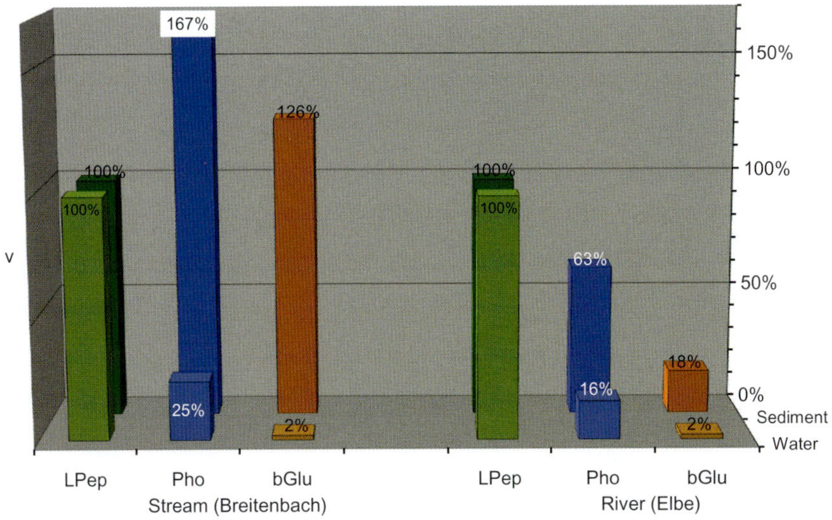

Figure 6.22 Relative leucine-aminopeptidase (LPep), phosphatase (Pho) and β-glucosidase (bGlu) activity in water and sandy sediments from the Breitenbach (Marxsen and Witzel, 1990, 1991; Marxsen, unpublished data) and the Elbe River (Wilczek *et al.*, 2005).

ments of the Breitenbach is also of much less relative importance than in the sediments of the River Elbe and marine environments (Hoppe *et al.*, 2002). The reasons might be:

1) The sediments of a small stream are more stable than is the case in the Elbe, where shifting sediments dominate (Wilczek *et al.*, 2005). Therefore much less fresh protein is supplied into the Breitenbach deposits, and the microorganisms must attack more carbohydrates, like cellulose, to satisfy their energy and carbon demands.

2) Another reason is probably the high inflow of groundwater into a stream like the Breitenbach. The groundwater contains almost only refractory, dissolved organic matter, such as humic substances (e.g., Marxsen and Fiebig, 1993). Groundwater enters the stream by diffuse perfusion through the streambed and is an important carbon source for streambed microbial communities (Fiebig and Marxsen, 1992).

Thus, with respect to the organic carbon resources, these communities are comparable to those in the water column of polyhumic lakes where Münster (1991) observed a similar relationship to that in the Breitenbach sediments, that is, high β-glucosidase and low peptidase activity.

6.3.2.5 Environmental Controls and Regulation

Extracellular enzyme activity in natural environments may be controlled via enzyme synthesis or through the changing activity of already available enzymes. Whereas constitutive enzymes are produced independently of an activator, ensuring a constant enzyme concentration, inducible enzymes are synthesized in the presence of a suitable substrate. Similarly enzyme synthesis can be repressed when a specific compound (often the end product of the reaction) turns off enzyme synthesis. Vice versa, enzyme synthesis can be de-repressed when such a specific compound is removed, thus turning on synthesis. Inhibition of enzyme activity occurs when a compound reacts with an enzyme and prevents its activity, for example, with PO_4^{3-} acting on phosphatase, but is also possible through other factors, such as temperature or pH (Bisswanger, 1994).

It is generally accepted that enzyme activity on macromolecular organic compounds (carbohydrates, proteins, or lipids) reflects the relative importance of those compounds. Thus there should be an increase in the respective enzyme activity with increasing amounts of a polymer source, if there are no more readily available resources. But there are few experimental data that confirm this assumption (Chappell and Goulder, 1994; Findlay *et al.*, 1997; Boschker and Cappenberg, 1998; Burns and Ryder, 2001).

During summer 1988 (May–September) a series of samples from sandy Breitenbach sediment was taken for measuring β-glucosidase activity (Thomas, 1990). These surface sediments (0–3 cm) contained particulate organic matter, between about 3 and 25 mg ml^{-1} sediment. Neumann (1995) found that FPOM in the Breitenbach deposits is very homogeneous (throughout time and space) in its proportions of macromolecules including carbohydrates and cellulose. He determined that the carbohydrate fraction only fluctuated between 30–35% of FPOM. Thus we can conclude that cellulose increases in the same way as total particulate organic matter. Thomas (1990) observed an increase in enzyme activity in the lower organic matter concentration range, up to about 10–15 mg OM ml^{-1} sediment (Figure 6.23). With further increasing organic matter content some fluctuations in enzyme activity occurred, but no permanent increase. Thus there is evidence that the availability of substrate induces enzyme synthesis or favors organisms producing the respective enzymes. However, some kind of saturation seems to occur which, in this case, prevents further increase in β-glucosidase activity above about 15 mg POM ml^{-1} sediment. This corresponds to about 5 mg ml^{-1} of carbohydrates and roughly 2–3 mg ml^{-1} of cellulose (calculated after Neumann, 1995).

But β-glucosidase activity might also be regulated by direct inhibition of enzyme activity, or through repression of enzyme synthesis by free glucose. Glucose (the reaction product of the hydrolysis of the artificial substrate MUF-β-glucoside as well as of cellobiose, the natural substrate of β-glucosidase) was added experimentally to Breitenbach sediments at increasing concentrations, up to 100 mmol l^{-1} (Marxsen and Fiebig, 1993). This resulted in a typical competitive inhibition pattern of β-glucosidase activity (Figure 6.24a), which is demonstrated by the Lineweaver–Burk plot (Figure 6.24b). The maximum hydrolytic activity (V_{max}) remains constant (in this case V_{max} is achieved in the inhibition experiment at concentrations much higher than 200 μmol l^{-1}), but the affinity of the enzyme for the substrate decreases, which is visible from the increasing K_m value. Such an effect is typical for competitive product inhibition (Pahlich, 1984). It has also often

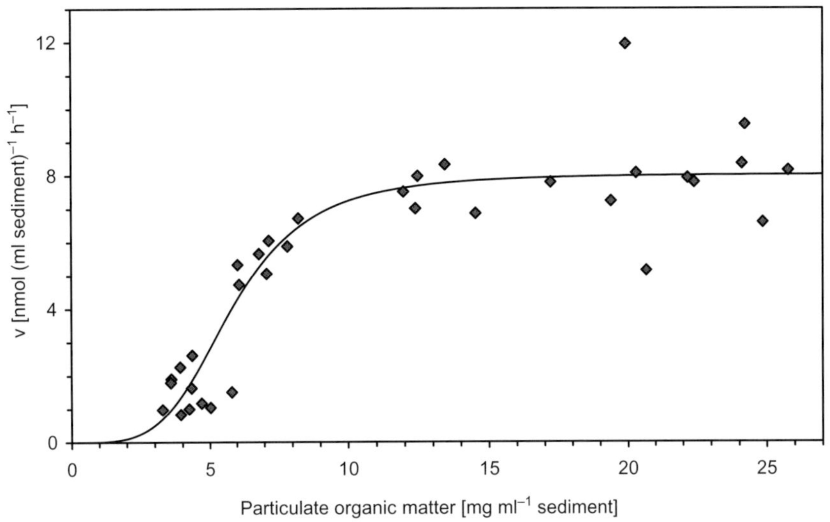

Figure 6.23 β-Glucosidase activity vs. particulate organic matter concentration in Breitenbach sediments during summer 1988 (May–September). The relationship was best described by the equation $v = 8.0\ \mathrm{POM}^{4.1}/(1363 + \mathrm{POM}^{4.1})$, where v is the enzymatic activity and POM the particulate organic matter concentration ($r^2 = 0.86$, $P < 0.01$). Data from Thomas (1990).

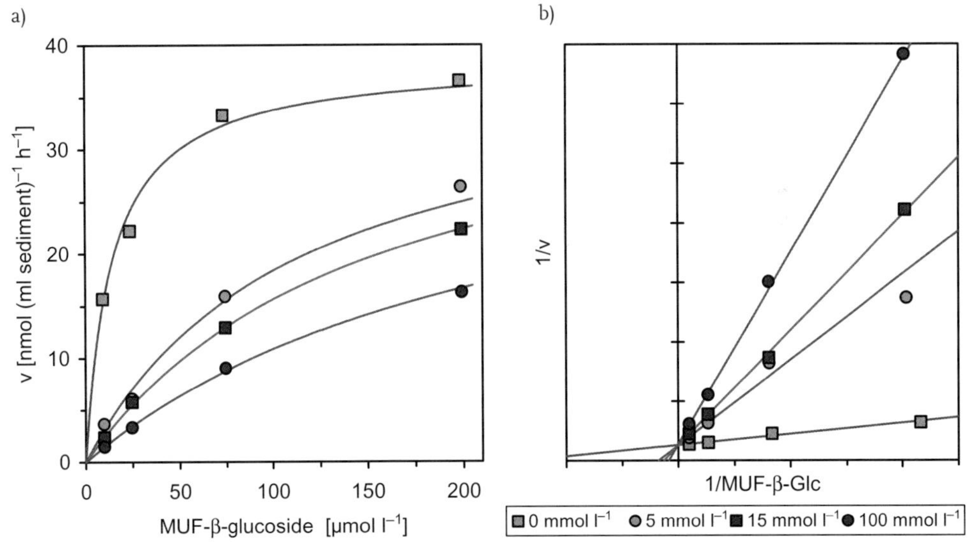

Figure 6.24 β-Glucosidase activity in Breitenbach sediments enriched with glucose. (a) Substrate saturation curves of MUF-β-glucoside (MUF-β-Glc). Increasing concentrations of the hydrolysis product glucose (concentrations up to 100 mmol l^{-1}) result in increasing inhibition of enzymatic hydrolysis. (b) Lineweaver–Burk plots. The straight lines intersect on the y-axis which is characteristic for the competitive type of inhibition. The lines of the Lineweaver–Burk plots were drawn with V_{max} and K_m values taken from a nonlinear regression of the untransformed data. Data from Marxsen and Fiebig (1993).

been reported in investigations on extracellular enzyme activity measurements in planktonic environments (e.g., Chróst and Overbeck, 1987; Chróst, 1989, 1990, 1991; Chróst *et al.*, 1986). However, the glucose concentrations which had to be applied to observe product inhibition were far above the natural glucose concentrations occurring in the interstitial waters of the sediments (see Section 4.3). Even the highest values determined in situ were below 1 μmol l^{-1}. Although the high activities of β-glucosidases can theoretically release high concentrations of glucose, its microbial lability ensures that free concentrations in interstitial, as well as in stream, water remain low. Thus product inhibition is probably not important for regulation of β-glucosidase activity in a stream like the Breitenbach. But whether glucose affects enzyme synthesis by repression at elevated glucose concentrations, or de-repression if glucose is withdrawn, could not be verified by this approach.

In the Breitenbach both ground and stream water discharge through the streambed sediments (Fiebig, 1995). These two types of water were perfused experimentally through cores from the streambed (Marxsen and Fiebig, 1993; Figures 6.25 and 6.26). Whereas similar values were achieved for V_{max} (Figure 6.25a), K_m was substantially higher with groundwater (a slower rate of increase in activity as substrate concentration increased). The linearization of the data in a Lineweaver–Burk plot (Figure 6.25b) suggests that competitive inhibition occurred

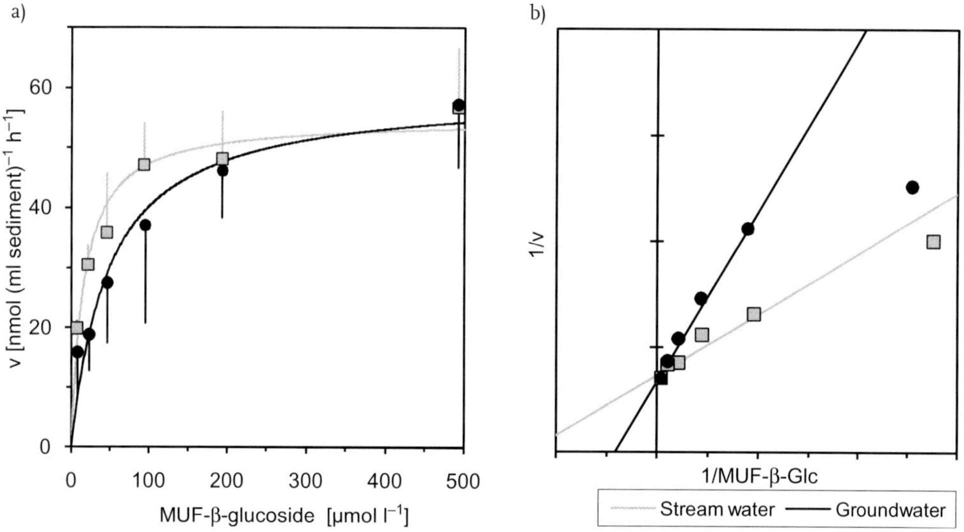

Figure 6.25 MUF-β-glucoside (MUF-β-Glc) hydrolytic cleavage in sediments of the Breitenbach during percolation with stream water and groundwater). (a) Substrate saturation curves. (b) Lineweaver–Burk reciprocal diagrams. Average values with standard deviations [in (a) only] are given. The lines of the Lineweaver–Burk plots were drawn with V_{max} and K_m values taken from a nonlinear regression of the untransformed data. Data from Marxsen and Fiebig (1993).

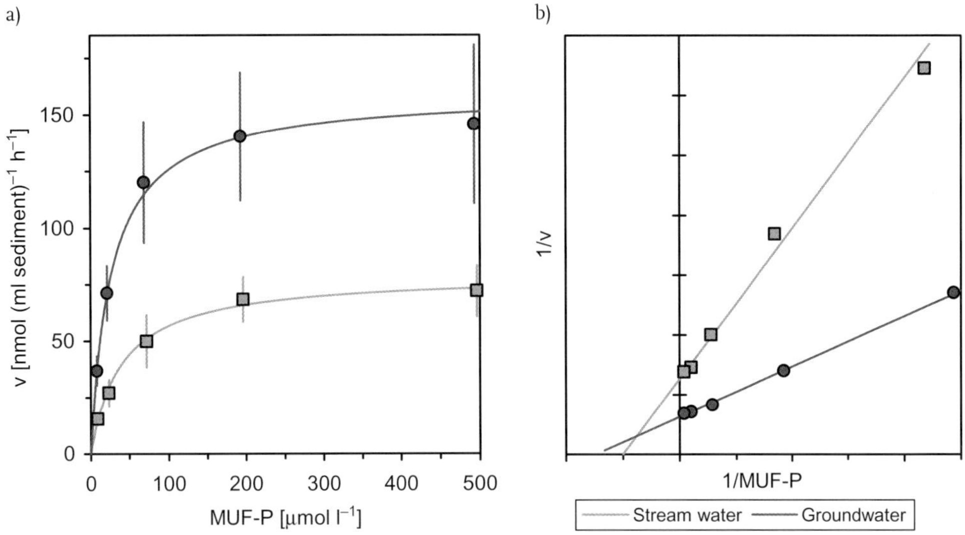

Figure 6.26 MUF-phosphate (MUF-P) hydrolysis in sediments of the Breitenbach perfused with stream water and groundwater. (a) Substrate saturation curves. (b) Lineweaver–Burk plots. Average values with 95% confidence limits [in (a) only] are given. The lines of the Lineweaver–Burk plots were drawn with V_{max} and K_m values taken from a nonlinear regression of the untransformed data. Data from Marxsen and Schmidt (1993).

when the cores were percolated with groundwater, probably due to humic material. This effect is well documented for limnic systems (Wetzel, 1991). The brown color of groundwater entering the Breitenbach suggests that it contains substantially higher concentrations of humic matter than the stream water.

The same approach was used to investigate phosphatase activity in the Breitenbach streambed sediments (Marxsen and Schmidt, 1993). Here both V_{max} and substrate affinity, K_m, were higher with groundwater (Figure 6.26). The contrasting trend for phosphatase compared to β-glucosidase is not surprising, because inorganic phosphorus concentrations were much lower in groundwater than in stream water (Table 6.15). Thus, if exposed to groundwater the phosphorus demand of the microbial community was higher. This probably accounts for the differences between kinetic parameters in sediments exposed to the different types of water influencing the Breitenbach streambed microbial communities.

6.3.2.6 **The Molecular Biology Perspective**

Over the past decades, methods in molecular biology have reached a stage of development that allows their application in ecology to answer previously intractable questions (Muyzer, 1999; Zwart *et al.*, 2002; Pernthaler *et al.*, 2004; Lozupone and Knight, 2007; Amann and Fuchs, 2008). Thus many of the old limitations of bacteriological investigations in microbial ecology have been overcome (also see Section 6.2.2). The importance of extracellular enzymatic degradation of

Table 6.15 Concentrations of soluble reactive phosphorus (SRP) and nonreactive phosphorus (NRP) in different types of water from the Breitenbach, used for a perfusion experiment on extracellular phosphatase activity in streambed sediments (Figure 6.26).

Type of water	SRP	NRP
	($\mu g\,l^{-1}$ P)	
Natural water		
Stream water	41	10
Interstitial water	24	12
Filtered and boiled water for experimental use		
Stream water	38	12
Groundwater	23	16

From Marxsen and Schmidt (1993).

macromolecular organic matter as a key process for organic matter decomposition and utilization in the Breitenbach, as well as in streams in general, had often been demonstrated (e.g., Breitenbach: Marxsen and Fiebig, 1993; Scholz and Marxsen, 1996; Wolf *et al.*, 1997; Hendel and Marxsen, 2000; Romaní and Marxsen, 2002; other streams: Freeman *et al.*, 1990; Chappell and Goulder, 1994; Tank *et al.*, 1998; Romaní and Sabater, 2001). Thus attempts were made to develop techniques for the detection of genes encoding extracellular enzymes that are involved in these degradation processes, and to analyze the occurrence of functional genes for enzymes involved in polymer degradation in stream environments (Marxsen and Witzel, 2005).

The most common organic compound in the Breitenbach, but also worldwide, is cellulose (Figure 6.18; Fritsche, 1985; Gilbert and Hazlewood, 1993; Béguin and Aubert, 1994; Bergquist *et al.*, 1999). A lot of information is available on genes encoding for the different enzymes involved in its degradation, but cellulose sequences are highly heterogeneous and grouped into many distinct families (Béguin, 1990; Gilkes *et al.*, 1991; Tomme *et al.*, 1995). Published primer sets could not be found for the distinct enzyme groups, but it was possible to gather primers that amplify DNA sequences known to encode for extracellular enzymes acting on other polymers (Williamson *et al.*, 2000; Bach *et al.*, 2001), including primers for chitinase genes. These were used to initiate the detection and analysis of functional genes for polymer degradation in a stream environment. Although chitin is not the most important polymer in such a stream, it is of relevance, occurring in many invertebrates and fungi, and it is the second most important polymer on a global scale (Gooday, 1990; Schrempf, 2001).

Therefore DNA was extracted from various Breitenbach habitats (Beier *et al.*, 2008). Amplification was performed using a nested PCR approach (Williamson *et al.*, 2000). The codehop primers GA1F and GA1R were used in the first step, specifically to amplify family 18A chitinase DNA fragments, with the consensus primers GASQF and GASQR in a second round, to produce DNA fragments for DGGE (degenerative gradient gel electrophoresis). Following this approach it was possible to demonstrate the occurrence of family 18 group A chitinase gene sequences in stream and spring water, in different sediments, in leaf litter, and in enrichment cultures from the stream. The sequences all belonged to the catalytic domain of subgroup 18A of one of the two chitinase families. Chitinases (EC 3.2.1.14) are classified, on the basis of sequences and structure, into two evolutionarily unrelated groups, family 18 and 19 of glycosyl hydrolases. Family 18 includes enzymes from bacteria, eukaryotes, and viruses. Family 19 contains most of the plant chitinases, but also chitinases from viruses, bacteria, nematodes, and protozoans. Bacterial family 19 chitinases were mainly detected in Actinobacteria (Cohen-Kupiec and Chet, 1998; Kawase *et al.*, 2004).

DGGE was used to observe the variability of chitinases in the different stream habitats. Cluster analysis of DGGE gels showed that different habitats exhibit different gene compositions: high similarities in water, and also to, and within, sediments, but larger deviations to decaying leaves and an laboratory enrichment culture (Figure 6.27).

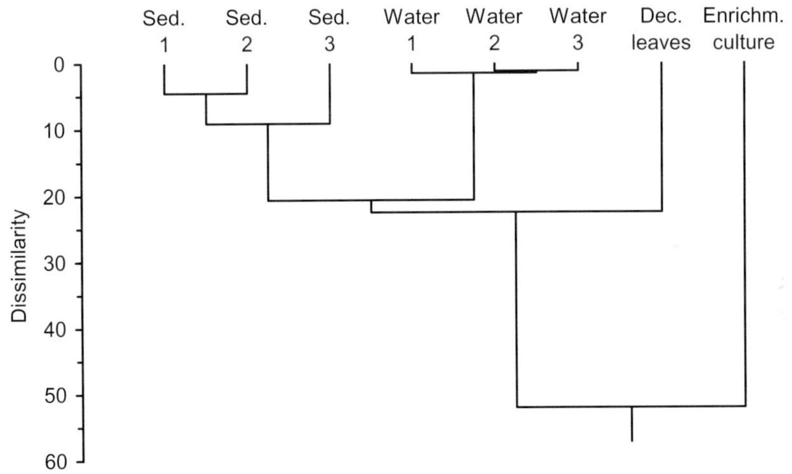

Figure 6.27 Cluster analysis on the similarity of DGGE profiles prepared with family 18A chitinase gene fragments obtained from eight different sites in the Breitenbach. The relative intensities of the bands were calculated for each lane and used for further statistical treatment. Clustering was performed by the Ward method based on Bray–Curtis similarities, calculated for each pair of lanes. Stream water samples showed high similarities. Similarities were also found within sediment samples and between stream water and sediment, but larger deviations to decaying leaves and most notably the laboratory enrichment culture occurred. Sed. 1 = coarse sediment, Sed. 2 = fine sediment with FPOM, Sed. 3 = fine sediment with much FPOM (all sediment samples from the middle reach), Water 1 = stream water upper reach, Water 2 = stream water middle reach, Water 3 = stream water lower reach, Dec. leaves = decaying leaves, Enrichm. culture = laboratory enrichment culture.

Table 6.16 Most similar sequences to Breitenbach family 18A chitinases from NCBI GenBank. Comparison of sequences was performed via a NCBI BLASTN nucleic acid search in Nucleotide collection (nr/nt) using Megablast (Zhang *et al.*, 2000) in August 2009.

Betaproteobacteria	*Chromobacterium, Doohwaniella, Janthinobacterium*	$n = 8$
Gammaproteobacteria	*Lysobacter, Saccharophagus, Stenotrophomonas*	$n = 4$
Firmicutes	*Bacillus, Paenibacillus*	$n = 4$
Actinobacteria	*Streptomyces*	$n = 34$
Bacteroidetes	*Rhodothermus*	$n = 1$

In a next step clones were obtained from the amplified DNA (Beier *et al.*, 2008), allowing the evaluation of 51 DNA sequences from Breitenbach water and sediment. The sequences were translated to proteins by the tool provided by http://www.in-silico.com (Bikandi *et al.*, 2004). The protein sequences were then compared by protein–protein BLAST (blastp) search (http://www.ncbi.nlm.nih.gov/BLAST/) to NCBI databases (Altschul *et al.*, 1997). Simultaneously conserved domains in the sequences were identified by the Conserved Domain Database (CDD) search approach (rpsblast at http://www.ncbi.nlm.nih.gov/Structure/cdd/wrpsb.cgi; Marchler-Bauer and Bryant, 2004).

All protein sequences were identified by the CDD search as potential family 18 glycosyl hydrolases. These putative chitinases appear to be diverse in origin, as the most similar GenBank sequences were obtained from different groups of bacteria (Table 6.16). The majority of the closest relatives, 34 from 51, were from *Actinobacteria,* eight from *Betaproteobacteria,* four each from *Gammaproteobacteria* and *Firmicutes,* and one from *Bacteroidetes.* Similarities were found between the Breitenbach sequences and sequences from other freshwaters (rivers and lakes) and from soil. Most were closely related to sequences of soil *Actinobacteria,* but no close relationships were detected with any sequences from marine and inland saline environments.

Besides cellulose, which is the most important polymer, globally as well as in the Breitenbach, chitin is among the other relevant polymers in streams. These first results on chitinase genes demonstrate the potential of molecular biology methods for detecting and analyzing functional genes involved in the degradation of polymers in a stream like the Breitenbach, providing a basis for further progress in investigating the role of extracellular enzymes.

6.3.3 Bacterial Production

6.3.3.1 The Importance of Bacterial Carbon Production in Streams

When the first studies on the ecological role of bacteria and other organisms in the Breitenbach (and other natural streams of the region) were begun (Marxsen, 1976, 1980a, b, c), it was generally accepted that bacteria were the main decomposers of organic matter in streams, as in other ecosystems (Thienemann, 1955). They were considered organisms with high metabolic activity (Zeuthen, 1953), but occurring at only very low quantities of biomass in aquatic systems (Odum 1956b, 1957a). However, in the 1970s, ideas began to grow that the significance of "bacterial secondary production" (Sorokin, 1965) was an important source of biomass for higher trophic levels (Kusnetsov and Romanenko, 1966, Overbeck, 1972, 1973; Pomeroy, 1974). This concept was further developed and is now well known as "the microbial loop" (Azam *et al.*, 1983), or "microbial food chain" (Pomeroy, 1994; Cotner and Biddanda, 2002; cf. Section 6.1).

Whereas most of these investigations on the importance of the microbial food chain were performed in pelagic environments of lakes and oceans, a first rough attempt was made to estimate bacterial production in the Breitenbach (Marxsen 1988a). This resulted in a somewhat speculative calculation of about 750 g C m^{-2} $year^{-1}$ for bacterial production in this stream. Stimulated by this finding, detailed investigations were performed on the quantity of bacterial production and its organic matter basis over the following decade (e.g., Fiebig and Marxsen, 1992; Marxsen 1996, 1999, 2001, 2006). The estimated high bacterial production made it particularly necessary to quantify its organic carbon basis (Fiebig 1992, 1995, 1997; Marxsen *et al.*, 1997; cf. Chapter 4), to ensure a consistent picture of the role of bacteria within the organic matter balance of the Breitenbach. From this detailed long-term study on the Breitenbach significant major advances have been made in our knowledge on the role of bacteria in streams (Marxsen, 2006).

- In contrast to the classical view, bacterial biomass was found to be, on an annual average, about 36% of total heterotrophic biomass in the Breitenbach (cf. Section 6.1).
- Total bacterial production was about 200 g C m^{-2} $year^{-1}$. This is of similar magnitude to net primary production.
- Also, in contrast to the classical view, bacteria accounted for 71%, that is, the major fraction, of heterotrophic production.
- But bacteria were also demonstrated to be the main decomposers in the Breitenbach, accounting for 78% of heterotrophic respiration and 59% of total respiration, thereby confirming their classical ecological role.

6.3.3.2 Methodological Approach

Several methods were used to determine bacterial production. For the investigations on the Breitenbach, leucine incorporation into protein, developed by Kirchman *et al.* (1985) for marine pelagic environments, was used after being optimized for measurements in streambed sediments (Marxsen, 1996). This technique provides a direct quantification of protein production. Because total bacterial cell dry weight is usually 50% or more protein, this essential part of bacterial production can be directly quantified. This reduces potential inaccuracies during the calculation of total production. Particularly in food web studies, biomass production is more relevant than DNA production, which is estimated via thymidine uptake (Fuhrman and Azam, 1980; Findlay *et al.*, 1984). Furthermore, the leucine technique is about one order of magnitude more sensitive than the thymidine method

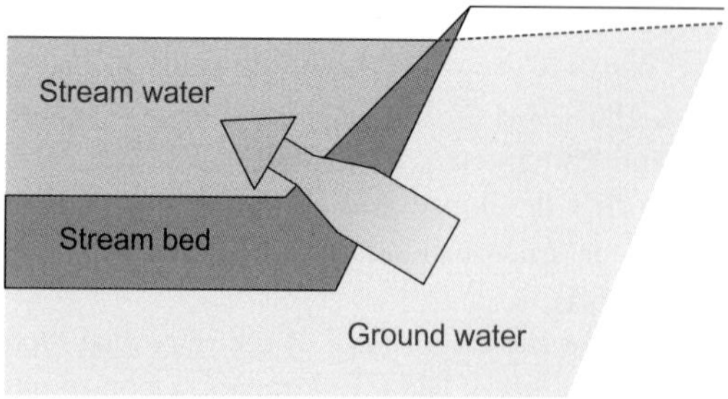

Figure 6.28 The perfused core technique (to investigate microbial activity in streambed sediments) is based on the observation that in the Breitenbach, as in many other streams, groundwater enters the stream by diffuse perfusion through the streambed. This natural process is simulated in the laboratory (Figure 6.29).

(Simon and Azam, 1989; Riemann and Bell, 1990), and nonspecific incorporation of leucine into macromolecules other than protein is less of a problem than the nonspecific incorporation of thymidine into molecules other than DNA (Riemann and Bell, 1990). Another advantage of the leucine technique is that longer incubation times can be used than with the thymidine technique (Marxsen, 1996).

To avoid experimental disturbance of the sandy streambed sediments before and during incorporation measurements, a technique using perfused sediment cores was developed (Fiebig and Marxsen, 1992). This simulates the natural process occurring in the Breitenbach, as well as in many other streams, whereby groundwater enters the stream by diffuse perfusion through the streambed sediments (Figure 6.28). To perform the experiments, cores were taken with specially prepared polypropylene syringes, 2 cm in diameter, with the lower part cut off. The cores were brought into the laboratory, where they were slowly percolated with water from below using a multichannel peristaltic pump (Figures 6.29–6.31). ^{14}C-labeled leucine (50 μmol l^{-1}) was added to the perfusion water, which was pumped through the cores for 24 h. At the end of the experiments, the columns were divided into separate layers from which protein was extracted (Marxsen, 1996). Leucine incorporation at gravelly sites was measured simultaneously with the experiments with sandy cores, but using a different procedure. Small stones from the stream bed (excluding stones exposed to light) were placed into Erlenmeyer flasks containing 10 ml of filtered and boiled stream water and were incubated in a shaking water bath. Incorporation of ^{14}C-labeled leucine into protein was determined over 4 h. The stones from the flasks were processed for protein extraction in a similar way to the sandy sediment. Monthly investigations using these two approaches were performed over a complete year, 1994, considering sediment from 0–7 cm deep (Marxsen 1999, 2001).

6.3.3.3 **Seasonal Fluctuation**

Bacterial production in the Breitenbach showed a clear seasonal pattern, with minimum values in winter and maximum values in summer, in a manner typical

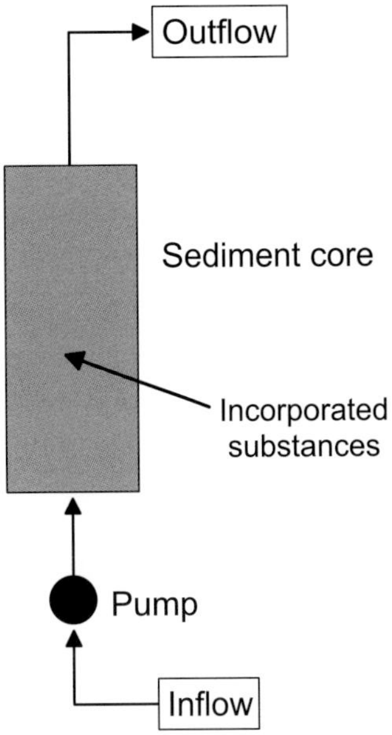

Figure 6.29 Small cores are taken from the sediment and placed in a temperature-regulated incubator set to ambient stream temperature. The cores are percolated slowly from below with water at perfusion rates similar to the natural ones. For investigation of the microbial activity in the cores, different substances (e.g., ^{3}H- or ^{14}C-labeled leucine, or methylumbelliferyl compounds; cf. Section 6.3.2) can be added to the perfusion water. During the experiment water can be sampled before and after perfusion through the cores (e.g., for determining liberated methylumbelliferone). At the end of the experiments the columns can be divided into several layers for detailed analysis (e.g., for quantifying incorporated leucine). Schematic diagram after Marxsen (1996).

Figure 6.30 Photo of a sediment core in the laboratory simulation of diffuse groundwater discharge through the streambed.

Figure 6.31 Overall view of the perfused core technique. The cores are percolated slowly from below with water using a multichannel peristaltic pump.

for temperate regions (Marxsen, 2001). This was especially clear on the gravel and stone surfaces (Figure 6.32b), but there was one exception to this general pattern. After an increase in production rates during spring a sharp decrease was measured, which had been caused by a large spate in April shortly before sampling. This event not only cleared the streambed of most of the sandy deposits, it also appeared to have abraded the biofilm from the gravelly and stony deposits. The high bacterial production rates measured just one month later indicate rapid recovery of the biofilm from this major disturbance.

A less pronounced seasonal pattern was found for bacterial production rates in the sandy sediments of the streambed (Figure 6.32c). Whereas about tenfold differences were measured in the coarse sediments, only somewhat more than twofold differences were found in the sandy deposits. The April sandy sediment samples did not show such a sharp decrease in production, but there were extremely large differences between the cores, as indicated by the high standard deviation. However, it must be stressed that the quantity of sandy deposits had been very much reduced by the storm, so the total production of the sandy deposits

Figure 6.32 Seasonal pattern of bacterial carbon production (BCP) in the Breitenbach during 1994 (after Marxsen, 2001). (a) Mean daily temperature (range 5–12 °C) and mean daily discharge (generally under 80 l s^{-1}, but with occasional high spikes), recorded continuously at gage GT2 about 300 m above the investigation area (data from H.-H. Schmidt; cf. Chapter 3). The black squares indicate the sampling dates. (b) Seasonal pattern of monthly BCP on the surfaces of small stones in the middle reach of the Breitenbach (below T4). Data are given in ng C cm^{-2} stone surface area h^{-1}. (c) Seasonal pattern of monthly BCP in sandy sediments (0–7 cm). (d) Seasonal pattern of monthly BCP per stream surface area (0–7 cm). In (b–d) data are means with standard deviations, and the dotted lines indicate the annual averages.

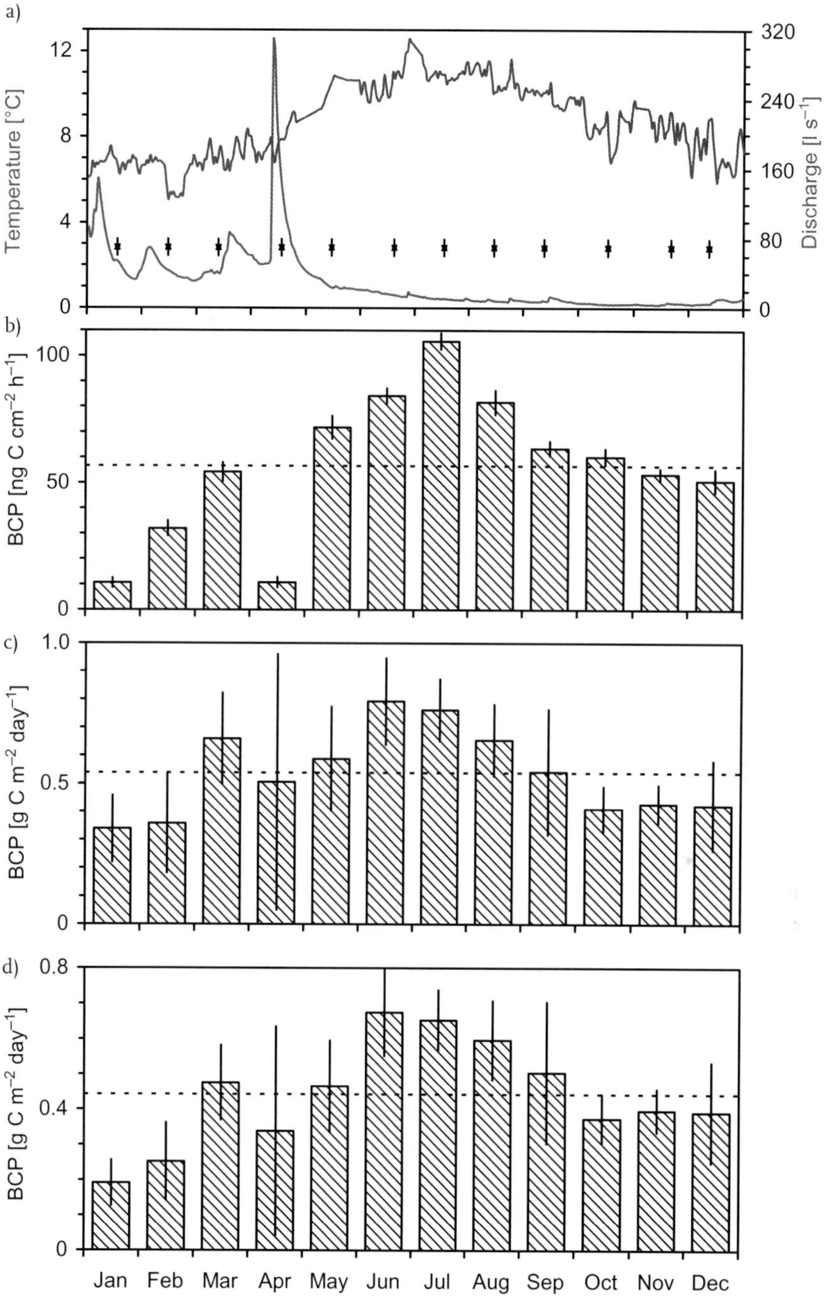
a)
Temperature [°C]
Discharge [l s−1]
0
4
8
12
0
80
160
240
320
b)
BCP [ng C cm−2 h−1]
0
50
100
c)
BCP [g C m−2 day−1]
0
0.5
1.0
d)
BCP [g C m−2 day−1]
0
0.4
0.8
Jan
Feb
Mar
Apr
May
Jun
Jul
Aug
Sep
Oct
Nov
Dec

was reduced by far more than is shown by the data from the few remaining sandy sediments.

If the total streambed (both types of sediments) is considered, the effects of the spate, as well as the seasonal trends, are intermediate to those on the single types, corresponding to the percentage of both types throughout the year (Figure 6.32d).

6.3.3.4 **Small-Scale Variability and Particulate Organic Matter Content**

Small scale variability of the structure of streambed sediments is typical for small streams like the Breitenbach. Thus, for most time of the year there is large variability in BCP over short distances, both vertically and horizontally. Figure 6.33a

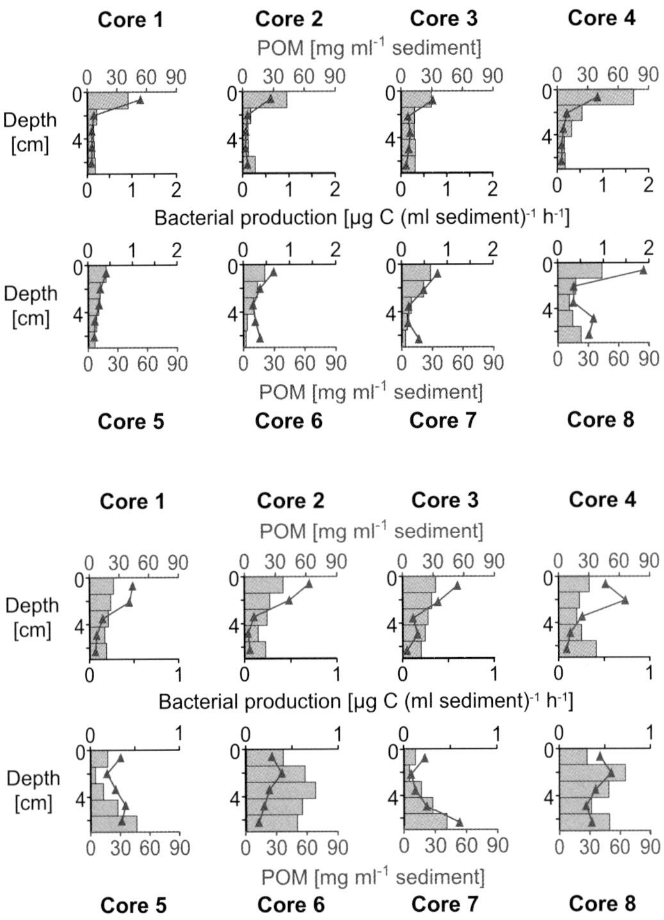

Figure 6.33 Vertical distribution of bacterial carbon production (gray bars) and particulate organic matter (black triangles) in eight cores of Breitenbach sandy sediment (Marxsen, unpublished data) in May 1994 (upper graph) and September 1994 (lower graph).

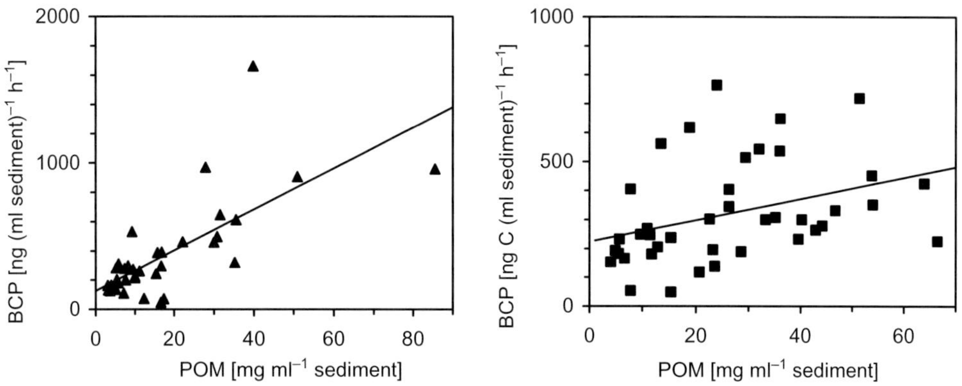

Figure 6.34 Relationship between particulate organic matter (POM) and bacterial carbon production (BCP) in layers from 0–7 cm depth of sediment cores from the Breitenbach (Marxsen, unpublished data) in: (a) May 1994 ($r = 0.73$, $p < 0.001$), (b) September 1994 ($r = 0.36$, $P < 0.05$).

shows the distribution of bacterial production over depth in sandy sediments from May 1994. All cores (0–7 cm depth) were taken close to each other (within 1 m distance) at a site some 50 m below trap T4. At that time the vertical distribution of BCP was similar for all cores. The highest values were always measured in the upper layers, where the particulate organic matter (POM) content was usually highest too. During May 1994 fresh sediments were deposited after the spate in April. However, for most of the year, the distribution of BCP and particulate organic matter content was more uneven, as demonstrated by the data from September 1994 (Figure 6.33b). Some cores showed similar distributions of BCP and POM content to those from May. But there were also samples where the highest BCP rates were found in the deepest layers, or distribution was fairly even over depth.

Thus, there seemed to be a relationship between bacterial production and particulate organic matter content in the Breitenbach sediments. Correlation analysis showed high correlation during May (Figure 6.34a), but the relationship was not very distinct for the September samples, although it was still significant (Figure 6.34b). Thus some relationship between bacterial production and particulate organic matter content was obvious, but this was not the only factor that influenced variability in bacterial production. The POM content of sandy sediments has often been found to be important for the distribution of bacterial production (e.g., Hendricks, 1993; Fischer *et al.*, 1996; Findlay and Sobczak, 2000), but large residual variances usually remained (Cole *et al.*, 1988; Marxsen, 1996; Sobczak *et al.*, 1998). These showed that other factors need to be considered to explain the distribution of BCP, not only temperature (Sander and Kalff, 1993; Marxsen, 2001), but also discharge and hyporheic flow within the sediments (Hendricks 1993, 1996; Findlay, 1995; Brunke and Gonser, 1997; Battin, 2000) as well as sediment stability (Fischer *et al.*, 2002).

6.3.3.5 Influence of Temperature and Discharge

Two main physical factors were recognized as influencing BCP (Marxsen, 2001):

1) Temperature explained most of the seasonal variation of bacterial production in sandy, as well as in coarse deposits (Figure 6.35). This was not surprising and has often been found in other studies of streams and rivers (Kaplan and Bott, 1989; Sander and Kalff, 1993; Fuss and Smock, 1996).

2) Discharge is the second important physical factor to influence BCP in running waters (Allan. 1995). Data from the Breitenbach shows its importance for BCP, especially on the surfaces of coarse substrata (Figure 6.36). The best

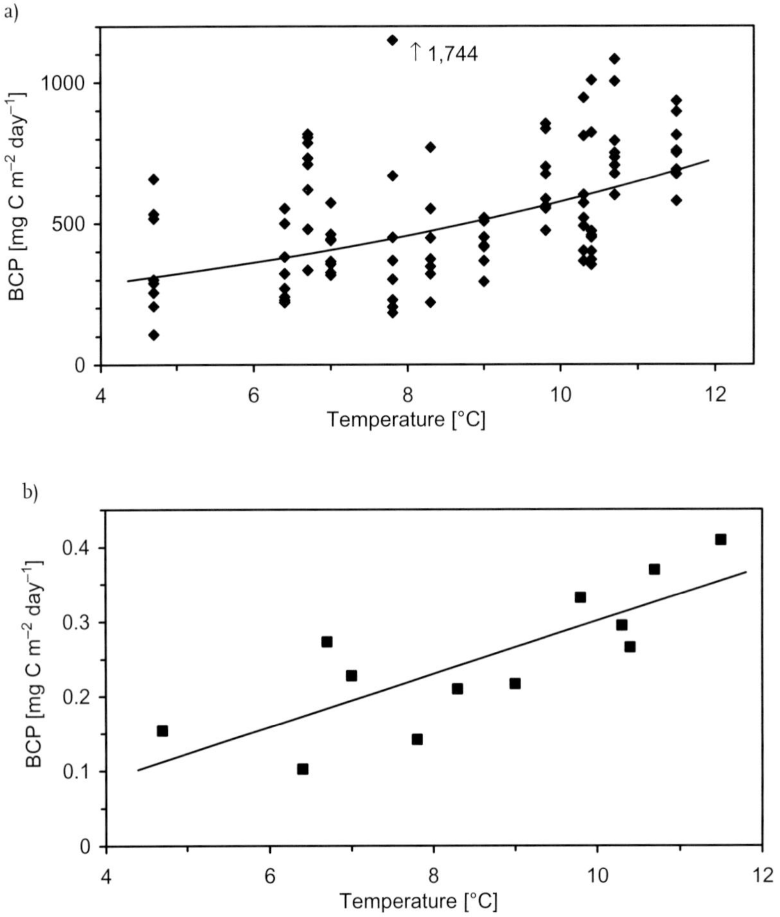

Figure 6.35 Relationship between water temperature and bacterial carbon production (BCP) in the Breitenbach (measurements in 1994, after Marxsen, 2001). (a) Within sandy sediment (0–7 cm depth, $n = 96$ for BCP, $n = 12$ for temperature, $r = 0.50$, $P < 0.001$). (b) Per stream surface area of coarse sediments (0–7 cm depth, $n = 12$, $r = 0.80$, $P < 0.001$).

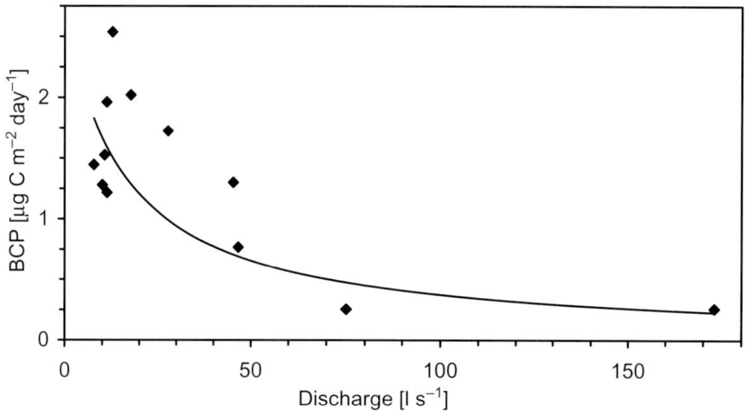

Figure 6.36 Relationship between discharge and bacterial carbon production (0–7 cm depth) on the surfaces of small stones from the Breitenbach ($n = 12$, $r = 0.86$, $P < 0.001$; measurements in 1994, after Marxsen, 2001).

fitted model (BCP = $1/[a + b^{Disch}]$) suggested a continuous decrease of production with increasing discharge, although the data plot casts some doubt on this interpretation (Marxsen, 2001).

Nevertheless the importance of both factors was validated by multiple regression analysis. This showed that, for the BCP measurements performed in 1994 in the Breitenbach, 77% of the variation in BCP was explained by temperature and discharge.

6.3.3.6 Comparing BCP between Different Streams and Rivers

From the monthly measurements in 1994, the annual bacterial carbon production for the upper 7 cm of the Breitenbach streambed was calculated to 162 g C m^{-2} (Marxsen, 2001). BCP measurements in sediments from other streams and rivers are available from only a few studies (Table 6.17), most of which used the thymidine technique, except that by Fischer and Pusch (2001), where leucine was used. Complete annual measurements were only available from Black Creek and Ogeechee River (Ga., USA; Findlay *et al.*, 1986; Meyer, 1988). Another problem that made comparison difficult was that different sediment depths had generally been investigated. Thus it was necessary to recalculate the data as production per 1-cm layer for comparison (Marxsen, 2001).

Bacterial production rates per 1 m^2 of streambed surface in the Breitenbach fluctuated between 190 and 670 mg C m^{-2} day^{-1} (annual average 440 mg C m^{-2} day^{-1}). This equates to 27–96 mg C m^{-2} day^{-1} (average 63 mg C m^{-2} day^{-1}) for a 1-cm layer and is well within the range of other, mostly North American, studies. Only Hudson *et al.* (1992) presented distinctly higher results, from southern Ontario streams during the summer. Measurements from the river Spree in eastern Germany (Fischer and Pusch, 2001), which were conducted on four occasions

Table 6.17 Bacterial carbon production (BCP) in rivers and streams (based on Marxsen, 2001). If available, mean annual data (ave) are given.

Site	Months covered	Depth of sediment in measurement	BCP (original data)	BCP (per 1-cm layer)	Reference
		(cm)	($mg\,C\,m^{-2}\,day^{-1}$)	($mg\,C\,m^{-2}\,day^{-1}$)	
Ogeechee River, Ga., USA	Jan–Dec	0–1	0.2–15	0.2–15	Findlay *et al.*, 1986
Black Creek, Ga., USA	Jan–Dec	0–1	0.2–26	0.2–26	Findlay *et al.*, 1986
Ogeechee River, Black Creek backwaters, Ga., USA	Jan–Dec	0–1	3.1–530	3.1–530	Findlay *et al.*, 1986
Ogeechee River, Ga., USA	Jan–Dec	0–20	60–2200 (ave: 980)	3–110 (av: 49)	Meyer, 1988
Southern Ontario streams, Canada	May–Aug	0–2	430–3400	215–1700	Hudson *et al.*, 1992
East Branch, Maple River, Mich., USA	Nov, Feb, May	9.5–10.5	1–7600	1–7600	Hendricks, 1996
Spree River, Germany	Aug, Oct, Mar, Jun	0–2	624–2600	312–1300	Fischer and Pusch, 2001
Breitenbach, Germany	Jan–Dec	0–7	190–670 (ave: 440)	27–96 (ave: 63)	Marxsen, 2001

Original data were recalculated to production rates per 1-cm layer, if necessary.

throughout the year, also exceeded the values from the Breitenbach. Furthermore, Hendricks (1996) data from a 10 cm layer (measured during November, February, and May) covered a broad, but very low range, from 0.001 to 7.6 mg C m^{-2} day^{-1}. Thus, although only a few measurements are available, bacterial carbon production data varies temporally and spatially by more than four orders of magnitude, with the data from the Breitenbach of a similar magnitude to that from the North American studies, towards the upper medium range.

6.3.3.7 **Annual Bacterial Production and Respiration**

Annual bacterial production of 162 g C m^{-2} had been measured in the Breitenbach streambed (Marxsen, 2001). However, some habitats were not considered in that study. Pelagic bacterial production is known to be quantitatively unimportant, below 1 g C m^{-2} $year^{-1}$ (Marxsen, unpublished data), but the surfaces of epiphytes are more important. This habitat was investigated in detail by Fischer and Pusch (2001) in the Spree River (eastern Germany). Here the bacterial production in the sediments was found to surpass the combined epiphytic and pelagic production by factors of 12 to 25. Very few data have been published on the bacterial production in the phototrophic biofilms of stone surfaces, but a comparison between the bacterial production in sandy sediments and on rocky surface biofilms is available from southern Ontario streams, but for summer only (Hudson *et al.*, 1992). From these data an average relationship of 17% between rocky surface and sediment production was calculated (range 1–80%; Marxsen, 2006). In the Breitenbach, rocks and small stones compose a considerable percentage of the streambed surface (1994: 14–62%, average 32%), but the measurements from 1994 do not include the phototrophic biofilms on these surfaces (Marxsen 2001, 2006). Thus, a few grams of BCP must be added for total production in this habitat. There were some investigations on the bacterial production on decaying leaves and macrophytes from streams in North America, but these streams contained different species from those growing in the Breitenbach. The data range from 0.3–22 (Newell *et al.*, 1995) to 4–32 (Kuehn *et al.*, 2000) and 7–149 μg C g^{-1} AFDM day^{-1} (Weyers and Suberkropp, 1996). Assuming an average bacterial production rate of 10–50 μg C g^{-1} AFDM day^{-1} and inputs of CPOM (allochthonous and macrophyte sources) of about 850 g m^{-2} $year^{-1}$, an additional 3–15 g C m^{-2} $year^{-1}$ of BCP must be taken into account for the Breitenbach. Summing up the estimates on the different Breitenbach habitats, which were not, or only scarcely, investigated, Marxsen (2006) assumed total bacterial production of about 200 g C m^{-2} $year^{-1}$ for this stream, equivalent to total net primary production (cf. Chapter 5).

del Giorgio and Cole (1998) reviewed the data on bacterial growth efficiency (BGE) on organic matter in aquatic environments and found a large range of data, from less than 10% to more than 60%. They calculated 30% as a reasonable medium value for vascular plants. This is in good accordance with the study of Benner *et al.* (1988), who determined average conversion efficiencies of 30% for bacterial growth on lignocellulose for estuarine and salt water. This also agrees with the range of efficiencies for DOC (summarized by Jahnke and Craven, 1997), from about 5% to more than 80%. Thus, if 30% is assumed as bacterial growth efficiency, bacterial respiration for the Breitenbach can be calculated to 470 g C m^{-2} $year^{-1}$ (Marxsen, 2006). This finding is in good agreement with the few studies on microbial or bacterial respiration in other streams. Fuss and Smock (1996) determined total microbial respiration as 2 g C m^{-2} day^{-1}, and Naegeli and Uehlinger (1997) measured between 1.7 and 2.5 g C m^{-2} day^{-1}. These data are somewhat higher, but not far from the daily average of 1.3 g C m^{-2} for bacterial respiration calculated for the Breitenbach (Marxsen, 2006). Only Fischer *et al.* (1996) published data on bacterial respiration in streambed sediments, for the Steina (Black Forest, Germany), of 0.62 g C m^{-2} day^{-1} during October.

6.3.4
Fungal Production

No special investigations were performed on the production of fungi in the Breitenbach, and very few data are available from other aquatic environments. Sinsabaugh and Findlay (1995) investigated the fungal production in the sediments of the Hudson River estuary (New York, USA). If their turnover times for fungal biomass (13–72 days) are applied to the Breitenbach, fungal production can be estimated as between 5 and 28 g C m^{-2} $year^{-1}$. Similar results are obtained if the study of Baldy *et al.* (2002) is used. From their P/B relationships of 0.004 to 0.1 day^{-1} in decomposing leaves in the Garonne River (France), fungal production in the Breitenbach corresponds to about 2–36 g C m^{-2} $year^{-1}$. The range is much higher when using the data of Kuehn *et al.* (2000) on microbial production associated with the degradation of decaying litter from *Juncus effusus* l., found in a freshwater wetland ecosystem in Alabama, USA. Based on between 73 and 2386 μg fungal biomass production g^{-1} AFDM day^{-1}, the annual fungal production for the Breitenbach is calculated as between 8 and 270 g C m^{-2}. Studies giving lower production rates seem more closely related to the Breitenbach. Therefore, Marxsen (2006) assumed the fungal part of the microbial production to be 20 g C m^{-2} $year^{-1}$, about 10% of the bacterial production (Table 8.31). However, this estimate is based on a weak dataset, but is close to fungal productions of 17 g C m^{-2} $year^{-1}$ from Walker Branch (Tenn., USA; Suberkropp, 1997), and 13 and 16 g C m^{-2} $year^{-1}$ from two woodland streams in Alabama, USA (Methvin and Suberkropp, 2003). Suberkropp (1991) reported a fungal growth efficiency of 35%. On the basis of 20 g C m^{-2} $year^{-1}$ fungal biomass production, this equates to about 40 g C m^{-2} $year^{-1}$ fungal respiration in the Breitenbach (Table 8.30).

7
The Fauna of the Breitenbach

Peter Zwick, Georg Becker, Rüdiger Wagner, and Joachim Reidelbach

More than 1000 species of Metazoa have been documented for the Breitenbach fauna. A variety of sampling methods were employed, for example, mud samples for microfauna, kick or Surber samples of the benthos, traps for emerging insects, and electrofishing. Much smaller species numbers have been reported from large rivers, like the Turiec in Slovakia ($n = 616$) or the Austrian Traun ($n = 442$; Zwick, 1998a). Species numbers as high as in the Breitenbach have only been reported for the Austrian section of the Danube (Moog, Humpesch and Konar, 1995). However, the difference in faunal diversity is apparent rather than real and is probably due to less comprehensive sampling in other studies. When only aquatic insects taken in emergence traps are compared for 20 studies of central European streams, the Breitenbach is average (Zwick, 1998a).

Attempts to find published information on final body size of the Breitenbach species were successful for only about 80% of the taxa. Missing species were congeners or close relatives of others of known size and presumably not very different. Therefore, although based on only 80% of the species present, Figure 7.1 gives a good idea of the small size of most of the animals in the stream. It is unclear whether this is an adapation to the food supply in streams, which is dominated by very fine particles (Section 4.2), or whether protection in small crevices and reduced impact of current speed are the driving forces.

It was only possible to deal continuously and in detail with most of the larger animals, the so-called macrozoobenthos, provided the taxa were sufficiently abundant. Where we lacked taxonomic competence, visiting colleagues helped, for example, with the Mollusca and macroscopic Annelida. However, such studies were brief, and the results are only presented in the general overview of the fauna and in tables in the Appendix[1]. For most very small forms, only isolated studies were made, to which reference is made in the taxon lists in the Appendix.

1) Supplementary material is available at www.wiley-vch.de/home/breitenbach and http://pubman.mpdl.mpg.de/pubman/item/escidoc:969687.

Central European Stream Ecosystems: The Long Term Study of the Breitenbach, First Edition. Edited by Rüdiger Wagner, Jürgen Marxsen, Peter Zwick, Eileen J. Cox.

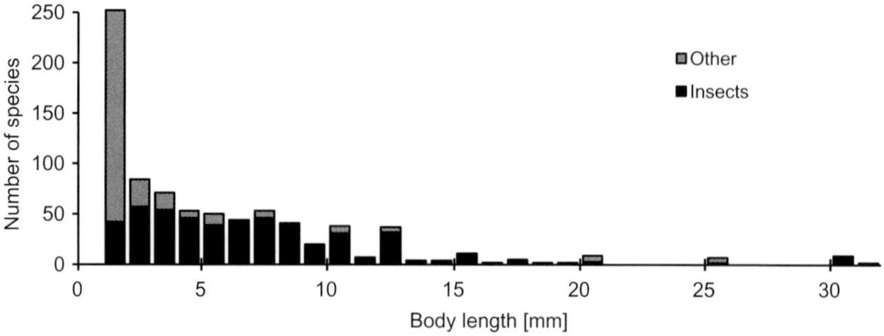

Figure 7.1 Size distribution of the metazoan fauna in the Breitenbach. The data cover 80% of the species present, data for the remaining taxa not available. Modified from Zwick (1998a).

7.1
Sampling Methods

Peter Zwick

7.1.1
Collecting the Macrozoobenthos

Quantitative samples of benthos were taken to study growth and development of species in the macrozoobenthos and to estimate their secondary production. Samples were taken with a modified Surber sampler (Figure 7.2) supplied with lateral sheet metal screens which directed all drifting material from the 20 × 25 cm sample area into the 0.5 mm mesh net bag. The underside of the heavy metal base surrounding the sampling area was fitted with 5 cm rubber foam to ensure a tight seal, even on a rough, stony surface while the person sampling stood on the metal frame.

Five samples were taken in 50 m subunits of the 300 m study stretch in Reach B, downstream from Trap T2. A different subunit was sampled at each occasion. Two parallels were taken in Reach A, just downstream from Trap T1, and two more in Reach G, upstream from the Kreisstrasse 60 to Pfordt. Samples were taken in seemingly similar riffles with gravelly bottom substrate, in Reaches A, B, and C of the Breitenbach. The nine parallels were normally taken within one week. Sampling was approximately monthly. Samples were coded by year and week number, for example, sample 8727 was taken in 1987, week 27; actual sampling dates were 25 June in reach B, and 29 June in Reaches A and G.

Substrate in the sampling area was removed to a depth of 5–7 cm with a shovel. The then exposed stream bottom was disturbed to ensure drift of remaining specimens into the sampler net. All material was kept in aerated water until it was processed the same day. Mineral and organic material were separated by washing, decantating, and fractionated sieving, down to a minimum mesh size of 200 μm. All live organisms were picked, the remaining detritus was scanned for dead

Figure 7.2 Oblique anterior view of the modified Surber sampler used in the present study; the sampling area in the base plate is 20 × 25 cm.

individuals at 6× magnification. Large stones were gently scrubbed with a toothbrush, gravel and sand were scanned for cased caddis and beetles at 6× magnification. Specimens were preserved in 75% ethanol.

7.1.1.1 Study of Specimens

Identified specimens were measured to the nearest 0.5 mm body length with a digitizing tablet or ocular micrometers at 6–40× magnifications (Ephemeroptera, Plecoptera). Coleoptera and Trichoptera were assigned to instar by size, usually head capsule width (HCW) or head capsule length (HCL). Damaged individuals and exuviae were assigned to size classes by comparing head capsule, labrum width, or other appropriate measures with intact specimens.

However, the larvae of most insects are much more difficult and less reliably identified than adults of the same species. Even in well studied groups, specific identification of larvae is not generally possible. However, the biocenosis in the Breitenbach is exceptionally well known. It is diverse, but includes limited numbers of species in each family or genus. Therefore, in the Breitenbach, specific larval identifications are often possible to the taxonomic specialist. However, even under such favorable conditions most young specimens cannot be identified to species because specific characters form successively at moults and are simply not yet developed in small individuals.

7.1.1.2 Secondary Production

Quantitative sampling of stream benthos is a very difficult task. The relevant manual of the International Biological Program called sampling "*extremely difficult*" (Hynes, 1971). A more recent evaluation of samplers and methods of sampling concludes "*eine vollständige Erhebung der Bestände ist mit den bisher bekannten Methoden jedoch nur bedingt möglich*" (Tittizer, 1999).

Estimates of secondary production of stream fauna are normally based on quantitative samples of aquatic life stages, the benthic larvae in the case of insects. Problems encountered during benthos sampling would be avoided if production estimates could be based on quantitative trap collections of the (usually reliably identifiable) emerging adult insects as Illies (1971) proposed. However, over the years doubts arose if emergence trap catches can be related to a defined area of stream bottom. We eventually concluded that this is not the case and after 1982 no longer used trap collections for production estimates. Malicky (1989, 2002b) arrived at similar conclusions.

Emerging adults represent only the biomass of the final cohort, not total secondary production. It was long doubtful how important the contribution of the very numerous very small specimens to total production might be. Probably this relation varies between taxa, streams, and years. Statzner and Resh (1993) used published data to evaluate "whether any relationship exists between the emergence biomass and the secondary production of stream insects. . . . Only because such a relationship was discovered did we go on to a detailed analysis of emergence". The said relationship is illustrated in a scatter plot with a regression line (Statzner and Resh, 1993, see their Figure 3), the parameters of which are not given. In the studies from which the data were taken a variety of methods were employed. For example, emergence was collected by very different traps, collecting efficiency must have varied widely. For example, some data were obtained with pyramidal traps covering ca. 700 cm^2 of stream bottom (Ritter, 1985; after Ritter, 1990), others with a 12 m long greenhouse (Benedetto Castro, 1975). Speir and Anderson (1974) did not directly measure emergence but equated the blackfly pupae in their benthos samples with emergence. Benthic sampling techniques varied also much between the studies in question.

Not surprisingly, estimates of the relation between production and emergence range from 25 : 1 for the univoltine caddisfly, *Agapetus fuscipes*, in the Breitenbach (Benedetto Castro, 1975) to approximately 10 : 1 or 4 : 1 in various Simuliidae and Chironomidae, some of which were plurivoltine (Neveu, 1973; Speir and Anderson, 1974).

To test the validity of methods and improve this unsatisfactory situation, approximately monthly benthos samples were taken from February 1986 to May 1988, following some preliminary tests in earlier years. Sampling focused on rheophilous taxa in lotic sites. Silted habitats were avoided, sampling of marginal habitats, especially under overhanging banks, was impractical in the narrow stream. Sampler operation in the Breitenbach was difficult and encountered most of the problems described in the literature (see above); almost every second sampling attempt failed.

The data analysis to estimate production can be performed in different ways (e.g., from instantaneous growth rates, Allen curves, average cohorts; see Benke, 1984) that deliver generally similar results but perform differently well (Morin, Mousseau and Roff, 1987) under particular conditions. In every method, production is estimated from differences in abundance and size of fauna between successive samples.

In view of the known problems of benthic sampling sampling accuracy should be carefully documented. In a classical study, Needham and Usinger (1956) found that 73 parallel samples from a seemingly uniform stretch of a river were required to estimate mean population density with 95% confidence levels. Chutter (1972) re-examined their data and concluded that the earlier estimate was optimistic: actually 448 parallel samples would have been needed. Only much smaller numbers of samples can be handled, and very many studies operate with 3–10 parallels. Variance of specimen numbers in the benthos were "much greater than the mean" in three parallel benthos samples (Clifford, 1972). Unfortunately, estimates of benthic production are typically not supported by statistical analysis.

The same problems were encountered in the present study. Among five parallel Surber samples taken in Reach B every month numbers of specimens differed very much, errors bars of successive samples often overlapped widely and reversed temporal trends of abundance outside periods of recruitment were also not uncommon. Meaningful estimates of differences between successive monthly samples which are needed for production estimates were often not possible. A few taxa that can be safely identified in all life stages and raise no sampling or identification problems were selected to illustrate these problems (Figure 7.3) which ultimately precluded estimates of secondary insect production from benthos samples.

Distribution in the stream is even less homogenous in taxa not simply dwelling on the bed surface or inhabiting shallow top layers of the sediment. Insects which, for example, dwell also on macrophytes, exploit all three dimensions of space in the stream. Reimo Lieske's data on the spatial distribution of *Nemurella pictetii* illustrate the problem particularly clearly (R. Lieske, personal communication).

Nemurella was studied in its preferred habitat, the short outlet of the Georgsborn, in February 2003. Not to destroy the small site, only three parallel samples were taken in each available microhabitat (sand, detritus, stands of *Galium palustre* and *Agrostis stolonifera*). Figures 7.4 and 7.5 illustrate exteme differences in density between microhabitats as well as between parallels. Especially within the plant stands, larval density differed between samples by up to four orders of magnitude. Distribution of different larval sizes differed also somewhat between microhabitats.

The high larval densities in the plant stands are explained by the feeding preference of *Nemurella* for epiphytic algae (Lieske, 2005). Macrophyte surface provides a suitable substratum for microalgae to grow on. Biermann (1990) compared data by different authors on the specific surface of submerged macrophytes. Surface area varies from 500 to 1500 cm^2 per gram dry weight. However, both inter- and intraspecific variation are high. The plant surface area in the Breitenbach therefore

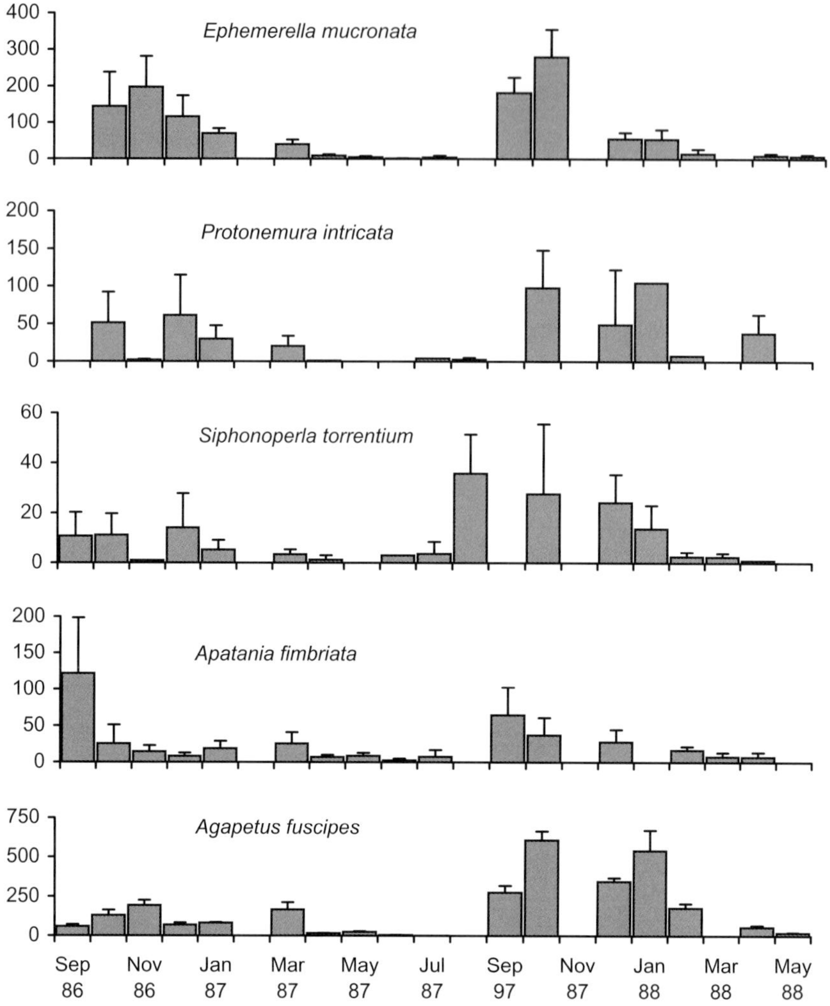

Figure 7.3 Abundance of selected taxa per sample area (0.05 m^2) in Reach B of the Breitenbach over time. Means + 1 SD.

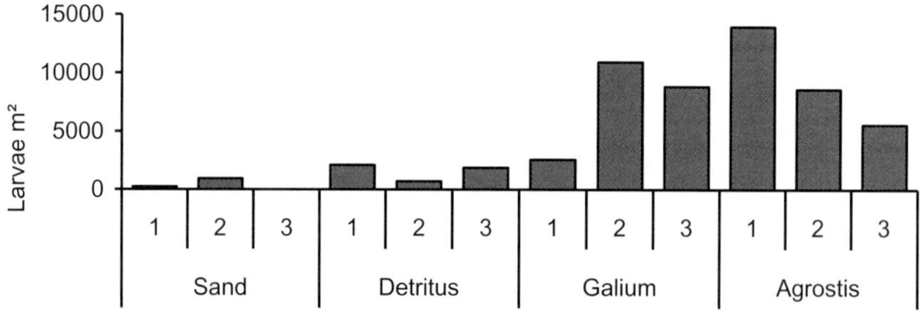

Figure 7.4 Number of *Nemurella pictetii* larvae m^{-2} stream bed in three parallel samples taken in each of four different microhabitats (R. Lieske).

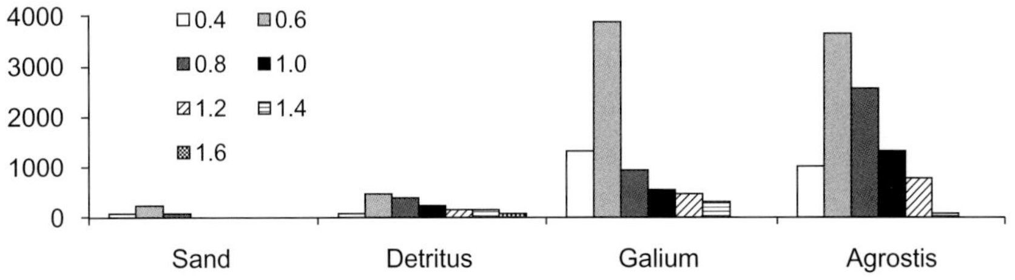

Figure 7.5 Number of *Nemurella pictetii* larvae of different sizes (HCW, mm) m^{-2} stream bed offering different microhabitats, including stands of vegetation in the spring outlet of the Georgsborn; means of three parallel samples (R. Lieske, modified data). Legend shows upper size class limits.

cannot be accurately estimated but it is clear that it is many times larger than the sampled streambed area suggests (R. Lieske, personal communication).

Unsolved additional problems concern taxa with simultaneously occurring species which cannot be identified while small. For example, the abundant tiny larvae of Nemouridae can only with difficulty (and accordingly high risk of error) be identified to genus. The same is true of specific identifications of *Protonemura* and *Leuctra* spp. Small *Baetis* larvae cannot be identified to species. Attempts to recur to auxiliary criteria and assign specimens to species by inferences based on the different modes of egg development of the two *Baetis* species concerned were unreliable in view of the irregular patterns of adult emergence.

Heterogeneity of the stream bottom substrate on a small scale was reported by many previous authors in the literature and was the obvious reason for failure also in the present study. The sampled gravel sometimes concealed a large underlying stone, or alternatively deep layers of fine sand. Substrate data were recorded but were insufficient for an analysis of specific substrate preferences. Larger numbers of monthly samples would have been destructive to the small stream, and could also not have been handled.

7.1.1.3 **Life Cycle Analyses**

In contrast, the population structure of the included species was well documented, size class distributions agreed closely between simultaneous parallel samples, and even between simultaneous samples in different reaches. Data on several dominant species are presented below.

7.1.1.4 **Effect of an Insecticide Poisoning**

In May 1986 an accidental cypermethrin insecticide poisoning occurred in the Aspegraben, near a locality called Morddelle. It reset almost all arthropod populations in the study stretch to zero (Zwick, 1992). Apparently, only a few large *Limnius* larvae with very hard cuticle living in sand layers of the stream bottom survived, at least for some time; see also under *Sericostoma*. Recolonization was by drift from upstream sites along the Aspegraben (e.g., *Baetis* mayflies and Diptera Simuliidae), by oviposition of adults which had emerged from the stream before the accident, and, in the case of shared species, probably also by ovipositing females originating from the River Fulda.

While samples taken soon after the insecticide accident were almost empty, later samples taken in 1986 documented the recolonization of the stream. Normal population structures were restored in 1987 and studied until May 1988.

7.1.2
Collecting Insects in Emergence Traps

Because aquatic insects have been the main focus since the start of the Breitenbach survey, we give a short outline of the type of emergence trap used, its function, strengths and weaknesses for the questions of interest (e.g., community assemblage, biodiversity, secondary production, life histories).

7.1.2.1 Goals

The Breitenbach and its community are assumed to be an example of an average stream of the Central European highland area underlain by Bunter Sandstone. The original intention of installing emergence traps was to obtain knowledge on secondary production in the stream (e.g., Illies, 1971, 1972). Later, additional questions were posed which eventually replaced the focus on production biology.

7.1.2.2 Types of Trap Used and Years of Operation

From April 1969, insects emerging from the Breitenbach were collected at one or more sites (compare Table 7.1). The number of traps in operation over a particular period depended on specific questions, for example, along-stream distribution patterns, effects of sediment quality on emerging insects, effects of specimen removal on subsequent generations.

Trap T2 was run from 1969 to 2006 and provided information on long-term variability that can now also be used to check for possible effects of global change. Additional traps were used for a limited time to analyze particular problems: traps T2 and T3, together with "intermediate" traps I and III were used to answer the question whether the removal of "so many" specimens in individual traps affects specimen numbers in downstream traps (1974–1982) and harms the entire stream community. The effects of artificial stream bottom siltation, or the addition of stones, were investigated with "intermediate trap III", T3/sand and T3/stone traps from 1980 to 1984. Community assemblage and distribution patterns of species along the stream were examined for an 11 year period (1983–1993) at four sites (T1, T2, T4, T6). Individual traps were used for closer inspection of species emergence or for experimental purposes.

At the beginning, an emergence trap was an empty greenhouse placed over the stream (Figure 7.6). White plastic covered the ground along the banks, and flexible plastic skirts "closed" the stream in- and outflow. In that way only insects that developed in that stream section were expected to occur inside the house. No specimens were supposed to escape from the house or to enter it from the outside.

Table 7.1 Overview of emergence traps run along the Breitenbach, 1969–2006. For T1–T6, refer to Table 2.2. GS = trap at the Georgsborn, I and III = "intermediate" traps run in combination with T2 and T3 for limited periods.

Years	**Emergence trap**										
		GS	T1	I	T2	III	T3, sand	T3, stone	T4	T5	T6
	Distance from source (m)	7	105	580	660	757	810	935	110	1475	1875
	Distance from mouth (m)	2068	1970	1495	1415	1318	1265	1140	975	600	200
1969–1973					X						
1974–1975		X		X	X	X	X				
1976				X	X	X	X				
1977–1980				X	X	X	X		X		
1981				X	X	X	X		X	X	
1982				X	X	X	X		X	X	X
1983			X		X		X	X	X	X	X
1984			X		X		X		X		X
1985			X		X				X		X
1986	22 May insecticide		X		X		X		X		X
1987–1993			X		X				X		X
1994–1996			X		X						X
Since 1997					X						

In the early years of study, insects were collected daily from the traps using forceps and aspirators, in later years with vacuum suction (Figure 7.7). In summer, it took more than 30 min for two persons to empty a 12 m long greenhouse. The procedure was quite effective; however, the condition of some of the collected specimens left a great deal to be desired. The collecting time was usually in the morning. After 1986, the operating traps were modified to collect continuously by installing an internal tent in the greenhouse, as had first been proposed by a then doctoral student, J. Reidelbach (Marten and Zwick, 1999). Collecting became far less time-consuming and more time was available to handle (sort, determine, count, measure, weigh) the collected material.

7.1.2.3 **Fundamentals of Emergence Trap Function**

Unconstrained free movement of water and aquatic biota are characteristic of running waters. To conform with the requirement that no study should itself

Figures 7.6–7.7 The first emergence trap on the Breitenbach (early 1970s) at site T2. Collecting emerging insects with vacuum suction (1970–1986); from left.

impact and alter the studied object, emergence traps were built in such a way that aquatic life was allowed unhindered passage, up- and downstream. It was therefore clear that specimens taken in a trap had not necessarily spent all their life in the area covered by the trap. The method was effectively non-invasive because, even when seven traps were operated simultaneously (1982–1983; Table 7.1), the sampled stream surface corresponded to only 2% of the total habitat area.

Emergence traps are a particular type of eclector employed over water bodies. Like other eclectors, the mode of operation relies on innate insect behavior. Positive phototaxis and negative geotaxis make most insects emerging from the water surface move upwards and towards light. Together both behaviours ensure that, once inside a transparent trap, most insects are retained in it for some time. There are exceptional examples of aquatic insects that behave differently, that are therefore not easily caught in emergence traps, for example, the micropterous males of the caddisfly, *Anomalopterygella chauviniana,* which remain on the banks waiting for females to emerge. Only larvae were found in the Breitenbach.

7.1.2.4 **Errors during Trap Operation**

At least some insects that are actively trying to leave a trap will succeed in doing so. In the late 1960s and early 1970s traps were sampled manually once a day. However, occasional tests during which traps were sampled several times a day showed that numbers of captured insects rose with sampling frequency, substantially in some groups (Ringe, 1974). Because of important natural variation in the numbers emerging per day, precise estimates of the share lost when sampling only once per day were not possible. Also, in the early years, when most of the trap surface consisted of glass panes, the air temperature within the traps increased so much on sunny days that numbers of specimens dried, died and were lost (Sandrock, 1978). Therefore, many panes of glass on the sides of the traps were replaced by fine-meshed screens to permit ventilation.

Entry of predators that consumed specimens in the traps was another source of error. Occasional frogs, toads or errant birds (*Erithacus, Troglodytes*) were rare and their impact of little importance. However, spider impact must have been much more important (Briers *et al.*, 2005; Collier, Bury and Gibbs, 2002) although it could not be quantified. Moderate numbers of Tetragnathidae were present in the emergent and littoral vegetation all year. Araneidae posed an obvious problem in late summer and autumn. These spiders were removed by the personnel operating the traps.

When an inner tent-like structure was installed inside the greenhouses, directing emerging insects into a tray with preservative (water, some detergent, and very little formaldehyde to prevent rapid decay of the collected specimens) collecting became continuous, and the above-mentioned problems were largely, if not completely, overcome. Another benefit was the undamaged condition of the collected material. The known attraction of insects to formaldehyde (Adis and Kramer, 1975; Zulka, 1996) did not bias the collections. Its attraction may have worked at a short distance, when specimens were next to the preservation trays, but could have affected insects prior to emergence, while they were still under the water.

7.1.2.5 Emergence Traps as Activity Traps and Terrestrial By-Catch

Most terrestrial insects developing in the grassy banks under traps were easily recognized as such and excluded from analysis. However, certain groups of Diptera including aquatic, amphibious, as well as terrestrial species, whose preferred habitats were not generally known caused problems. Therefore, all terrestrial vegetation was eventually suppressed under black plastic sheets and wooden boardwalks.

Despite their name, emergence traps do not function exclusively as such, but are at the same time activity traps, collecting insects moving over the stream. Animals that happen to fall onto the water surface drift for some distance before they drown or rise again from the stream surface. If they rise while under a trap, they are retained in it by the same mechanisms as emerging individuals. This is not a failure of trap function; rather, by-catch due to insect activity is an obligatory component of a functional emergence trap over a stream. An emergence trap can be regarded as a habitat island, and the intrusion of taxa from outside is in agreement with the theory of island biogeography (MacArthur and Wilson, 1967), as is the elevated proportion of intruders in those traps closest to another stream (Figure 7.8). However, if non-emerging specimens caught because of their activity around the trap are not recognized as such by the observer, this may lead to serious errors in data analysis.

Two groups of taxa were caught due to their activity around or over the stream. The largest component of by-catch was terrestrial insects without any specific relationship to the stream. For example, seasonal migrations between winter quarters and food plants account for the abundant occurrence of a beech (*Fagus silvatica*) leaf-mining weevil (*Euthoron fagi*, Curculionidae). Numerous longhorn beetles (Cerambycidae: Lepturini) were trapped when *Filipendula* flowered along the stream. In early years, when limited farming still occurred in the Breitenbach

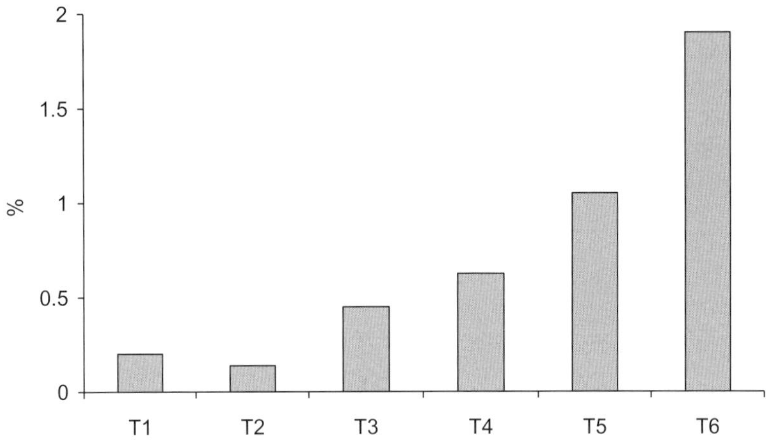

Figure 7.8 Share of "alien" caddisfly specimens (%; 1982–1985) in six emergence traps along the Breitenbach. For site locations refer to Figure 2.3 and Table 2.2 (Chapter 2).

valley, manure was occasionally spread on the surrounding grassland and attracted numerous terrestrial beetles that were also taken in the emergence traps. *Aphodius* spp. (Scarabaeidae) dominated those genera interested in the dung per se, the rove beetles *Philonthus, Oxytelus, Tachinus* and other were abundant among taxa preying on fly maggots in dung and other decomposing organic material.

The activity trap function of the emergence traps caught about 500 species of terrestrial beetles as by-catch, corresponding to ca. 8% of the German beetle fauna, including several rarities (data available on request from P. Zwick). This figure still excludes many difficult and species-rich groups (Nitidulinae, *Atheta* group, Ceutorrhynchini, Apionini) that were not, or only exceptionally, identified. If continued long enough, the by-catch of terrestrial beetles would eventually approach an inventory of the regional fauna.

Less numerous, but relevant to our study, were taxa with some systematic relationship to the stream, or a preference for cool and moist habitats along it. The rove beetle, *Lesteva longoelytrata* (Staphylinidae), provides an outstanding example of seasonal migrations during which much time is spent alongside streams (Burkart and Zwick, 2006). Preference for cool, moist sites, found around trap in- and outlets, probably lured numbers of terrestrial hoverflies (Diptera: Syrphidae) into the emergence traps (Löhr, 1988, 1991, 2004a, 2004b).

The mode in which most Plecoptera oviposit (Section 7.2) may also bias emergence trap collections towards old, ovipositing, stonefly females (Zwick, 1977). The probability for such bias varies locally. It is greatest at sites just below attractive oviposition sites with turbulent flow. However, because of the need of Nemouroidea (i.e., all Breitenbach species, except *Isoperla* and *Siphonoperla*) to feed before ovipositing, eggs are usually laid towards the end, or even after, the actual emergence period. This may lead to a striking excess of females in late collections (Figure R3). Old ovipositing or spent plecopteran females can also be recognized by certain physical characteristics (Zwick, 1990) and were accounted for in our analysis.

Females of the short-lived mayflies usually die shortly after oviposition, many without ever rising again from water. Accordingly, spent female mayflies did not appear in traps. With special effort, caddifly specimen age can be estimated from individual condition (Petersson and Hasselrot, 1994). However, an average female accidentally caught in an emergence trap after ovipositing would not normally be recognized. Among the Diptera, female blackflies (Simuliidae) with developed eggs had clearly not emerged freshly, but were several days old and had been caught after feeding on blood from lifestock or wildlife, far from the trap (Section 7.6.4.6).

7.1.2.6 **Emergence Trap Collections and Estimates of Secondary Production**

We were always aware that specimens taken in traps had not necessarily developed in the area under the trap, because water flow and animal movement in the water were deliberately not constrained by the traps. When Malicky (1989, 2002b) covered the stream bottom under an emergence trap with a metal sheet and swept it clean every day he nevertheless collected large numbers of specimens – exactly as expected. Nevertheless, if faunal movements along the stream were in balance or steady state, collections could be regarded as samples from random stream sections of the same length as the traps. Therefore, as proposed by Illies (e.g., Illies, 1971, 1972, 1974) production estimates using emergence trap collections appeared possible.

However, when several traps were placed on a short stretch of the Breitenbach (1974–1982; Table 1), which appeared to offer similar conditions to the fauna, unexpected differences were nevertheless observed between traps (unpublished data of the Flussstation). Artificial siltation of one trap (Section 8.2) highlighted the importance of stream bottom structure. As shown above, benthic sampling revealed important differences in colonization on a small scale, even when the visible sediment surface appeared comparable between sites. The initially assumed relationship between emergence trap collection and area of stream bottom under the trap was not confirmed, nor were some of the presumed rules governing insect emergence (Illies, 1982, 1983). Therefore we abandoned production estimates from emergence trap collections on the Breitenbach after 1982. Malicky (2002a) later arrived at the same conclusion from his own data on Austrian streams. However, Statzner and Resh (1993) held a more positive view (see above) and used various published emergence trap data to test several ecological theories.

7.1.2.7 **Evaluation – Statistical Treatment of Data**

Rüdiger Wagner

Studies on emerging aquatic insects were used to compare specimen numbers (size, weight) between years (temporally), and from more than one trap along the Breitenbach (spatially). These questions can be subsumed under comparative "mensurative experiments" in the sense of Hurlbert (1984). Four traps arranged along a stream invoke the criticism of spatial pseudo-replication because all are placed on an individual stream, and if traps are run for a number of years on the same place, temporal pseudo-replication might be assumed.

Among others, Hargrove and Pickering (1992) and O'Neill (2001) expressed the view that landscapes and ecosystems *per se* are not replicable. Nevertheless, it is possible to obtain reliable results about the combination or dependencies of environmental variables and communities. However, ecological and statistical conclusions have to be separated. Unreplicated studies may nevertheless produce viable ecological conclusions. The repetition of an experiment may provide better insight into a problem than increasing the replicates in a single experiment.

Four traps along the Breitenbach were handled as four "treatments", because sites differ in individual years and in several environmental variables, that is, most variables, in particular water temperature, have strong and generally predictable spatial gradients while others, like discharge, water depth and channel width, probably vary with time.

Four traps are replicates in the sense that they provide information about the population variation along the stream. The same traps are also replicates in the sense that they provide data on population variation over time.

Also, the same site in subsequent years does not constitute pseudo-replication because stream bottom sediments are constantly subject to the variable discharge. Thus, although it is the same site, environmental conditions at any time differ from other periods, and subsequent years will describe the variability of the community over a certain stream area.

Thus, we conclude that replicate samples or measurements are dispersed in space (and/or time) in a manner appropriate to Hurlbert's demands.

For statistical analysis the Statistical Package for Social Sciences (SPSS 2007, and subsequent versions) was normally used. Additional programs and packages used are indicated in the respective sections. CANOCO (Ter Braak, 1988) was the preferred program for community and environmental data analysis.

Results (in the strict sense) are valid only for the Breitenbach, but commonsense, biological knowledge, and the use of intuition are also permissible. Hence, if we carefully treat, analyze, and interpret the data, bearing in mind the possibility of pseudoreplication, many questions (along stream distribution, temporal variation, and more) may be answered with comparative and concluding statistics without fundamental or serious errors. For further generalization additional expert knowledge or commonsense can be employed.

7.2
Protozoa: Ciliophora

Peter Zwick

Of the nonmetazoan, unicellular or acellular animals in the Breitenbach, only the free-living ciliates were studied from 1990 to 1992 by G. Packroff (Packroff, 1994) and published in part (Packroff and Zwick, 1996, 1998; see these for details and additional references). Epizoic ciliates were occasionally seen on larger beetles and the like, and symphoriontic Ciliophora species possibly occur in the stream but these were not studied.

While there is an important body of literature on ciliates as bioindicators (e.g., Berger, Foissner and Kohmann, 1997; Foissner *et al.*, 1991, 1995), there are only a few studies of unpolluted streams (e.g., Foissner, 1997). The aim of our study was to provide a species inventory and investigate the possible role of Ciliophora as consumers of the important bacterial biomass in the Breitenbach, a first-order stream.

The Breitenbach Ciliophora were sampled in two ways. Artificial substrata (glass microscope slides) were exposed in several habitats to obtain inhabitants of the periphyton. Glass slides were found to support the same fauna as various natural substrata in the stream. Ciliates living in fine-grained sediments rich in organic material were examined in sediment cores. Cores (2.6 cm diameter) were subdivided into 2 cm slices to determine depth distributions. Identifications were based on observation of live specimens on the day of sampling, and on silver impregnated, preserved material.

Live animals on microscope slides could be counted directly. Sediment samples were first diluted 10-fold with stream water and gently stirred to obtain a homogeneous suspension. Four 250 µl subsamples were pipetted into sedimentation chambers and live animals observed and counted under an inverted microscope. An additional 10 ml of the suspension was preserved with $HgCl_2$ solution. Unfortunately, the condition of preserved specimens in sediments deteriorated quickly.

Altogether, 143 ciliate taxa were recorded in the clean, cool, Breitenbach, many more than the literature suggested might occur in this type of habitat. Only *Pseudochlamydonella rheophila* is restricted to running waters, while most other ciliates observed in the Breitenbach are ubiquitous species that also occur in standing, and even in organically polluted waters. A total of 113 taxa (including some sessile species) occurred in the periphyton, but 39 of these were also found in sediments where 69 taxa (all vagile) were observed. Diversity was lowest in the Georgsborn where only 23 periphytic taxa were found, while at the other five sampling sites along the stream 75–91 taxa were found in the periphyton. Apart from the spring, species composition and diversity differed little between sites along the stream.

The mean annual abundance of ciliates in the periphyton was 30–140 specimens cm^{-2}, corresponding to 0.1–52.0 µg wet weight. Peak abundance was 1400 ciliates cm^{-2}, or 430 µg wet weight. Table 7.2 presents the species most regularly found in periphyton samples.

Up to 4500 ciliates were found in 1 ml of sediment, but yearly mean abundance ranged from 850 to 1500 individuals ml^{-1} sediment. Along-stream differences were not significant. Most taxa were bacterivores, but at some sites, increasing abundances in spring resulted from a larger share of grazing ciliates containing diatom frustules, at the time of maximum diatom abundance (Cox, 1990b). Most specimens were found in the top 2 cm of the sediment. However, the genera *Metopus, Brachonella,* and *Caenomorpha,* which are usually considered to prefer sapropelic habitats, occurred deeper in the sediment, except during periods of minimal flow when oxygen depletion probably occurred in the deeper sediments.

Small to very small ciliates (20–60 µm long) contributed 75–95% of the community. Larger species were never numerically dominant and predatory taxa >200 µm (*Monilicaryon monilatus, Litonotus cygnus*) occurred at low abundances. In the top 2 cm of sediment biomass usually ranged from 5–60 µg wet weight ml^{-1} sediment.

Annual production was estimated to be up to 2.5 g C m^{-2} in the periphyton, and 6 g C m^{-2} in the sediment. Packroff (1994) concluded that feeding by ciliates has

Table 7.2 Ciliate species observed in 70% or more of all periphyton samples. Species with a mean dominance of ≥ 5% in bold script.

Aspidisca cicada	***Glaucoma scintillans***	*Stylonychia mytilus*
A. lynceus	***Holosticha monilata***	***Tachysoma pellionellum***
Chlamydonella minuta	*H. pullaster*	*Trochilia minuta*
Cinetochilum margaritaceum	***Litonotus* spp.**	*Vorticella campanula*
Frontonia acuminata	***Platycola decumbens***	

only limited impact on the available bacterial biomass, probably less than their impact on algal biomass. Similarly, in a laboratory study feeding selected ciliate species on fluorescently labeled bacteria, Hecht (1992) estimated that Ciliophora probably consume less than 1% of bacterial production in the Breitenbach.

7.3 Platyhelminthes and Annelida

Peter Zwick

From September 1976 to October 1979, Platyhelminthes and Annelida were studied from all types of limnic habitats in upper Hesse, between the Vogelsberg and the Rhön mountains, by Peter Schwank. His important studies (Schwank, 1979, 1980, 1981a, 1981b, 1981c, 1981d) provide much information on these groups in the Central German highlands. The Breitenbach was more intensely studied than other watercourses and a short account of P. Schwank's core results is presented.

Altogether 94 species of free-living Platyhelminthes and 81 Annelida were recorded in the study, 49 and 55, respectively, from the Breitenbach (Schwank, 1981a). At the time of publication (Schwank, 1979, 1980, 1981a, 1981e), 29 species of Platyhelminthes and 8 species of Oligochaeta were new records for Germany. Of the flatworms, two were the first European records, and 13 species of Platyhelminthes were new to science and first described by P. Schwank. Species lists are provided in the Appendix.

7.3.1 Platyhelminthes

Of the Platyhelminthes, Catenulida were the dominant group, contributing 18% of the species and about 40% of the specimens. The evident success of Catenulida in running waters may be related to their high rate of asexual reproduction. Members of the Dalyellidae (particularly dominant in lotic mosses) made up 13% of species and up to 25% of specimens; in contrast, the Typhloplanidae were not abundant but represented more than 40% of species.

A quantitative study of the distribution of the large carnivorous, lithorheobiontic triclad *Dugesia gonocephala* in the middle reach of the Breitenbach, from about 100 m upstream of T2 down to T4, recorded 0.2 to 6.0 specimens m^{-2}, depending on stream reach and season. Estimates indicated that the density and biomass of *D. gonocephala* were of the same magnitude as those of an abundant carnivorous caddisfly, *Rhyacophila fasciata*.

The temporal abundance of *D. gonocephala* varied considerably. Mean density was highest [~5 specimens per running meter (= rm^{-1}) stream bottom] in July 1978, but decreased to November 1978 (~3 specimens rm^{-1}) and even further to May 1979 (~1 specimen rm^{-1}). The decline towards November 1978 was a result of both the death of the parent generation and a massive sedimentation of fine inorganic and organic material, which filled interstitial spaces and thereby reduced available habitat. A further decrease in specimen numbers until May 1979 was related to heavy winter and spring spates that year, which may have seriously disturbed the necessary microhabitats. By autumn 1979, densities were again approximately the same as in the preceding year.

The distribution along the stream barely changed with time. In the stretch between T2 and T4 the abundance of *D. gonocephala* was uniform and low. Densities increased significantly only about 100 m upstream from T4. The species was also rare in the downstream section with its large areas of fine sediments.

7.3.2
Annelida

Most species of Annelida belonged to the Enchytraeidae (27) and Naididae (24), but four tubificid species, *Tubifex tubifex, Aulodrilus pluriseta, Rhyacodrilus coccineus* and *Limnodrilus hoffmeisteri*, represented up to 60% of the Oligochaeta.

Based on numerous ecological observations in several streams in East Hesse, Schwank (1981c) recognized characteristic taxocoenoses in springs (crenon), streams (rhithron), and small rivers (potamon). He further distinguished several microhabitats in each of these, for example, lotic and hygrophilic mosses, and hydropelic, macrophytic and interstitial habitats in springs. Seven and three taxocoenoses were differentiated in streams and rivers, respectively, inhabiting lotic and hygrophilic mosses, macrophytes, wet borders without mosses, lenitic mud areas and the hyporheic zones. Coenoses characteristic of springs and small streams were very diverse and specialized, but those in lotic mosses of montane streams were highly distinct. Another very characteristic coenosis occurred in the hyporheic zone, in which Catenulida dominated.

Platyhelminthes taxocoenoses were generally more complex than those of Annelida. Species of small Platyhelminthes are particularly stenoecious. Schwank therefore assumed that changes in species distribution and abundance of the worm taxa were related to spatial and temporal variation of microhabitat availability, caused by floods but also by periods of low flow, when sedimentation is particularly important.

From comparison with individual collections elsewhere in Central Europe, Schwank (1981d) concluded that most of the stream taxocoenoses identified in East Hesse are widely distributed, occurring also in other low mountain ranges in Europe.

7.4
Nematoda

Peter Zwick and Heino Christl

The Nematoda of the Breitenbach were studied during Peter Schwank's ecological study of worm-like metazoa in the stream (Schwank, 1985). There is also a paper on nematode parasites of insects in the Breitenbach (Kaiser and Schwank, 1985). However, the main study was from 1992 until 1995, during Heino Christl's doctoral study (Christl, 2008).

To obtain material from bed sediments, Schwank (1985) relied mainly on the method of habitat deterioration. He also used available keys to identify the specimens. Schwank's material was not available for revision using the strict taxonomical criteria subsequently established by Christl (2008). However, except for about 15 taxa, the species recorded by Schwank were also encountered in Christl's study. If, in a worst case scenario, those 15 species were misidentified, the possible error in species number in the combined taxonomic list (see Appendix) remains minor.

Christl (2008) tested several extraction methods. A flotation technique using diluted Ludox as the extraction medium proved to be the most suitable. However, even this method is not rigorously quantitative, because the tiny animals may use adhesive glands to attach themselves to mineral particles, or may remain concealed among FPOM, which has a similar density to the Nematoda themselves.

Available identification keys proved to be unsuitable for determining the majority of adult Monhysterida. The dominant *Eumonhystera* species was new to science, but even with specimens that apparently belonged to known species, the identification process was cumbersome and prone to errors, particularly since there are inconsistencies within published keys and disagreements between keys. In order to obtain more reliable and reproducible results Christl (2008) developed a computerized polytomous identification system as an Excel macro linked to a digitizing tablet and microscope. This allowed instant comparison of all measured morphometric parameters with published data. The system was extended by an index system, assuming a unimodal distribution of morphometric characters, which enables the user to examine entire data bases automatically.

The Breitenbach sediments were inhabited by an abundant and diverse nematode community. 241 taxa were recorded in Christl's study, of which 53 were generally aquatic species, another 98 are considered to be amphibious, and the remaining 90 are classified as terrestrial (see Appendix). However, the aquatic species were generally the consistently present and dominant ones. Mean abundance was 51 individuals per milliliter sediment (3–787 individuals ml^{-1}), corresponding to a density of ca. $2.55 \times 10^6 m^{-2}$ stream bed. Abundances sometimes varied by an order of magnitude between replicate samples. No clear seasonality was detected, but there were consistent differences between sampling areas along the stream.

Abundance and composition of the nematode community were affected by a winter flood to different degrees in different stream reaches. In particular, the abundance of monhysterids decreased after the flood, whereas large Triplonchida and Enoplida remained more constant, and hence their relative contribution increased. These changes partly coincided with distinct changes in the abiotic sediment parameters.

Ordination analysis revealed patterns in the community and allowed the effects to be visualized and the influence of measured environmental variables to be quantified. 34–60% of the total variance could be captured in the first four axes (indirect methods). When latent variables were constrained to linear combinations of measured environmental variables (direct, constrained method) 17–30% of the total variance was captured as species–environment interaction (and used in the model), whereas 29–41% of the variance was allotted to the covariable, "sampling date". These results, which were in part highly significant, show that there are links between abiotic parameters and community composition, but also that there must be additional, unmeasured, environmental variables that have an effect on the nematode community. Among the measured environmental variables sediment grain size (median) showed the strongest link with the latent variables, followed by water content, and content of organic matter. Also the amount of very fine mineral particles (less than 5 μm) and the microbial activity in the sediment affected nematode community composition. However these links were less pronounced than expected.

R-strategic bacterial feeders dominated in the Breitenbach in terms of abundance, whereas in terms of biomass larger, K-strategy species were dominant. Generation times varied between 23 and 365 days, some longer periods were observed directly, but the majority was estimated allometrically. This corresponds to 1–17 generations $year^{-1}$, although there is considerable uncertainty at the upper margin of this range. Nematode biomass (standing stock) was approximately 0.45 g fresh weight m^{-2}, and the production was estimated to be 0.6 g carbon m^{-2} $year^{-1}$.

Nematodes incorporate bacterial biomass. It was therefore assumed that they might act as an important link between bacterial production and higher trophic levels. However, the data on biomass and production indicate that this pathway is only of minor importance in the trophic net of the sediment. Bacterial and macrozoobenthos biomass and production were more than an order of magnitude higher than those of the nematodes.

Other potential functions of nematodes in the benthic community, such as generating hot spots of increased bacterial production or producing rare exoenzymes, are discussed, although this study did not aim to test these hypotheses. However, observations from this study indicate that links between nematodes and other biota tend to be bidirectional, that is, nematodes can predate on ciliates, but at the same time become the prey of ciliates. Bidirectional interactions are possibly the rule for the majority of interactions between nematodes and other organisms.

7.5
Mollusca

Peter Zwick

Molluscs play a subordinate role in the Breitenbach and were only examined over a short period by a visiting specialist, A. Piechocki, from Łodz (Piechocki, 1985).

The following brief comments are also based on other published information, mainly Gittenberger and Janssen (1998) and Glöer (2002).

In attempts in the Vogelsberg and Rhön Mountains to save the dwindling populations of the freshwater pearl mussel, *Margaritifera margaritifera*, by transferring specimens to other, seemingly suitable, streams some specimens were introduced into the Breitenbach in the 1950s by K. Müller (J. H. Jungbluth, personal communication). The mussels did not survive and, apart from a single shell fragment that turned up in 1987 on a mole heap, a few meters from the stream, just upstream from the Kreisstrasse 60, left no trace.

The autochthonous bivalve fauna of the Breitenbach consists of only two species of tiny pea clams that are found in fine sand and soft sediments. While *Pisidium casertanum* occurs in many different types of standing and flowing waters, including those that regularly fall dry, *Pisidium personatum* is a cold-stenothermous inhabitant of wells, springs, and spring runs, including intermittent ones. The life history of both hermaphroditic species is flexible and varies with conditions in particular habitats. The birth period begins in spring and may continue over summer, unless the water body dries up, something the mussels may survive for several months. The total life span of the small mussels is believed to be one to two years. Piechocki (1985) only found numbers of both species far upstream from the main study area, in the upper section of the Aspegraben. Very few were taken in benthic samples, downstream from the Georgsborn, between 1986–1988. Their life history in the Breitenbach is unknown.

Prosobranch snails are represented by the small bisexual *Bythinella compressa*, a characteristic member of the fauna of calcium-poor springs over a restricted geographical area, from the Vogelsberg to the Rhön Mountains (Piechocki, 1985; Glöer, 2002). The species is regularly found on hard substrata in the Georgsborn where it is abundant. However, it does not normally occur elsewhere in the stream. The Breitenbach population was not specifically studied. Typically, one capsule containing a single egg is laid per month. Young snails hatch after about four months and graze on epilithic algae, including blue-greens. Reproduction starts when the shell is about 2.2 mm high. Species of *Bythinella* are generally cold stenotherms. However, Tertiary fossils suggest that the present relict distribution in springs may result, not from a preference for cool temperatures, but rather from avoidance of winter cold. Indeed, a close relative, *B. dunkeri* (v. Frauenfeld, 1857) has been shown to thrive, grow larger and live longer at 20 °C in the laboratory than in the field (Oswald *et al.*, 1991).

Ancylus fluviatilis, a basommatophoran pulmonate, was dominant during a particular formation phase of the Baltic Sea. Today it is a characteristic inhabitant of streams. Its ecology has been well studied (e.g., Calow, 1973; Schwenk and Schwoerbel, 1973; Streit, 1975). The snail grazes on periphyton but blue-green algae are not digested. A muscular stomach with cuticular ridges grinds the food with the aid of ingested small sand grains. *Ancylus fluviatilis* is common in many streams in East Hesse, but its abundance in the Breitenbach is very low, except in the lowermost section, downstream from Kreisstrasse 60. It is commoner in stony sections of the River Fulda. Perhaps the relatively coarse-grained sandstone is not particularly suitable for the animal which keeps a resting place on a rock clean of periphyton, to which it returns after grazing expeditions. Temperature in the Breitenbach may also be limiting because egg formation occurs only at 13 °C or more. Size and longevity vary with temperature, the maximal size is about 8 mm. The snails are polyploid and hermaphroditic, and selfing dominates (Städler, Lowe and Streit, 1995, 1996). Normally, the snails reproduce only once in their lifetime.

Two species of Lymnaeidae were found by Piechocki. The large snail identified as *Lymnaea (Radix) peregra forma typica* must now be called *Radix balthica* (Linnaeus, 1758) (Glöer, 2002). It is typically an inhabitant of standing waters and also occurs in the fish ponds in the Breitenbach valley. Piechocki (1985) found moderate numbers far upstream from the main study area in the Aspegraben, few further downstream. Between 1986 and 1988, *R. balthica* was not found in benthic samples downstream from the Georgsborn.

The other lymnaeid, *Galba truncatula*, host of the liver fluke, was caught in small numbers by Piechocki (1985) and only exceptionally in benthic samples. It is amphibious and often dwells outside water. Its main habitat in the valley is the manmade ponds where it is found at the edges.

Although *Carychium minimum*, a basommatophoran snail, is entirely terrestrial (Kerney *et al.*, 1983), it has been recorded from the Breitenbach along with a few Stylommatophora (Piechocki, 1985). They were: *Succinea putris* (Linné, 1758), *S. oblonga* Draparnaud, 1801, *Cochlicopa lubrica* (Müller, 1774), *Nesovitrea hammonis* (Ström, 1765), *Euconulus fulvus* (Müller, 1774). The species of *Succinea* in particular favor moist habitats and are often knocked into the stream from littoral vegetation during field work.

7.6 Arthropoda

Peter Zwick

A segmented exoskeleton, formed by articulated plates of sclerotized cuticle, and segmented legs are part of the groundplan characters of arthropods. The exoskeleton is flexible but not extensible. In order to grow, arthropods must sporadically shed and replace the hard outer layers of their cuticle. This is a waste of material, and moults per se are stressful for the organism. During, and shortly after, the process the body is unprotected and vulnerable. On the whole, protection by this armor-like exoskeleton probably explains the outstanding evolutionary success of the group, the largest in the animal kingdom and occurring in all habitats.

7.6.1 Chelicerata

The spider clade is generally carnivorous. Migratory wolf spiders (Lycosidae) occur on the ground alongside the Breitenbach, sometimes moving across quiet water surfaces, and web builders (mainly Tetragnathidae and, in late summer, Araneae) among littoral and emergent vegetation catch emerging insects. Both probably contribute to high loss rates of adult water insects (Sections 7.6.4.2 and 7.6.4.6). Fishing spiders (*Dolomedes* spp.) and the submerged *Argyroneta aquatica* occur beside, or in, ponds in the Breitenbach valley (Zwick, unpublished observations) but play no role in the stream itself.

7.6.2 Hydrachnidia

Twenty-three species of water mite (see Appendix) have been recorded from the stream proper. However, several of these prefer standing waters and were probably washed into the stream from connected ponds and pools. The complex life cycle of water mites was only recently elucidated (Böttger, 1977; Gerecke, 1994). It includes instars which remain within the cuticle of the previous stage, which led to confusion in stadal nomenclature. Contary to earlier assumptions, water mites are generally predators (Böttger, 1970). Late developmental instars are ectoparasites, sucking body fluids from adult water insects. They either attach themselves to pre-imaginal instars and remain on them during metamorphosis, either actively moving from one life stage to the next at moults, or being dragged through the cuticle. Alternatively, mites wait on the water surface to attack emerging adult flies. When satiated, they simply let go, regardless of whether their host flies near or far from water (Böttger, 1976; Ullrich, 1978). Losses must therefore be high, but mite egg numbers are nevertheless quite small. Aspects of life cycles, preferred host midges, and number of attached mites were studied in the Breitenbach (Ullrich, 1978).

In his dissertation, Ullrich (1976) mentioned 21 Hydrachnidia species for the Breitenbach. However, he speculated that even more species may occur in the drainage area. The intensely studied *Sperchon setiger* produced egg masses with up to 70 eggs (other species 8–40), mean reproduction rate was 21 eggs per female. Embryogenesis lasted 4–5 weeks, but up to seven weeks at approximately 13 °C in *Protzia eximia*. Eggs of *S. glandulosus* deposited in late summer entered dormancy. Development proceeded after eggs had been stored for at least two months at 6 °C in the laboratory. To develop further, larvae must undergo a parasitic phase on aquatic insects. Host detection is random.

More than 6000 larvae were found on emerging aquatic insects (phoresy, parasitism). Most (~1650) occurred on Chironomidae, 716 on Simuliidae, and 14 on Trichoptera. *Protzia eximia* was detected on all three groups, *Sperchon setiger* exclusively on Simuliidae. The rest were only found on Chironomidae, of which Orthocladiini and Tanytarsini were the preferred hosts. Tanypodinae were rarely infected and no specimen was ever found on Diamesini.

Infestation of Simuliidae was higher (usually 3–6, but up to 21 larvae per specimen) than Chironomidae (1–3, up to 15). Minimum duration of the parasitic phase is 2–3 days, but may last up to almost two weeks. Postlarval quiescense at the protonymphal stage then occurs.

The developing deutonymphs feed preferably on small chironomid and simuliid larvae. Duration of this instar is species-specific lasting between six weeks and seven or even eight months, and is followed by a second quiescent period (tritonymphal stage).

Laboratory observations indicated that *P. eximia* adults do not feed. Adults of the other species fed on larvae of Chironomidae and Simuliidae, some species even capturing Ephemeroptera and Plecoptera. The life cycle usually lasts four to six months.

Males produce spermatophores, but there is no direct contact between male and female (copulation). The uptake of a spermatophore was observed only once in the laboratory. The great number of spermatophores available in the field probably assures female insemination.

7.6.3
Crustacea

Most Crustacea grow and molt throughout their lifetime and female receptiveness is correlated with molting. The class is of marine origin, its primitive members, as well as the majority of the numerous species living in the sea, where ontogenesis usually involves planktonic larvae. A few clades were able to colonize freshwater and terrestrial habitats. Limnic crustaceans have no planktonic larvae.

The fauna of the Breitenbach is relatively poor and comprises only small species. The large Astacidae used to be common in streams in the area (the late Dr. Gothe, personal communication) until an introduced fungal disease (*Aphanomyces astaci*, Phycomycetes) wiped them out. Although there are no records of its presence, Astacidae may also have been present in the Breitenbach. The extant fauna comprises a number of small species (see Appendix), which are never observed in large numbers, plus two amphipods, *Gammarus fossarum* and *G. pulex*, which grow to over 1 cm in length. *Gammarus pulex* also occurs in the River Fulda, where *G. roeselii* and the isopod *Asellus aquaticus* also abound. Neither of them has been recorded in the Breitenbach, like *A. coxalis meridionalis* Herbst (Williams, 1975), which lives in puddles on the floodplain along the Fulda (P. Zwick, personal observation).

The two *Gammarus* species were very abundant and extensively studied in the 1960s and 1970s (see the Flussstation publication list at http://edoc.mpg.de/display.epl?mode=doc&id=475487&col=22&grp=399). Gammarids were strongly affected by an accidental stream poisoning in 1986 (Zwick, 1992). Unlike most insects they never recovered to their previous abundance levels.

The smallest representatives in the stream are minute water fleas and Copepoda (see Appendix), all of them bottom-dwelling or restricted to protected microhabitats, such as mosses, where they feed on microparticles. When abundant, Harpacticida may graze heavily on bacterial populations (Perlmutter and Meyer, 1991). However, they are not at all numerous in the Breitenbach (Klemp, 1979). Nevertheless, most species were only found in very small numbers (Figure 7.9), suggesting that the fauna is incompletely known.

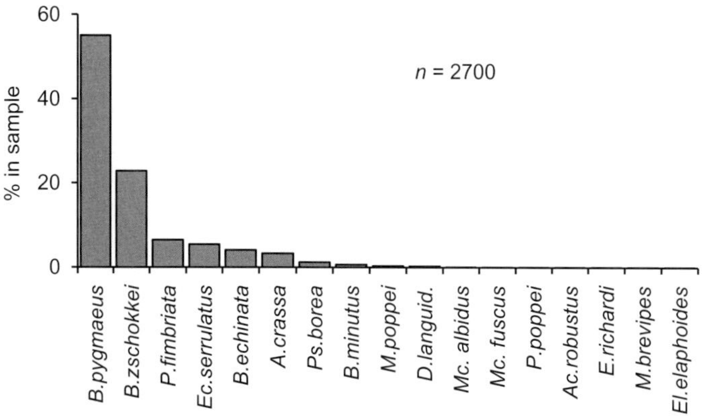

Figure 7.9 Relative abundance of the different Copepoda in the Breitenbach; full names in Appendix. After Klemp (1979).

7.6.4 Insecta

Peter Zwick

7.6.4.1 General and Minor Taxa

In contrast to the originally aquatic and marine Crustacea, insects appear to be primarily terrestrial. This is strongly supported by their tracheal respiration. Because of its build and efficiency, tracheal respiration limits the maximum possible size of insects (Wigglesworth, 1972; Seifert, 1999), and tracheae filled with water cannot function at all. However, insects have solved the problem in various ways and are the dominant animal class on land and in freshwater. Their almost complete absence from the marine environment remains unexplained. Several modes of respiration are observed in the Breitenbach (Table 7.3). The ability to osmoregulate is another prerequisite for life in dilute freshwaters and has been achieved in characteristic ways by different taxa (Table 7.3).

The term "minor taxa" refers to groups that are not dominant in the Breitenbach and therefore were not within the focus of our studies. The present chapter provides brief outlines to supplement the taxon lists in the Appendix. The dominant orders, Ephemeroptera, Plecoptera, Coleoptera, Trichoptera, and Diptera, are treated in separate sections below.

7.6.4.1.1 Collembola

The name Insecta is used here in a comprehensive sense, to include the primitive, primarily wingless, insects. Of these, only the springtails comprise a fair number of water-associated species, although none of them leads a submerged life. However, several species, not only the small (~1 mm) *Podura aquatica* Linné, 1758, are epineustic and have a strongly hydrophobic cuticle. They are fine particle feeders that often occur in large numbers on the water surface. An up to date

Table 7.3 Modes of respiration and osmoregulation in groups of aquatic insects present in the Breitenbach. Some cutaneous respiration is inevitably involved and may be sufficient in small specimens of taxa whose larger representatives rely on other modes of respiration. Species not occurring in the Breitenbach may be different. PBF = ingestion of prey body fluids; Pu = pupae.

	Respiration					Osmoregulation			
	Atmospheric oxygen	Cutaneous	Tracheal gills	Rectal gills	Plastron	Chloride cells	Chloride epithelia	Rectal papillae	Closed surface and/or PBF
Ephemeroptera			X			X			
Odonata				X				X	X
Plecoptera			X			X			
Heteroptera	X					X			X
Megaloptera			X						X
Coleoptera									
Dytiscidae	X								X
Gyrinidae	X								X
Haliplidae	X								X
Hydraenidae					X				X
Elmidae					X				
Scirtidae	X		X						
Trichoptera			X				X	X	
Diptera									
Tipulidae	X								
Limoniidae	X								
Ceratopogonidae		X							
Chironomidae		X							
Dixidae	X								
Thaumaleidae	X								
Simuliidae		X						X	
Ptychopteridae	X							X	
Psychodidae	X		X						
Tabanidae	X		X						
Rhagionidae	X		X						
Stratiomyidae	X		X						
Empididae: Hemerodom.	X	X	x(Pu)						
Dolichopodidae	X								
Ephydridae	X								

synopsis of collembolan biology, with keys to water-associated taxa is provided by Palissa (2000). The species in the Breitenbach remain to be studied.

7.6.4.1.2 **Pterygota**

Unlike the Crustacea and the primarily wingless Collembola, all winged insects grow to a particular final size, reproducing only in the last life stage, the adult or imago, which does not molt. Orders are often divided between Hemimetabola and Holometabola, but only the latter is a monophyletic clade. Hemimetabola is simply a convenient collective term for insects at an early evolutionary stage, whose larvae generally resemble the adults. They have compound eyes, true cerci and develop wings in several steps at potentially numerous moults. This provides students with excellent means of analyzing life cycles and relating environmental factors to particular steps in metamorphosis.

Further, the Holometabola have a secondary type of larva which does not resemble the imago. Holometabolous larvae usually grow without a major change in body shape over fewer moults, and visible metamorphosis occurs in a single step, the pupal stage.

Interestingly, the three most primitive pterygote orders, the Odonata, the Ephemeroptera, and the Plecoptera, are all entirely aquatic as larvae. There has been speculation on a possible connection between an aquatic life and the origin of insect wings from gills (see Wichard, 1997; Zwick, 2009).

In the literature, the term "aquatic insects" is sometimes restricted to taxa in which both adults and larvae live in water. For example, beetles whose long-lived larvae are completely aquatic, but eventually develop into short-lived terrestrial adults, are termed "false water beetles" (Jäch, 1998). However, adults of many "true" water beetles (*sensu* Jäch, 1998) depend on atmospheric air, are potentially terrestrial, and most are ready to fly away whenever conditions in the aquatic habitat deteriorate.

In contrast, we use the term "aquatic insects" in a broad sense, including all those that live in water for part of their life. Very often, aquatic insect life cycles involve a fundamental change of habitat, with reproductive, winged adults emerging from water to live on land and breathe air. We have taken advantage of this migration between habitats for quantitative collections of adults in emergence traps. However, there are other types of life cycle, with one or more life stage occurring outside water (Table 7.4).

7.6.4.1.3 **Odonata**

Odonata are a very distinct group of predatory insects whose beauty, amazing flight abilities and interesting biology (Corbet, 1999) attract exceptional interest; there are journals specifically dedicated to Odonata. The indirect mode of sperm transfer, and especially the active removal of a predecessor's sperm by mating males (e.g., Waage, 1984; Michiels, 1989; Miller, 1995), served as outstanding example of sperm competition and "the egoistic gene".

The adult thorax is strongly oblique, so that the bristly legs are positioned directly below the head and are used to catch prey during flight. Resting adults

Table 7.4 Aquatic insects of the Breitenbach, distribution of the different life stages over aquatic and terrestrial habitats. Entries are in parentheses if individual species differ from main type of respective group. A = aquatic; H = hygropetric; T = terrestrial; n.a. = not applicable, hemimetabolous; V = variable.

	Life stages			
	Egg	**Larva**	**Pupa**	**Imago**
Ephemeroptera	A	A	n.a.	T
Odonata	A	A	n.a.	T
Plecoptera	A	A	n.a.	T
Heteroptera	A	A	n.a.	T
Megaloptera	T	A	T	T
Coleoptera				
Dytiscidae	(A)	A	T	T
Gyrinidae	A	A	T	T
Haliplidae	V	A	T	T
Hydraenidae	?	T	T	A
Elmidae	A	A	T	A
Scirtidae	T	A	T	T
Trichoptera	A	A	A	T
Diptera				
Tipulidae	V	V	V	T
Limoniidae	(A)	V	V	T
Ceratopogonidae	V	(A)	(A)	T
Chironomidae	(A)	(A)	(A)	T
Dixidae	A	H	H	T
Thaumaleidae	A	H	H	T
Simuliidae	A	A	A	T
Ptychopteridae	A	A	A	T
Psychodidae	(A)	(A)	(A)	T
Tabanidae	(A)	(A)	(A)	T
Rhagionidae	(A)	(A)	(A)	T
Stratiomyidae	(A)	(A)	(A)	T
Empid.: Hemerodromiinae	A	A	A	T
Dolichopodidae	?	?	?	T
Ephydridae	(A)	(A)	(A)	T

either hold their wings widely spread (Anisoptera) or erect over the abdomen (Zygoptera). Because of the strong obliqueness of the thorax this looks as if the wings are folded back over the abdomen, which, however, Odonata are unable to do. The modified larval labium functions as an extensible grab to rapidly seize distant prey. Respiratory and osmoregulatory functions are mainly performed by the rectum, which is actively ventilated, with three gill-like anal lamellae in zygopteran larvae (Wichard and Komnick, 1974a, 1974b).

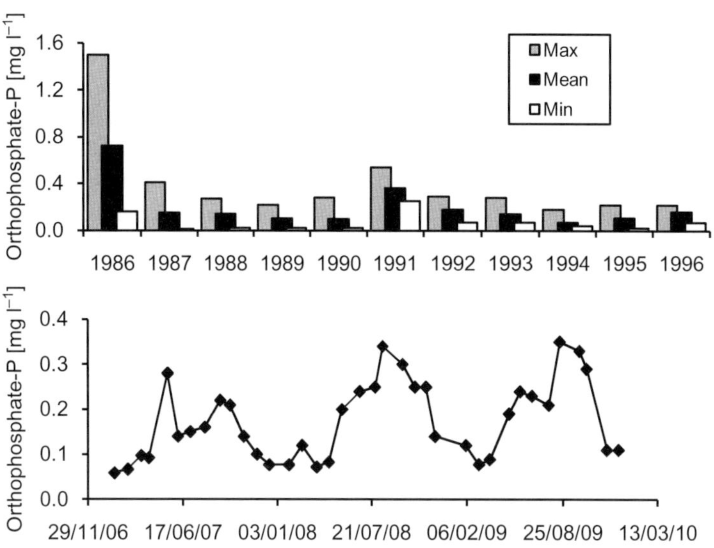

Figure 7.10 Orthophosphate-phosphorus (P) in the water of River Fulda. Top: annual maximal, mean, and minimal concentration at Fulda Kämmerzell (modified from Siebert, 1998). Bottom: seasonal variation of concentration at Schlitz-Pfordt; data points connected by auxiliary line. Courtesy of Hessisches Landesamt für Umwelt und Geologie.

The most readily observed representatives of the order by the Breitenbach are the metallic blue males of *Calopteryx splendens* (Harris, 1782). However, they are alien to the stream, and one of the most spectacular examples of potential faunal exchange between the Breitenbach and other running waters. Larvae abound in the River Fulda and, especially in June, large swarms of adults are seen. Males usually disperse widely. They migrate up the Breitenbach valley and perch on vegetation overhanging the stream. However, *Calopteryx* never oviposits in the Breitenbach and there are no larvae. A similar situation occurs in *Gomphus pulchellus* Selys, 1840, which also inhabits the River Fulda since its water quality has improved (Figure 7.10). In summer, occasional specimens are seen basking in the sun on the bare soil of mole hills along the Breitenbach (P. Zwick, unpublished data).

Cordulegaster boltoni (Donovan, 1807), a large dragonfly with black body striped in bright yellow, is the largest insect in the Breitenbach, but is not often observed. Ovipositing females were occasionally seen but no larvae were taken in benthic samples. This is partly due to the rarity of the animal, partly to its hidden life. Cool lowland and hill streams are the typical habitat (Donath, 1989), where larvae hide in stream-bed sediments, leaving just the eyes exposed. Prey is located visually and specialized sensory setae on the dragonfly's head also perceive fine-scale current patterns caused by moving prey (Verschuren, 1987). Larval life lasts several years.

7.6.4.1.4 Heteroptera

The Heteroptera, or true bugs, are advanced hemimetabolous insects with a fairly ordered ontogeny leading to adulthood via a fixed number (usally five) of larval instars. The characteristic transformation of the originally biting insect mouthparts into a flexible, plurisegmented sucking rostrum is one of the diagnostic apomorphies of the order. Others are, for example, the hardening of only the basal part of the front wings to form hemielytra protecting the membraneous hind wings, and the reduced number of antennal segments (usually five). The order includes terrestrial and two, very different, aquatic superfamilies.

The Nepomorpha are distinguished by short antennae held under the head, and resting on an air film underneath the body. They perceive shifts in the air film caused by gravity when the animal changes position. The air film is replenished during brief surfacing of active taxa, or via caudal respiratory siphons in ambush predators. Most Nepomorpha inhabit standing waters and are common in ponds and pools in the Breitenbach valley. Water scorpions and needle bugs (*Nepa, Ranatra*) are too sedentary to be washed into the stream from the pools. In contrast, the agile Corixidae, Notonectidae and Naucoridae quickly escape in flight.

The normally flightless (Hoffmann, 2008), plastron-breathing, ground bug *Aphelocheirus aestivalis* (Fabricius, 1794), and the equally flightless *Plea minutissima* Leach, 1817 occur in the River Fulda, where in the past few years the former has become locally common (Siebert, 1998; P. Zwick, unpublished data). Neither was found in the Breitenbach.

The second aquatic bug superfamily, Gerromorpha, unites specialized dwellers on the surface film of waters, stagnant little pools, large lakes, rapidly flowing streams, and the open ocean alike (Møller-Andersen, 1982). Only a very small selection was observed, but not specifically studied, at the Breitenbach. The slow-moving Hydrometridae walk on quiet surfaces among emergent vegetation, while the Veliidae race on riffles, or rest on solid supports near the shoreline when hunting is interrupted. *Velia* feeds on drifting prey, probably both emerging and ovipositing aquatic insects, as well as terrestrial taxa that fall onto the water (Erlandsson *et al.*, 1988).

7.6.4.1.5 Megaloptera

The holometabolous order is poorly represented in Europe where only few species of *Sialis* (Sialidae) occur. *Sialis fuliginosa*, a typical inhabitant of cool streams with dense emergent vegetation, appeared in emergence traps, and occasional larvae were found in the benthos of the Breitenbach (Zwick, unpublished data). In the upper section of the nearby Rohrwiesenbach adults emerged from April to June, most in May (Sandrock, 1978). Some of the adults resting on vegetation, or flying near the Breitenbach in May and June, were *S. lutaria* (L.), common in ponds in the valley and in soft sediments of the River Fulda (Zwick, unpublished data). Between April and July, Sandrock (1978) trapped considerable numbers in the more sluggish Rohrwiesenbach.

The present general account of their biology is mainly based on Hölzel (2002). The dark adults do not feed. They are about 1–2 cm long, with thick heads and long filiform antennae and are seen, mainly in May and June, flying near water bodies. At rest they hold their dark, coarsely veined, wings in a roof-like position over the abdomen. Sexual attraction over longer distances is by pheromones, at short distances sexes communicate by vibrational signals (Rupprecht, 1975). Eggs are closely packed, one by one in flat batches of up to 1000 eggs. Batches are found on the underside of foliage, branches or bridges, always directly above water. The two local species differ in egg incubation, mean hatch and optimal hatching conditions. Hatching success occurs at lower temperatures in the stream-dwelling *S. fuliginosa* than in the still-water species (Elliott, 1995).

Hatching larvae fall into water, initially drifting or swimming freely before disappearing in bottom sediments, leading a nocturnal life. Larvae are unmistakable with their paired filamentous abdominal gills and a single long terminal filament. They catch prey with large, strongly toothed mandibles. When sampled, even large larvae are often trapped in the surface film. They progress through approximately 10 instars over 1–3 years and then move onto land to build a pupal chamber in shallow soil.

7.6.4.1.6 **Neuroptera**

The graceful adults (more than 2 cm long) of the lacewing, *Osmylus fulvicephalus*, were regularly seen fluttering between emergent vegetation in spring and early summer, and are the only representative of the order in the Breitenbach. They are mainly carnivorous and feed on other insects (Stitz, 1931; Hölzel and Weißmair, 2002). Large numbers may gather in culverts, performing complex courtship and mating behavior. Eggs are laid singly or in small groups on vegetation next to water. Larvae have spear-like, gently outwardly curved, mouthparts and multi-segmented antennae of similar length, unusual for the Holometabola. The body is soft with only small areas of hard cuticle bearing stiff, erect setae. Two retractile tubes at the tip of the abdomen with strong recurved hooks are used to anchor the larvae on land while they hold their foreparts underwater to hunt aquatic prey. Prey is approached very slowly and punctured with the mouthparts, formed of the mandibles and maxillae, with a channel between them. Initially strong poison, and then digestive enzymes are injected into the prey, which is completely digested pre-orally; the midgut is closed. The anus functions only as an outlet for silk produced by the Malpighian tubules when the third instar larva spins a loose cocoon on land, the following spring.

Although all its life stages are terrestrial we include *Osmylus* among the aquatic insects because its larvae depend on submerged prey. They were not abundant in traps over the Breitenbach, but considerable numbers emerged under a trap on the nearby Rohrwiesenbach, mainly in June (Sandrock, 1978). Their impact on the limnic fauna remains to be established.

7.6.4.1.7 **Hymenoptera**

The holometabolous Hymenoptera are an insect order that is generally associated with dryness and sun, not water. Yet, parasitic wasps exploit the limnic fauna, although their larvae and pupae are never exposed to water. Emerging adults also move immediately into the air and to land, only females return to water to oviposit. The Breitenbach aquatic Hymenoptera are the least known insects in the stream, very few have ever been caught.

Several families are involved, many parasitize insect eggs and are accordingly very small, often only about 0.3 mm long. The wings of these minute wasps lack veins but have marginal hair fringes. Underwater wings are used to swim. The largest aquatic wasp, the approximately 5 mm long *Agriotypus armatus* (Curtis, 1832) does occur in the Schlitz area, but not in the Breitenbach. Emerging adults are recognized by their large thoracal spine (hence the name *armatus*), and larvae concealed inside the cases of Goeridae and other Trichoptera are easily recognized by the floating silk band that serves as a gill (Messner, 1965; Bürgis, 1993).

7.6.4.2 Ephemeroptera

In recent years it became fashionable to refer to the Ephemeroptera, Plecoptera and Trichoptera as EPT. Sometimes, the Coleoptera are also included, expanding the abbreviation to EPTC. This is fine where it is simply used as a handy short term, but sometimes, even distribution, seasonal change, diversity, overall abundance of "EPT" is presented, as if the included orders represented an ecological unit. In fact, differences outweigh similarities. Mayflies and stoneflies share a hemimetabolous mode of development, they are similar in this particular respect. The shared habitat also selects for adaptation to similar conditions, but each of the included orders has developed its own characteristic traits. In fact, the EPT orders are mainly those aquatic insects that are most readily routinely identified, larvae in the benthos as well as adults. However, the ecological requirements and properties of these orders diverge widely. Therefore, they are here presented separately, each in its own right.

7.6.4.2.1 General and Species List

Diagnostic Structures and Position in the Phylogenetic System

Ephemeroptera are easily recognized. At rest, mayflies hold their usually richly veined wings erect over the abdomen and cannot fold them back. The contour of each wing is roughly triangular. Hind wings are strikingly smaller than the front wings, or are even missing. Mayflies have two long and delicate cerci, and usually an additional median terminal filament, resulting in three tails. However, the median filament is absent from the males of several families. In a few species and genera it is even absent in the larvae.

Larvae have conspicuous, often foliate lateral gills on up to seven abdominal segments, facetted eyes, and compared to the adults, mostly long, multi-segmented antennae. Two long cerci are alway present, most also have a long terminal filament. In doubtful cases, they can be recognized by one-segmented tarsi with a single claw.

Ephemeroptera share direct development from a primary larva with compound eyes to the winged adult with several other hemimetabolous orders. However, mayflies are the most ancient winged insects and the sister group to all others. They are the only insects with a sexually immature winged stage, the subimago, which must molt to attain adulthood and sexual maturity. A large number of families is recognized but their phylogenetic relations are the object of much controversy; details cannot be presented here.

Subimaginal and imaginal mayflies do not feed, their mouthparts are nonfunctional although drinking water has been documented (Takemon, 1993). During their adult life mayflies are fueled solely by materials ingested during the larval stage. Adult life is accordingly very short, the Breitenbach species seem to live for 1–3 days, and most is spent in flight, walking is of little importance. Some species even shed their legs at the molt to adulthood, or retain nonfunctional rudiments, apart from the male forelegs, which hold the female during aerial copulation. *Oligoneuriella rhenana* (Imhoff, 1852) in the River Fulda is an example of this, but no Breitenbach species exhibits leg abscission. Another distinctive, primitive trait of the Ephemeroptera is the possession of paired separate penes and female genital openings. This becomes obvious in females carrying paired egg masses.

Larval Functional Feeding Groups

While the short-lived adults are rather uniform their larvae are structurally very diverse and belong to a wealth of different life forms, for example different functional feeding groups. The Central European mayflies include grazers and scrapers feeding on periphyton, such as *Baetis vernus* and *B. rhodani*. On average, these two species together account for 95% of all mayfly specimens in the Breitenbach. The Heptageniidae are also grazers. Gatherers and collectors consuming detritus are numerous, several species in the Breitenbach fall in this category, or are mixed gatherer–grazers like the Ephemerellidae (Willoughby and Mappin, 1988; McShaffrey and McCafferty, 1990, 1991; Riaño, Basaguren and Pozo, 1997). In the Breitenbach the Leptophlebiidae and *Centroptilum luteolum* are detritivores (Bauernfeind and Humpesch, 2001). Carnivores are exceptional and there are none in the Breitenbach. Shredders are not known among mayflies.

Filtering mayflies collect material from flowing water by holding setal brushes on their legs against the current, like *Oligoneuriella* (Elpers and Tomka, 1992), or build tunnels in which the beat of paired abdominal gills creates a current from which material is filtered, for example, *Ephemera* (Ladle and Radke, 1990). Both genera are common in the River Fulda, but only the latter occurs sporadically in the Breitenbach. Perhaps particle drift in the Breitenbach is insufficient to feed these large insects. Also, *Ephemera* may be unable to build its tunnels in the normally loose, floating sand of the Breitenbach, but is common in the similarly small Rohrwiesenbach near Schlitz, which has a loamy bottom.

Species List of the Breitenbach

Mayflies are generally not particularly well represented in cool, spring-fed runs, like the Breitenbach (Brittain, 1982, 1990). Jacob (1986) provided an overview of material collected in 1984. Today, the faunal list of the Breitenbach (see Appendix) includes 18 species, about 13% of the German list. Only *Ephemerella mucronata* is almost restricted to epirhithral streams like the Breitenbach (Bauernfeind and Humpesch, 2001). The other species have much less strict habitat preferences, especially *C. luteolum*. In the Breitenbach, only few species are common all the time. *Centroptilum luteolum* and *Paraleptophlebia submarginata* are abundant only at some sites in some years.

Large populations of the Breitenbach species also occur in the River Fulda (Table 7.5). Given the fact that both mate-finding and oviposition may involve long distance flights, some exchange of mating partners, and especially of ovipositing females, between the neighboring populations is inevitable, although it cannot be documented in detail. The fauna of the River Fulda was impoverished by pollution in the 1960s and 1970s (Zwick, 1969; Marten, 1983) but has since recovered (Siebert, 1998), especially after tertiary sewage treatment was introduced (Figure 7.10). Now, the River Fulda is again a viable pathway for downstream drift from its montaineous headwaters, and even a habitat for selected species.

In contrast, the Breitenbach was not impacted by anthropogenic pollution. Nevertheless until the early 1980s, at best, single individuals of *Ecdyonurus subalpinus, Epeorus assimilis,* and *Centroptilum luteolum* were caught. Now they are regularly found and are sometimes abundant in the middle and lower sections of the Breitenbach. Neither *Epeorus* nor *Ecdyonurus* are found in other small streams around Schlitz, for example, the nearby Rohrwiesenbach (Gümbel, 1976; Sandrock, 1978) which drains into the River Schlitz.

7.6.4.2.2 **Mayfly Larvae in the Benthos**

Fundamental Differences between the EPT Orders

Development of Ephemeroptera and Plecoptera on the one hand and of Trichoptera on the other hand offer very different opportunities for life history analyses which are related to the different numbers of instars through which larvae grow before metamorphosis or pupation.

Mayflies and stoneflies grow through 10 or more larval instars that differ in linear measures of cuticular structures, for example head or pronotum width, length of individual abdominal tergites, and ultimately in total body length. Sexual size differences are marked during at least the second half of the development period. Except during early larval life, size distributions within given instars are therefore bimodal. In consequence, sizes of instars overlap and cannot be used to identify instar numbers. The advantage is that size of cuticular structures is proportional to mean body mass of specimens. Size can therefore be used for a detailed analysis of stepwise biomass gain.

Many mayflies seem to grow through particularly large numbers of instars but no detailed analysis is available. Difficulties are caused by allometric growth which occurs during late instars when the strong flight muscles develop and sexual differences in eye- and head-structure form (Beer-Stiller & Zwick, 1995).

These problems do not occur in stoneflies which are otherwise similar and accordingly much better known. The large *Dictyogenus fontium* (Plecoptera: Perlodidae; data underlying Zwick and Zwick, 2010) provides a typical example. Each of the 16 larval instars of *Dictyogenus* experiences an intrastadial gain of about 40% dry weight. The final instar is about 1.7 times heavier than a mature penultimate

Table 7.5 The mayfly faunas of the Breitenbach and River Fulda, in the section into which the Breitenbach drains. Black: common (and at all sites in the Breitenbach). Gray: moderately common in the River Fulda (occasionally or locally common in the Breitenbach). u = common at upstream sites; x = sporadic occurrence.

	Fulda (Hartershausen – Hutzdorf; Siebert, 1998)	**Breitenbach (present study)**
Siphonoperla torrentium		
Baetis fuscatus		x
B. muticus	x	x
B. rhodani		
B. scambus	x	x
B. vardarensis		
B. vernus		
Centroptilum luteolum		
Cloeon dipterum	x	X
Oligoneuriella rhenana		
Rhithrogena semicolorata	x	x (sp.)
Ecdyonurus dispar		
E. macani		x
E. subalpinus		x
E. torrentis	x	x
E. venosus group	x	
Electrogena affinis	x	
Epeorus assimilis	u	
Heptagenia flava	x	
H. sulphurea		
Ephemerella (S.) ignita		
E. mucronata	x	
E. (T.) major		
Caenis luctuosa		
Paraleptophlebia submarginata		
Habrophlebia fusca	x	x
H. lauta		x
Habroleptoides confusa	x	
Ephemera danica		x
Potamanthus luteus		

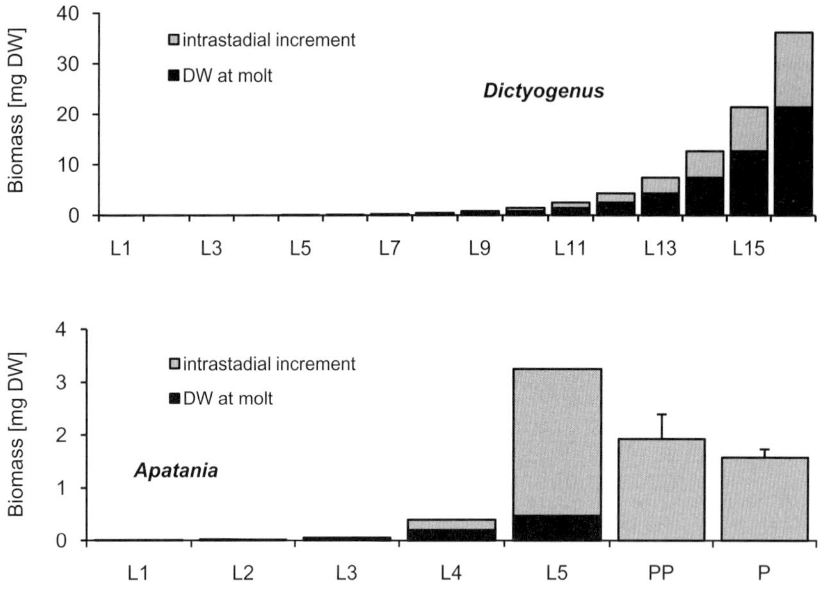

Figure 7.11 Diagram of growth and biomass gain of the large stonefly *Dictyogenus fontium* (top; Zwick and Zwick, 2010 and unpublished data) and of the caddisfly *Apatania fimbriata* (bottom; data from Aurich, 1992). Larval instars (L) on the abscissa are numbered. PP and P are prepupae and pupae, respectively.

instar (Figure 7.11). The relation between body length and dry weight is linear on logarithmic scales. Sexes differ much in linear measures of cuticular parts and in biomass; males possibly grow through one instar less than females. When the final larval instar is attained, adult emergence follows shortly, usually within 2–4 weeks. The larva turns directly into the winged adult because metamorphosis is prepared stepwise, during several instars that can externally be distinguished by their wing pads. Only cuticular biomass is lost which is a minimal amount (Zwick, 1982a).

In contrast most of the holometabolous Trichoptera pass through only five larval instars which are often easily recognized by eye. Sizes of cuticular measures do normally not overlap between instars (Table 7.23) and do not differ prominently between sexes. Even in the last instar the size distribution remains unimodal (Figure 7.12) although sexual differences in biomass are important (e.g., Wagner, 1990a, 2002a).

Aurich (1992) documented the intrastadial weight gain of *Apatania fimbriata* (Figure 7.11), his data agree with summary information on *Chaetopteryx villosa* (Wagner, 1990a). During each early instar weight increases up to threefold. However, during the fifth instar weight increases by a factor of 7–9. A mature last instar *Apatania* larva is about eight times heavier than a larva just molted to last instar, and even 10 times heavier than an average fourth instar larva. This weight

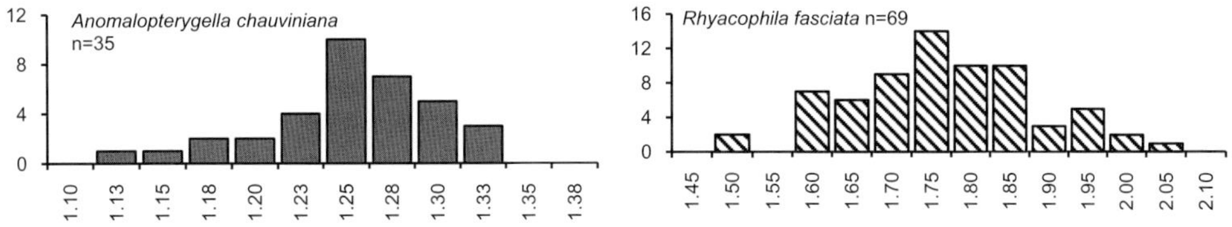

Figure 7.12 Unimodal distribution of head capsule size in the last instar of *Anomalopterygella chauviniana* (head width) and *Rhyacophila dorsalis* (head length).

gain takes often long, several months may elapse between entry into the last instar and adult emergence. Added biomass is accommodated mainly in the poorly sclerotized larval abdomen which is concealed in a case.

Observation of tube case structure suggests successive lengthening of the case by the growing larva may be related to mass gain. This would agree with data on saddle cases of *Agapetus* (Becker, 2005). When maximum weight is attained metamorphosis is initiated. This process consumes considerable biomass, resulting in weight losses during the prepupal and pupal phases.

Ephemeroptera Life Histories

Not all of the mayfly species recorded in emergence traps on the Breitenbach were taken in benthos samples. Species preferring habitats with slow flowing water and sandy or muddy bottom were not taken at all (*Centroptilum luteolum*), or in small numbers permitting no analysis of development (Leptophlebiidae). The latter is also true of Heptageniidae which were still much rarer in the stream during the benthic studies in 1986–1988 than in later years. Larval growth and seasonal change of population structure is described in graphs showing percentage abundance of half millimeter classes of body length (front of head to tip of abdomen, excluding appendages) in samples of the respective species.

Sometimes it is informative to convert size distributions into biomass. Equations to calculate biomass of the taxa in the Breitenbach from equations offered in the literature for closely similar taxa were checked for plausibility of results. Using the Plecoptera as an example, several different formulas offered for German Nemouridae (which are too similar to be identified by shape!) yield strongly divergent results: the dry weight of an 8 mm long specimen (Meyer, 1989) would range from 0.46 mg (*Nemoura*) to 3.05 mg (*Protonemura* spp.), which cannot be. The cause of such errors lies in the limited number of readings over too small a size range from which regressions were calculated. When extrapolated to the natural size range of animals any accidental small error in slope leads to gross over- or underestimates. After careful comparison of literature data with our own supplementary data the regressions in Table 7.6 were adopted for mayflies and stoneflies.

The weight of each size class was multiplied with its percentage share in the sample. The sum of all size classes corresponds to the standing crop of a population of 100 larvae representative of the momentary size distribution.

Table 7.6 Regression parameters used in biomass estimates of larvae from the Breitenbach with regressions of the general form: ln DW = ln a + b lnBL; where BL is body length (in mm) and DW is dry weight (in mg). Values of intercept a are not logarithmically transformed.

Applied to:	*Siphonoperla*	*Isoperla*	*Leuctra* spp.	Nemouridae	*Ephemerella mucronata*	*Ephemerella ignita*	*Baetis* spp.
Source	*Chloroperla*, after Burgherr and Meyer, 1997	*Dictyogenus*, after Zwick and Zwick, 2010	Modified after Burgherr and Meyer, 1997, and Benke *et al.*, 1999	McNeely, Finlay and Power, 2007	*Ephemerella* sp. 2, after Benke *et al.*, 1999	*Serratella* sp. 1, after Benke *et al.*, 1999	Wallace and Gurtz, 1986, after Benke *et al.*, 1999
Intercept a	−5.800	−5.087	−5.620	−5.185	−4.358	−4.733	−4.893
±1 SE	±0.5	±0.0689	±0.32	±0.001	±0.0014	±0.0015	Not given
Slope b	2.670	2.710	2.665	2.762	2.444	2.584	2.423
±1 SE	±0.24	±0.0314	±0.21	±0.107	±0.266	±0.479	Not given
r^2	0.800	0.985	0.910	not given	0.810	0.760	0.940

Larval densities per square meter stream bottom in figure legends of size frequency diagrams are only intended to give an idea of sample sizes from which percentage distributions of body lengths were calculated. These density estimates are too crude to estimate absolute biomass in the field, or even to infer in-site production by comparison of data on successive sampling dates!

Ephemerella (E.) mucronata* and *E. (Serratella) ignita The two Ephemerellidae can be identified to species from the first instar on by the paired tergal processes of *E. ignita*. Emerging adults are temporally vicariant but both species oviposit in summer. However, because eggs of *E. ignita* pass through diapause and hatch only in spring, the larval populations are also almost completely separate timewise (Figure 7.13).

When data are converted as explained above and presented as biomass gain of a representative average population of 100 larvae over time, both Ephemerellidae are seen to experience maximum growth and biomass gain during the warm season (Figure 7.14), although not at the same time.

Baetis rhodani* and *B. vernus Of the dominant genus *Baetis* only *B. rhodani* and *B. vernus* were taken. Distinctive larval species traits, for example the marginal spines on gills of *B. rhodani,* first appear when larvae are about 2 mm long (Werneke, 1997). Young larvae can therefore not normally be assigned to species which is an important hindrance to *Baetis* life history analyses. However, the Cypermethrin accident in May 1986 had occurred before *B. vernus* emerged. No adult *B. vernus* were taken in emergence traps in 1986, and very few of them in 1987. Therefore, all small larvae collected in 1986 and almost all taken in 1987 must have been *B. rhodani* (Figure 7.15). Only in 1988 considerable numbers of *B. vernus* emerged again from the stream. Larvae taken early in 1988 must have included both species but were too small to be identified.

B. rhodani is frequently described as a bivoltine species with an early spring generation of large adults giving rise to a more numerous summer population of usually smaller adults. However, other types of life history involving plurivoltinism apparently also occur (Benech, 1972b; Clifford, 1982). Egg development of *B. rhodani* is direct and the incubation period short (Bohle, 1969). The presumed single brief phase of vernal oviposition cannot be reconciled with long phases of recruitment and adult emergence observed in the Breitenbach.

The number of overwintering larvae is at best disproportionally small when compared to the suggested daughter generation. Repeatedly, there was no spring emergence from the Breitenbach at all, but summer populations were nevertheless numerous, see below. Immigration of ovipositing females from elsewhere appears therefore probable but the situation seen in the benthos of the Breitenbach additionally suggests that *B. rhodani* is actually polyvoltine and continuously reproducing during much of the warm season, with sometimes a small number of late larvae continuing growth and emerging as above-average-sized adults after winter. *B. rhodani* exploits the Breitenbach as a supplementary summer feeding and breeding plot, same as the blackfly, *Simulium (Odagmia) ornatum* (Section 7.6.4.6).

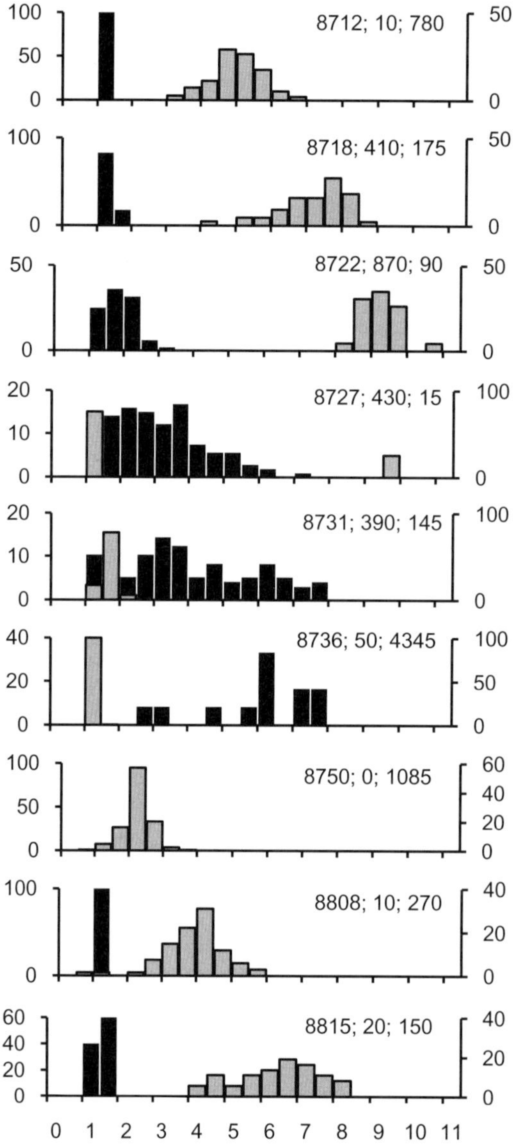

Figure 7.13 *Ephemerella ignita* (dark bars, left ordinate) and *E. mucronata* (light bars, right ordinate) in the Breitenbach. Frequency (percentages in samples; ordinate) of body length classes (abscissa; in mm). Figures in legends are calendar weeks in 1986–1988 followed by approximate specimen numbers m^{-2} streambed. Data are for Reach B for *E. mucronata*, while those for *E. ignita* were pooled from all sampling sites.

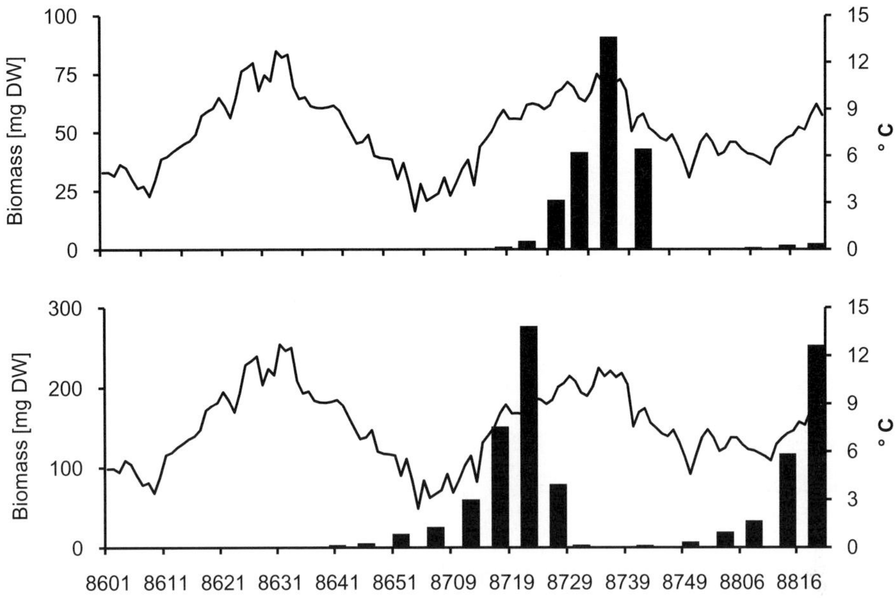

Figure 7.14 *Ephemerella ignita* (top) and *E. mucronata* (bottom), standing crop of average cohorts of 100 larvae at different times of the year, plotted against mean weekly water temperature at Measuring Station 2. Units on abscissa are calendar weeks in 1986–1988.

B. vernus was almost missed during the present benthos study. Therefore, the data of U. Werneke (1997) for *B. vernus* are shown in Figure 7.16, for comparison. Egg dormancy guarantees the larvulae hatch only after winter (Bohle, 1969; Schmidt, 1984). The recruitment period was relatively short, but adult emergence nevertheless continued for several months. However, no larvae survived during winter.

The most common mayflies in the Breitenbach strongly support the view that Ephemeroptera generally have a higher thermal demand than the Plecoptera which is also reflected by patterns of global biodiversity of the two hemimetabolous orders (Brittain, 1990).

Because adult emergence was extended and occurred over a wide size spectrum, and because in *B. rhodani* larval recruitment also continued for a long time, conversion of size distributions to biomass values (as done for *Ephemerella*, Figure 7.14) gives no precise picture of periods of maximal growth during the warm season.

7.6.4.2.3 **Mayfly Emergence**

Identification

Mature larvae of most Breitenbach species swim to the water surface, where the subimago quickly emerges from the larval skin. Only poor swimmers like the

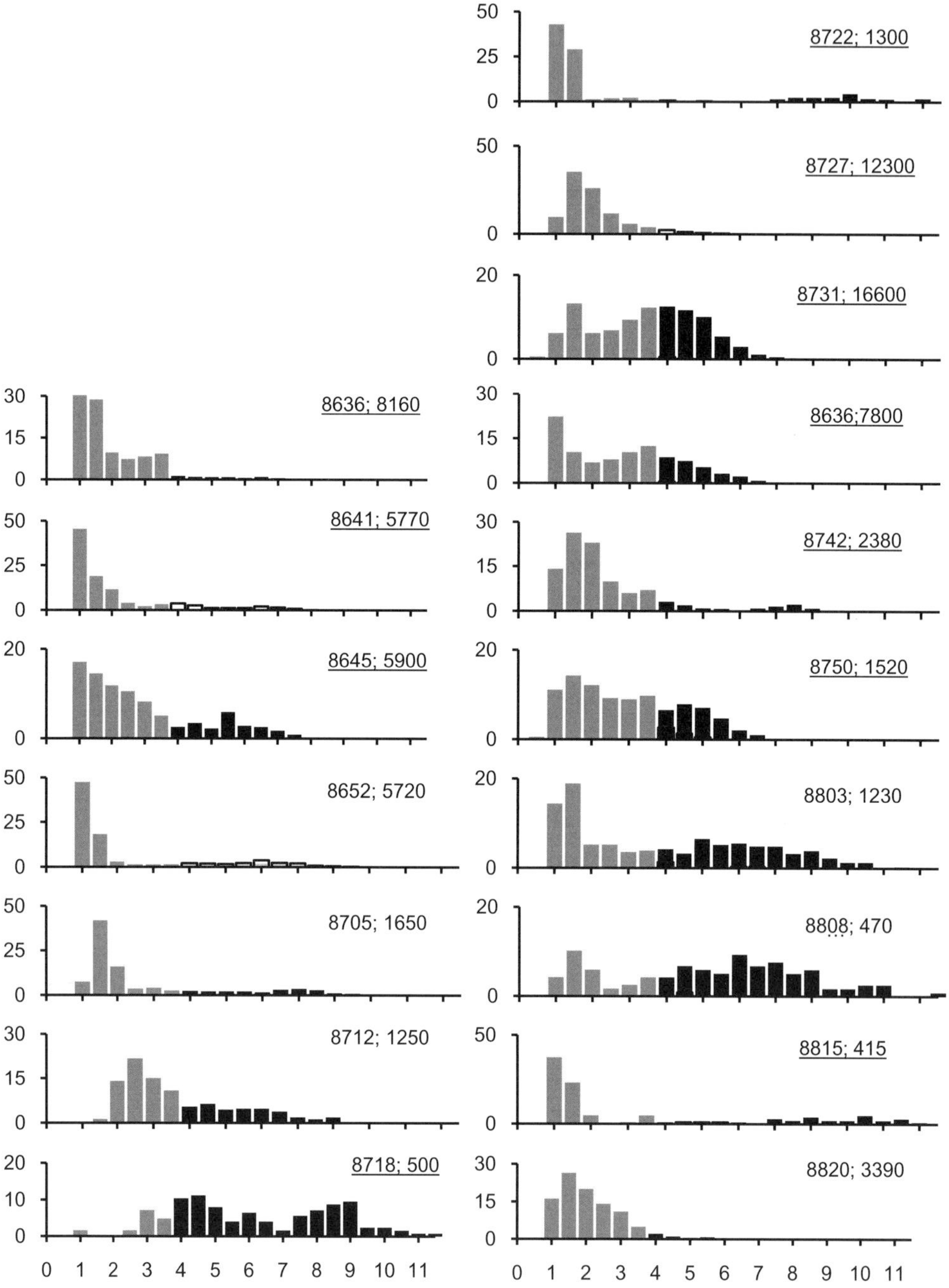
8722; 1300
8727; 12300
8731; 16600
8636; 8160
8636;7800
8641; 5770
8742; 2380
8645; 5900
8750; 1520
8652; 5720
8803; 1230
8705; 1650
8808; 470
8712; 1250
8815; 415
8718; 500
8820; 3390
0 1 2 3 4 5 6 7 8 9 10 11

Figure 7.15 *Baetis* spp., growth and development of larval populations in Reach B of in the Breitenbach. Frequency (% in samples; ordinate) of body length classes (abscissa; in mm). Dark columns represent *B. rhodani*, light ones indicate small larvae that cannot be identified to species. However, in 1986 and 1987, they must also have been *B. rhodani* (see text), while small larvae early in 1988 were probably largely of *B. vernus*. White bars show the few identified *B. vernus*. Figures in legends are calendar weeks in 1986–1988 followed by approximate specimen numbers m^{-2} streambed. Underlining indicates adult emergence in T2.

Heptageniidae require emergence substrata. Subimagines flutter away immediately and rest at protected sites near the stream for up to 2–3 days, until the molt to the very short-lived adult occurs. Mayfly emergence periods are therefore practically identical with flight periods.

Emergence traps collect subimagines whose specific identification is problematic (Jacob, 1986). The same is also true of most adult females, keys do not cover them, or at best cursorily (Kimmins, 1954; Studemann *et al.*, 1992; Bauernfeind and Humpesch, 2001). Subimaginal wings have a seam of fine microtrichia and are translucent rather than glass clear. In a few taxa they are distinctly patterned (Studemann *et al.*, 1992). Genera that are represented by only one species in the Breitenbach posed no problem. Similarly the two Breitenbach Ephemerellidae species have separate flight periods and females could be confidently assigned to the same species as simultaneously emerging males.

The two most common Breitenbach species, *Baetis vernus* and *B. rhodani,* are best separated by the pattern on the metatergum of *B. rhodani* (Kimmins, 1954; Figure 7.17). However, this pattern is often pale and poorly defined, and a similar very vague pattern is often seen in *B. vernus,* and in other additional species (Engblom, 1996). Therefore, significant numbers of *Baetis* were sometimes not identified to species, or the identifications were not beyond doubt, until G. Stüber discovered an additional distinctive character. *Baetis vernus* has two dark spots on sternite 9, as opposed to pale ones in *B. rhodani* subimagines. Identifications were validated by dissecting and mounting the concealed, barely pigmented internal penis sclerites of subimaginal males, which resemble those of adult males (Figure 7.17). The pointed sclerite of *B. rhodani* also distinguishes it from *B. fuscatus,* which is similar to *B. vernus.* Jacob (1986) described eye characters that distinguish male subimagines of *B. rhodani* and *B. vernus.*

Sex Ratios

Both sexes started to emerge at the same time and emergence was essentially synchronous, as would be expected for short-lived animals. A regular stepwise shift from excess males to excess females was rarely observed (Figure 7.20, top). For no apparent reason the sex-ratio oscillated during long emergence periods. To avoid accidental errors in sex ratio estimates, if only very few specimens emerged (<20, or <50 in the common taxa), sex ratios were not considered. Males outnumbered females in the annual catch of most species. An approximately balanced sex

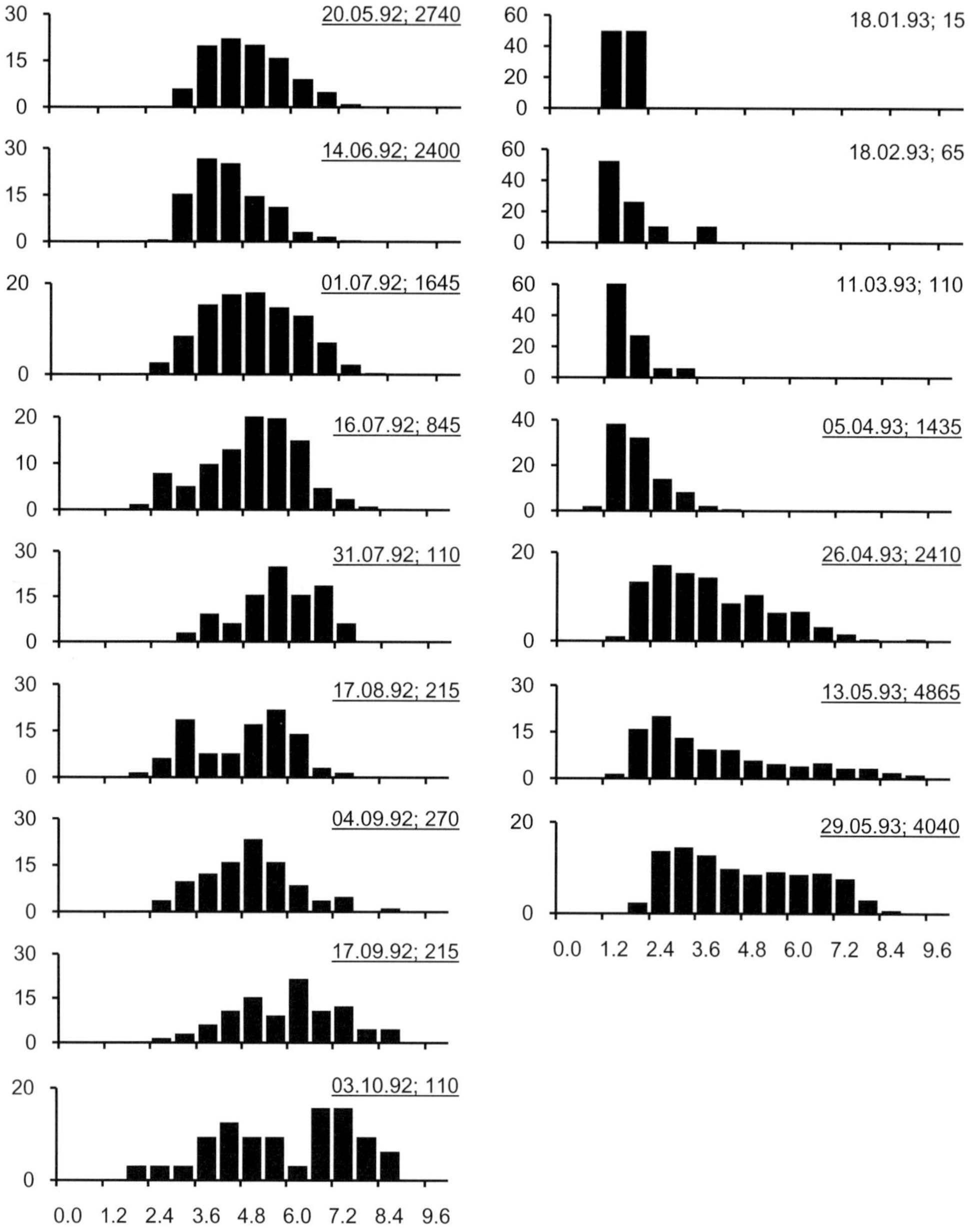

Figure 7.16 *Baetis vernus*, population structure in Reach B of the Breitenbach. Frequency (% in samples; ordinate) of body length classes (abscissa; in mm). Figures in legends are sampling dates followed by approximate number of specimens m^{-2}. Underlining indicates adult emergence in T2. After data of Werneke (1997).

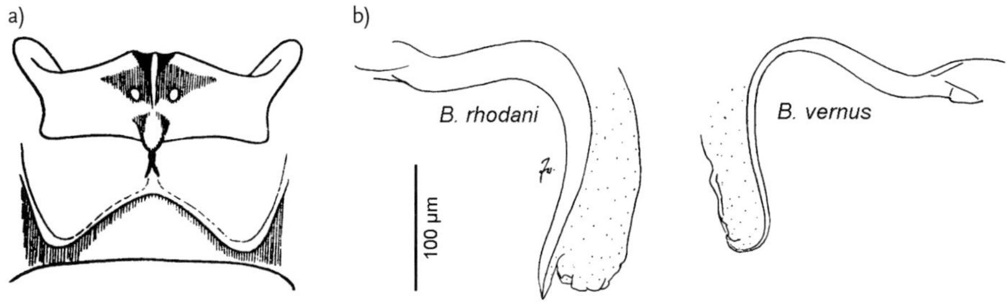

Figure 7.17 Subimaginal *Baetis* spp. Left: *B. rhodani*, pattern of metatergite (from Kimmins, 1954). Right: left and right side penis sclerites of *B. rhodani* and *B. vernus*, respectively. Original artwork.

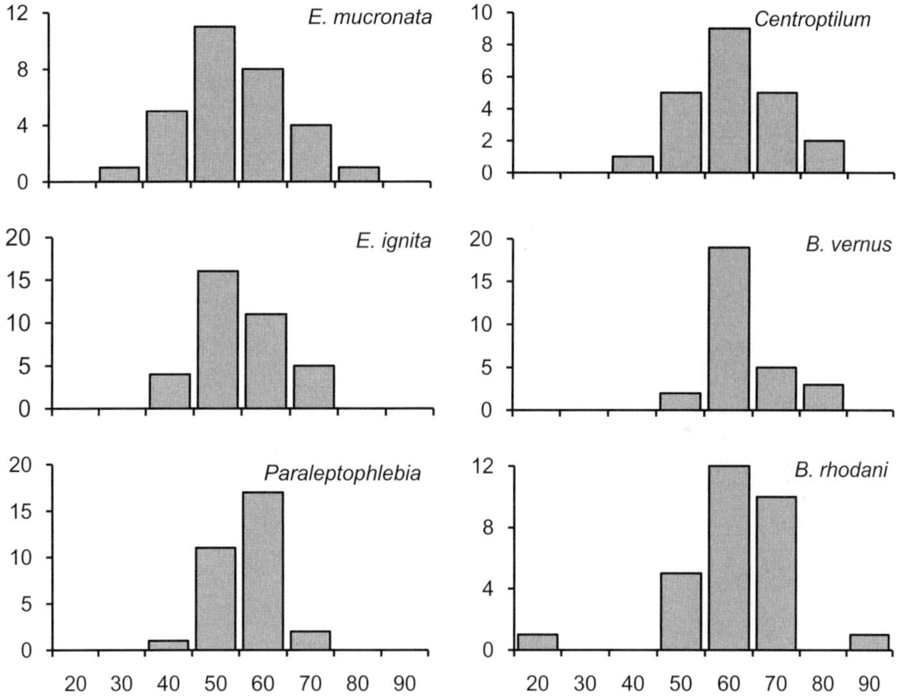

Figure 7.18 Frequency of males (%) in the annual catch of several mayflies from the Breitenbach. Minimum sample size was 100 specimens. Ordinate: frequency. Abscissa: percentage of males in annual catch.

ratio was usual only for the two *Ephemerella* species (Figure 7.18). A male bias appeared most clearly in *B. rhodani*.

Standard Emergence Patterns

There are several convenient descriptors of insect emergence. Length of emergence period is defined as the interval during which 5–95% of the cumulative emergence occurs. This avoids bias through individual early or late outliers. For a given species, emergence period is extended little by increasing specimen numbers. Peak emergence is an important landmark, not necessarily coinciding with median emergence, when 50% of all emerging specimens emerged.

Emergence patterns are characteristic species traits. Attempts to identify environmental cues triggering emergence of particular species are below discussed for some Plecoptera. Minor modifications, such as short term interruptions of emergence in response to spates, extreme weather, and anthropogenic disturbance or other singular events, affect taxa with similar modes of development similarly. Examples among mayflies and Plecoptera are presented. Long-term shifts of emergence timing in response to climatic change are addressed separately (Section 8.8).

With the exceptions below, periods and patterns of emergence of common Breitenbach mayflies varied little between years. As a rule patterns were also congruent between different traps along the stream, those in Trap T2 being representative of the entire study stretch. In contrast, abundance along the stream was variable, and several species had changing core areas of distribution along the stream (Section 8.3). Standard emergence patterns and periods were extrapolated from several decades of trap collections and are presented as bar charts (Figure 7.19). The patterns are usually skewed, numbers of specimens rising fast until peak emergence, but declining more slowly afterwards.

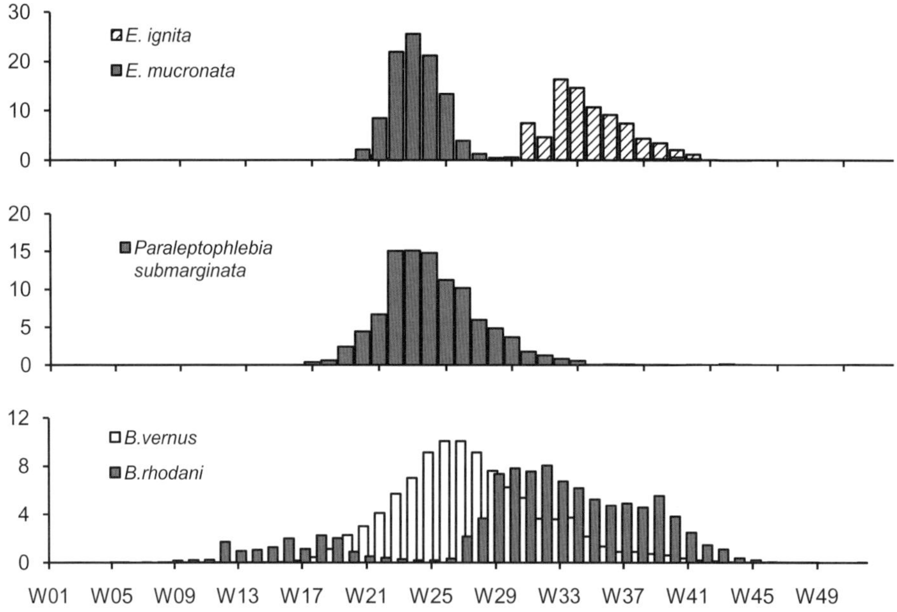

Figure 7.19 Standard emergence patterns and periods of the common Breitenbach mayflies based on collections in Trap 2, over 30 years. Bar charts show average weekly percentage of annual emergence.

Ephemerella mucronata, the only typically epirhithral mayfly in the Breitenbach, emerges over only eight weeks, which is the shortest emergence period of all. It overlaps completely with *Paraleptophlebia* which emerges over almost four months. Peak emergence of both is before the summer solstice, a little earlier for *Paraleptophlebia* than for *E. mucronata*. The two ephemerellid species are temporally vicariant (Figure 7.19), also during their larval stages in the benthos (Figure 7.13).

Centroptilum luteolum is bivoltine, one generation overwinters as the larva, the other develops in summer (Bauernfeind and Humpesch, 2001). Emergence patterns and times varied much (Figure 7.20). Distinctly bimodal emergence would be expected and did in fact occur, but emergence could also continue during most of the warm season. In other years, specimen numbers in the two generations differed so much that emergence was essentially unimodal, either in spring or autumn. Spring emergence was sometimes so insignificant that it was doubtful that the first cohort was the parent generation of the second cohort.

Baetis vernus emerged for several months, peaking in weeks 21–25, but in some years as late as week 29. In most years, *B. vernus* peak emergence occurred between the two generations of *B. rhodani*, but both species showed many exceptions from the standard patterns discussed below. Specimen numbers along the stream varied much, and the core area of distribution along the stream also varied between years.

Aberrant Emergence Patterns of *Baetis* spp.

Jacob (1986) noticed that *Baetis* emergence patterns from the Breitenbach only partly agreed with previous data. An additional, largely separate second cohort of *B. vernus* emerged in several years (1975, 1976, 1980, 1983, 1986), peaking around week 35 or later (Figure 7.21). This conflicts with the repeatedly recorded univoltine summer life cycle (Clifford, 1982). In the Breitenbach, *B. vernus* passes through an egg diapause ensuring the normal univoltine pattern (Bohle, 1969; Schmidt, 1984). The possible cause for the aberrant emergence patterns seen in *B. rhodani* may perhaps also apply, see below.

Based on the literature (summarized by Clifford, 1982) small numbers of large-bodied adults from overwintering larvae sometimes emerge early from central European populations of *B. rhodani*. They lay directly developing eggs from which a summer generation of usually much smaller adults arises, mostly in larger numbers. Summer emergence lasts long and may extend into autumn, but no separate autumnal cohort occurs. This standard pattern (Figure 7.19) was only rarely observed in the Breitenbach (e.g., 1974, 1987, 1994, 1995; Figure 7.22).

Breitenbach emergence patterns of *B. rhodani* are unusually varied. Sometimes, only some, not all of the traps along the stream, were involved, especially when the species was relatively rare (Figure 7.23). Emergence was sometimes distinctly trimodal (Figure 7.24) (1975–1978; 1985; also 1990, Traps 4 and 6 in Figure 7.23), a pattern conflicting with the development outlined above. Whenever this pattern occurred, the second and third peaks tended to start earlier at downstream than upstream sites (Figure 7.23). In contrast, from 1979 to 1984, there was significant unimodal summer emergence, without the preceding emergence of a hibernating generation (Figure 7.25).

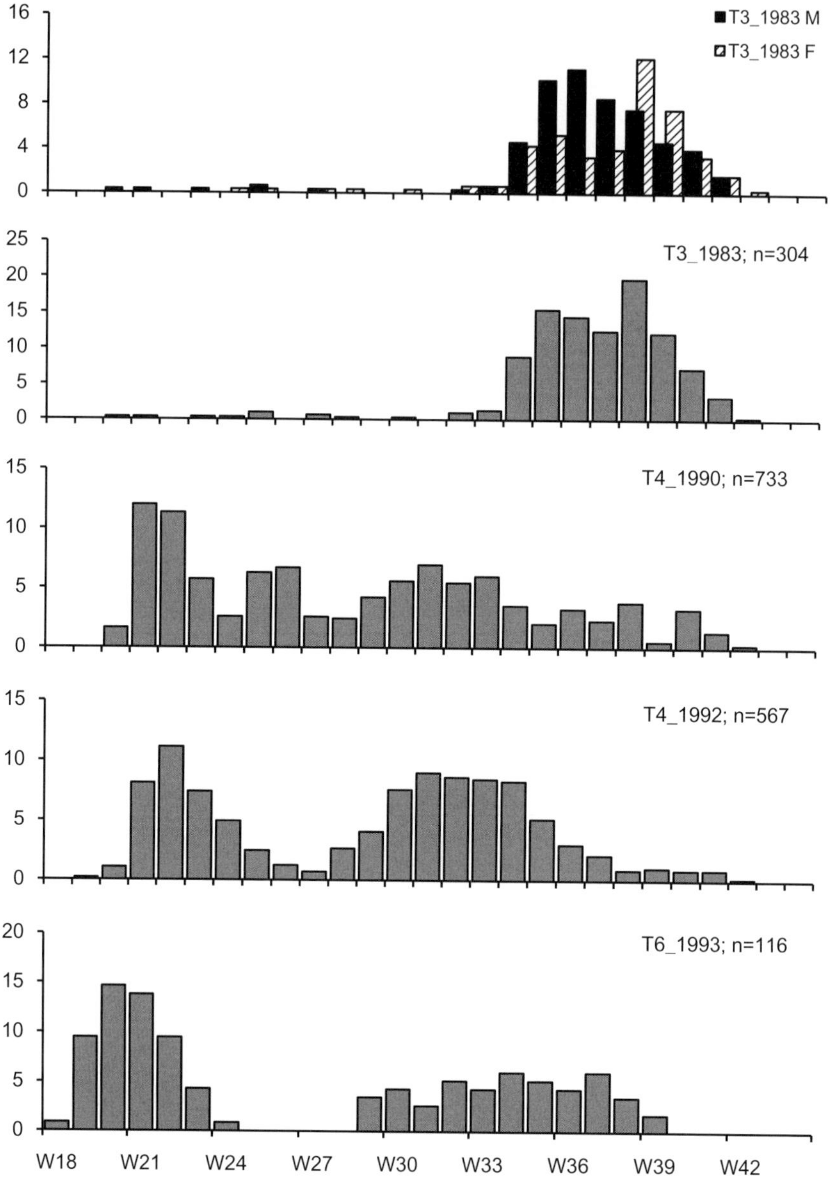

Figure 7.20 *Centroptilum luteolum*, examples of different emergence patterns from the Breitenbach at several sites and in different years. Ordinate: percentage of annual catch. Abscissa: calendar weeks.

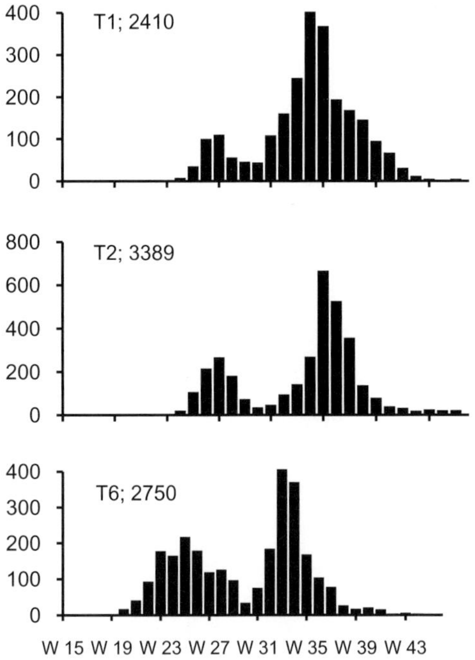

Figure 7.21 *Baetis vernus*, 1980, anomalous bimodal emergence patterns along the Breitenbach. Trap and specimen numbers indicated. Ordinate: numbers emerged per week. Abscissa: calendar weeks.

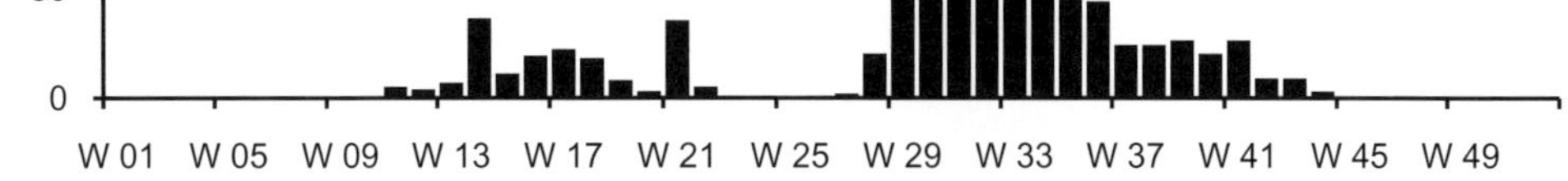

Figure 7.22 *Baetis rhodani*, emergence of 1158 specimens in Trap 6, 1995. Ordinate: numbers emerged per week. Abscissa: calendar weeks.

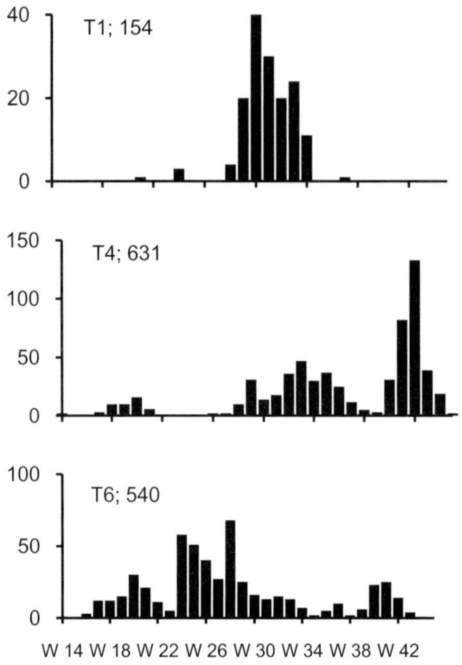

Figure 7.23 *Baetis rhodani*, 1990, different patterns of emergence along the Breitenbach. Trap and specimen numbers indicated. Ordinate: numbers emerged per week. Abscissa: calendar weeks.

Possible Explanations of Aberrant Patterns

Behavioral Constraints In the Breitenbach oviposition of all *Baetis* spp. occurs under water (Bauernfeind and Humpesch, 2001) on selected oviposition substrata. Females alight on stones or logs protruding from turbulent flow and crawl underwater to the lower face of the object, or up to 50 cm on the stream bed, in search for a suitable oviposition site (Bengtsson, 1988). Each female attaches a single batch of eggs. Batches resemble flat plates in which eggs are densely packed, usually in a single layer, about 56–60 eggs mm^{-2} (Werneke, 1994). *Baetis* oviposition is behaviorally constrained because availability of oviposition substrata acceptable to females depends on local bed structure and stream discharge. Peckarsky *et al.* (2000) observed the uncoupling of adult emergence and oviposition in the nearctic *Baetis bicaudatus*. Oviposition sometimes occurred in streams from which no adult had yet emerged, whereas streams producing many females might receive no oviposition if adequate oviposition sites were unavailable.

Female dispersal presumably also plays a role in the *Baetis* species in the Fulda-Breitenbach system where local recruits are not necessarily the offspring of resident females. Turbulent flow of the Breitenbach must be attractive for females coming from the slow-flowing calm section of the River Fulda into which the

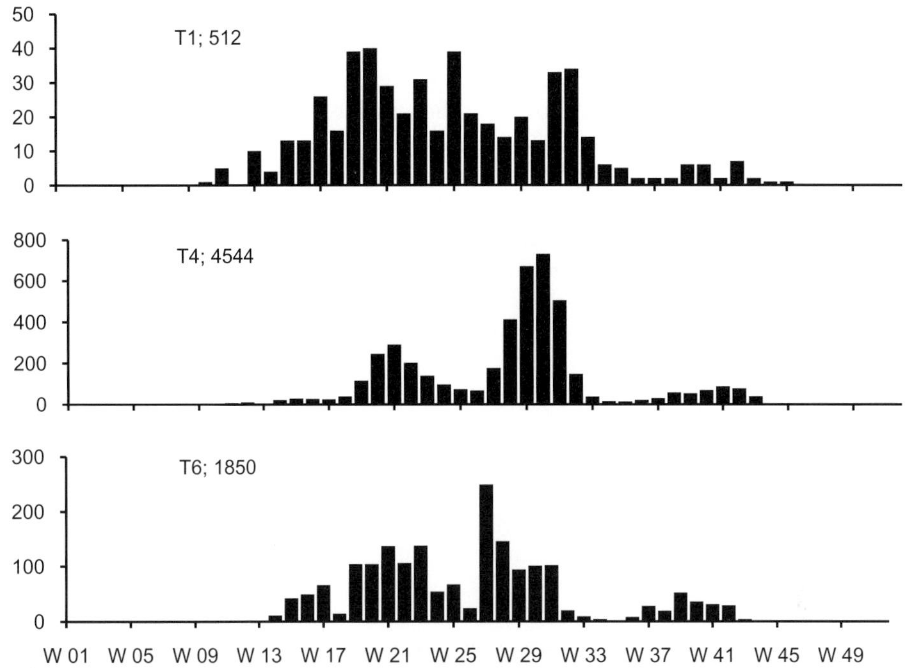

Figure 7.24 *Baetis rhodani*, 1985, variable emergence from the Breitenbach, including trimodal patterns. Ordinate: percentage of annual catch. Abscissa: calendar weeks.

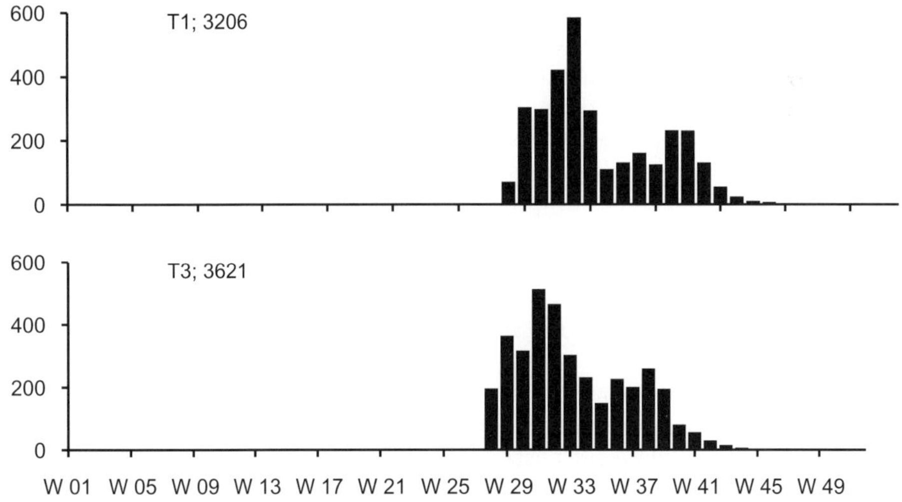

Figure 7.25 *Baetis rhodani*, 1982, examples of strong summer emergence despite absence of an overwintering parent generation. Ordinate: numbers emerged per week. Abscissa: calendar weeks.

Breitenbach drains. Timing of populations differs in the two water courses. For example, in 1992/93, the Fulda population of *B. rhodani* was about three weeks ahead of the Breitenbach population (Werneke, 1997, p. 87). Immigration of ovipositing females may therefore alter the structure of larval populations, and ultimately the adult emergence patterns in the adjacent Breitenbach.

The dependence on particular oviposition sites invited for experimental manipulation of their availability and permitted direction of females to controlled artificial sites. Eggs on these were counted and compared with numbers of adults of the preceding and following generations. It turned out that during the long larval life, 92–97% of the population were lost, and almost 99% of the survivors were again lost to the terrestrial environment during the few days of adult life (Werneke and Zwick, 1992).

Cohort Splitting There is some evidence that, in some years, larvae developing late in summer may not complete their development but instead continue to grow over winter and emerge as large individuals the next spring, together with offspring of their own, faster growing siblings (Bengtsson, 1973, 1988). The same was found in Breitenbach populations of the stonefly, *Nemurella pictetii* (Nesterovitch and Zwick, 2003).

Larval Drift *Baetis* larvae are strong active swimmers able to migrate upstream against the current, but also particularly liable to drift with the current when searching for rich feeding plots, or when disturbed. In early studies of *Baetis* drift, diel drift patterns were mainly assigned to specific behavioral traits and the effects of latitude (Müller, 1970). Whatever the ultimate cause of drift, it may affect *Baetis* spp. population structure and production estimates so strongly that models were developed to compensate for this in production estimates (Ciborowski, 1987).

After accidental insecticide poisoning in May 1986, at the Morddelle site (white bar in Figure 7.26; Zwick, 1992), large numbers of blackfly and *Baetis rhodani* larvae were observed drifting into the devastated study stretch from unaffected upstream sites. Specimens grew while the front of drifting larvae advanced downstream. Many eventually emerged under Trap 4. At Trap 6, some specimens had survived poisoning and emergence continued normally. The late summer generation at this site was probably a mix of indrifting specimens and offspring of local parents.

7.6.4.2.4 Mayfly Reproduction and Development

Mate Finding and Mating

Swarming of males to attract females is typical of mayflies. Swarms form at encounter sites which differ between species (Bauernfeind and Humpesch, 2001), as does the mode of swarming. Specific swarming is important for the reproductive isolation of related taxa (Savolainen *et al.*, 1993). In particularly short-lived adults, like the Nearctic *Dolania*, swarming and mate encounter lasts a few minutes, on two or three days, at some particular phase of moonlight and at some

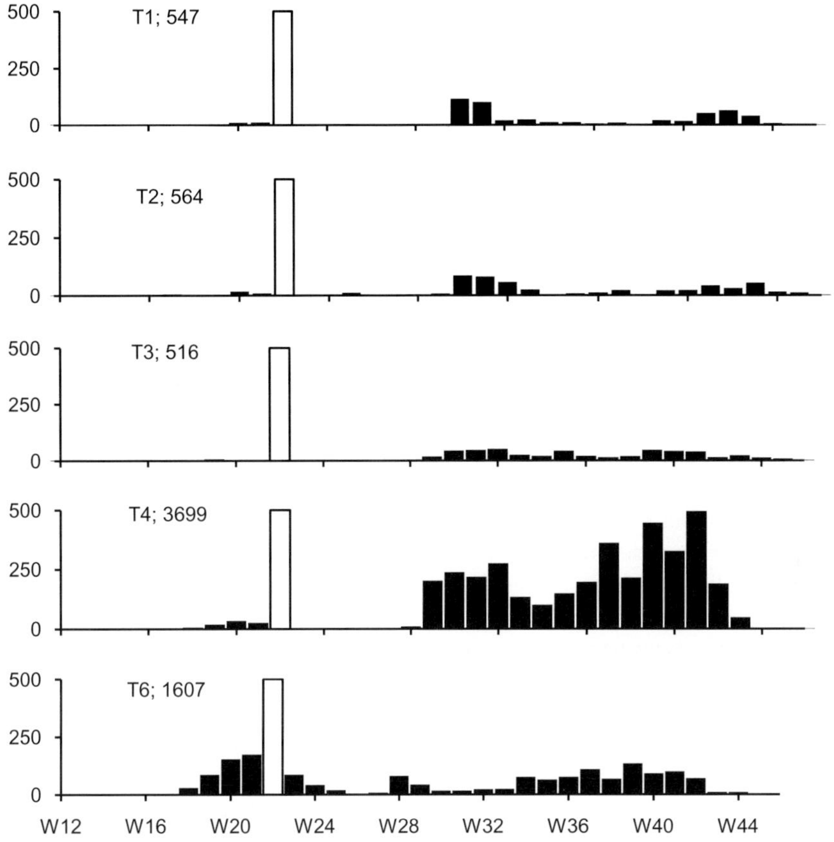

Figure 7.26 Emergence of *Baetis rhodani* from traps along the Breitenbach in 1986, after insecticide poisoning in May (white bar), upstream from the study area. Numbers in legends are specimens emerging. Ordinate: numbers emerged per week. Abscissa: calendar weeks.

particular temperature (Peters *et al.*, 1986). At the other extreme, for example, in the largest European mayfly, *Palingenia longicauda* (Olivier, 1791), spectacular swarms may form, which are widely known as "Theissblüte" (see, e.g., video by Preisig and Leumann, 2008). Ovigerous females of *Ephoron virgo* (Olivier, 1791) gathering at traffic lights may become a traffic hazard (Kureck, 1992; Butz and Schuwerk, 1993; Tobias, 1996).

Mayfly swarms along the Breitenbach are unspectacular; the species involved are easily overlooked, being small and not strikingly pigmented. They also typically swarm in small groups, often far from water (Studemann *et al.*, 1992, p. 24; Bauernfeind and Humpesch, 2001). Small swarms of *Ecdyonurus* spp. are sometimes seen along the upper River Fulda, with males rising and dropping abruptly by half a meter or more, but swarms of the Breitenbach population have escaped observation.

Swarming males spend all their energy reserves while females are more economical, flying only briefly (Sartori *et al.*, 1992). When entering swarms females are immediately mated in flight. Large male size enhances mating success (Harker, 1992). The male flies beneath the female, holding her by her wing bases with his long frontlegs. The male abdomen curves up until the genitalia meet. Mating is usually short and often completed within a few seconds, but may last up to 2 min (Illies, 1968; Brittain, 1982; Fischer, 1984). Males may mate several times before they die (Flecker, Allan and McClintock, 1988) while females mate once, oviposit, and die. Spent females do not rise again and (unlike some stoneflies) do not appear in emergence traps.

Early records of parthenogenetic development of mayfly eggs (Degrange, 1954) have since been confirmed in many species (e.g., Harker, 1997). Study of the temperature dependence of egg development is often performed on material obtained by dissection of the last instar nymphs (e.g., Humpesch, 1980). Indeed, some mayflies, especially species inhabiting intermittent water courses, reproduce mainly or entirely parthenogenetically (Jazdzewska, 1973; Sweeney and Vannote, 1984; Gillies and Knowles, 1990).

In view of the widespread potential for parthenogenetic development, release of eggs from large larvae mutilated during predator attacks was proposed as a possible explanation for the occasional appearance of individual larvae outside the normal recruitment periods of the respective species (Harker, 1997). However, in the case of individual out of season larvae of *E. ignita* which are sometimes observed in the Breitenbach a simpler explanation probably applies, see below.

Fecundity

Formation of mayfly gametes is generally complete in the last larval instar (Brittain and Sartori, 2003), egg mass being derived entirely from maternal biomass. Fertility ranges from around 100 in *Dolania*, to over 12 000 eggs in *Palingenia* (Brittain and Sartori, 2003). The relatively small Breitenbach species lay smaller numbers of eggs (Table 7.7).

Large female body size is a distinct fitness advantage (Sweeney and Vannote, 1981), decline of fecundity with female body size is well documented. At the

Table 7.7 Fertility of some Ephemeroptera species inhabiting the Breitenbach.

Species	Egg number	References
Baetis rhodani	200–4500	Benech, 1972; Werneke and Zwick, 1992
B. vernus	300–2000	Werneke, 1991
Centroptilum luteolum	<1500	Bauernfeind and Humpesch, 2001
Ephemerella (S.) ignita	150–1170	Degrange, 1960, Elliott, 1978
E. mucronata	1230–1670	Degrange, 1960
Paraleptophlebia submarginata	<1200	Bauernfeind and Humpesch, 2001
Habrophlebia fusca	<1200	Bauernfeind and Humpesch, 2001
H. lauta	<1200	Bauernfeind and Humpesch, 2001

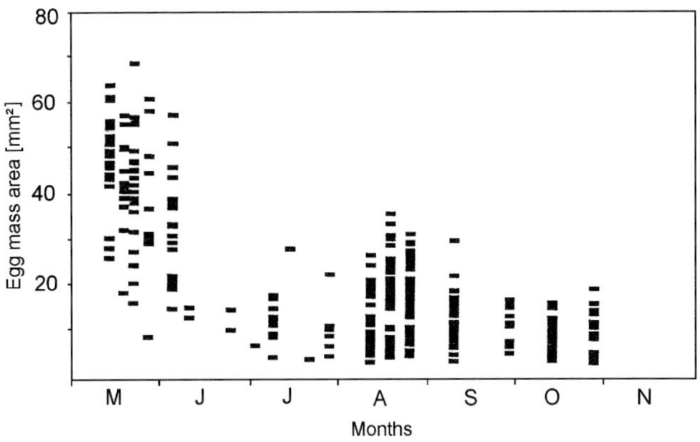

Figure 7.27 Seasonal decline of *Baetis rhodani* egg mass area in the Breitenbach in 1990 (from Werneke, 1991).

Breitenbach, *Baetis* batch size declined with seasonally decreasing adult size, especially between the two generations of *B. rhodani* (Figure 7.27; 1991, 1994). Size of individual eggs may also decline with maternal size (Benech, 1972a; Fink *et al.*, 1991; Corkum *et al.*, 1997). Eggs may also differ in biochemical quality (Benton, 1989) but there is no information on species in the Breitenbach.

Egg morphology differs much between taxonomic groups, especially in the number and location of micropyles, presence, arrangement and specific structure of hapteres. Many species also have specific chorionic patterns (e.g., Kopelke and Müller-Liebenau, 1981; Studemann and Landolt, 1997; Koss, 1968), but we did not study these aspects.

Oviposition

Females recognize suitable oviposition sites visually, by the reflection of polarized light, mainly UV (Schwind, 1991). They are sometimes misled by technical products with similar optical properties (Kriska *et al.*, 1998). Records of upstream oviposition flights by mayflies have been published (Illies, 1968; Flecker and Allan, 1988; Bauernfeind and Humpesch, 2001), but there is no evidence of it at the Breitenbach (Werneke, 1994). However, Müller (1982) postulated that upstream flights that compensate for larval downstream drift occur generally, as part of a so-called colonization cycle.

Mode of oviposition varies between taxa (Bauernfeind and Humpesch, 2001), many fly down and dip into the water surface, releasing a few eggs each time. Ephemerellidae and Baetidae in the Breitenbach lay all their eggs at once, but in different ways. Females of *Ephemerella* extrude the eggs and carry a dark green spherical egg mass for some time. They are attracted to lights (Bohle, 1972b) and eggs can easily be collected for studies. *Ephemerella* egg masses remain intact in water and the silt-coated balls, kept together by the entangled hapteres, were occasionally found in benthos samples (Zwick, personal observation).

The characteristic mode of oviposition of *Baetis* species and its possible significance for the unusual, often irregular patterns of *Baetis* emergence from the Breitenbach were discussed above.

Egg Development

Both direct and indirect development involving an egg diapause, occur in the Ephemeroptera. The dominant Breitenbach genera, *Baetis* and *Ephemerella*, each include one directly developing species (*B. rhodani*, *E. mucronata*) and another which passes through diapause (*B. vernus*, *E. ignita*; Bohle, 1969, 1972b; Sowa, 1975; Studemann *et al.*, 1992; Bauernfeind and Humpesch, 2001; Schmidt, 1984). The Leptophlebiidae, which are rarely important in the Breitenbach, all have univoltine winter cycles and must therefore develop directly (Bauernfeind and Humpesch, 2001).

The temperature dependence of direct egg development is usually studied at various constant temperatures in the laboratory. Mean incubation period (MIP; the time over which half of all eggs that eventually hatch have hatched), hatch success (%) and hatching period (as a measure of synchrony) are suitable descriptors of egg development.

Baetis rhodani can serve as example. Mean incubation period and hatching period both rise strongly with falling temperatures (Benech and Vignes, 1972; Elliott, 1972; Figure 7.28). At Breitenbach summer temperatures, eggs hatch fast and synchronously, but at low temperatures hatching is very extended. Hatching success also varies with temperature. Generally, success is often lower in mayflies than in stoneflies, and in most no hatching occurs at 5 °C or below (Brittain, 1990). *Baetis rhodani* has an optimal range, approximately 5–15 °C, beyond which success declined in both directions (Elliott, 1972; Figure 7.29). In *Baetis* mortality also increases with extreme length of diapause (Bohle, 1969). Low pH may also negatively affect hatching success (Rowe, Hudson and Berrill, 1988), in line with the generally low tolerance of mayflies to acid waters (Guerold, Vein and Jacquemin, 1991; Baltes, 2000; Brittain and Sartori, 2003).

Mayfly egg diapauses typically interrupt embryonic development at some advanced stage, for example in *Baetis vernus* and *Ephemerella ignita*. In the range normally encountered in rhithral streams both pre- and post-diapause development are accelerated by warmth but slow down at higher temperatures. The temperature shortly before the onset of diapause influences its length in *B. vernus*. Warmth extends, low temperatures shorten the egg rest, and are also required for diapause termination. *Baetis vernus* may start to develop directly if it experiences 0 °C before it would normally enter diapause. At the height of diapause, *Baetis* eggs are cold-resistant to −14 °C (Bohle, 1969, 1972a, 1972b). To explain variable emergence patterns of the species, Illies and Masteller (1977) suggested that the mode of development changed with winter temperatures.

Egg development may take a very long time because of diapause but egg rest helps synchronize changes in population structure (Schmidt, 1984). This is the case in *E. ignita* and *B. vernus* which both oviposit in summer but overwinter as dormant eggs. Massive recruitment occurs in spring when temperatures rise again

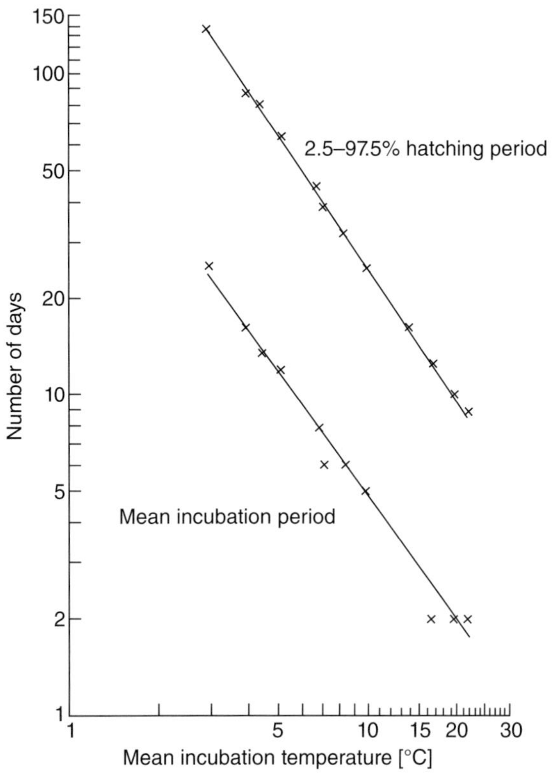

Figure 7.28 *Baethis rhodani*, negative relationship of mean incubation period and hatching period with mean incubation temperature. Modified from Elliott (1972).

(Bohle, 1969, 1972a, 1972b). While *B. vernus* diapause appears to be obligatory, populations of *E. ignita* differ. Unlike northern ones, western and southern European populations may develop directly, in part or completely (Elliott, 1978; Thibault, 1971). Indeed, single *E. ignita* larvae that must have hatched from directly developing eggs were occasionally encountered in the Breitenbach (Zwick, unpublished data), which may simply reflect the species' wide potential for different types of life cycle (Clifford, 1982).

7.6.4.2.5 **Morphological Change during Mayfly Growth and Development**

First Instar Larva

Newly hatched mayfly larvae differ considerably from older ones. Gills and the terminal filament are often lacking, instead of facetted eyes only five ommatidia are present, and most neonate larvae have a double row of setae on the tarsal claw, which disappears in later instars (Illies, 1968). Some of the initially missing structures are distinctive characters of the order Ephemeroptera, which are only added

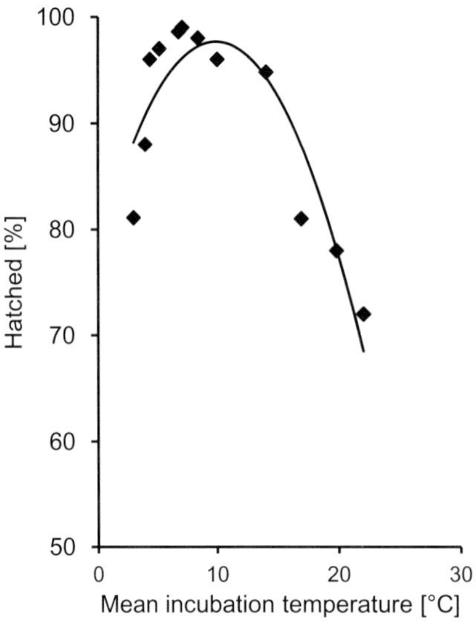

Figure 7.29 *Baethis rhodani*, dependence of hatch success on mean incubation period (polynomial trendline fitted by Excel). After data given by Elliott (1972).

successively during early moults. Gills appear on anterior segments before posterior ones. When all are present, the set of second lamellae is added in species with bifoliate gills (Grandi, 1941). Once structurally complete, only microscopic structures, such as the number of hair sensilla, increases at every molt. It is a problem for benthic studies in the Breitenbach that denticles on the gill margin of *B. rhodani,* which are needed to separate them reliably from *B. vernus,* only appear when larvae are over 2 mm long. In contrast, the distinctive tergal processes of the two *Ephemerella* species are present from the beginning (Werneke, 1997). However, on the whole, larvae grow without morphological change until they begin to prepare for metamorphosis (Clifford, 1982; Campbell, 1991).

Instar Number

The number of instars in Ephemeroptera is apparently variable, ranging from 12 to over 40 (Ide, 1935; Grandi, 1941; Fink, 1980; Brittain, 1982; Brittain and Sartori, 2003). Illies and Masteller (1977) estimated 18 instars for *B. vernus* from the Breitenbach. Individual rearing and counting of exuviae, or estimates from growth ratios at moults (Benton and Pritchard, 1988) is one reliable method to estimate instar number, study of the "Palmén body" is another (Benech, 1972b; Ruffieux *et al.*, 1996). Topology prevents the extraction of the shed cuticle at moults from a particular crossing of cephalic tracheae, which accumulate forming the multi-layered Palmén body. Its true structure was only recently recognized, but it remains possible to determine the number of moults from counts of its layers (Fink, 2008).

Only 10–17 larval instars were observed in a fairly large species of *Siphlonurus*, and numbers differed between pond and stream habitats (Ruffieux *et al.*, 1997). *Baetis rhodani* was found to have 9–14 larval instars (Benech, 1972b). The authors believe that the relative thickness of the shed cuticle rises with temperature. However, their figure (Ruffieux *et al.*, 1997; see Figure 5) suggests that there may also be a relationship with instar number, larger specimens shedding a relatively thicker cuticle.

Generally, the number of moults appears to be higher in mayflies than in stoneflies. This may be related to the very complex and fragile cuticular mouthparts of Ephemeroptera, which are quickly worn and need replacement. Arens (1989, 1990) performed experiments showing that wear of mouthparts can induce moults without somatic growth. He was even able to induce moults for structural repair in unfed specimens, which grew smaller at each molt.

Normally, growth rates at moults decline with increasing specimen size, but some growth and weight gain occur even at the last molt. Brook's or Dyar's rule (Dyar, 1890, Crosby, 1973) of a constant growth rate at moults was not confirmed in the mayfly, *Ameletus celer* (Benton and Pritchard, 1988).

Structural Change and Metamorphosis

The primary metamorphic process, development and maturation of gonads, is difficult to study. Last instar Ephemeroptera (Brittain, 1982) contain mature gametes; testes even start regressing before the adult molt (Harker, 1992).

As seen in published figures (Clifford, 1982), where distinctly different wing pads reveal that instars were wrongly counted because of sexual size differences, body size is unsuitable for instar determination. The biomass of emerging female mayflies is much larger than male biomass (e.g., Sartori *et al.*, 1992; Sweeney *et al.*, 1995).

Appearance of external wing pads is a clear sign of approaching metamorphosis. Wing pad shape appears to change in a characteristic way at moults, which would permit the counting of pre-imaginal instars. To quantify this change some other linear body measurement is needed as reference, but this is difficult because, (unlike in stoneflies) several structures grow allometrically in late instar mayflies (Clifford, 1970; Campbell, 1991; Rösch, in press). Also, in convex-bodied active swimmers like the Baetidae, an increase of strong indirect flight muscles alters the thorax shape. Measurements of the pronotum are very useful in the Plecoptera, but in the Ephemeroptera the prothorax is small and often positioned under the convex pterothorax. Shape and size of late instar larvae heads are variously affected by sexual dimorphism (Gupta, *et al.*, 2000), much more in Baetidae males with turbinate eyes than in, for example, the Ephemerellidae.

For all mayflies living in the Breitenbach, the eventually emerging subimago moults again while sitting on some support. None of the local species becomes sexually mature after a partial molt, or even without a molt of the female, as occurs in several other taxa (Illies, 1968), nor do any molt in flight like the male *Oligoneuriella* in the River Fulda (Zwick, personal observation).

Early Metamorphosis – A Strategy?

The interplay between ontogeny and certain ecological aspects has led to speculation about mayfly strategies. For example, it was suggested that in the presence of predatory fishes mayflies may sacrifice the advantage of large body size and high potential fertility in order to emerge early, despite being small. This would shorten the period of exposure risk, which involves the danger of not reproducing at all (Peckarsky *et al.*, 2000).

However, there is insufficient information on the basic ontogenetic processes in Ephemeroptera. It is unclear whether mayfly ontogeny really permits implementation of the proposed strategy, or whether developmental constraints require reconsideration. Our own presently incomplete data (Rösch, in preparation) suggest that at least two instars with morphologically defined shapes must precede metamorphosis. Therefore, spontaneous metamorphosis in response to the appearance of predators can only occur after at least two more molt intervals. Also, some minimum size is clearly required before a larva can prepare for metamorphosis. On the whole, mayfly development appears somewhat more flexible than that of stoneflies.

The risk of mayflies becoming sexually mature late in the season, when climatic conditions become unfavorable, has been suggested as a stimulus for metamorphosis at a fairly small size. The induction of metamorphosis by season is suggested by the frequently observed, so-called seasonal size decline. Some mayfly examples are discussed together with Plecoptera, see below (see also Figure 7.27).

7.6.4.3 Plecoptera

7.6.4.3.1 General and Species List

Plecoptera are commonly known as stoneflies or springflies. The origin of the frequently used first name is doubtful, the second alludes to the early appearance of many species. Adults of some species are found walking on snow alongside streams, like *Leuctra prima* at the Breitenbach. In comparison with other aquatic insects, Plecoptera in general are exceptionally rheophilous and cold stenothermous (Brittain, 1990). However, the order also comprises a few limnic and many tropical species (Zwick, 2003a). Plecoptera are hemimetabolous, eggs and larvae are aquatic, adults terrestrial.

Diagnostic Structures and Position in the Phylogenetic System

Stoneflies belong to the Neoptera and fold their four soft wings back; resting flat on the abdomen, sometimes slightly rolled around on the sides. This, together with long, multi-segmented antennae, a large, freely moveable prothorax, and three-segmented tarsi with large arolium, makes them unmistakable. Many also have two long cerci, but there is never a terminal filament. The free-running larvae possess facetted eyes and long, multisegmented antennae and cerci, and three-segmented tarsi with paired claws, which also readily distinguishes them. In the European fauna, gills (if present) are filamentous bushels near the leg bases, or inconspicuous on the ventral side (neck of some Nemouridae, on the coxae in *Taeniopteryx*), or on the paraprocts. Foliate gills never occur.

The position of stoneflies within the Neoptera is uncertain (Zwick, 2009), but conventionally the Plecoptera are regarded as close relatives, or members, of the orthopteroids (Polyneoptera). The classification of the Plecoptera themselves is better known, morphologically based proposals (Zwick, 1973, 2000) receiving overall support from DNA studies (Terry and Whiting, 2003, 2005). The split into suborders reflects the break up of Gondwanaland, the four families of the Antarctoperlaria being restricted to Australia, New Zealand and South America while most of the 12 families of the Arctoperlaria occur only in the Holarctic and Oriental regions. Six of the eight European families are represented in the Breitenbach.

Species List for the Breitenbach

Seventeen species inhabit the Breitenbach (see Appendix), 14% of the German fauna, and one third of that in the entire catchment of River Fulda. All Breitenbach species are common widespread taxa found in many unpolluted foothill streams in Central Europe.

Some additional species found in the Breitenbach do not normally live there. They are:

- *Nemoura sciurus* Aubert occurs in most small streams west of the township of Schlitz, but for unknown reasons is absent from the Breitenbach and other streams east of the Schlitz and Fulda rivers. A single specimen found among Breitenbach material was probably due to contamination of the equipment after sampling traps on the Rohrwiesenbach, where the species occurs.
- *Protonemura nitida* (Pictet) was recorded once. The specimen must have been displaced as the species only occurs in montane streams in the Vogelsberg and Rhön mountains, not around Schlitz.
- *Leuctra inermis* Kempny was intentionally introduced in June 1986 to stock the stream after it was cleared of Plecoptera, due to insecticide poisoning (Zwick, 1992). Four hundred egg masses collected from females trying to oviposit on a polished car roof near the Birxbach/Rhön (Zwick, unpublished data) developed sucessfully, equivalent to approximately 120 000 hatchlings. In the following years a few adults were collected, but the species soon disappeared again. *Leuctra inermis* is common above 400 m NN, but does not occur at lower elevations.
- *Perlodes dispar* (Rambur) is again common in the River Fulda since its water quality improved (Siebert, 1998; Zwick and Zwick, unpublished data). It occurred sporadically in emergence traps. A larva taken in the Breitenbach benthos was probably the offspring of a wandering female of this large insect.
- *Perlodes microcephalus* (Pictet) does not occur in the area. The single specimen taken in a trap must have originated from an egg lost during incubation experiments in the Georgsborn in the late 1980s.

There are no Perlidae in the Breitenbach, nor in any other small streams near Schlitz. *Perla marginata* (Panzer) was common and widespread in the nearby Rhön Mts (Illies, 1955), where it is a rarity today. The reason for this large-scale decline is unknown but the species still occurs in small spring runs in the Harz Mts and the Schwarzwald (Zwick, personal observations). The extended sandy areas in the stream would appear to be suitable habitat for *Isoptena serricornis* (Pictet), which may be common in large continuous pleistocene sand areas, but was also found in streams where sand-beds are derived from the breakdown of Bunter Sandstone (Otto and Westermann, 2003; Dorfer, 2008). However, *Isoptena* requires permanently moving, well ventilated sand (Zwick and Hohmann, 2005) which is not continuously available in the Breitenbach.

Stoneflies generally have fairly strict habitat requirements. Ranges along major stream continua, for example along the River Fulda, are more restricted than those of many mayflies (Illies, 1953a, 1953b; Marten, 1983; Siebert, 1998). Along-stream patterns at the Breitenbach agree with known preferences. For example, *Brachyptera* spp., *Leuctra prima*, and *Protonemura auberti* prefer upstream sites, *Nemoura flexuosa* downstream ones. *Nemoura flexuosa* has only occurred in the Breitenbach since 1977, when pollution of the River Fulda began to lessen, but it is not common in the river.

The stonefly fauna in the Breitenbach is almost entirely different from that in the adjacent part of the River Fulda (Table 7.8), vicariance is pronounced. However, most species in the Breitenbach, plus several more, occur in distant upstream rhithral sections of the River Fulda.

The distinctness of the two neighboring faunas is better explained by individual species preferences for particular habitats than by limited dispersal capacities. Estimates of adult dispersal away from streams are controversial. In some studies, adults were caught at most a few metres away from streams (e.g., F. Wagner, 2003; Collier and Smith, 1995; Petersen *et al.*, 1999; Briers *et al.*, 2002). The genetic structure of a North American peltoperlid population suggests very limited dispersal and interpopulation exchange (Schultheis *et al.*, 2002). Other authors record numbers of Nemouroidea walking or flying 60–200 m from streams (Thomas, 1966; Mendl and Müller, 1970; Kuusela and Husko, 1996). Considerable dispersal along stream corridors has also been documented for *Leuctra ferruginea*, a rather poor flier (Macneale *et al.*, 2005).

In the Breitenbach valley, all taxa were frequently seen flying up into trees on the sides of the valley. This is in accord with records by Kjeldsen (1991), see her Table 7). These specimens presumably move into the canopy to forage because the food requirements of the long-lived Nemouridae and Leuctridae (Table 7.11) cannot be satisfied in the riparian stretch alone. The other occasions when stoneflies fly longer distances are oviposition flights.

Trophic Relations and Functional Feeding Groups

Plecoptera larvae fall into only a few different functional feeding groups and have only three major types of trophic relationships.

- Larvae of most families are detritivores consuming dead plant material. Their bulky mandibles have large molar areas to grind food. The huge Pteronarcyidae in North America and East Siberia are important shredders (Stewart and Harper, 1996). Most Nemouroidea are also detritivores but because they are much smaller, many only nibble on dead leaves, or are gatherers. Some detritivorous Nemouridae are known to prefer epilithic algae over detritus (Ledger and Hildrew, 2001; Lieske and Zwick, 2007), but lack corresponding mouthpart specializations. All the delicate Leuctridae gather fine detritus. Whole mounts of larvae from the Breitenbach show guts densely filled with very fine material,

Table 7.8 The largely vicariant Plecoptera faunas of the Breitenbach and the adjacent section of River Fulda. Black = common; gray = moderately common; x = sporadic occurrence.

	Fulda (Hartershausen–Hutzdorf; Siebert, 1998)	**Breitenbach (present study)**
Brachyptera seticornis		gray
B. risi	x	gray
Protonemura auberti		black
P. intricata		black
P. meyeri		gray
P. nitida		x
Nemoura cambrica		black
N. cinerea		gray
N. flexuosa	x	gray
N. marginata		black
Nemurella pictetii	x	gray
Leuctra digitata		black
L. fusca	black	x
L. inermis		x (introduced, Zwick)
L. nigra		black
L. prima		black
Isoperla goertzi		black
I. grammatica	black	x
Perlodes dispar	black	x
P. microcephalus		x (introduced, Marten)
Siphonoperla torrentium		black

including mineral particles. All are opportunistic feeders and for extended times the Nemouroidea can survive on fine particulate organic matter. This is an important safeguard against significant resource limitation.

- *Brachyptera* and allies graze biofilms with a rasp on their galea (Wichard *et al.*, 1995, 2002) but also consume some detritus.
- Perloidea (= Perlodidae, Perlidae + Chloroperlidae) larvae are predators with modified mouthparts suited to grasp live prey. The mandibles have sharp teeth, but no mola. When young and small, most larvae also eat detritus, or have a mixed diet. There is no information on how important accidental or deliberate consumption of microbes or microcrustaceans may be. *Isoperla goertzi* is the largest carnivorous stonefly in the stream. It tears prey to bits or swallows small prey whole. The latter feeding mode is typical of the Chloroperlidae, *Siphonoperla* larvae often contain undamaged blackfly or chironomid larvae that completely fill the gut (Zwick, unpublished data).

Unlike adult Ephemeroptera and Trichoptera, most adult Plecoptera also consume particulate food. However, the opposite is sometimes stated in general literature because the mouthparts of large adult Systellognatha are nonfunctional, the mandibles being reduced to soft lappets (Zwick, 1980), of which *Perlodes* is an example. However, the opposite is the case in the smaller Perlodidae, and generally in the Chlorperlidae. Adult mandibles resemble those of the carnivorous larvae, being slender without mola, and more or less sclerotised. The smaller the animal, the more functional the mandibles. The presumed interplay between size and reserves stored during larval life needs study. All adult Nemouroidea and Antarctoperlaria have functional mandibles with large molae and many are known to feed (Zwick, 1980).

7.6.4.3.2 **Plecoptera Larvae in the Benthos**

The present section deals with the age structure and biomass gain of larval populations of stoneflies. Methods used to illustrate mass accrual are the same as for mayflies.

Except *Brachyptera seticornis* and *B. risi,* larvae of all stonefly species recorded from the Breitenbach were found in the benthos. A single larva of *Perlodes* sp. was found confirming that members of the genus can occasionally develop in the small stream. Structure of mandibles and maxillae easily separates the two superfamilies occurring in Europe and also in the Breitenbach. The generally detritivorous Nemouroidea are represented by Taeniopterygidae (*Brachyptera*), Nemouridae and Leuctridae, the Perloidea by Perlodidae and Chloroperlidae.

Generic identification of large larvae is no problem (Zwick, 2004) but small specimens and specific identification generally pose important problems. However, in regular quantitative benthos samples very small larvae by far dominate, and their correct taxon assignment is decisive for analyzing and comparing their life histories. Problems with identification of small larvae of Nemouridae and Leuctridae were overcome to different degrees.

Nemouridae

The most diverse stonefly family in the Breitenbach is represented by all four European genera. Once about 2 mm long larvae can easily be identified to genus (Zwick, 2004). *Protonemura* larvae can be identified to species with the key by Raušer (1956), but specific identification of *Nemoura* larvae is possible only for some (Zwick, 2004). Nemouridae are characteristic inhabitants of springs and rhithral streams. They are generally shredders of dead leaves and gatherers of detritus, but also grazers to a certain degree (Steiner, 1991), which is best documented in *Nemurella* (Ledger and Hildrew, 2000; Lieske and Zwick, 2007).

Genus *Nemoura* Of the four species present in the Breitenbach, only larvae of *N. cinerea* can be identified. *N. cinerea* is uncommon in the Breitenbach except in a short weedy stretch downstream from the Georgsborn, before the spring run enters the main course. No benthos samples were taken in this small section, elsewhere in the stream larvae of *N. cinerea* played no role.

N. marginata, *N. flexuosa*, and *N. cambrica* are closely related, even adult females can be distinguished only with difficulty; there is no published key. According to Raušer (1980) specific identification may seem possible but is actually not; only pharate males can be separated. Therefore species of the marginata group must be dealt with together, as *Nemoura* spp. The gill-less larvae can at all sizes be safely distinguished from the equally gill-less *Nemurella*, but separation from very small larvae of *Protonemura* and *Amphinemura* was possible only after mounting on microscope slides and special study; see below.

The life cycles and growth patterns of the three closely related species overlap completely (Figure 7.30). Together, the three species present a more synchronous univoltine picture than many single univoltine species do. Plecoptera in general are said to continue growth over winter, while other taxa experience a winterly growth stop; see, for example, Figure 7.14. *Nemoura* species provide an excellent example of winter growth and enormous weight gain in very early spring (Figure 7.31).

Nemurella pictetii This widespread monotypic genus is restricted to springs and spring runs, even isolated small spring seeps may support populations. Adults are therefore occasionally found almost everywhere which is probably why in general literature *N. pictetii* is sometimes considered to be ubiquitous. In samples from Reaches B–E of the Breitenbach only occasional specimens were taken. *Nemurella* is abundant on vegetation and on the stream bed of the small ditch connecting the Georgsborn to the main course of stream, and in the outflow of the spring-fed ponds in Reach C. For the semiterrestrial egg masses see Figure 7.61.

Larvae can be identified from the first instar on, by the long tibial setae and other details (Zwick, 2004) which faciliated analysis of the unusual life history of *N. pictetii*. Wolf and Zwick (1989) documented that some larvae can develop from egg to adult in about four months while other siblings from the same egg mass follow normal univoltine development. Provided the stream warms during summer, *Nemurella* exhibits cohort-splitting also in the field (Figure 7.32;

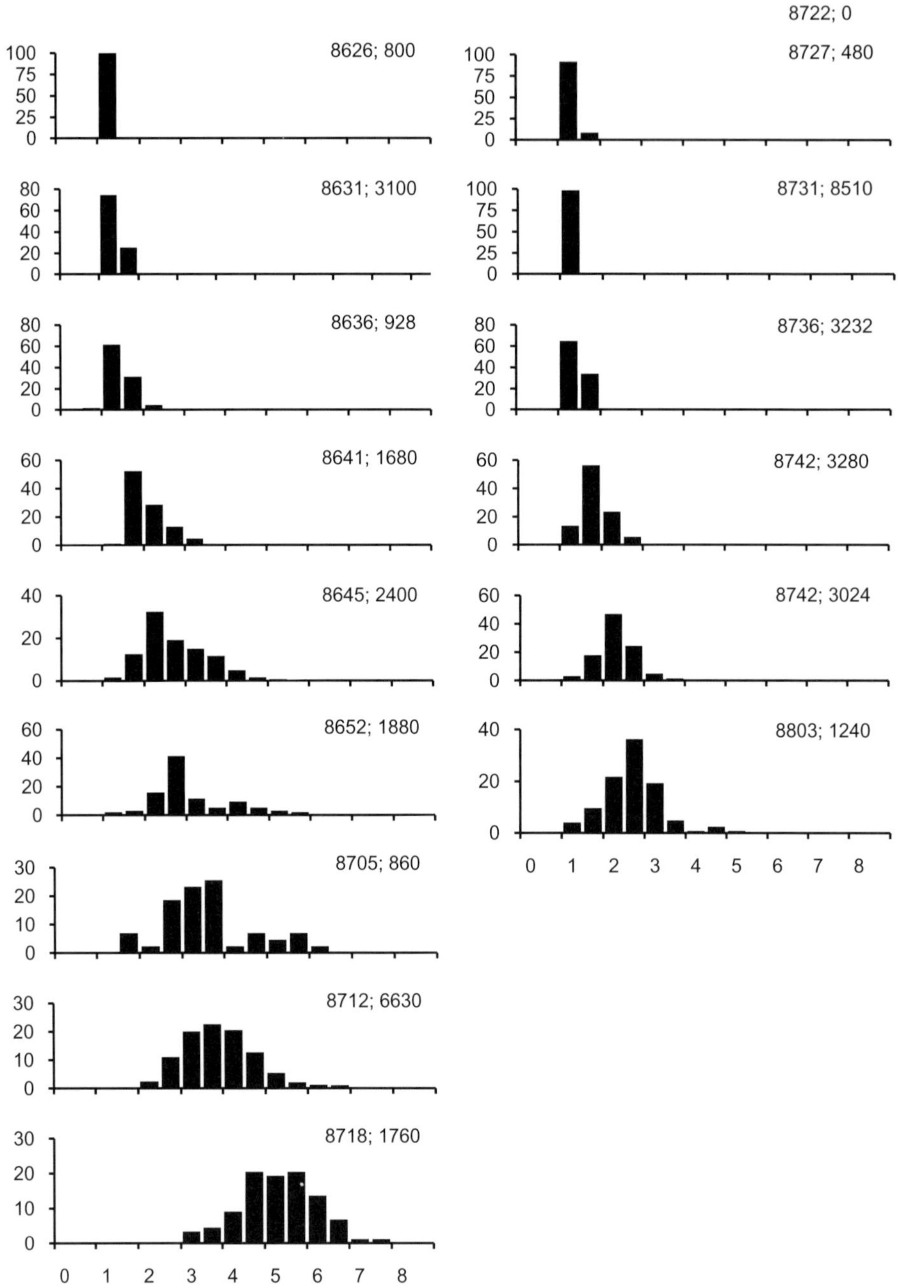

Figure 7.30 *Nemoura* spp. of the marginata-group (*N. cambrica*, *N. flexuosa*, *N. marginata*). Frequency (percentages in samples; ordinate) of body length classes (abscissa; in mm). Figures in legends are calendar weeks in 1986–1988 followed by approximate specimen numbers m^{-2} streambed.

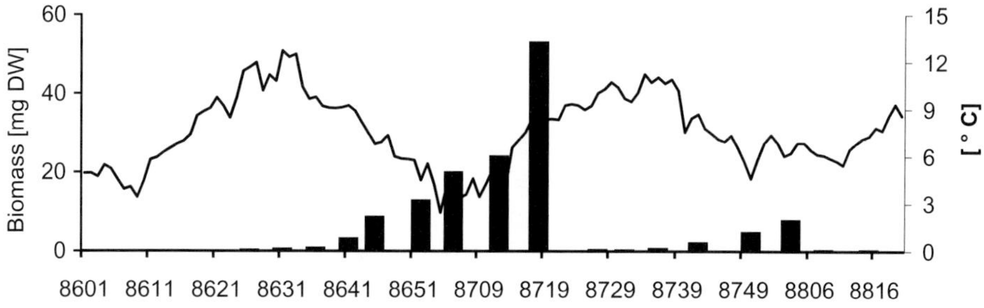

Figure 7.31 *Nemoura* spp. of the *marginata* group, invernal biomass gain. Standing stock of an average population of 100 specimens of the momentary size distribution (bars) and mean weekly water temperature at Messstation 2. Units on abscissa are calendar weeks in 1986–1988.

Nesterovitch and Zwick, 2003). Recruits of specimens emerging in spring appear around July when late specimens of the parent generation are still numerous. Some of the small larvae grow to adulthood by late summer or autumn of the same year. Others grow slowly over winter and emerge the next spring, together with slow-growing siblings of their parents. At summer-cool sites, for example the spring of the River Fulda at 800 m a.s.l., development is simple and univoltine. In contrast, British and North Skandinavian populations are semivoltine (Brittain, 1978; Elliott, 1984).

Seasonal size decline during summer emergence from the Breitenbach is pronounced, but the size reduction does not occur shortly before metamorphosis but much earlier. Size decline was observed in the antepenultimate instar (Nesterovitch and Zwick, 2003) and may actually happen even earlier. However, because younger specimens cannot be identified this remains putative.

Genus *Protonemura* Life histories of the three species in the Breitenbach differ distinctly. The large *P. meyeri* is widespread over most of Europe and, together with *P. praecox* (not in the Breitenbach) the earliest Central European representative of the family. It exhibits continuous growth over winter (Figure 7.33).

In contrast, the small *P. intricata* is an expansive species of Mediterranean origin (Zwick, 1981). Its recruitment begins long after the extended adult flight period but is reasonably synchronous because length of egg diapause depends on oviposition date (mechanism unknown; Marten and Zwick, 1989). Growth during the cold season is very limited but becomes fast as water warms. Although larvae of the two species coexist much of the time their size spectra do overlap only when they are very small but soon become separated through differential growth (Figure 7.33).

The pattern of biomass gain of *P. meyeri* is very similar to that of *Nemoura* spp., maximum growth is even one month earlier. The small *P. intricata* also attains maximum standing stocks distinctly before water temperatures peak (Figure 7.34).

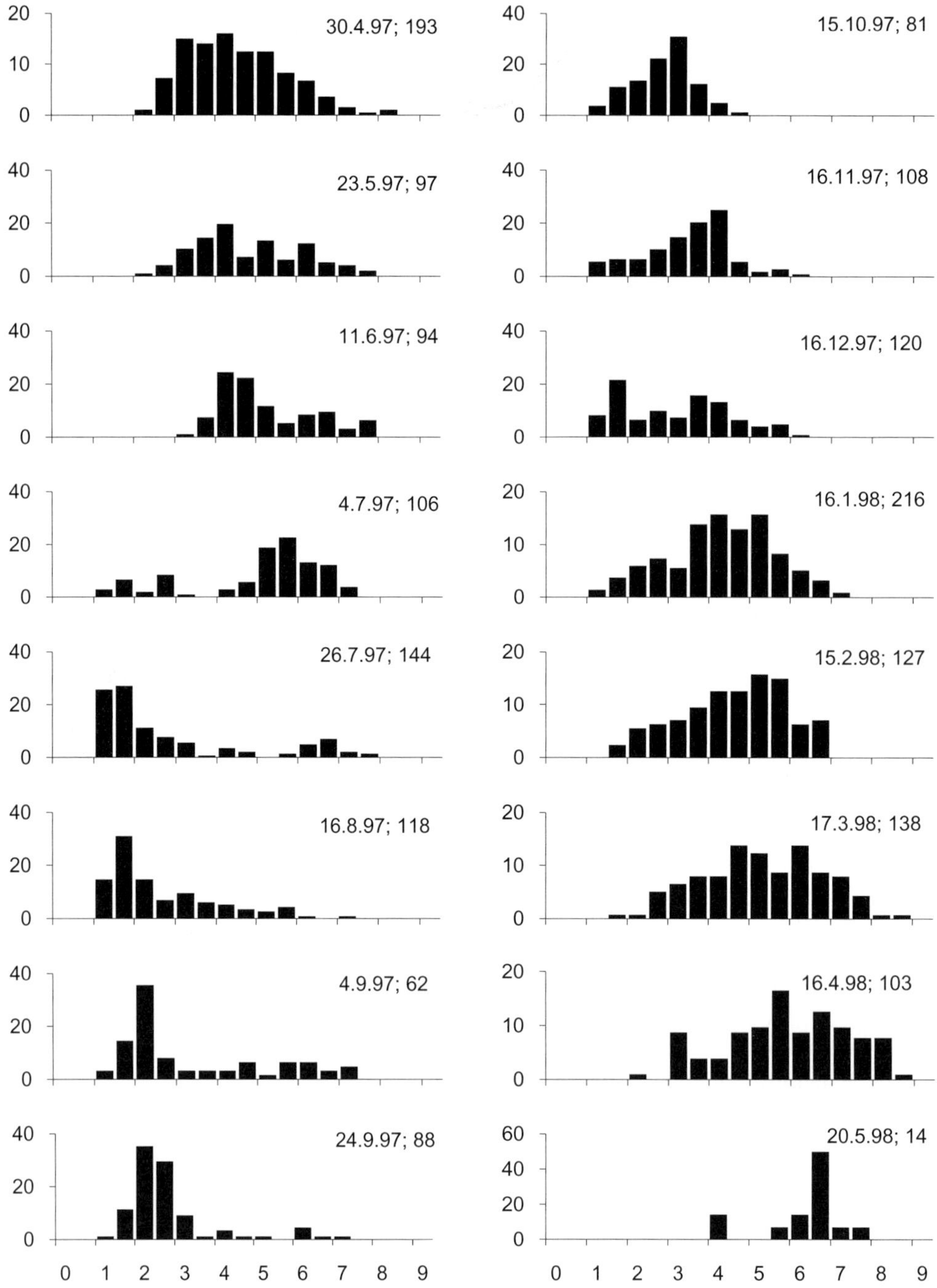

Figure 7.32 *Nemurella pictetii* in the spring run of the Georgsborn, 1997–1998, based on the data of A. Nesterovitch (Nesterovitch and Zwick, 2003). Frequency (percentages in samples; ordinate) of body length classes (abscissa; in mm). Figures in legends are calendar weeks in 1986–1988 followed by approximate specimen numbers m^{-2} streambed. Adults emerged from March to October, except in early July.

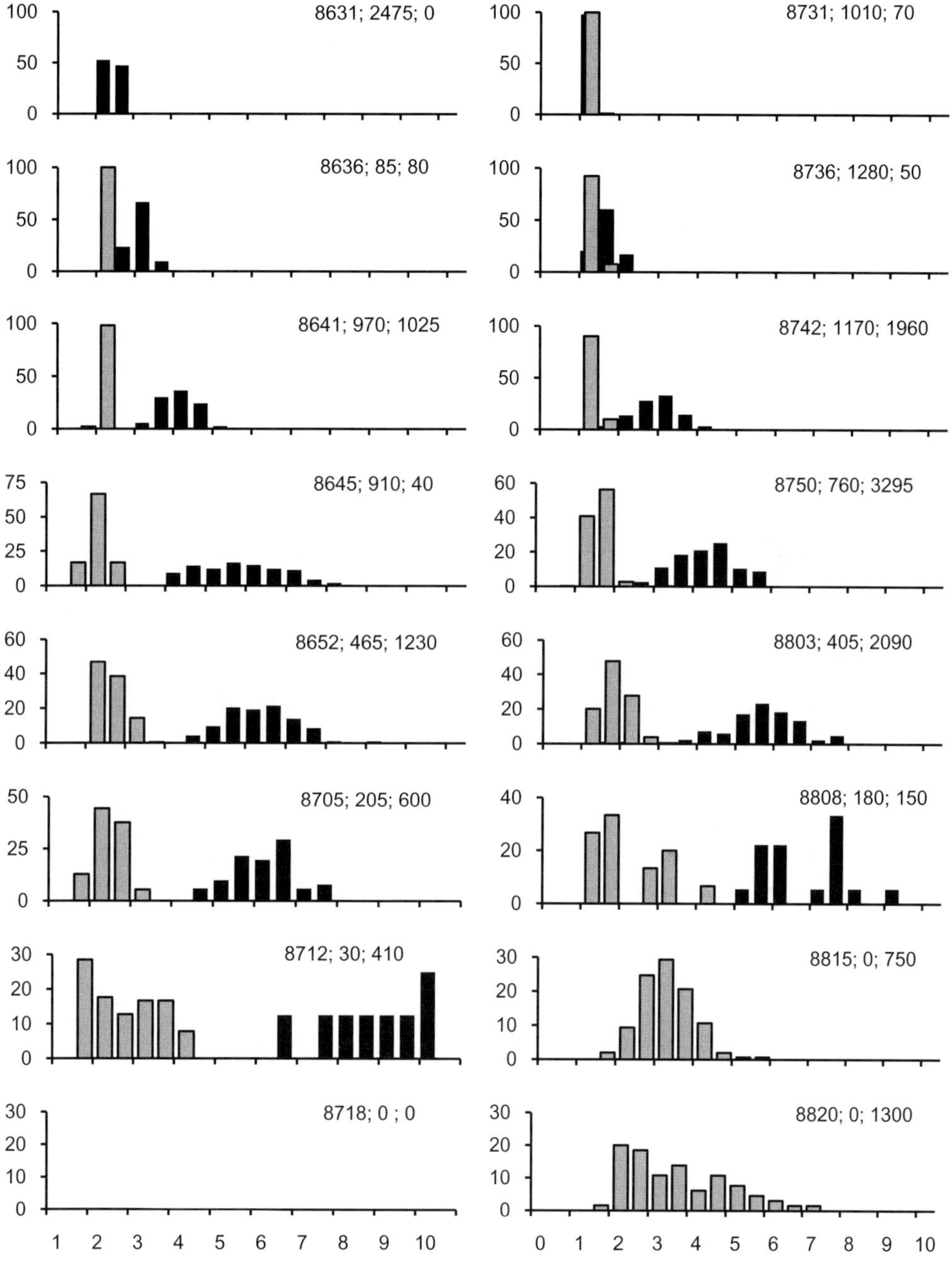

Figure 7.33 *Protonemura meyeri* (dark bars) and *P. intricata* (light bars) in Reach B of the Breitenbach. Frequency (percentages in samples; ordinate) of body length classes (abscissa; in mm). Figures in legends are calendar weeks in 1986–1988 followed by approximate specimen numbers of *P. meyeri* and *P. intricata* m^{-2} streambed.

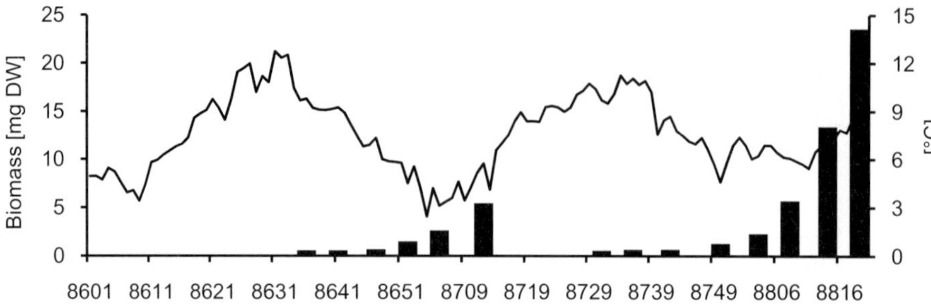

Figure 7.34 *Protonemura intricata*, standing stock of an average population of 100 specimens of the momentary size distribution (bars), and mean weekly water temperature at Messstation 2. Units on abscissa are calendar weeks in 1986–1988.

Protonemura auberti belongs to a species complex apparently representing an example of speciation during Pleistocene separation, and subsequent postglacial spread. *P. auberti* was named from the nearby Wasserkuppe. Less than 100 km further west and southwest (as far south as the Pyrenees) it is replaced by *P. risi*. In a population in Lower Austria, the normally distinctive male genitalia vary from typical *P. auberti* to what looks like typical *P. risi*. However, the very peculiar female genitalia reveal that all Austrian specimens are *P. auberti*. Eastward follows a group of closely related species (or possibly subspecies) in the Balkans and Carpathians, distinction from *P. auberti* requires confirmation. The habitat preference of the insect probably adds to isolation of local populations and character drift: *P.auberti* inhabits only springs and immediately adjacent spring runs.

The preference for upstream Breitenbach sites is pronounced. The largest population lives in the Georgsborn where larval population structure (Marten, 1987) and presence of adults suggested almost continuous reproduction during the warm season. The species seems to opportunistically exploit the equitable temperature in the spring. Possibly, some specimens there follow univoltine, others semivoltine development. This “untidy life cycle”, as H.B.N. Hynes called this, is still apparent in Reach B, where the species is only moderately common (Figure 7.35) and sesonally more restricted. The life cycle becomes more regular further downstream where the animal is rare (Figure 7.48). The temporal pattern of biomass gain is irregular and disturbed by long periods of adult emergence and larval recruitment.

Amphinemura standfussi The genus is of holarctic distribution, Europe has the lowest species diversity of all continents. Only three species are widespread: *A. triangularis* in the hyprhithron, *A. sulcicollis* in the entire rhithron, and *A. standfussi* in springs and epirhithral streams. It is the only European *Amphinemura* found also in lowland streams, for example in Schleswig-Holstein. In the Breitenbach, *A. standfussi* is occasionally abundant, in Reaches A and B only. However, during the present benthic survey no larvae were encountered. In the Breitenbach, eggs enter a diapause, see below. *Amphinemura* larvae are generally found less easily

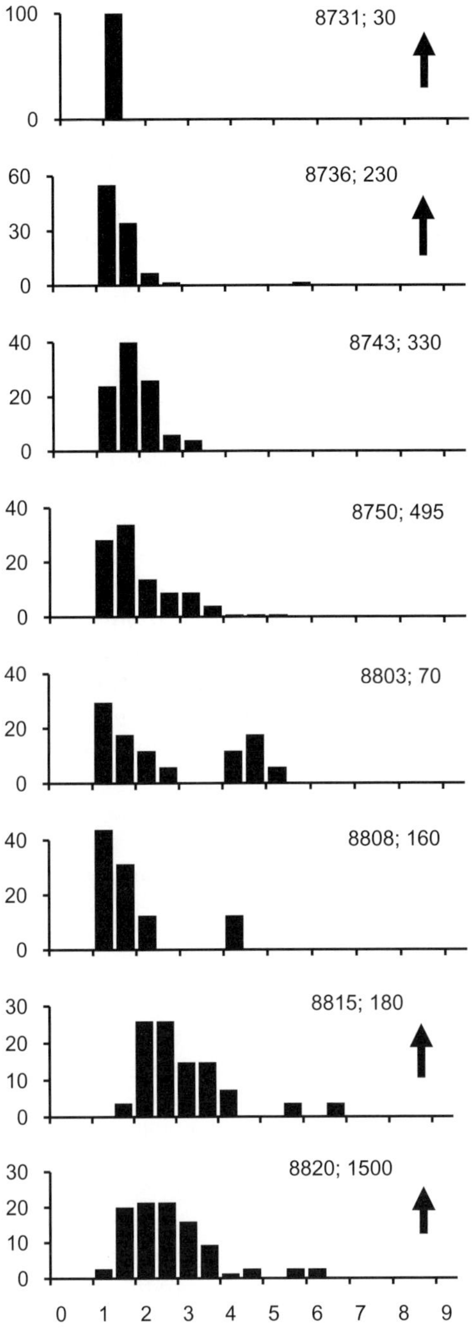

Figure 7.35 *Protonemura auberti* in Reach B of in the Breitenbach. Frequency (percentages in samples; ordinate) of body length classes (abscissa; in mm). Figures in legends are calendar weeks in 1986–1988 followed by approximate specimen numbers m^{-2} streambed. Arrows indicate adult emergence.

than the often extremely abundant adults, especially of *A. sulcicollis*. Habitat preferences and conditions leading to mass development are unknown.

Leuctridae: Genus *Leuctra*

Some of the very many European *Leuctra* species emerge all the time during the frost-free season, and this is so with the three species in the Breitenbach: *L. prima* in late winter and early spring, *L. nigra* in late spring to late summer, *L. digitata* in autumn, until the first frosts. This seasonal separation is probably an expression of traits typical of the different phylogenetic groups to which the species belong. Competition avoidance through temporal separation cannot be excluded but appears unlikely in view of the abundant supply of fine particulate organic matter (FPOM) on which the larvae feed.

Specific identification of *Leuctra* larvae is not generally possible (Zwick, 2004), but was successful for the distantly related indigenous species in the Breitenbach. *L. nigra* holds an isolated position in the genus, its only close relative is *L. bidula* Aubert from Spain. *L. nigra* larvae can be distinguished by a setal brush on a bump on the lower side of the head. In dorsal view it is seen between eye and antenna (Figure 7.36). Already the first instar larva has a single strong bristle in this place. As soon as large larve of the other species develop wing pads, *L. digitata* which belongs to the *fusca* group (Zwick, 2004) strongly differs from *L. prima* (*prima* group) in the shape of the rear margins of meso- and metanotum (Figure 7.36). Smaller larvae can be distinguished by gradual differences in setation, with difficulty, but the different phenology helped assign larvae correctly.

Introduction of *L. inermis* (which also belongs to the *prima* group) after the insecticide spill caused problems of identification and some small *L. inermis* may have been wrongly associated. They were abundant and seemed to grow slowly all the time. Large mature larvae were missed, but small numbers of the species appeared in emergence traps for a few years before it disappeared again completely (Table 7.9).

In the field, *L. inermis* is not found at elevations below 400 m a.s.l., perhaps the reason of its failure in the Breitenbach lies there. Competitive exclusion by the other species is improbable because in the upper River Fulda and elsewhere in the field *L. inermis* coexists with exactly the same species as in the Breitenbach plus several more and is usually more common than any of them.

The life histories of *L. prima* and *L.digitata* are univoltine and synchronous. Initially, size distributions of their larvae overlap but then rapidly diverge. *L. digitata* grows during summer, *L. prima* mainly over winter (Figures 7.37 and 7.38).

Hidden life in layers of sand had protected some of the *L. nigra* larvae during the cypermethrin poisoning in May 1986, and some adults emerged after the accident. The benthic samples document generally very slow growth of their offspring which experienced cohort splitting (Figure 7.29). Some adults emerged from this cohort in spring 1987. Many other specimens grew more slowly and emerged much later, at a time when there were already recruits from their fast-growing siblings. Distinctly bimodal adult emergence sometimes occurs (Figure

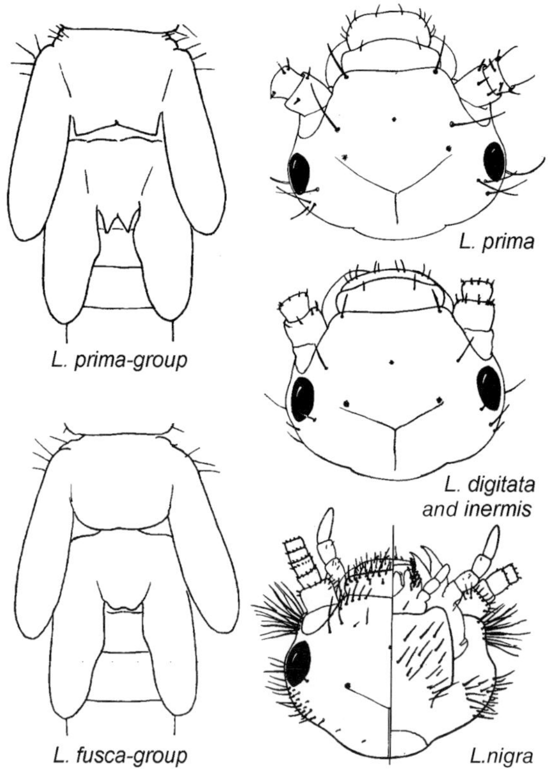

Figure 7.36 Left, the characteristic contours of pterothoracic nota of last instar larvae from two major species groups of *Leuctra*. Right, late instar heads of the *Leuctra* spp. in the Breitenbach, *L. nigra* in combined dorsal and ventral views (modified from Zwick, 2004).

Table 7.9 Numbers of adults of the introduced species *Leuctra inermis* taken in Trap 2, 1988–1992; the species disappeared again from the stream.

1988	1989	1990	1991	1992
17	18	20	4	9

7.49). Under such circumstances, no clear seasonal pattern of biomass accrual is observed (Figure 7.39).

Perlodidae: *Isoperla goertzi*

Larvae of Perlodidae can be recognized as members of the family by their distinctive habitus. The bases of the maxillae which are the main organs of prey capture extend sideways beyond the limits of the actual head capsule and are visible in dorsal view making the head appear exceptionally large. Even the smallest larvae

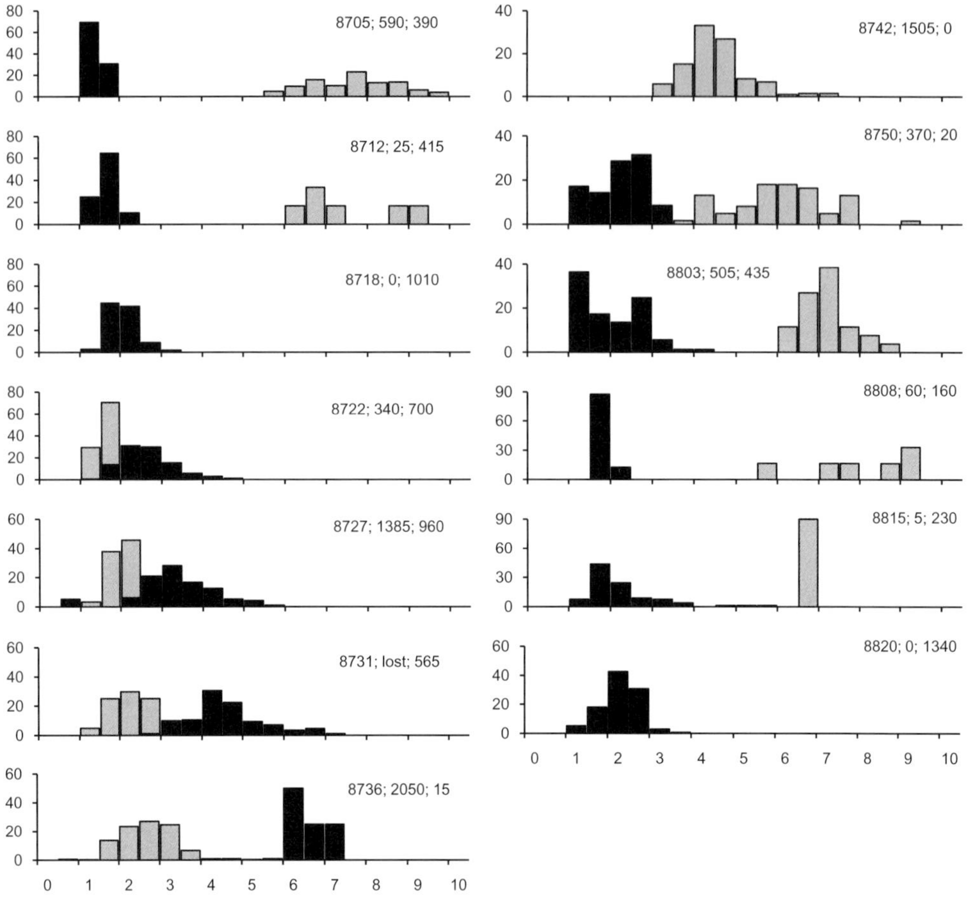

Figure 7.37 *Leuctra prima* (light bars) and *L. digitata* (dark bars) in Reach B of the Breitenbach. Frequency (percentages in samples; ordinate) of body length classes (abscissa; in mm). Figures in legends are calendar weeks in 1987–1988 followed by approximate specimen numbers m^{-2} streambed.

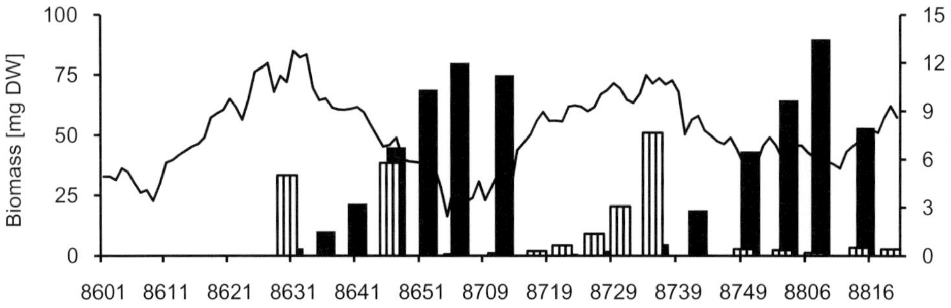

Figure 7.38 *Leuctra prima* (gray bars) and *L. digitata* (striped bars), standing stock of average populations of 100 specimens of the momentary size distribution, and mean weekly water temperature at Messstation 2. Units on abscissa are calendar weeks in 1986–1988.

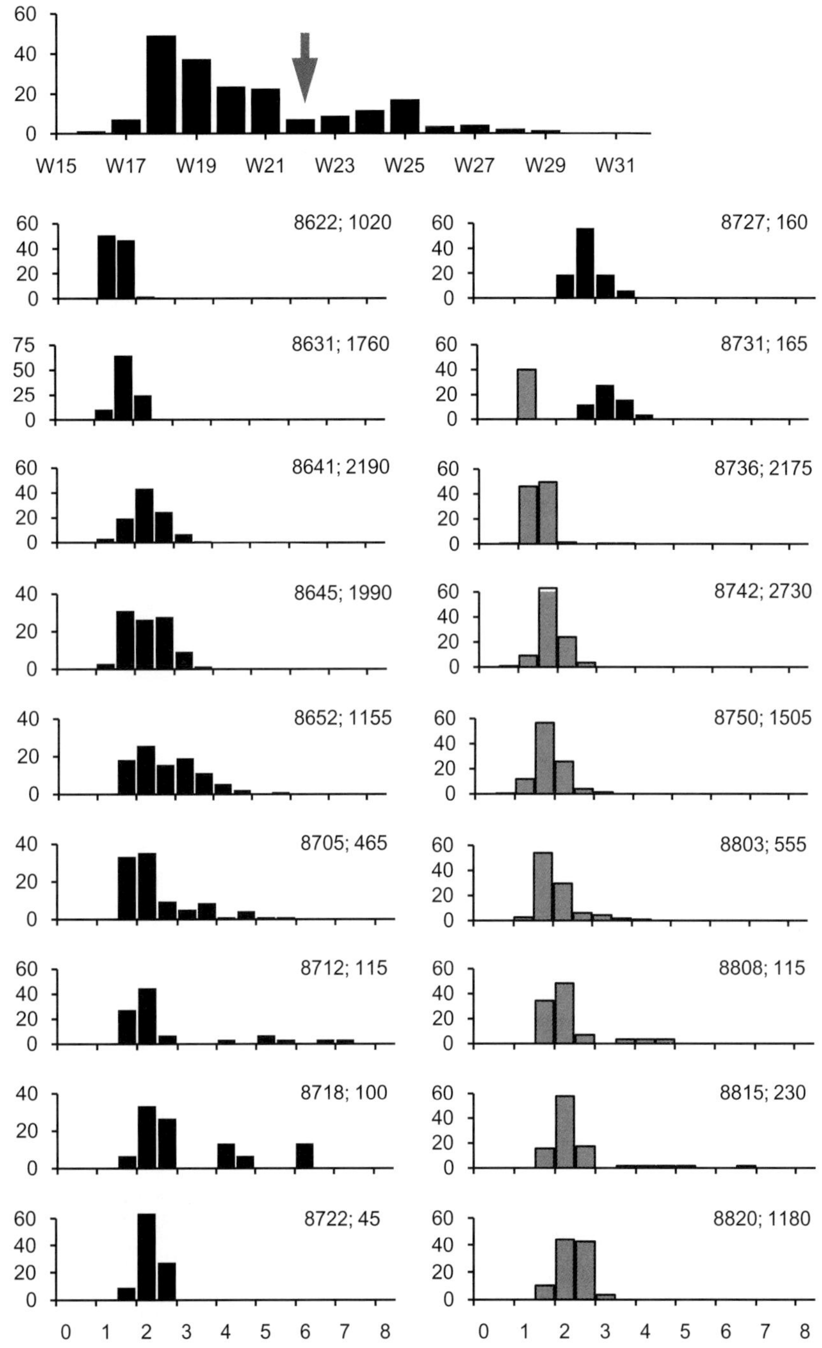

Figure 7.39 *Leuctra nigra* in Reach B of the Breitenbach. Top, adult emergence (percent of annual emergence) continued after the insecticide application (arrow). Bottom: frequency (percentages in samples; ordinate) of body length classes (abscissa; in mm). Figures in legends are calendar weeks in 1986–1988 followed by approximate specimen numbers m^{-2} streambed.

of *Perlodes* can by their bidentate laciniae be separated from small *Isoperla* larvae which have more but smaller lacinial teeth. Of the diverse genus *Isoperla* only a single species, *I. goertzi* which is typically found in montane spring runs and epirhithral streams occurs in the Breitenbach. This is fortunate because meristic characters proposed for specific identification of late instar larvae (Illies, 1952a, 1952b) are doubtful (Zwick, 2004). They definitely change with specimen size (Hynes and MacPhee, 1970) so that smaller larvae which do not yet show the sometimes characteristic cephalic color pattern are unseparable.

During the first year of sampling *I. goertzi* followed a simple univoltine life history (Figure 7.40, left). The relatevely long recruitment period reflected extended hatching from eggs (Schwarz, 1970). However, population structure was atypical because the insecticide accident had removed all larvae surviving from previous generations. In 1987–1988, the typical irregular structure with extended irregular recruitment from dormant eggs and the accordingly heterogenous larval population were restored.

Accordingly, the conversion of body length distributions into biomass values gave a regular pattern only in the first study year (Figure 7.41). Significant standing stocks were found only during half a year, biomass accrual increased when the cold season ended. Maximum standing crop was attained in early May, immediately before adult emergence. The wider and irregular seasonal distribution of standing stocks observed the following year is probably more typical of the species (Figure 7.41).

Growth during mainly the warm season is in accordance with the central to south European distribution of the species and the overall pattern of diversity of *Isoperla* species in Europe (www.faunaeur.org).

Chloroperlidae: *Siphonoperla torrentium*

Larvae of Chloroperlidae are unmistakable because of the minute asymmetrically inserted terminal segment of the maxillary palpus and the compact conical cerci, especially in small specimens. Larvae of the family are engulfers that swallow prey whole. All Chloroperlidae have a tendency to subterranean life, the subfamily Paraperlinae lives exclusively in very deep hyporheos (Stanford and Gaufin, 1974). The German *Isoptena serricornis* is only found in deep shifting river sand and has ballast stones in its gut (Zwick and Hohmann, 2005). It preys mainly, perhaps exclusively, on Oligochaeta. *S. torrentium* is the only widespread and common representative of the family in Central Europe. It is absent from lowlands and does not extend north beyond the Harz mountains and the Sauerland. It dwells mainly in shallow sandy bottom layers which led to survival of a few larvae after the insecticide application in May 1986. Several Breitenbach specimens that turned transparent in permanent slide mounts contained extended larvae of Chironomidae or Simuliidae. The life history is straightforward univoltine, with short recruitment and emergence periods. Standing crops increase slowly during the cold season but the increase to vernal standing crops is tremendous (Figures 7.42 and 7.43).

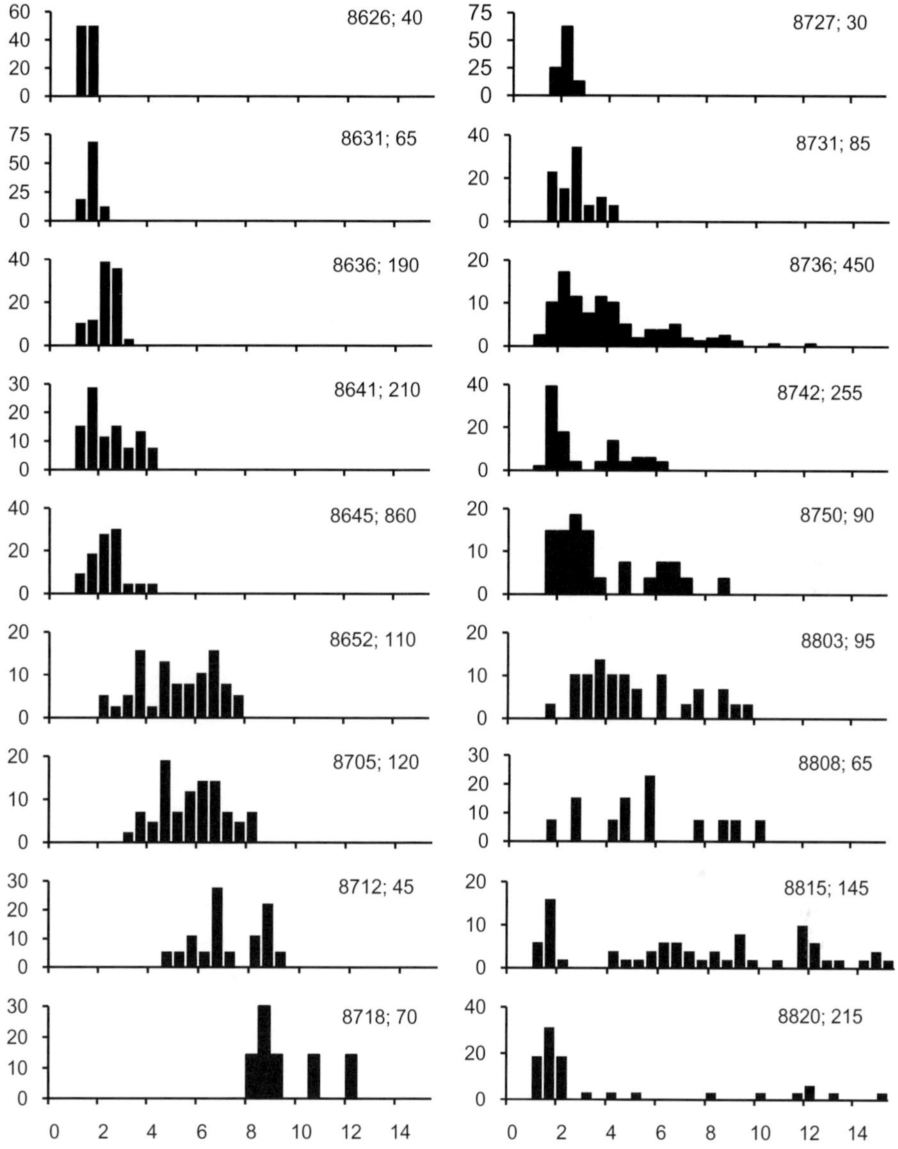

Figure 7.40 *Isoperla goertzi* in the Breitenbach, Reaches A–C combined. Frequency (percentages in samples; ordinate) of body length classes (abscissa; in mm). Figures in legends are calendar weeks in 1986–1988 followed by approximate specimen numbers m^{-2} streambed.

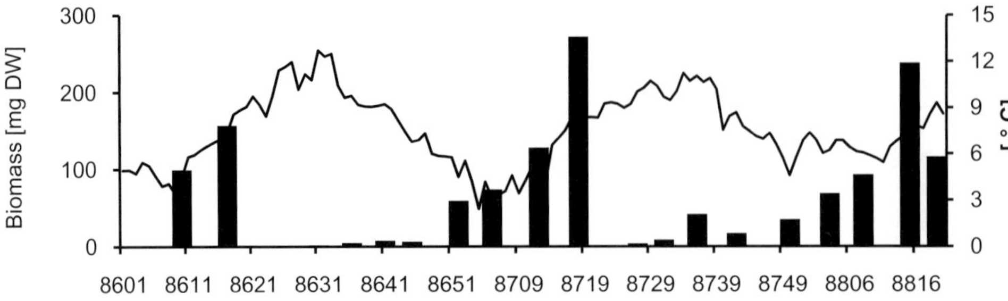

Figure 7.41 Standing stocks of a representative population of 100 larvae of *Isoperla goertzi* during the sampling period, and mean weekly temperature at Messstation 2. The graph includes samples before the insecticide accident that are not covered by Figure 7.40. Units on abscissa are calendar weeks in 1986–1988.

7.6.4.3.3 **Plecoptera Emergence**

Threshold Emergence Temperatures and Emergence Cues

Some Plecoptera species emerge from the Breitenbach throughout the entire frost-free period between, on average, Weeks 8–42 (Figure 7.46, Table 7.10). The greatest diversity is attained shortly before the middle of year, when up to 11 species occur simultaneously in traps. Emergence periods may be distinctly shorter than flight periods which include the total lifespan of the adults. Especially females may be quite long-lived (Table 7.12)

Plecoptera are suggested to maintain their specific life cycles and flight periods particularly rigidly, and to emerge at precisely the same time every year. In the literature emergence periods are described as synchronous and short, and diagrams depict the regular stepwise sequence of different species emerging in temporal succession (e.g., Aubert, 1946; Brinck, 1949; Illies, 1955; Kerst and Anderson, 1974). The Plecoptera are regarded as prime examples of competition avoidance through temporal vicariance, especially between congeners, as postulated by Monard's (or Gause's) principle (Illies, 1952a, 1952b). Flight periods were believed to be short, and to occur at the same time each year (Illies, 1953a, 1953b, 1955). Our data contradict some of the beliefs. In particular, emergence periods are much longer than the presumed flight periods, and the closest relatives occur at the same time (Figure 7.46).

Emergence dates of particular species shift along altitudinal (Kühtreiber, 1934; Brinck, 1949; Ravizza, 1974; Lillehammer, 1975) or latitudinal gradients (Brinck, 1949) suggesting that temperature may be a decisive emergence cue. If they are flexible in this respect, with global warming, Plecoptera flight periods might begin earlier. Otherwise species replacement would be expected. The Breitenbach material invited for tests, short-term aspects are addressed here.

Control of emergence by low temperature appears probable for species emerging so early in the spring that adults might be confronted with lethal air temperatures. In the Breitenbach, *Leuctra prima* and *Protonemura meyeri* at the end of

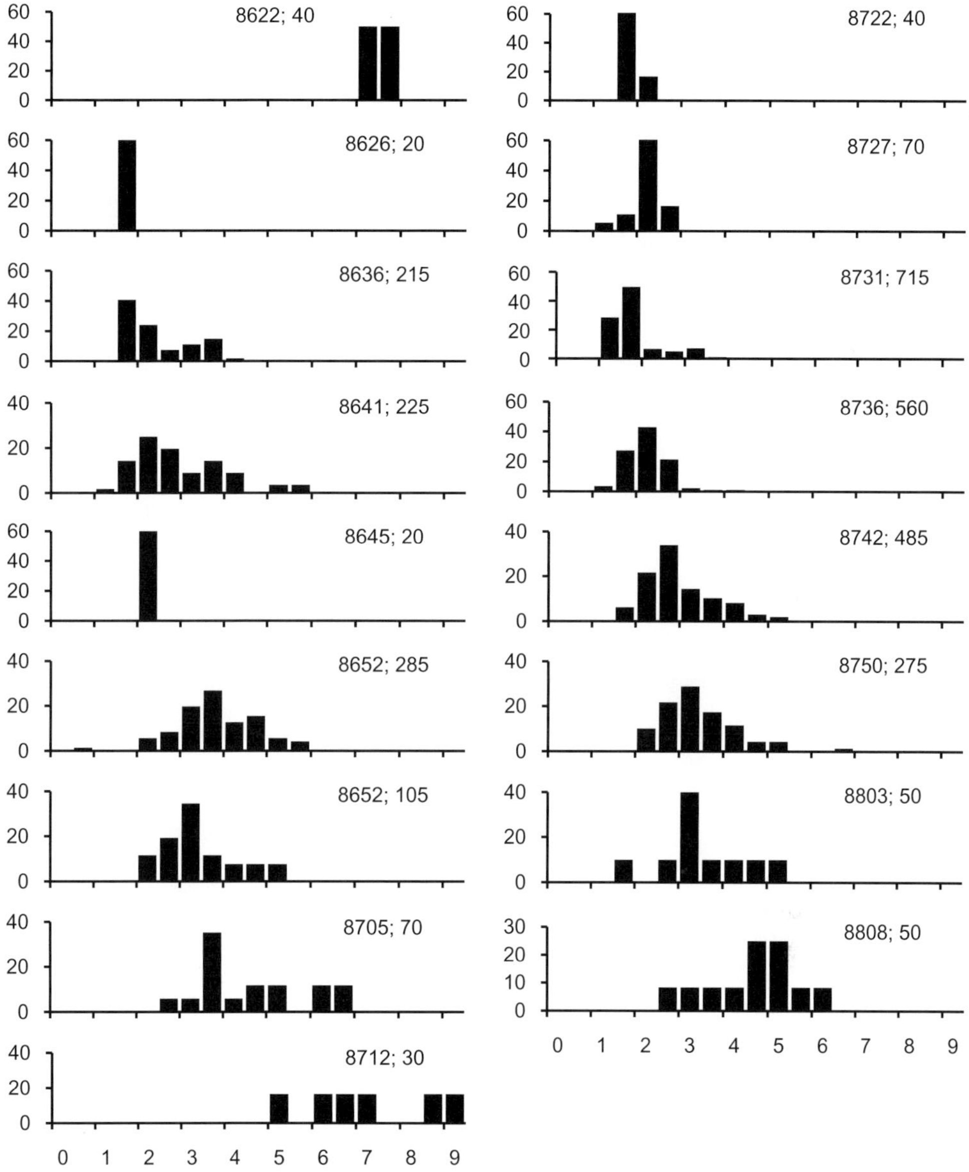

Figure 7.42 *Siphonoperla torrentium* in Reach B of in the Breitenbach. Frequency (percentages in samples; ordinate) of body length classes (abscissa; in mm). Figures in legends are calendar weeks in 1986–1988 followed by approximate specimen numbers m^{-2} streambed.

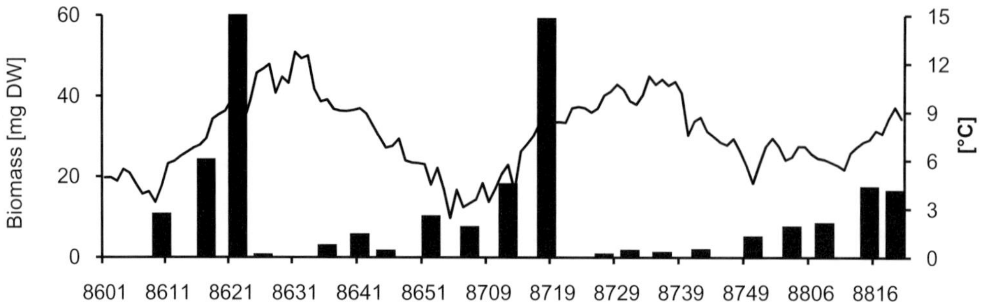

Figure 7.43 Standing stocks of a representative population of 100 larvae of *Siphonoperla torrentium* during the sampling period in Reach B, and mean weekly temperature at Messstation 2. Units on abscissa are calendar weeks in 1986–1988.

Table 7.10 Synchronous emergence of abundant univoltine Breitenbach Plecoptera, with week of mean emergence, number of weeks required for 5–95% of the population to emerge, and rank of average abundance. Thirty-year averages.

	Calendar week of mean emergence	Length of 90% emergence period (weeks)	Rank of average abundance
L. prima	9	7	5
P. meyeri	13	9	11
N. flexuosa	16	9	12
N. marginata	18	9	8
N. cambrica	19	5	6
L. nigra	20	9	1
I. goertzi	22	14	10
P. intricata	22	10	2
S. torrentium	22	8	7
A. standfussi	26	13	9
P. auberti	31	17	4
L. digitata	39	5	3

winter, and *L. digitata* in late autumn are relevant (Figure 7.44). Small numbers of *L. prima* occasionally emerged at around 2 °C which is the lowest mean weakly temperature at Trap 2, which occurred on less than 0.3% of all occasions. To avoid being misled by single aberrant specimens emergence < 5% was not considered. Regular emergence began at 4 °C, a weekly mean temperature attained on only 2% of all occasions. Emergence of *L. prima* was complete by the time mean water temperature attained 8 °C. However, low air temperature may temporarily delay emergence (Figure 7.51). *Protonemura meyeri* also began to emerge regularly at 4 °C, although most specimens appeared between 5–9 °C. Most *Leuctra digitata*

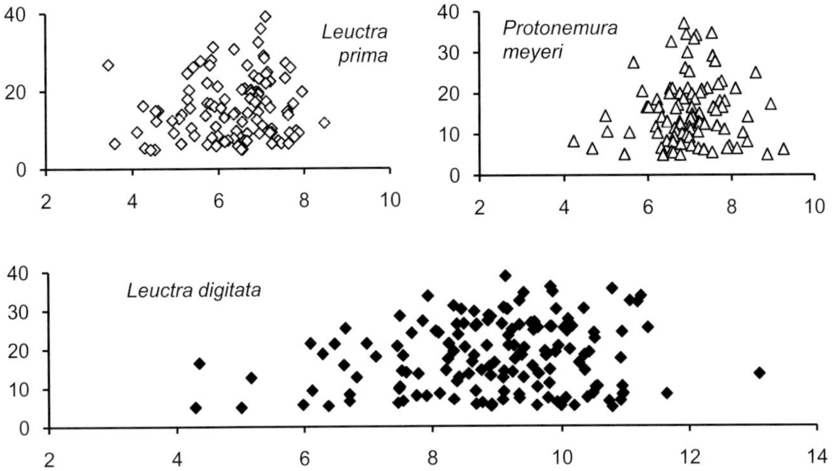

Figure 7.44 Weekly share of annual emergence in T2 (ordinates; percentages > 5%) of the earliest and latest emerging Plecoptera in the Breitenbach at the mean temperatures (abscissa) recorded during the respective weeks in 1974–2004.

emerged while the stream was still warm, although small numbers hatched just before freezing. However, all larvae emerged, none were left in the benthos.

There were evidently no temperature constraints on the emergence of these species in the Breitenbach. Peak values in Figure 7.44 are not necessarily preferred emergence temperatures, additional different factors, in particular daylength, may affect emergence.

Some late summer larvae of *Nemurella pictetii* overwinter, keep growing and emerge the next spring (Nesterovitch and Zwick, 2003). Temperature proved to be of decisive importance for developmental pathways of *N. pictetii* when the Breitenbach was compared with the upper section of River Fulda which experiences a less benign regime. However, cohort splitting is caused by the acceleration of growth of some larvae by summer warmth, rather than by autumnal cold retarding growth (Wolf and Zwick, 1989; Nesterovitch and Zwick, 2003).

Mean temperature and daylength are correlated and change seasonally but not synchronously in the Breitenbach, which on long-term averages, is warmest around calendar week 33, when daylength is already declining rapidly. Daylength does not change between years and is therefore the more reliable predictor of seasonal change. The two separate annual cohorts of *Nemurella pictetii* were analyzed for the effects of daylength and temperature. While the vernal cohort develops under the rising arm of the seasonal curves of both daylength and temperature, the estival cohort emerges under declining daylength while temperature still rises but drops after Week 34.

Temperature appeared to control emergence. Emergence of both cohorts started or ended, respectively, at about 8 °C. The quite different daylengths in spring and late summer had no recognizeable effect (Figure 7.45).

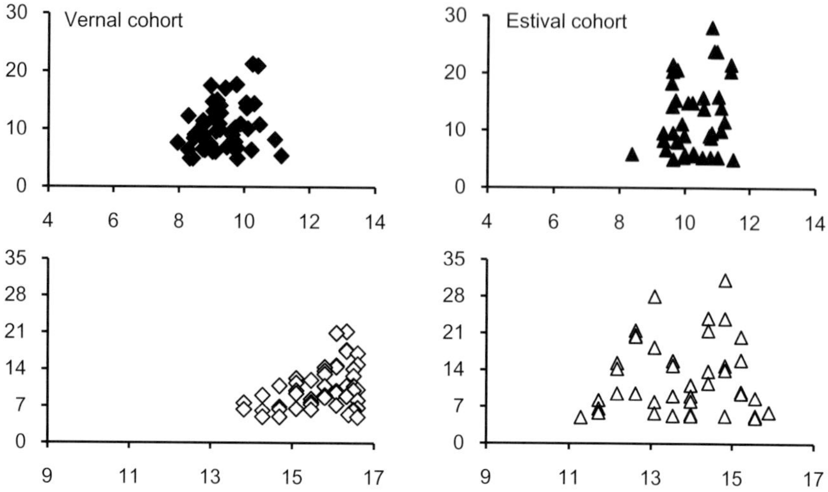

Figure 7.45 Weekly share of annual emergence (ordinate; percentages > 5%) of the two seasonal groups of *Nemurella pictetii* in the Breitenbach plotted against mean temperature (°C; top) and daylength (h; bottom). Data from Trap 2, 1991–1995 and 1998.

Patterns of Stonefly Emergence from the Breitenbach

Unimodal Patterns Most species exhibit synchronous unimodal emergence at the same time every year (Figure 7.46). The two rare *Brachyptera* species follow the same pattern.

Emergence of individual species continues for much longer than the literature suggests. Five to 14 weeks are required for the central 90% of populations to emerge, most frequently nine weeks (Table 7.10). In interspecific comparisons, length of emergence period depends neither on the number of specimens emerging (expressed by mean rank of abundance) nor does it matter during which season emergence occurs.

Plurimodal Emergence Stoneflies are either univoltine or semivoltine, there are no regularly plurivoltine species. Bimodal emergence of several stonefly species in the Breitenbach nevertheless occurs, at least in certain years or at some sites, because of different modes of cohort splitting.

In *Isoperla goertzi*, the split occurs because eggs may develop directly or enter diapause. When low numbers of *I. goertzi* emerge, emergence tends to be synchronous and unimodal. When the pattern is bimodal, *I. goertzi* is usually more abundant, with overlapping groups of univoltine and semivoltine specimens. The first emergence peak is at the same time in all traps, but the second is earlier and ends sooner at the downstream sites (Figure 7.47).

In Central Europe, *Protonemura auberti* is common in springs and epirhithral spring runs, but extends downstream into metarhithral reaches. Under the almost

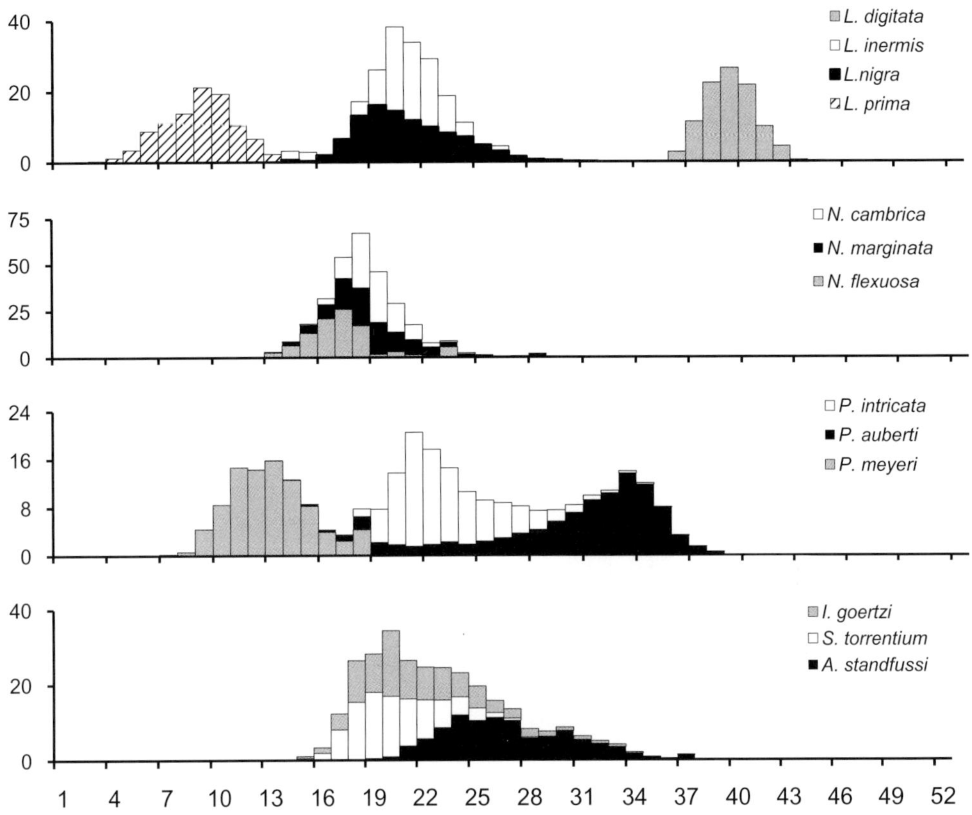

Figure 7.46 Standard patterns of unimodal emergence of the abundant Plecoptera in the Breitenbach. Ordinates are percentages of annual emergence in Trap 2, averages during 30 years of trap collections. The abscissa shows calendar weeks. A = *Amphinemura*, I = *Isoperla*, L = *Leuctra*, N = *Nemoura*, P = *Protonemura*, S = *Siphonoperla*.

aseasonal temperature regime in springs, *P. auberti* emerges over many months, for example in Trap 1 (Figure 7.48). Further downstream, in Trap 2 and the others with a distinctly seasonal regime, the species is less abundant and exhibits synchronous summer emergence.

The sand-dwelling *Leuctra nigra* took two years to complete its development under a seasonal temperature regime fairly similar to that of the Breitenbach, but grew faster and became univoltine under higher constant temperatures (Elliott, 1987b). In our own laboratory, developmental mode of siblings differed under different constant temperature regimes, and cohort splitting occurred. Emergence of *L. nigra* from the Breitenbach is variable. In Trap 1 emergence is extended but unimodal. An occasional second peak of abundance is not caused by emergence but by late ovipositing females. In Trap 2, bimodal emergence occurred in some years, but not in others (Figure 7.49). This probably reflected cohort splitting but

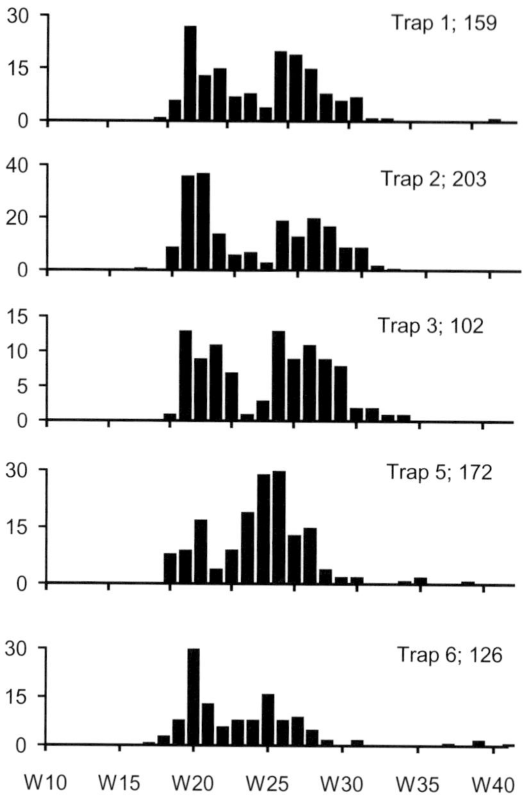

Figure 7.47 Bimodal emergence of *Isoperla goertzi* resulting from overlap of uni- and semivoltine cohorts. Ordinate: numbers emerging; abscissa: calendar weeks. Legends list traps and numbers caught in 1981.

mean weekly water temperatures during the respective and proceeding years suggest no apparent cause for the switch between developmental modes.

Nemurella pictetii follows several developmental pathways and is semivoltine, univoltine, or partially bivoltine. It is the only stonefly that produces more than one generation per year. Central European populations were studied in detail (Zwick, 1977; Wolf and Zwick, 1989; Nesterovitch and Zwick, 2003). Emergence is bimodal or trimodal, but may be unimodal in West and North Europe (Iversen, 1978; Hildrew, Townsend and Henderson, 1980; Fiałkowski, 1986). The Breitenbach pattern is bimodal (Figure 7.50), a mix of uni- and bivoltine cohorts. There is no stable precedence of one sex over the other during emergence. The population includes normally growing univoltine specimens, plus fast-growing individuals that develop to adulthood in about 100 days (Wolf and Zwick, 1989). The spring peak is a mixture of the two groups. The summer peak includes only the fast-growing kind that hatch from eggs laid in spring. Siblings from a single egg mass may behave in either way (Wolf and Zwick, 1989; Lieske, 2005).

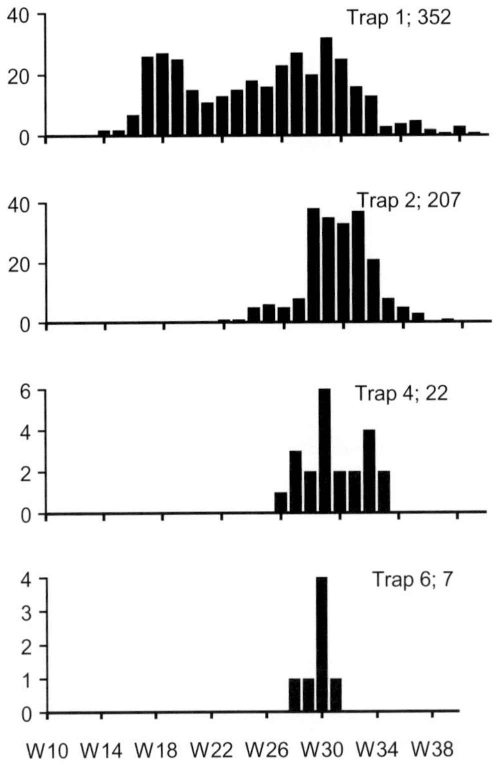

Figure 7.48 Bimodal and unimodal emergence patterns of *Protonemura auberti* along the Breitenbach. Ordinate: numbers emerging; abscissa: calendar weeks. Legends list traps and numbers caught in 1983.

Short-Term Control of Emergence by Abiotic Factors

Water and Air Temperature Early emerging *Leuctra prima* may encounter air temperatures that are markedly below water temperature and run a high risk of death from nocturnal frost. Adults normally descend into snow at night, but when the snow surface freezes many are found dead (Zwick, unpublished data). In a laboratory experiment, mature last instar larvae kept at 4 °C water temperature but 2 °C air temperature were repeatedly observed leaving the water briefly, but none emerged. As soon as the air was allowed to warm to the water temperature, numerous individuals emerged. The significance of low air temperatures was occasionally also apparent in the field when emergence was reduced or stopped during cold spells even though water temperature remained in the range supporting emergence (Figure 7.51).

In other species, falling temperature may also reduce emergence, even though the temperatures are not dangerous to the insects. It is not known if specimens do not emerge or whether emerged juvenile adults remain hidden between littoral debris and vegetation.

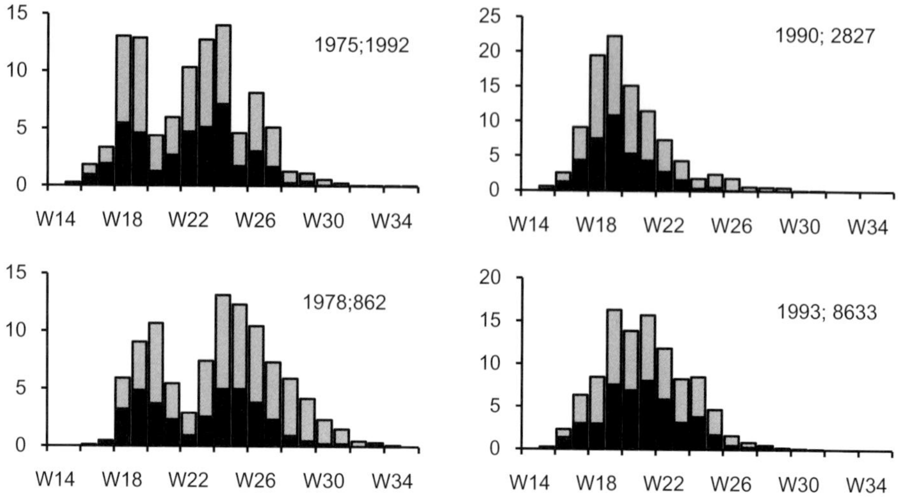

Figure 7.49 *Leuctra nigra*, Trap 2, bimodal and unimodal emergence patterns; males dark, females light bars. Ordinate: percentage emerging; abscissa: calendar weeks. Legends list years and numbers caught in Trap 2.

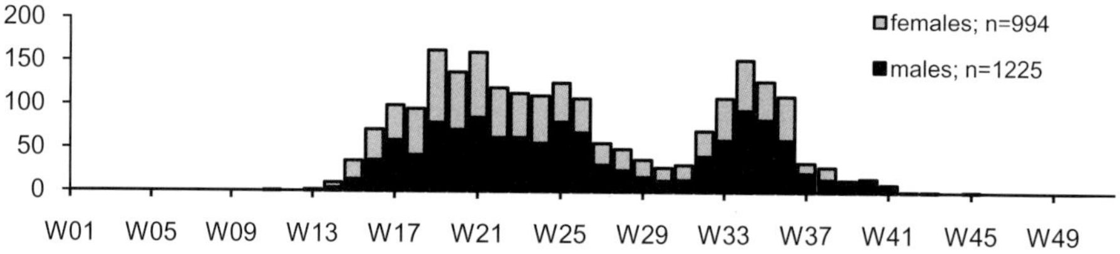

Figure 7.50 Bimodal emergence of *Nemurella pictetii* (Trap 1, 1983) resulting from coexistence of univoltine and bivoltine cohorts. Ordinate: numbers emerging per week. Abscissa: calendar weeks.

Adverse weather, extreme pH, discharge, lack of food, and other factors must also affect younger specimens and probably delay growth, development, and metamorphosis. Such delays cannot be recognized in emergence patterns because the time requirements of the aforementioned processes are unknown in the hemimetabolous mayflies and stoneflies.

Episodic Temperature Effects Episodic environmental impacts cause temporary lows in emergence but do not completely prevent emergence. Such effects can only be distinguished from normal variations in numbers emerging during massive emergence, when numbers drop simultaneously in all traps, or at the same time in several species. However, it was generally rare for different species to emerge in sufficiently large numbers at the same time, so the number of documented environmental impacts on emergence is not large.

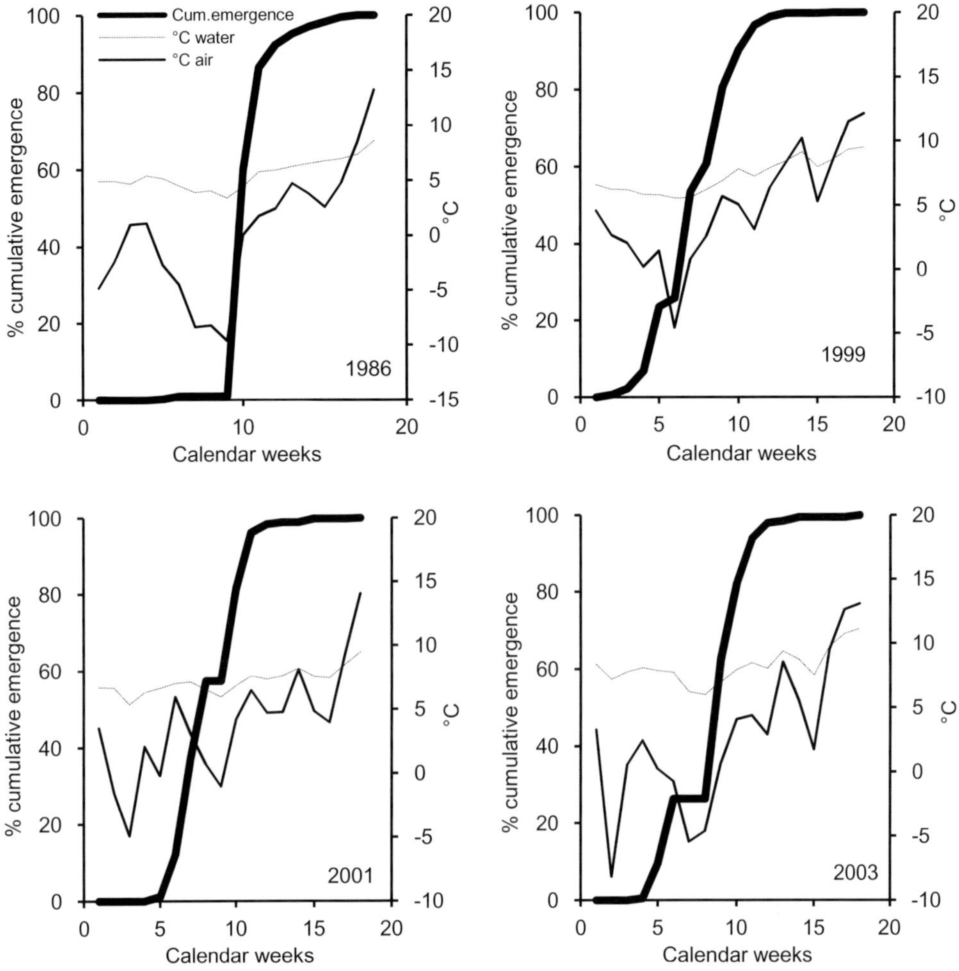

Figure 7.51 Cumulative emergence of *Leuctra prima* (Trap 2) delayed by frosty air at times when water temperatures would have supported emergence.

Sudden low emergence, apparently caused by a drop in water temperature, was most frequently observed in *Leuctra prima* ($n = 6$), *Paraleptophlebia submarginata* ($n = 4$), *Protonemura intricata, Siphonoperla torrentium* (each $n = 2$), *Ephemerella mucronata, Amphinemura standfussi, L. nigra,* and *Isoperla goertzi* (one case each; Figure 7.52). Effective drops in temperatures were on average 0.9 °C lower than the preceding week. This may seem little but is actually as much as the two-week difference in long-term average temperatures in the Breitenbach. The difference also corresponds to about 10% of the annual mean temperature in Central Europe, or to ca. 130 m difference in altitude.

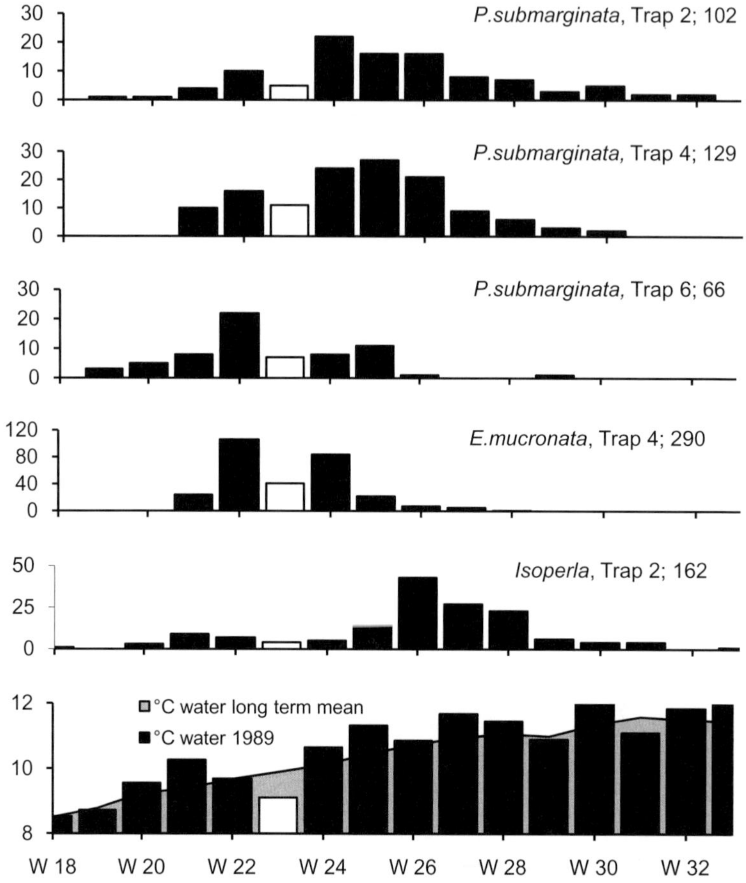

Figure 7.52 A sudden drop of water temperature below long-term mean (bottom) depressed emergence of several Ephemeroptera and Plecoptera (top). *Paraleptophlebia submarginata*, *Ephemerella mucronata*, and *Isoperla goertzi* provide examples. Ordinate: numbers emerged; and mean weekly water temperature, respectively. Abscissa: calendar weeks in 1989. White bars highlight exceptional week.

Discharge Discharge is the most important habitat engineer in streams, through sediment re-arrangement. An impact of abruptly rising discharge on emergence was rarely seen, best in *Leuctra prima* (Figure 7.53). A second spate three weeks later had less impact, probably because most of the population had already emerged by then.

At the times when most other Plecoptera emerge, spates were uncommon, but one summer spate affected several species in all traps in the same way. After the spate, all emerged in numbers similar to those before the event, there were no apparent losses from the populations (Figure 7.54).

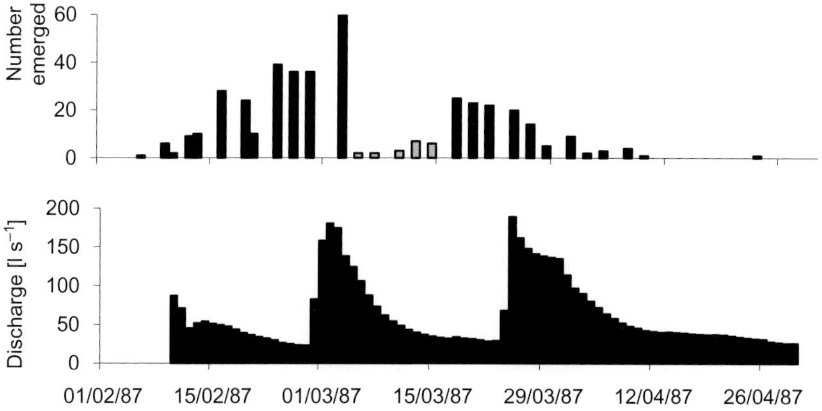

Figure 7.53 Reduced emergence of *Leuctra prima* (top) following a spring spate.

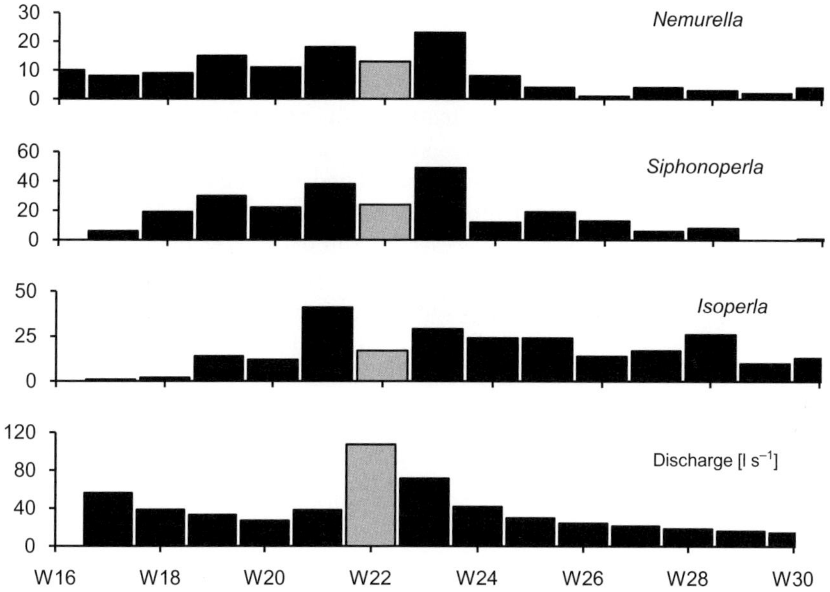

Figure 7.54 Decreased emergence of several stoneflies during a summer spate. Ordinates are numbers emerged (top three) and discharge (l s^{-1}; bottom); the abscissa shows calendar weeks in 1995.

Global Irradiance and pH In one case each, *N. flexuosa, P. intricata, L. prima,* and *S. torrentium* exhibited distinctly reduced emergence when total global irradiance dropped abruptly, with a simultaneous rise of air humidity, during bad weather. This is plausible but, because not all abiotic parameters were available for comparison, the influence of fluctuating light intensity on emergence remains uncertain.

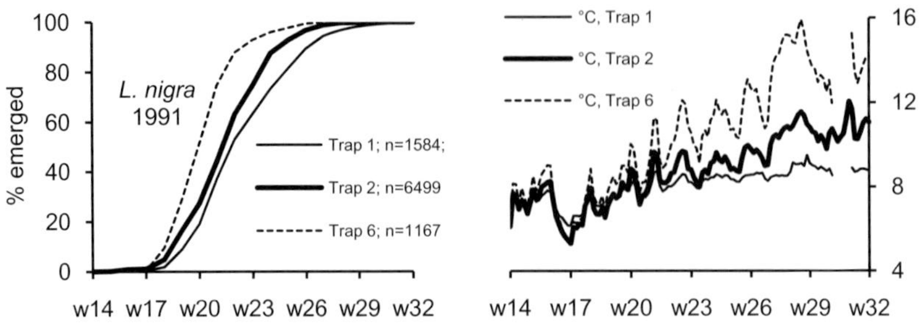

Figure 7.55 *Leuctra nigra*: when cumulative emergence starts in spring water temperature is the same along the stream. Emergence ends first in the trap with the highest, last in the trap with the lowest summer temperature. The abscissa shows calendar weeks.

The acidity of the Breitenbach water usually varied more strongly over a single day (Section 3.3.2) than between successive days. Accordingly, pH never appeared to impact emergence.

Shift of Emergence along the Stream Emergence periods in Trap 1 under strong thermal influence from the spring frequently differ from those in other traps. To avoid the possible influence of numbers emerging on the length of the emergence period, species that are more or less equally abundant along the Breitenbach were checked for the importance of temperature. Emergence starts synchronously along the stream early in the season. However, later in the year, when there are differences in mean daily summer temperatures between sites, emergence tends to end sooner at downstream than at upstream sites for several stoneflies and mayflies (Figure 7.55; e.g., also *Isoperla*, Figure 7.47, and Ephemerellidae). However, a causal relationship remains unproven. The alternative assumption, that lifetime accumulation of thermal energy at the different sites might explain the phenomenon, was not supported by our data.

7.6.4.3.4 **Plecoptera Reproduction and Development**

Sex Ratios

Over many years of emergence trapping, the sex ratios of species from the Breitenbach varied greatly between years. In the longer perspective, none of the species has a well balanced sex ratio. Female bias in several species is mainly due to the appearance of old, ovipositing females in traps (Figure 7.60), but the Nemouridae and *Leuctra prima* consistently have some excess males. Bias in sex ratios is probably related to both differential behavior and longevity of the sexes, as well as the trapping method. Petersen *et al.* (1999) reported a strong female bias in several species, including *L. nigra* and *Nemurella pictetii* (Figure 7.56).

As soon as they hatch Perloidea males actively search for females. *Dictyogenus* males were shown to leave traps before they could be collected (Zwick, 1982a), and the same probably applies to *Isoperla* and *Siphonoperla*. In some years, addi-

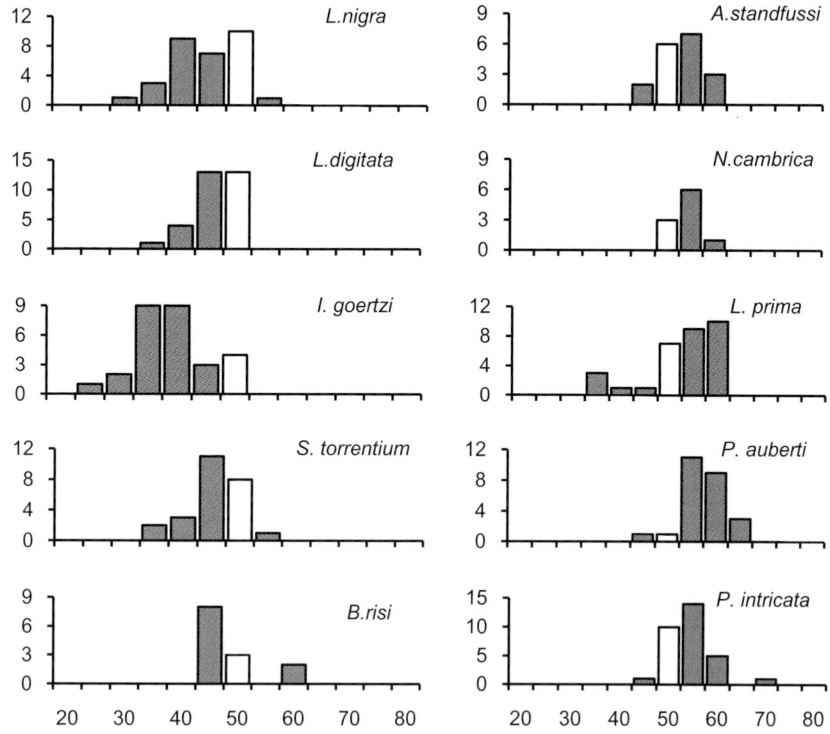

Figure 7.56 Frequency (ordinate) of certain sex ratios (percentage of males) in annual emergence trap collections (minimum 100 specimens) of several Plecoptera from the Breitenbach. The 50% value is highlighted in white. Taxa in the left column are female, those in the right column are male-biased.

tional females of both species arrived after the emergence season, when males were still alive. Presumably, they had returned to the stream for oviposition and were trapped in the process. These errors could be avoided by sexing large numbers of pre-emergent, late instar, larvae during the entire emergence season. However, the numbers taken in Breitenbach benthos samples were too low to do this.

Protandry

Except for individual outliers, sexes start emerging at the same time. However, there is regularly a distinct protandry in the sense that males are initially more abundant than females, while the reverse is true late in the emergence season. Figure 7.57 presents several examples.

Mating Encounters and Mating

Long distance attraction of sexual partners by swarming or pheromones is unknown in the Plecoptera. Accidental encounter of the sexes is aided by the

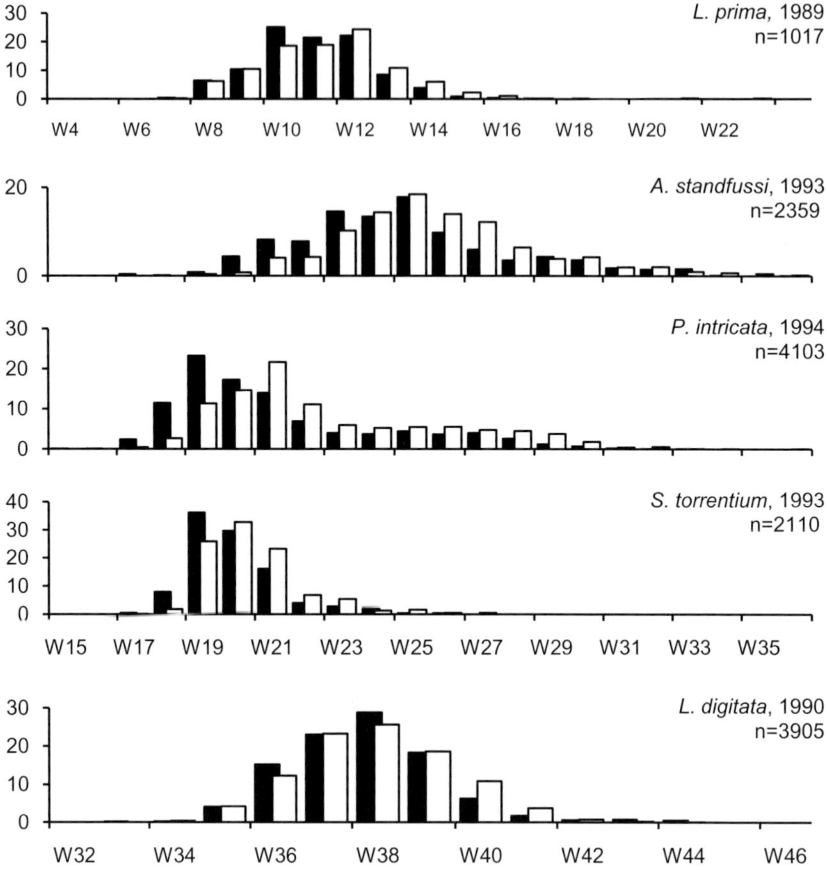

Figure 7.57 Examples of protandric emergence of several Plecoptera in Trap 2 of the Breitenbach. Ordinate: percentage of annual emergence of males (black bars) and females (white bars). Abscissa: calendar weeks.

animals' inherent tendency to climb to the tops of bushes, rocks, or other substrata ("hilltopping"; Stewart, 1994). Plecoptera emerging from streams in open terrain, without dense riparian growth, leave the banks and fly away, often uphill, to optical marks which become mass aggregation sites (Harz, 1966). Species behave differently. For example, many hundred *Nemoura marginata, Protonemura auberti, Nemurella pictetii,* and sometimes *Brachyptera* spp. gather every year on trees and walls behind the source of the River Fulda in the Rhön. They arrive from downhill springs and spring runs, but *N. cinerea* and *Leuctra nigra,* which also develop there in large numbers, never appear (Zwick, unpublished observations).

Drumming signals produced by males have long been known (Newport, 1851), and are characteristic of the order. Male abdominal beats are not audible but cause substratum vibrations which are perceived by female genual receptors, mostly over a range of a few decimeters, but up to 8 m on dead wooden branches (Stewart and Zeigler, 1984). Male Chloroperlidae do not even hit the substratum, vibrations are transmitted to the substrata via their legs (tremulation; Rupprecht, 1981). In the Taeniopterygidae, Nemouridae, and Leuctridae drumming is also widespread but not easily observed and was therefore discovered much later than the drumming of Systellognatha. On sunny days, male *Isoperla goertzi* are the only, easily observed, drummers along the Breitenbach.

Nemouridae and *Brachyptera* spp. must feed before they mate (Hynes, 1942). In contrast, *Leuctra* spp. mate as soon as partners meet, often within emergence traps, as shown by paint particles in their guts. Copulation occurs on solid substrata. Males mount females obliquely, curving the abdominal tip under the female until their parts meet. Similarity of mating position with the Embioptera was used to suggest a close relationship between the orders, but disagreement in all details contradicts this assumption (Zwick, 2009). Modes of mating and of sperm transfer are extremely varied in Plecoptera (Zwick, 1973, 1980). *Leuctra* and *Protonemura* spp. may remain paired for hours while the female feeds. Males mate repeatedly, but second matings of females are uncommon.

Fecundity, Egg Size, and Origin of Biomass Invested into Offspring

Fecundity Adult body size is certainly related to potential fecundity in the Plecoptera, which makes large size a distinct fitness advantage. However, this has not been documented in stoneflies because accurate estimates of fecundity are difficult. In many taxa eggs are not laid all at once, but in several batches (Elliott, 1987a), sometimes at long time intervals during which additional ovarian cycles could possibly occur (Harper and Hynes, 1972). Data in Table 7.11 are estimates of average specimens. Very low egg numbers are characteristic of the Chloroperlidae (Zwick, 1980; Yoshimura, 2003, Tierno de Figueroa and Marfil-Saza, 2004).

Table 7.11 Estimated fecundity of some stoneflies inhabiting the Breitenbach.

Isoperla goertzi and *I. grammatica*	250–350	Zwick, 1980
Siphonoperla torrentium	ca 60	Zwick, 1980
Brachyptera risi	~1500	Khoo, 1968
Leuctra spp.	600 to >2000	Zwick, 1980
Leuctra nigra	~250	Elliott, 1987a
Amphinemura standfussi and *A. sulcicollis*	800–1200	Zwick, 1980
A. standfussi	~500	Elliott, 1988a
Nemoura cambrica	~800	Elliott, 1988b
N. cinerea	~1200	Elliott, 1988b
Nemurella pictetii	1100	Zwick, 1980
N. pictetii	~500	Elliott, 1988b
Protonemura meyeri	~1500	Elliott, 1988b
P. meyeri	~570	Marten, 1990
P. auberti	~1200	Marten, 1990

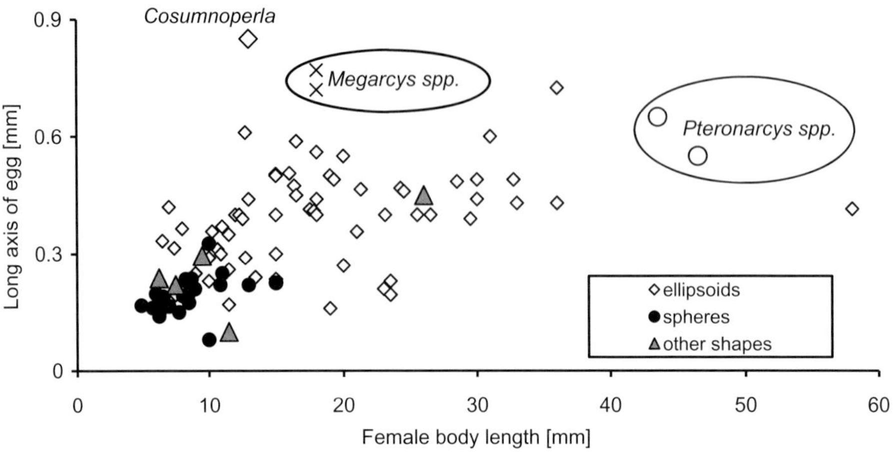

Figure 7.58 Relationship between maternal body length and long axis of eggs. Black spheres are Nemouroidea. Original Zwick, sources available upon request. Ordinate: specimen numbers. Abscissa: calendar weeks.

Egg Size Maternal size does not affect individual egg size in the Plecoptera (Zwick, 1999a). In interspecific comparisons, stonefly egg size is weakly positively correlated with maternal size, irrespective of shape and structural details (Figure 7.58). Some taxa with exceptionally large eggs are striking outliers, but the ecological significance is unknown. A similar situation occurs in mayflies (Fink *et al.*, 1991).

Spherical eggs are the most compact geometrical form. The Pteronarcyidae are the only large species with ball-shaped eggs, which have a rigid, multilayered chorion. In contrast, the almost spherical eggs of the Nemouroidea save on chorionic materials; their egg shell is thin, often described as a membrane. Its complex structure was only recently discovered (Rosciszewska, 1996). The soft eggs of the Nemouroidea are tightly packed within the small female bodies and are the smallest examples in the order. Pre-oviposition eggs of *Isoperla* spp. are concave on one side while eggs of many large Perlodidae are triangular in cross section. Those of *Perlodes microcephalus* and *P. dispar* are tetraedric, which permits optimal packing inside the maternal body.

Origin of Egg Biomass Only a few Plecoptera have reduced mouthparts that are unsuitable for consuming particulate food. In the autochthonous Breitenbach fauna, only adult *Isoperla* do not feed. Although egg mass must be derived from maternal biomass, *I. goertzi* emerges with only partly developed eggs. Eggs are yolked and mature within approximately one week, at the expense of a set of larval swimming muscles which the adults no longer need (Zwick, 1973).

Females of all other Breitenbach Plecoptera feed and are sufficiently long-lived (Table 7.12) to build up considerable biomass (Figure 7.59). They consume various

Table 7.12 Female longevity (in weeks) of selected Plecoptera from the Breitenbach estimated from average age of ovipositing females appearing in the upstream Trap 1 after the normal emergence period ended. The number of years in which sufficiently many females were available is indicated for each species.

	L. prima	*P. meyeri*	*N. cambrica*	*N. marginata*	*N. flexuosa*	*S. torrentium*	*I. goertzi*	*L. nigra*
Weeks	5.4	6.3	5.3	5.0	6.2	3.4	3.0	7.7
Years of observation	7	2	3	4	3	3	1	1

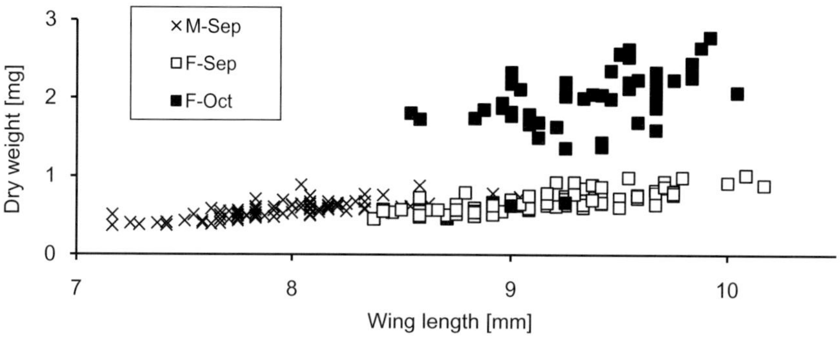

Figure 7.59 Leuctra digitata, wing length and dry weight of freshly emerged males (M) and females (F) taken in September, and of aged females taken in October, when males were no longer taken.

terrestrial detritus, unicellular algae, Cyanobacteria, fungal hyphae, and the like, grazed off bark or stones. The Chloroperlidae typically consume large amounts of conifer pollen which they gather from leaf surfaces, gaining considerable weight (Zwick, 1990). Adult Plecopteran food consumption was recently documented in a number of taxa (e.g., Smith and Collier, 2000; Tierno de Figueroa and Fochetti, 2001; Tierno de Figueroa and Sánchez-Ortega, 2000).

Female Longevity, Upstream Oviposition Flights, and Oviposition

Female Longevity Older Plecoptera are visibly darker than young or even freshly emerged ones, and additionally, the abdomen of old females is either swollen, because it is full of eggs, or empty and shriveled after oviposition (spent females). It can be assumed that the first emerging females are also the first to oviposit, while the latest emerging ones oviposit last. Based on this, patterns such as those shown in Figure 7.60 provide estimates of female longevity, using the time difference between first, peak and last emergence, and first, peak and last oviposition flights, respectively. Female Leuctridae and Nemouridae clearly live for over a month, female *Siphonoperla* for more than three weeks (Table 7.12).

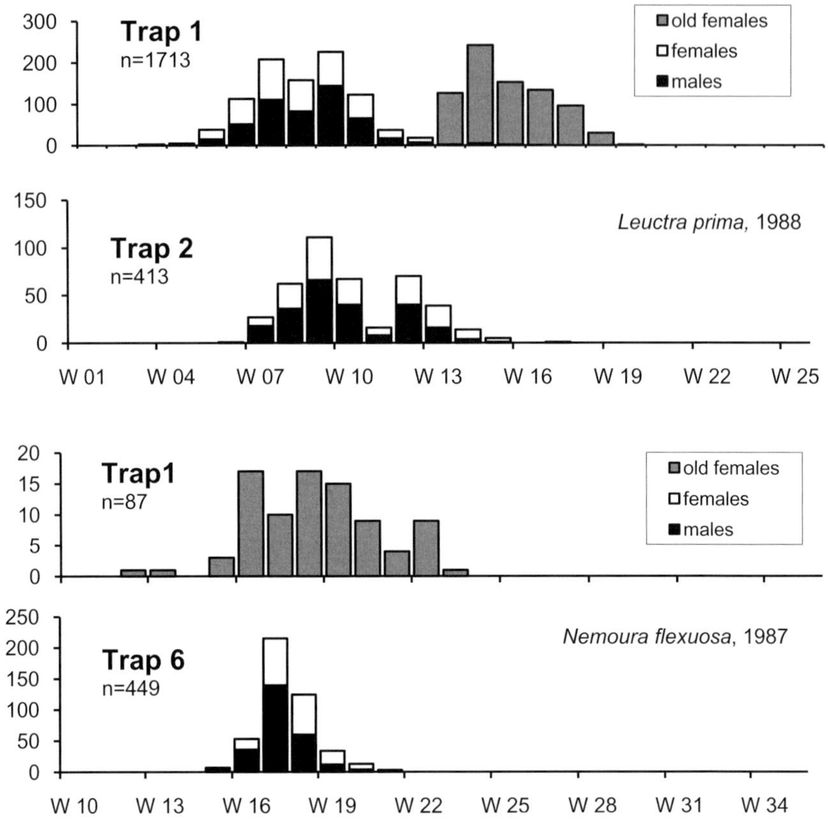

Figure 7.60 Upstream oviposition flights suggested by large numbers of old females of *Leuctra prima* appearing after actual emergence, but only in upstream Trap 1. Old females of *Nemoura flexuosa* appeared in the uppermost trap where the species never emerges. Modified after Zwick (1990). Ordinate: specimen numbers; abscissa: calendar weeks.

Upstream Oviposition Flights Females probably locate streams for oviposition visually (Schwind, 1991; Kriska *et al.*, 1998). Circumstantial evidence, especially the occasional occurrence of hyporhithral and potamal species (*Hydropsyche* spp., *Rhyacophila nubila, Perlodes dispar*) in the lower Breitenbach suggests that females sometimes oviposit into waters that are atypical for the species concerned.

The Plecoptera of the Breitenbach provided direct and circumstantial evidence for upstream oviposition flights. Sticky traps placed over the stream at various sites trapped females with egg masses almost exclusively next to Trap 1 (Zwick, 1990). Upstream oviposition flights were also inferred from large numbers of old females, either spent or with ready-to-lay eggs (Zwick, 1990), which appeared in some years at the end of, or after, emergence, but only in the upstream trap, Trap 1 (Figure 7.60). Females of *Nemoura flexuosa* carrying egg masses were caught on sticky traps next to Trap 1, even though *N. flexuosa* actually emerges only at more downstream sites (Zwick, 1990).

Upstream oviposition flights also apparently lead to a shift in size distribution along the stream, between emergence and oviposition of *Leuctra prima*. The species emerges after frost, usually in late winter. After mating, females forage at some distance from the stream to which they later return (in flight) to oviposit. While wing lengths of emerging females increased in a downstream direction, wing lengths of old ovipositing females caught in upstream traps increased in an upstream direction. Evidently, large females had flown upstream from downstream sites. Their winglengths exceeded those of females that had emerged at the upstream sites (Zwick, 1990).

Modes of Oviposition No Breitenbach stonefly emerges with ready to lay eggs; oviposition commences at least one or two weeks after emergence (*Isoperla, Siphonoperla*). Plecoptera preferentially oviposit on overcast, warm, humid afternoons. Females extrude egg masses, which differ in shape between families, and all local species carry them to water in flight (Zwick, 1980). Stoneflies as well as mayflies may mistake glass plates, dark car roofs or plastic foils for water because these materials reflect polarized UV light in a similar way (Schwind, 1991; Kriska *et al.*, 1998). Egg masses for experiments could therefore be obtained by collecting females that were trying to oviposit on the glass roofs of emergence traps. *Leuctra* spp. can release egg masses in the air to drop into the stream. Most other taxa fly down to wash their eggs into the water, when the current may carry the females away for some distance before they rise again from the water, sometimes under an emergence trap. At preferred oviposition sites, hundreds of spent females, but hardly any males were trapped (Zwick, 1977).

Egg masses disintegrate in water, eggs of the Nemouridae in an explosive manner. Individual eggs sink and are apparently washed into the interstices of the stream bed before, after a short time in water, their surface coating or special attachment disks, the so-called anchors (Perloidea), become sticky (Zwick, 1980).

Egg-carrying females of *Nemurella pictetii* were not seen in flight, they must walk to oviposition sites. Females seem to oviposit preferentially on stones protruding above the water surface, but other oviposition sites are also used (Nesterovitch and Zwick, 2003). Stones at the Georgsborn are also used for oviposition by other merolimnic insects, for example, the caddisfly, *Apatania fimbriata* (Enders and Wagner, 1996). Egg clutches of both species consist of a gelatinous mass and can, at first sight, be confused. Egg masses of *N. pictetii* are fairly flat, often irregular and laid close together, neighboring masses may overlap and become confluent, and eggs are white (Figure 7.61, left). In contrast, egg masses of *Apatania fimbriata* are spherical, always deposited separately from one another, and eggs are yellow (Figure 7.61, right; R. Lieske, personal communication).

Oviposition of *Leuctra nigra*, the most common stonefly in the Breitenbach, is enigmatic. On suitable afternoons hundreds of pregnant females suddenly climb up grass stems on the pastures some metres from the stream. They do not fly but soon simply disappear. Over several decades of study, only three flying females

Figure 7.61 Egg masses of *Nemurella pictetii* (left; in the foreground at least two egg masses are visible) and *Apatania fimbriata* (right) from the spring of the Breitenbach (photo by R. Lieske).

with eggs were ever taken. Captive females died without laying eggs (Zwick, unpublished data).

Direct Egg Development

Immediately after oviposition direct development starts with a significant uptake of water and swelling of eggs (Zwick, 1999a). Parthenogenesis has only been observed in experiments, in a small number of eggs (Degrange, 1958), but does not seem to be significant in nature.

Temperature and Photoperiod Temperature dependence of egg development is similar to the Ephemeroptera. However, the Plecoptera have the lowest developmental threshold temperature (near 1 °C) of several insect orders (Honek and Kocourek, 1990).

Temperature dependence is usually studied at constant temperatures in the laboratory, for example Marten (1990). However, development under fluctuating temperature is often thought to be faster than at the same constant mean temperature (Section 7.9). In principle, the relationship of incubation period with temperature is not linear (van t'Hoff's rule), acceleration during warm periods must be expected to cause faster development than at the arithmetic mean daily temperature. Different plecopteran species behaved differently in experiments (Lillehammer *et al.*, 1991); there was no uniform reaction to fluctuating temperature.

Acceleration of development by warmth, indicated by a negative slope of the linear regression between incubation period and temperature (logarithmic scales), is normally observed under a Central European climate (Table 7.13). However, in other climatic regions there are also species that do not react to temperature (regression slope near zero), or whose egg development is retarded by rising temperature (positive slope). Pritchard *et al.* (1996) proposed that all species with

Table 7.13 Regression parameters to calculate mean incubation period (MIP, in days) as function of constant incubation temperature (*T*, in °C) for selected Central European Plecoptera with direct development as: ln MIP = ln a + b*T. All regressions are highly significant. Species from the Breitenbach in bold script.

Species	ln a	b	Reference and origin of non-German material
Taeniopteryx nebulosa	4.634	−0.896	Zwick and Hohmann, 2003
T. auberti	5.911	−1.089	T. Schmidt, 1991
Amphinemura sulcicollis	5.820	−1.020	Brock, 1985 (Harz Mts)
A. sulcicollis	5.829	−1.056	Zwick, unpublished data (Rhoen Mts)
Nemoura avicularis	5.596	−0.951	Zwick, unpublished data
N. cambrica	5.927	−1.117	Zwick, unpublished data
N. cinerea	5.476	−0.850	Brittain and Lillehammer 1987 (Norway)
N. flexuosa	6.066	−1.190	Zwick, unpublished data
N. marginata	5.927	−1.107	Zwick, unpublished data
N. sciurus	5.905	−1.162	Zwick, unpublished data
N. uncinata	5.849	−1.079	Zwick, unpublished data
Nemurella pictetii	5.268	−0.930	Wolf and Zwick, 1989
Protonemura auberti	5.497	−0.884	Marten, 1990
P. hrabei	5.347	−0.713	Marten, 1990
P. meyeri	5.649	−1.009	Marten, 1990
P. nitida	5.247	−0.723	Marten, 1990
P. praecox	5.537	−1.016	Marten, 1990
Leuctra albida	6.127	−1.032	Zwick, unpublished data
L. aurita	6.321	−1.047	Zwick, unpublished data
L. braueri	6.120	−0.943	Zwick, unpublished data
L. digitata	5.746	−0.902	Zwick, unpublished data
L. fusca	5.609	−0.789	Elliott, 1987a (England)
L. fusca	5.663	−0.815	Lillehammer, 1985 (Norway)
L. geniculata	5.911	−1.035	Elliott, 1987a (England)
L. hippopus	6.143	−1.135	Zwick, unpublished data
L. inermis	5.576	−0.885	Zwick, unpublished data
L. nigra	6.319	−1.148	Zwick, unpublished data
Leuctra prima	5.922	−1.019	Zwick, unpublished data
Capnia atra	5.220	−0.930	Brittain and Lillehammer, 1987 (Norway)
C. bifrons	5.756	−0.999	Elliott, 1987a (England)
Isoptena serricornis	5.524	−0.875	Zwick and Hohmann, 2003
Siphonoperla torrentium	7.142	−1.367	Zwick, unpublished data
Isoperla difformis	5.538	−0.865	Zwick, unpublished data
I. grammatica	6.019	−0.860	Saltveit and Lillehammer 1984 (Norway)

negative slope be classified as warm-adapted, which would include all species in Table 7.13. However, several of these occur exclusively in permanently cool, montane, habitats and are generally regarded as cold stenotherms. Hatching success at different temperatures must be included in the classification. High hatching success is only achieved over a rather narrow temperature range by most species (e.g., Lillehammer, 1985, 1987; Elliott, 1987a, 1988a, 1988b, 1991a, 1991b; Marten, 1990). Outside this range success is lower on both sides of the peak, rendering the taxa competitively inferior. For the Breitenbach species, peak success is achieved at 5–10 °C, and their classification as "warm adapted" cannot be upheld. Where species lists overlap, incubation parameters of British and Norwegian populations agree closely with our own data on the same species. Photoperiod has no noticeable influence on stonefly mean incubation periods; these were the same under total darkness (experiments by Lillehammer and colleagues) and under short- and long-day regimes.

Photoperiod has no noticeable influence on stonefly mean incubation periods;these were the same under total darkness (experiments by Lillehammer and colleagues) and under short- and long-day regimes (Zwick's present data) in all species compared (Table 7.13).

Comparisons between Taxa

Differences between taxa are best appreciated when the mean incubation periods are compared at average temperatures encountered by the insects in the Breitenbach. Data in Table 7.13 and the long-term data on stonefly emergence were used for this purpose. Oviposition was assumed to begin around mean emergence time of a given species, and average mean incubation period was estimated from long-term mean water temperatures in the Breitenbach during the six weeks after oviposition. Table 7.14 lists the average sequence of emergence and expected peak recruitment for the common Breitenbach species.

The restricted size spectrum of species living in the Breitenbach does not allow a test of whether larger eggs generally exhibit longer mean incubation periods than smaller ones (Gillooly and Dodson, 2000), but data on large Central European

Table 7.14 Long-term (1985–2005) average of weeks of mean emergence, average temperature (°C) during the following six weeks, resulting mean incubation periods (MIP), and weeks of expected peak recruitment of dominant Plecoptera with direct development in the Breitenbach.

Species	Week of mean emergence	Mean temperature during the next six weeks	Average length of MIP (days)	Expected week of peak recruitment
Leuctra prima	9	6.7	54	17
Protonemura meyeri	13	8.0	35	18
Nemoura flexuosa	17	9.4	30	21
N. marginata	18	9.6	31	22
N. cambrica	19	9.9	29	23
Leuctra nigra	20	10.2	39	26
Siphonoperla torrentium	21	10.4	51	28
Protonemura auberti	31	10.8	30	35
L. digitata	38	8.1	47	45

Perloidea support the hypothesis (Marten, 1990; Zwick, 1982a; Zwick and Zwick, 2010).

Congeners Elliott (1987a, 1988b) pointed out that hatching from eggs was spread over the year between the British species of Leuctridae and Nemouridae and contributed to resource partitioning. Whether temporal vicariance at the time of oviposition was maintained through larval life was not addressed. The numerous species of the large genera *Leuctra* and *Protonemura* fall into a number of distinct species groups. Species in the Breitenbach belong to different species groups and are only distantly related. In contrast, the *Nemoura* species from the Breitenbach are siblings in the *Nemoura marginata*-group.

Tables 7.13 and 7.14 show that the closest relatives are most similar in egg biology. For example, peak larval hatch of the extremely closely related *Nemoura* spp. differs by only one week, larval life overlap must be almost complete. In the seemingly widely separate life cycles of *Leuctra* spp. significant overlap occurs through cohort splitting in *L. nigra*. Despite different times of peak recruitment, temporal segregation of *Protonemura* spp. is insignificant because *P. auberti* has an extremely long emergence and recruitment period, which also coincides with peak recruitment of the diapausing species, *P. intricata*.

Differences between Families There are clear differences at family level (Figure 7.62). Directly developing Nemouridae are faster than Leuctridae, which are in turn faster than the chloroperlids. *Siphonoperla* has a longer incubation period than others at the same temperature and lays very few eggs. Data on additional species suggest that this is typical for the Chloroperlidae (Zwick, unpublished data).

Hatching

Stoneflies with direct development usually hatch synchronously over the entire favorable temperature range. Massive hatch occurs within 2–3 days of first hatch. At less favorable temperatures, extended hatching occurs. For example, at 4 °C it

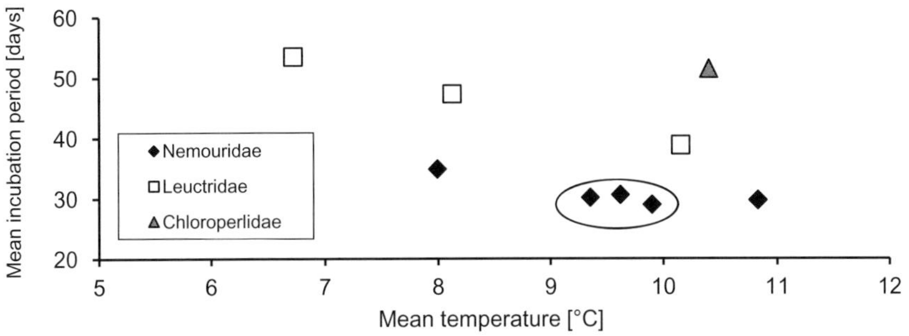

Figure 7.62 Mean incubation periods of common Plecoptera at the mean temperatures under which they normally develop in the Breitenbach. Siblings in the genus *Nemoura* surrounded by an ellipse.

Table 7.15 Plecoptera in the Breitenbach whose eggs undergo diapause.

Species	Reference
Brachyptera risi	Neumann, 1989
Brachyptera seticornis	Neumann, 1989
Amphinemura standfussi	Zwick, 1999a
Protonemura intricata	Zwick, 1981; Marten and Zwick, 1989
Isoperla goertzi	Schwarz, 1970

took 2–4 weeks, at 16 °C it took 1–2 weeks until 90% of all Leuctridae and Nemouridae larvae with direct development hatched (Elliott, 1987a, 1988b). Extended hatching periods are often observed for diapausing eggs.

Indirect Egg Development Involving Diapause

Several Plecoptera in the Breitenbach have indirect development involving a diapause (Table 7.15). Plecopteran egg diapause precedes embryogenesis. Most eggs rest during the warm season and larvae hatch in the autumn. Distinct egg swelling indicates termination of diapause and onset of morphogenesis (Zwick, 1999a). An additional type of diapause occurs in some Perloidea where ready to hatch embryos may remain in the egg for some time, factors triggering ultimate eclosion are not known. It is not clear if the development of *Isoperla goertzi* is of the latter type (Schwarz, 1970).

Egg diapause may be facultative or obligatory. *Isoperla goertzi* eggs undergo facultative diapause, some eggs developing directly, others after diapause (Schwarz, 1970). *Perlodes* eggs in the nearby River Fulda provide extreme examples of facultative diapause. Some siblings develop directly, others remain in diapause for one or two years (Marten, 1991).

The facultative diapause of *Protonemura intricata* is exceptional in that its duration varies with date of oviposition. Early eggs rest for long periods, very late eggs may develop directly (Marten and Zwick, 1989). Eggs of both *Brachyptera* species hatch in autumn, after obligatory diapause lasting for 80–110 days, diapause duration rising with temperature (Neumann, 1989). In Northern Europe, *Amphinemura standfussi* develops directly (Lillehammer *et al.*, 1989), spring emerging populations in Holstein (Zwick, 1967) may be similar. The Breitenbach population emerges in the second half of year, its eggs spend the summer in diapause (Zwick, 1999a). Unfortunately, the distinction between diapause and active development is not always made. For example, Elliott does not mention diapause, but his data on overall mean egg incubation periods in *Brachyptera risi* (Elliott, 1988a) and *A. standfussi* (Elliott, 1988b) suggest that British populations behave like those in the Breitenbach.

A diapause of up to three months occurs in soft-shelled eggs of the Nemouridae and Taeniopterygidae, as well as in eggs of the Perloidea with a hard chorion. Egg rests of more than a year may lead to a long-lived supply of dormant propagules

(Zwick, 1996a, 1996b, 2003c). Such long rests occur only in taxa with hard-shelled eggs. No Breitenbach species is concerned.

Ecological Significance of Egg Diapause The diapause of *P. intricata* eggs appears to be reminiscent of safeguards against summer-dry Mediterranean streams from whence the species immigrated to Central Europe (Zwick, 1981). The same probably applies to *B. risi*. In the Mediterranean, close relatives of both species survive months of complete summer drought in intermittent streams. Even if kept continuously in water, eggs do not hatch before autumn (Malicky, 1982).

Possible advantages of egg rest include:

- Extended hatching periods after diapause spread risk.
- Autumnal hatching may synchronize shredders with seasonal leaf fall (e.g., Cummins and Klug, 1979; Lamberti and Moore, 1984).
- Avoidance of unsuitably high summer temperatures. This may play a role in *A. standfussi*, which seems to have immigrated from Scandinavia and does not hatch at 16 °C (Lillehammer *et al.*, 1989). However, excess heat is not an issue in the cool Breitenbach.
- Competition avoidance through temporal vicariance of ecologically similar, close relatives. Breitenbach Plecoptera offer no examples, but *Ephemerella* species do.

Morphoplogical Change during Plecopteran Growth and Development

During hemimetabolous development, somatic growth, imaginipetal change, and eventually maturation and metamorphosis, are not separated. Instead quantitative and qualitative changes occur simultaneously. As far as can be judged from the external appearance of the animals, for the greatest part of larval life, somatic growth and biomass gain, that is, the quantitative aspects, prevail. Changes of population structure over time and seasonal patterns of growth and final size attained by the various species were documented above. The present section addresses the qualitative changes of the growing individual and the number of instars through which it passes.

Very precise measurements of size are needed for this. Body length (BL) is affected by segment telescoping in life, as well as due to preservation artifacts. Other measures were therefore used. Width of head capsule (HCW) and labrum (LW) are linearly correlated and provide excellent measures of total size. LW of exuviae can be used to back-calculate HCW of specimens before moults. Thus, the complete growth trajectory of individual specimens in captivity can be traced.

Sexual dimorphism of stoneflies is apparent only in the secondary genital characters and in absolute size, the biomass of virgin females being approximately twice that of males. Other than in the Ephemeroptera, there is no significant allometric growth. Shape and overall body proportions are maintained during growth. Even when pterothecae appear and increase stepwise at late instar moults, the other body proportions remain unchanged.

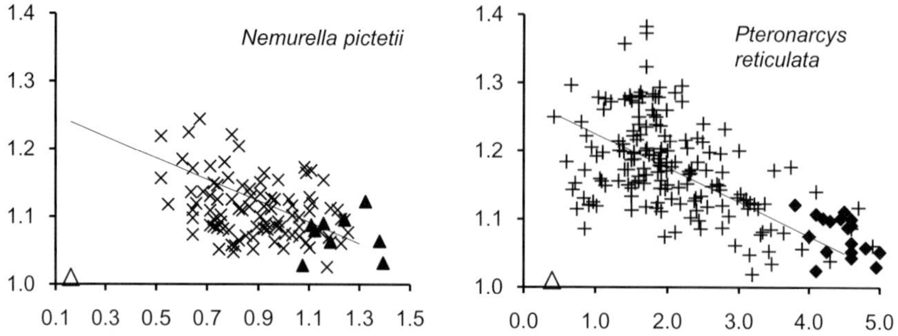

Figure 7.63 Increment at moults (ordinate: growth ratio, i.e., times larger after a molt) in larvae of different size (abscissa, HCW, in mm) of the genera *Nemurella* and *Pteronarcys*. Molt to last instar indicated by full symbols. Triangles on abscissa give the HCW of first instar larvae. Lines mark the hypothetic median growth track (Zwick, unpublished data; and modified from Zwick and Teslenko, 2002, respectively).

During laboratory maintenance of fresh, field-collected larvae as well as during long term laboratory breeding of several species, size of exuviae and the corresponding molted larvae were used to estimate increments at individual moults. Overall growth rates decline with increasing specimen size but, even at the molt to last instar, some growth and weight gain occurs. Individual variation is important and, together with sexual size differences, leads to a continuum of sizes in which instars cannot be distinguished (Figure 7.63).

Brook's or Dyar's rule (Dyar 1890; Crosby, 1973) of constant growth rates at moults was not confirmed; see also Frutiger (1987), or Benton and Pritchard (1988) on the mayfly *Ameletus celer*. However, in other Plecopteran species the assumption of a constant percentage increase at moults seems to hold (Sephton and Hynes, 1982a; Elliott, 1984, 1987a, 1987c, 2009, for 24 British species).

Different growth rates might be expected in species at different trophic levels because assimilation from different types of food varies from about 10% in detritus, 30% in algae, to ca. 70% in animal food (Benke and Wallace, 1980). However, there is no unambiguous relationship, probably because detritivores feed more or less continuously while feeding intervals of predators vary with hunting success.

For example, detritivorous *Leuctra* spp. and the carnivorous *Siphonoperla* are of similar sizes, and both grow up over only 10–11 months. Grazing *Brachyptera* larvae require less time (approximately September–April) to grow to a slightly larger size than the detritivorous Nemouridae. The predators *Perlodes* and *Dictyogenus* grow at very low temperatures, during winter or in perennially cool streams, respectively. The first requires six, *Dictyogenus* 33 months to attain 2.5 cm body length (Zwick and Zwick, 2010). Both are Perlodidae but belong to different tribes which apparently have different metabolic efficiencies. However, there is hardly any information on the physiology of the Plecoptera.

Despite the relatively numerous moults, freshly molted pale stonefly larvae are rarely found. Plecopteran larvae seem to molt mainly at night, subsequent pigmentation is rapid.

Males and females but also specimens of the same sex can attain adulthood and emerge at very different body sizes. Therefore, size alone provides no measure of degree of maturity or proximity of larvae to adulthood but such information is needed to understand key processes in life history regulation, such as frequency of moults, initiation of sexual differentiation and metamorphosis, causes for seasonal size decline, influence of predator or competitor presence on size at emergence, etc. Knowledge of instar number and of morphological change, especially in preparation of metamorphosis, can provide the necessary supplementary information.

Body size, numbers of hair sensilla, cercus and antennal segment numbers, all increase at moults. Gills, taxon-specific length relationships of tarsus segments, specific types of pilosity or color patterns also develop step by step at moults. However, the magnitude of changes varies too much to separate instars, except occasionally the few first ones (Zwick, 1982a, 1982b; Zwick and Teslenko, 2002). The criticism (Fink, 1980, 2008) of the conventional methods for estimating instar numbers also applies here.

In Plecoptera, instar number cannot be counted directly from some sclerotized structure as is possible with the Palmén body of the Ephemeroptera. Instead, indirect approaches are needed that relate growth to moults and instar numbers. In this connection, knowledge of structure and size of the first and the final wing-bearing instars are important. Data on several species from the Breitenbach are here supplemented by information on the slow-growing *Dictyogenus fontium* (Ris, 1896), whose large size and hairless smooth cuticle allowed measurements and observations (Zwick and Zwick, 2010; and unpublished data) that were not possible for Breitenbach taxa, and also not with the large *Perlodes* from the River Fulda. The latter grow rapidly to final size between September and March, but are barely accessible for most of this time.

First Instar Newly hatched Plecoptera larvae leave the egg burster on their embryonal cuticle in the empty egg shell. First instar larvae of the suborder Arctoperlaria agree in the following features (Zwick, 1980): specimens are unpigmented; the number of antennal segments is nine; the number of cercus segments is three in species whose adults have multisegmented cerci, but four if adults have one-segmented cerci; eyes are represented by four ommatidia; gills are lacking; single chloride cells are present. First instar larvae of the suborder Antarctoperlaria differ in segment counts, and some paired abdominal gills may be present (Sephton and Hynes, 1982b).

In the local fauna, first instar larvae of Nemouroidea start feeding immediately, while those of the Perloidea live for several days on yolk supplies in the gut. They only develop functional mouthparts and start feeding at the first molt, in the second instar.

Sexual size differences of adults are important. Some authors ascribed them to sexually different growth rates, while others believed that there were fewer instars in males (Elliott, 2009; see also references in Stewart and Stark, 2002; Zwick and Teslenko, 2002), but whether sexes already differ in size at birth was not tested.

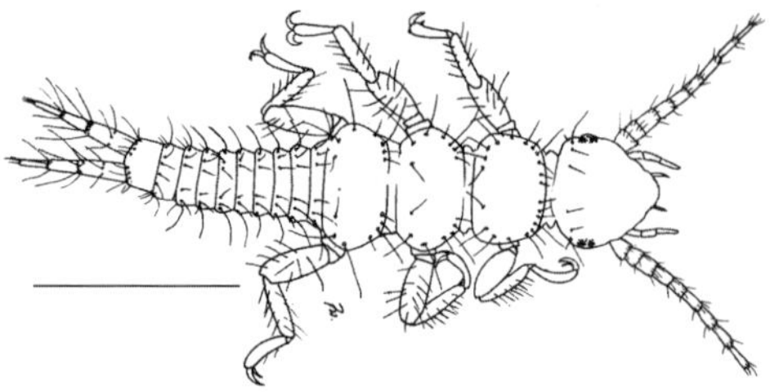

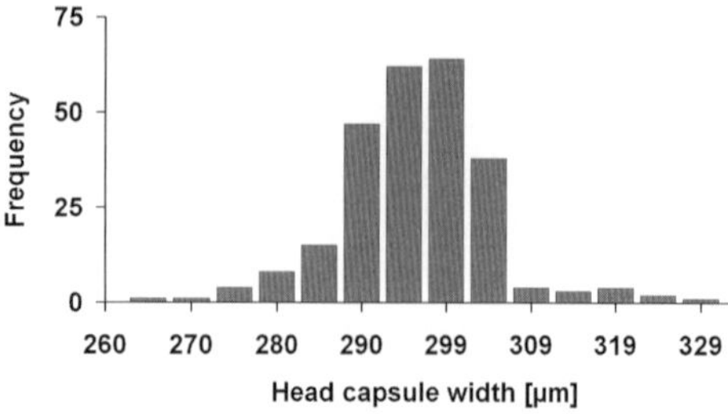

Figure 7.64 Dictyogenus fontium, a typical first instar stonefly larva (left; scale bar is 0.5 mm; from Zwick, 1982b), and frequency distribution of first instar head widths (right; from Zwick and Zwick, 2010).

The range of head widths of many freshly hatched *Dicytogenus fontium* was unimodal, not sexually dimorphic. Size differences between eggs were never described, and in our study first instar HCW did not differ between egg masses. In view of the apparently uniform egg sizes, the great range of observed HCWs was particularly unexpected (Figure 7.64).

Metamorphosis Apart from chloride cells, larval tissues and structures, even gills, are carried over into the adult stage. Gills persist in reduced form, but are no longer functional. Internal changes, especially gonad formation, growth, function, and degeneration were not observed in the present study. Gonads are known to attain maximum size in late larval instars, but are no longer active in the adult (references in Zwick, 1980). Here, the focus is on exoskeleton changes.

Morphological change at metamorphosis is generally rather limited and involves only the following:

1) Spiracles and gonoduct open, secondary sexual structures become exposed.
2) An arolium appears between tarsal claws.
3) Meso- and metathoracic nota have complex sculpturing to allow longitudinal muscles to operate as indirect flight muscles.
4) Wings become freely exposed and articulated.
5) Adults of many families shorten the long larval cerci to one or a few segments.

The molt to the adult insect occurs on land, mainly at dusk. About 15–20 min elapse until the wings are fully extended, and once that is achieved, the animal is able to fly. However, complete hardening of the exoskeleton takes several hours. Full pigmentation is only attained in old individuals.

A study of genital thecae or scars and of pterothecae allows the identification of several late pre-metamorphic instars, which provides a handle for life history analysis.

Genital Characters and Sexual Dimorphism The first indication of a vaginal scar in *Dictyogenus fontium* appears when gonad differentiation begins, after approximately the first third of larval life. Early sexual differentiation, as seen in *Dictyogenus,* probably also occurs in other Plecoptera. Vaginal scars of *D. fontium* adopt their characteristic shapes over the next five instars, until female emergence (Zwick, 1982a; Zwick and Zwick, 2010). Males cannot be recognized externally until the last instar. Many other taxa differ in this respect, but only rarely are thecae of secondary male genitalia large enough for instar analysis. In males of *Leuctra* and *Protonemura* paraprocts differ in shape between late instars. In *Nemurella,* the last three instars can safely be distinguished by the huge male paraprocts (Zwick, 1991).

When gonad and vaginal scar formation begin the previously unimodal size distribution of the larval population becomes bimodal, the size difference increasing at molts (Zwick, 1982a, 1982b; Zwick and Zwick, 2010). The final sexual size difference usually corresponds to the increase at a molt. Whether males indeed go through one or a few instars fewer than females needs checking (Elliott, 1984, 1987a, 1987c, 2009, Khoo, 1964; Oberndorfer and Stewart, 1977; Zwick and Teslenko, 2002; see also references in Stewart and Stark, 2002). Females of species which must still accumulate biomass for yolking the eggs during adult life (e.g., all Nemouroidea) have empty storage space, being distinctly larger but relatively lighter than males (Figure 7.59).

Pterothecae When wing sheaths or wing pads (pterothecae) of *Dictyogenus fontium* start to develop, gonads are already quite advanced (Zwick and Zwick, 2010). Enlargement of wingpads at moults suggests exceptionally rapid growth before adult emergence. However, the relative growth of cuticular structures (Figure 7.63) and specific biomass gain (mg gained mg^{-1} previously existing) are not greater than before, although the absolute weight gain is (Figure 7.11).

Absolute size is unsuitable to identify instars, size of last instar males and females, and of specimens in different instars, overlap completely. *Leuctra prima* (Table 7.16) provides a typical example.

Although the shape of wing pads differs between families, in almost all families (including all European taxa) the development of pterothecae occurs in three steps. Each instar is distinguished by a characteristic shape, proportions between body and wing pad size only vary over a narrow range in each instar (Zwick, 2003b). Often, the last three instars are readily recognized (Figure 7.65), and are coded as L_n for the last, L_{n-1} and L_{n-2}, for the penultimate and antepenultimate instars, respectively.

Table 7.16 *Leuctra prima*, range of head capsule widths (μm) in the three pre-metamorphic, wing-bearing larval instars (data from Zwick, 1991).

Ln-2, unsexed	Ln-1, unsexed	Ln male	Ln female
552–724	582–1002	721–1017	746–1160

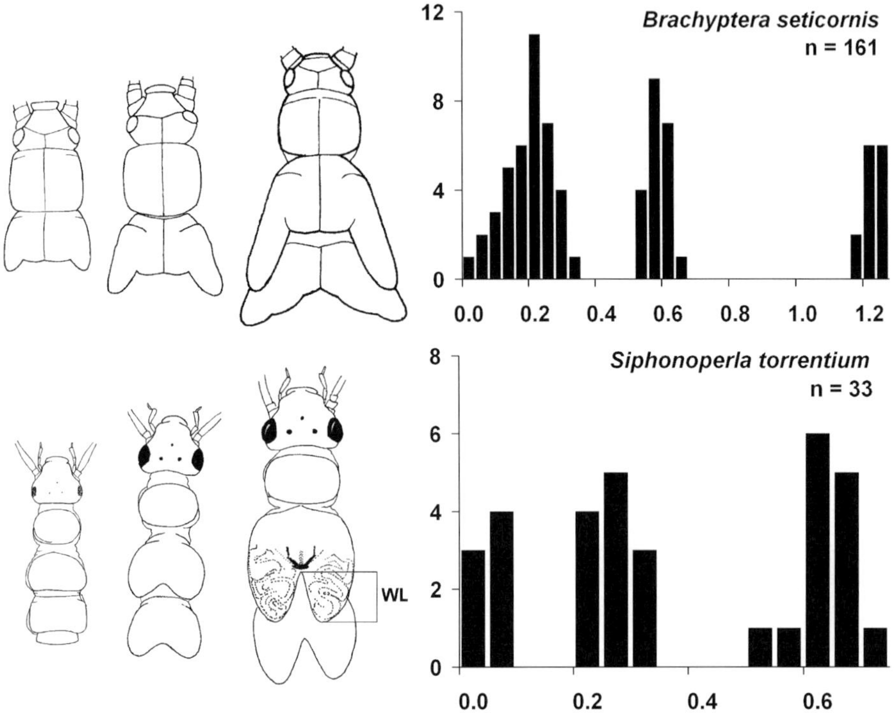

Figure 7.65 Wingpads of two very differently shaped stonefly species from the Breitenbach in the three last larval instars.The way in which wing length (WL) was measured is indicated. Head capsule width (HCW) was measured across and including the eyes. Bar charts show the frequency of quotients WL/HCW; *n* is the number of specimens measured (modified from Zwick, 2003b).

Changes during the Last Instar There is no consensus on the way in which wings form (Zwick, 2009). Morphological change of wing pads during the final instar, as shown in *Leuctra* (Figure 7.66) also occurs in the other species in the Breitenbach. Immediately after the molt to last instar, the completely transparent pterothecae curl up. They soon straighten and become translucent until the corrugated structure of the folded wings appears. Around this time, the gut is evacuated, as revealed by a sudden drop of the share of ash in dry weight. When wing pads appear black to the naked eye, apolysis has already occurred, the animal has become a pharate adult and normally needs to emerge within less than a day. *Nemoura cinerea* and *Nemurella pictetii* spend almost one month in the last instar (Zwick, in press), larger species may take markedly longer. At the extreme, 6–9 months are spent in the final instar by *Pteronarcys* species (references in Nesterovitch and Zwick, 2003).

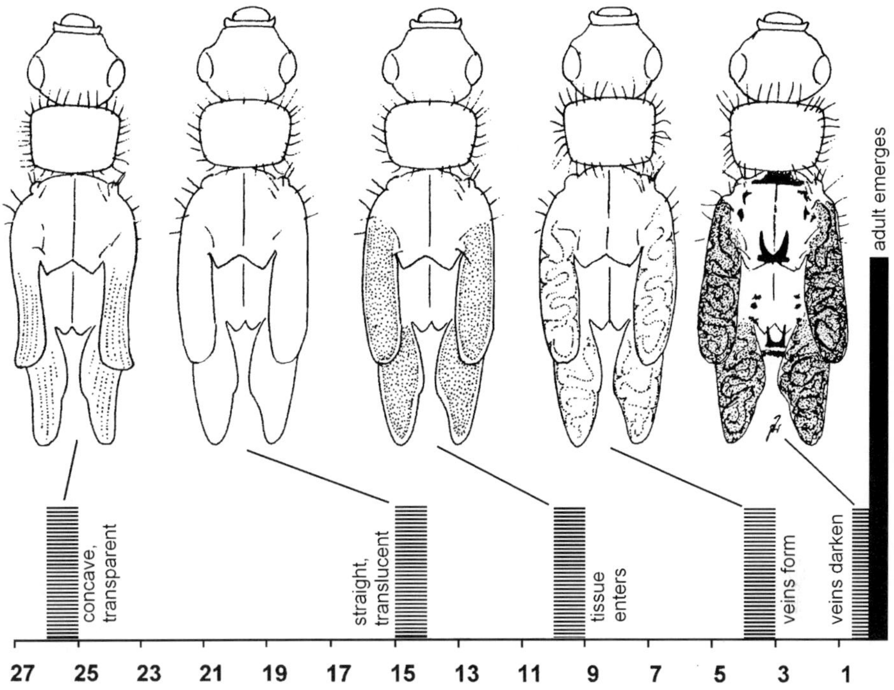

Figure 7.66 Phases of wing development in last instar Nemouroidea. Top: aspect in *Leuctra* sp. (modified from Zwick, 1991). Bottom: temporal sequence (abscissa: number of days before adult emergence) of the same developmental phases in *Nemoura cinerea* and *Nemurella pictetii*. Bars mark onset of the respective condition (from Zwick, in press).

Instar Numbers

Size of the first, and the last three instars was used to estimate instar number in a model describing a hypothetical specimen whose growth declines regularly along a median track (trend lines in Figure 7.63). The estimate was 16 instars for the huge *Pteronarcys* species, which agrees with counts during laboratory rearing (Zwick and Teslenko, 2002). Estimates for the slightly smaller *Dinocras cephalotes* (Perlidae; Frutiger, 1987) are similar. Similar or lower instar numbers seem to apply to the small stoneflies in the Breitenbach. Twelve to 15 instars were recorded in *Leuctra nigra* and *L. geniculata* (Elliott, 1987a, 1987c); laboratory breeding of *L. digitata* suggested the same numbers (Zwick, unpublished data). Even in some of the largest Plecoptera (Antarctoperlaria) instar numbers range from 10 to 14 (Sephton and Hynes, 1982a). Much higher instar numbers published in the older literature are clearly incorrect.

The above-mentioned model required 19 instars for small *Nemurella*, but Elliott (1984) published 15–17 instars. However, it is not known to which developmental type (bi- or univoltine) the tested individuals belonged, the sample may also have been heterogenous.

Seasonal Size Decline

The literature provides many examples of the earliest emerging insect adults being large, those of the same species that appear later being successively smaller. As in mayflies, large size provides a fitness advantage. Yoshimura *et al.* (2003) reported greater reproductive success in early- than in late-emerging specimens of Japanese *Isoperla* and *Sweltsa*. However, ultimately each trait was a function of body size. The term, seasonal size decline, is commonly used, but because some species increase in size in autumn, seasonal size change would be a better term. Autumnal size increase was first reported in blackflies (Friederichs, 1921; Rühm, 1971, 1995), is known in the Plecoptera (Zwick, 1977) and was also observed in several Ephemeroptera (M. Rösch, unpublished data) and Plecoptera in the Breitenbach.

The sizes of mayflies and stoneflies emerging from the Breitenbach were usually recorded as average dry weight of individuals in pooled biweekly, or monthly, samples in 1982–1984. Collective samples veil individual variation, and there is also some risk of bias through old females arriving in traps, after gaining weight from extensive terrestrial feeding. Individual linear body measurements (wing length in adults, head width in larvae) avoid these problems, but such labor-intensive work is only rarely practical. Seasonal size change was not pronounced in all Ephemeroptera and Plecoptera in the Breitenbach. It was not observed in species with very short emergence periods, for example *Ephemerella mucronata* and *Leuctra digitata*. Changes in size were minor over the four months in which *Ephemerella ignita* was trapped. Even over the much longer emergence period of *Protonemura auberti*, mean specimen size did not change significantly.

Size declined distinctly during emergence of other stoneflies and mayflies emerging from the Breitenbach. Figure 7.67 presents some examples of seasonal

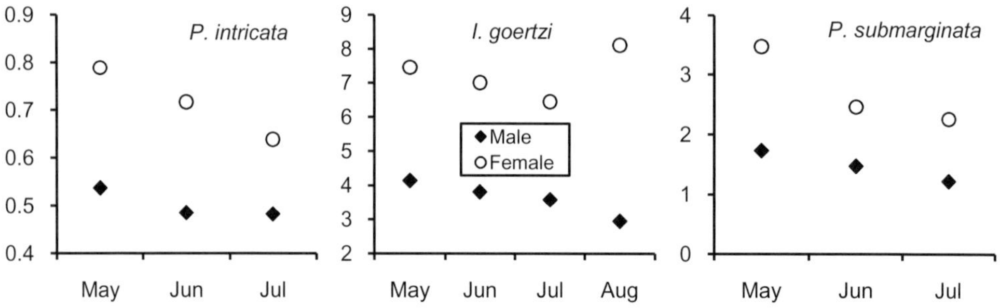

Figure 7.67 Seasonal decline of average specimen dry weight (ordinate; in mg) of *Protonemura intricata*, *Isoperla goertzi*, and the mayfly *Paraleptophlebia submarginata* emerging from the Breitenbach, 1982–1984.

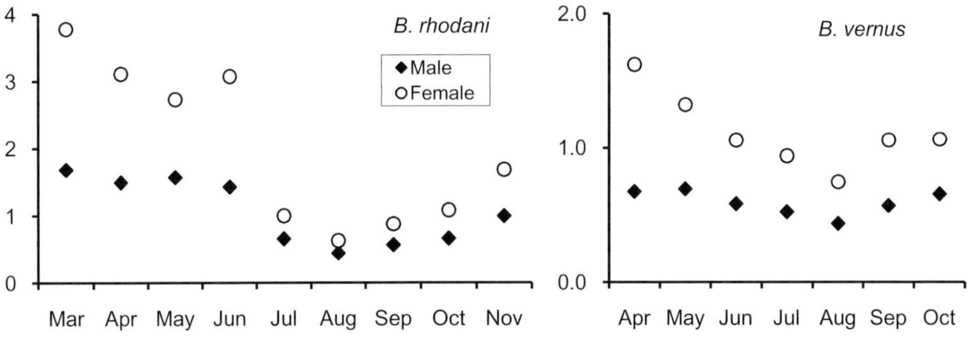

Figure 7.68 Seasonal changes in mean individual dry weight (ordinate; in mg) of the mayflies, *Baetis vernus* and *B. rhodani* emerging from the Breitenbach in 1983.

size decline. In other species, for example some mayflies, variable patterns of changing body size were observed (Figure 7.68).

For example, *Baetis vernus* exhibited gradually increasing body weight when it emerged abundantly from May to October 1982. In 1983, emergence occurred in three peaks between April and October, average specimen size declined until August, and then increased again. In 1984 the emergence period was late and only lasted from August to October, size declined all the time. The spring generation of *Baetis rhodani* included very few specimens in 1982–1984. Summer emergence was bimodal over the first two years. Size initially declined gradually but dropped sharply between the generations. Towards autumn, size increased again (Figure 7.68). However, in 1984, *B. rhodani* size declined continuously, from February to late October. Emerging from the artificially sanded trap in 1983 *Centroptilum luteolum* also displayed a slight autumnal rise in body weight, but specimen numbers were too low to be entirely certain of the effect.

Wing length of all female *Leuctra prima* was recorded in 1987, 1989, 1990, and 1992. In each year, size declined as emergence progressed, and absolute size differed between years. Differences did not reflect different degrees of intraspecific competition because average size and number emerging were not correlated. In contrast, the negative relationship between individual wing length (WL; mm) and calendar week of emergence (CW) (Figure 7.69) was described by a highly significant regression:

$$WL = -0.178 \pm 0.0051 * CW + 9.402 \pm 0.0474; R = 0.618; n = 1026$$

Possible Causes of Seasonal Size Changes Seasonal changes in food supply may lead to seasonal differences in adult body size (Chapter 7.9). This is improbable for the Ephemeroptera and Plecoptera in the Breitenbach, which are grazers or feed mainly on fine particulate organic material (FPOM), which is rarely in short supply (Section 4.2.3). However, negative relationships between temperature and body size have been observed in a bacterium, in protists, many plants and animals (Atkinson and Sibley, 1997). Opinions on the underlying causes diverge and some

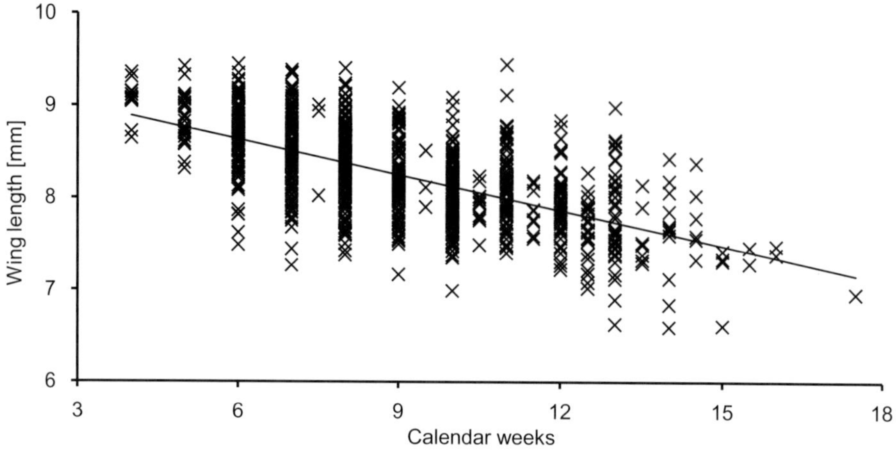

Figure 7.69 *Leuctra prima*, wing length of females taken in Trap 2 on the Breitenbach during 1987, 1989, 1990, and 1992 plotted against the calendar week in which each specimen emerged. Linear trend line of size decline fitted by Excel.

are contradictory. Most studies consider the effects of temperature on juvenile growth of poikilotherms, but it is also possible that cell size is affected (Atkinson and Sibley, 1997).

While Lonsdale and Levinton (1985) propose that development at the lowest tolerable temperature always results in the largest body size, Makarieva, Gorshkov and Li (2005) suggest the opposite: maximum body mass of terrestrial poikilotherms should approximately double per 10 °C temperature increase. However, this would not be the case in aquatic poikilotherms because of constraints resulting from the low concentrations of oxygen in water, together with a largely temperature-independent diffusion flux rate of oxygen per unit body surface. The temperature response of basic physiological processes, such as anabolism and catabolism, might differ in magnitude. Differential effects of temperature on ecdysone and neotenine are sometimes proposed to explain seasonal size declines in arthropods. Growth and development both occur faster at higher temperatures, but development, for example insect molting, increases more than somatic growth. As a result, specimens would normally grow to a larger size at relatively low temperatures than their conspecifics at higher temperatures (Blanckenhorn, 1997; Ratte, 1985; Gilbert and Raworth, 1996).

Hogue and Hawkins (1991) found a negative correlation between adult body size and temperature in 75 of about 90 published observations, most under variable temperatures. When, on which sensitive stage, and for how long raised temperatures acted is not clear. Even comparisons between different temperature regimes are not always precise, sometimes simply "warm" and "cold" conditions, prevailed, in individual cases only around the time of adult emergence. Seasonal change of adult size is generally perceived as seasonal size decline, while autumnal rises of body size are not discussed.

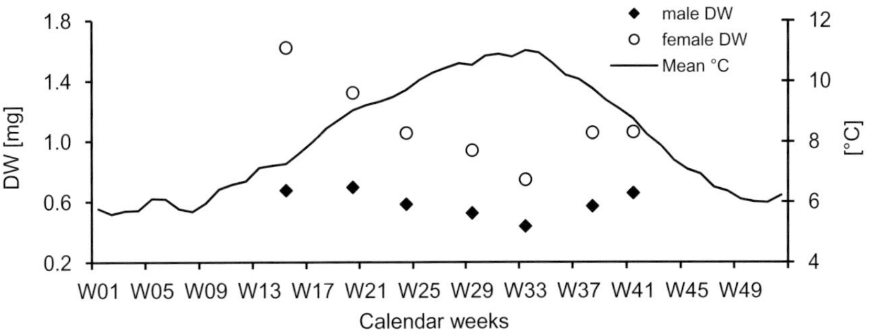

Figure 7.70 Mean dry weight (DW, mg) of *B. vernus* (1983) and long-term (1974–2005) mean weekly temperature in the Breitenbach.

Long-term average Breitenbach temperatures present a mirror image of size changes in *Baetis vernus* (Figure 7.70) which may support the suggested causal relationship. However, insect size changes only at molts and changes of temperature can become effective only with some delay, at least one molting interval. For the small stoneflies in the Breitenbach this is almost one month, less in many mayflies. The temporal resolution of the present data permits no analysis.

Suggested size declines during hurried development released by some environmental cue, for example predator kairomones (Peckarsky *et al.*, 2000), seem improbable in view of morphological constraints during wing development. More than 90% of all stonefly species are fully winged, at least the females which need to fly to disperse and oviposit. Rare exceptions occur in species living under extreme habitat conditions at the highest altitudes, latitudes, or among Capniidae emerging in winter. If development were shortened by one or more instars adults with short wings and unable to fly would certainly result. However, this never occurs in populations under temporate climate. Generally, any change of prospective body size appears improbable once wing pad formation has started. If late during development vital requirements such as food supply are not met, specimens probably die, instead of attaining adulthood at some smaller size.

Indeed, the prospective size of specimens is pre-determined several instars before adulthood, at a much earlier stage than generally assumed. Seasonally declining size of *Nemurella pictetii* from the Breitenbach and the spring of the River Fulda was shown to affect already the earliest identifiable instar, L_{n-2} (Nesterovitch and Zwick, 2003; Figure 7.71). Because younger specimens can presently not be assigned to instars there is no way to check if the effect occurs possibly even earlier.

Different causes for seasonal size change can be imagined:

- Successive emergence of initially large and then successively smaller individuals might be based on genetic constitution and reflect different degrees of overall fitness leading to characteristic developmental careers (Klingenberg and Spence, 1997).

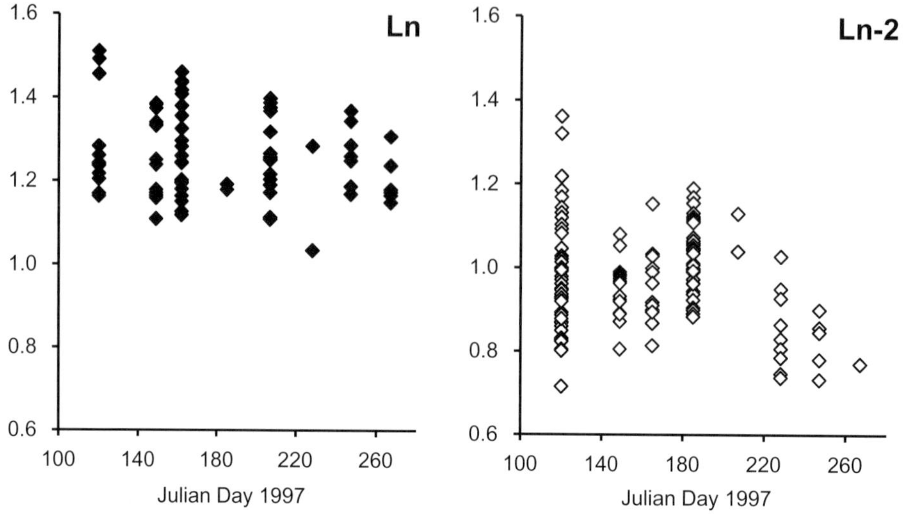

Figure 7.71 Size reduction of larval *Nemurella pictetii* in the last and the antepenultimate instar in the Breitenbach, 1997. Ordinate: HCW (in mm; after the data of Nesterovitch and Zwick, 2003).

- A variety of factors, such as available food, competition, age, temperature, photoperiod and possibly others influence specimen size. Once specimens have attained a specific minimum size they become sensitive to some environmental signal directing metabolism, from somatic growth towards increasing differentiation and maturation. Adult size would then depend on the size attained by the time the environmental signal was released or perceived.

The last point was made by Sweeney and Vannote (1981) who stated "the decrease in adult .. size . . . seems to result from the sequential metamorphosis of progressively smaller larvae". Tierno de Figuero and Sanchez-Ortega (2004) reached the same conclusion. I agree and also presume that daylength plays an important role in the environmental signal releasing metamorphosis. This would explain seasonal size change as well as the stable stonefly phenology which varies little between years. Daylength and temperature are related, but daylength does not vary inter-annually, which makes it a better indicator of season. Indeed, Nesterovitch and Zwick (2003) found better correlation of *Nemurella* specimen size with time (number of days since winter solstice) than with temperature. They also showed an inversion of temperature effects with different directions of photoperiodic change. At 6 °C or lower, growth of *Nemurella* stopped under declining daylength but resumed at the same temperatures when daylength was increasing.

At present, we do not know what factor or combinations of factors initiate metamorphosis and thereby determine the size of prospective adults.

7.6.4.4 **Coleoptera**

Beetles are the largest insect order with over 300 000 named (and possibly some 30 million existing) species worldwide (Erwin, 1982). Adult size ranges from 0.5 to 180.0 mm. Beetles are extremely diverse in structure, function, and all aspects of their ecology, but do not include any endoparasites. Coleoptera are present in all climates and in every kind of habitat. Some even live in the marine littoral which is colonized by very few insects (Lawrence and Brittain, 1991; Cheng, 2003). There are numerous freshwater representatives of the suborders Myxophaga, Adephaga, and Polyphaga. Klausnitzer (1996) provides a general overview of water-associated Coleoptera.

The beetles in the Breitenbach were studied in benthic samples in 1986–1988, over several years of emergence trapping, in drift samples (mainly at the mouth of the Aspegraben in 1984), and from occasional samples taken over many years. The present data include P. Neumann's unpublished records (1989) from the Aspegraben. Several ponds connected with the Breitenbach contributed indrifting specimens which are recorded below. Had the ponds themselves been sampled, more species would have been added to the list (see Appendix) which presently includes 83 aquatic or water-dependent species.

7.6.4.4.1 **Suborder Myxophaga**

Adults and larvae of all three families live submerged or in water-saturated littoral habitats; larvae have tracheal gills. Only the Microsporidae, represented by the single genus *Microsporus* (formerly *Sphaerius*), occurs in Germany. *Microsporus acaroides* (Waltl, 1858), the single Central European species, has been sifted out of flood debris of the Fulda and Schlitz rivers (coll. Zwick), but was not recorded from the Breitenbach, where it may nevertheless occur. Adults are only 0.7 mm long and the insect may have been overlooked.

7.6.4.4.2 **Suborder Adephaga**

The Adephaga include the ground beetles (Carabidae) and a suite of aquatic families, most in lentic waters. Several aquatic families occur in the Breitenbach and provide the majority of entries in the taxonomic list in the Appendix. The number of larval instars is three.

Most Adephaga are predators, as are usually their larvae. In most taxa, the pupae are the only terrestrial life stage. However, larvae and adults breathe atmospheric air and have to surface at short intervals. Adults, especially of the smaller species, often carry air bubbles as physical gills. Yet, these do not last long because, under hydrostatic pressure, nitrogen in the air supply diffuses quickly into the surrounding water making the bubble shrink.

From an ecological point of view, most of the many aquatic Adephaga play a minor role in fast-flowing, headwater streams like the Breitenbach, at least as these are presently structured. In the original dense riverine forest of the River Fulda, backwaters, floodplain puddles, pools formed by trees fallen across streams, quiet areas behind snags and, possibly, beaver ponds, were all intricately connected with the main courses of running water. Unfortunately, a recent study suggesting that lotic dispersal through such networks of water bodies is an established strategy (van der Meurrer *et al.*, 2006) does not cover beetles. In the meantime, man has largely separated standing water bodies from streams, ending the former intimate coexistence of lentic and lotic organisms. However, groundwater- or spring-fed ponds continue to drain into the stream.

In the Breitenbach valley, three sets of ponds exist, or existed. The uppermost one beside the Aspegraben was a single pool, intended to hold water against possible forest fires, but was sometimes used as a recreational fish pond. Three very small ponds situated 150 m upstream from Trap 2 served the same purpose. A set of four larger ponds fed by strong springs near Trap 4 was built for fish farming but was left unattended for years. Fish stocking was resumed about 1996.

Dytiscidae

Swimming beetles are usually strong flyers. Flights from one habitat to another are often nocturnal and the large spectacular *Dytiscus* are known to land, by mistake, on asphalt roads, glass roofs and other structures that reflect polarized UV light in the same way as water surfaces do (Schwind, 1991). Most Dytiscidae recorded from the stream are ubiquitous freshwater inhabitants, with a preference for standing water. In particular, single specimens of the genera *Colymbetes, Acilius,* and *Dytiscus,* taken in emergence traps or drift samples, do not normally occur in streams.

Swimming beetles oviposit between algal mats, or directly into living plant tissue (*Dytiscus*) where there is a good oxygen supply, or in algal masses just above water surface. All larvae are carnivorous, the prey spectrum being a function of predator size, and for the group as a whole, ranging from microscopic Crustacea to little fish. Larvae practice pre-oral digestion by injecting digestive fluids into the prey via a mandibular canal, through which the liquified prey body is eventually ingested. The mouth proper is firmly closed. Larvae cannot swim and must regularly crawl up, or use their positive buoyancy to reach the surface, where they breathe atmospheric air. Pupation occurs in soil chambers on land.

Agabus bipustulatus and *A. biguttatus nitidus* are typical inhabitants of small spring runs in hilly country. They also inhabit intermittent streams in which adults estivate and hibernate, under stones or pieces of wood on the dry bed. In the Breitenbach, they are mainly encountered in small numbers in upstream sections, that is, the Aspegraben and just downstream from the Georgsborn. Another Agabini, *Platambus maculatus,* is a characteristic inhabitant of quieter streams, preferring slow ditches, quiet bays, or protected near-shore macrophyte stands in larger water courses. Small numbers are encountered near the mouth of the Breitenbach but the species is common in reedy shores of the River Fulda.

The Hydroporinae are the most diverse group; most species are small. *Oreodytes sanmarkii* is a common inhabitant of metarhithral and larger streams in the Rhön and Vogelsberg Mountains. Because it is vividly striped in black and yellow it is not easily overlooked. Yet, only a single adult (but no larvae) were found in the Breitenbach. Only a few adults of the generally rarer running water species *Nebri-*

oporus elegans and *Stictotarsus duodecimpustulatus* were collected. All three also occur in the River Fulda.

Among the small Hydroporinae, *Hydroporus ferrugineus* is a relatively large, very slender and partly depigmented, rufous insect. Outside the lowlands, it is widely distributed but sporadic and uncommon in Germany. It is found in groundwater wells, mine galleries and is occasionally flushed into springs from subterranean waters. A few specimens were collected in the uppermost section of the Breitenbach, near the mouth of the Aspegraben. In this area, there always seems to be some underground flow through gravel beds accumulated at the fault near the valley bend, where the large Georgsborn is situated.

Haliplidae

Most Haliplidae prefer standing or slowly flowing waters. The adults are active swimming predators which surface to breathe and can fly. The extremely slow-moving larvae are bottom dwellers, or move between algal vegetation. They are mainly vegetarians, feeding on algae whose cell contents they suck through a canal in their mandibles. Larvae of several subgenera of *Haliplus* have special tibial and tarsal structures enabling them to grasp and hold filamentous algae, which are then emptied cell by cell. The hard, dark, larval cuticle carries numerous tiny tubular processes which, by their fine structure, are gills. These tubular gills are extremely long in the genus *Brychius*, giving the larvae an unmistakable feathery appearance. Larvae pupate on land.

No haliplid larvae were found in the Breitenbach, but they occur along the shores of the ponds adjacent to the Breitenbach, in which adults also abound. Without doubt, the numerous *Haliplus* spp. taken in the Breitenbach came from standing waters, most were probably only in transit. However, several of the same species live and reproduce in quiet habitats of the River Fulda.

The only exclusively stream-living haliplid, *Brychius elevatus*, is known to prefer hyporhithral and potamal sites. However, only a single specimen was caught in the Breitenbach, although the species is regularly, though not abundantly, found in the River Fulda.

Gyrinidae

Most whirligig beetles inhabit the surface of lentic waters, where schools of shiny, silver beetles parade. Their highly specialized locomotory system allows almost frictionless movement on the surface film, but renders steering difficult, so that the beetles move in circles or spirals on water surfaces. They are predators with divided eyes, able to simultaneously watch the surface film and underwater environments for prey. When capturing prey or escaping from threat at the surface underwater movement is straight, because the extruded genitalia are used to steer. Oviposition is between algal mats and the like.

The larvae are gilled predators on the stream bottom, to which they attach using the four strong hooks of the pygopodium. If larvae come into contact with the surface film, for example during collecting, they are trapped in it. Mature larvae move onto land and climb up trees, carrying soil on their backs to build cocoons (Rupprecht, 1971). Given the lack of natural bodies of standing water in the area, *Gyrinus* is not commonly seen; the only specimen caught in the Breitenbach was apparently in transit. No gyrinid larvae were observed in the Breitenbach.

In contrast, *Orectochilus villosus* with its dull gray, finely pilose matt surface, is a river inhabitant. It is nocturnal and hides during the day under undercut banks or among littoral debris, where it is easily collected. Larvae are regularly found in the River Fulda. *Orectochilus villosus* is common in the region but was never found in the Breitenbach.

7.6.4.4.3 **Suborder Polyphaga**

The largest beetle suborder includes several water-bound families of different habits.

Staphylinidae

The large family of rove beetles includes hygrophilous genera found only in the immediate vicinity of water bodies, especially streams, including along the Breitenbach. Several of the colorful *Paederus* belong here. *Dianous coeruleus* (Steninae) is exclusively found in water-soaked mosses, on rocks or between coarse debris in streams, but is not aquatic. It is not common at the Breitenbach.

The sister genus, *Stenus*, contains many water-bound species. Some walk on sandy beaches in search for prey, others move between emergent aquatic vegetation and cross small areas of open water, walking on the surface film. When washed into water by waves, many are able to skim on water using secretions of the pygidial gland as a propellent (Jenkins, 1960; Schildknecht *et al.*, 1975). They steer towards emergent rocks, vegetation or other firm ground, which they detect with their huge eyes. However, no *Stenus* is aquatic or preys on aquatic fauna. Instead, they hunt for Collembola which are caught with the glutinous, long labium that can be thrust far forward, much like the tongue of a chamaeleon.

The genus *Lesteva* includes a number of extremely hygrophilous species. To find them, collectors' manuals direct students toward wet mosses on weirs, mill races, and so on. *Lesteva longoelytrata* is common in the Breitenbach, and large numbers appear in emergence traps twice each year. The spring collections comprise immature specimens with undeveloped gonads; the autumnal catch includes only dark individuals with mature gonads. It was therefore assumed that the species spends the summer along the stream, feasting on riparian and small emerging fauna, for example, Chironomidae, growing and maturing its gonads at the expense of benthic production.

However, we have found that the beetle is an active scavenger for only a short period in spring, when it experiences some weight gain. Most of the year is spent at rest, adjacent to the stream. In autumn, mature beetles fly to distant terrestrial habitats, with extensive bare soil, such as fields with young winter cereals, where they reproduce during winter. The young generation returns to water courses the following spring (Burkart and Zwick, 2006).

Helophoridae

The Helophoridae and the Hydrophilidae are the two families of the Hydrophiloidea present in the Breitenbach. The superfamily Hydrophiloidea is mainly lentic and, like the Dytiscidae, most species enter the Breitenbach from the ponds, but do not normally, or permanently, live in the stream itself. An obvious, common characteristic of its families is the lengthening of the maxillary palpi, which function like antennae. The real antennae are short and often concealed beneath the head. They have one pointed flagellar segment that is used to break through the surface film. The pilose club is bent back under the head and forms a bridge to transfer atmospheric air into the air film covering much of the ventral side and extending into subelytral air supplies.

Another common trait is that the long-lived adults are phytophagous, mainly algal feeders or detritivores. In contrast, larvae are normally carnivorous and aquatic, and most pass through three instars in a very short time. Larvae are found between vegetation and debris at the surface, where they breathe atmospheric air. Apart from the Berosini which do not occur near Schlitz, gill respiration plays only a suppplementary role.

Silk production to spin egg cocoons with the ovipositor is common in the group. Cocoons are amazingly elaborate in the largest species, and once completed, the female leaves them alone. However, several smaller taxa, for example *Enochrus*, practice brood care, and females carry their cocoon on the ventral side of the abdomen, holding it with the hind legs and bringing it into a more protected position if disturbed. The hatching larvae disperse immediately.

The family Helophoridae comprises the single European genus, *Helophorus*. The name refers to a light mud cover on older beetles. With few exceptions, adults are aquatic but the carnivorous larvae are terrestrial, living in water-saturated soil and mud. The larvae of some *Helophorus* species feed on decaying plant matter and are terrestrial, several are agricultural pests on beet, grain seedlings, and other crops (Angus, 1992). Most pass through three instars, but a few species have only two larval instars.

The about 100 European *Helophorus* species are very similar and difficult to identify, but the specific status of currently recognized taxa has been confirmed by chromosome studies and/or hybridization experiments (Angus, 1992). Identification by morphological characters alone can sometimes remain problematical, for example, among the largest species in the subgenus *Helophorus* s. str.

Emergence trap catches included large numbers of several aquatic *Helophorus* species. All live close to the surface, among vegetation and along the shoreline, never in the benthos. They are well known invaders that arrive immediately in almost any water body anywhere, for example, a full washing bowl in a domestic garden (Jäch, 1994). Apparently there are numerous beetles in the air, which are no doubt attracted by the polarized light reflected from water surfaces (Schwind, 1991), but also from several technical products, for example car roofs, glass plates or plastic foils. Both the stream surface and the glass panes of the traps attract such beetles, but those trapped in the preservation trays of the emergence traps

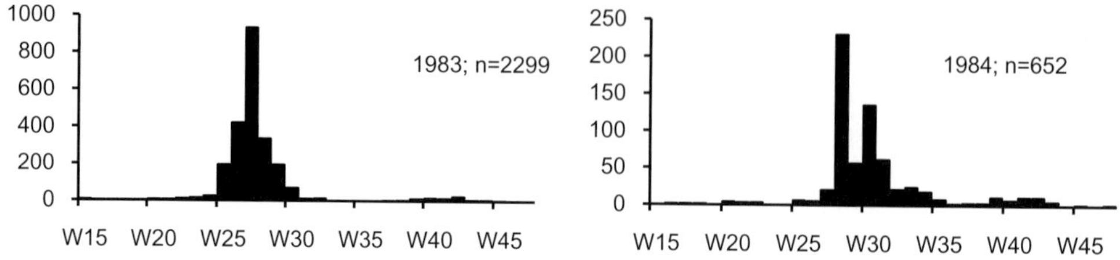

Figure 7.72 *Helophorus brevipalpis* in emergence traps along the Breitenbach, all traps pooled. Seven traps were operated in 1983, only five in 1984. Ordinate: numbers caught. Abscissa: calendar weeks.

had clearly attempted to leave again in search of more suitable habitats, or were freshly hatched from pupae along the stream edge.

The largest species, *Helophorus grandis,* is lentic and occurred sparingly in emergence traps. The slightly smaller sibling species, *H. aquaticus* and *H. aequalis,* are more ubiquitous and also recorded from flowing waters. They have sometimes been treated as a single species, but were separated after chromosome studies. Yet, morphological distinction remains difficult because of their variability. The taxa agree in their phenology. Single specimens were taken as early as week 13, and as late as weeks 40–43, while the majority appeared in traps during weeks 21–29.

Figure 7.72 shows two examples of trap catches of the common *H. brevipalpis,* a reliably identified small species. Single specimens appeared at various times, but most were found during an early summer peak, and comprised clean, freshly hatched specimens. A second, shallow pulse was observed in autumn. Numbers caught (3–7 traps pooled) per year ranged from as few as 27–68 specimens (1980, 1981, 1986, 1988) to over 2000 in 1983. Regardless of numbers, phenology was always similar. In 1983 and 1984 the main peak coincided with sudden significant warming, the low in 1984, with a marked temporary drop in mean weekly temperature.

Hydrophilidae

The list of Hydrophilidae includes widespread and common species. The greater proportion lives in lentic waters, also abundantly in pools in the Breitenbach valley from whence some drifted into the Breitenbach and were taken in drift samples. As in comparable cases, the same species were not caught in stream benthos samples, nor were larvae found. Most of the species are good flyers, with good dispersal capacity.

Hydrobius fuscipes has several morphs. The one in the Breitenbach is typical of small streams and regularly found in protected sites among vegetation or debris. Several *Laccobius* species were caught, as single specimens or in small numbers. Only *L. striatulus* is regularly encountered in the Breitenbach, although too rarely to describe its phenology. It is common in vegetation-rich marginal zones of both standing and running waters.

Anacaena species occur in marginal stream habitats, for example in water films on the underside of pebbles or pieces of bark and similar objects, in spring-fed pools, on banks or among detritus at the water's edge. Disturbed beetles cannot swim but walk along the underside of the surface film, the air-covered ventral side up, until they reach some firm support along which they descend into deeper water. The two recorded species (see Appendix) are widespread and generally common, but not along the Breitenbach. Species coexist along the entire stream, being least rare in Reach A. *Anacaena lutescens* is the commoner species, but altogether numbers are low, up to 48 specimens were found in traps. Phenology is only partly revealed by the limited trap samples. Occasional over-wintering specimens appear in very early spring (Week 12), more are trapped in early summer (Weeks 19–26), and single ones again in late summer (Week 47).

Elmidae

Adult Elmidae are small (Breitenbach species 1–3 mm) hard, black, slow beetles that have characteristic ventral patterns of silver, sometimes extending onto the sides of the elytra. The silver sheen is caused by total reflection of the light by an incompressible air film, a plastron, held between special hair-like structures (Wichard, Arens and Eisenbeis, 2002). Plastron respiration relieves the insects of needing to leave the water to breathe. It works best in agitated water and elmids therefore typically inhabit running waters. The vernacular name, riffle beetles, highlights this habitat choice. Of the Breitenbach species, *Oulimnius tuberculatus* is also found on wave-swept stony lake shores, for example in North Germany. Adults cling to hard bottom substrata with their enlarged tarsi and claws, whence the names Haken- and Klauenkäfer. These structural modifications are not impressive among the Breitenbach fauna, but several exotic taxa have exceptional shapes with hypertrophied legs, especially tarsi (e.g., Jäch, 2003). For spelling of the scientific name, refer to ICZN (1995). Nomenclature follows Löbl and Smetana (2003).

Early taxonomists attempted to identify species from the adult habitus and surface structure. Students from the Flussstation were the first to base the taxonomy on genital structures (Illies, 1953a, 1953b; Steffan, 1958, 1961). Most species prefer some particular section of stream and several Elmidae are excellent bioindicators (Knie, 1977; Moog and Jäch, 1995). Larvae have characteristic shapes, genera are immediately recognized (Figure 7.73).

Today, accurate descriptions and keys to adults and larvae are available (Berthélemy and Riols, 1965; Berthélemy, 1979; Steffan, 1979; Jäch, 1992; Klausnitzer and Richoux, 1996), but the two most common species cannot be consistently separated. Sainz-Cantero, Sanchez-Ortega and Alba-Tercedor (1985) therefore suggested that they are simply varieties of a single species. However, males can be identified after dissection and mounting of genitalia. The widespread *E. aenea* and *E. maugetii* coexist in large numbers in the Breitenbach, in fact almost everywhere. Most studies suggest that *E. aenea* occurs mainly upstream of *E. maugetii*, which is more pollution-tolerant and even occurs under mesosaprobic conditions (Grasser, 1994). Random samples from the Breitenbach revealed no distributional

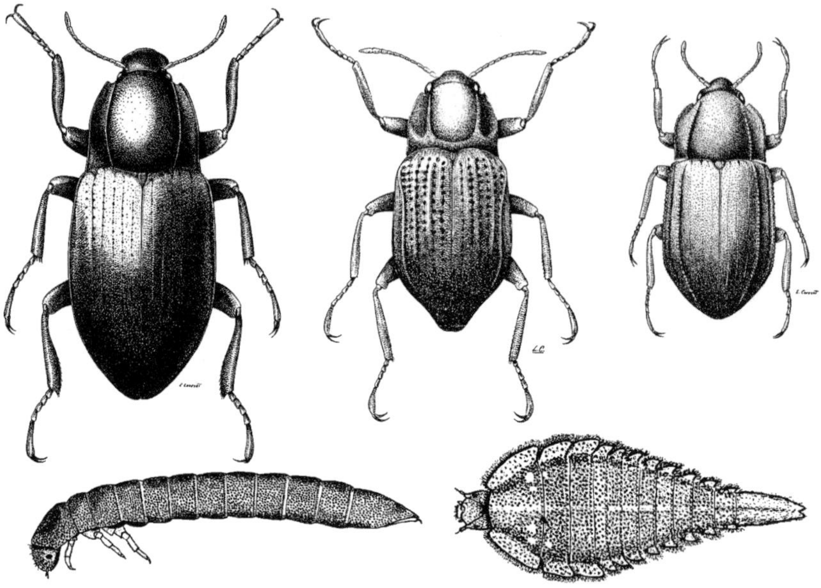

Figure 7.73 Elmidae, habitus. Top, from left: adult *Limnius volckmari* (Panzer), *Elmis maugetii*, and *Oulimnius tuberculatus* (from Olmi, 1976). Bottom, from left: larval *Limnius* sp., *Elmis* sp. (from Holland, 1972). Not to scale.

trend in the study stretch. Females cannot be identified (Schulte, 1989)[1] but Berthélemy and Riols (1965) succeeded in separating late instar larvae of the two sibling species. However, the differences are minor and specimens in instars 1–3 cannot be distinguished.

The two *Elmis* species are here treated as though they were only one. Information loss through combining the two sibling species is probably unimportant because distinct habitat segregation among the Elmidae has so far only been observed between genera (LeSage and Harper, 1976). However, even between genera, dietary overlap is complete (LeSage and Harper, 1976; Tavares and Williams, 1990). Furthermore there is no temporal segregation, because life cycles last two or more years, with most larval instars and adults being simultaneously present all year. The record of a third species, *Elmis coiffaiti* Berth. and Clav., 1961 (Reuter, 1974) from the Breitenbach was a misidentification (det. Berthélémy, personal communication).

Limnius perrisi, *Elmis aenea*, and *E. maugetii* form the core of the Breitenbach elmid fauna, and all are typical inhabitants of epirhithral streams (Moog and Jäch, 1995). *Oulimnius tuberculatus* was regularly collected in the benthos, but only a single larva was ever found in the main study stretch. The species' actual habitat

1) Dr. Jäch, the recognized taxonomic expert on the group, initially offered to identify sample specimens but eventually abstained, to avoid being "speculative".

is probably somewhere along the upper stream section in the Aspegraben, where two more species also once drifted in.

Riolus subviolaceus is titanophilous and a characteristic inhabitant of limestone habitats (Steffan, 1979, Garcia Criado and Fernandez Alaez, 1995). Fewer than 10 adults (but three or four of the strikingly bristly larvae) were caught, most in drift samples at the mouth of the Aspegraben, but once also in the benthos at Site 2. Only four adults of *Esolus parallelepipedus* were taken, in Aspegraben drift samples.

The occurrence of a limestone indicator species in the very poorly buffered Breitenbach is surprising because the distributional capacity of the local Elmidae is limited. Elmidae are slow-moving during all their life stages. Adults may be macropterous or short-winged, including mixed populations. Immediately after eclosion, winged adults may engage in a dispersal flight, usually at dusk (Brown, 1987; Takahashi and Sato, 1988). Records by Jäch (1997) also concern some of the Elmidae which occur in the Breitenbach, but we have few data of our own. Fewer than 10 adult *Elmis* were found during 30 years of emergence trap collections, several with open wings. *Limnius* and *Oulimnius* never occurred in traps. In view of the restricted elmid colonization potential, the impact of two events affecting the entire biocenosis had a more lasting and adverse effect on Elmidae than on many other taxa.

Late in 1983 (by 7 November) the lower Aspegraben fell completely dry. After bed sediments had been dry for 43 days, and frozen for about 20 days, 25 live *Limnius* larvae representing all stages, and occasional *Elmis* larvae were recovered from one liter of sediment. The drought occurred shortly after leaf fall in the surrounding deciduous forest and equipment had been set up to estimate the amount of leaf material drifting into the study stretch from upstream. The equipment was also used to record drifting insects (mesh size 0.5 mm).

Flow resumed on 14 January 1984, but discharge was barely $1\,l\,s^{-1}$ until 31 January. Large numbers of Elmidae appeared immediately (Table 7.17). After 10 days the numbers of beetles declined to a few per day, and remained at that level

Table 7.17 Numbers of Elmidae and Hydraena drifting out of the Aspegraben during low flow ($\sim 1\,l\,s^{-1}$) after a long drought (Period 1), and estimates of numbers from short term samples during the first two days of quickly rising discharge (Period 2). Sampling was discontinued when discharge exceeded $300\,l\,s^{-1}$. ad. = adult; la. = larva.

	Period 1	Period 2
Elmis ad.	3400	1800
Elmis la.	690	430
Limnius ad.	970	300
Limnius la	310	3420
Oulimnius tub.	740	45
Hydraena spp.	670	55

for the rest of the period of low discharge, until 31 January. Apparently, beetles and smaller numbers of larvae had relocated after the drought.

Large numbers of Elmidae drifted again (Table 7.17) when sudden snowmelt and strong rain increased the flow at the beginning of February. On 7 February the stream flooded, and for some time the entire, 15 m wide, valley became a roaring stream. Sampling had to be discontinued, but eight, short term (1 min), drift samples were taken daily until discharge exceeded $300\,l\,s^{-1}$. During that period, before the flow had even peaked, many thousand Elmidae were washed downstream. Drift samples taken later, once a week at the same site, contained an average of 32 adult *Elmis* per day. In 1986–1988, population densities in the stream section between Traps 1 and 6 (specimens m^{-2}; means of nine samples $month^{-1}$) were 59 (21–110) adult *Elmis*, 56 (11–90) adult *Limnius*, 1.5 adult *Oulimnius*.

Elmidae were also extremely numerous in the catastrophic drift following an accidental insecticide poisoning in the Aspegraben in May 1986, but accurate counts were not possible. Samples taken in 1996, 1997, and 2000 suggest that the populations of *Elmis* and *Limnius* recovered, reaching densities similar to those before the disturbances. However, *Oulimnius* remains much rarer today than it used to be in the past.

The low proportion of drifting larvae can be only partly explained by loss from the drift net through the mesh, mainly of the sausage-shaped *Limnius* larvae. *Limnius* larvae often live deeper in sandy substrata and are more protected against displacement by spates and poisoning than adult *Limnius*, and more than adult and larval *Elmis*.

Bionomics and Life History Elmid pupae rest for one to two weeks in small pockets in riparian soil, the other life stages are aquatic. Young adults can be recognized by their brownish clean body surface, as opposed to black, often with some dirt crust in aged specimens. Young adults enter streams in autumn and early winter (Beier, 1948; Holland, 1972); they are long-lived, surviving more than 10 years in the laboratory (Kodada and Jäch, 2005). They reproduce for much of the warm season, but whether the local species survive to the next spring and reproduce again is unknown.

Small groups of eggs are attached [either spun (Illies, 1953a, 1953b), or glued (Brown, 1987)] to moss leaf axils, stone crevices, and other protected sites. Egg development takes about two weeks (Beier, 1948; Brown, 1987), which is in agreement with Breitenbach life cycle data. The number of larval instars differs between genera, ranging from five to eight (Beier, 1948; Holland, 1972; LeSage and Harper, 1976; Tavares and Williams, 1990).

In *Elmis*, several body dimensions can be used to establish that there are six instars. Head capsule width (HCW) of instars 1 and 2 is identical, but they differ in other measures (Figure 7.74, left) and also in the setation of the paratergal lobes of their flat bodies (Berthélémy and Riols, 1965). The diameter of the tubular body clearly separates all seven instars of *Limnius* (Figure 7.74, right). The cuticle is smooth in the first instar, later it is covered with denticles and carries sharp teeth

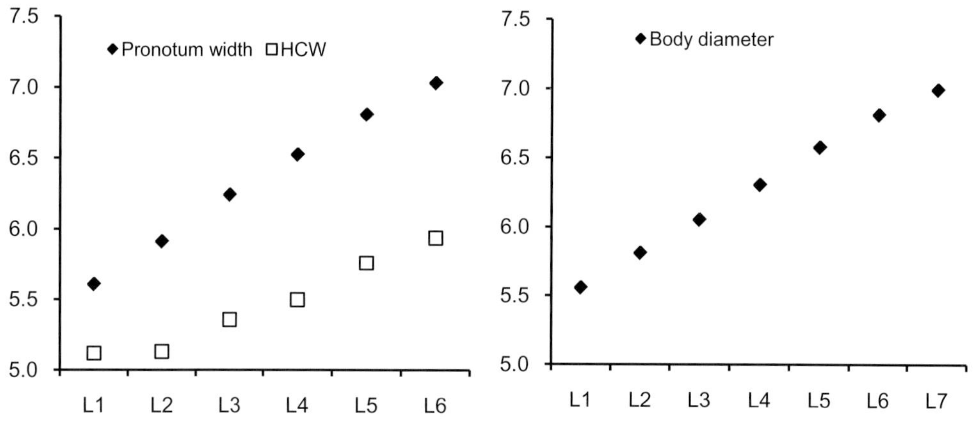

Figure 7.74 Natural logarithms of mean body dimensions (ordinate) for the larval instars (abscissa) of *Elmis* spp. (left) and *Limnius perrisi* (right) from the Breitenbach. Each mean is based on 10–17 individual readings; error bars are smaller than symbol sizes.

along the rear edges of the segments. In both genera, size differences between instars conform well with Brook's (or Dyar's) rule of a constant percentage increase at molts, that is, characteristic dimensions are linearly arranged in logarithmic plots (Figure 7.74).

Larvae of both genera have gastric mills composed of a ring of massive, longitudinally oriented, teeth, followed by a longer section with irregular smaller denticles (Beier, 1948). The crushed gut contents contain fine detritus, as recorded for other taxa (LeSage and Harper, 1976; Tavares and Williams, 1990) but Beier (1948) suggests that diatoms and Desmidiaceae are their main food.

The last instar larvae possess nine open spiracles on the mesonotum and the first eight abdominal tergites. Each spiracle is a small cone situated at the base of the paratergites and is only slightly larger than the pearl-shaped cuticular ornaments. A small tracheal canal is visible although Beier (1948) stated that the spiracles are only used after larvae have gone on land to pupate. Only last instar specimens possess more or less spherical tracheal extensions, which occupy much of the abdominal space (Beier, 1948). They can be compressed by muscle action, provide active control of buoyancy, and facilitate the drift of mature larvae to pupation sites (Brown, 1987).

The exoskeleton of adults and larvae is strong enough to resist mechanical impact during sediment movement. In view of this hard cuticle, cutaneous respiration cannot play a major role. The tiny first instar larvae already have some tracheal gill filaments at the abdominal tip, which can be withdrawn and protected under a sclerotized valve. Larvae dwell on the underside of stones, among gravel and sand, but also in clumps of moss. Dislodged flat *Elmis* larvae roll up, while the wire-like *Limnius* larvae curl firmly around a sand grain.

Opinions on life cycle lengths of Elmidae diverge. *Stenelmis crenata* in a summer-warm prairie stream in the United States was univoltine (Stagliano and Whiles,

2008). Berthélemy and Riols (1965) and Berthélemy and Ductor (1965) suggest univoltine cycles for Pyrenean populations of three genera, including *Elmis maugetii*. In contrast, Holland (1972), LeSage and Harper (1976) and Tavares and Williams (1990) estimate semivoltine life cycles between two and three or more years, which agrees with our own observations below. Because of the longevity and simultaneous occurrence of most larval instars and adults, interpretations of population structure are very difficult (LeSage and Harper, 1976). A recent synopsis (Elliott, 2009) states that *Limnius volckmari* (Panzer), a species of mainly hyporhithral distribution also completes larval development in one year. Adults of all species concerned here are said to take a full year until sexual maturity (Elliott, 2009).

The stream poisoning in week 21 of 1986 removed most insects and also most of the elmid populations, but some middle and late instar larvae appeared in the benthos a short time afterwards. They probably survived hidden in sediments, as did numerous *Sericostoma* larvae (Trichoptera), another sediment dweller (Zwick, unpublished data). Most adults were probably young individuals entering the stream after the poison had been flushed out.

First instar *Elmis* larvae were found in midsummer 1986 (Week 31), their growth could be followed during the next month. Young adults of this generation appeared only in spring 1988, after almost two years (Figure 7.75, left). First instar larvae of another generation appeared exactly one year later, 1987, in Week 31, recruitment continued for a long time. Larvae rapidly molted to second instar but when sampling was discontinued almost one year later, instar 2 was still dominant. At that time, the previous cohort was still present, including few larvae in instars 5 and 6 that seemed to be delayed siblings of the adults which appeared at that time. The *Elmis* spp. in the Breitenbach take about two years to complete their development.

The case of *Limnius perrisi* (Figure 7.75, right) is less clear, the species seems to develop extremely slowly. The population structure remained almost unchanged for one year, from the summer of 1986 to the summer of 1987 when the population density declined. It appears that last instar larvae moved on land to pupate, but because late instar larvae and adults were never abundant this cannot be documented. In late summer 1987, numerous first instar larvae hatched and a population structure formed which was very similar to the one previuosly observed and persisted until the end of benthic sampling in 1988. Much effort was spent to establish safe criteria for instar distinction (Figure 7.74) because surprisingly the third larval instar was rare, except around the middle of 2007. The life cycle length of *L. perrisi* cannot be safely inferred from the present data, but development may take even longer than in the smaller *Elmis*.

Dryopidae

In the older literature the Dryopidae included the Elmidae. True Dryopidae are distinguished in the adult stage by a very large, bowl-shaped pedicellus, which contains and conceals the short, club-like flagellum. This resembles the antennae of adult whirligig beetles, but while the modified antennae of the latter serve as

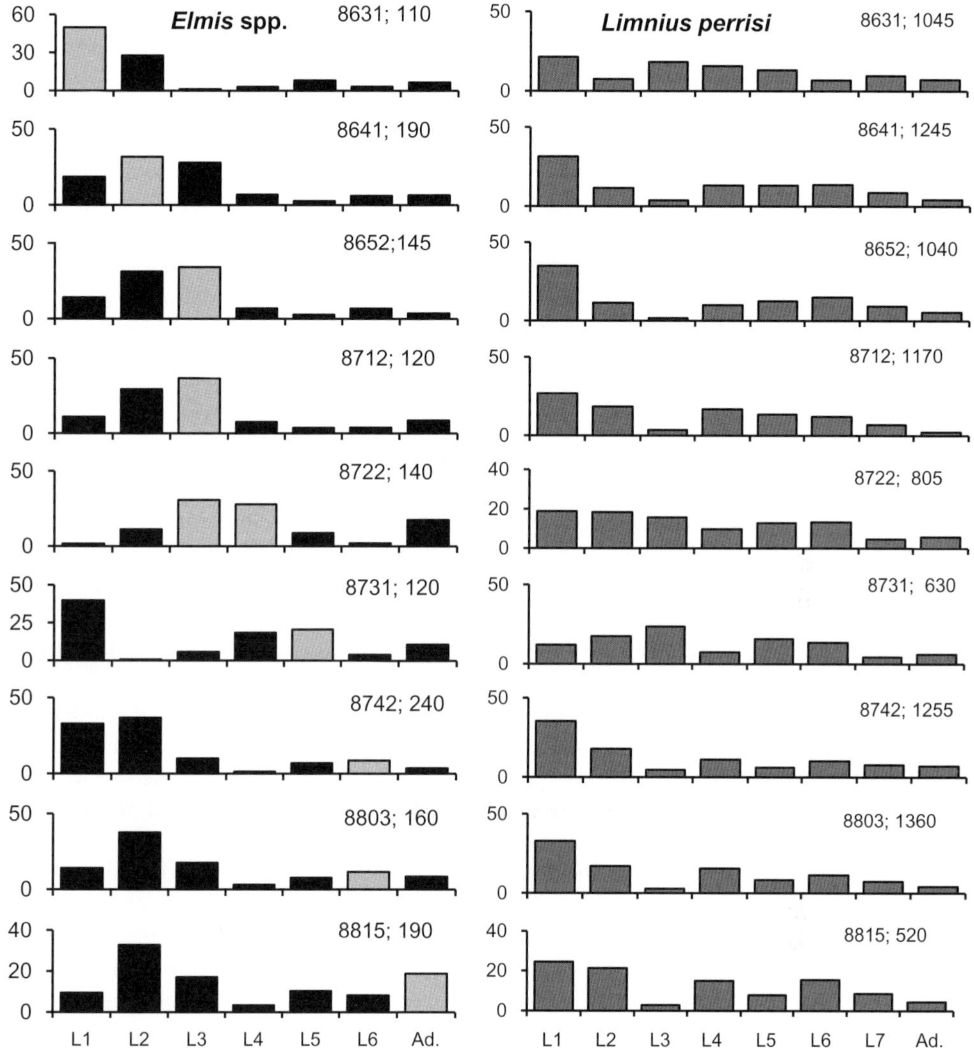

Figure 7.75 *Elmis* spp. and *Limnius perrisi* population structure in the Breitenbach during almost two years of observations. Ordinate: percentage of sample. Abscissa: the larval instars and adults. In *Elmis*, light gray bars illustrate an attempt to follow the new recruits of Week 31, 1986, throughout the two-year sampling period. Figures in legends are calendar weeks in 1986–1988 followed by approximate specimen numbers m^{-2} streambed.

sense organs, analyzing water movement during swimming or surface waves for prey detection, a similar function is out of the question in the slow moving semi-terrestrial Dryopidae. The significance of this striking structure is unknown.

The family mainly includes amphibious taxa (Brown, 1987). Adults have a dense, hydrophobic, hair covering over the entire body, holding an air bubble that serves as a physical gill when beetles are submerged. Females deposit eggs with their sable-shaped ovipositor in moist substrata. Fecundity and number of larval instars are unknown. Olmi (1976) described adult and larval *Dryops* digging superficial galleries into decaying woody material. Captive larvae consumed decaying plant matter, such as lettuce leaves. Olmi also quoted a study of a large population of *D. ernesti* adults and larvae in moist, agricultural soil, as damaging maize and grain seedlings by eating their rootlets. Pupation may occur in galleries in rotting wood, or in riparian soil.

Along the Breitenbach, the widespread, common, *Dryops ernesti* is a stream-margin inhabitant that hides under little stones or pieces of wood. Disturbed adults may escape in flight. We did not observe it in water, and never encountered any larvae. As a riparian species, no special attention was paid to it. The number of records is too limited to provide insight into the species' voltinism or bionomics.

The ecology of *Pomatinus substriatus* (formerly known as *Helichus* s.) seems to agree with *Dryops* (Olmi, 1976). A single specimen occurred in an Aspegraben drift sample after the winter drought in 1983/84. This record is interesting because *P. substriatus* is a thermophilus species, common along streams in the Mediterranean area. It is widespread in central Europe, but very rare in Austria and Germany. The only previous record in Hesse ("Hanau, Wetterau") was in 1845 (Horion, 1955). There are several recent records from Baden-Württemberg (Alf, 1991).

Scirtidae

The Scirtidae or marsh beetles were long known as the Helodidae or Elodidae. Most Scirtidae have aquatic larvae but terrestrial adults; only adult *Hydrocyphon* may spend time under water. Several genera are common members of lentic and lotic communities. In the Breitenbach, the family is represented by the genera *Cyphon*, *Elodes*, and *Odeles*.

There are several species of small *Cyphon* in Germany, but only the ubiquitous and common *C. coarctatus* appeared in small numbers in our traps. Only the 1983 and 1984 catches were analyzed. These consisted predominantly of males (86 and 89%, respectively). Collections of *Cyphon* usually tend to be male-biased, even though the sex ratio is balanced. Males are simply more active than females, and are therefore more readily collected (Nyholm, 1972). The species overwinters in the larval stage, whereas most other German species overwinter as adults, or as adults and larvae (Klausnitzer, 1971; Nyholm, 1972). In the Breitenbach, *C. coarctatus* begins to emerge at the same time as *Odeles marginata*, which would be late for overwintering adults. The life history more probably resembles that of *O. marginata*, see below. In 1984, emergence was even later, with a late peak (Figure 7.76, right). The larvae were never collected in the Breitenbach. They possibly live in peripheral marshy habitats or very deep in the substratum; many larvae of some other *Cyphon* species were recorded from groundwater (Klausnitzer, 2009).

The larger *Elodes minuta* and *Odeles marginata* are typical of rhithral streams. The range of *E. minuta* extends further downstream, but in the Breitenbach both

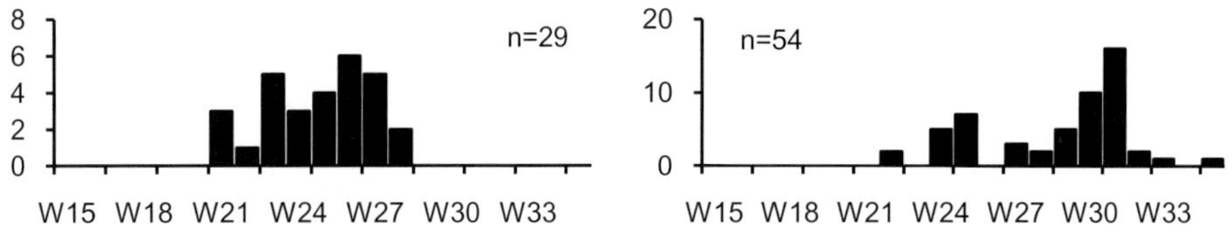

Figure 7.76 Numbers of *Cyphon coarctatus* emerging from the Breitenbach in Weeks 15–35 in 1983 and 1984. All traps pooled.

coexist everywhere and are common. *Elodes tricuspis*, of which a single male was once trapped, may be a pond-outlet specialist (Zwick, 2007). Adults of the two common species are 4.5–6.0 mm long and easily identified by their different colouration. Larvae have distinctively shaped heads, particularly the labrum, and also differ in general body shape and surface texture (Hannappel and Paulus, 1994; Klausnitzer, 2009; Zwick and Zwick, 2008a).

Scirtid Biology Apart from habitat preferences, seasonal aspects of their life cycle and the amazing feeding mechanism of marsh beetle larvae (Beier, 1952), knowledge of scirtid biology is incomplete. Some errors dating back to respected early pioneer studies were many times repeated in the literature (e.g., Klausnitzer, 1996). Our recent studies corrected several aspects and most were included in the latest summary (Klausnitzer, 2009).

Adults are mainly observed on shaded littoral vegetation, often flying for a short distance. Activity often has no obvious purpose. For example, dozens of *O. marginata* moved restlessly over about 1 m^2 of dead beech leaves along a tiny stream in the Odenwald, without any interaction between individuals (Zwick, unpublished observation). Adult *Cyphon* may feed on fine terrestrial detritus or fungal spores. Adult *Elodes* and *Odeles* were presumed to be carnivorous (Klausnitzer, 1996), but this is precluded because their sharply pointed mandibles are inserted too far apart to interact. Guts never contain particles; some greenish fluid is present upon hatching from pupae (Zwick, 2007).

During copulation, female *Cyphon* reach into the male and grasp the spermatophore with sclerotized structures in their bursa, which were therefore called prehensors (Nyholm, 1969, 1972). The same term was applied to sclerites in female *Elodes* (Klausnitzer, 1996) but the latter simply provide a support for the male parts during copulation (Zwick, 2007).

Eggs and oviposition are generally unknown in the family. Hoof (2001) presented indirect evidence of *Microcara testacea* egg survival in moist soil. The animal lives in intermittent, summer-dry pools and may be specially adapted. Eggs of the limnic *Scirtes* are laid in gaps in damaged duckweed and develop directly. The first instar larva has neither functional mouthparts nor spiracles (Zwick and Zwick, 2008b) but is nevertheless active.

Larval Biology General structure and morphology were described by Benick (1924, 1925) and especially by Beier (1949). Larvae depend on atmospheric oxygen.

The main tracheal trunks are transformed into spacious air sacs without taenidium (Beier, 1949), and end in a pair of open caudal spiracles. Larvae are usually strongly overcompensated and they need to cling to substrata to avoid floating. They are unable to swim and the running-water taxa roll up when carried away by the current, until they are deposited at stream edges, or can grasp some support. Finger-like anal protrusions, first described as gills (Beier, 1949), are osmoregulatory organs absorbing ions from the dilute environment against the concentration gradient, but may have some supplementary respiratory function (Wichard and Komnick, 1974a, 1974b). Indeed, larvae have survived experiments during which they were deprived of access to air for several hours (M. Hecht, personal communication).

Beier's account of larval mouthpart function (Beier, 1949, 1952) remains valid. The extremely complex structures attracted much attention, particularly for phylogenetic analysis and taxonomic purposes (e.g., Hannappel, 1991). The distal part of the soft, flat mandibles, and the galea of the maxilla, bear large, dense hair brushes, with which detrital deposits are brushed into the oral cavity, 70–80 times in a minute (Beier, 1949, 1952). The comb-like mandibular prostheca and mola interact with an amazing armature of elaborate spines, combs, claw-like teeth and brushes on the dorsal side of hypopharynx. Together, these components form a grinding and filtering apparatus for the fine material which is eventually compressed to form food balls, and sucked into the pharynx. Benick (1925) observed that larvae chewed and ate fresh leaves of *Berula* when fine detritus was in short supply. Paradise and Kuhn (1999) also classify the North American *Sacodes pulchella* (Guerin, 1843) and *Prionocyphon discoideus* (Say, 1825) which both live in phytotelmata as shredders skeletonizing leaf litter.

Much low quality detrital food is consumed and large amounts of feces are produced. There is no particle selection, markers such as soot or dye powder are indiscriminately ingested. Gut passage time has been estimated from the appearance of colored feces in larvae provided with labeled food (Figure 7.77).

Life Cycle and Larval Population Structure Adult *Odeles marginata* display simple monophasic emergence between Weeks 17 and 28. In five of the nine study years, peak emergence was in Week 23, but it may be as early as Week 20, and single specimens may appear at other times during the warm period of the year. Adult

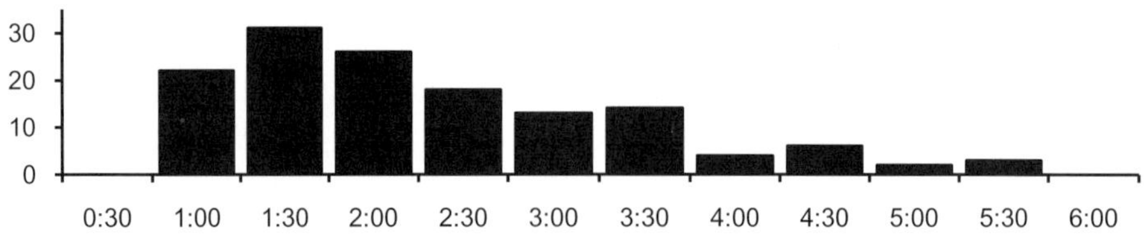

Figure 7.77 *Elodes minuta*, gut passage times in laboratory experiments using food labeled with karmesin powder ($n = 123$; M. Hecht, unpublished data). Ordinate: number of observations. Abscissa: time in hours.

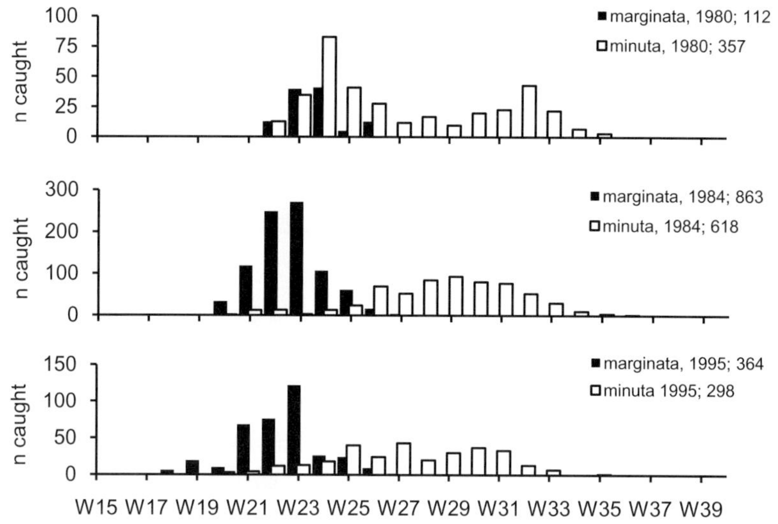

Figure 7.78 Different patterns of emergence of *Odeles marginata* and *Elodes minuta* in Trap 2 of the Breitenbach. Legends give year and number caught; the abscissa shows calendar weeks.

Elodes minuta emergence continues for longer. The pattern varies from unimodal with an early peak, or unimodal with a central peak, to distinctly bimodal between years (Figure 7.78). Attempts to relate the different patterns to water temperature failed because the sensitive period of the semivoltine life cycle is unknown. The life cycle is also uncoupled from stream temperature for approximately one month, when the prepupal and pupal phases are spent on land.

After adult emergence two different groups of larvae coexisted in the Breitenbach. The first group comprised many half grown *Odeles* larvae remaining in the benthos. By October many, by November all, attained last instar and over-wintered. The second cohort consisted of small larvae appearing in July. They were half grown in October/November, when growth stopped. Development was resumed in April (Figure 7.79, left). Our data on the semivoltine life cycle of *O. marginata* confirm the findings of Rasmussen in Denmark (Rasmussen 1978).

The situation in *E. minuta* was similar but population structure was less regular, due to cohort splitting in summer. Larval size classes at the end of winter could not be assigned to separate cohorts. Around June, the largest specimens left to pupate; they were possibly univoltine. Slightly smaller individuals attained the last instar soon after the first group but overwintered a second time. They were semivoltine and emerged in the following early summer (Figure 7.79, right). Fresh recruits were first noticed in July and grew very rapidly. By October the semivoltine and new groups merged, which restored the wide range of sizes in spring, when cohorts cannot be distinguished. Graphs of a population in Jutland (Rasmussen, 1978) suggest the same happens there. However, speed of larval growth, and whether some individuals are ready to pupate after the first winter may vary with

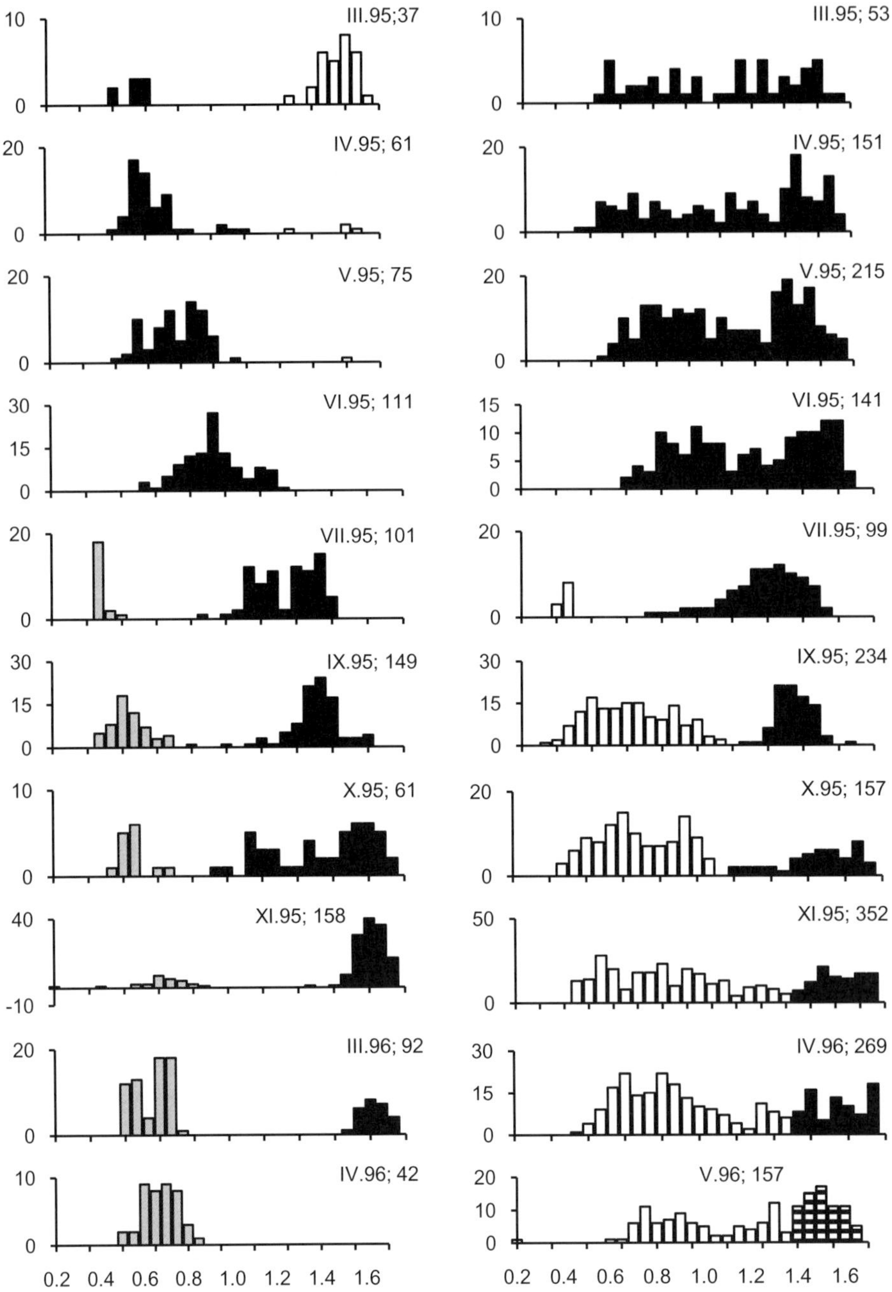

Figure 7.79 Size class frequency distribution (ordinate: percentage in sample; abscissa: head capsule width in mm) of *Odeles marginata* (left) and *Elodes minuta* larvae (right) in the Breitenbach, 1985–1996, all sites pooled; number of specimens indicated (data by M. Hecht). Different shading identifies separate cohorts, last instar of *E. minuta* in the final diagram cross-hatched.

conditions in given years, lead to different larval size distributions, and ultimately to variation in emergence patterns.

Only the first and the last instar can morphologically be distinguished. The first instar has remarkably long and strong caudal seate which disappear at the first molt. In *Elodes, Odeles and Cyphon* first instar larvae have functional mouthparts and spiracles (Zwick and Zwick, in preparation). The last instar is distinguished by a sharp apical tooth on the mandible, the so-called incisivus which is used to construct a spherical pupal chamber in shallow soil (Zwick and Zwick, 2008b). Other genera, especially *Scirtes* and *Hydrocyphon,* differ in these respects (Klausnitzer, 2009; Zwick and Zwick, 2008a, and in preparation).

At molts, the number of antennal annuli increases. Benick (1924, 1925) thought the number of annuli combined with body length separated five larval instars. His view was almost universally accepted in the literature. However, analysis of the published supporting data (Benick, 1924, 1925; Beier, 1949) revealed that they are actually inconclusive (Zwick and Hecht, 2008). By rearing separately kept individuals from egg to last instar the number of instars is now known to be 9–12 in *Cyphon, Elodes, Odeles, Prionocyphon,* and *Scirtes* (Zwick and Zwick, 2008a, and in preparation).

7.6.4.4.4 **Terrestrial By-Catch**

Traps on the Breitenbach are designed to catch animals emerging from the stream area under the trap, and largely do so. However, there is always some by-catch of terrestrial organisms, including about 550 species of terrestrial beetles (list available on demand). *Lesteva longoelytrata* (see above) belongs in this category. The analysis of by-catch is of interest mainly with regard to trap methodology (Section 7.1).

7.6.4.5 **Trichoptera**

Rüdiger Wagner

7.6.4.5.1 **General and Species List**

Trichoptera (caddis flies) are holometabolous insects (with pupal stage), they are the phylogenetic sister group of the Lepidoptera (moths and butterflies), and together both orders form the superorder Amphiesmenoptera. It is remarkable that the scientific term Trichoptera is derived from the terrestrial adults, that is, wing (Greek: *pteron*) covered with hairs (Greek: *trichos*), whereas the common name in many countries very often applies to the "house" of the larvae. Due to their close phylogenetic relation to Lepidoptera (scales on wings), adult caddis flies are often confused with "moths". However, the presence of trichia on wings is distinctive, as well as the long antennae and the comparatively well, and diversely developed, mouthparts (de Moor and Ivanov, 2008).

Caddis flies are known from the middle and late Triassic, but the main differentiation processes occurred in the Jurassic and early Cretaceous, with the breakup of the supercontinent Pangaea and the southern Gondwana continent. Two suborders, the Annulipalpia and Integripalpia, are usually distinguished. A third group the "Spicipalpia" (including the families Glossossomatidae and Rhyacophilidae) is not generally accepted (Kjer, Blahnik and Holzenthal, 2002).

The larvae are remarkable in that they construct fixed shelters, some in relation to filtering nets, or transportable cases. However, the larvae of a restricted number of species are free living. Several functions have been ascribed to the cases: camouflage, shelter, breathing device, and in streams and rivers, ballast against dislocation by running water (e.g., Statzner and Holm, 1989; Waringer, 1989a, 1989b; Williams *et al.*, 1987). Larvae are apneustic and depend on dissolved oxygen for respiration. They have found their way into almost all aquatic habitats, in standing as well as in running waters, springs, streams and rivers, ponds, lakes, and bogs. Many larvae have a hygropetric way of life, mainly petrimadicolous (*Tinodes*, *Bereaea*); limimadicolous and bryomadicolous taxa are of comparatively little importance. Several taxa from different families have found their way back into the terrestrial environment, including *Enoicyla pusilla* found by the Breitenbach. Caddis fly larvae cover all functional feeding groups and thus are therefore important processors of organic matter in freshwater food chains. They are also a keystone taxon in the development of the River Continuum Concept (Vannote *et al.*, 1980).

Final instar larvae built a pupal case and spin a cocoon in which the transformation of the last aquatic larva to the terrestrial adult takes place. The pupa is exarate (with free appendages) and covered by a semi-transparent pupal integument (de Moor and Ivanov, 2008). The cocoon of Rhyacophilidae species lies free in the case, whereas in the other families it is connected to the case. At the end of pupal development the cocoon contains the "pharate" imago. Using its specialized mouthparts the pharate adult disengages from the cocoon and the case. With the help of broadened and often setose legs it then swims to the water surface and goes on land. In most cases, the molt occurs on land, although sometimes already on the water surface. The adult leaves the exuvia behind.

To date more than 12 600 caddis flies, in 610 genera and 46 families, have been described worldwide (de Moor and Ivanov, 2008). However, the discovery of new species is far from over.

The west and east Palearctic have quite similar faunas (Sørensen similarity index 0.49), but less similarity to Neartic (0.35) and Oriental (0.34) regions. The extant fauna of the west Palearctic region contains the greatest number of species of the families Limnephildae, Sericostomatidae and Beraeidae – all Integripalpia. At present, it contains 149 genera and about 1520 species (de Moor and Ivanov, 2008). Further species await discovery and description, particularly based on molecular studies. In Europe, endemism most often occurs in the mountainous areas, the Pyrenees, Alps, Balkans, Carpathian Mountains or the Caucasus. Although caddis fly larvae are not in direct contact with the weather pronounced altitudinal distribution patterns and spatial patterns exist along running waters (Vannote *et al.*, 1980; Pauls *et al.*, 2008).

Species are usually distinguished by morphological features of the adults or larvae (Malicky, 2004; Waringer and Graf, 1997; Tobias and Tobias, 1981). However, even in the Breitenbach, at least two taxa are still not unambiguously named. This is partly due to the re-examination of species with modern molecular methods, and why use of the names "*Wormaldia occipitalis*" and "*Sericostoma personatum*" here remains provisional (Leese and Wagner, 2005).

Twenty-nine families and several hundred caddis fly species have recently been reported from Europe (Malicky, 2004); 74 species in 44 genera from 15 families have been identified from the collections on the Breitenbach (Appendix). Several species are widespread and euryecious, and occur not only in streams but also in small rivers of the area. Thus, the influence of the River Fulda may be noticed in numbers of specimens and species collected in the downstream area of the Breitenbach in particular (Figure 7.8). Furthermore, small ponds in the valley (partly used for private fish farming, Chapter 2) affect the stream community because specimens from standing waters (e.g., *Limnephilus* spp.) seem to invade the stream permanently, including the emergence traps.

The remaining species are typical of small streams in low mountain areas as long as they remain more or less unaffected by man. However, even in this thinly populated part of Germany, the great majority of small streams have undergone periods of changes (due to "land melioration"), by straightening water courses for agricultural use, or structuring ponds for private fish farming (Chapter 2), and are inhabited by more or less heavily altered communities. The Breitenbach however, remains one of the rare, only slightly altered, small (first- to second-order) streams in Central Germany.

The life cycle of the holometabolous Trichoptera comprises egg, usually five (up to seven) larval instars, pupa and the terrestrial adult. At the Breitenbach, the life cycle of the most abundant species is univoltine, but development of *Rhyacophila fasciata*, *Wormaldia occipitalis* and *Drusus annulatus* requires between one and two years. For *Sericostoma personatum* a life cycle lasting between one and up to four years was calculated, based on laboratory treatments at constant temperatures (Wagner, 1990b). For the species concerned the number of emergence peaks per year provides no indication of life cycle duration.

The seasonally variable temperature patterns along the stream, and the emergence patterns of individual species at different sites, indicate that univoltine life cycles (with one distinct emergence peak) may be less definite near the source. Under the seasonally comparatively constant temperature, life cycles of *A. fuscipes* and *A. fimbriata*, and probably *W. occipitalis*, become desynchronized, and the emergence period is extended because specimens are ready to pupate at almost any time of the year, and adults emerge when conditions in the terrestrial environment are adequate (Figure 7.80). Emerging *A. fimbriata* were collected from February to August in the spring (Georgsborn; Gümbel, 1976). Emergence at T2 and further downstream is restricted to about 10 weeks in May, June, and July.

Several species were examined in detail, that is, reared from the egg in the laboratory and in the field. The temperature dependence of embryogenesis, instar duration, food consumption and the existence of diapause or dormancy were studied. Results of laboratory studies were compared with data from the field survey. Field experiments were also conducted to confirm or falsify the reliability

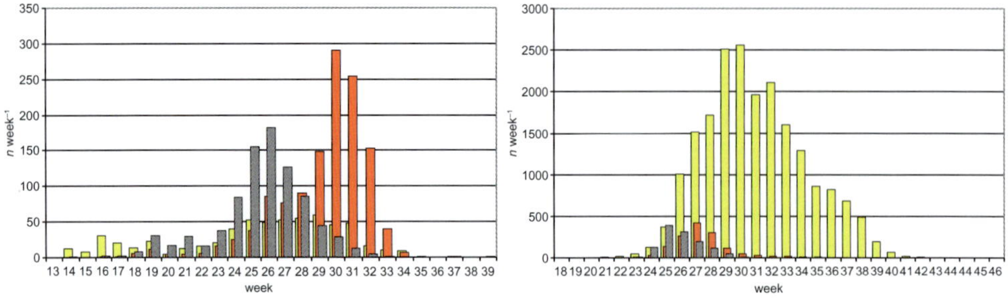

Figure 7.80 Weekly sum of emerging *Apatania fimbriata* (1985, left) and *Agapetus fuscipes* (1989, right) in traps T1 (yellow), T2 (red), and T4 (gray).

Table 7.18 Usual emergence periods of the major Trichoptera at the Breitenbach (site T2, 5–95% yearly emergence 1969–2005). Emergence may differ by more than one week at more upstream or downstream sites.

Species	J	F	M	A	M	J	J	A	S	O	N	D
Agapetus fuscipes						■	■	■				
Apatania fimbriata				■	■	■	■	■				
Drusus annulatus					■	■	■	■	■	■		
Chaetopteryx villosa										■	■	■
Potamophylax cingulatus								■	■	■		
P. luctuosus					■	■	■	■				
Plectrocnemia conspersa					■	■	■	■	■	■		
Rhyacophilda fasciata					■	■	■	■	■	■	■	
Silo pallipes					■	■	■	■	■			
Sericostoma personatum						■	■	■				
Tinodes rostocki					■	■	■	■				
Wormalidia occipitalis						■	■	■	■	■	■	

of laboratory data. Size–weight–sex relations provided further data to test various hypotheses (Section 7.9).

In individual years emergence usually starts in late March with *A. fimbriata* and lasts until December, when adults of the short winged *C. villosa* may be found walking on the snow cover (Table 7.18). The emergence period of most species is four weeks or longer, several taxa are on the wing for up to five, or even six, months. Along the stream emergence periods of the same species may vary drastically, for *A. fuscipes* and *A. fimbriata* it may be about twice as long at

Table 7.19 Mean male share (%) of six abundant caddis flies (1983–1991) from four collecting sites (pooled data), and determined individually for four traps (T1, T2, T4, T6; sd = standard deviation [mean all years and traps; *n* = 36]; T1–T6 mean percentage in individual traps in nine years [*n* = 9]; / = insufficient data).

Species	Pooled	sd	*n*	T1	T2	T4	T6	*n*
Ag. fuscipes	60	8	36	51	63	60	65	9
Ap. fimbriata	45	19	36	30	48	58	/	9
C. villosa	64	6	36	68	64	64	60	9
D. annulatus	63	13	36	53	65	71	/	9
S. personatum	65	6	36	64	62	68	66	9
T. rostocki	45	6	36	45	41	48	45	9

T1 as at T2 (Fischer *et al.*, 1998). In the same year, emergence maxima between traps only 400 m apart may differ by four weeks or more, for example *A. fimbriata* (Figure 7.80). Furthermore, as mentioned above, for several species the number of "emergence peaks" does not necessarily correspond to the "number of generations".

7.6.4.5.2 **Reproduction**

Sex Ratio

Under ideal conditons the sex ratio (♂/♀) in populations should be 1 : 1, however this was not met in the abundant caddis flies, as shown by results from a nine year period (1983–1991) with four traps along the stream. *Ap. fimbriata* and *T. rostocki* have sex ratios with slightly less males than females. In the other species mentioned, the proportion of males is 60–65%. Differences between years are larger in *Ap. fimbriata* and *D. annulatus* but comparatively small in the other species (Table 7.19).

In four species the percentage of males changed along the stream, increasing steadily with distance from the source in *A. fuscipes*, *A. fimbriata* and *D. annulatus*, but decreasing in *C. villosa*. No changes were evident in *S. personatum* and *T. rostocki*. We have no indication as to whether these changes are related to specific or sex-related behavioral patterns, to collecting methods or to any other reason.

Adult Feeding

Knowledge about the terrestrial part of the caddis fly life cycle is still limited. Until quite recently it was disputed whether adult caddis flies were active feeders, although the uptake of water or dewdrops had been documented. Adult Trichoptera feed on dewdrops, plant liquids and solutions of many substances on leaf surfaces. Syrnikov *et al.* (2005) experimented with adult caddis flies and discovered that food quality affected longevity, fitness and sexual attractiveness of both sexes. Longevity of the Breitenbach taxa is species-specific, and may last from 1–2 days

Table 7.20 "Pheromone agents" from some Breitenbach Trichoptera identified by J. Bergman (University of Hamburg): *Apatania fimbriata*, *Chaetopteryx villosa*, *Potamophylax luctuosus*; *Tinodes rostocki*. f = female; m = male.

Species	Agent 1	Agent 2	Agent 3
A. fim f	3-Octanon	6-Methyl-3-octanon	Eucalyptol
C. vil m	4-Methyl-3-hexanon	4-Methyl-3-heptanon	
C. vil f	4-Methyl-3-hexanol	4-Methyl-3-heptanol	
P. luc m	4-Methyl-3-heptanon		Sordidin
P. luc f	4-Methyl-3-heptanon	4-Methyl-3-heptanol	Sordidin
T. ros f	4-Nonen-2-on	2-Nonanon	2-Heptanon

(female *A. fimbriata*) to more than 3–4 weeks (*C. villosa*). Mass swarming was never observed in the typical Breitenbach species. However, *Rhyacophila* and *Potamophylax* specimens are good fliers that disperse along the Breitenbach valley with directed flight. Upstream migration was not observed, so doubts remain about the validity of the colonization cycle theory (Müller, 1973), at least for the Breitenbach caddis flies. In particular, "typical" species were repeatedly observed hiding in the dense bank vegetation under increasing wind velocity or rain.

Pheromones

Pheromones play an important role in mating by many caddis flies. The presence of such substances was demonstrated for *Apatania fimbriata* by Solem and Solem (1991), and later for other Breitenbach species. It is now well known that males are attracted by alcohols and ketones produced and released by females (Bergmann *et al.*, 2002). In cooperation with Jan Bergmann (University of Hamburg) pheromones of several Breitenbach caddis flies were identified (Table 7.20) and those of female *A. fimbriata* were successfully tested for their effectiveness to attract males in the field (Wagner, unpublished data).

Oviposition

Most emerging females have more or less fully developed eggs in their abdomens, however, Syrnikov, Melnitzky and Ivanov (2005) showed experimentally that the quality of adult food may affect sexual behavior, number of sexual partners, and probably reproductive success.

Trichoptera have very diverse oviposition modes and different shapes of eggs and egg masses. Size and shape of egg masses of the Breitenbach species are so different that in many cases they can be attributed to genus or even species.

Three basic oviposition modes are usually distinguished: individual or small groups of eggs (e.g., Rhyacophilidae, Psychomyidae); eggs in single layered masses with small amounts of mucous (e.g., Hydropsychidae, Polycentropodidae); spherical, elliptical and irregularly shaped masses with a larger mucous matrix that may swell considerably in water (e.g., families Beraeidae, Limnephilidae, Phryganeidae, Sericostomatidae).

Some species, mainly Limnephilidae, oviposit in the terrestrial environment beside flowing or standing waters, whereas *Hydropsyche* females dive, even into deep waters, to deposit their eggs on river bottoms (Deutsch, 1984). However, females of the Breitenbach species mainly oviposit on hard substrata under the shallow stream water. In most cases females produce only a single egg mass (but compare Becker, 2005; Section 7.8). Oviposition seems to occur preferably at dawn or twilight.

Egg masses of *Plectrocnemia conspersa* and *Hydropsyche* species do not swell; they consist of only a single layer of more elongate eggs, and were only found under water, on larger stones or on pieces of wood. *Agapetus fuscipes* females deposit small numbers of eggs in a single layer on stones, and cover them with a sand grain (cap stone; Anderson, 1973; Becker, 2005; Section 7.8).

Oviposition site preferences of some other species are as follows: *Apatania fimbriata* lays small (4–8 mm diameter) hemispherical egg masses (Figure 7.61) on stones just above the water surface and in dark cavities in the stream bank. Eggs are at risk from various predators in such places (Enders and Wagner, 1996). *Drusus annulatus* fastens 2–4 cm long egg masses 3–6 cm deep under water on vertical stony structures and in crevices underneath stones at low current speed (2–6 cm s^{-1}). These egg masses are remarkable because they consist of two gelatinous matrices, an inner spherical one, typical of many Limnephilidae, containing the eggs, and an outer one in which the inner matrix lies, as if in a bag. Whether the outer matrix acts a protection against desiccation in case of falling water level or against underwater bacterial or fungal attacks is not clear. In other streams of the region egg masses of the related *Ecclisopteryx madida* appeared almost identical but were found outside the water in damp hollows along streams (Enders and Wagner, 1996; Wagner and Höchst, 2001).

Most other Limnephilidae produce mainly spherical egg masses some pinhead-sized with densely packed eggs. Many of them swell to 1 cm^3 or more, by taking up water for some hours, thereby losing their color. The distance between the eggs increases during the swelling process and, in the end, individual eggs are hardly visible. Egg masses of *Potamophylax* species are even larger; they appear translucent and of irregular shape. A few egg masses of *P. nigricornis* were observed in the terrestrial environment above the Georgsborn. They were positioned so that hatching larvae will fall into the water below (Sternberg-Holfeld unpublished data).

Egg masses of *Sericostoma personatum* (Sericostomatidae) were rarely found in the field. In the laboratory, females deposited egg masses under water on stones, but mainly on vertical structures 5–10 cm below the water surface (Zwick, unpublished data).

Fecundity

For a given caddis fly species, fecundity varies with female body size; larger females contain more eggs. This was evident in *Agapetus fuscipes* from two adjacent springs. Specimens from the "warmer" spring (yearly mean 9 °C) were larger than those from the "cold" spring (yearly mean 8 °C; Figure 7.81).

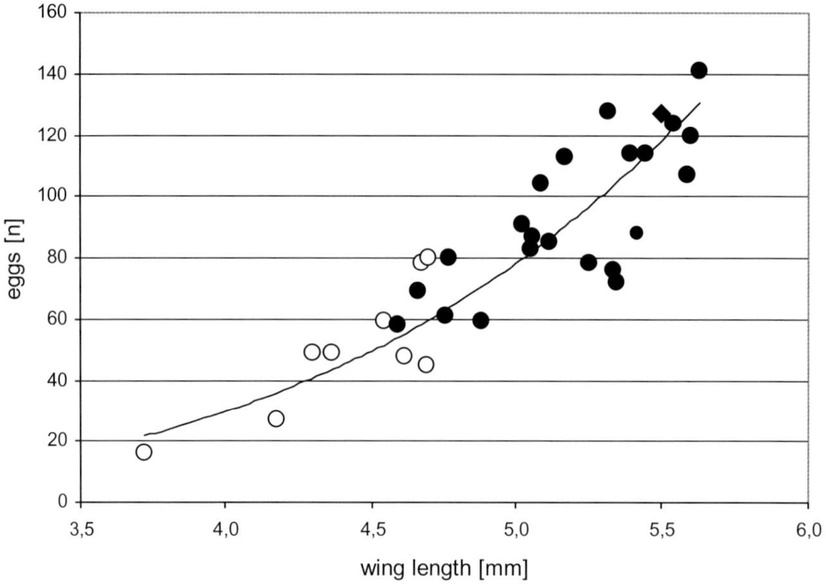

Figure 7.81 Size (forewing length) and egg number of *Agapetus fuscipes* females from a cold (open circles) and a warm (full circles) source (regression $y = 0.0722x$ exp 4.342, $r^2 = 0.83$; Sternberg, unpublished data).

Table 7.21 Egg numbers (mean ± SD, or min–max) per female of abundant caddis flies from the Breitenbach.

Species	Mean	SD	min–max	Reference
C. villosa	150		50–290	Wagner, 1986
P. conspersa	461		220–797	Hildrew and Wagner, 1992
Ap. fimbriata	204	83		Enders, 1994
D. annulatus	200		150–250	Enders, 1994
A. fuscipes	82	31	16–141	Sternberg (unpublished data)

In addition, egg number per dissected female varied depending on whether females had oviposited or not. This partly depends on oviposition mode, because the Glossosomatidae, like *A. fuscipes* and some other taxa, deposit several egg masses, whereas most other species produce only a single egg mass.

Depending upon taxon, ecological studies of the Breitenbach species revealed fewer than 100, up to about 800, eggs per female (Table 7.21). It is clear that females of larger species contain more eggs.

Egg size differed between species. Eggs of the two *Potamophylax* species were of the same size (Figure 7.82). Mean egg size and mean female dry weight (as a

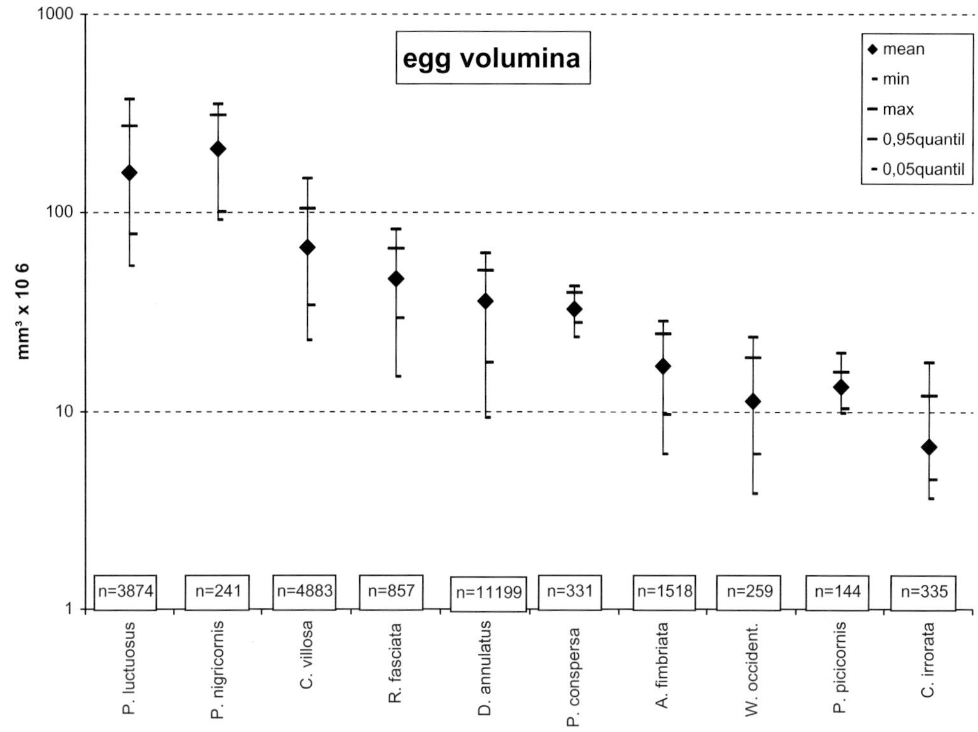

Figure 7.82 Egg volumes ($mm^3 \times 10^{-6}$) of some Breitenbach caddis flies (mean, 0.95 quantile, minimum, and maximum), n = number of eggs measured (Sternberg, unpublished data).

measure of size) were highly significantly correlated, at least for the Breitenbach species; larger species have larger eggs ($P < 0.0001$). The relationships between egg size and higher taxonomic group, feeding mode or functional feeding groups of the larvae were not significant (Figure 7.83).

7.6.4.5.3 **Egg Development**

Water temperature affects duration of embryogenesis and hatching success of caddis fly eggs. These studies were predominantly carried out in the laboratory, at different constant temperatures; only a few experiments were conducted with diurnally fluctuating temperatures. In addition, comparative field studies with egg masses, and on egg development, were carried out on some species. We were lucky to observe oviposition and successful embryogenesis of *Rhyacophila fasciata* at different temperatures in the laboratory (Sternberg, unpublished data).

Egg development in caddis flies is usually direct, without diapause. However, the externally firm, brown-green egg masses of *Sericostoma* spp. may enter a diapause, probably as a strategy to withstand dry conditions. Embryogenesis starts immediately after oviposition; in many species a gelatinous matrix between the eggs takes up water from the environment so that the egg masses swell. An

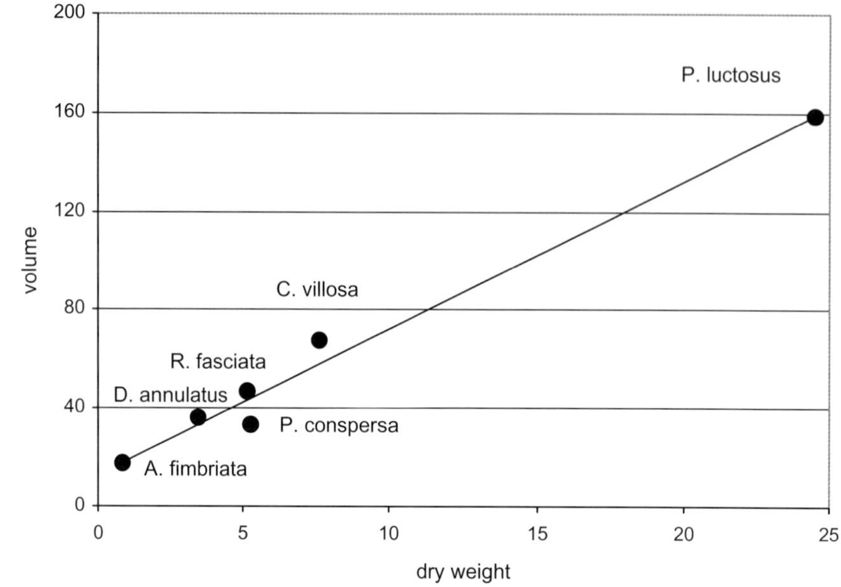

Figure 7.83 Mean egg volume (10^{-6} mm^3) and mean female dry weight (mg) of some Breitenbach caddisflies; (linear regression $y = 5.967x + 12.87$, $r^2 = 0.982$). Egg volumina of the congeners *Potamophylax nigricornis* and *P. luctuosus* are the same.

increase in egg diameter during embryogenesis has been described for several species (Enders and Wagner, 1996; Hildrew and Wagner, 1992; Sternberg, unpublished data).

Incubation periods of caddis fly species were defined as the time interval between deposition of an egg mass and hatching of the larvae in the outermost layer of the spherical egg masses. Comparison of field and laboratory results revealed that, in the field, almost all larvae usually hatched within one, or at most two, days from the spherical egg masses, whereas in the laboratory hatching at the center of an egg mass was delayed for some days, probably due to a suboptimal oxygen supply.

Mean incubation period decreased with rising temperature. The best general model to describe the dependence of incubation period on temperature was a power function, $y = a{*}x^b$. Figure 7.84 and Table 7.22 provide all the relevant information on the Breitenbach species and an additional species from the Vogelsberg Mts (the parthenogenetic *Apatania muliebris*). Some species were tested by different authors in subsequent years, however without significant differences.

Laboratory and Field Studies

The upper temperature limits for successful development were discovered for two species. Larvae of *Apatania fimbriata* and *Drusus annulatus* did not hatch at constant temperatures above 18 °C, because developing larvae had consumed all the available yolk (energy reserve). By the end of embryogenesis only the empty scle-

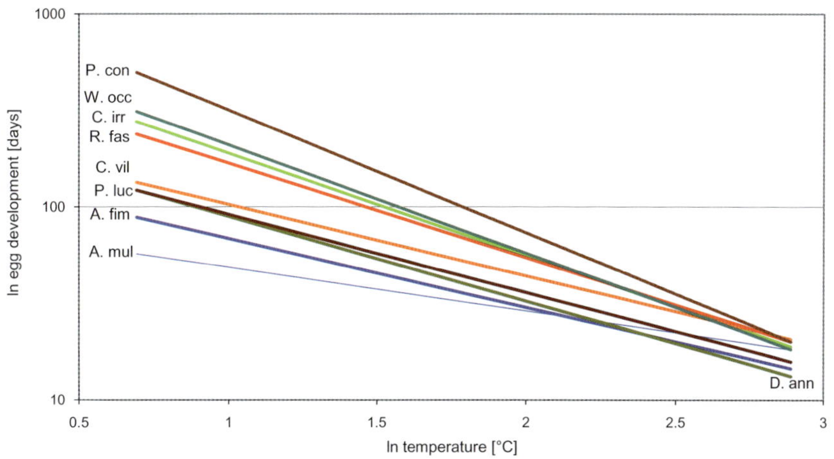

Figure 7.84 Incubation period of nine caddis fly species (*Apatania muliebris* from the Vogelsberg mountains) as a function of constant incubation temperature in the laboratory (*Apatania fimbriata, A. muliebris, Crunoecia irrorata, Chaetopteryx villosa, Drusus annulatus, Potamophylax luctuosus, Rhyacophila fasciata, Wormaldia occipitalis, Plectrocnemia conspersa*). Compare Table 7.22.

Table 7.22 Dependence of incubation period (y) of several Trichoptera species from the Breitenbach on constant water temperatures in the laboratory ($y = a{*}x^{exp}b$).

Species	**a**	**b**	**r^2**	**Reference**
A. fimbriata	156.0	−0.82	0.98	Enders and Wagner, 1996
A. fimbriata	152.2	−0.84	0.89	Sternberg, unpublished data.
A. fimbriata	165.5	−0.89	0.98	Aurich, 1989
A. muliebris	82.1	−0.52	0.94	Wagner, unpublished data.
C. irrorata	644.0	−1.22	0.97	Sternberg, unpublished data
C. villosa	255.0	−1.15	0.97	Wagner, 1986
C. villosa	241.0	−0.85	0.93	Sternberg, unpublished data
D. annulatus	245.7	−1.01	0.98	Sternberg, unpublished data
D. annulatus	222.0	−0.94	0.96	Enders, 1994
P. luctuosus	232.2	−0.93	0.99	Sternberg, unpublished data
R. fasciata	518.8	−1.125	0.79	Sternberg, unpublished data
W. occidentalis	762.5	−1.29	0.86	Sternberg, unpublished data
P. conspera	1364.8	−1.46	0.97	Hildrew and Wagner, 1992
S. personatum	?	?	?	Wagner, unpublished data

rotized remains of the larvae were visible in the eggs (egg masses were not attacked by bacteria or fungi).

The results of these laboratory experiments raise the question whether egg development in those reaches of the stream where temperature is usually above 18 °C is also unsuccessful. This might be the reason why both species are rare, or do not occur, in the downstream sections (D and E).

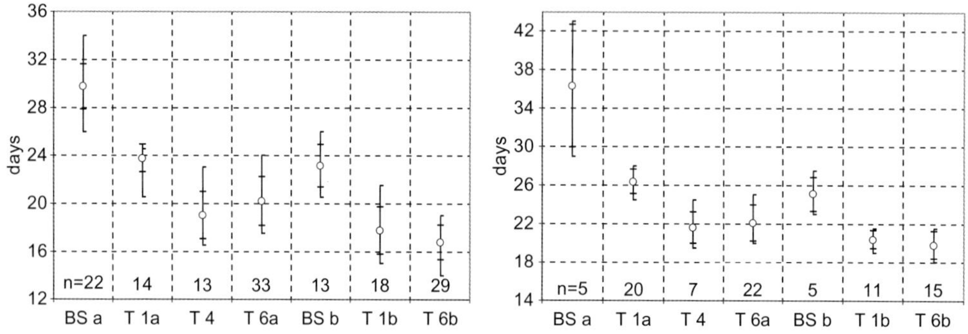

Figure 7.85 Successful egg development of *A. fimbriata* and *D. annulatus* at different field sites. BS = source; T1, T4, T6 = trap sites; a = under water; b = 1 cm above water surface; *n* = number of egg masses per site. Mean = open circle; sd = elongate bars; max–min = short bars. (Sternberg, unpublished data)

However, field experiments demonstrated that egg development in both species was also successful downstream, under higher, but diurnally fluctuating, water or air temperatures. Egg masses positioned approx. 1 cm above the water surface developed successfully, and even faster than those under water (Figure 7.85). They are absent for some other reason.

In 1983/84 egg masses of *Chaetopteryx villosa* were kept at different sites in the Breitenbach under natural water temperatures, almost constant at T1, with increasing diurnal amplitude downstream (T2, T3, T4). Those exposed under almost constant temperature (sections A and B) hatched at the time predicted from laboratory treatments. Larvae from egg masses with increased daily temperature fluctuation (section D) hatched earlier than predicted, probably due to the increased diurnal temperature fluctuations (Wagner, 1986).

Results of a few laboratory experiments with limnephilid egg masses showed similar effects. Development of *C. villosa* and *Anabolia nervosa* egg masses kept under sinus-shaped diurnally fluctuating temperature regimes (8–12 °C) was up to 40% faster than in those kept at constant 10 °C (Wagner, unpublished data).

7.6.4.5.4 **Larval Development**

Freshly hatched larvae immediately started case construction. Houses of one-day-old larvae of *Drusus annulatus, Apatania fimbriata* and *Sericostoma personatum* were well constructed and shaped like those of older larvae, but smaller. In contrast, the cases of young *Chaetopteryx villosa* resembled the accidental accrual of various organic and inorganic particles, and only after some days was a more or less "proper" case constructed.

Unlike the Ephemeroptera and Plecoptera, Trichoptera are holometabolous, that is, with a low number of instars, and a pupal stage in which all the morphological and physiological changes between aquatic larvae and terrestrial adults occur. The low instar number is one reason for the very unequal accrual of biomass between

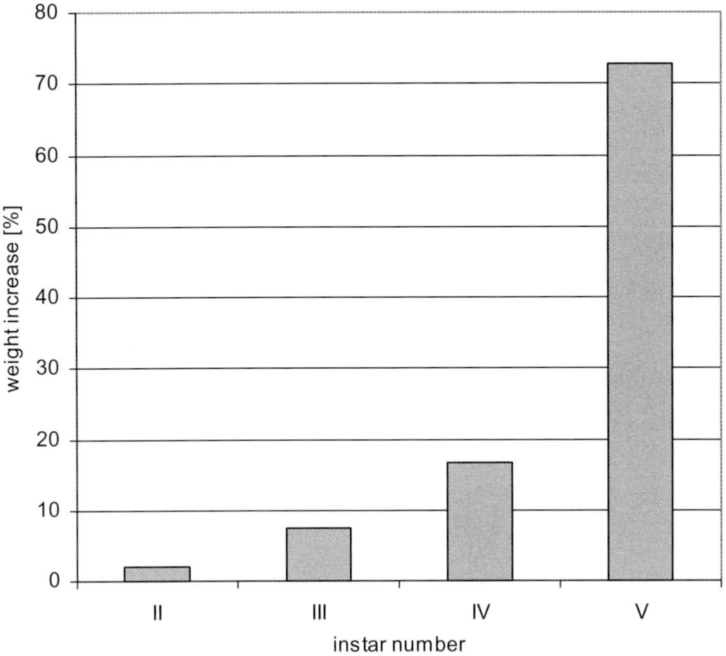

Figure 7.86 Mean weight increase (% maximal larval weight) in instars II to V of ten caddis fly species of all major functional feeding groups.

instars. The final larval instar is evidently "weight decisive", up to 80% or more of the adult biomass is accumulated during this instar in all species (Figure 7.86). Environmental variables, mainly temperature, food quality and quantity, are also major determinants for adult weight.

All trophic categories and functional feeding groups are represented by the caddis fly larvae in the Breitenbach. The majority are shredders, feeding on coarse particulate organic matter (CPOM). However, most shedders are rather opportunistic feeders, sometimes even feeding on green bank vegetation outside the water, fine particulate organic matter (FPOM), small live or dead animals of their own, and other, species (Anderson, 1976; Wagner, 1986, 2002a). The Glossossomatidae and some Limnephilidae (*Apatania, Drusus*) are the dominant scrapers in the upper Breitenbach reaches (T1–T3) whereas the gallery constructing *Tinodes rostocki* (Psychomyiidae) is very abundant in stream reaches at sites T3, T4 and further downstream. Larvae of the *Rhyacophilidae, Polycentropodidae* and *Odontocerum albicorne* are predators.

Laboratory experiments with shredders at constant temperatures, and with an excess of green alder leaves as food, confirmed the strong effect of water temperature on larval growth. Field experiments on the effects of additional limiting environmental factors (food quality, temperature pattern, sex) on larval growth and adult size were mainly carried out with *Chaetopteryx villosa*. Periods of main weight gain were easily recognized from life cycles (Figure 7.86; Wagner, 2005a, 2005b). One exception was the larvae of *Sericostoma personatum*, which only feed at night (CPOM) on the sediment surface and rest during the day a few cm deep in the sediment. Defecation effected an increase in organic content in the upper sediment layer (Wagner, 1991).

Winter dormancy was observed in the life cycle of *Apatania fimbriata* in the field, and was verified in laboratory treatments at different temperatures (Aurich, 1992). Its larvae were the first shown to possess defensive secretions (mixture of fatty acids) effective against small invertebrate predators, but not against larger invertebrates or fish (Wagner *et al.*, 1990).

Larval habitat preferences were easily obtained by: (i) inspection of substrata in the field and (ii) substratum manipulation under emergence traps over three years (Section 8.2). Availability of specific habitats is crucial for the success of populations in streams. The seasonal and annual variable discharge simultaneously creates and disturbs habitats. Only if the specific traits and preconditions in the stream match well can a species be successful (in terms of abundance and biomass). Due to long term low flow conditions, sand was deposited on the stream bottom causing changes in the relative abundance of species. However, no species recorded in 1969 was lost until recently, and no "new" species was successful.

Autecological aspects include the study of weight and size of adults collected in subsequent years along the Breitenbach. Spatial (along the stream) and temporal (between years) differences in size and weight were found to be significant (Wagner, 2005a, 2005b). The effects of intra- and interspecific competition and limited food source were proven statistically, relating individual size and weight of emerged specimens and abundance in individual traps (Wagner, 2005a, 2005b).

7.6.4.5.5 **Distribution of Adult Caddis Flies in Traps along the Breitenbach**

The distribution of caddis flies (and other aquatic insects) along the Breitenbach is uneven. Several of the abundant species prefer the vicinity of the source and T1, while others have their highest abundances in the lower stream section (T4–T6). Distribution centers along the stream are in good agreement with the generally known habitat preferences.

For example, *A. fuscipes*, *C. villosa*, and *P. cingulatus* are typical inhabitants of the stream section in the vicinity of T1, whereas *R. fasciata*, *S. pallipes*, and *T. rostocki* prefer the lower stream sections, T4–T6 (Figure 7.87).

Although these patterns appear more or less stable, in some species differences may occur between years. From 1983 to 1993, *T. rostocki* was absent around T1, relative abundance declined at T2, but increased at T4 and T6. In some years *A. fimbriata* had its highest abundance at T2, but its mean abundance declined slightly from T1 to T4; it was almost absent from the most downstream site, T6. No species preferred the area at T2, so this section is an along-stream transition zone, at least for caddis flies.

7.6.4.5.6 **Spatial and Temporal Weight Patterns of Adult Caddis Flies**

Size (forewing length) and weight are two variables often used to distinguish animals of different sexes, from different years, collecting sites or generations.

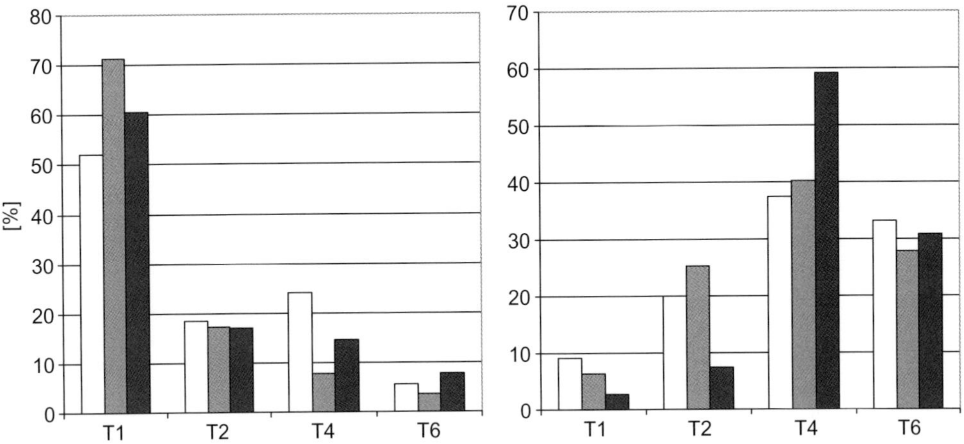

Figure 7.87 Left: mean share (%) of *A. fuscipes* (white), *P. cingulatus* (gray), and *C. villosa* (black) at four sites along the Breitenbach. Right: mean share (%) of *R. fasciata* (white), *T. rostocki* (gray), and *S. pallipes* (black) at four sites along the Breitenbach.

Both measures are quite easy to determine and mostly highly correlated. Females and males from all traps were weighed in monthly, 10-day, or weekly intervals, in different years to determine spatial and temporal weight variation. Monthly intervals provided insight to the principal temporal and spatial patterns. Weekly intervals gave information on temporal variation within a year, and spatial variation if data from several traps were compared. In some respects temporal weight patterns are different from those described for the Ephemeroptera and Plecoptera.

The first caddis fly species found to show obvious size differences between sexes and sites along the stream was *Chaetopteryx villosa*. Even in the field and by eye it was evident that females were larger than males at any site, and the size of both sexes also increased with distance from the source. Food shortage, due to high larval densities, and probably to competition by other shredders, was suspected to be the most important cause. However, through comparative laboratory and field studies, it became clear that the weight and size increase of both sexes of *C. villosa* along the Breitenbach did not depend on increased competition for food but on the site-specific seasonal water temperature patterns (Wagner, 1992, 2002b, 2002c, 2002d, 2005a, 2005b). Specimens from downstream sites (T2, T4, T6) grew to a larger size than those from the spring region, because higher water temperature downstream favored growth during the period of most effective weight gain (fourth and fifth instars) in the life cycle. Food effects only became evident if poorest (FPOM) and best (CPOM) food qualities were compared experimentally. The same applies for another abundant shredder, the crustacean *Gammarus fossarum* (Section 7.9).

Nine Years Data

For the period 1983 to 1991, substantial measures of weight and size of emerged caddis flies, elucidated additional interesting spatial and temporal patterns. Data

Table 7.23 Analysis of weights of eleven caddisfly species from four traps on the Breitenbach (1983–1991). Females heavier than males (significance); wcds = weight change down stream (↑ = increase, ↓ = decrease); wdiff ♀, wdiff ♂ = significance of weight difference on four sites for sexes. (Kruskal–Wallis and Wilcoxon rank tests).

Species	♀ > ♂	wcds	wdiff ♀	wdiff ♂
D. annulatus	$P < 0.0001$	♀♂↓	A = B = E>G	A = B = E>G
R. fasciata	$P < 0.0001$	♀↓	A = B = E>G	n.s.
A. fimbriata	$P < 0.009$	n.s.	n.s.	n.s.
P. luctuosus	$P < 0.013$	♀♂↑	A = G<B = E	A<B = E = G
S. pallipes	$P < 0.0001$	n.s.	n.s.	n.s.
S. personatum	$P < 0.0001$	♀♂↑	A = B = E<G	A<B = E<G
T. rostocki	$P < 0.0001$	n.s.	n.s.	n.s.
A. fuscipes	$P < 0.0001$	♂↑	n.s.	A<B = E = G
P. conspersa	$P < 0.0001$	n.s.	n.s.	n.s.
P. cingulatus	n.s.	♀↓	A = B = E>G	n.s.
C. villosa	$P < 0.0001$	♀♂↑	A<B = E = G	A<B = E = G

are presented for the eleven most abundant species. Comparison of annual mean weights per site for the entire period showed that females are heavier than males, except in *P. cingulatus* (Table 7.23).

Weights decreased downstream in *D. annulatus*, *R. fasciata*, and *P. cingulatus*, but increased in *C. villosa P. luctuosus*, *S. personatum* and male *A. fuscipes*. No weight difference occurred between sites for *T. rostocki*, *A. fimbriata*, and *P. conspersa* (Wagner, 2005a, 2005b). These species have some peculiarities that probably make them more independent of limitations (particularly food) than others. In *T. rostocki*, the construction of galleries probably provide shelter and constant food quality and quantity throughout the year (Becker, 1993, 1994; Section 7.8). *A. fimbriata* is dormant from November to February (low temperature) and thus not competing for food in the stream when biofilm production is restricted by low water temperature and short day-length.

Downstream weight increase in *C. villosa* depended on water temperature, and this is probably the same for other species, at least for shredders (below). Upstream weight increase, in particular for *S. personatum*, probably depended on its variable life cycle that may last between one and several years; it remains unknown how many years specimens live in the stream before emerging. The data for *P. luctuosus* were too few, and why, in *A. fuscipes*, only males should benefit from better conditions remains unclear.

Weights in Traps along the Stream in 1987

In 1987, specimens of most species were individually measured, weighed and sexed. The differences between sexes in the six species remain obvious at all sites and are highly significant (Figure 7.88). Confidence limits of the data on females are usually wider than for males. Mean values for the nine years corresponded

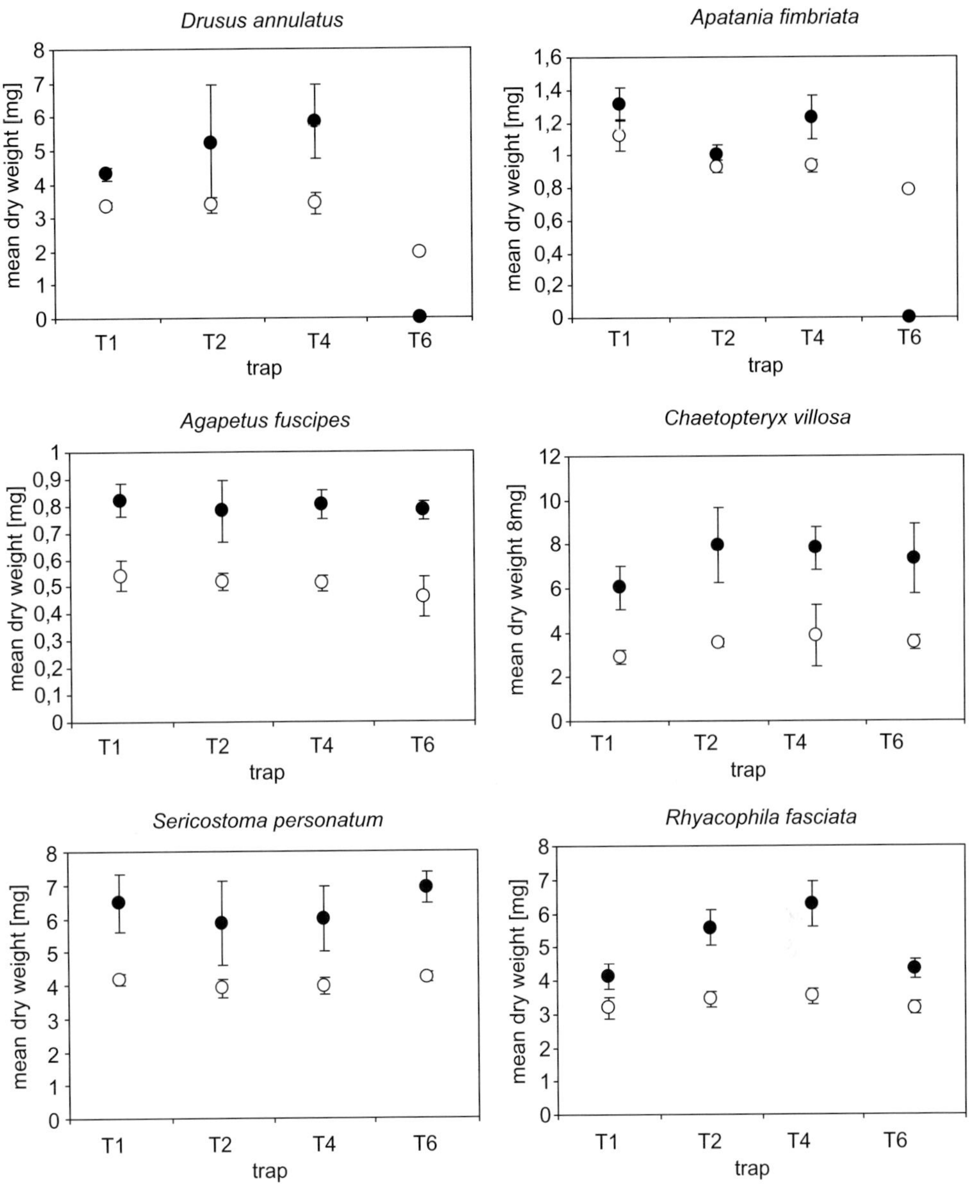

Figure 7.88 Mean dry weight (mg; $x \pm 95\%$CL) of both sexes for six abundant caddis flies at four sites along the Breitenbach 1987 (data based on individual weights; females = black circles; males = white circles).

with the 1987 data. There were no differences in females and males of *P. conspersa* along the stream.

If, as rule of thumb, confidence limits enclose the mean value of the compared variables, then differences are not considered significant, and in most cases only rough weight trends along the Breitenbach can be deduced. The lightest specimens of *S. personatum* and the heaviest specimens of *D. annulatus* and *R. fasciata* were found in the mid-stream section (T2, T4). Lighter specimens of *C. villosa* occurred at T1 compared with the remaining sites. *A. fuscipes* and *A. fimbriata* tended to be lighter downstream (Figure 7.88). The reasons for this were temperature for *C. villosa*, and food shortage for the grazers (see below).

Trends of the long-term series and the data from an individual year corresponded roughly; however, results from several subsequent years improve the quality of the results.

Seasonal Weight Variation in Caddis Flies

Results for the Trichoptera are restricted to those years and collecting sites, when and where species were sufficiently abundant and results were interpretable. Surprising differences were discovered between species, years, stream site, and functional feeding group.

In *Apatania fimbriata*, trends of monthly mean weight did not match between collecting sites and years. In 1985 at site T1, mean weight decreased to August ($r^2 = 0.88$; linear regression) but remained almost constant at T2. At T4 weights remained almost constant from April to July, but decreased in August and September ($r^2 = 0.48$).

There were different results for 1988. At T1, female and male weights showed a tendency to increase over the course of the year ($r^2 = 0.41$ female; $r^2 = 0.57$ male), but at the more downstream sites (T2, T4) there was a steady, significant (T2: $r^2 = 0.75$ f, $r^2 = 0.83$ m; T4: $r^2 = 0.87$ f, $r^2 = 0.27$ m) trend to lighter specimens later in the year. Two opposing tendencies of weight development occurred together in the same year, depending on stream site. In 1991, although the emergence period was considerably longer at site T1, female weights tended to decrease ($r^2 \sim 0.6$) in all three traps (apart from three heavy specimens at T2 at the end of the emergence period; Figure 7.89).

In *Agapetus fuscipes* weight increased towards the end of the year in 1983 and 1985, but decreased in 1984 (monthly data). In 1987, the seasonal weight decrease was significant only at T1 (weekly data; T1 $r^2 = 0.9$; T2, T4, T6 $r^2 < 0.2$; Figure 7.90, left). In 1988, no significant trend was detectable (weekly data, low abundance; Figure 7.90, right).

However, the years with the highest abundances (1989, 1991), particularly at site T1, provided remarkable results. It was evident that females were heavier than males at all sites during the entire flight period (T1 as example). In one year female dw was different along the Breitenbach, with specimens at sites T2–T6 being heavier than upstream, where (T1) the weekly mean weight increased in the second half of the emergence period. A significant seasonal trend was not detected at sites T2 and T4 ($r^2 < 0.1$), and weight decreased at T6 ($r^2 = 0.40$).

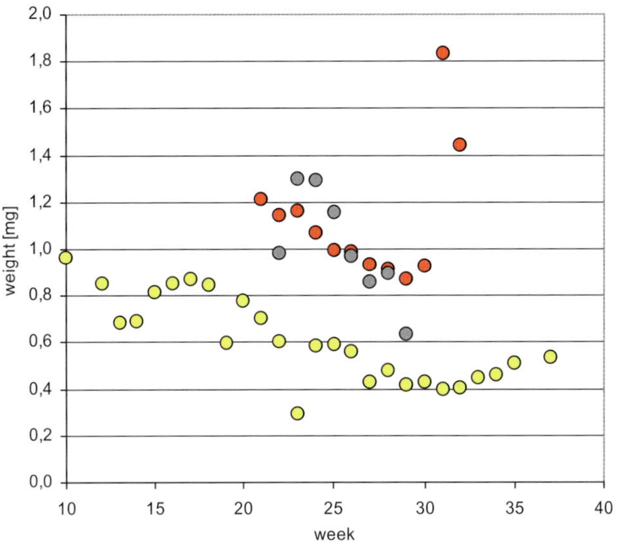

Figure 7.89 Weekly mean weight of *A. fimbriata* females in three traps along the Breitenbach in 1991 (T1 = yellow; T2 = red; T4 = gray).

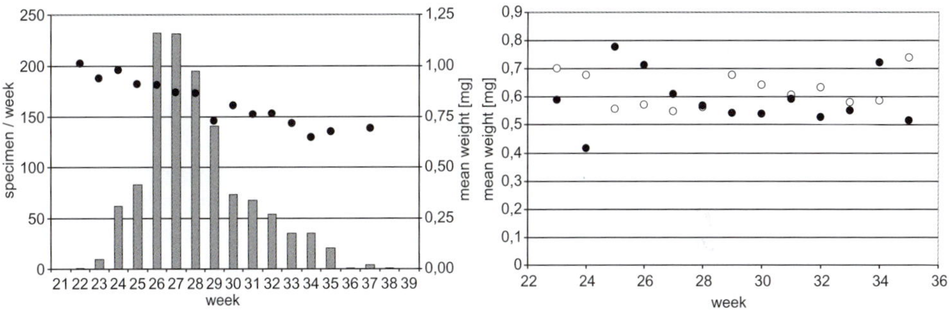

Figure 7.90 Left: weekly mean female weight (dots) and emergence (columns) of *A. fuscipes* at T1 in 1987 (n = 1250). Right: weekly mean female weights in 1988 in T1 (white dots; n = 1674) and T2 (black dots; n = 1166).

At T1 weekly weights of females and males showed clear, parallel trends, with the lowest values occurring in week 28 (Figure 7.91, left). For 1989 this pattern was best described by a quadratic regression: $y = ax^2 + bx + c$ (females: $y = 0.0029x^2 - 0.1824x + 3.06$, $r^2 = 0.88$; males: $y = 0.0021x^2 - 0.1288x + 2.1235$, $r^2 = 0.89$). 1991 presented a similar situation to 1989. In traps T2, T4 and T6, weights decreased (T2: $r^2 \sim 0.8$; T4: $r^2 \sim 0.4$; T6: $r^2 \sim 0.96$) but at T1, female and male weights decreased until week 31 but increased again thereafter. The quadratic regression for both

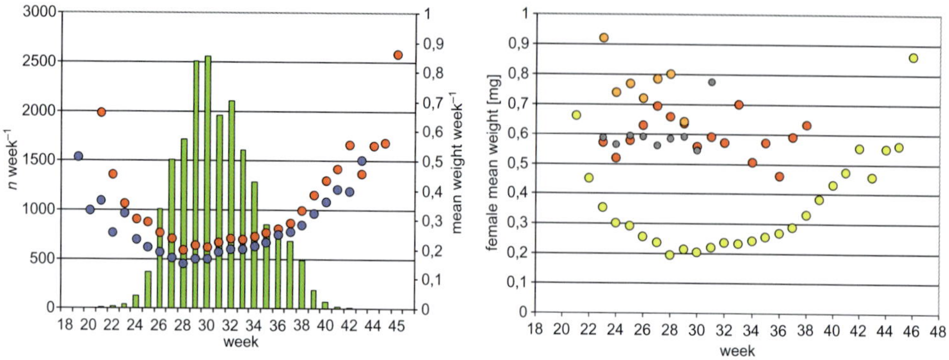

Figure 7.91 Left: weekly mean weights of males and females and emergence (green) of *A. fuscipes* in T1 (female = red dots; male = blue dots). Right: weekly mean female weights in T1 (yellow; $n = 19969$), T2 (red; $n = 1405$), T4 (gray; $n = 1217$), and T6 (orange; $n = 182$) in 1989.

was again significant (females: $y = 0.0019x^2 - 0.0476x + 0.6681$, $r^2 = 0.79$; males: $y = 0.0012x^2 - 0.0238x + 0.4111$, $r^2 = 0.67$).

The decline of mean weekly body mass until maximum emergence (and increasing mass with decreasing numbers of emerging adults) was regarded as an effect of the nutritional conditions and intra-specific competition acting on the population in the period before emergence. With pupation and emergence, an increasing number of specimens became inactive competitors for food, and younger larvae prefer different feeding sites (Becker, 1994, 2005; Section 7.8). Thus better feeding conditions are available to the remaining older larvae that may then grow to greater weights (weight decisive last instars). This is seen as increased mean DW towards the end of the emergence period. Temperature (low and more or less constant at T1) then plays a subordinate role, whereas inter- and intraspecific (biological) competition of scrapers for a limited food (biofilm) dominates.

At the other sites, low winter temperature may create a bottleneck for the population. However downstream, due to the much lower densities (although there may be additional grazing competitors), optimal feeding will be easier for individuals. In the intervening year, 1990, no seasonal weight decline occurred at T1, but considerable decline (without final increase) occurred at the other sites.

Results differed for other functional feeding groups. Weights of females and males of *Chaetopteryx villosa* (omnivorous, shredder) decreased with increasing length of emergence period (Figure 7.92). However, the trend was unequal in different years. While the trend was similar in 1987 and 1990 at T1, female dry weights were disparate at the other sites. They decreased in parallel in 1987, but appeared undirected in 1990, increasing at T6, and finally decreasing at T4.

In *Tinodes rostocki*, mean female weight tended to decrease synchronously in all years. Trends were low but significant every year, clear in 1985 and 1988, less so in 1989.

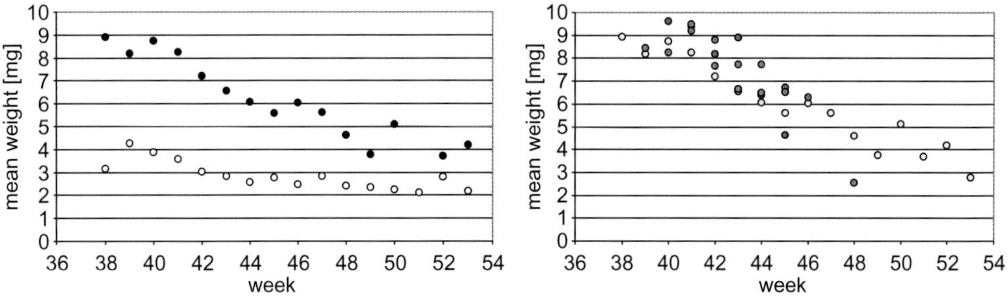

Figure 7.92 Left: Weekly mean female (black) and male (white) weights of *C. villosa* in T1 in 1987. Right: weekly mean female weights in T1 (yellow), T2 (red), T4 (gray), and T6 (orange) in 1987.

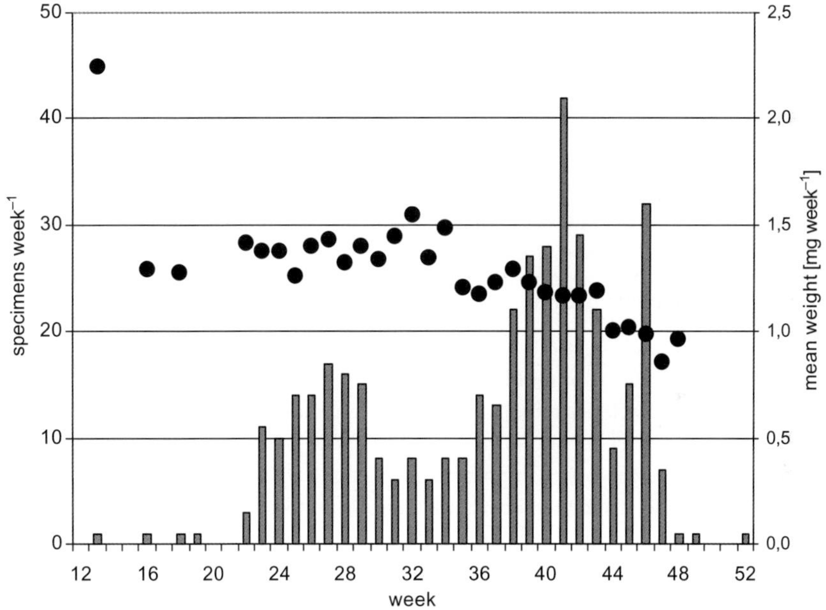

Figure 7.93 Weekly female weight (black dots) and sum of emerging *W. occipitalis* adults (columns) at T1 in 1991.

Wormaldia occipitalis was sufficiently abundant only in 1991. On a weekly basis males were lighter than females; weight did not differ between traps T1 and T2. If female weights are compared with the weekly sum of emerging adults three emergence peaks (weeks 28, 42 and 46) occur; during these periods females weighed more than 1.25–1.50 mg, about 1.25 mg and about 1 mg respectively (Figure 7.93).

However, there was no correlation between numbers and weight of emerging specimens. This means a seasonal, specimen weight decline is also usual for caddis flies, but site and time dependent exceptions may occur in several species. Effects of temporal food shortage in years with extremely high abundance affected *Agapetus fuscipes* weights at T1, and competing grazers, such as *Drusus annulatus* and *Apatania fimbriata*. The *Baetis* species were apparently not affected by this situation, either abundance was too low, emergence started or was finished before the maximum competition period and, most probably, the way an *Agapetus* larva "defends" its feeding area is relevant for this species (compare Becker, 1987a, 1987b; Section 7.8). A similar effect was not observed for the other abundant scraper, *Tinodes rostocki*. This may indicate optimal foraging inside and around the movable retreats. A size decline was usually evident in most years in the shredder *Chaetopteryx villosa*, but as in 1990, weight changes over time may differ between sites.

7.6.4.5.7 **Trichoptera in the Benthos**

Peter Zwick

Most caddisfly genera in the Breitenbach are represented by single species which greatly facilitated larval identification using Waringer and Graf (1997) and Wallace, Wallace and Phillipson (2003). British keys are not normally applicable to central European fauna, but at the generic level the present one is. The habitus of retreat- and saddle-case-making taxa is characteristic of individual families and/or genera and even the smallest specimens are easily recognized. In contrast, identification of very small tube-bearing larvae, especially first instar, remains difficult. Not all those in the Breitenbach can be identified using Siltala (1907), Nielsen (1942) and Hetrick, Morse and West (1998). Where appropriate, notes are added.

During the two years of regular benthic sampling we did not collect larvae of all genera or even species taken as adults in emergence traps during 30 years. This is due to different collecting efforts but the discrepancy between the lists partly also results from the often important adult flight activity. Individual specimens originating in the River Fulda certainly contribute to the wealth of species in the emergence traps. At times, such erratic species may even oviposit and develop in the Breitenbach without being regular members of its fauna. *Hydropsyche* spp. are probable examples. On the other hand, two of the most abundant caddisflies in the Breitenbach are rather sedentary. *Agapetus fuscipes* can fly but does not often do so. Unlike most other caddisflies it is also not attracted to light traps. *Chaetopteryx villosa* has short-winged males but fully winged females. However, adults occur in winter, at temperatures not normally supporting insect flight.

The present chapter focuses on the dominant taxa in the Breitenbach of which sufficiently many larvae were found during the 1986–1988 benthos sampling campaign to permit analysis of the population structure. Life cycles and various other aspects of the dominant species have been previously described but our own study contributes supplementary information. Arrangement is by numerical dominance in the stream community, and by functional feeding group/trophic category.

Table 7.24 Head capsule width (HCW) or length (HCL) of the larval instars of several Trichoptera from the Breitenbach.

		L1	L2	L3	L4	L5
Plectrocnemia conspersa	HCL (mm)	0.280	0.454	0.851	1.397	2.072
	STABW	0.035	0.015	0.043	0.166	0.239
	n	3	4	7	8	4
Odontocerum albicorne	HCW (mm)		0.350	0.800	1.133	1.556
	STABW		0.058		0.058	0.088
	n		4	1	3	9
Micrasema longulum	HCW (mm)	0.189	0.249	0.340	0.464	0.636
	STABW		0.009	0.020	0.034	0.048
	n		10	21	32	58
Rhyacophila fasciata	HCL (mm)	0.330	0.505	0.769	1.193	1.746
	STABW	0.018	0.027	0.047	0.088	0.117
	n	40	35	52	46	69
Anomalopterygella chauviniana	HCW (mm)			0.560	0.795	1.252
	STABW				0.029	0.048
	n			1	42	35

Body length of Ephemeroptera and Plecoptera larvae in the benthos samples were measured. The same is impractical with Trichoptera larvae concealed in cases but even for the free-living taxa the effort would be unrewarding. Trichoptera grow through only few instars, normally five, and these can in most cases be immediately recognized by eye because the characteristic head capsule dimensions are neither sexually dimorphic nor do they overlap between instars (Table 7.24). This is convenient, but has also disadvantages because 85% or more of all biomass is gained and stored in the mostly concealed abdomen of the long-lived last instar (Figure 7.11). In all Breitenbach Trichoptera, maximum biomass gain occurs shortly before adult emergence, in the warm season; winter growth plays no role.

Agapetus fuscipes

The most abundant caddisfly of the Breitenbach is a saddle case maker (Weaver, 1984). It is a rather indiscriminate feeder of epilithon ingesting also fair amounts of fine detritus. It is exceptional in passing through seven larval instars (Becker, 2005; Section 7.8). The species was studied repeatedly (e.g., Benedetto Castro, 1975; Becker, 2005; Becker and Menzler, 1998), only a brief summary is appropriate. *A. fuscipes* prefers the uppermost sites in the stream.

At first sight, *Agapetus* seemed unaffected by the cypermethrin accident because its cases remained on the stones. However, the poisoned larvae had left them and drifted naked. The cases were attached to stones by silk. Tiny young larvae subsequently appeared at the same time as normally, but in distinctly lower numbers

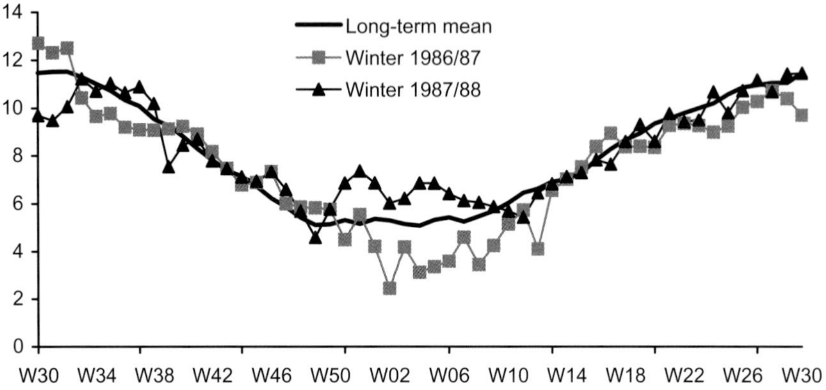

Figure 7.94 Mean weekly temperatures during winters 1986/87 and 1987/88, compared to the long-term mean (1969–2005).

than in the subsequent year. Development during the warm season was very similar between the two years but winters differed. In 1986/87 mean winter temperature was below long-term average (Figure 7.94). Population structure hardly changed after Week 45, until Week 12, in 1987 (Figure 7.95). The winter 1987/88 was warmer than average, and growth of *Agapetus* was less affected. Becker (2005) whose data were pooled across several years with quite different winter temperatures reported no winterly growth.

Apatania fimbriata

This rhithral species is most abundant at upstream sites. A. *fimbriata* was previously studied for its life cycle (Aurich, 1992; Enders and Wagner, 1996), its chemical defense (Aurich *et al.*, 1991; Wagner *et al.*, 1990, 1993) and its larval feeding (Becker, 1990).

The anteriorly slightly widened tube with its downcurved opening is distinctive in all instars. Larvae tend to leave their tubes during preservation. Aurich (1992) proposed the horseshoe-shaped mesonotal sclerites for distinction from first instar larvae of *Drusus annulatus*. Also, the pronotum of *Apatania* is almost twice as wide as long, with straight rear edge while that of *Drusus* is about 1.5 times wider than long, with U-shaped, medially narrow rear margin.

There was no difference in development between Reaches A and B, data were pooled. Massive recruitment occurred from the end of July to the beginning of August (Figure 7.96). It was followed by rapid growth up to third, exceptionally to fourth instar. After end of September population structure did not change until March/April. Larvae attached themselves by silk in protected sites under stones and spent winter in dormancy (see also Aurich, 1992). In the first study year, mainly instars two and three overwintered, the latter dominated. The next year, roughly equal shares of instars three and four overwintered. The difference must be attributed to a period of cold during six weeks in August–September 1987,

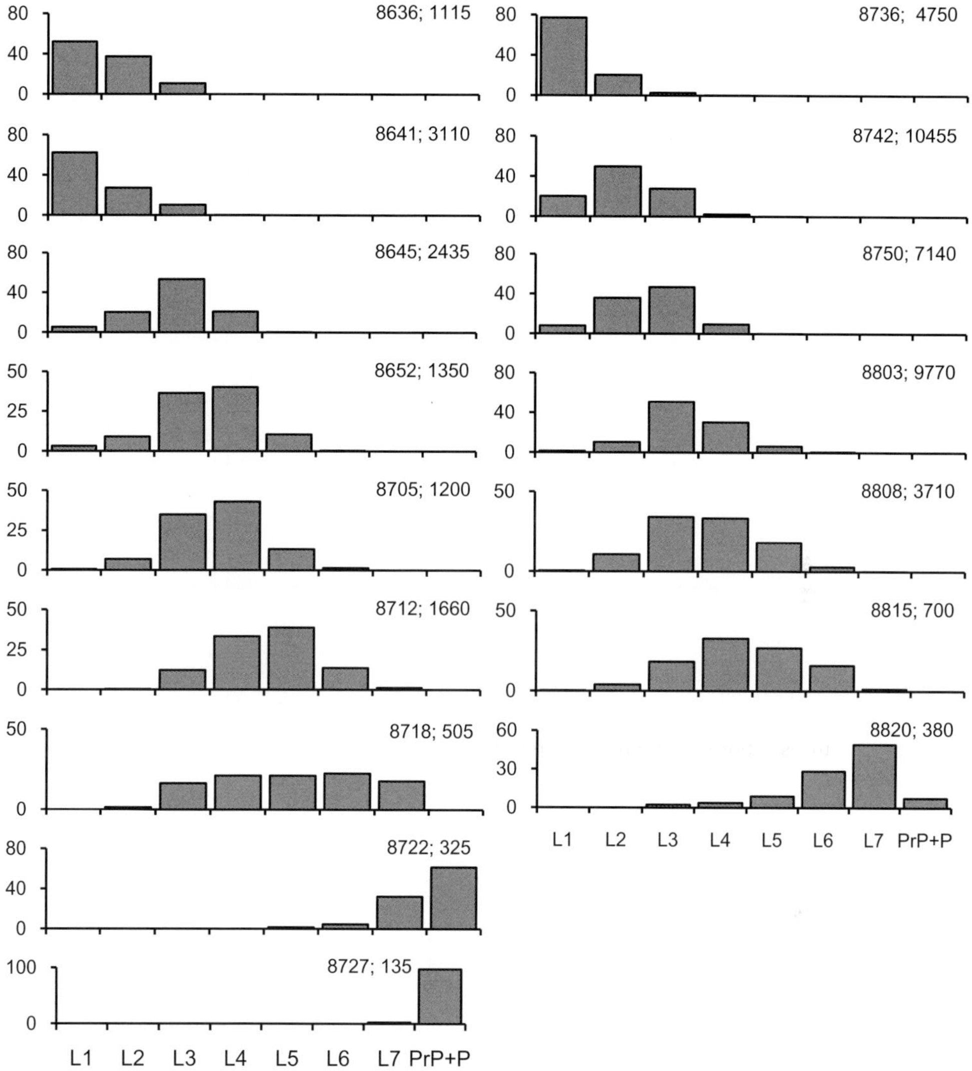

Figure 7.95 Population structure of *Agapetus fuscipes* in Reaches A (left) and B (right) of the Breitenbach. Ordinates show percentages of population, the abscissa larval instars, prepupae and pupae. The legends give calendar weeks in 1986–1988, followed by approximate number of specimens m^{-2} streambed.

during early larval growth, when the temperature was clearly below the long-term average (Figure 7.94).

The stand-still in larval growth was noticed soon after the autumnal equinox, when temperature dropped from 8–9 °C to 6–7 °C. In 1988, growth was again recorded two weeks after the spring equinox, at 7.5 °C. This may indicate a threshold temperature for larval growth at around 7 °C, and a possible additional

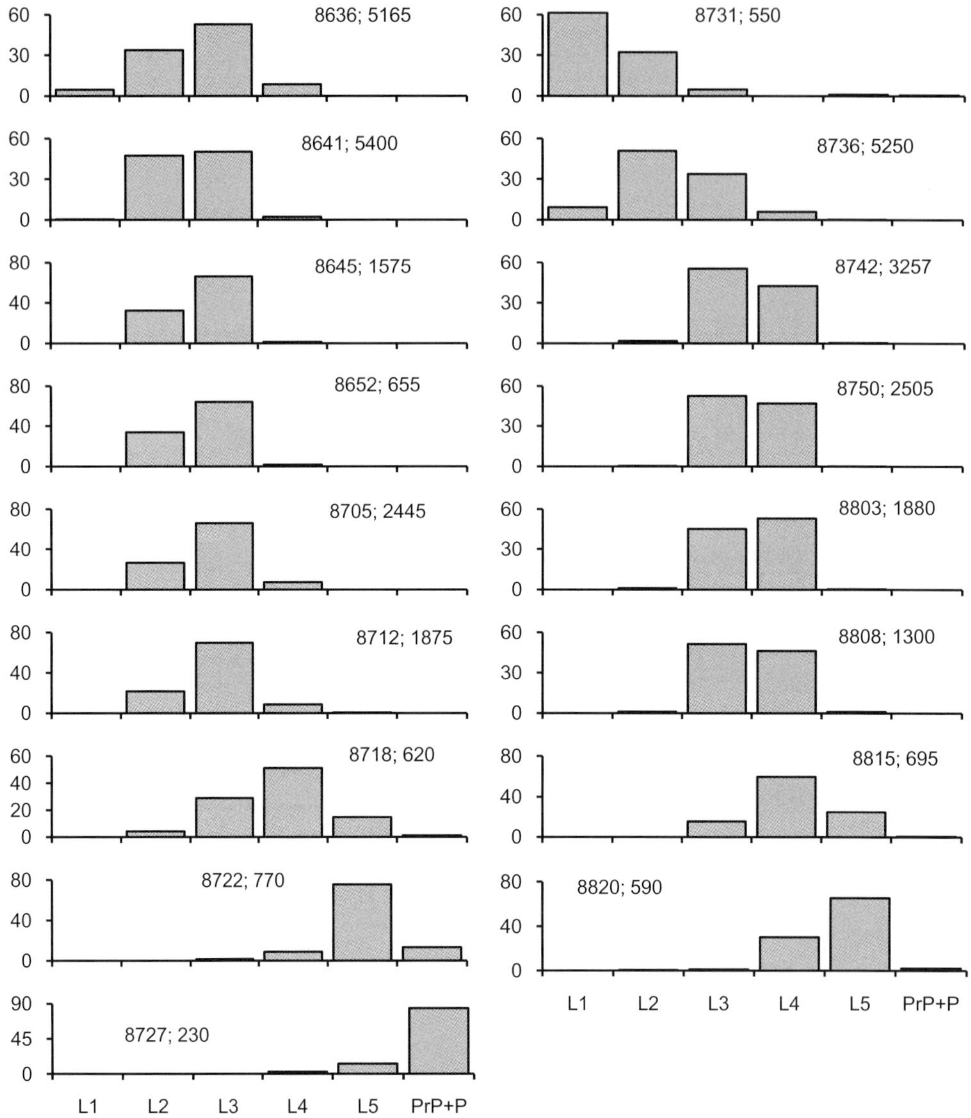

Figure 7.96 Population structure of *Apatania fimbriata* in Reaches A and B of the Breitenbach. Ordinates show percentages of population, the abscissa larval instar, prepupae and pupae. The legends give calendar weeks in 1986–1988, followed by approximate number of specimens m^{-2} streambed.

influence of daylength. However, in the laboratory slow growth occurred even at constant 6 °C and although photoperiod was constantly 14/10 h. Larvae in instars three and four started dormancy in November, even at 10 and 14 °C (Aurich, 1992). Aurich (1992) concluded that dormancy was endogenous. Yet, in the field, also the second instar was affected in the cool year 1987, but not in 1988.

Drusus annulatus

The largest grazer in the stream (Becker, 1990) requires rewarding feeding plots. This leads to its (partial) exclusion by large numbers of grazing *Agapetus fuscipes* (Becker and Menzler, 1998). The rough chestnut-brown integument distinguishes older larvae in their straight sand tubes at first glance. The rear part of pronotum becomes more raised as larvae grow and takes a characteristic shape (Waringer and Graf, 1997).

The *Drusus* life cycle in the Breitenbach is straightforward univoltine. Emergence is extended, adults are usually found all the times from May to September. Accordingly, the recruitment period of *D. annulatus* is very long but appears even longer because the overwintering population included a strong share of first instar larvae. Only by the end of March had all molted to second instar. Whether these were all overwintering specimens or whether eggs also overwintered to hatch after winter is unknown (Figure 7.97). The winterly stop of growth is distinctly shorter than in *Apatania*.

Micrasema longulum

The slender, gently curved tubular cases are unmistakable because of the cloverleaf-shaped window in the caudal closing membrane, in older instars. Young instars build from fine mineral material. As they grow, the brown silken inner tube becomes successively more exposed at the anterior end (Bohle, 1974). From instar 4 on, the tube consists usually entirely of translucent organic material because the early mineral-covered tube sections are cut as the larva grows (Zwick, 1998b). The delicate tubes were mostly found on stones in fast current, often among mosses. The strong tendency of large larvae to aggregate and pupate in groups lead to an extremely clumped distribution often preventing realistic estimates of density. The species is not common, in general, density was less than 100 specimens m^{-2} stream bed. *M. longulum* ingested more moss than other caddis larvae but since moss was not recognizably digested it may have been no more than a carrier for diatoms which are the preferred food of the larvae (Becker, 1990).

It was generally assumed that tubecase makers pupate in their larval tubes. The sudden lengthening and straightening of the tube unveiled that in *Micrasema* pupation does not take place in the larval tube but in a special parallel-sided pupal chamber (Zwick, 1998b). The slightly conical larval tube is removed before the case is attached to rocks. Since, similar behavior has been recorded in additional taxa (Bohle, 2004; Malicky, 2000; Wiggins, 2001).

In 1986–1988, *M. longulum* was moderately common, mainly in the lower stream section but was not found in Reach A; Becker (1990) observed the same. More than 60 third instar larvae were found after the insecticide accident but the

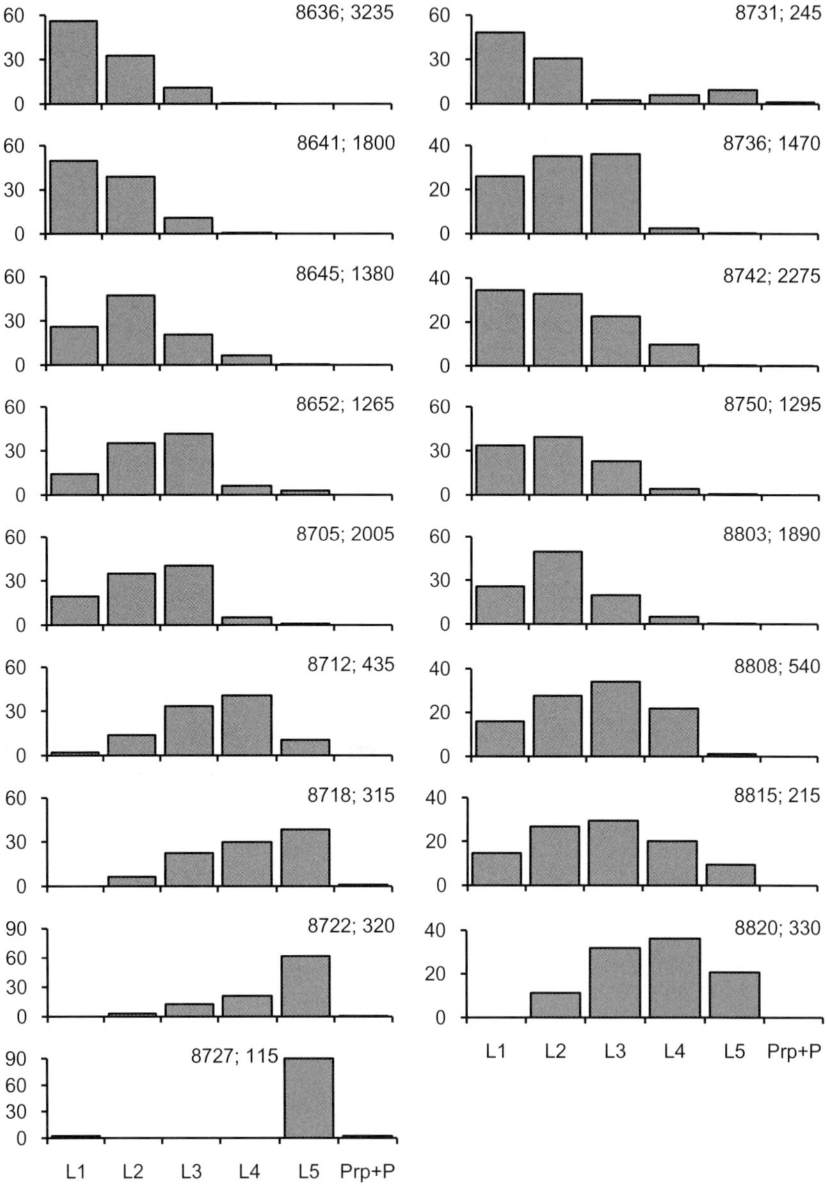

Figure 7.97 Population structure of *Drusus annulatus* in Reaches A and B of the Breitenbach. Ordinates show percentages of population, the abscissa larval instars, praepupae and pupae. The legends give calendar weeks in 1986–1988, folllowed by approximate number of specimens m^{-2} streambed.

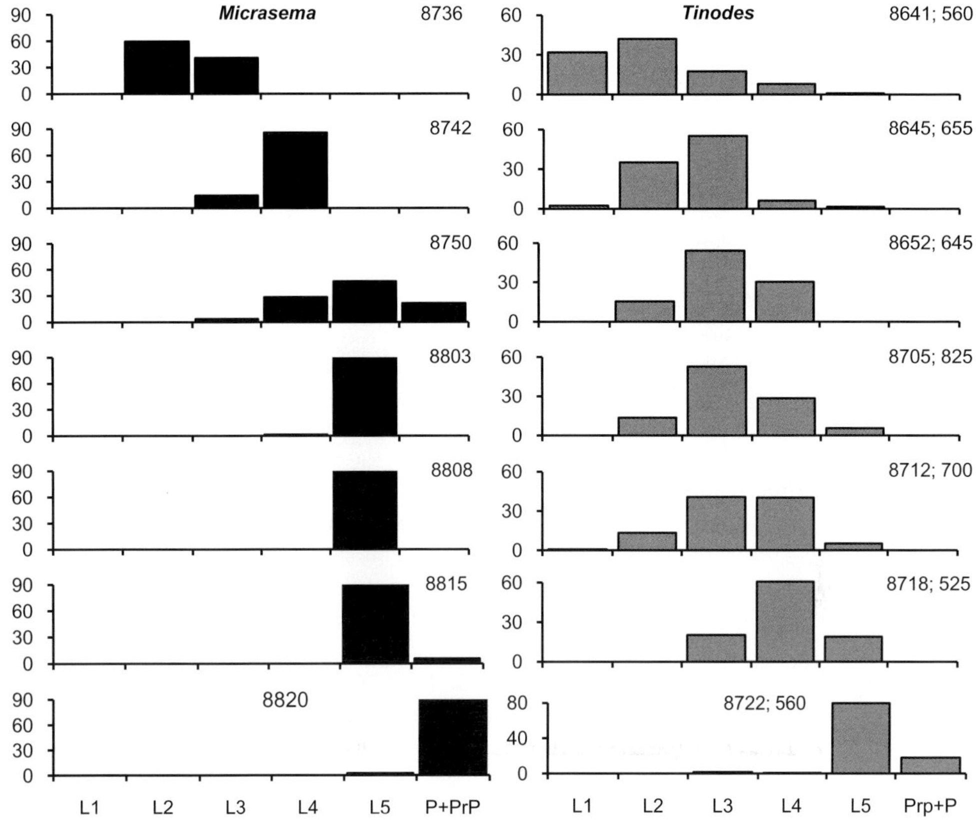

Figure 7.98 Population structure of *Micrasema longulum* (left; Reaches C and E pooled) and *Tinodes rostocki* (right; Reaches A, C, and E pooled) in the Breitenbach. Ordinates show percentages of population, the abscissa larval instars, praepupae and pupae. The legends give calendar weeks in 1986–1988, followed (right only) by approximate number of *T. rostocki* specimens m^{-2} streambed. *Micrasema* was mostly rare and too clumped for realistic estimates of density.

population then disappeared until a year later when second and third instar larvae were collected; the first instar was not found. Growth over summer was rapid (Figure 7.98, left side), the population overwintered in the last instar. Occasional specimens attached cases and entered the prepupal stage in December, but most pupated only in May.

Tinodes rostocki

The life history of this dominant tunnel-building grazer and collector in 1987 was described by Becker (1993); the present data agree with his. Details of biofilm structuring, tube wall properties and chemical and microbial microzonation were described by Stief and Becker (2005). Clearly, *Tinodes* larvae use their gallery walls in some way as feeding plots but the actual mode of feeding is not known.

The winterly interruption of development lasted from November to March, instars three and four being dominant. Like in *Apatania fimbriata,* and probably for the same reason, population development was more advanced in 1987 than the year before when instar three clearly dominated in the wintering population (Figure 7.98, right side).

Rhyacophila fasciata

The Trichoptera include three of the few major predators in the Breitenbach, the free-ranging *Rhyacophila fasciata* and the catchnet-building *Plectrocnemia conspersa* and *Hydropsyche* spp. The last occur only occasionally in the lower Breitenbach but are abundant in the River Fulda.

Rhyacophila fasciata was abundant. Mean increase of linear measures at moults was 52% (Table 7.24). Almost all larval instars were present all the time. Apparently, mainly some late instar larvae survived the insecticide accident. For several months after the incident, until March the next year, the last instar dominated. Some new recruits were collected over winter, and for a short period only the smallest and largest instars were present. After major recruitment in August 1987 all instars were present all the time. Several additional bursts of recruitment from October to December and again in March and April contributed to the confused population structure to which the evident longevity of the last instar added (Figure 7.99).

The adult emergence period is bimodal, with peaks in spring and again after a summer break (Illies, 1971; Sandrock, 1978). The length of the life cycle remains unclear. Knöfel (1983a, 1983b) had similar difficulty interpreting the voltinism of *Rh. fasciata.* She showed that a minimum of about 2700 degree-days was required to complete larval development, that is, 11 months at the central European mean temperature of 8 °C. She apparently anticipated that part of the population may require longer and suggested uni- to semivoltine development. Probably, univoltine and semivoltine cohorts overlap, much as in *Rh. dorsalis* and *Rh. munda* (Elliott, 1968).

Hydropsyche spp.

Although the larvae build relatively fine-meshed catch nets and are able to filter fine particulate material from the water body, they are essentially predators that discard FPOM as long as animal food is available (Petersen, 1985). No more than 57 specimens were taken in the Breitenbach at all sites together from spring 1986 to spring 1988. First and second instar larvae (altogether 30) were encountered in spring and summer. Instars 3–5 overwintered, 10 pupae or prepupae were found in early summer (Weeks 31–36), all but one in Reach E. Specimens large enough for specific identification were *H. siltalai.* The small *Hydropsyche* population possibly results from oviposition of females coming from the River Fula.

Anomalopterygella chauviniana

Larvae of *Anomalopterygella chauviniana* are unmistakable by their amazing, saddle shaped pronotum. Five fifth instar larvae were taken in Reach C at the end of July 1987, and three fourth, one fifth instar specimen in April and May 1988. The

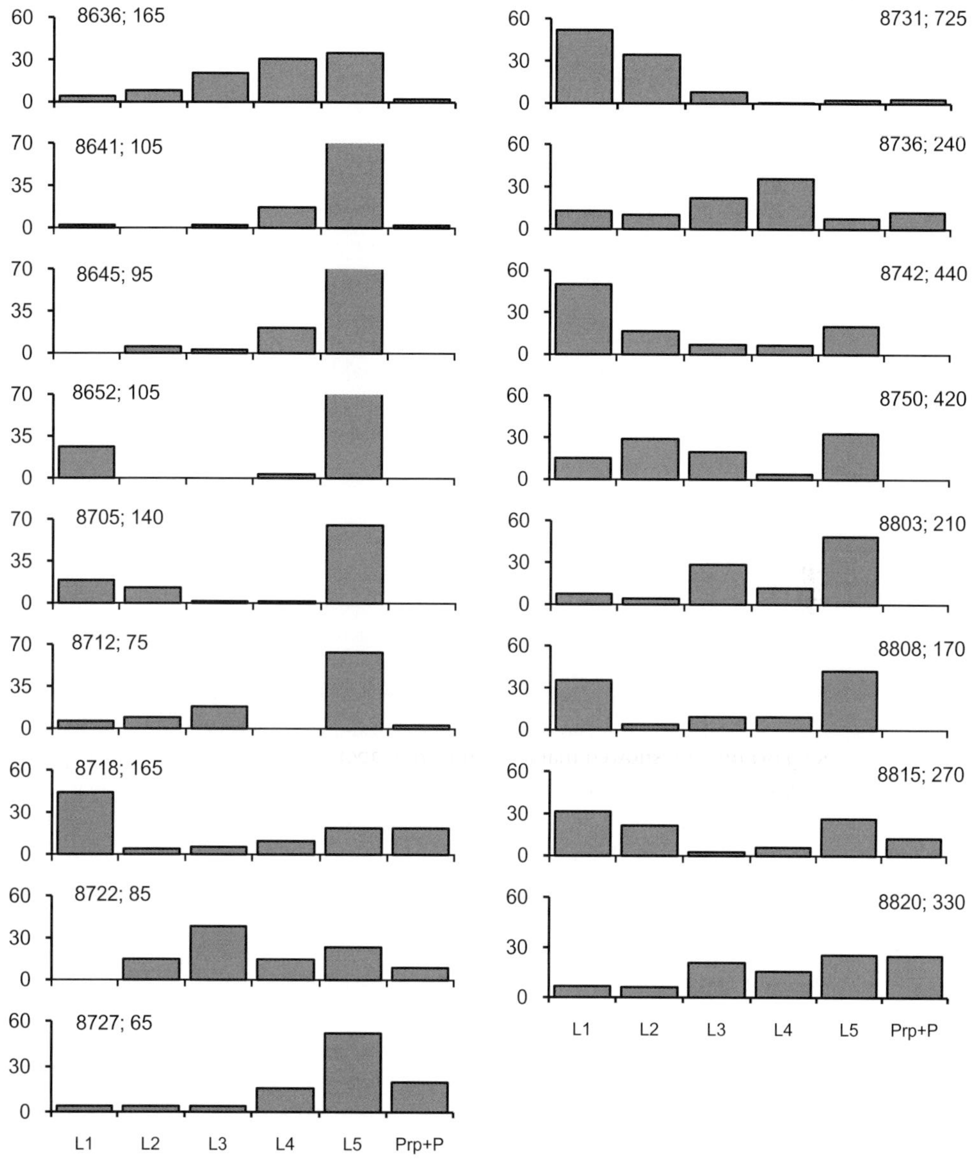

Figure 7.99 Population structure of *Rhyacophila fasciata* in the Breitenbach benthos. All sites pooled. Ordinates show percentages of population, the abscissa larval instars, praepupae and pupae. The legends give calendar weeks in 1986–1988, followed by approximate number of specimens m^{-2} streambed.

strongly brachypterous, long-legged males resting on emergent rocks or on the banks are extremely swift runners and excellent skimmers. They wait for emergence of the fully winged females which apparently do not move far from the water's edge. Because of this behavior, adults were not taken at the Breitenbach. Many were observed along the River Fulda where *Anomalopterygella* was the most abundant case bearing caddisfly in the hyporhithron of the River Fulda (Zwick and Zwick, in press). The head capsule dimensions in Table 7.24 were taken from these specimens (Figure 7.100).

Odontocerum albicorne

Only 24 larvae were taken (Table 7.24) They are mentioned to suggest additional study. Cases of large larvae are always inhabited by numerous bdelloid Rotatoria. Also, from the neck membrane of one last instar *Odontocerum* projected paired membraneous tubes, possibly a defensive gland like in *Apatania*.

7.6.4.5.8 **Life Cycle Studies**

Rüdiger Wagner

The following text summarizes the life cycle studies on some caddisflies. The studies were performed to understand how species survive in the variable environment of a stream. It is necessary to identify the limits for successful development of all instars, the limiting variables for the growth of individuals in a generation, and the success of a population over several years. This is also important for understanding and interpreting the interdependencies of species and environmental variables based on statistical analyses of long-term data. Numbers must be interpretable, ecologically relevant, and sound. The knowledge of species ecology links field and laboratory research with mathematical models and can help corroborate or disprove theories.

The Life Cycle of *Apatania fimbriata* in the Breitenbach

Until recently the genus *Apatania* was part of the subfamily Apataniinae (family Limnephilidae) but has now been placed in the family Apataniidae. In some years it was sufficiently abundant to study its life cycle in more detail.

Based on its preferred water temperature, the species is cold stenothermous, and larvae are grazers. In the Breitenbach the univoltine (Figure 7.80) species occurs from the spring down to T4. Although mean abundance between T1, T2, and T4 differed little from 1983 to 1993, in individual years maximum abundance occurred around T2 and T4. Further downstream *A. fimbriata* was always very rare.

With respect to the yearly discharge pattern *A. fimbriata* attained its highest emergence abundance when there was good flow with low variation during larval development. Under such conditions, mean abundance was around 500 specimens, whereas under dry or snowy winter conditions, it was around 200 specimens per trap and year. From the long-term data set (T2, 1969–2006) it is evident that flow patterns with unpredictable floods harm the population and abundance in emergence traps drops.

The emergence period of *A. fimbriata* differs markedly between collecting sites. For the year 1985 it was from the beginning of April (week 14) to the end of August (week 39) (Figure 7.80). Emergence maximum at T4 was in week 26, in week 30 further upstream at T2, but without a distinct maximum at T1, showing more or less similar emergence between weeks 24 to 31. Over the course of the day, maximum emergence occurred around midnight ±2 h (Aurich, 1989), and minimum emergence around noon.

The sex ratio in field collections was surprising. Hand-net collections contained almost exclusively males (80% and more). This means that either the females prefer different microhabitats, or the adult life time differs between sexes. Laboratory tests showed that females mated only once and, in most cases, soon or immediately after emergence. Oviposition occurred in the next 1 to 3 days, and females died shortly after. However, males lived for up to 3 weeks and many copulated several times. This explains the observed field sex ratio.

In many caddisflies, pheromones play a major role in the communication between sexes. Solem and Solem (1991) first demonstrated the presence of female pheromones in the Breitenbach population. Three substances were detected in the glands of *A. fimbriata* females (Table 7.20, and http://pherobase.org/database/species/species-Apatania-fimbriata.php). Two substances tested in the field successfully attracted males (unpublished).

Females deposit a single, small translucent, spherical egg mass at the water's edge, usually not more than 1–2 cm above, and rarely below, the water surface. Maximum flow velocity beside the egg masses was always below 8, usually $3 \pm 2\,cm\,s^{-1}$. Deposition of a second egg mass was never observed. Mean egg number per mass was 208 ± 83 (Enders and Wagner, 1996). Immediately after oviposition, egg masses appear bright yellow because eggs are densely packed. By absorbing water, egg masses swell over the next hours to 5.6 ± 1.1 mm diameter, distances between eggs increase, and egg masses become almost invisible. There was no significant relationship between egg mass size and egg number (Enders, 1994).

During egg development (embryogenesis, Table 7.22.) the yellowish color disappears and finally, shortly before the larvae hatch, eggs appear whitish. During embryogenesis egg masses and individual eggs are endangered due to several abiotic and biotic environmental variables. Entire egg masses may be displaced by the current or by animals living along the banks (e.g., Amphibia, water shrew (*Neomys fodiens* [Pennenat, 1771]). Of 198 egg masses monitored in the field, 6% were not seen to hatch. Of the remaining 186 masses, 52.2% successfully finished their development, while 25.2% were lost, most for unknown reasons, but 3 were infected by fungi.

Of the remaining egg masses (22.6%), some were lost due to mechanical damage (4.6%) but most due to occupation by egg predators, particularly Chironomid larvae (*Thienemaniella gracilis*), and first instar larvae of the neuropteran *Osmylus fulvicephalus* (Enders and Wagner, 1996). Chironomids are matrix intruders, but first instar *Osymylus* larvae eat individual eggs from the outer layer thanks to their

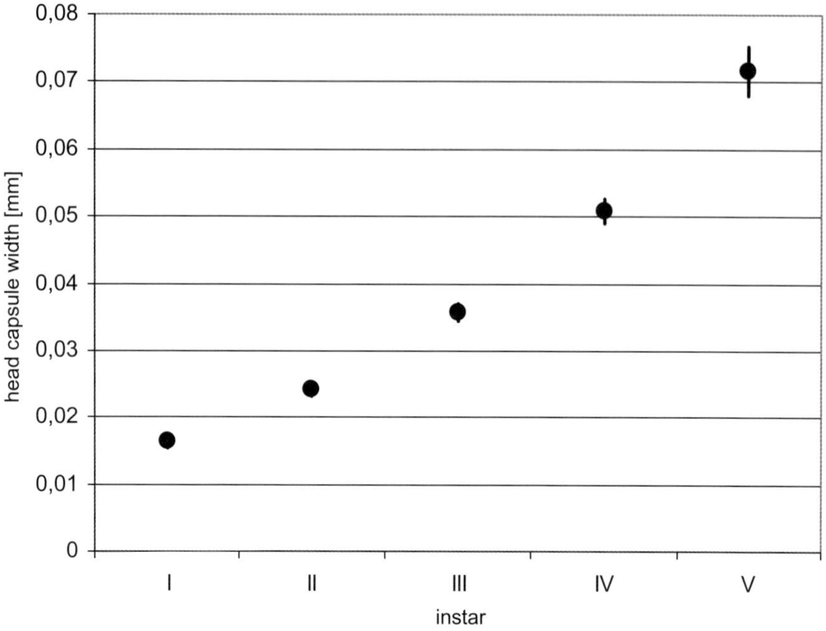

Figure 7.100 Head capsule width of the five larval instars of *A. fimbriata* (mean ± SD; regression highly significant).

elongate mouth parts. Temperature dependence of egg incubation in the laboratory was already described.

Larval Development and Predators in the Field Five larval instars were distinguished (Aurich, 1992; Figure 7.100), and there was no overlap in head capsule width between subsequent instars. The relative increase in head capsule width between instars was about 1.47 (instars 1/2 and 2/3), 1.42 (instars 3/4), and 1.41 (instars 4/5), close to the expected value of about 1.5 following Dyar's rule.

Four instars occurred together only at the beginning of the life cycle, in September. Most larvae had developed to instars 3 or 4 by November, and the population was roughly synchronous until February. Over that time, larvae rest in large aggregations on the undersides of large stones, preferably those with crevices (larval dormancy). This dormancy synchronizes the development of the population. Increased temperatures in the laboratory had no effect on occurrence and duration of dormancy, which is presumably endogenously controlled. During dormancy cases are fastened to the substratum by a few silk threads, and this is a period of very low activity. However, if larvae are disturbed (e.g., by higher discharge or other calamities), they may move and try to refasten their case elsewhere. Mean weights of overwintering third and fourth instar larvae were not significantly different between November and the following February, even though they did not feed. Final growth occurs from February onwards; larvae thus take advantage of the first algal bloom in the stream every year. According to Aurich (1989), instar 5 larvae were 3 to 4 times heavier than instar 4 larvae (~0.4 or ~1.7 mg). Final instar larvae develop quickly in March, April, and May. Pre-pupae and pupae occur from April (Figure 7.96). These instars were again usually found in aggregations of many to several hundred larvae underneath large stones that lie with some space between them and the underlying stream substratum. Emergence usually starts in April and May.

Larval Predators and Defensive Strategies Larvae of *A. fimbriata* have antipredatory strategies. If they are endangered by predators, particularly *Rhyacophila fasciata* larvae, they withdraw deeper into their cases. If the enemy invades the case, the larva ducks its head down and secretes drops of a fluid on the hind part of the head capsule. A gland in the prothorax produces the secretions that are released through two pairs of openings on the lateral sides of an eversible, sac-shaped structure on the front of the prothorax. If the invader touches the drops, it immediately leaves the case and seems to be paralyzed for some time (Aurich, 1989).

Gas chromatography–mass spectrometry analyses (Wagner *et al.*, 1990a, 1990b) of the secretion revealed about 30 fatty acids (6 to 14 carbon atoms long), the major components being 12–14 carbon atoms long. Laboratory and field tests with saturated fatty acids and predators (*Rhyacophila* sp., *Plectrocnemia* sp. *Isoperla* sp.) demonstrated paralyzing effects. In laboratory choice experiments, *Rhyacophila* sp. larvae significantly avoided *A. fimbriata* larvae, compared with "chemically unprotected" *Agapetus fuscipes* and *Drusus annulatus* larvae. Large predators, such as fish or large stone-flies (Perlidae), were not affected. The fatty acids therefore appear to be effective repellents against the most important predator in *Apatania* larvae's natural environment, that is, *Rhyacophila* sp.

Loss Rate over the Life Cycle This was estimated for *A. fimbriata* in the Breitenbach for 1993, based on data from the T1 area. The loss was 82% at the adult stage; only 14 of 72 females emerging per meter stream produced an egg mass (~200 eggs per mass ~2.800 eggs per meter). During embryogenesis (about 4 weeks) the loss rate was 35% of the remaining eggs, that is, 1800 first instar larvae hatched successfully from 2800 eggs. These losses occur within about 4 weeks (~250 specimens lost per week). Assuming that about 150 specimens will emerge in the following year (50% females), weekly loss falls to about 35 specimens per week, presumably showing higher rates for the younger instars. Based on mean emergence over several years, loss rate is about 98% between egg and adult, which means that 2 adults will emerge from about 200 eggs (Enders and Wagner, 1996). Loss rates, both in the terrestrial and in the aquatic environment, are of similar magnitude, but those on land occur over a much shorter time than those in water.

The Life Cycle of *Chaetopteryx villosa* in the Breitenbach

Larvae of the caddisfly *Chaetopteryx villosa* are one of the dominant shredders in the Breitenbach (Wagner, 1986). This species belongs to the tribe Chaetopterygini, within the subfamily Limnephilidae, whose species preferentially emerge in late

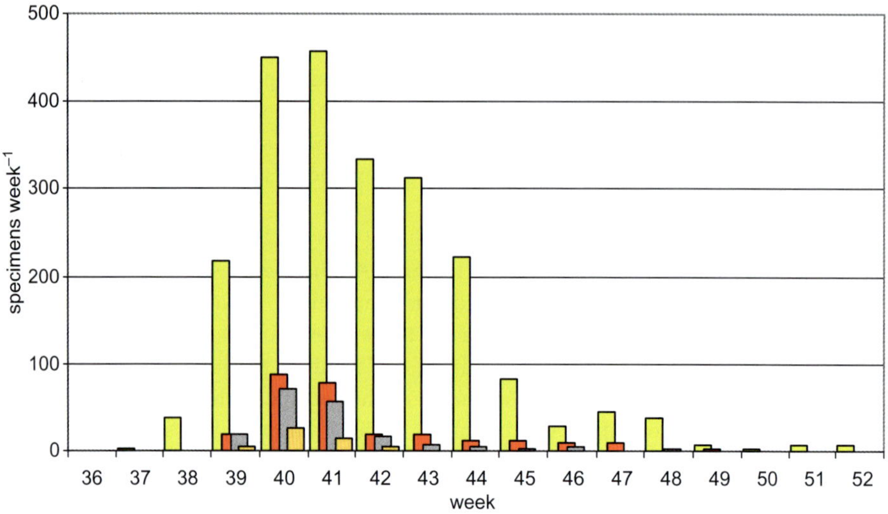

Figure 7.101 Weekly emergence of *C. villosa* at four sites along the Breitenbach (T1 yellow; T2 red; T4 gray; T6 orange) in 1983.

summer and autumn. Adult *C. villosa* were repeatedly found crawling on snow cover in December, or even January. Adults are brachypterous; the larger, heavier females seem to be totally unable to fly, whereas the smaller, lighter males may flutter over short distances. The sex ratio in the Breitenbach population was approximately 2 males to 1 female.

C. villosa is found along small streams, but it also occurs in mid-sized rivers, such as the Fulda. The species is considered eurythermous and ecologically flexible. At the Breitenbach emergence starts in mid-September at site T1, while the start of emergence at T2 and further downstream is delayed by approximately two weeks (Figure 7.101). Emergence of both sexes is simultaneous (Figure 7.102).

Females are mated soon after emergence, pheromones probably playing a significant role in recognition. Bergmann *et al.* (2002) identified methylhexane and methylheptane compounds, 4-methylhexan-3-one being found in both sexes, and classified them as pheromones (http://pherobase.org/database/species/species-Chaetopteryx-villosa.php). In the afternoon unmated females usually climb available vegetation and seem to actively "scent". Depending on the temperature, males either flutter toward the females or climb the vegetation. Males crawl from behind under the wings on one side of the female and try to contact the female genitalia by bending the abdomen. Once successful, the couple sits in the vegetation in the typical position for caddisflies. Copulation persists for days at least in the field and for more than three weeks in the laboratory. It is not clear whether one male produces one or several spermatophores. Unmated males are able to separate pairs in order to mate.

Usually only a single egg mass is produced and deposited, but oviposition has, so far, only been observed in the laboratory. Freshly laid egg masses are pinhead-

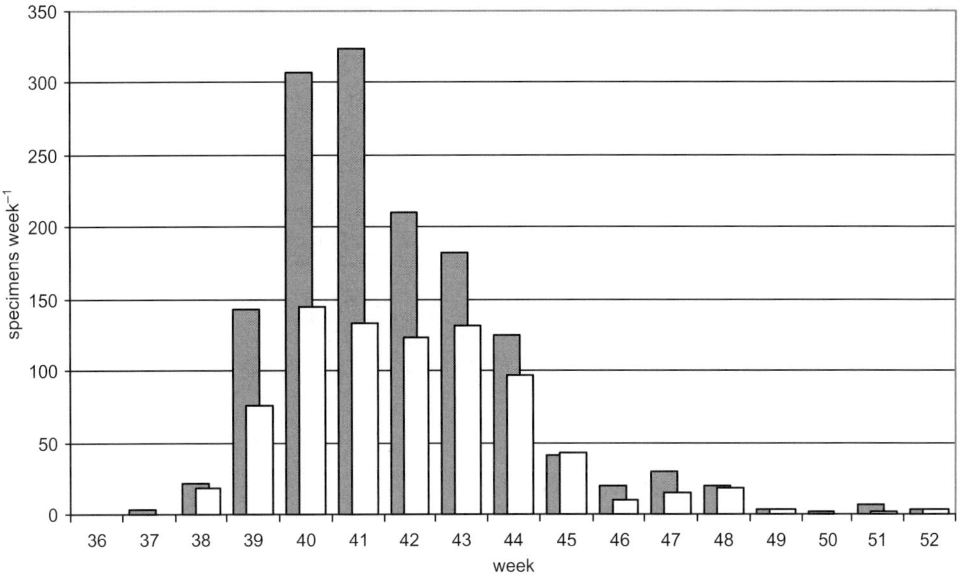

Figure 7.102 Simultaneous emergence of females (n = 815; white) and males (n = 1439; gray) of *C. villosa* at T1, in 1983.

sized, brownish to orange, and may absorb up to 1 cm^3 of water over an hour, becoming almost translucent, the distance between the eggs increasing. Egg masses have not been found in the field, probably due to high discharge. However, from the large number of egg masses obtained in the laboratory, several were taken to the field in glass tubes to study egg development. Duration of egg development in the laboratory was strongly temperature dependent (Figure 7.84, Table 7.22; Wagner, 1986). Egg masses in the field developed in line with laboratory predictions at T1 (~constant temperature) but, with increasing diurnal temperature fluctuation, development was accelerated at T2 and T4 (Wagner, 1986).

The life cycle comprises egg, five larval instars, pupa, and adult. Instars are easily distinguished by head capsule width (Figure 7.103), with relative width increases between instars of 1.5 (I–II), 1.33 (II–III), 1.45 (III–IV), and 1.56 (IV–V). Increase in mean weight between instars was between 2.9 (instars I–II) and 3.5 (IV–late instar V; Figure 7.104).

There is a clear along-stream distribution in *C. villosa*. Every year, maximum abundance occurred at T1 with reduced specimen numbers in traps T2 to T6. No significant differences in abundance were found between years of different discharge patterns, probably as a result of the species' ecological flexibility.

Size and weight of females and males varied along the stream; the highest abundance at T1 was accompanied by the smallest and lightest specimens; size and weight, particularly of females, increased downstream. These differences were at first puzzling because larvae fed for several weeks longer on CPOM at site T1

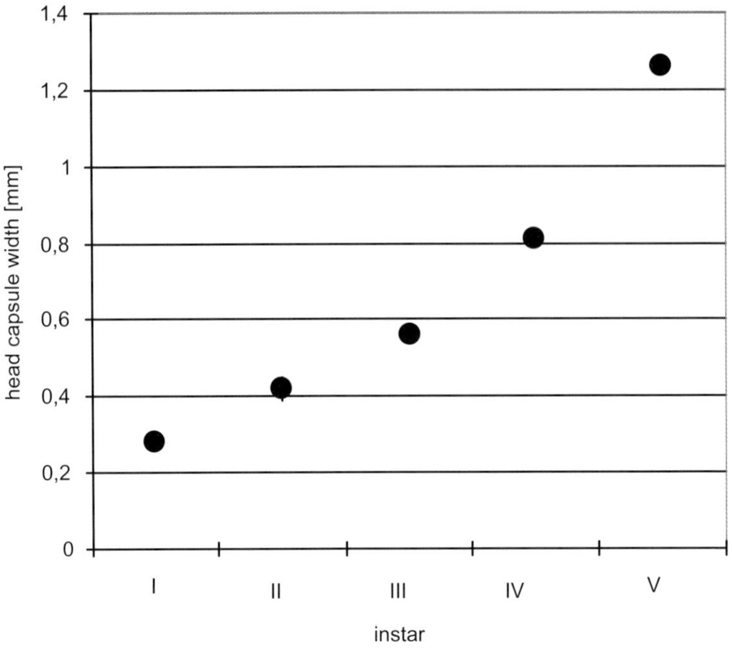

Figure 7.103 Head capsule widths of *C. villosa* larval instars I–V (mean ± 95 CI; mm; CI smaller than the circles' diameter).

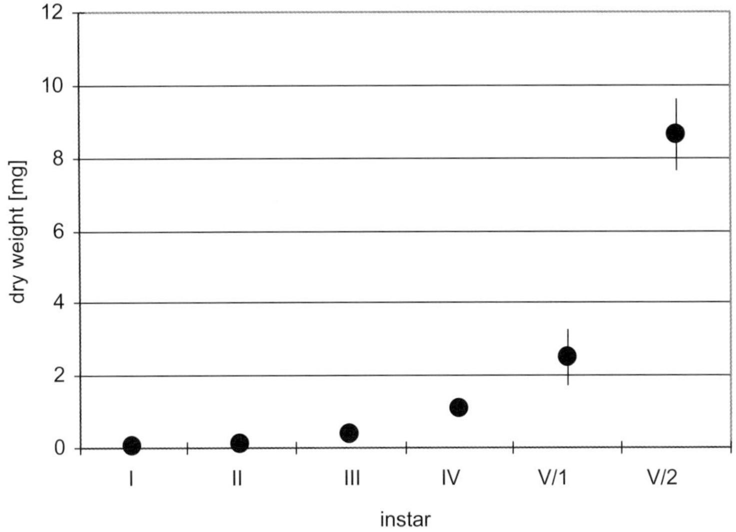

Figure 7.104 Mean weight of *C. villosa* larval instars I–V (mean ± 95 CI; mm) (V/1 early instar, June–July; V/2 late instar August–September).

than further downstream, seemingly without any effect on size. The hypothesis, food limitation, was rejected, because even the major shredders *Gammarus* spp. did not seem able to measurably limit the food sources. Furthermore, like several other caddisfly shredders, *C. villosa* larvae may leave the water and feed on grass and macrophytes in the terrestrial environment (Wagner, 1992).

Captive laboratory breeding and field experiments finally revealed size limitation by temperature linked to the life cycle. Because most pupa and adult weight (~80%) in caddisflies is accrued at the final instar, larvae at this stage are "weight decisive". Under unlimited food resources, temperature will limit growth. Low summer temperature at T1, but increased temperature downstream, leads to larger and heavier adults emerging at T4, even though the total larval life was 6 to 8 weeks shorter downstream (Wagner, 2002c; see Section 7.9).

Size and weight differences between females and males seem to depend on slightly different life cycles between the sexes. Prospective females seem to pass through larval instars 1 to 4 rather quickly, so that they live and feed for a longer time as large specimens. Males, however, remain in the early instars for longer and have a shorter period in which to feed as maximum-sized larva. Although time spent feeding under water is the same for both, it will still result in different final weights and sizes. This was confirmed by rearing individuals, and a simple growth model, including information about temperature-dependent growth rates and duration of instars, was developed (Wagner, 2002c).

The Life Cycle of *Sericostoma personatum* in the Breitenbach

It is more difficult to deal with the life cycle of *Sericostoma personatum* than other species. The taxonomic status of the westpalaearctic *Sericostoma* species is still the subject of debate. The Central European "species-pair", *S. personatum* (from Breitenbach and other nearby streams) and *S. schneideri* Kolenati, 1848 (from the River Fulda), have been investigated using various molecular methods (Leese, 2004).

The separation of *S. personatum* and *S. schneideri* into two distinct groups, or potentially sibling species, was not supported. Mitochondrial haplotypes exhibited a unique geographical pattern, with one group primarily found in Eastern Europe (including Finland), another in Central Europe (including Norway, Sweden), and a third predominantly located in the Alps (Leese, 2004). Morphological characters usually used for the identification of species may be polymorphisms, or the result of phenotypic plasticity, and were therefore considered unhelpful for resolving systematic questions (Leese, 2004).

Both *Sericostoma* species (from the Breitenbach and River Fulda) showed differences in emergence period and diurnal activity. Compared with *S. personatum*, emergence of *S. schneideri* (Fulda) was delayed by 2–3 weeks. The latter emerged preferably at night, whereas *S. personatum* emerged actively throughout the day. *S. schneideri* mainly mated between 16.00 and 20.00, and *S. personatum* between 6.00 and 10.00 (Leese, 2004).

Most laboratory-deposited egg masses remained partially developed in diapause at various incubation temperatures (Leese, 2004; Wagner 1990b). The induction of maternal diapause due to unfavorable environmental conditions, simulated in the laboratory during development of the last instar larva and pupa, seems to be the most likely explanation (Leese, 2004). In small laboratory flight cages, females deposited egg masses several centimeters deep underwater on the glass walls of the aquaria or other solid structures. It was not clear whether they developed with or without diapause. In the field, oviposition occurs in summer during low flow; the strong outer layer may be an adaptation to avoid desiccation (Leese, 2004).

The number of larval instars is still uncertain. Authors previously reported 5 (Nielsen, 1942) or 6 (Elliott, 1969) instars for "*Sericostoma personatum*", and 10 instars were reported for a South European species (Denis, 1981). In the Breitenbach, seven instars have been discriminated for the Breitenbach population (Wagner, 1990b), in agreement with Iversen (1973) and Rennerich and Schuhmacher (1984) (Figure 7.105). In addition, specimens were kept individually and molts observed, but clear distinction between subsequent instars was impossible. Assuming seven instars, the increase in head capsule width was between 1.5 and 1.7 until instar 4, dropping to 1.2 to 1.3 in instars 5 to 7.

Life cycle duration was strongly temperature dependent. In the laboratory, four groups kept at 10°C and 14°C (light dark 8/16 or 14/10) completed their life cycle

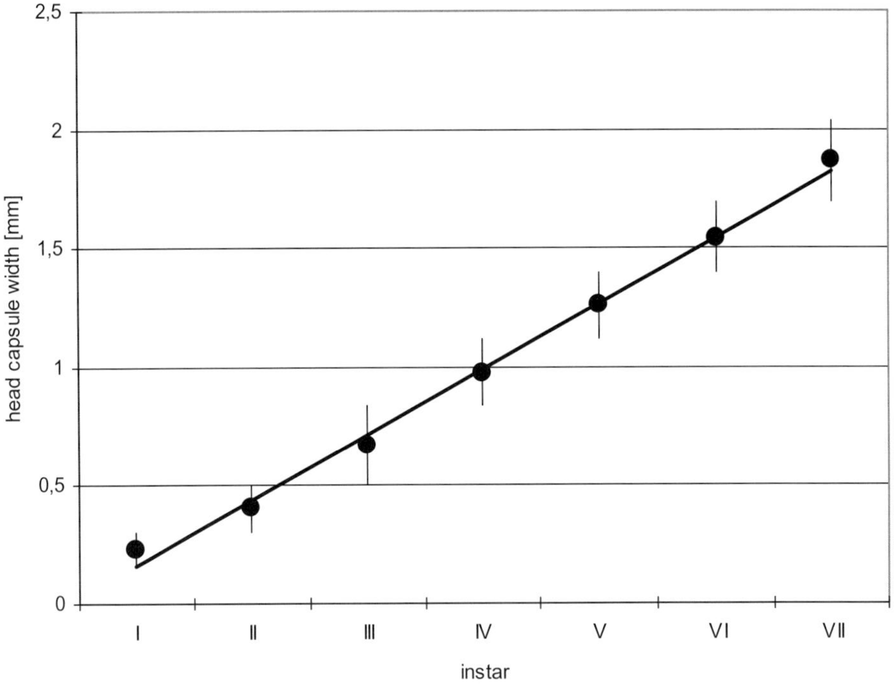

Figure 7.105 Head capsule width ($x \pm 95\%$ CI) of (presumed) seven *Sericostoma personatum* instars from the Breitenbach (linear regression, $y = 0.277x - 0.116$, highly significant $r^2 = 0.995$).

in about 1 year, but at least 1.5 years are required at constant 6°C (Wagner, 1990b). With a potential egg diapause, the life cycle in the Breitenbach may last between 1 and 2 or even 3 years.

Larvae of this species are shredders that feed, preferably at night, on CPOM on the substratum surface, resting some centimeters deep in the substratum during the day. They do not feed on organic matter in the substratum (Wagner, 1990b). Larval guts are therefore full at the end of the night but empty at the end of the day. Laboratory tests demonstrated that the daily consumption of the last instar larvae is significantly related to the light–dark rhythm (Wagner, 1990b; Figure 7.106). In two experiments, run over a 90-day period, the presence of a fifth instar larva increased the sediment organic content by 42.9 or 59.8 mg per 16 cm^3, an increase of 75.8, or 185.6%, compared to a control without larvae (Wagner, 1991). FPOM excreted into the sediment can be subsequently used by collectors, for example, *Ptychoptera paludosa* larvae (compare Diptera). The export of organic material from streams may also be delayed, although only on a small scale.

Emergence lasts from June to September (Figure 7.107), ending five weeks later at T1 than at the downstream sites. Females are larger and heavier than males, but weight differences between sites were insignificant in 1987 (Figure 7.108). Over the period from 1983 to 1993, both sexes were found to be significantly heavier at the most downstream site (T6) compared to more upstream sites (Wagner, 2005b).

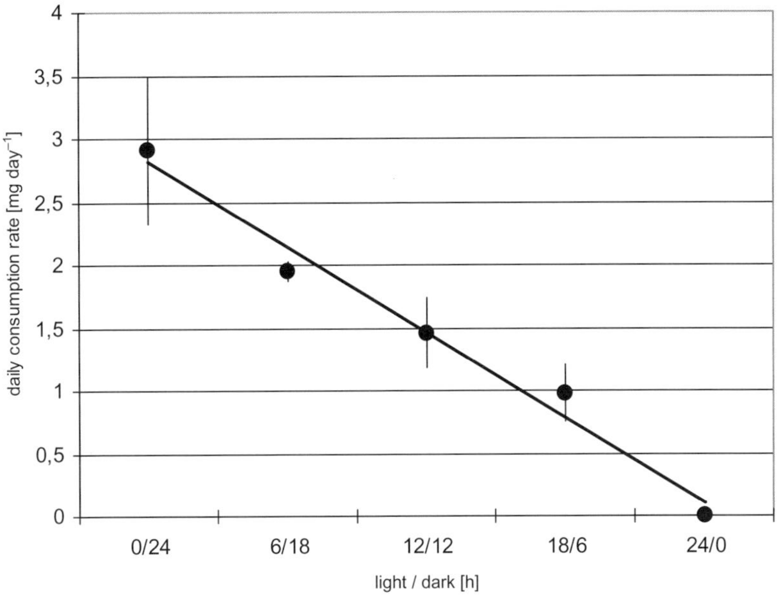

Figure 7.106 Daily consumption rate (mg day^{-1}) of last instar of *Sericostoma personatum* larvae under different day lengths (light/dark) at 10°C water temperature ($y = -0.679x + 3.497$; $r^2 = 0.98$).

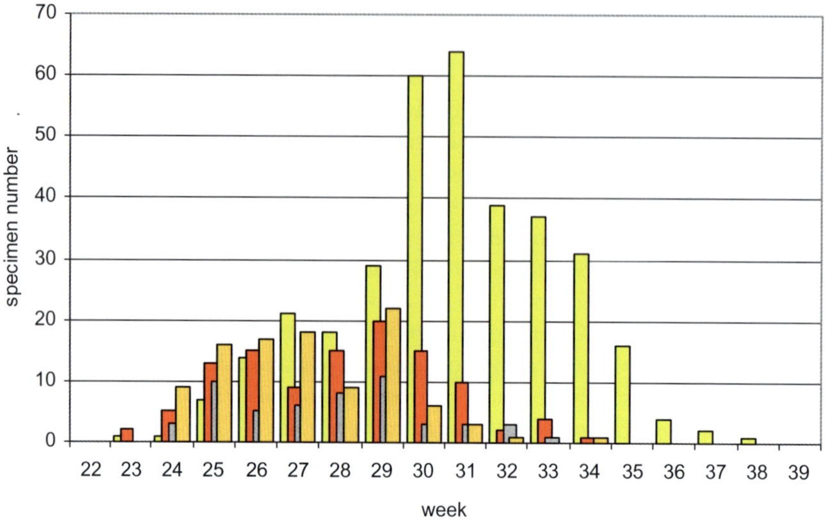

Figure 7.107 Weekly emergence of *Sericostoma personatum* at four collecting sites along the Breitenbach in 1984 (T1 yellow; T2 red; T4 gray; T6 orange).

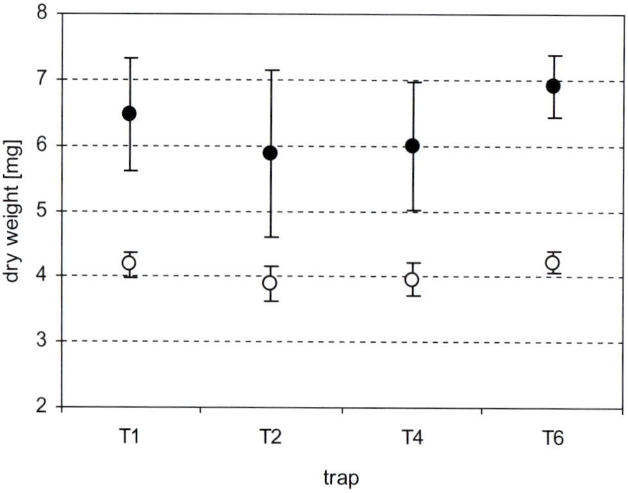

Figure 7.108 Female and male dry weights ($x \pm 95\%$ CI) of *Sericostoma personatum* in four traps along the Breitenbach in 1987.

The Briefly Colonial Life of *Plectrocnemia conspersa* (Curtis) Hatchlings

Research into the egg development time of the caddisfly *Plectrocnemia conspersa* (Curtis) revealed an interesting behavior of newly emerged hatchlings that had not been previously observed in caddisflies. Unlike most other freshly hatched caddisflies, larvae of *P. conspersa* flock together in a kind of "colonial web", but only under running water conditions (Hildrew and Wagner, 1992).

The egg masses are typical of the family Polycentropodidae, single-layered, with only small amounts of spumaline between the elongate eggs, containing about 440 eggs on average (minimum 150, maximum 800). Freshly deposited egg masses were light brown, the color deepening to brown within a day of deposition. In the laboratory, egg incubation period was significantly related to temperature (Table 7.22), between 93 days (constant 6°C) and 20 days (constant 18°C). During embryogenesis egg length increased by 17%, hatching took about half an hour, and hatchlings still had yolk in their guts for about another day. The next day they started feeding.

There was a marked difference between larvae kept in standing or in running water. In still water, larvae spin individual, long silken ducts, and depart from the egg mass in all directions. However, in running water, a large, single net, up to 10 cm^2 area, was formed by the larvae around the egg mass. As in standing water, larvae spent most of the time spinning, but the abdomens showed no ventilation movement and turning movements were less frequent (Hildrew and Wagner, 1992). This "colonial net" persisted for about seven days, and during that time larval density in the net declined from more than 30 larvae cm^{-2} to about 3–4 larvae cm^{-2}. The behavior of larvae in the net changed after a few days.

During an initial period (3–4 days) larvae either ignored or "explored" (scanned) each other without aggression. From day 4 onwards "aggressive" behavior between larvae was noted. Although one half-eaten larva was seen, cannibalism was seemingly unusual. Typically, defeated larvae left, sometimes being expelled from the net. A few larvae were seen actively entering the drift (Hildrew and Wagner, 1992).

Feeding started on the second day after hatching and larvae only took live prey. In the laboratory they were mainly fed on brine shrimp larvae, sometimes with small mayfly larvae (placed on the net). The net could hold prey larger than an individual *Plectrocnemia* larva. On one occasion up to six *Plectrocnemia* larvae (1.5 mm long) were seen feeding successfully on a *Baetis* larva (3.0 mm long). This particular behavior of young *P. conspersa* larvae had been unknown because field collecting immediately destroys the nets that are fastened between a larger "cap stone" and the stream substratum. Several authors (e.g., Nielsen, 1942) described finding young *Plectrocnemia* larva in "mud nettings" beneath stones, presumably remnants of the colonial net.

A colonial net may increase the survival rate of young larvae, probably helping to avoid accidental drift. Silk sharing may be the main advantage, probably linked to the likelihood of obtaining a first meal, even if the prey is much larger than the individual predator (Hildrew and Wagner, 1992).

7.6.4.6 **Aquatic Diptera**

Rüdiger Wagner

Quite a number of people participated in the Diptera section (Nematocera and Brachycera); many colleagues have determined specimens from different years and collecting sites over the last decades. Because only a few groups were dealt with over the entire study period (1969–2005) coverage remains fragmentary and this is also the case for groups with low specimen numbers. In all these cases, comments are restricted to certain species and general ecological remarks.

However some families (e.g., Simuliidae, Dolichopodidae), are described more extensively, because the respective groups are abundant and important in, or along, running water courses, and information on these groups is scarce. Lists of identified species are presented in the floral and faunal documentation (Appendix).

7.6.4.6.1 **Tipulidae**

Rüdiger Wagner

The family Tipulidae (crane flies), together with the Limoniidae and Cylindrotomidae, form the superfamily Tipuloidea. More than 4000 Tipulidae species are described worldwide and more than 1200 occur in the Palearctic Region (De Jong *et al.*, 2008). Tipulid larvae, leatherjackets, are found in a wide range of habitats, mostly terrestrial or semi-aquatic. Only a few are truly aquatic, including the dominant species in the Breitenbach. Some may be locally abundant and of great ecological importance, for example, as food for a variety of animals, or as plant pests. Larvae of other taxa are destructive feeders of grasses and crops and may cause severe commercial losses (Oosterbroek and Theowald, 1992). Adults seem to feed on plant liquids to a limited extent.

About 30 Tipulidae species have been collected in the emergence traps along the Breitenbach. The relationship of several species to aquatic habitats is more or less close. Larval development of approximately half of them is aquatic, according to Theowald (1978). However, he did not mention *Nephrotoma* species and *Tipula* species of the subgenera *Pterelachisus, Schummelia, Lunatipula, Vestiplex,* and *Beringotipula* as aquatic. It remains difficult to relate crane fly larvae to aquatic or non-aquatic environments, because they live alongside streams in habitats that may be subject to floods or droughts. Oosterbroek (1978, 1979a, 1979b) provided some information for the genus *Nephrotoma,* and Brinkmann (1991) presented an extensive study on crane flies and related taxa of a North German lowland stream, indicating the probable aquatic development of some more species.

A tipulid life cycle typically lasts one year for most Breitenbach species; it consists of an egg stage (1–2 weeks), four larval instars and a pupal stage of 1–2 weeks. Summer and winter diapause is often noted (Oosterbroek and Theowald, 1992). Final instar larvae may be very large (up to finger size in *T. maxima*). Several species had only short emergence periods, for example, *T. pagana* in autumn.

Tipula maxima and *T. pagana* are dominant at the Breitenbach, with frequent presence and abundance along the stream and between years. Larvae are shredders, that is, they feed on any kind of decaying material, preferably coarse particulate organic matter (CPOM). They live in the fine sandy and gravel sediments, with high organic matter content, and along the banks of the stream, preferably in habitats with high amounts of CPOM.

In individual years, size–weight relationships were significant for both sexes (dw = wing-length $^{\text{exp b}}$; male: b = 1.194, $r^2 = 0.995$, $n = 173$/female: b = 1.35, $r^2 = 0.987$, $n = 92$ for 1983). With dry weights of up to 70 mg for females, and up to 40 mg for males, *T. maxima* has the highest individual weight of all aquatic insects in the Breitenbach. The weight of one *T. maxima* female equals the weight of about 60–70 females of the caddisfly *Agapetus fuscipes*, or twice the weight of the heaviest caddisfly *Potamophylax cingulatus*.

Information on the emergence period comes from data from four sites along the stream over 11 years (1983–1993). Every year *T. maxima* emerged in May and June, without any evidence, even on a weekly basis, that emergence started earlier at an upstream or downstream site.

The time series of the abundance in four traps along the stream over 11 years indicate very strict parallel temporal development. Statistical tests showed that species abundance was not related to any pattern along the stream or to any discharge pattern. This is exceptional compared with most other, particularly EPT, species, which either have clear abundance patterns along the stream, or are related to one discharge pattern, or both (Section 8.4). This is interpreted as a clear indication that larvae of *T. maxima* (and presumably other Tipulidae species) are dependent on very local and short-term effects of food availability and quality, moisture and temperature conditions, which are scattered along the stream. Larvae seem to be less susceptible to environmental variation in the stream probably due to their individual size and mobility.

Specimen numbers of the remaining species are comparatively low, so emergence period and other facts are omitted.

7.6.4.6.2 **Limoniidae and Pediciidae**

Rüdiger Wagner

The Limoniidae (in the new sense) and Pediciidae were separated from one another by Starý (1992). Adults are small to medium-sized flies. Together with the Tipulidae, Trichoceridae (terrestrial) and Cylindrotomidae, both families form the infra-order Tipulomorpha. The Limoniidae are often confused with the Tipulidae, but adults of most Limoniidae have a short rostrum. About 11 000 species of Limoniidae and Pediciidae are known worldwide (De Jong *et al.*, 2008), and 286 Limoniidae, and 38 Pediciidae have been recorded from Germany (compare Reusch and Schrankel, 2006).

Ninety-seven Limoniidae, nine Pediciidae, and two Trichoceridae species were collected in traps at the Breitenbach (Appendix). Adults of some species, for example, the abundant *Molophilus ater*, are brachypterous. Larvae of most Limoniidae and Pediciidae are semi-aquatic or aquatic, those of the Trichoceridae and the limoniid subfamily Ulinae are terrestrial.

A typical limoniid life cycle comprises egg, four larval instars and pupa; egg and pupal stage are normally short (1–2 weeks), larval development is longer. As usual, environmental conditions (temperature, humidity) affect the duration of a life cycle that may be as short as six weeks, or as long as four years (Reusch and Oosterbroek, 1997). At the Breitenbach most species seem to have only one generation per year (univoltine), but a few are bivoltine (e.g., *M. griseus*). However two emergence peaks do not necessarily indicate two generations because, from rearing in the laboratory, it became evident that retarded hatching from oviposition in the spring, or even in the autumn of the previous year including a winter diapause (Reusch, personal communication). In genera with several species (*Molophilus, Ormosia*) there is a seasonal sequence of taxa (Figure 7.109). For instance, the short-winged *M. ater* is a typical spring species, congeners emerge during the summer months. In *Ormosia* a few species emerge in the spring and summer, but the most abundant taxon, *O. hederae,* emerges in autumn.

Adults are usually found in deciduous vegetation along water courses. Many species may be very abundant and are extensively preyed upon by many vertebrates. The morphological variability of limoniid larvae is generally high; they are hemicephalic, elongate and metapneustic. Many have unpaired creeping welts others (*Dicranota*) have paired prolegs on abdominal segments 3 to 7. Larvae occur in all kinds of semi-aquatic or aquatic environments; they seem to prefer wet soils with large amounts of decaying organic matter (CPOM), and may sometimes be abundant in moss carpets (hygropetric). Some are found almost exclusively in decaying wood where they feed on fungal mycelia (*Lipsothrix*).

Larvae of the Pediciidae and most Limnophilinae are carnivorous; large pediciine may even successfully attack predaceous caddisfly larvae of the genus, *Rhyacophila* (W. Graf, personal communication). Pediciidae larvae are aquatic, but those of the Limnophilinae prefer semi-aquatic conditions. A few Chioneinae larvae are also aquatic, but most are semi-aquatic; all are saprophageous. Most larvae of the Breitenbach Limoniinae species are semi-aquatic and phytosaprophageous, they include xylophageous, fungivorous, and even omnivorous species (Reusch personal communication; Reusch and Oosterbroek, 1997, with further references). Species of the subfamily Ulinae live in, and feed on, fungi. The pupal stage is fairly short, pupae are also found in larval habitats, but under drier conditions.

Results from the few years when data were available from all traps along the stream indicate that at least the most abundant taxa have distinct distribution patterns along the stream, with decreasing [*M. crassipygus* (a synonym of *ochrescens*; see Savchenko, Oosterbroek and Starý, 1992), *M. propinquus, M. bifidus*] or increasing (*O. depilata*) downstream abundance, or with highest specimen numbers in the mid-stream section (*M. ater, O. hederae*; Figure 7.110). These distribution patterns are reminiscent of those of other undoubtedly aquatic taxa, for example, Plecoptera and Trichoptera, but differ from that of the closely related Tipulidae (here *T. maxima*). These patterns can be interpreted as showing closer connection of the Limoniidae to the truly aquatic Breitenbach habitats, compared with several other groups of aquatic or semi-aquatic Diptera.

7.6.4.6.3 **Chironomidae**

Rüdiger Wagner

The Chironomidae are an ecologically diverse, and taxonomically difficult, dipteran family of predominantly aquatic species. However, they are frequently neglected in studies on streams and rivers because species richness, abundance are high and taxonomy is difficult.

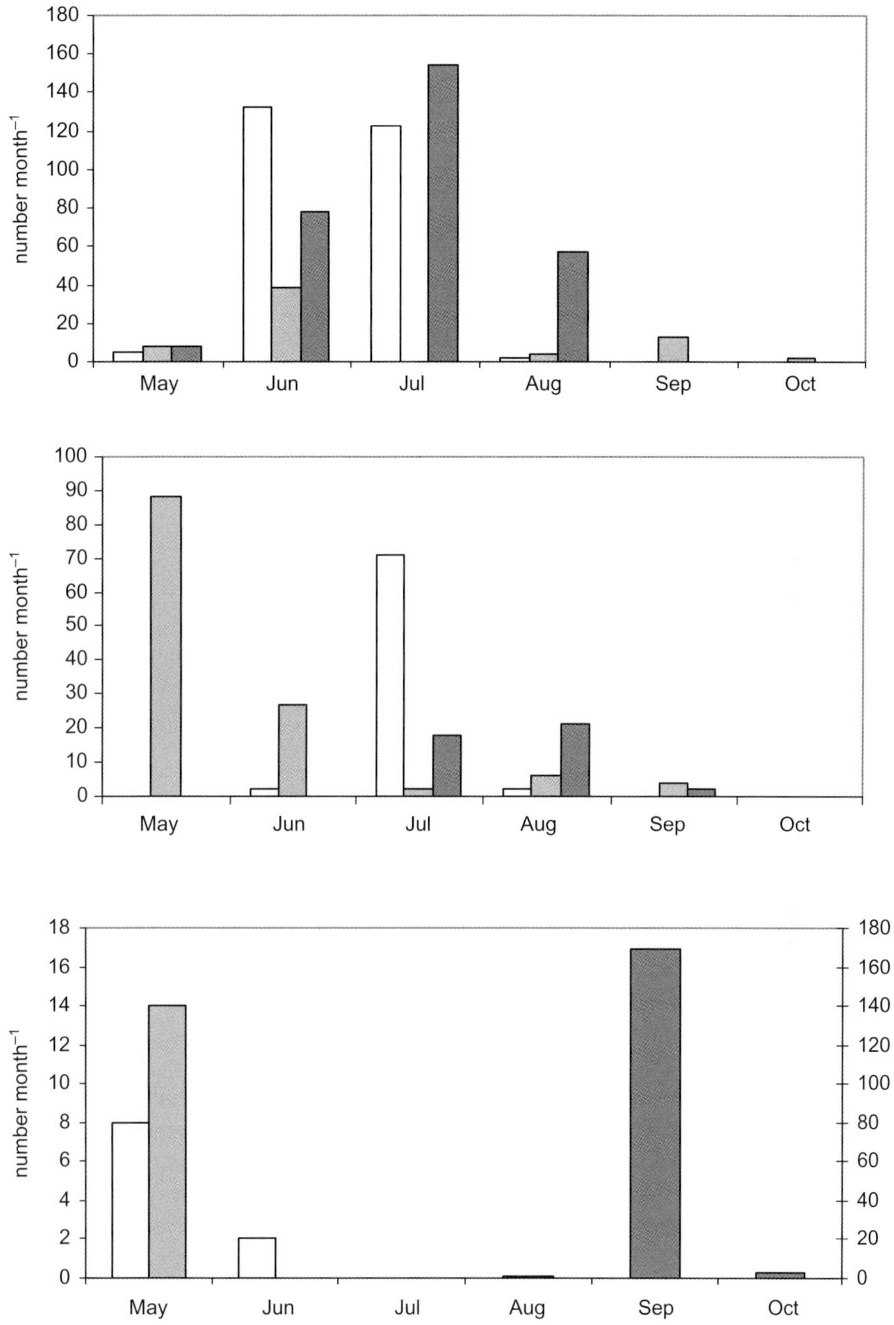

Figure 7.109 Seasonal emergence sequence of dominant species of the genera *Molophilus* and *Ormosia*, 1984 in trap T2. Above: *M. bifidus* (white), *M. griseus* (gray), *M. medius* (dark). Middle: *M. ochrescens* (white), *M. propinquus* (gray), *M. undulates* (dark). Below: *O. depilate* (white), *O. lineate* (gray), and *O. hederae* (dark). Data are specimen numbers per month; *O. hederae* on second y-axis respectively.

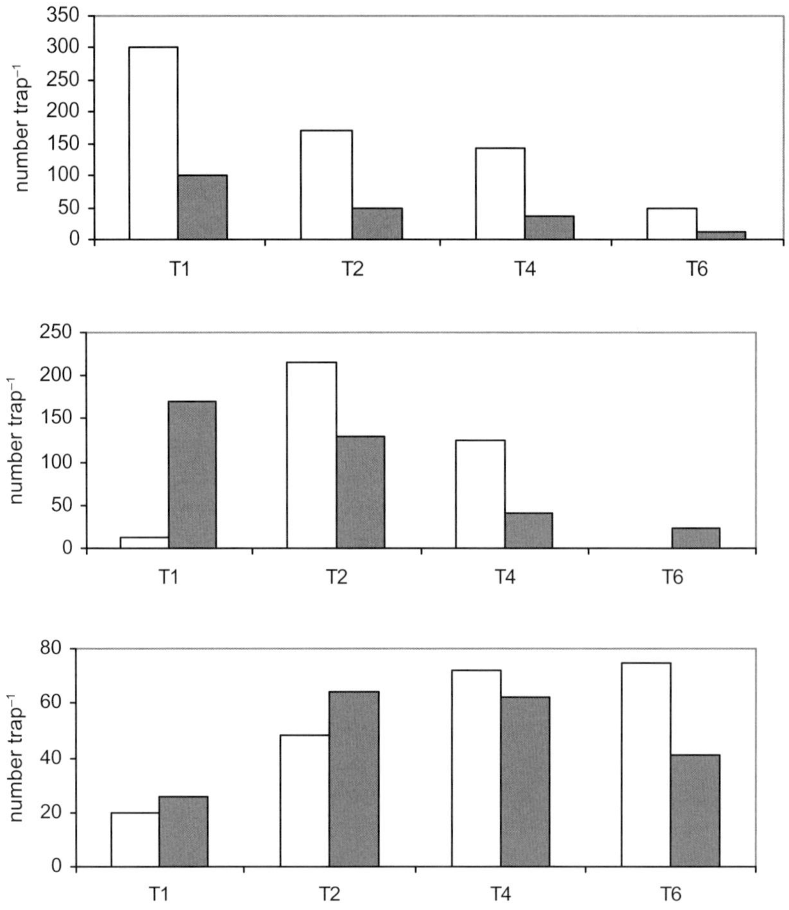

Figure 7.110 Abundance (specimens per year) of *Molophilus* and *Ormosia* spp. in four traps along the Breitenbach in 1984. Above: *M. ochrescens* (white), *M. propinquus* (gray). Middle: *M. ater* (white), *M. bifidus* (gray). Below: *O. depilate* (white), *O. hederae* (gray).

Adult Chironomidae are small to medium-sized flies. The apneustic larvae are elongate, cylindrical with a strongly sclerotized, nonretractile head capsule. They occur in almost every wet or aquatic habitat. The ecological requirements of the larvae of many species are known empirically. Most of the life cycle is spent in (usually) four larval instars; pupal and adult stages are comparatively short, lasting at most a few weeks (Oliver, 1981; Lindegaard, 1997).

The larvae of most groups construct cases of fine substratum particles glued by salivary secretions. They are microphagous, that is, algal grazers or detritus collectors. Larvae of the subfamilies Tanypodinae, some Orthocladiinae and Chironominae, are carnivorous and thus free-living (without cases) (Oliver, 1981; Lindegaard, 1997). But, as usual in a diverse family like the Chironomidae, more

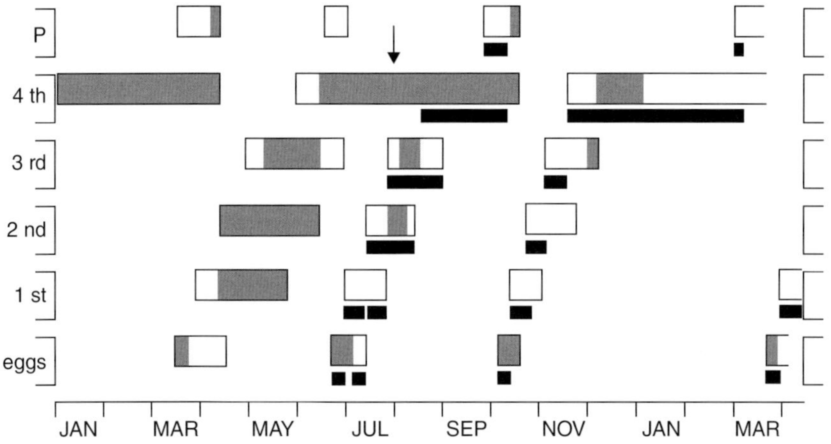

Figure 7.111 Life cycle of *P. branickii* in the Breitenbach 1988/1989. Open bars indicate minimum time spans during which instars occurred in the stream. Shaded segments indicate occurrence of instars in field samples, white segments indicate presence deduced from associated (laboratory) observations. Arrow indicates presence of larvae from the spring and summer generations. Closed bars show rearing of *P. branickii* during the summer (laboratory) and from winter to early spring (*in situ*); from Nolte and Hoffmann (1992a, with kind permission of Springer).

types of feeding mode occur. Most pupae are sedentary, again often protected by a case.

In the Breitenbach adults were found to emerge almost all the year round. Species and specimen numbers are maximum in summer, lowest from October to March. The life cycle duration is between a few days and several months. Emergence patterns were used to indicate generation numbers per year. However, after the life cycle studies of Nolte and Hoffmann (1992a, 1992b) the assumption that emergence peak and generation numbers per year are comparable must be carefully re-examined.

At constant 8 °C water temperature, a short-day photoperiod (compared with long-day) increased the generation time of *Pseudodiamesa branickii* from about 80–90 to 140–150 days. However, it is possible that two "strains" of the species were represented in a single population, one bivoltine and the other trivoltine (Figure 7.111). In contrast, for *Diamesa incallida* the occurrence of eight to ten generations per year was deduced based on field and laboratory studies (Figure 7.112).

Studies of larval microdistribution demonstrated that mosses are crucial in the life cycles of many chironomids; egg masses are laid there. The fauna of the moss *Platyhypnidium* sp. was dominated by *Diamesa* (Diamesinae) and *Eukiefferiella* (Orthocladiinae; Nolte, 1990). Near the source the abundance of these genera was about twice as high as at 1 km downstream.

Metriocnemus hygropetricus larvae were found in the spray zone but never submerged. Smaller larvae of *Micropsectra atrofasciata* occurred in submerged zones

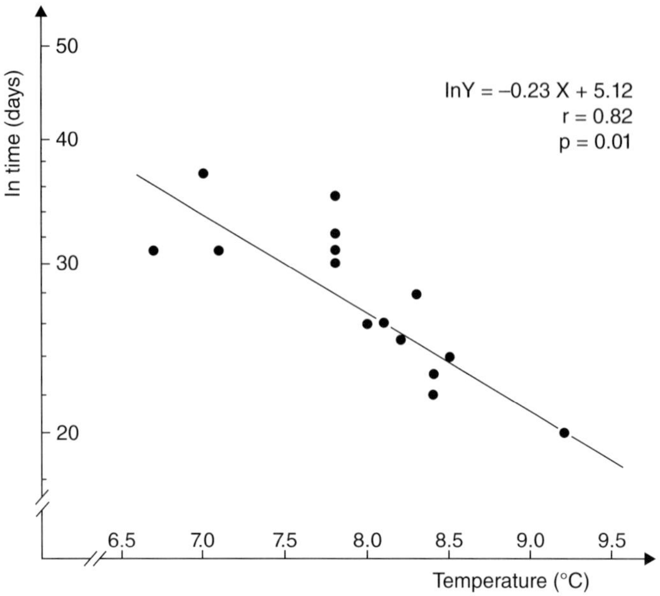

Figure 7.112 Regression of development time of *D. incallida* against ambient water temperature; from Nolte and Hoffmann (1992b, with kind permission of John Wiley & Sons).

shielded from the current, the larger instars seem to move into other habitats (mud, sand). A similar behavior is suspected for many other species (Nolte, 1991).

More than 210 chironomid species have been reported from emergence traps at the Breitenbach (Appendix), several remain unnamed because of taxonomic difficulties (Ringe, 1974, 1976; Siebert, 1979a, 1979b, 1980; Becker and Wagner, 2004; Hergersberg, unpublished). Specimen numbers per trap may vary between about 10 000 and more than 150 000 specimens $year^{-1}$. Some new species have been described from studies on sources and streams in the vicinity of Schlitz, two of these from the Breitenbach (Siebert, 1979a, 1979b).

Data are only available for a limited number of collecting sites and years because of the high species and specimen numbers, and taxonomic difficulties. Nevertheless, community composition at various sites in different years provides interesting information.

The characterization of chironomid communities is based on the representation of species and higher taxa in subfamilies, and on the hypothesis that systematically close relatives have similar ecological requirements (Lehmann, 1971; Caspers, 1983). Based on the data of Fittkau and Reiss (1978) these are:

Tanypodinae, predominantly in lacustrine or slowly flowing waters rich in macrophytes (predators).

Diamesinae and Prodiamesinae, in nutrient-poor, fast-flowing and cold streams, larvae feeding on algae and detritus (collectors, scrapers).

Orthocladiinae, Orthocladiini, characteristic species of running waters, larvae feeding on algae (scrapers).
Orthocladiinae, Metriocnemini, crenophilous, semi-aquatic, hygrophilous, or terrestrial.
Chironominae, Chironomini, in nutrient-rich standing waters.
Chironominae, Tanytarsini, in ± nutrient-rich standing or slow-flowing waters.

Data from the Georgsborn are available for 1974 (33 species det. Ringe, in Gümbel, 1976). Here the Metriocnemini were dominant and contributed 56% of specimens, Orthocladiini made up 34%, and Diamesinae 8% of the total specimen numbers. It is evident that the subfamily Orthocladiinae is the dominant chironomid subfamily at every site. The Metriocnemini are more abundant than the Orthocladiini at the source, but this is reversed at the other stream sites. The Diamesinae prefer the region just below the source.

For 1987, Hergersberg (unpublished) provided data on chironomids from sites T1 (67 species) and T2 (111 species). As usual, the Orthocladiini were the most abundant tribe with respect to species and specimen numbers. They made up about 65% of all specimens at T1, and 53% at T2. At T1 the Diamesinae contributed about 25% of specimens, only 5% at T2; the Metriocnemini contributed 9% at T1, but 36% at T2. These differences indicate that the Diamesinae were most abundant and diverse at T1, about 100 m downstream the source, where seasonal and diurnal temperature variation was low compared with the downstream sites (Figure 7.113).

In 1983, about 156 000 Chironomidae, from 109 taxa, were recorded from two emergence traps (T3), one exposed above a sandy substratum and another 150 m downstream above large, moss-covered sandstones, the clearest substratum difference in the stream (Becker and Wagner, 2004). Ninety-nine species were recorded from the sand and 86 from the stone trap. Proportions of Prodiamesinae (~8%) and Chironomini (~12%) were higher in the sand trap. The relative abundance of the Tanytarsini was higher in the stone trap (~27%; Figure 7.114).

Two species, *Odontomesa fulva* and *Paracladopelma camptolabis,* were found exclusively in the sand trap. Total specimen number was 24% lower in the stone trap. The similarity between these, and some other stream sites and years, was always below 0.50; high similarity was even observed between the contrasting sediments in that year (1983).

Thienemann's rule, stating that communities from increasingly monotonous uniform habitats were less diverse than from diverse habitats, was not confirmed. However, this may be partly due to a methodological weakness of emergence traps: a reduced current speed underneath the sand trap improved the chances of successful emergence from drifting pupae.

7.6.4.6.4 Ceratopogonidae (Biting Midges)

Rüdiger Wagner

The Ceratopogonidae are phylogenetically closely related to the Chironomidae, Thaumaleidae and Simuliidae, all placed in the superfamily Chironomoidea, the

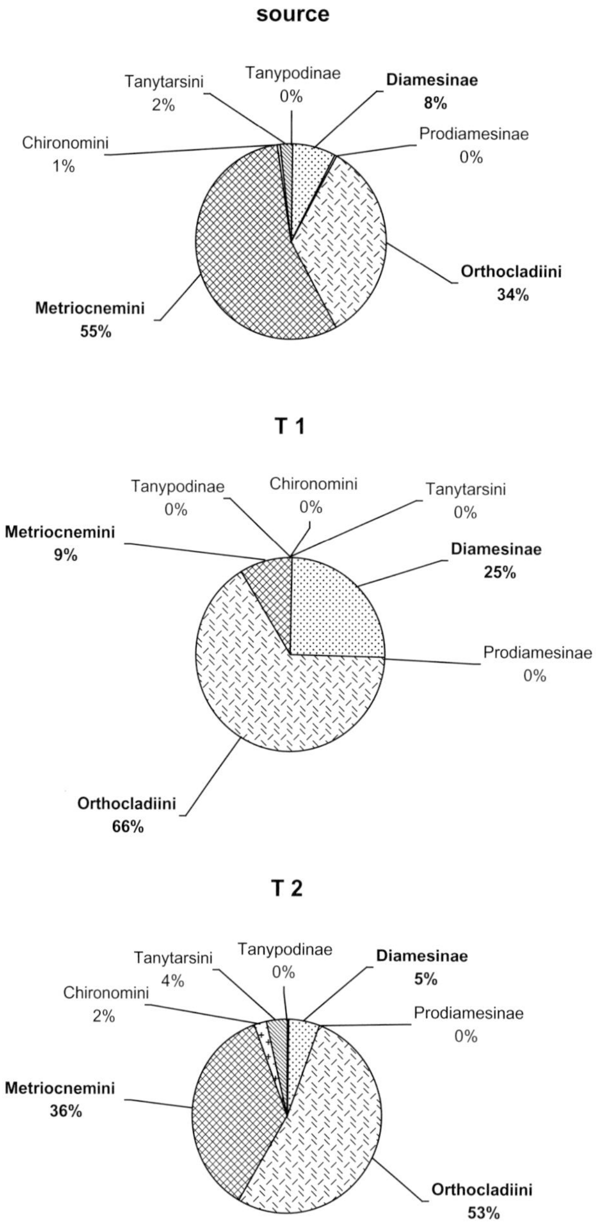

Figure 7.113 Share (%) of specimens in subfamilies and tribes of Chironomidae at the source (1974; Gümbel, det. Ringe) and sites T1 and T2 (1987; Hergersberg, unpublished).

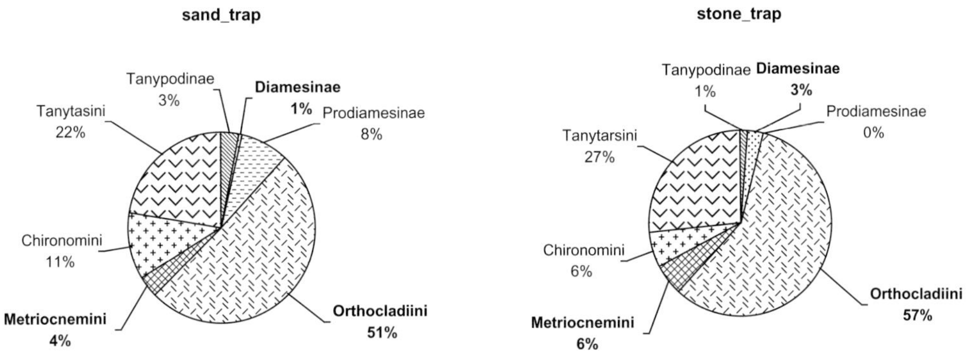

Figure 7.114 Share (%) of specimens in subfamilies and tribes of chironomidae over stoney and sandy stream segments at the Breitenbach.

sister group of the Culicoidea in the infraorder Culicomorpha. About 6000 species are currently known (Wagner *et al.*, 2008).

Adult Ceratopogonidae are small, slender to robust flies, the females having biting mouth-parts. They are blood suckers of vertebrates and insects (e.g., Odonates), or predators (Downes and Wirth, 1981). They live in moist places around larval habitats. Males and females may visit flowers, and females of some species lay the first egg batch without a blood meal (autogeny). However, they need to feed on blood to undergo another ovarian cycle. In Europe the bites of some species can cause considerable nuisance, and others are important because they transmit animal pathogens (filarial worms, viruses).

The larvae are characteristic of damp habitats, some may be found in semi-aquatic habitats or among mosses. They vary considerably in shape between genera (Boorman, 1997).

Eggs are usually deposited on moist substrata and embryogenesis lasts several days. Larvae pass through four instars and the pupal life span is several days. The life cycle of some species kept under the laboratory conditions lasts only 3–4 weeks, however, most European species seem to be univoltine in the field.

A clear ecological classification is difficult, many inhabit a wide range of habitats, from truly terrestrial to truly aquatic. Larvae of the subfamily Ceratopogoninae are mostly predators; they feed on all animals that they can overpower, and are even cannibalistic. Other taxa feed on algae, bacteria or collect fine particulate organic matter (FPOM).

The taxon lists from the Breitenbach and Rohrwiesenbach include 78 taxa in 10 genera (Havelka, 1975; Appendix). Species numbers at stream sites were about twice those at the source. Seven of 31 species occurred exclusively at the source, and 34 exclusively at stream sites in the Breitenbach. There were few taxon differences between the streams, and these were probably accidental, because many taxa only occurred in low abundances. Adults usually emerge from mid-April to mid-October, with two emergence peaks in summer (June, August).

7.6.4.6.5 **Dixidae (Meniscus Midges)**

Rüdiger Wagner

Adult Dixidae are fragile Diptera that resemble the phylogenetically related Culicidae. The family is notable for its remarkable larvae that rest in a U-shaped position associated with emergent structures (vegetation, dead wood) and the water meniscus along unpolluted streams and stream sources (Disney, 1999). Seven, of about 30 European Dixidae species have been reported from the Breitenbach (Appendix); all are widespread across Europe; *D. nebulosa, D. puberula, D. aestivalis,* and *D. serotina* are even known from North Africa, and *D. obscura* has spread down to Israel. The Breitenbach dixids are typical of many lower elevation small streams in Central European highlands.

The eucephalic dixid larvae have ambulatory combs on the ventral side of abdominal segments, and in the genus *Dixa* there are "crowns of setae" on the dorsal side of several abdominal segments. The terminal complex around the caudal spiracles is heavily sclerotized with lateral "paddles" and elongate hairs. Larvae feed on microorganisms and decaying plant or animal material trapped in the water or on the surface film. The genus *Dixa* is characteristic of running waters; *Dixella* prefers still waters. Dixid larvae are sometimes a significant component of invertebrate drift in streams (Elliott and Minshall, 1968; Elliott and Tullett, 1977). *Dixa* larvae are sensitive indicators of pollutants in streams (Thomas, 1979).

Dixid pupae resemble Culicidae pupae but are less mobile, resting in a U-shaped position near the larval habitats. Adults may have larval water-mites attached, usually on the abdomen.

The life cycle covers egg, four larval instars, pupa, and adult. Peach and Fowler (1986) described all instars of *D. autumnalis* and reported a life cycle duration of about 60 days. Emergence patterns from sources indicate at least two generations per year. Depending on species, temperature and nutrition, more generations probably develop.

As long as environmental factors, particularly temperature, in the Breitenbach are suitable, adults of *D. submaculata* emerge. *Dixella aestivalis* is restricted to summer and autumn. If the above life cycle duration data are roughly applicable to Dixidae in the Breitenbach, one to three generations per year can be expected. Figure 7.115 shows monthly sums of specimens emerging at two sites in 1991: at T2, two emergence peaks for *D. submaculata* and *D. nebulosa*, and one for *D. puberula*, indicate the probable generations. However, in the same year just 1000 m downstream conditions are different. At T6, *D. nebulosa* has three emergence peaks, *D. puberula* has one. The abundance of *D. submaculata* decreased dramatically so that peaks like those at T1 are not as clear at T6.

Individual species varied in abundance along the stream (spatial) and between discharge patterns (temporal). The abundance of *D. nebulosa* and *D. submaculata* decreased downstream; that of *D. puberula* was correlated with the discharge pattern. Because *D. puberula* is the most abundant dixid, variation in the sum of all Dixidae varies with discharge (compare Section 8.4).

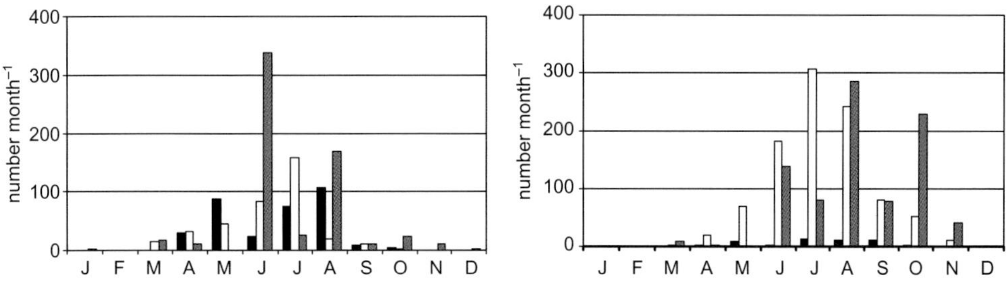

Figure 7.115 Emergence (monthly sum) of *Dixa nebulosa* (gray), *D. puberula* (white), and *D. submaculata* (black) at sites T2 (left) and T6 (right) in 1991.

7.6.4.6.6 **Thaumaleidae**

Rüdiger Wagner

Thaumaleidae (trickle midges) are rare in emergence traps. Adults resemble black flies (Simuliidae), but they lack the large anal lobe at the base of the wing. Three species were recorded in low numbers in the Breitenbach. The elongate larvae are associated with thin water layers in sources and spray-zones along cold and clean streams from sea level to about 3000 m a.s.l.. Vaillant (1955) demonstrated that they live **on** rather than below the water surface. Larvae are eucephalic, with a nonretractile head. They feed on the biofilm on stones and decaying leaves. The prothorax carries a pair of respiratory tubes and a proleg, and thorax and abdominal segments have varying patterns of setae. The eighth abdominal segment has a pair of stigmata surrounded by transverse groups of setae and flanked by a pair of setose finger-shaped processes. Pupae are subcylindrical without a cocoon; the prothorax has a pair of respiratory horns.

Life cycles of the Thaumaleidae are almost unknown. It is assumed that the life cycles of the widespread species are poorly synchronized. Adults emerge during the entire year, when temperature in the terrestrial environment is suitable. The life cycle of *T. cebennica* probably includes a diapause, from mid-summer to next spring, and lasts one year (Wagner, 2002).

7.6.4.6.7 **Simuliidae**

Joachim Reidelbach

The Simuliidae (blackflies) are a large family of nematoceran Diptera, comprising more than 2000 valid species worldwide (Adler and Crosskey, 2009). The Chironomidae, Ceratopogonidae, and Thaumaleidae are considered to be most closely related families (Crosskey, 1990).

European blackflies are rather uniform in size, morphology and general ecology. The dark-colored adults are 1.5 to 5.0 mm long (with wing length in the same range or slightly shorter; e.g., Rivosecchi, 1978) and their dry weight ranges from about 0.3 to 2.0 mg per individual (Kazimírová, 1981; Stamer, 1987). Their shape is compact with a large thorax and arched scutum, which may partly hide the small head of females. Size and morphology of the males are very similar to females, except for the genitalia and the large compound eyes, which completely cover the dorsal side/surface of the head. The eyes are horizontally divided into a larger dorsal part with large ommatidia and a ventro-lateral part with small ommatidia. Functionally, the dorsal part of the male eye is adapted to detect small, fast moving objects, such as females flying over a swarm of males (Kirschfeld and Wenk, 1976).

Adult males live for several days and feed on nectar and pollen, often on flowers of the Apiaceae, whereas females may live for several weeks and suck ythe blood of hosts several times with consecutive gonotrophic cycles and oviposition between blood meals. Females of most species feed mainly on mammals (including man) or birds, and may thereby transmit various human and animal pathogens/diseases. Adult blackflies are agile, fast and persistent flyers. They can fly long distances looking for hosts and their obligate breeding habitat, running waters of all kinds and sizes (Garms and Walsh, 1988; Crosskey, 1990).

Modes of oviposition are surprisingly diverse, from dropping single eggs above the water surface during flight, to laying large composite egg masses on a variety of species-specific types of submersed, marginal or moist terrestrial substrata (Golini and Davies, 1988; Timm, 1988; Zwick and Zwick, 1990).

However, larvae of most species exhibit similar modes of attachment and nutrition. Using a sticky gland secretion ("silk", Kiel and Röder, 2002; Barr, 1982, Kiel *et al.*, 1989) and an abdominal hook circlet, they attach themselves to plant or stone surfaces which are exposed to considerable currents, with small scale substratum-near velocities from about 5 to 80 cm s^{-1} (Malmqvist and Sackmann, 1996; Lacoursière and Craig, 1993). It is noteworthy that the long-term attachment of larvae does not require muscle activity and thus seems to be very efficient with respect to energy consumption. The secretion is also used for a looping step by step movement on the substratum and as an anchor thread ("lifeline") during drift (Kiel *et al.*, 1989; Reidelbach and Kiel, 1990). This enables larvae to colonize new trailing natural and artificial substrata (Rühm and Pegel, 1986a, 1986b; Reidelbach, 2002). They disperse from egg masses and drift to downstream sections to reach more suitable substratum or flow conditions, or to escape from predators and other disturbances (Rühm, 1998; Kiel, 1996; Kiel, Dickmann and Rühm, 1998).

Larvae are up to 10 mm long and open their paired, fan-shaped cephalic filter structures into the current. These fans protrude from the anterior part of the head capsule and capture fine particulate matter from the water using different types of adhesion for very fine particles (10 nm to about 50 µm) and sieve mechanisms for large or filamentous particles (Ross and Craig, 1980; Wotton, 1976; Crosskey, 1990).

Unlike many other running water macroinvertebrates blackfly larvae have no specialized external respiratory surfaces, but depend on gaseous exchange through their unsclerotized thoracic and abdominal skin. Together with their mode of food particle capture, this results in complete dependence of larval development on a steady and considerable (micro-)current, and places blackflies among the few, truly rheobiontic, taxa.

As a result of their "sophisticated" and unique reversible attachment, movement and landing after drift, the substratum surfaces preferred by larval blackflies are quite clearly distinguished from those microhabitats inhabited by the majority of coexisting macroinvertebrates; relatively clean surfaces in moderate to very fast mainstream currents, when on trailing vegetation often distant from the benthos, where most biological activity is concentrated.

This makes the qualitative sampling of blackfly larvae and pupae from natural and artificial substrata quite easy and selective, and minimizes sorting effort. On the other hand, quantitative sampling is very difficult and time-consuming, due to the clumped and irregular distribution of larvae and pupae. Blackfly abundances are therefore often underestimated by common quantitative benthos sampling methods.

In most species, larvae pass through seven instars until they pupate in a cocoon, fixed to the same types of substrata as the larvae, but in slower currents, for example, on the lower faces of stones. In all species, the pupa is immobile and protected by a thick, stiff cuticle. It is shielded against the current and other influences by a bag-, slipper-, or shoe-shaped cocoon made of silk, so that an external respiratory organ is needed. These striking, paired structures protrude as a group of four to forty branched cuticle tubes from both sides of the pupa's thorax, forming long and slender filaments in most European species. They project from the cocoon's opening, which is always oriented downstream, and function as a plastron. They are of great importance for species identification.

Pupal development and metamorphosis takes four to ten days, the duration being strongly dependent on water temperature (Prügel, 1988). The pupa darkens considerably, from pale yellow to dark brown, or almost black, and finally gas is formed between the pupal and the adult cuticle. During eclosion the adult leaves the pupal exuvia in a gas bubble and is swept to the water surface, where it is able to fly away immediately.

Because of the immobile pupae and their specific preferences for certain substrata and flow conditions, the spatial and temporal distribution of adult emergence is mainly determined by the respective distribution of pupation, and the availability of adequate substratum conditions. Temperature and diurnal rhythms provide short-term controls of emergence patterns.

Larvae, pupae, and adults are parasitized by a number of taxa, the most important being mermithid nematodes, Sporozoa, water mites, and Trichomycetes. A great number and variety of taxa have been observed to feed on all life stages of blackflies (Crosskey, 1990; Werner, 2004; Werner and Pont, 2003). Eggs are ingested by net-spinning trichopteran larvae and chironomid larvae. Larvae are preyed on by flatworms, many taxa of predaceous water insects, fish and even adult terrestrial Diptera. Pupae may be attacked by fish and large insect larvae with strong mouthparts. Numerous adult terrestrial insects and birds prey on blackfly imagines.

As a family, the ecological niche of European blackflies in general and their role within the running water community are quite well defined: Larvae and pupae live attached to substrata, exclusively in steadily running water with considerable flow velocity. Larvae are semi-sessile, passive filter feeders that use their mouthparts as filter organs, being able to filter and ingest very small particles, well below 1 µm. No other group of lotic macroinvertebrates efficiently utilizes this spectrum of small organic particles drifting in the water body. Larvae convert this material to compact feces that are much larger than the ingested particles. These are added to the stream sediment by egestion and may serve detritus collectors as food, and reducers as organic substrata.

The taxonomic situation of the family is characterized by persistent difficulties in the reliable morphological separation of closely related species and members of species complexes. This is especially true for larvae and females because of a lack of characters and, particularly because there is great variability in continuous characters. The application of cytogenetic methods has led to the recognition of many more cytospecies and cytotypes. But these methods can only be used for larvae and are not practicable for large numbers of specimens. Thus, the taxonomic resolution for ecological studies or population dynamics of blackflies is often limited to the level of genus or species group.

The Blackfly Species of the Breitenbach

Joachim Reidelbach

The blackfly fauna of the Breitenbach consists of eight species (Appendix) that are regularly or repeatedly found in the stream and in the emergence traps. Four of them (*Prosimulium tomosvaryi, Simulium vernum, S. cryophilum, S. ornatum*) are the dominant species and occur in considerable numbers. The total number of adult blackflies emerging from the complete length of the section T1 to T4 was calculated for two years as 555 000 (1984) and 488 000 (1988) individuals, representing a total dry mass of 293 g and 299 g respectively, that leaves the stream per year (Reidelbach and Christl, 2002).

Only single adults of four additional pecies [*S. (N.) angustitarse, S. (N.) lundstromi, S. (E.) aureum, S. (E.) angustipes*] were recorded from the traps, but it is unknown whether their aquatic stages may live in very low numbers in the Breitenbach. They may also have reached the traps by flying in from adjacent running waters (Reidelbach and Christl, 2002). Moreover, the traps yielded three species that have been shown to have originated from the adjacent potamal section of the River Fulda (where they are typical inhabitants), and one species (*S. noelleri*) that is typically found on the weirs of the ponds in the Breitenbach valley. Two more species [*S. (N.) brevidens, S. (N.) urbanum*] were only found as single pupae in the headwater section U.

Species Distribution in Space

Despite marked variations from year to year, a longitudinal zonation is recognized for the dominant species, *S. cryophilum*, which is a typical inhabitant of springs and small headwaters, and thus has its highest proportions near the springs in both parts (sections U, A, and B) of the Breitenbach. Larvae could be observed a few decimetres below the mouth of rheocrene springs. With increasing distance from the springs the very closely related species, *S. vernum*, gradually “takes over”, but both species disappear towards T6. Highest absolute numbers occur in the region of T1 and T2. Only very low numbers of the third species of the *S. vernum* group, *S. costatum*, were found in different traps.

The blackfly biocenosis of the upper part of the headwater (part of section U) was clearly dominated by species of the *S. vernum* group. The valley is quite narrow in this section and most parts are shaded by forest. Trailing marginal and submersed macrophytes are rare and the main colonization substrata for blackfly larvae are stones, pebbles, and leaf litter in autumn.

The abundance of *S. (S.) ornatum* increases towards the mouth of the stream and is influenced by the population in the River Fulda (see below). In years with large populations of *S. ornatum*, this species clearly dominates the lower 1.5 km of the stream during summer, while it is rare, or even absent, in the uppermost headwater sections. This longitudinal zonation is probably also connected to the availability of suitable ovipositon substrata: marginal grass leaves that hang into the water, so that the female can land on the dry surface and lay its eggs on the leaf just below the water surface. This type of substratum is predominantly found in unshaded stream sections and is therefore rare in the shaded parts of section U (Timm, 1994; Reidelbach, 1994). In addition, larvae of *S. ornatum* mainly colonize trailing leaves and stems of marginal and submersed grasses.

P. tomosvaryi is distributed rather evenly along the stream, with an indistinct peak from the lower part of section U down to the reach of T1 and T2.

Species Distribution in Time and Life Cycles

There are marked abundance variations of all stages during the year, ranging from zero to an estimated 30 000 larval individuals m^{-2} stream length (Reidelbach, 1994). A high degree of life cycle synchrony is found among the abundant species in winter and spring. They all overwinter as the larval stage, pupate in spring and emerge in late April and May. Low numbers of adults of the *S. vernum* group also appear from March to June. This leads to the situation that the stream is almost abandoned by larvae from the end of May to mid June. In some years and sections of the stream, this colonization gap is very striking and may last for several weeks (Figure 7.116, compare spring 1984 and 1985). Based on the low numbers of pupae and adults, this pattern also seems to apply to the rare species in the stream.

Prosimulium tomosvaryi Temporal and life cycle differentiations are distinct only for *P. tomosvaryi*. The eggs of this species are laid in May and June on wet mosses just above the water level and undergo a dormancy which lasts for several months until October or even January (Zwick and Zwick, 1990; Timm, 1993b). This means that *P. tomosvaryi* is virtually absent from the stream throughout the summer. Larvae develop quite synchronously over the winter (Figure 7.117) until they gather on stable plant substrata to pupate in April and May. The average growth rates of larval populations of *P. tomosvaryi* during December and January were calculated to 1.4–2.2% body length day^{-1} and were considerably higher than those of the syntopic *S. vernum* group larvae (0.1–0.9% day^{-1}, Reidelbach, 1994).

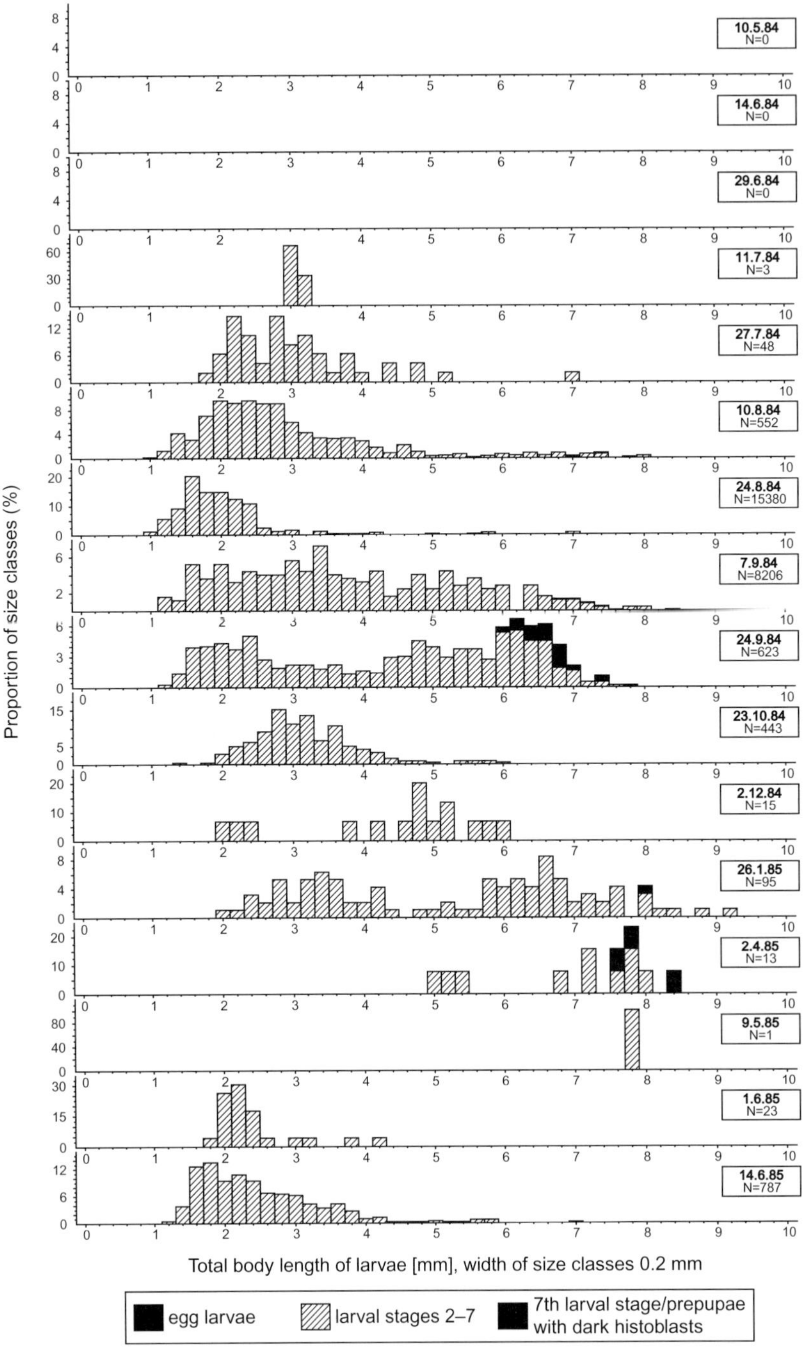

Figure 7.116 Larval growth (in terms of frequency distributions of body length) of *Prosimulium tomosvaryi* developing at the downstream end of stream section U from hatch (October) to final instar (May, June). Larvae were sampled using standardized artificial substrata.

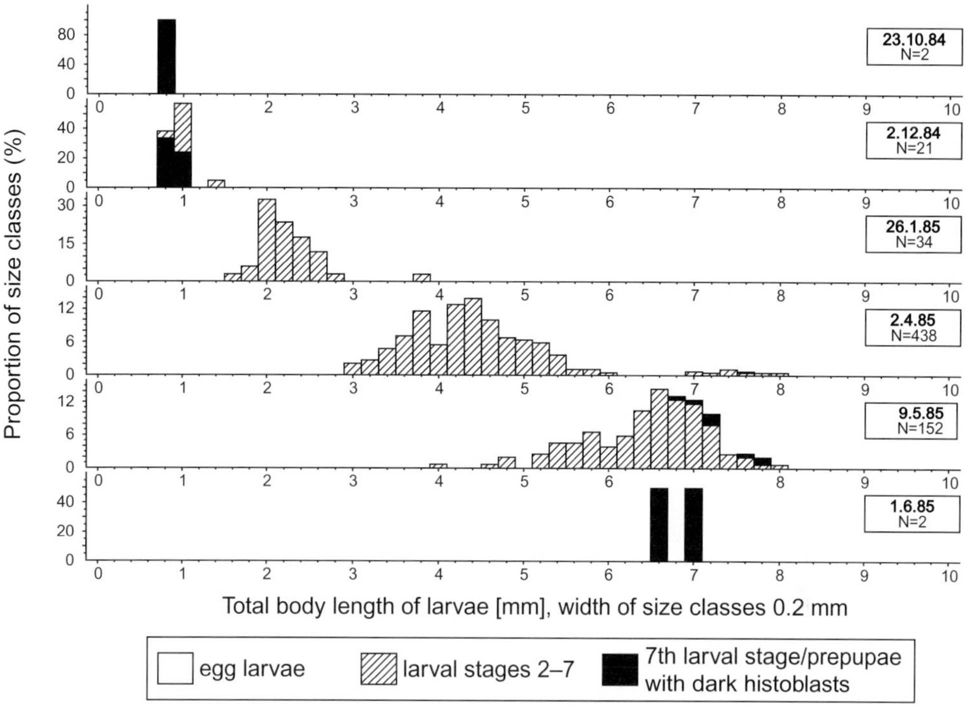

Figure 7.117 Larval growth (in terms of frequency distributions of body length) of the *Simulium ornatum* group developing in stream section C (downstream T2) over the course of one year; first instar larvae and some other very small larvae could only be identified to genus level and are thus not included in the diagram; note the occurrence gap in early summer 1984 and the widely varying numbers of specimens in the samples.

In winter, *P. tomosvaryi* is very abundant in the lower part of section U and replaces *S. ornatum* (see also below). Here the stream is mainly unshaded, and the trailing, durable, leaves of *Juncus* sp. are frequent and used as a preferred substratum by *Prosimulium* larvae (Reidelbach, 1994).

Although the life cycles of the rare species are not completely known, *P. tomosvaryi* is probably the only univoltine blackfly species in the Breitenbach, and the only one with a dormant period in its life cycle (eggs of *S. noelleri* are dormant during winter and thus withstand freezing and desiccation. In the Breitenbach valley this species exclusively inhabits artificial habitats and substrata such as weirs and pond outlets).

The life history of *P. tomosvaryi* is relatively stable, adapted to low water temperatures and avoiding desiccation periods (late summer and autumn), as well as parasitism of theadults by water mites, which do not occur before July, but affect all other species (Christl, 1991; Reidelbach, 1994).

The remaining abundant species have a second generation per year, with larval development during summer and emergence in September and October.

***Simulium (Nevermannia) vernum* Group (*S. vernum, S. cryophilum, S. costatum*)** The life cycles of *S. cryophilum* and *S. vernum* seem to be very similar with very extended flight and oviposition periods, and possibly retarded embryonic development (compare Timm, 1993a), resulting in highly desynchronized larval development.

In most cases, the autumn numbers of emerging adults were considerably higher than in spring. This might be due to higher losses in the larval population during winter, which could be a result of competition with *P. tomosvaryi* larvae, which are absent in the summer. This is particularly true for the stream sections A and B where the abundances of *P. tomosvaryi* and the *S. vernum* group are high but those of *S. ornatum* low. Numbers of emerging individuals were similar for both generations in trap P2, located about 1400 m upstream the Georgsborn, where *P. tomosvaryi* is rare.

Desynchronization is so marked that individual adults of *S. cryophilum* even appear in winter, between November and February. It is likely that these individuals do not reproduce. The life cycles of *S. vernum* and *S. cryophilum* probably represent an opportunistic, risk-spreading strategy, suitable for stream sections disturbed by temporal desiccation (parts of section U) or spates.

***Simulium (Simulium) ornatum* Group (*S. ornatum, S. trifasciatum*)** The population dynamics of *S. ornatum* in the Breitenbach are a quite intricate. From June or July in many years, very large numbers of young larvae occur in the section below trap T2 (Figure 7.117). This desynchronized larval cohort leads to heavy pupation and emergence in September and October (Figure 7.118). The first occurrence and the abundance peaks shifted with time, from the mouth of the stream towards the upper sections. This was shown for all stages (egg masses, larvae, pupae, emerging adults) in 1984. Together with the presence of many young and gravid females in the emergence traps without corresponding numbers of males or pupae in the stream, it wase clear that females originating from the River Fulda invaded the Breitenbach valley in large numbers and oviposited on trailing vegetation in the stream.

S. ornatum was the only species that was found in both the Breitenbach and the Fulda. Interestingly, the large numbers of adults in the autumn developing from the allochthonous egg masses only produce small numbers of eggs, and consequently a small population of overwintering larvae. Adults emerging in the spring produce an early summer larval cohort, mainly pupating in July and emerging in July and August. This small, early summer cohort and its possible offspring overlap to a great extent with the invading large cohort, so that the fates of the two cohorts over the following winter cannot be distinguished.

It is therefore unclear whether there is a small autochthonous population of *S. ornatum* with two generations per year, to which a large allochthonous cohort is added in the summer. Eggs and young larvae originating from this latter cohort

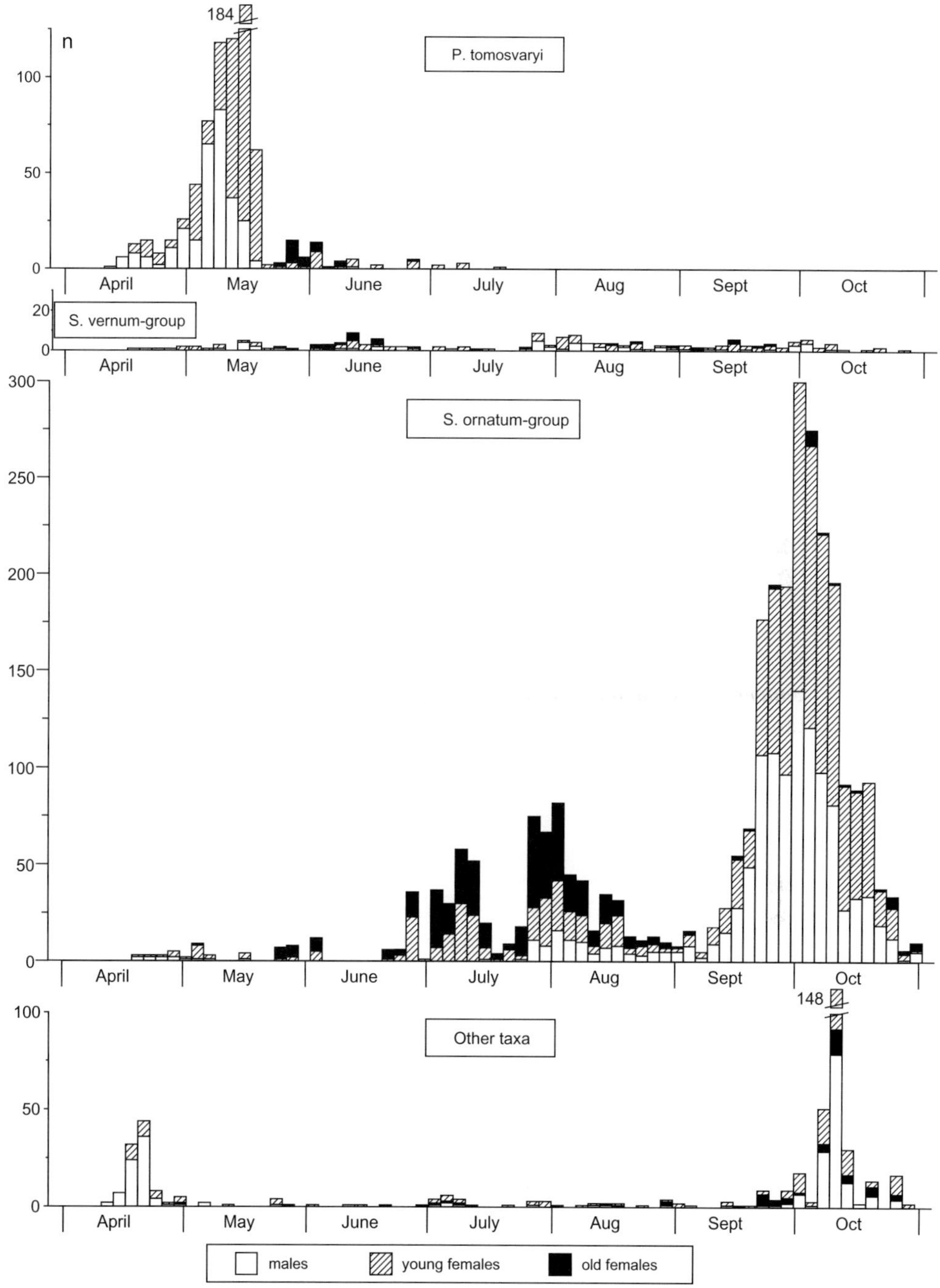

Figure 7.118 Numbers (*n*) of four taxa of adult blackflies caught in emergence T6 in 1984, summarized by three-day periods. "Old females" (black) are females that did not hatch under the traps but emerged elsewhere and flew into the traps as adults. *S. vernum* group males: hatched = *S. vernum*; open/white = *S. cryophilum*.

in late autumn could suffer from high mortality due to low temperatures, competition or other reasons, so that the presence of descendants of the Fulda population remains episodic. Larvae of the autochthonous population would have to live for seven to eight months over winter, longer than for larvae of *P. tomosvaryi* or the *S. vernum* group, which hatch considerably later.

It seems more likely that the overwintering larvae originate mainly or completely from the large allochthonous summer cohort. This could mean that the occurrence of *S. ornatum* in the Breitenbach is completely dependent on regular input from the River Fulda.

The life cycle of *S. ornatum* may be interpreted as an *R*-strategy with strong tendency to disperse and tolerate differing habitat conditions.

The second species of the *S. ornatum* group, *S. trifasciatum*, was rarely found as pupae and adults, Except once at T1, when *S. trifasciatum* contributed more than fifty percent of the total *S. ornatum* group adult males, its numbers were very low compared to *S. ornatum*. This is a striking example of the greatly varying numbers of individuals and proportions of a species between years, in the absence of any apparent reason. It is likely that the life cycle of *S. trifasciatum* in the Breitenbach closely resembles that of *S. ornatum*, with emergence in April/May and August to October, but *S. trifasciatum* can be assumed to be truly autochthonous in the Breitenbach. It should be noted that *S. ornatum* and *S. trifasciatum* regularly occur together in streams and no consistent ecological differentiation has yet been found between the two species (e.g., Jedlicka, 1978) except that *S. ornatum* is more tolerant to pollution and occurs not only in small streams but also in large rivers.

No consistent evaluation of the life cycles or strategies is possible for the remaining species (*S. argyreatum*, *S. monticola*, *S. aureum* group, *S. ruficorne* group) because of the low numbers of specimens found.

Diurnal Emergence Periodicity

The eclosion of adult black flies exhibits a pronounced diurnal rhythm with emergence during daylight (Reidelbach, 1994). This periodicity, as well as the number of individuals emerging on consecutive days, seems to be influenced by abiotic factors, the most of which is presumably the intensity of global radiation. This is the only parameter that mature pupae can perceive which reflects weather conditions in their future terrestrial environment (Christl, 1991; Reidelbach, 1994). Water temperature is only weakly linked, and with considerable delay, to terrestrial variables such as air temperature, and so on.

Upstream Movements of Adults and Dispersion

In different years it was shown that many gravid and host-seeking females, and some males, flew for considerable distances upstream along the Breitenbach (Figures 7.118 and 7.119). In particular, conspicuous numbers of the three potamal species that undoubtedly emerged exclusively from the River Fulda were captured along the uppermost section of the Breitenbach, about 3.5 km from the River Fulda, by handnet or in trap P2 about 1400 m upstream the Georgsborn (Reidel-

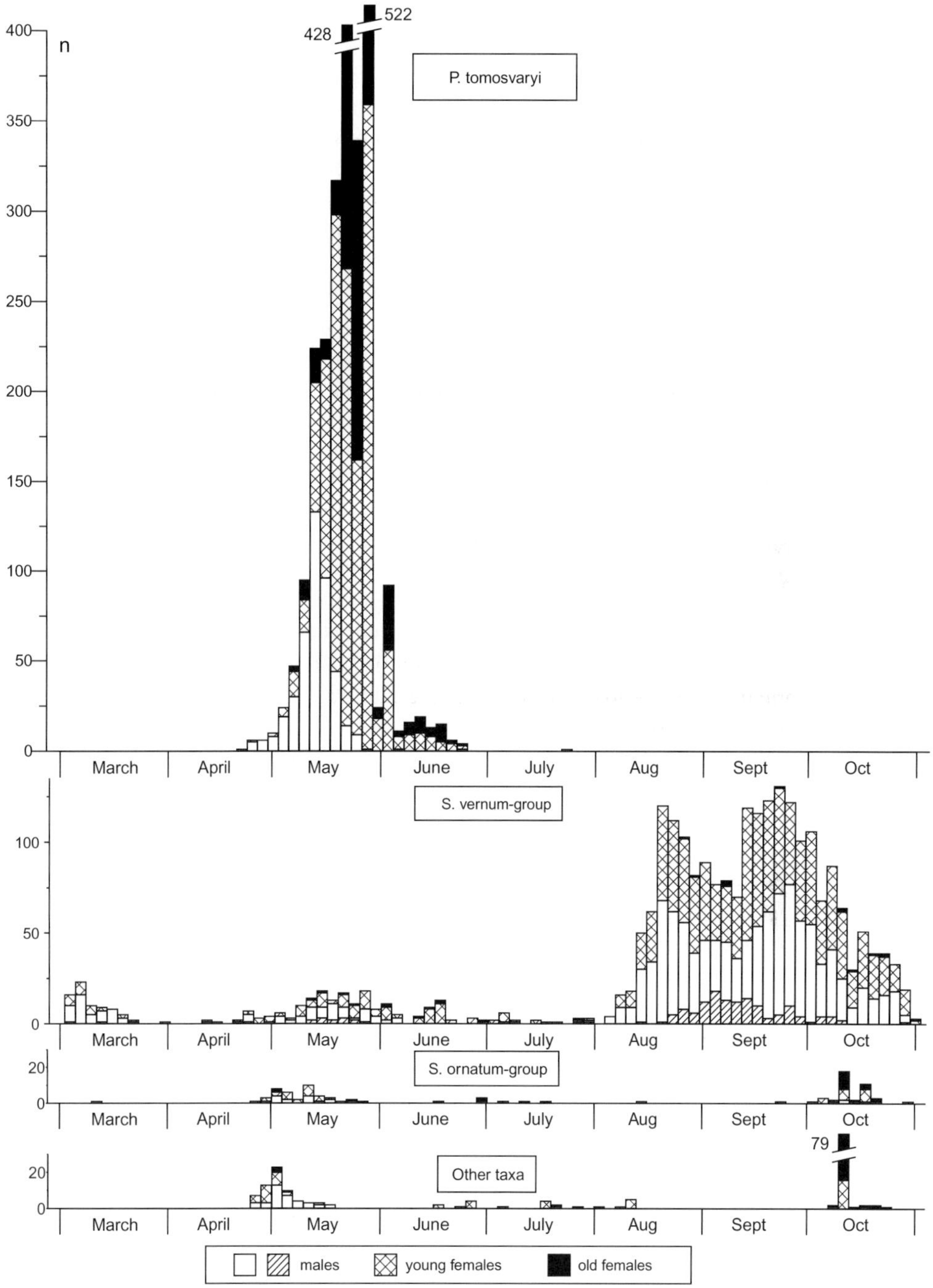

Figure 7.119 Numbers (*n*) of four taxa of adult blackflies caught in emergence T1 in 1986, summarized bythree-day periods. "Old females" (black) are females that did not hatch under the traps but emerged elsewhere and flew into the traps as adults. *S. vernum* group males: hatched = *S. vernum*; open/white = *S. cryophilum*.

bach and Christl, 2002). Egg masses of *S. ornatum*, found in the lower part of section U on natural and artificial substrata in July and August, were probably oviposited by females originating from the River Fulda.

Remarkably high numbers of engorged, gravid and post-oviposition females of *P. tomosvaryi* were mainly found in T1 (Figure 7.119). With respect to the temporal shifts of emergence in different traps, this indicates that females from lower sections of the Breitenbach flew upstream and apparently gathered around T1, where the valley narrows and bends (compare Plecoptera; Zwick, 1990).

Being active long-distance fliers, it may be a common or even general strategy of blackflies to disperse from their original habitat and colonize other running waters where substrata and oviposition sites may be adequate, but conditions for larval and pupal development may not be optimal. This could lead to the temporary occurrence of a species, for one or a few generations or seasons, in a certain water body.

Due to above-average mortality from suboptimal conditions the species possibly vanishes after some time. Disturbances (spates, droughts, extreme temperature events, human impact) may play important roles in this extinction process, as well as in the creation of unoccupied ecological niches.

Such an opportunistic colonization strategy is indicated by the Breitenbach emergence data and by several other long-term emergence studies (Kazimírová, 1981; Chmielewski and Hall, 1993; McCreadie *et al.*, 1994). Collections of aquatic stages are generally unsuitable for such interpretations because rare species are easily overlooked due to the patchy distribution of larvae and pupae and discontinuous sampling.

7.6.4.6.8 **Psychodidae (Moth Flies)**

Rüdiger Wagner

Adult Psychodidae (*sensu lato*) are small to medium-sized Nematocera. Their relationship to other families remains unclear. Members of three subfamilies occur in the Breitenbach: the Trichomyiinae, Sycoracinae and Psychodinae. The Trichomyiinae and Sycoracinae that have occasionally been found in emergence traps appear "naked", that is, with only few setae on body and wings. Trichomyiinae larvae live in decaying soft wood (xylophageous), a rare substratum in the Breitenbach. The asselliform Sycoracinae larvae usually prefer calcareous water and adults are restricted to biotopes with Amphibia and Reptilia; females must suck the blood of these vertebrates to develop eggs.

The Psychodinae are the "typical" moth flies; their body is covered by a dense covering of black, brown and white hairs and setae, forming specific color patterns. Larvae are recognizable by their elongate heavily sclerotized body, usually divided into 26 ring-shaped pseudosegments (annuli), each with one or more "tergal plates". Two rings form the three thorax and the first abdominal segments, respectively, three rings the second to seventh abdominal segments, and the siphonal segment contains the remnants of the remaining abdominal segments. Due to the typical number, shape and position of setae on annuli and segments, most larvae can be determined to species level and may therefore be useful for the ecological assessment of biotopes (Vaillant, 1971, 1982; Wagner, 1979, 1980).

Larvae of most Breitenbach species are shredders, or they collect fine particulate organic matter deposited along stream banks or in moss carpets. Larvae of the tribe Psychodini prefer fouling material, and some are abundant in sewage treatment works. Larvae of the Pericomini only occur in the presence of sufficient oxygen. In particular larvae of the genus *Berdeniella* occur exclusively in moss carpets of cold mountain streams. The Telmatoscopini usually have intermediate oxygen demands and prefer leaf packs (habitat and food). Pupae are indistinguishable; they live near the larval habitats but in drier situations.

Sixty-one species from 20 genera have been reported from the Breitenbach (Appendix), but only about a dozen are of ecological importance, due to high abundance, either in particular years, at individual sites, or in certain microhabitats. Several species new to science were discovered from the Breitenbach (Wagner, 1973, 1975, 1978).

The life cycle is univoltine for most species, in particular for the abundant *Berdeniella* taxa, the most characteristic moth flies inhabiting the stream. However, *B. manicata*, which prefers the downstream section (T6) and is characteristic for larger streams, is bivoltine. *Berdeniella* larvae are preferably found in the partly or totally inundated moss carpets on stones at the shore line, which provide a moderate transition from terrestrial to aquatic conditions under particular discharge regimes. Species are members of different groups and can be distinguished by the length of the body setae even in the field. Although respiration is accomplished via the terminal open stigmata, larvae survive inundation for several days or even weeks. If submerged, four elongate prolongations with elongate hairs (flabellar apparatus) hold an air bubble that acts as respiration device (physical gill).

Other species with two generations per year are *Pneumia trivialis* and *Pericoma blandula*. As in other bivoltinous aquatic insects size and "fitness" (= number of offspring per female) of the autumn generation of *P. trivialis* appeared reduced compared with the spring generation (Wagner, 1973).

All Psychodini species are probably multivoltine with generation times between one and several weeks, depending on species and environmental conditions. Some Psychodini, such as the polyvoltinous *Psychoda parthenogenetica* were collected in every month of the year. Emergence of the Telmatoscopini and Pericomini is more or less restricted to a particular time of year; most species are on the wing from May to July.

Data from four sites and 11 years showed that variation along the stream (spatial) and between discharge patterns (temporal; cross references) occurred. In the source and the upper reach the taxocoenose is dominated by *Pneumia mutua* and *Ulomyia fuliginosa*; in moss carpets along the entire stream the genus *Berdeniella* is dominant, with *B. illiesi* in the T1–T2 region, *B. unispinosa* mainly in the T2–T4 region, and *B. manicata* in the downstream section (T4–T6). *Pneumia trivialis* occurs along the entire stream.

Preference for particular discharge patterns is low, however permanent low flow was favored by most species (Section 8.4).

7.6.4.6.9 Ptychopteridae (False Crane Flies)

Peter Zwick

The general habitus of the Ptychopteridae resembles that of crane flies (Tipulidae and Limoniidae *sensu lato*), but they are easily distinguished. The Ptychopteridae have an appendage, the prehalter at the base of the halter, and longitudinal folds in the wing membrane resembling veins. Unlike in crane flies, their legs are firmly attached to the thorax.

Adults are up to 12 mm long, with dark bands or spots on their wings. The body may have a green or blue metallic sheen with yellow marks, or be patterned in black, orange and yellow. The running-water species are essentially black. The head lacks ocelli and the firm antennae are elongate. The large labellum is equipped with pseudotracheae which are used to absorb sugary Homoptera secretions from leaf surfaces (Shcherbakov and Lukashevich, 2005). The sugar is metabolized; our unpublished tests with C^{14}-labeled glucose retrieved respired CO_2 (Kyrieleis, 2000; Zwick, unpublished data). The abdomen is elongate, usually club-shaped in males; females have a slender blade-like ovipositor.

Of the eight German species (Zwick, 1999b) several that develop in lentic biotopes or are euryecious have larval and pupal respiratory organs that are longer than their bodies, while the same organs are less than the length of the body in running-water species. *Ptychoptera albimana* is an opportunistic species occasionally found in the Breitenbach, but *P. paludosa* is a common and widespread inhabitant of small streams. The life cycle is univoltine (Wolf and Zwick, 2001), *P. paludosa* mainly emerges from May to July.

In shallow sections of the stream, the eucephalous larvae shift considerable amounts of fine organic material in soft fine sediments, from several centimetres deep to the stream surface. Larvae stand vertically, head down, breathing atmospheric oxygen through stigmata at the tip of a retractable abdominal siphon. With their specialized mouthparts they extract fine particulate organic material from the sediments (Wolf, Zwick and Marxsen, 1997). After a short gut passage time considerable amounts of organically rich feces are released on the sediment surface, the FPOM becoming available to other members of the stream community. Due to water flow, such habitat engineering is not obvious because the feces do not accumulate.

Moults occur on the sediment surface. Larvae overwinter completely submerged and are found in Surber samples from the deep stream bottom (Zwick, unpublished data). In spring, larvae attain their fourth instar, feed extensively and pupate on the sediment surface, into which the pupae then dig. They also stand vertically in the sediment, head up, the spiracle at the tip of the much lengthened, right prothoracal horn floating in the surface film.

7.6.4.6.10 Brachystomatidae

Rüdiger Wagner

Only recently changes in the higher-level phylogeny and classification of the Empidoidea resulted in a change in the systematic position of genera and tribes of the former Empididae. In particular, the position has changed for some genera that were provisionally grouped in the tribe Trichopezini, later subfamily Trichopezinae (Sinclair, 1995) in the Empididae. These genera and species are now placed in the subfamily Trichopezinae of the family Brachystomatidae (Sinclair and Cumming, 2006).

The Trichopezinae are of variable size but, because they are rare, very little is known of their biology. Adults have been repeatedly found around sources and running waters, and adults repeatedly seen to feed on adult insects, holding their prey with their front femura and tibiae (Ulrich, 1981). Male genitalia are bean-shaped in genus *Heleodromia*, those of *Trichopeza* species asymmetrical.

From the adult habitats it can be deduced that larvae occur along springs and streams, most probably in the water–land transition zone. The univoltine life cycle includes an egg stage, most probably three larval instars, pupa, and adult. Adults are on the wing in spring and summer. A list of the Breitenbach species is provided in the Appendix.

7.6.4.6.11 **Empididae (Dance Flies)**

Rüdiger Wagner

The Empididae include the subfamilies Hemerodromiinae and Clinocerinae (Sinclair, 1995) with aquatic larval development, and the terrestrial Empidinae. Twenty-seven aquatic species have been found in the Breitenbach, several of which were new to the German fauna at the time of discovery. Many terrestrial dance flies were collected in the traps but are not dealt with here.

Larvae and adult Empididae are predators. Adult Hemerodromiinae have a mantoid habit and catch prey (e.g., Protura, Chironomidae, Mycetophilidae) with their enlarged front legs, adopting a sit and wait strategy. Males also use their front legs for competition fights for the best mating places (Joost, 1980). The front legs of adult Clinocerinae are narrower but also covered with spines and bristles. Flies are active hunters, catching emerging aquatic insects, or fishing larvae from just below the water surface, from moss carpets or biofilms.

Empididae larvae are usually inconspicuous, pale, elongate, with an incomplete head capsule and 11 body segments. Clinocerinae larvae are larger than the Hemerodromiinae. Prolegs are weakly developed in *Chelipoda* and *Dolichocephala*, stronger in *Clinocera* and *Wiedemannia*. The ecological demands of most larvae are not precisely known, but they occur in a variety of stream habitats, in gravel beds as well as in moss carpets. Larvae have frequently been found in crowds of simuliid larvae on which they prey, but they also feed on chironomid larvae and other aquatic invertebrates. Empididae larvae may invade the pupal cases of caddis flies. They eat the pupa and develop in the case. Hemerodromiinae pupae have remarkable, extremely elongate, spiracular processes that are used to anchor pupae between substratum particles of stream sediments, and for respiration.

The life cycle includes an egg stage, three larval instars, pupa, and adult (Sinclair and Harkrider, 2004). Adults are on the wing from May to October. Most taxa are univoltine but *Chelifera pyrenaica* has two generations per year, which differ in individual size and weight (Wagner and Gathmann, 1996). Two emergence peaks, for example, of *Chelifera precabunda* and *C. precatoria*, may indicate two distinct generations. Life cycles of *Clinocera stagnalis* and *C. wesmaeli* appear desynchronized; adults were collected all the year round, even in winter.

The Breitenbach empidid community is typical of undisturbed small streams in low mountain areas in Europe (Wagner, 1983). Some species are spread all over Europe; the range of *C. concinnicauda* includes Mongolia, *Clinocera stagnalis* and *D. irrorata* even occur in the Nearctic region.

A large variation in adult number and size in emergence traps along the stream and in different years is usual. At most, in 1982 about 3000 specimens of *C. trapezina* were collected in T5, but fewer than 130 in the other traps. Adult abundance varied along the stream for several species: *C. pyrenaica* was most abundant at T1, *W. bohemani* and *C. trapezina* prefered T2. *W. lamellata, Dolichocephala irrorata, D. oblongoguttata, Clinocera madicola, Chelifera precabunda* and *C. precatoria* had their highest abundance at the downstream sites T4 and T6. Species with low variability along the Breitenbach have much larger distribution ranges than the size of the stream, which probably explains why distribution patterns along the stream are rare. Interannual variation in abundance was not affected by discharge for most species. Only numbers of *D. irrorata* increased significantly under low flow conditions; further details are given in Wagner and Gathmann (1996).

7.6.4.6.12 **Dolichopodidae**

Rüdiger Wagner

Ten subfamilies, and more than 130 species in 24 genera, were determined from the emergence collections along the Breitenbach since 1980. In addition, about 12 more species from the subfamily Medeterinae, the larvae of which are undoubtedly terrestrial were found, about one-third of the extant German fauna (Bellstedt, Stark and Meyer, 1999). Because the larval ecology of most European Dolichopodidae remains almost unknown, an empirical habitat compilation of most species is given in subfamiliar sequence. Information provided here is mainly based on approximately 60 000 specimens collected in emergence traps (Dr. Hans Meyer, Kiel, personal observation) and on empirical information from Caspers and Wagner (1982), Meyer and Heydemann (1990), Negrobov (1991), Pollet (2000) and Vaillant (1978).

It is presumed that larvae and adults are predators. Many adults have raptorial front, and sometimes middle, legs covered with elongate spines or small knob-shaped setae. Prey is caught and held with the front legs. Dolichopodidae larvae are usually pale, with a reduced head capsule and eleven body segments. There are no prolegs but creeping welts on the ventral side, and the distal end of the body is slanted and possesses a pair of spiracular openings; a unique combination of morphological features. The ecological demands of larvae remain unknown. However, some have been found in aquatic environments, for example, in moss-carpets in streams, in bank sediments of streams and rivers, or in the bank sediments of great lakes such as Lake Constanz (Wantzen, personal communication).

Dolichopodinae

Twenty-nine *Dolichopus* species have been recorded from emergence traps along the Breitenbach. *D. nigricornis, D. popularis,* and *D. ungulatus* were the most abundant and were classified as eurytopic, euriecious, and sivicolous species. Vaillant (1978) included *D. popularis* and *D. ungulatus* as "aquatic species", with larvae occurring along rivers, in swamps, and in moist soil. However, the most abundant species in the green houses, *D. nigricornis,* was not mentioned as aquatic. Of the remaining species, *D. arbustorum, D. argyrotarsis, D. campestris, D. latilimbatus, D. linearis, D. mediicornis, D. nitidus, D. nubilus, D. plumipes, D. signatus, D. subpennatus, D. trivialis,* and *D. urbanus* are classified as ripicolous, or associated with aquatic habitats (Pollet, 2000; Vaillant, 1978). *D. atripes, D. notatus,* and *D. vitripennis* are tyrphopilous, and the remainder are eurytopic, euriecious, silvicolous, or occurring in marshes. *N. nodicornis,* another comparatively rare species, occurs in riparian habitats, swamps, and moist soil (Vaillant, 1978). The majority of *Hercostomus* species are associated with aquatic habitats: *H. aerosus* (Vaillant, 1978), *H. celer, H. chalybeus, H. chetifer, H. longiventris,* and *H. nanus. H. chrysozygus* and *H. fulvicaudis* are also found in marshlands. The others are associated with forest and classified as silvicolous. *Poecilobothrus* species are either silvicolous (*P. comitialis*), or eurytopic, and ripicolous (*P. nobilitatus*). All *Sybistroma* species are associated with stream banks and ponds: *S. crinipes, S. disciples,* and *S. sphenopterus* are ripicolous; *S. obscurellus* is known from swampy areas, moist soil, and forests. Most *Dolichopus* species are on the wing in early and high summer, *Hercostomus* spp. fly some weeks later. *Sybistroma* spp. seem to prefer high and late summer; some specimens were found in September.

Sciapodinae

Although *Sciapus platypterus* was the most abundant Sciapodinae in a small forest stream (Caspers and Wagner, 1982), both species collected on the Breitenbach are classified as silvicolous, with an early summer flight period.

Xanthochlorinae

X. ornatus and *X. tenellus* are eurytopic, euryiecious, and silvicolous.

Rhaphiinae

Several species are regularly found along springs, streams, and ponds: *Rhaphium brevicorne, R. elegantulum, R. ensicorne, R. fasciatum, R. fissum, R. gravipes, R. micans, R. penicillatum, R. quadrispinosum,* and *R. riparium.* However, *R. fascipes, R. micans,* and *R. riparium* prefer transition habitats (ecotones) between aquatic and terrestrial environments, the remaining taxa are euriecious, and silvicolous. Nematoproctus distendens is often found in riparian habitats.

Peloropeodinae

These are ripicolous and silvicolous (*A. flaviventris*), or eurytopic and silvicolous (*C. molliculus*); both emerge in June and July.

Enliniinae

Achalcus cinereus and *A. flavicollis* are related to aquatic habitats, reed marshes, swamps, or moist soil. *A. flavicollis* was classified as aquatic (Vaillant, 1978).

Hydrophorinae

Two genera and two species were collected. *Liancalus virens* occurs in the environment of springs, on stream and river banks, and in hygropetric zones (Vaillant, 1978). *Hydrophorus litoreus* is regularly found in riparian habitats.

Neurigoninae

These are characteristic forest insects (silvicolous), their larvae are not aquatic.

Sympycninae

Sympycnus spp. are either euriecious (*S. pulicarius*), or found in marshland and on the banks of streams and rivers (*S. aeneicoxa*). Most *Syntormon* spp. are also frequently found on banks of springs, streams and rivers: S. *aulicus, S. denticulatus, S. monilis, S. pallipes, S. pumilus, S. sulcipes,* and *S. tarsatus. Syntormon metathesis* and *S. monilis* occur in marshlands; *S. filiger* prefers saltmarshes (halotopophilous) and *S. sulcipes* is associated with heathland and moors; *S. tarsatus* is pratinicolous. *Lamprochromus bifasciatus* is rare; its preferred habitat is described as ripicolous.

Campsicnemus adults are also rare along streams; they are conspicuous in early spring and autumn, when other Dolichopodids are rare. *C. curvipes* is eurecious; it was observed in temporary small water bodies (Vaillant, 1978). *C. scambus* is similarly classified and was found in marshes (Vaillant, 1978). *C. loripes* is classified as eurecious, but is also associated with swamps, and wet soil (Vaillant, 1978). *C. lumbatus, C. marginatus, C. picticornis,* and *C. pumilio* are mainly ripicolous, but show transitions to swampy and marsh habitats.

Teucophorus species were also rarely collected; they are termed silvicolous, although adults may frequently be found in the environment of swampy springs and small streams.

Diaphorinae

Diaphorinae are ecologically flexible. Several *Argyra* species seem to be restricted to more terrestrial habitats: *A. argentina, A. atriceps, A. auricollis, A. grata, A. hoffmeisteri,* and partly *A. diaphana* and *A. leucocephala.* All other species are ripicolous. *A. diaphana, A.* sp. nov., *A. ilonae, A. leucocephala,* and *A. setimana* are eurytopical, preferring the banks of streams, rivers, or standing waters (Vaillant, 1978). *A. discedens, A. elongate,* and *A. vestita* are restricted to reed marshes and small temporary water bodies (Vaillant, 1978).

Diaphorus hoffmannseggi and *D. winthemi* prefer riparian habitats; *D. nigricans* inhabits heathlands and moors (tyrphophilous); *D. oculatus* is a forest species.

Few *Chrysotus* spp. occur in aquatic habitats; *C. cilipes, C. gramineus, C. laesus, C. viridifemoratus* live in swamps and moist soil (Vaillant, 1978), *C. femoratus* and *C. obscuripes* in heathlands and moors. *C. melampodius, C. viridifemoratus,* and *C. suavis* are ripicolous. *C. blepharoceles* and *C. cilipes* occur in grasslands; *C. femoratus, C. gramineus,* and *C. neglectus* are euriecious; *C. collini, C. cupreus,* and *C. laesus* prefer forests; *C. palustris* and *C. pulchellus* are known from sandy habitats, such as coastal dunes.

The life cycle includes an egg stage, probably three larval instars, pupa, and adult. Most adults are on the wing from May to October, but some *Campsicnemus* species may be found particularly in late autumn and winter. Although almost nothing is known about life cycle duration and larval habitat preference, most taxa seem to be univoltine. However, emergence patterns of some abundant species indicate a bimodal emergence pattern that may indicate two or more generations per year.

The long-legged fly community of the Breitenbach is typical of undisturbed, small, stream catchments in European low mountain areas. Most species are distributed all over Europe and the range of some species covers large areas, even including the Nearctic region.

Variation in numbers and size of adults in emergence traps along the stream and between years may provide further evidence for the probably aquatic development of individual species if they show similar correlations between environmental variables and abundance as EPT species that are known to be aquatic. Sixty-eight abundant species were analyzed for significant relationships between abundance, year of collection, site and discharge pattern using non-parametric correlation (Spearman's rho), Kruskal–Wallis and U-tests.

Correlation analysis showed a significant ($P < 0.05$) relationship of 22 species with "year", 23 species with "site", and 21 species with "discharge pattern". At first sight the same relationship of 20 species to year and discharge pattern is remarkable, but this is due to the "years" being coded by "discharge pattern"; both are temporal variables.

Part of the the Dolichopodidae showed mainly along stream (spatial) variation, whereas another group of species showed predominantly temporal variation. Five species, *Rhaphium monile* and *R. zetterstedti,* varied simultaneously in space and time (year, trap), and three (*Dolichopus argyrotasus, D. campestris, Hercopstomus cupreus*) varied between years, traps, and discharge patterns.

The species with highest abundance in the upstream region (T1) were *Sympycnus aeneicoxa, Dolichopus signatus, Hercostomus vivax,* and *Xanthochlorus ornatus. Rhaphium riparium* and *Teucophorus monacanthus* had their highest abundance in the mid-stream region (T2, T4). However, 18 species had their highest abundance at the most downstream site (T6): *Dolichopus argyrotarsus, D. campstris, D. linearis, D. longicornis, D. pennatus, D. plumis, D. trivialis, D. ungulatus, Hercostomus celer, H. cupreus, H. nanus, Sybistroma disciples, Syntormon pumilum, Syntormon* sp., *Raphium monotrichum, R. zetterstedti,* and *Anepsiomyia flaviventris.* Only 31 of 68 species showed significant and long-term variation in abundance along the stream.

Twenty significant correlations of species abundance and discharge pattern were detected (17 with $P < 0.05$, 3 with $P < 0.1$) . The abundance of Dolichopodidae was mainly related to "good flow" or "low flow" patterns, both with low discharge variation. Five species (*Syntormon fuscipes, Teucophorus monacanthus, Chrysotus collini, Sybistroma sphenopterus, Raphium commune*) were positively correlated with "good flow". *Achalcus cinereus, Neurigonia quadrifasciata, Syntormon fuscipes, Campsicnemus scambus, Dolichopus picipes, Poecilobothrus nobilitatus, Sybistroma crinipes,* and *Anepsiomyia flaviventris* were positively related to long-term low flow. *Syntormon denticulatum* and *S.monile* were negatively correlated with spring floods after snowy winters (Chapter 8.4).

7.6.4.6.13 **Stratiomyidae**

Rüdiger Wagner

The Stratiomyidae (soldier flies) are slender to robust, usually brown, yellow or orange, sometimes shiny, metallic Diptera. The distinctive features are a closed discal cell in the wing center and an "armed" scutellum, often with one to three pairs of strong spines. Larvae are usually elongate, flattened, comprising a head capsule and 11 body segments. They feed on decaying plant or animal material; some terrestrial larvae are predators. Larvae of subfamily Stratiomyinae are aquatic and some can even survive high salinity. The ecological demands of other taxa are not precisely known (Roskošný, 1973, 2000).

Five species of Stratiomyidae have been recorded from emergence traps on the Breitenbach (Appendix). The subfamily Clitellariinae includes most of the "typical" aquatic taxa, the larvae have an apical chaplet of pinnate setae that keep the open apical stigmata dry. *Oxycera pardalina* is distributed in Europe from Great Britain and Scandinavia to the Caucasus area. Body coloration is variable, and color "forms" have been described as different species. The larvae are hygropetric, preferring moss carpets in clear springs, pristine streams and karst areas, up to more than 2000 m a.s.l. The flight period is May to July, one generation per year.

There is controversy over whether all species of the genus *Beris* Latreille, 1802 are aquatic. There is evidence that *B. clavipes,* and *B. vallata* larvae develop in wet moss and under decaying plant material along springs, streams and rivers. Their distribution covers almost all of Europe. Larvae of *B. chalybata* have been bred in moss; adults are quite common along streams, as are those of *B. strobli* (larva unknown).

7.6.4.6.14 **Rhagionidae**

Rüdiger Wagner

The Rhagionidae (snipe flies) are slender flies with elongate legs and elongate conical abdomen. The body is pilose, light to dark brown. Larvae are hemicephalic, pale elongate, cylindrical, with differently shaped distal lobes. Larval biology is almost unknown, but they are probably predators of other insect larvae and even feed on mucus covered Oligochaeta (Thomas, 1997). The larvae of *Rhagio* species are terricolous and thus not included here. *R. lineola* is most abundant in wood soils; *R. scolopaceus* and *R. tingarius* are also common taxa.

Chrysopilus larvae develop in cold streams and damp soils in stream banks. Larvae and pupae of *C. erythrophthalmus* were found in aquatic moss carpets and in gravel sediments of cold streams (water temperature permanently below 13–14 °C; Thomas, 1978). *C. cristatus* larvae prefer water soaked soils at the edge of streams; *C. asiliformis* is thermophilous.

The flight period lasts from May to July, and the life cycle probably takes one year. The first larval instar has an "ruptor ovi", that is, is nonfeeding. Growth is slow and the life cycle may even last two years. Pupation occurs out of water, on bank areas protected from dehydration (hygrophilous bryophytes, or wet gravel for *C. erythrophthalmus*).

7.6.4.6.15 **Tabanidae**

Rüdiger Wagner

Twenty Tabanidae species (horse flies) have been collected over the years in the traps at the Breitenbach. About 3500 species of these mid-sized to large hematophagous Brachycera are currently known (Kniepert, 2000). The outstanding feature of tabanids is that females bite to drink the blood of wild animals, cattle, and man. However males and females drink honey or eat pollen from flowering plants. Larvae found their way into many different biotopes. Immature stages of several species are still unknown, but most larvae prefer or live exclusively in the vicinity of moist or aquatic habitats.

Species are classified ecologically into four categories (Kniepert, 2000): hydrophilous (*Chrysops* and some *Hybomitra* species), semi-hydrophilous (early larval instars are aquatic, later instars, particularly pre-pupal larvae occur on dried water-beds; this group includes many *Hybomitra* species), edaphic (entire life cycle in soil) and xerophilous species (larvae develop in dry soil). All Breitenbach species are hydrophilous or semi-hydrophilous. *Hybomitra* larvae prefer organic substratum, *Tabanus* larvae mineral soil (Kniepert, 2000).

All Tabanidae species are univoltine in Central Europe; cold climate and poor nutrition may extend the life cycle to up to four years. Large egg masses are deposited on plants, and may be parasitized by some Hymenoptera and Diptera. The first instar larvae molt on the (terrestrial) egg mass; instars 1 and 2 nourish their yolk, instar 3 enters the water. The following instars (up to eleven recorded as adaptation to low temperatures and poor nutrition) prey on insect larvae, molluscs, annelids, and even congeners. Aquatic larvae feed during winter.

Pupation and emergence occur at night. Different horse flies are present at the stream from late May to August. *Chrysops caecutiens* is the first noticeable and abundant tabanid in the year. *Haematopota* species are quite abundant in July and particularly in August. Most other species are comparatively rare.

One blood meal is obligatory for successful egg development (unautogenous species), although some species may develop a first egg batch without a blood meal (autogenous, *Haematopota pluvialis, Heptatoma pellucens*). However, blood meals are obligatory for a second or even third egg batch (Kniepert, 2000).

7.6.4.6.16 Syrphidae

Rüdiger Wagner

The Syrphidae (hover flies) are slender to robust Cyclorrhaphan flies with a characteristic wing venation, that is, the cubital vein ends at the apex or vein R_{4+5}, and an unattached longitudinal vein runs throughout the wing. Adults are often abundant and occur in a wide range of habitats; they are frequent flower visitors and important pollinators (Thompson and Rotheray, 1998).

Larvae are diverse in shape and habit, mainly saprophagous, but also predatory, parasitic and phytophagous. There are three larval stages; the third being distinguished by the posterior breathing organs that mostly project as a short to long, sclerotized, tube-like structure (Rotheray and Gilbert, 1999).

Approximately 6000 Syrphidae species are known worldwide, belonging to three subfamilies: the Microdontinae, Syrphinae, and Eristalinae (Thompson and Rotheray, 1998). Most Eristalinae, but only a few Syrphinae, are aquatic. The Microdontinae are ant-associated and do not seem to include aquatic species (Rotheray and Gilbert, 1999; Stahls *et al.*, 2003).

Most larvae of aquatic syrphids are saprophagous. Larval morphology is modified in many ways for an aquatic life, for example, with extended interspiracular setae to anchor the breathing tube at the water surface, or with retractile breathing tubes (Hartley, 1958, 1961; Rotheray and Gilbert, 1999).

134 Syrphidae species of 51 genera have been found in emergence traps along the Breitenbach (Löhr, 1988, 2004a, 2004b; Appendix). However, most species occurred in low densities so that information on flight period or generation time remains fragmentary.

Surprisingly two species, *Xylota florum* and *X. segnis*, which both prefer forests and have only a low preference for wetlands, were most abundant in the emergence traps (Löhr, 1988). All specimens were several days old (fully sclerotized). Furthermore the percentage of females was extraordinarily high (~90% in both taxa) and most females had mated and were ready to oviposit. It was evident that both taxa had their highest abundance in the traps in years with dry summers, that is, specimens search out and visit the moistest habitats in the landscape.

Field observations and experiments revealed that this (probably artificial) finding depended on the territorial behavior of males and on female oviposition behavior (Löhr, 1991). Females prefer dark and damp hollows with wet ground (e.g., below fallen trees) for oviposition, and males position themselves near such places. Females try to enter such hollows directly. The entrances and exits of the Breitenbach into and out of the emergence traps (closed with swinging black plastic strips) mimic such biotopes and attract males (they take their territorial positions near the entrance) as well as females. Due to their direct flight many females enter the trap in the hope of finding an appropriate oviposition site, and all are captured. Males only enter the trap by chance. This is a clear evidence that a certain portion of particularly airworthy adult insects intentionally enters the traps (compare Empididae, Dolichopodidae). However, in this particular case, the invasion of non-aquatic insects was revealed by careful observation and field experiments (Löhr, 1988, 1991).

7.6.4.6.17 **Lonchopteridae**

Peter Zwick

The Lonchopteridae are a distinctive Diptera group with uncertain relationships to other families. It is usually placed at the margin between orthorrhaphan and cylclorrhaphan Diptera. Characters of the wing venation make the family easy to identify. Wings are narrow and lanceolate with an acute tip, and transverse nerves are almost absent. In males the anal vein ends into the rear wing margin, but in females it returns into the distal part of vein m_{3+4}.

The life cycle of the Lonchopteridae is univoltine and comprises six larval instars, the last differing remarkably from the others in that it possesses six elongate projections, one pair on the prothorax and mesothorax respectively, and a pair close to the distal spiracles (Baud, 1973; Vaillant, 2002). Pupation proceeds within the sheath of the final larval instar, and the puparium resembles the larva. Adults are found in the vicinity of larval habitats along springs and streams and are common in emergence traps.

The ecological demands of most larvae are not precisely known. They are amphipneustic and occur in moist soil in woods and in pastures (Vaillant, 2002). Larvae are known from decaying organic matter, dead leaves, plant debris, under stones on the edges of water courses or semi-aquatic habitats. They are saprophagous, microphagous or, presumably, mycetophagous (Bartak, in Wagner *et al.*, 2008). Unambiguously aquatic habitats have only been documented for two species: *L. nigrociliata* (Duda) and *L. lutea* Panzer (Vaillant, 2002).

At least two of the nine Lonchopteridae known from Germany (Bährmann and Bellstedt, 1988) have been recorded from the traps at the Breitenbach, the most abundant being *Lonchoptera lutea* and *L. tristis*. Vaillant (2002) reported that, within *L. lutea*, two biologically distinct forms with differing larval ecological requirements can be distinguished. Larvae from North Africa and the Southern Alps occur on the banks of rheo- and limnocrenes, where they can be found among decaying leaves and in wet soil. Larvae of the same species from other parts of Europe seem to be “less aquatic”. However, they seem to tolerate short inundation periods.

7.6.4.6.18 **Ephydridae, Muscidae, Fanniidae**

Rüdiger Wagner

These families were poorly represented in the collections, or specialists were not available for further extensive studies. Thus the species lists remain fragmentary (Appendix).

7.7 Chordata

Peter Zwick

The nature conservation maintenance plan for the Breitenbach catchment reports 89 vertebrate species (Chapter 2, Table 2.3). Mammals, birds, and reptiles are not "aquatic" in a strict sense. However, many of them drink from the stream, bathe in it, or feed on plants or animals growing in or along the stream.

Of the five amphibians, *Rana temporaria* Linnaeus, *Bufo bufo* (Linnaeus), *Triturus vulgaris* (Linnaeus), and *T. alpestris* (Laurenti) prefer standing water (e.g., fish ponds, rain-filled puddles in the valley) for spawning and are not usually encountered in the Breitenbach. However, in some winters hibernating *R. temporaria* were occasionally caught under overhanging banks in the lower stream section (T4–T6), or recovered from the stream bottom during benthos samples in reaches A and B. Fire salamanders, *Salamandra salamandra* (Linnaeus), are regularly observed in the adjacent forests, mainly along the uppermost stream section. The viviparous females release their offspring into tiny trickles and spring seeps that often dry out at the end of summer. From there they drift downstream, and individuals were regularly seen in the pothole below gauging station GA. During benthic sampling only a very few were caught further downstream. Salamanders were not normally found in the more downstream area (T3 and below).

Three "fish" species, one member of the Cyclostomata and two Osteichthyes, were recorded from the Breitenbach. In the lower stream sections Becker and students repeatedly encountered young chub (*Leuciscus cephalus* Linnaeus) that swim in from the nearby River Fulda. The Breitenbach provides them with shelter and food.

The Breitenbach harbors a possibly indigenous population of trout (*Salmo trutta fario*). Measurements during early years of our study showed that specimens had a maximum total length of about 22 cm and a maximum body mass of about 130 g. During a spate, two 15 cm long specimens were taken in a drift net exposed at gauging station GA, but the upper limit of trout occurrence is not known. Single trout nests were inadvertantly disturbed during benthic sampling, one in Reach A, another in reach G. The late J. Illies once electrofished immediately downstream from the main road in the Fulda valley. The trout had full stomachs containing exclusively *Asellus aquaticus*, that is, the fish had foraged in the River Fulda. Rainbow trout that escaped from fish ponds in the valley could apparently not survive in the stream.

The lamprey (*Lampetra planeri* Bloch) remains hidden in sandy deposits throughout its lifetime, except during reproduction. A spawning pair was only observed once, in the area of T4, even though the species is not uncommon in the lowest stream reach. The eyeless larvae (ammocoetes) filter-feed on fine organic particles, diatoms, algae, protozoans and detritus for about 6.5 years. Metamorphosis occurs synchronously and is triggered by low temperatures, between June and August. Adults do not feed. By spring, their gonads are fully developed and spawning occurs at 10–11 °C, especially after sudden rises in temperature. The spent animals die, after about half a year of adult life (Holcik, 1986; Salewski, 1991; Maitland, 2003).

7.8
Ecophysiology, Behavior, and Life Cycles of Scrapers

Georg Becker

7.8.1
Crowding Problems and Feeding Strategy of *Tinodes rostocki* (Trichoptera: Psychomyiidae), a Dominant Epilithic Caddisfly in the Middle and Lower Reaches of the Breitenbach

7.8.1.1 Larval Age Structure, Colonization of Natural Substrata, and Intraspecific Competition for Space in *T. rostocki*

7.8.1.1.1 Species Distribution and Typical Larval Retreats

Tinodes rostocki McLachlan, 1878 (Trichoptera: Psychomyiidae) is a typical univoltine inhabitant of the rhithral environment (Botosaneanu and Malicky, 1978; Robert, 2001; Graf *et al.*, 2008). In contrast to Alderson (1969) and Jenkins (1977), who found *T. rostocki* mainly in small, shaded woodland streams, larvae are common in both shaded and unshaded sites of the Breitenbach, and one of the dominant macroinvertebrate species in the middle and lower reaches of this stream (e.g., Illies, 1971, 1978, 1983, Section 8.4). *Tinodes rostocki* larvae colonize the upper surfaces and sides of stones, and build retreats constructed of fragments of mineral or detritus held together with silk, fixed to the stones by silk. Larvae also produce a silken lining on the inner side of the wall.

There are several hypotheses about how larvae align their retreats with the surrounding conditions: they orient their retreats according to food availability (Danecker, 1961; Alderson, 1969), current direction (Tomaszewski, 1955) or intraspecific competitors (Hasselrot, 1993; Alecke, Spänhoff, and Meyer, 2005). Averaged across the year, the diets of *T. rostocki* larvae are composed of 31% epiphytic algae (diatoms 78%, cyanobacteria 19%, green algae 3%), 50% detritus, and a relatively high percentage of mineral particles (11%; Becker, 1990).

At several sites in the Breitenbach, the stones were densely covered with larval retreats throughout the year (Figure 7.120), which as a result of the crowding might induce intraspecific competition for space. Similar territorial conflicts have also been reported for other Psychomyiidae (Alderson, 1969; Hart, 1983; Hart and Robinson, 1990).

7.8.1.1.2 Larval Development

This study investigated larval development of *T. rostocki* from September 1987 to June 1988 at a site in the lower reaches of the Breitenbach, where the stream is about 2 m wide and shaded by trees (alder, beech, oak, pine). The Bunter sandstone substratum was heterogeneous, ranging in size from cobbles to fine sand. During

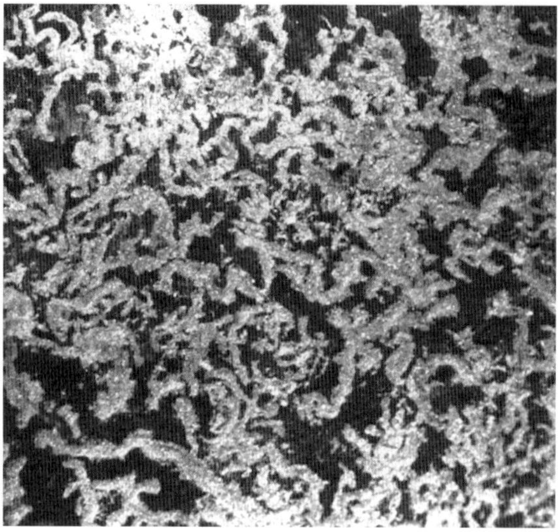

Figure 7.120 High density of *T. rostocki* retreats (instars I–III) on Bunter Sandstone.

the study period water depth ranged from 8 to 20 cm, and current velocity from 0.19 to 0.45 cm s^{-1}. Discharge (Figure 7.121a) was low in autumn (15 and 43 l s^{-1}), increased during winter (60–102 l s^{-1}), peaked during spring (max. 310 l s^{-1}) and decreased through to June (18 l s^{-1}). The mild winter resulted in relatively high water temperatures (Figure 7.121b), with means between 5.8 °C (December) and 6.6 °C (January). Means of 11.3 and 10.9 °C were recorded for September and June, respectively. A minimum diel mean of 3.1 °C was measured in December and a maximum diel mean of 12.4 °C in June.

Tinodes rostocki produced one generation per year in the Breitenbach (Figure 7.121c, Table 7.18). On the first sampling event (15.9.87) the population consisted of first, second, and third instars, with second instar larvae constituting 58% of the population. Fifth instars were found from November to May. The overwintering larval population was composed primarily of third and fourth instars. No developmental pause could be discerned during the mild winter of 1987/88. These observations are similar to those of Alderson (1969), and suggest that there is no photoperiodic control of winter diapause. In May 8% of the larvae had pupated, the remainder were all fifth instars, and by June all larvae had pupated.

7.8.1.1.3 **Larval Abundance**

Larval densities decreased fourfold, from 3100 ± 548 larvae m^{-2} in October 1987 to 762 ± 226 in May 1988 (mean ± SD, $n = 5$; Figure 7.122a). In contrast, biomass (DW) increased twofold from 405 ± 104 mg m^{-2} in October to 801 ± 196 in May (mean ± SD, $n = 5$; Figure 7.122b). During larval development stones were densely covered with retreats and the space requirement of larval retreats increased from

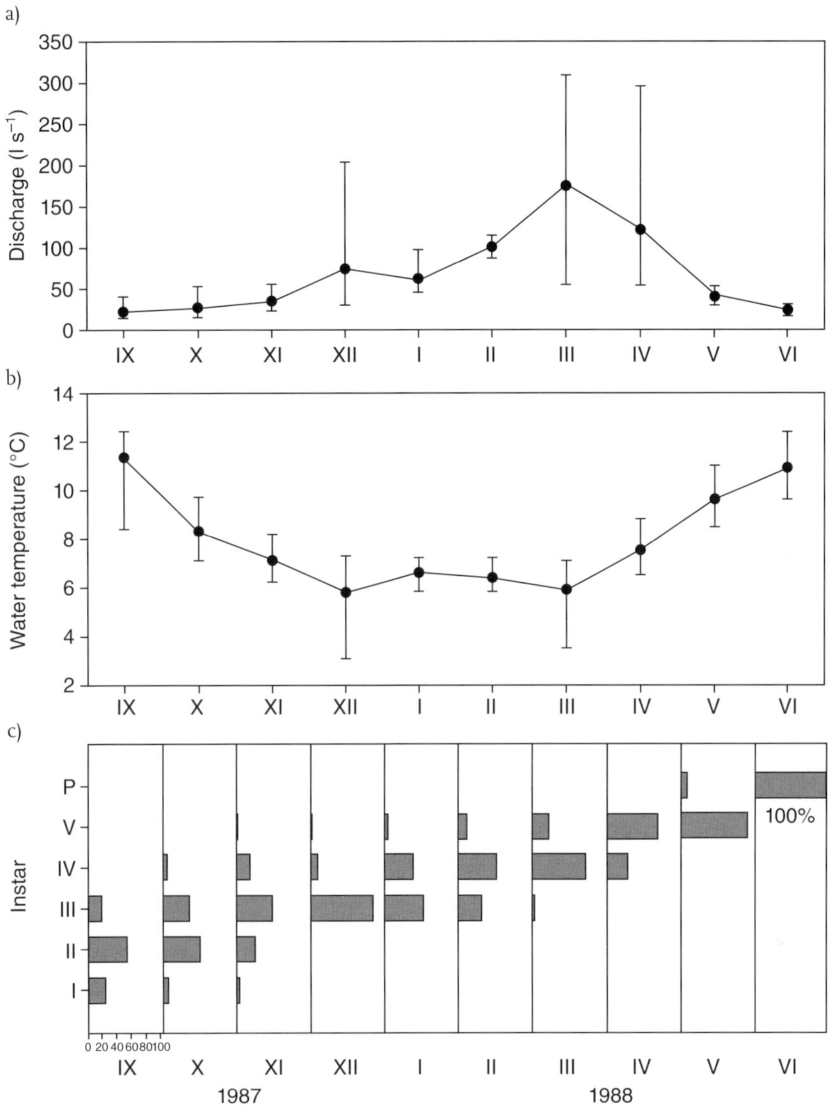

Figure 7.121 Mean values (with ranges) for monthly discharge (a), water temperature (b), and *T. rostocki* age structure (c) from September 1987 to June 1988 in the middle part of the Breitenbach.

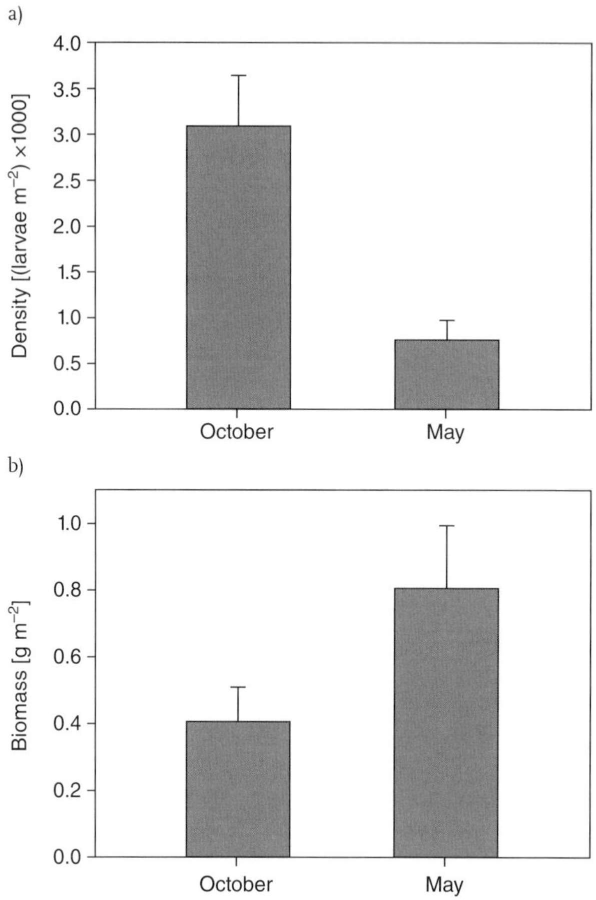

Figure 7.122 Density (a) and biomass (b) of *T. rostocki* larvae in October 1987 and May 1988. The lines show standard deviations.

4 mm^2 (average) for instar I larvae to 250 mm^2 for instar V (Becker, unpublished data). Although *T. rostocki* is a dominant species in the Breitenbach, larvae were found at much lower densities than has been previously reported in other studies of Psychomyiidae. For example, *Tinodes waeneri* reached an average density of 11 500 individuals m^{-2} under favorable lentic conditions (Dall *et al.*, 1984) and Hart and Robinson (1990) reported 2010 *Psychomyia flavida* individuals m^{-2} in a third-order stream towards the end of larval development.

7.8.1.1.4 **Larval Age Structure on Undisturbed and Manipulated Substrata**

To determine larval movements between stones field experiments were conducted at the same site in the Breitenbach. It was investigated whether larval age structure on undisturbed stones differed from that on stones whose natural population had been removed three weeks earlier. This study also determined whether manipulated stones directly adjacent to unmanipulated stones were more densely colonized after three weeks than manipulated stones a minimum of 30 cm from unmanipulated stones.

Observations of laboratory cultures of *Hydropsyche contubernalis* showed that optimal feeding positions were occupied by large larvae, while smaller larvae occupied suboptimal positions at the border of the territories of the larger larvae (Becker, 1987a, 1987b). Only when the larger larvae relinquished their positions to pupate could the smaller larvae occupy them (compare Englund and Olsson, 1990; Englund, 1992). Similarly, Hart (1983) reported that small larvae of *P. flavida* were unable to colonize optimal feeding positions when larger occupants were present. It was therefore expected that mainly smaller larvae searching for more favorable conditions would colonize the manipulated stones.

In September, six, similar-sized, mid-stream, sandstone stones were cleared of larvae, marked, and replaced in their original positions. Three stones were placed directly adjacent to colonized stones, while three others were placed at least 30 cm from colonized stones. The underlying substratum on which these stones were situated had a particle size of less than 1 cm. Every three weeks between October and May (12 occasions), the six marked stones, along with an unmanipulated control stone, were brushed to determine larval densities and age structure. The marked stones were then returned. For each occasion, homogeneity of the age structure on the manipulated stones and a single unmanipulated stone was tested using the chi^2 statistic.

On six of the 12 sampling occasions, significant differences were observed between the age structures of manipulated stones, and unmanipulated controls (Table 7.25). In each such case, contrary to expectation, the manipulated stones supported a higher proportion of larger larvae than the controls. These differences occurred between October and February, when water temperatures were lowest

Table 7.25 Comparison of age structure in *T. rostocki* from manipulated stones and unmanipulated control stones (chi square statistic). Y = Yates correction; ns = not significant.

Sampling date	D.F.	Chi-square	Yates	P
01.10.87	2	4.88		ns
13.10.87	2	11.93		$P < 0.01$
05.11.87	2	3.16		ns
26.11.87	2	31.45		$P < 0.001$
17.12.87	1	11.13	Y	$P < 0.001$
07.01.88	1	5.24	Y	$P < 0.05$
28.01.88	2	37.12		$P < 0.001$
19.02.88	2	38.85		$P < 0.001$
11.03.88	1	0.65	Y	ns
06.04.88	1	1.06	Y	ns
25.04.88	1	3.32	Y	ns
18.05.88	1	0.55	Y	ns

(Figure 7.121b). The average discharge during this period increased from $21\,l\,s^{-1}$ (first half of October) up to $101\,l\,s^{-1}$ (second half of February), reaching a maximum of $203\,l\,s^{-1}$ in December (Figure 7.121a). It can be assumed that stream discharge did not exert any marked influence on the age structure of manipulated stones because significant differences were found between manipulated and unmanipulated stones at both high and low discharges. Considering that, from April onwards, the population consisted principally of either fifth instar larvae or pupae, significant differences would not have been expected in the last three samples. No significant differences were found in the age structures of unmanipulated mid-stream stones, indicating that age structure differences between unmanipulated and manipulated stones were unlikely to be due to within-treatment variation.

7.8.1.1.5 Larval Movements between Substrata

Several reasons have been suggested for macroinvertebrates in running waters to drift actively (e.g., Wiley and Kohler, 1984; Siler *et al.*, 2001; McIntosh *et al.*, 2002). It is likely that larvae usually only leave their substrata when induced to do so by changes in biotic (e.g., competition for food or space, influence of predators) or abiotic factors (e.g., current velocity, drought). Otherwise they simply subject themselves to increased risk of mortality, due to drift-feeding predators (Allan, 1981) or increased energy expenditure, due to the necessity of building new retreats. Five 24 h drift catches between November (1989) and February (1990) were examined to determine the drift density of *T. rostocki*. Stream discharge rates were relatively low when the drift samples were collected ($4.5–14.7\,l\,s^{-1}$). *Tinodes rostocki* drift intensity was very low compared to the high abundance of these larvae at this site, only four larvae were caught in five days. Although colonization of the manipulated stones showed that larval movements occurred over the entire period of development, the mechanisms by which new substrata become colonized are unclear. *Tinodes rostocki* larvae have never been seen outside their retreats in daytime. It appears that Psychomyiidae larvae are generally rare in drift samples (Elliott, 1968; Bishop and Hynes, 1969; Dudgeon, 1983). Significantly higher larval densities on adjacent than distant stones, illustrate the importance of distance for the successful colonization of new substrata by *T. rostocki*. It seems improbable that larvae would crawl from one stone to another across areas of unstable fine sediment. These results indicate that *T. rostocki* larvae only drift over short distances. It is assumed that they utilize silk as safety filaments, as seen when other caseless and case-building caddisflies colonize new substrata [Hydropsychidae: *H. contubernalis*, *H. pellucidula*, *H. exocellata*, *H. angustipennis*, Becker (personal observation); Limnephilidae: *Drusus annulatus*, *Apatania fimbriata* (Menzler, 1995)].

The colonization of manipulated stones over each 3 week sampling interval was always sparser than on the control stones. Although the combined surface area of manipulated stones at a distance from colonized stones was larger ($600\,cm^2$) than that of the manipulated stones adjacent to colonized stones ($498\,cm^2$), the distant stones supported lower larval densities than the adjacent stones (Table 7.26). Assuming that the number of colonizing larvae was correlated with the surface

Table 7.26 Comparison of larval densities on stones placed adjacent to colonized stones and stones placed at least 0.3 m from colonized stones. The bracketed values show the number of larvae converted to 600 cm^2.

Sampling date	Larvae on adjacent stones	Larvae on distant stones
24.09.87	81 (98)	52
01.10.87	21 (25)	22
13.10.87	39 (47)	28
05.11.87	59 (71)	49
26.11.87	34 (41)	30
17.12.87	15 (18)	14
07.01.88	15 (18)	9
28.01.88	18 (22)	8
19.02.88	17 (20)	12
11.03.88	29 (35)	8
18.05.88	31 (37)	24

area of the stones, larval densities of the adjacent stones were converted to larvae per 600 cm^2 (Table 7.26, bracketed values). The comparison of larval densities (Wilcoxon matched pairs signed rank test) showed significant differences ($P < 0.05$) between adjacent and distant stones for converted and unconverted values.

7.8.1.1.6 **Intraspecific Competition for Space**

These results indicated that, as a consequence of crowding, the high larval densities in the middle and lower reaches of the Breitenbach induced intraspecific competition for space. But there is a question as to the cause of age structure variation between manipulated and unmanipulated stones. Two theoretical explanations are proposed.

On the one hand, because of territorial conflicts arising from crowding, larger rather than smaller larvae may have left their original positions on unmanipulated stones and colonized manipulated stones. Smaller larvae, with their lower space requirements, would have less difficulty in relocating themselves on the original stone because they could use the smaller spaces between the territories of larger larvae. Larger larvae, with a greater space requirement, would be compelled to relocate to other substrata via drifting.

On the other hand, populations colonizing manipulated stones could have similar age structures, but a lower density than control stones. With less crowding, larvae on manipulated stones could then grow faster, due to favorable factors such as higher food quantity/quality and less intraspecific competition. However, it is doubtful whether food conditions were really more favorable on manipulated stones, given that the latter were initially brushed clean of larvae and most of the periphyton.

Additionally, the newly built retreats on the manipulated stones may have contained less food than the older retreats on unmanipulated stones (Stief and Becker, 2005; see below). Furthermore, most of the significant differences between age

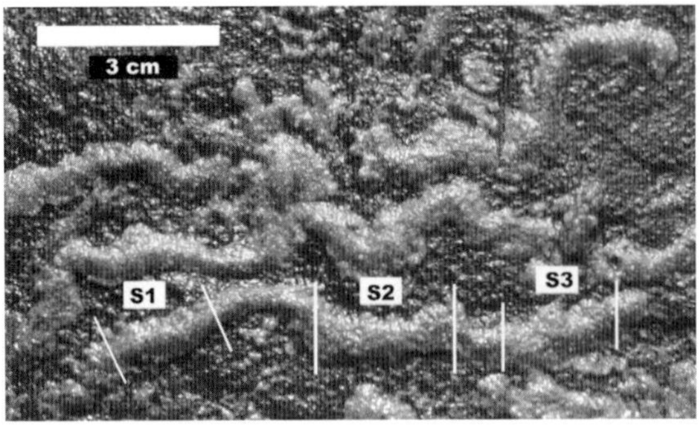

Figure 7.123 Retreats of *T. rostocki* with sections (S1–S3) used for pigment analysis.

structures were found during winter. Although larval development was not interrupted by a distinct developmental pause, larval growth rates were reduced on both types of substratum as a result of the low water temperature. The difference is therefore unlikely to have been due to different growth rates.

7.8.1.2 Retreats of *T. rostocki* as a Substratum for Larval Food Resources

7.8.1.2.1 Function of the Tinodes Retreat

Galleries of *T. rostocki* cover large proportions of submerged hard substrata and can be considered three-dimensional extensions of the stone surface (Figures 7.120 and 7.123). For the occupying larva the retreat functions as efficient protection against predators, desiccation, and accidental displacement by currents. Consequently, the larva spends most of its time inside the retreat and necessarily feeds on the biofilm on the retreat wall rather than on the surrounding epilithic biofilm (Hasselrot, 1993). Larvae of *T. rostocki* have never been observed scraping biofilms around their galleries (Hasselrot *et al.*, 1996), nor were feeding tracks found on stone surfaces close to the galleries (Becker, unpublished data).

Therefore, it was investigated whether the macroscopically visible gradient of pigmentation along the retreats mirror the longitudinal gradient of microphytobenthic activity and biomass, and whether the *T. rostocki* gallery biofilms differ significantly from the surrounding epilithic biofilms with respect to photosynthetic/respiratory activity and microphytobenthic community composition (Stief and Becker, 2005). Fifth instar larval retreats, identified by head capsule width (Becker, 1993), were sampled with their inhabitants in November and December 2002 from the middle reach of the Breitenbach, where stones are densely covered with larval retreats (Figures 7.129 and 7.123; Becker, 1993).

7.8.1.2.2 O_2 Production and Photopigment Composition of Undisturbed Retreat Biofilms and Surrounding Epilithic Biofilms

The photosynthetic/respiratory activities of non-manipulated larval retreat biofilms and the surrounding epilithic biofilms were measured within 48 h of sampling using O_2 microsensors at 7.7 °C in an aquarium (see below). O_2 microprofiles in the diffusive boundary layer (DBL) were measured above 14 randomly chosen sections of *Tinodes* retreats, and above 8 randomly chosen patches of stone biofilm, illuminated and in the dark. The positioning of the microsensor tip relative to the retreat or stone surface was controlled by viewing the sensor and sample through the aquarium wall under a dissection microscope. The 14 retreat sections were categorized as a light-brown, poorly developed biofilm (i.e., eight "poor" sections) and as dark-brown, rich biofilm (i.e., six "rich" sections). All sampled retreats had a "poor" and a "rich" section. The profiles were measured 3–5 times within each retreat section and patch of stone biofilm (Stief and Becker, 2005).

After microsensor measurement, retreats were photographed, cut into 2–4 longitudinal sections while still *in situ*, then cut off the surface with a scalpel and immediately frozen at –85 °C (Figure 7.123). Biofilm covering defined areas of stone surface was also scraped off with a scalpel and deep-frozen for further processing. HPLC analysis was performed with a Waters 2690 separations module equipped with a PDA 996 photodiode array detector (Waters, USA). Pigments were identified according to their retention time and absorption spectrum using a benthic microphytobenthos photopigment library. Digital *in situ* images of the retreats were used to determine the surface area of the retreat sections by image analysis (Nikon, Lucia G). The absolute pigment content was obtained from the peak area and the calibration curves of methanol-diluted pigment standards (DHI, Denmark). For more methodical details see Stief and Becker (2005).

O_2 production rates in the light (Figure 7.124a) and O_2 consumption rates in the dark (Figure 7.124b) were significantly higher in the "rich" sections of the *Tinodes* retreats compared to both "poor" sections and stone biofilms (ANOVA and Scheffé post hoc test, $P < 0.001$). There was no significant difference between "poor" retreat sections and stone biofilms. The chlorophyll-*a* content was significantly higher in the "rich" than the "poor" sections (Figure 7.124c, ANOVA and Scheffé post hoc test, $P < 0.01$). In contrast, the stone biofilm was similar to the "poor" sections ($P = 0.955$). Chlorophyll-*a* content and O_2 production per unit area of all sections at 200 μmol photons $m^{-2} s^{-1}$ were significantly correlated (Spearman, $r^2 = 0.832$, $P < 0.01$, $n = 18$). Fucoxanthin and β-carotene were also identified in the biofilm samples. In the retreat biofilms chlorophyll-*a* comprised 42.0%, fucoxanthin 56.5%, and β-carotene 1.5% of the total pigment by weight, irrespective of biofilm thickness (Figure 7.125). However, the stone biofilm comprised 84.5% chlorophyll-*a*, 10.5% fucoxanthin, and 5.0% β-carotene (Figure 7.125). The fucoxanthin to chlorophyll ratio of the stone biofilm was significantly lower than that for the retreats, that is, 0.15 compared to 1.27 (Mann–Whitney U-test, $P < 0.001$). The fucoxanthin to chlorophyll ratio did not vary significantly between "poor" and "rich" sections of the nontransplanted retreats ($P > 0.05$).

Thus, biofilms of *Tinodes* retreats and surrounding stone surfaces differed in microphytobenthic community metabolism and composition. The macroscopically visible gradient of pigmentation along the retreats mirrored the longitudinal gradient of microphytobenthic activity and biomass. All sampled retreats had a "poor" section with low photosynthetic/respiratory activity and microphytobenthic

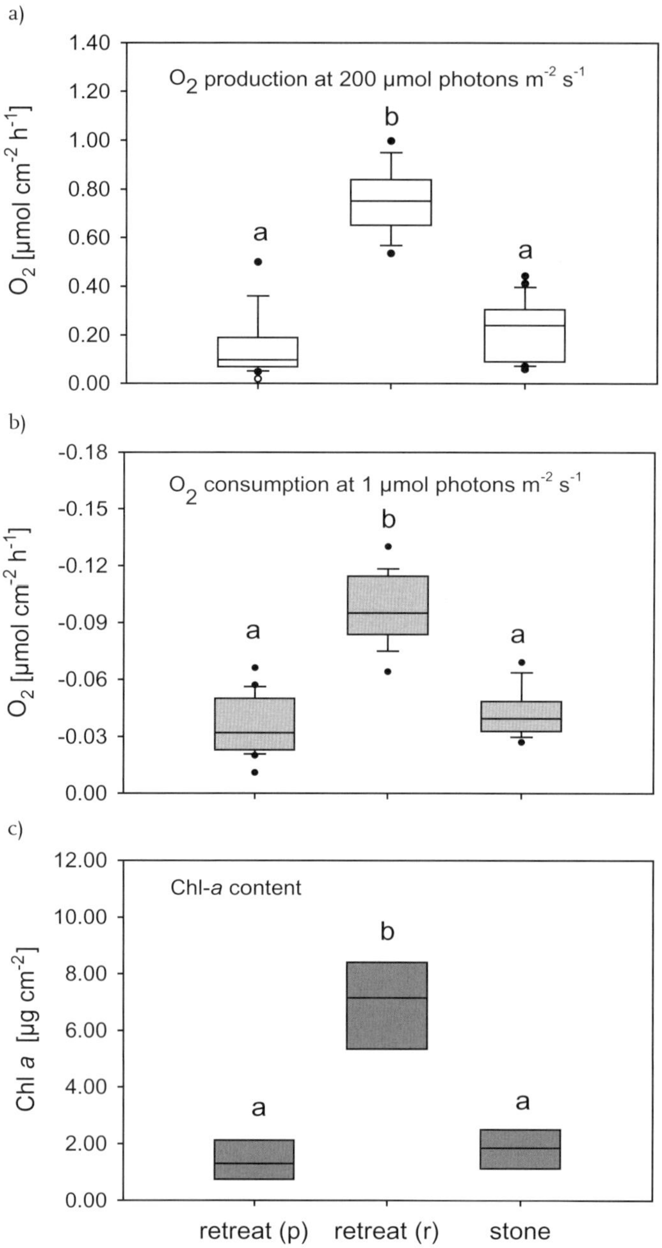

Figure 7.124 Net O_2 conversion rates (A + B) and chlorophyll-*a* content (C) of retreat and stone biofilms. Positive rates correspond to production, negative rates to O_2 consumption. Box plots give median and first and third quartiles; whiskers give tenth and ninetieth percentiles; circles are outliers. Letters indicate significant differences between the medians. Whiskers are not shown for the chlorophyll-*a* content of stone biofilms ($n = 4$). Retreat (p) = poor sections of the retreats; retreat (r) = rich sections of the retreats.

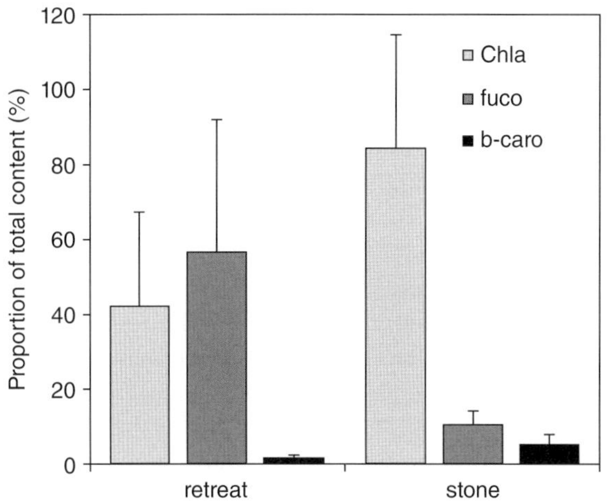

Figure 7.125 Relative photopigment composition of retreat and stone biofilms. Means + 1 SD of 14 pooled retreat sections and 4 stone biofilms are shown. Chl a = chlorophyll-*a*; fuco = fucoxanthin; b-caro = β-carotene.

biomass at one end, and a "rich" section with high photosynthetic/respiratory activity and microphytobenthic biomass at the opposite end. Only the "rich" sections of the retreat had higher photosynthetic/respiratory activity and microphytobenthic biomass than the stone biofilm. The structural polarity of the retreats was not due to differences in total community structure, since the fucoxanthin to chlorophyll ratio did not differ between the two ends of the retreats. The *Tinodes* retreat differed from the epilithic biofilm in microphytobenthic community composition, but only the dark-brown ends were distinct in terms of activity and biomass. This combined process and structure analysis confirmed earlier reports on microphytobenthic colonization of *Tinodes* retreats (*T. rostocki:* Hasselrot *et al.*, 1996; *T. waeneri:* Hasselrot, 1993; Kahlert and Baunsgaard, 1999).

The longitudinal gradient of microphytobenthic activity and biomass along the *Tinodes* retreat may be established as a result of larval retreat construction and degradation. Larvae elongate their retreat at one end by affixing new mineral particles, and break down the other end to feed on the biofilm attached to the detached particles (Hasselrot, 1993; Hasselrot *et al.*, 1996; Alecke, 1998; G. Becker, unpublished observations). This was confirmed by the dominance of diatoms in larval guts (Becker, 1990), which matches the diatom dominance in the retreat (Hasselrot *et al.*, 1996). The high proportion of mineral particles in the guts may also be a result of this larval behavior (Becker, 1990). As a consequence of polarized retreat elongation and degradation, an age gradient is established, which can be translated into a colonization gradient for microalgae and bacteria. This means that the "poor" section at one end has been recently constructed and is still poorly colonised by microorganisms, whereas the "rich", older section at the opposite end, represents a densely colonised and metabolically active site. The maximum chlorophyll-*a* content of "rich" sections of *T. rostocki* retreats ($7.0 \pm 1.5\,\mu g\,cm^{-2}$, mean $\pm$ 1 SD, $n = 6$) were distinctly higher than of epilithic biofilms reported by Cox, 1990b, Werneke (1997) and Becker (unpublished data) for the same stretch of the Breitenbach at the same time of year. These authors found average chlorophyll-*a* contents of between 0.5 and $2.8\,\mu g\,cm^{-2}$ in November and December. The maximum for "rich" sections of *T. rostocki* retreats also exceeded the chlorophyll-*a* content of *T. waeneri* retreats ($2.5 \pm 0.2\,\mu g\,cm^{-2}$) reported by Hasselrot (1993). The microphytobenthos standing stock and activity on the surrounding epilithic biofilm were at the lower end of values found on the retreats. This may either result from better growth conditions on the retreat (see above) or from the higher, overall grazing pressure on the undefended stone biofilm than on the *Tinodes* territory, that is, the retreat.

As fucoxanthin to chlorophyll ratios (fuco/chl-*a*) of 0.33–1.65 are indicative of diatom dominated microphytobenthic communities (Lucas and Holligan, 1999), the retreat biofilms were obviously dominated by diatoms (fuco/chl-*a*: 1.27), although the stone biofilms were not (fuco/chl-*a*: 0.15). This confirms earlier observations on diatom dominance on retreats of *T. rostocki*, compared to *Chamaesiphon* sp. (Cyanobacteria) dominance on the surrounding stone biofilm (Hasselrot *et al.*, 1996). High abundances of diatoms were also found on chironomid larval retreats (Pringle, 1985; Hershey *et al.*, 1988).

7.8.1.2.3 Microprofiles of O_2 and pH Outside, Inside, and in the Wall of Transplanted Retreats

The distinct microenvironment of the retreat wall and lumen was also investigated using retreats that were experimentally transplanted onto microscope slides, which allowed microsensor measurements in an otherwise hidden compartment (Stief and Becker, 2005). Under laboratory conditions larvae had completely attached their retreats to the slides after 7–14 days and were ready for experiments. Slides with transplanted retreats were fixed vertically against the front window of an aquarium, about 3 cm below the water surface (Hasselrot, 1993; Stief and Becker, 2005). Samples were illuminated with a fiber-optic halogen lamp, and aeration created a unidirectional water current of $1\text{–}2\,mm\,s^{-1}$ near the retreats. Photosynthetic/respiratory activity was measured above and within the biofilms using O_2 microsensors with a tip diameter of $15\,\mu m$ at different light intensities (Revsbech, 1989). The effect of photosynthetic/respiratory activity on pH was investigated using LIX-type pH microsensors with a tip diameter of $5\text{–}15\,\mu m$ (De Beer *et al.*, 1997). O_2 concentration profiles were recorded at a spatial resolution of 25 or $50\,\mu m$ using a computer-controlled micromanipulator. O_2 concentration and pH microprofiles were recorded in the central part of four transplanted *Tinodes* retreats. Three retreats were inhabited by a *Tinodes* larva, while one retreat had been vacated by the larva immediately prior to measurement. Profiles were repeated 3–5 times for each retreat. Average O_2 concentrations and pH were determined for the external water body, the retreat wall, and the retreat lumen. For more methodical details see Stief and Becker (2005).

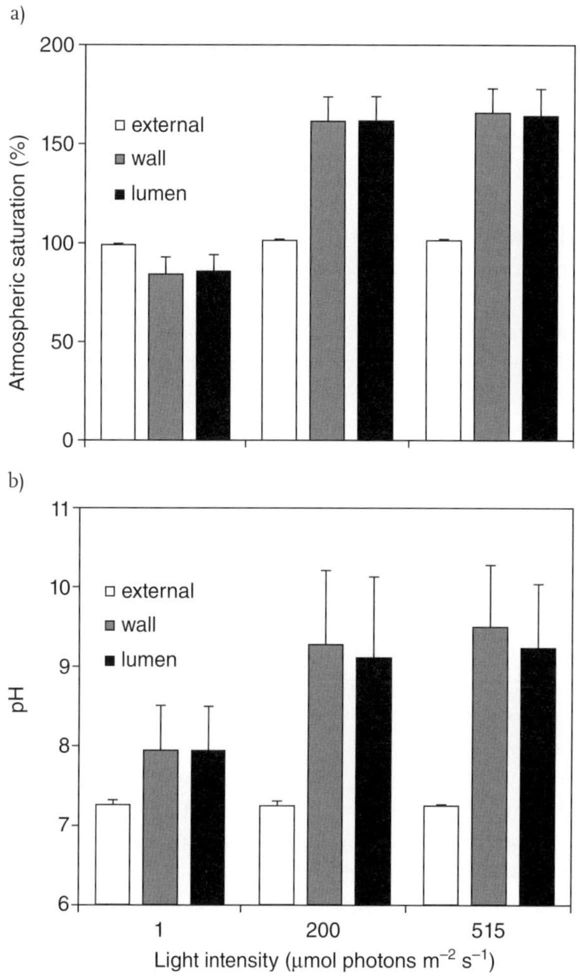

Figure 7.126 Oxygen saturation (a) and pH (b) in external medium, retreat wall, and retreat lumen measured at different light intensities in transplanted retreats. Means + 1 SD (*n* = 4).

When illuminated, O_2 concentration and pH increased strongly in the DBL surrounding the retreats, and were high within the retreat wall and lumen (Figure 7.126). Compared to the surrounding medium, O_2 concentrations were typically elevated by 200–300 $\mu mol\,l^{-1}$ giving internal oxygen saturations of 150–180% (Figure 7.126a). The pH typically increased from 7.2 in the surrounding medium to 10 within the retreat wall and lumen (Figure 7.126b). Average values for wall and lumen were almost identical, at 200 and 515 μmol photons $m^{-2}\,s^{-1}$. In the dark, O_2 concentration decreased slightly in the DBL and within the retreat wall, while pH continued to increase slightly. Within the retreat wall and lumen O_2 concentration was generally less than 100 $\mu mol\,l^{-1}$ lower than in the surrounding medium, resulting in internal oxygen saturation always being above 75% (Figure 7.126a). The pH increased from 7.2 in the surrounding medium up to pH 8.7 in the retreat (Figure 7.126b). Two-factorial ANOVA revealed significant effects of both light intensity and biofilm on oxygen saturation and pH in the retreat microenvironment ($P < 0.001$) and a significant interaction between light intensity and biofilm compartment (O_2: $P < 0.01$; pH: $P < 0.05$). Oxygen saturation and pH were significantly different between retreat wall and lumen, and the external medium (two-factorial ANOVA and Scheffé post hoc test, O_2: $P < 0.001$, pH: $p < 0.05$). No significant differences were found between retreat wall and lumen.

7.8.1.2.4 **Animal–Microbial Interactions**

Understanding animal–microbial interactions within stream epilithon is not only interesting with respect to the complex foraging strategies of species, but at high macroinvertebrate densities, animal-induced changes in microbial taxonomic composition, spatial organization and metabolic activity may have a significant influence on the energy and carbon flow through the epilithic food web.

Stief and Becker (2005) found evidence that the wall of the retreats represents a biofilm compartment exposed to a contrasting microenvironment to the external medium. When illuminated, both the retreat biofilm and inhabiting larva were exposed to substantially higher O_2 concentration and pH than in the water column. However, it is not known what conditions prevail within the epilithic biofilm, as this was too thin to be probed with microsensors. Stief and Becker (2005) could only speculate that nutrient concentrations may also have differed between the internal and external water.

The retreat encloses a small volume of water, whose exchange with the external water body remains largely unknown (Hasselrot, 1993; Kahlert and Baunsgaard, 1999). On the one hand, a degree of isolation can be expected due to the relatively thick wall of the retreat (500–750 μm). On the other hand, larvae have been observed to perform undulatory body movements that increase water exchange between the internal and external compartments (Hasselrot, 1993; G. Becker, unpublished observations). Our microsensor measurements indicated that, when illuminated, the water inside the retreat was supersaturated with O_2 (150–180% oxygen atmospheric saturation) and highly alkaline (pH 10). Comparative measurements in retreats with and without a larva revealed that water exchange was somewhat higher between internal and external water, when a larva was undulating in the retreat. Nevertheless Stief and Becker (2005) claimed that, at least under their laboratory conditions, the undulatory body movements of the inhabiting larva could not fully compensate for the photosynthesis-driven chemical changes inside the retreat. From this it can be speculated that some larval excretions may be retained within the retreat, which would elevate nutrient concentrations above those of the external water (Hershey *et al.*, 1988; Plaganyi and Branch, 2000). Bacteria and microalgae associated with *Tinodes* retreats may therefore efficiently recycle the nutrients excreted by the *Tinodes* larva (Kahlert and Baunsgaard, 1999) and growth may be stimulated in the attached microorganisms that eventually enrich the retreat. Preliminary studies showed high densities of the diatom

Achnanthidium sp. on both the inside and outside retreat walls of *T. rostocki* (Hasselrot *et al.*, 1996). In summary, the metabolism of the retreat biofilm and larval excretions may both shape the chemical microenvironment of the *Tinodes* retreat, and consequently the activity and biomass of the associated microorganisms.

Possible mechanisms of diatom enrichment include the deliberate removal and disposal of unsuitable microphytobenthic species (e.g., *Microcoleus vaginatus*, Cyanobacteria; Hart, 1985), the reduced ingestion of adnate diatoms compared to overstorey algae by scrapers (Hill and Knight, 1988), and the fast recolonization by immigrant diatom species (Bergey, 1995). Stief and Becker (2005) did not attempt to investigate the mechanism of diatom enrichment on the *Tinodes* retreat. Instead, it is assumed that the particular architecture and microenvironment of the retreats favored diatoms more than other groups of microphytobenthos. Owing to their unique motility, diatoms may benefit from both the elevated nutrient concentrations on the inside, and the optimal light conditions on the outside of the retreat walls. Macroinvertebrates that are able to graze on diatoms (e.g., trichopteran and ephemeropteran scrapers) benefit greatly from the dominance of these microalgae because of their high polyunsaturated fatty acid content (Mayer and Likens, 1987; Steinman *et al.*, 1987a, 1987b; Huryn and Wallace, 1988; Blomquist, Borgeson and Vundla, 1991; Dunstan *et al.*, 1994). In contrast, cyanobacteria often are unmanageable, or even toxic to grazers (Hart, 1985). In the case of *T. rostocki*, the high quality food source represented by diatoms on the larval retreat is exclusively at the disposal of the inhabiting larva, because it will obstinately defend the retreat against con- and heterospecific invaders (Hasselrot, 1993).

7.8.1.3 Feeding strategy of *T. rostocki*

7.8.1.3.1 Feeding Behavior of Larvae

To understand the foraging strategy of *T. rostocki* it is necessary to know the energy demand for gallery building and the energy content of the gallery. Typical behavioral patterns of larval feeding, spinning, and gallery construction were analyzed using video sequences. Most of the retreats contained both an older section with a visible biofilm and one or two new sections with a thin biofilm. Larvae cut large particles out of old sections of gallery wall, feed on the attached organic material, transport the particles to the new section, and fit them into the wall again (Becker, unpublished data). This behavior explains the slow change in retreat position on the stone surface. Alecke (1998) found that fifth instar larvae of *Tinodes unicolor* extended their retreats by up to 21 mm week^{-1}. Thus, the larval investment in time and energy to build, repair, lengthen, and break down the gallery seems to be very high.

7.8.1.3.2 Retreat Construction in the Course of the Ontogeny

Because of the substantial time and energy investment, *Tinodes* retreats are intensively contested resources. However, what is the larval feeding strategy, and what is the significance of the retreat size to larval growth and development? To answer these questions, retreat size and composition, and larval biomass and abundance were analyzed during larval ontogeny along the Breitenbach.

In the course of their ontogeny, average retreat size increased in the upper part of the stream by a factor of 62, from 4 mm^2 (first instar larvae) to 250 mm^2 (fifth instar larvae). Between November and May, the mean length and width of the retreats of fifth instar *T. rostocki* larvae were 4.5 and 0.4 cm, respectively ($n = 430$; Becker, unpublished data), considerably exceeding larval body length (about 0.8 cm). The gallery wall contained about 7570 mineral particles cm^{-2} (particles >33 μm), while the organic material (silk and biofilm) only constituted up to 3% of the dry mass of old sections and up to 2% of new sections (Becker, unpublished data). Organic content of the retreat wall increased with increasing larval stage, due to the thicker biofilms, linked to the longer occupation of the retreats by older larvae. The larvae showed different rates of development along the stream. Although first instar larvae hatched two weeks later in the upper than the lower part of the stream, the composition of the larval population was similar in both stretches of the stream in autumn (Gergs, 2004).

7.8.1.3.3 Significance of the Larval Retreat for Feeding and Growth of *T. rostocki* Larvae

Retreat size played a crucial role for the growth of fifth instar larvae between November and April at different sites along the stream (Becker, unpublished data). At an unshaded site in the middle part of the stream, the average retreat size increased significantly between November and February, from about 1.5 mm^2 to 2.5 mm^2. However, during this period average larval biomass increased slightly, but not significantly. Between February and April larval biomass increased significantly and was negatively correlated with decrease in retreat size. At a second unshaded site in the lower stretch of the stream, average retreat size increased significantly between November and March, while larval biomass did not increase significantly over this time. However, between March and April, that is, one month later than at the first site, there was a negative correlation between increasing larval biomass and decreasing retreat size. Thus the larvae significantly expand their retreats to increase their food resources on the retreat wall before they start to grow. When larvae grow significantly, retreat size is distinctly reduced. The results show the significance of retreats for feeding and growth of *T. rostocki* larvae.

7.8.1.3.4 Effect of *T. rostocki* Foraging on the Microbial Community of the Breitenbach

Throughout larval ontogeny, comparative HPLC analysis of photopigments of retreat biofilms and the surrounding epilithic biofilms showed a dominance of diatoms, a high-quality food, on retreats from both younger and older larvae, but not on stones (unpublished data from G. Becker with P. Stief and R. Schoon, both from the Max-Planck-Institute for Marine Microbioloy in Bremen). Significant differences in photopigment composition between retreat and the surrounding epilithic biofilms were regularly found in the upper, middle, and lower reach of the Breitenbach. These results demonstrate that the complex larval behavior has a significant influence on the quality of their food resources, the biofilms of their retreats, and demonstrates "gardening" activity by *Tinodes* larvae throughout their development. Substrata were densely covered by *T. rostocki* retreats along a 1 km stretch of the middle and lower reach, between 30 and 80% of the Bunter Sandstone (Figure 7.120). At these high larval densities, the microbial community of epilithic biofilms is significantly affected by the structuring impact of *T. rostocki* with respect to the larva-induced changes in taxonomic composition, spatial organization (increasing patchiness), and microorganismal metabolic activity. Therefore the specific feeding strategy of *T. rostocki* may have significant implications for total benthic primary production and its flow into the epilithic food web in small streams such as the Breitenbach. The strong effect of another *Tinodes* species, *T. waeneri,* on the epilithic community has been observed in the lake littoral (Harrison and Hildrew, 1998, 2001).

7.8.2 Life Cycle and Ovipositing Adaptations of *Agapetus fuscipes* (Trichoptera, Glossosomatidae) to a First-Order Stream

7.8.2.1 Life Cycle of Glossosomatidae

Knowledge of the number of immature stages and of the patterns of growth and development throughout the year is important basic information on a species' life history. Such information is essential to understanding a species' adaptation to its environment, as well as the functio-ning and interactions of biological communities (e.g., Butler, 1984; Roff, 1992; Stearns, 1992).

Glossosomatid species are typical scrapers in all rivers and streams (Wiggins, 1996), and generally have five larval instars (e.g., Anderson and Bourne, 1974; Irons, 1988; Sameshima and Sato, 1994; Houghton and Stewart, 1998; Alvarez and Pardo, 2005; Jin and Ward, 2007), which is also typical for most other Trichoptera. However, several authors have reported more than five larval instars in *Agapetus fuscipes* Curtis (Nielsen, 1942; Benedetto Castro, 1975; Iversen, 1976; Recasens and Murillo, 1986, and Sangpradub, Giller and O'connor, 1999; Becker, 2005). *Agapetus fuscipes* (Figure 7.127) is the dominant trichopteran species in the upper and middle reaches of the Breitenbach (e.g., Illies, 1978; Sandrock, 1978; Becker, 1990, 2005; Wagner and Schmidt, 2004), and is common in European upland and lowland streams (e.g., Pitsch, 1993; Robert, 2001; Fischer, 2003; Nijboer, 2004).

However, there are insufficient studies on the life cycle of this species and some are contradictory. Specific information about instar number and size are essential for understanding the seasonal timing of life cycle processes, population synchrony, distributional pattern, and persistence in the face of disturbance. Size and biomass are especially important with respect to risk factors such as predation and exposure to hydrodynamic forces.

Figure 7.127 Larvae of *Agapetus fuscipes* scraping a thin epilithic biofilm on Bunter Sandstone substratum.

7.8.2.2 Measurements of Pronotum Length, Larval Mass, and Case Length in *A. fuscipes*

Becker (2005) analyzed frequency histograms of pronotum length, which allows individual instars of *A. fuscipes* to be distinguished and data on the mean biomass of individual larval stages. Molt increments between larval instars and the presence of the individual larval stages in the upper reach of the Breitenbach were also investigated throughout the year.

To analyze pronotum length, body mass and case length, *A. fuscipes* larvae were sampled in the upper, perennial reach of the Breitenbach near trap T1, which has an annual mean temperature of about 8 °C, and an annual temperature amplitude of between 4 °C and 16 °C (see Section 3.2). All analyzes were undertaken using fresh larvae and cases. Pronotum length was measured along the mid–dorsal ecdysial line, and was the preferred measurement because differences between larval instars were clearer than when the more commonly measured head capsule width was used. To ensure that the youngest larval stage was not missed in the field, egg masses were sampled from the Breitenbach and cultured in laboratory streams at about 14 °C. After hatching, young larvae were fed with epilithic biofilms from the Breitenbach for six weeks, and pronotum lengths of the young larvae (instars I–III) were compared with those from larvae from the field samples. Because sex discrimination is not reliable at the last larval instar of Trichoptera, ranges of prepupal pronotum lengths from the upper reach of the Breitenbach were used to avoid counting smaller males and larger females as different instars. In addition, the length of the corresponding larval case was measured, and young larvae and their cases (instars I–III) were weighed in groups of up to 15 individuals, the older larvae (instars IV–VII) and their corresponding cases were analyzed individually.

7.8.2.3 Number of Larval Instars

The frequency histogram of pronotum lengths shows that *A. fuscipes* has seven larval stages in the upper reach of the Breitenbach (Figure 7.128, Becker, 2005). The pronotum lengths of instars I–V were distinct, whereas the peaks of instars VI and VII overlapped slightly. The larval pronotum lengths (instars I–III) from laboratory culture corresponded exactly with the field specimens, showing that all larval instars were present in the field samples (Becker, 2005). Furthermore, the range of pronotum lengths of instar VII larvae corresponded well with the prepupal pronotum lengths (Becker, 2005). The prepupae had reached the end of larval development, and were enclosed in their pupal cases shortly before pupation. The last two peaks therefore indicated two different larval instars rather than sexual dimorphism in the last larval instar. These observations agree with the findings of several authors, who have described seven larval instars in various parts of Europe (Nielsen, 1942; Iversen, 1976; Recasens and Murillo, 1986; Sangpradub *et al.*, 1999). From a Spanish stream (Catalonia, south-eastern Spain) Recasens and Murillo (1986) reported pronotum lengths that were somewhat smaller than in the Breitenbach population. However, they noted that *A. fuscipes* had a bivoltine life history in Catalonia, and the shorter generation time may explain the smaller size of their larvae.

Although glossosomatid larvae are common and often abundant (Wiggins, 1996), the life cycle and number of larval instars have only been documented for

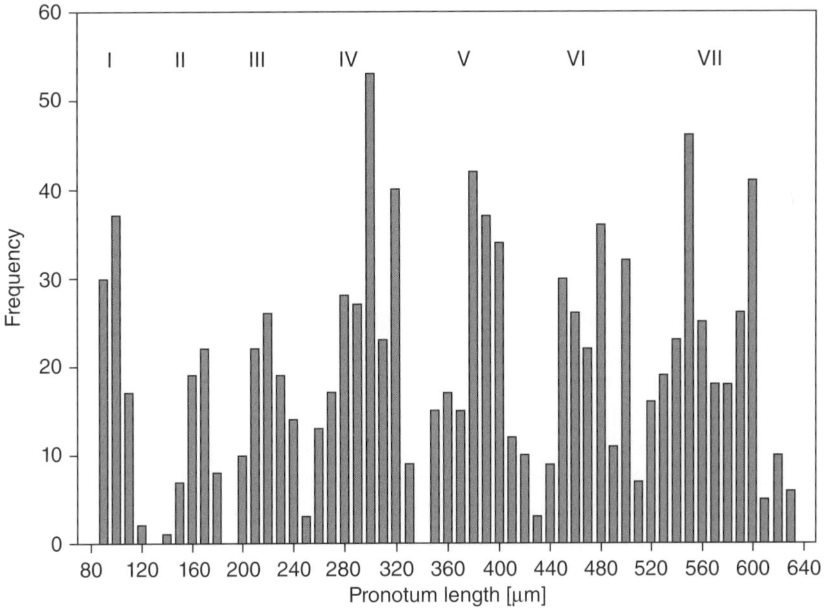

Figure 7.128 Frequency distributions of pronotum length of seven larval instars of *A. fuscipes* in the Breitenbach ($n = 1056$). I–VII = larval instars.

a few species. The number of instars is only known in 4 of 14 glossomatid species in Central Europe (Pitsch, 1993; Robert, 2001), that is, *A. fuscipes, Synagapetus iridipennis* (Otter, 1989), *Glossosoma conformis* (Sangpradub *et al.*, 1999), and *Glossosoma intermedia* (Fjellheim and Raddum, 1998). The last three species each have five larval instars, as do *Agapetus quadratus* (Spain; Alvares and Pardo, 2005), *Agapetus yasensis* and *Glossosoma inops* (Japan; Sameshima and Sato, 1994), and *Agapetus pontona* and *Agapetus monticolus* (Australia; Marchant and Hehir, 1999). In the Nearctic region, the life cycle of only nine of 80 known glossosomatid species (Wiggins, 1996) has been analyzed (Jin and Ward, 2007; for an overview see Houghton and Stewart, 1998). All nine were found to have five larval instars. Thus *A. fuscipes* is the only scraping caddisfly species in the Northern Hemisphere currently known to have more than five larval instars. This means that an important life cycle feature of this species, the instar number, differs markedly from the typical number for Trichoptera and all other glossosomatid species. This is of particular importance because major life cycle features, such as development and growth patterns, are under strong lineage-specific control and may be fixed within lineages (e.g., Butler, 1984; Stearns, 1992).

Within *A. fuscipes* seven larval instars seems to be fixed; only Benedetto Castro (1975) found eight larval instars in laboratory cultures of larvae from the Breitenbach, but not in the Breitenbach itself. However, those findings with respect to pronotum length do not agree with other measurements reported for larvae in the Breitenbach and other streams (Becker, 2005).

Few northern hemisphere trichopteran species are known to have more than five larval instars. For a Danish spring, Iversen (1976) reported six larval instars in the headwater species, *Berea maurus*, with a nonsynchronous life history. The number of larval stages is not clear for some *Sericostama* species, and it is not known whether the number of stages is fixed or variable (Nielsen, 1942; Elliott, 1969; Iversen, 1973; Denis, 1981; Resh *et al.*, 1981; Wagner, 1990a, 1990b). *Sericostama* species are detritivores, most of which have a semivoltine, or longer life cycle. Furthermore, in some southern hemisphere trichopteran species with a nonsynchronous life history, more than the typical five larval instars have been assumed (Winterbourn, 1978; Towns, 1981; Dean and Cartwright, 1987).

7.8.2.4 Larval Mass and Case Length throughout the Ontogeny

The case lengths, and especially the larval biomasses, of the instars partly overlapped (Becker, 2005). However, pairwise comparison of case length and biomass showed significant differences between the seven larval instars (Mann–Whitney U-test, all pairs with both parameters $P < 0.001$). When pronotum length was used to distinguish larval instars, the correlation between pronotum length, larval biomass, and case length (Figure 7.129) showed clear distinction between the four older larval instars (IV–VII). The values of instar VII larvae are widely distributed compared to the other instars because approximately 55% of mean larval biomass of instar VII larvae was accumulated during the development of this larval stage (see also Figure 7.130). Nielsen (1942) found that, in a Danish stream, case lengths of the first larval instars of *A. fuscipes* ranged over 1.33–1.85 mm, and case lengths

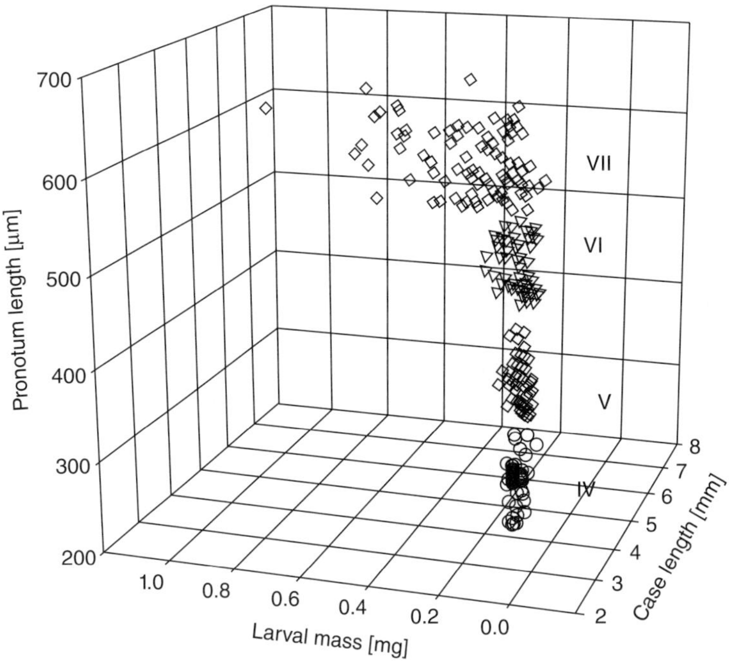

Figure 7.129 Relationship between pronotum length (µm), larval biomass (µg), and case length (mm) in older *A. fuscipes* larval instars. IV–VII = larval instars.

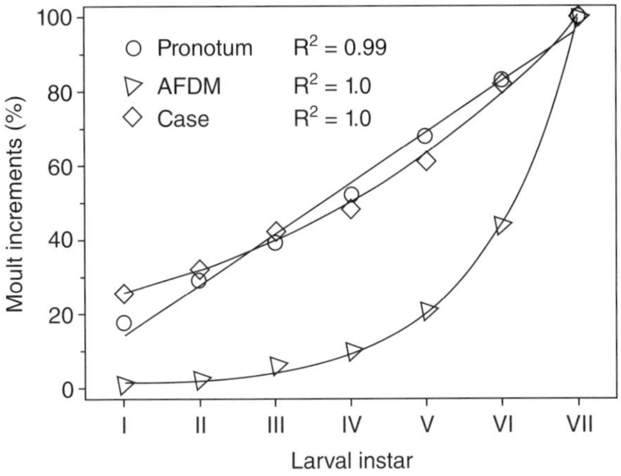

Figure 7.130 Regression lines and functions of molt increments of pronotum length, larval biomass, and case length in larval instars.

of the last larval instars were 4.6–8.0 mm; both agreeing with the case lengths for *A. fuscipes* in the Breitenbach (1.22–2.18 mm and 4.75–7.6 mm, respectively; Becker, 2005).

The AFDM values of the last instar larvae in the current study were relatively low (mean ± SD, 391 ± 84 μg) compared to the case mass (Becker, 2001). No comparable data for the AFDM of glossosomatid larvae could be found in the literature. Fjellheim and Raddum (1998) and Marchant and Hehir (1999) calculated the mean dry mass of preserved larvae, giving values of between 200 and 300 μg for last instar larvae of three different glossosomatid species, lower than for larvae of *A. fuscipes* in the Breitenbach (Becker, 2005). Alvarez and Pardo (2005) analyzed the mean dry mass of preserved larvae of *A. quadratus*, a trivoltine species from a temporary stream in South Europe (on the island of Majorca), and found also lower values, with a mean dry mass of 230 μg for last instar larvae.

Clear distinction between the larval instars in *A. fuscipes* using pronotum length, larval biomass, and case length (Figure 7.129) may be possible because of the step-like, post-molt increase, not only of pronotum length, but also of case length. In contrast to many other trichopteran larvae, glossosomatid larvae do not lengthen and widen their case continuously as they grow. Glossosomatid larvae build a new case after each molt and do not modify the size of the case during an intermoult period (Hansell, 1968a, 1968b; Anderson and Bourne, 1974; Bohle and Fischer, 1983).

7.8.2.5 **Molt Increments**

The molt increment between two larval stages (Dyar's rule) for hard cuticle parts of trichopteran species with five larval instars is about 1.5 (Nielsen, 1942). The molt increments in *A. fuscipes* were significantly lower than in other Glossosomatidae (Becker, 2005), with a mean molt increment of 1.35 for pronotum length over the seven larval stages in *A. fuscipes* from the Breitenbach, decreasing from 1.67 at the second/first instar molt to 1.20 at the seventh/sixth instar molt. The regression of pronotum length against larval instar was linear (Figure 7.130). In glossosomatid species with five larval instars, mean molt increments for head capsule lengths were distinctly higher, between 1.41 and 1.49 in four Japanese species (calculated from Sameshima and Sato, 1994), and 1.42 in *Glossosoma nigrior* (Jin and Ward, 2007). The mean molt increment of larval AFDM was 2.25 in *A. fuscipes*, decreasing from 2.64 at the second/first instar molt to 2.28 at the seventh/sixth instar molt (Becker, 2005). The regression showed a typical exponential growth curve (Figure 7.130). These mean molt increments were distinctly lower than the values of 2.79 and 2.92 for two Australian *Agapetus* species with five larval instars (calculated from data of Marchant and Hehir, 1999). The mean molt increment of case lengths in *A. fuscipes* was 1.26, and was relatively constant throughout ontogeny (Becker, 2005). The regression between case length and larval instar showed a slightly exponential function (Figure 7.130).

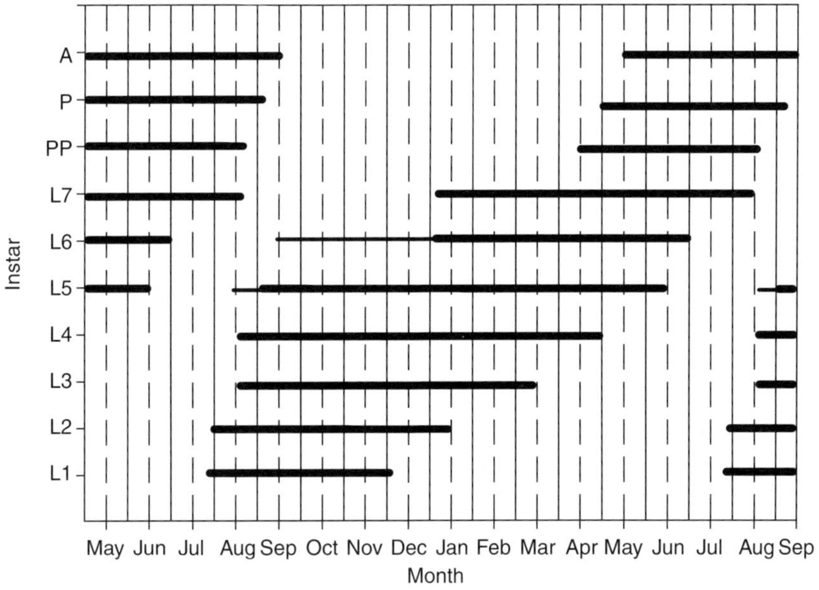

Figure 7.131 Life cycle of *A. fuscipes* in the upper reach of the Breitenbach. I–VII = larval instars; PP = prepupae; P = pupae; A = adults. Emergence data provided from Illies (1971, 1978), Sandrock (1978), Fischer *et al.* (1998), and Wagner and Schmidt (2004). Bold lines = larvae were found regularly; fine lines = larvae were found rarely.

7.8.2.6 The Presence of Developmental Stages in the Breitenbach throughout the Year

Qualitative samples were taken on 40 occasions between February 1997 and November 2003 in the upper reach of the stream and provided information on the occurrence of the different developmental stages of *A. fuscipes* in the Breitenbach.

The population structure of *A. fuscipes* showed a clearly univoltine life cycle in the upper reach of the Breitenbach (Figure 7.131, Becker, 2005), with little inter-annual variance for 6.5 years of collection (Figure 7.131 thin lines). Instar I larvae occurred in mid-July and were found until the beginning of December. Instar II larvae were found from mid-July to January, instars III and IV larvae from mid-August to mid-March and the end of April, respectively. Instar V larvae were found from the beginning of September (occasionally from the end of August), until mid-June. At the beginning of December the population consisted of instars I–V. Detailed observations showed that the larvae fed intensively throughout the winter, and that their development was not interrupted. Sixth- instar larvae were found until the end of June, but seldom before January. Instar VII larvae never occurred before January, and the last individuals were recorded at the end of August. Prepupae and pupae were found in April, with the last pupae being found at the beginning of September. However, most of the population pupated between May and July. At least five different developmental stages of *A. fuscipes*, with different ecological demands, coexisted in the Breitenbach at any given time (Figure 7.131).

7.8.2.7 Life Cycle Adaptation of *A. fuscipes* to a Small Uphill Stream

Life cycle duration can be highly variable in the Glossosomatidae, from 1.5 months for the summer generation of the trivoltine population of *Glossosoma inops* at water temperatures between 17.5 and 20.0 °C (Sameshima and Sato, 1994), to almost one year for most species in the northern hemisphere at lower temperatures (e.g., Fjellheim and Raddum, 1998, Sangpradub *et al.*, 1999; Becker, 2005). In the upper reach of the Breitenbach, *A. fuscipes* showed a clearly seasonal life cycle with distinct annual classes and slowly growing larvae due to the relatively constant water temperature (average 8 °C). However, the population was rather asynchronous. Thus individuals of several size classes occurred at all times of the year. Most of the larval biomass accumulated in late winter and spring. These characteristics correspond well with the slow seasonal life cycle in Hynes' (1970a, 1970b) classification scheme. The emergence period in the Breitenbach extends from May to September (Illies, 1971; Sandrock, 1978; Fischer *et al.*, 1998; Wagner and Schmidt, 2004), although most adults (88%) emerged in June and July (Sandrock, 1978). First instar larvae were present for almost 4.5 months and instar II larvae for 6.0 months, due to the extended emergence period. Individuals of at least five immature stages were present throughout the year. These characteristics of the immature stages do not fit easily within Hynes' (1970a, 1970b) slow seasonal life cycle classification.

The life history traits of *A. fuscipes* are more complex than the typical lineage-specific changes in Trichoptera, but the ecological advantage of this atypical number of immature stages is unclear. Two findings seem to be important in this context. On the one hand, larvae regularly feed on thin biofilms of low chlorophyll-*a* content, 0.2 $\mu g\,cm^{-2}$ (Becker, unpublished data). Due to feeding over the rough Bunter Sandstone substratum (Arens, 1990; Becker, 1990, 2001) high abrasion of the mouthparts may be compensated for by two extra molts. On the other hand, the wide range of larval size classes, which is extended by two additional larval instars, may enhance the ability of *A. fuscipes* to cope with unfavorable conditions in their abiotic and biotic environment, spreading the mortality risk (Den Boer, 1968; Danks, 1987). At least five immature stages, each with different ecological demands (Majecki *et al.*, 1997; Nijboer, 2004), for instance to risk factors such as predation and exposure to hydrodynamic forces and nutritional requirements, were always present. Between five and eight developmental stages occur at the same time, within and outside the stream, if egg and adult stages are included.

Because the number of glossosomatid species with well known life histories and larval stages is relatively small worldwide, it is not known whether a life cycle of more than five larval instars is within the normal phylogenetic variability of the Glossosomatidae. However, it is assumed that the large number of larval instars in *A. fuscipes* is a species-specific adaptive strategy to headwater streams like the

Breitenbach, indicating an adaptive response to long-term environmental effects (e.g., Butler, 1984; Sweeney, 1984).

7.8.2.8 Ovipositing Behavior of *Agapetus fuscipes*

When, like *A. fuscipes,* females of trichopteran species with a short adult lifetime emerge, at least one part of their eggs is mature and ready for laying (Jervis *et al.*, 2005). Different ovipositing strategies have been reported for caddisflies (Wallace and Anderson, 1996). Anderson (1973) described an interesting behavioral adaptation of the ovipositing behavior of *A. fuscipes* females. Females dive under the water, select a small stone (cap stone) and carry it to a larger stone (base stone). Eggs are laid in a gelatinous matrix on the larger stone and the small stone is attached as a cap while the cement is still soft (Figure 7.132). Observations in the Breitenbach have shown that cap stones always completely covered the egg masses

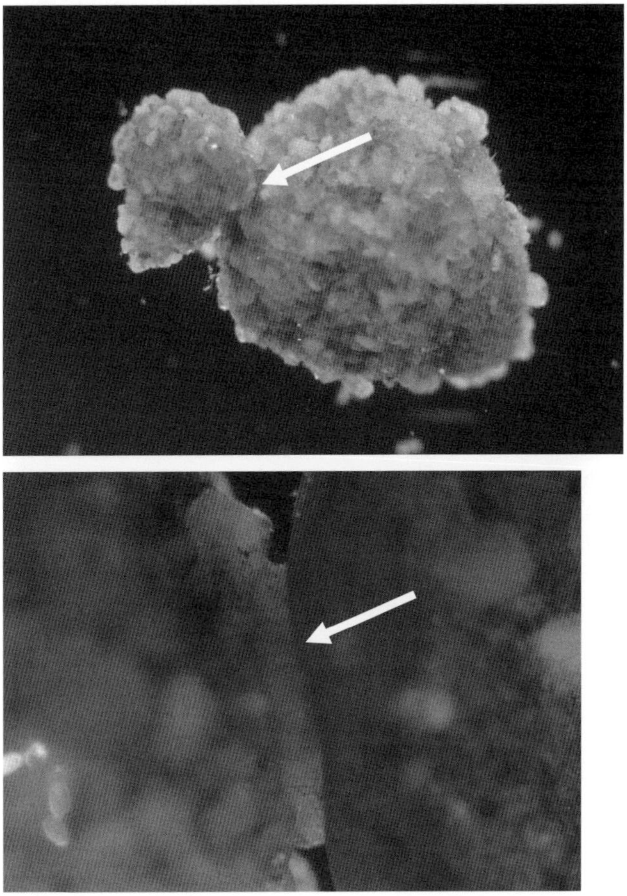

Figure 7.132 Photographs of eggs between cap and base stones.

of this species (Becker, 1991); eggs never protruded from beneath the cap stones. The following question therefore arises: Are females able to adjust the number of eggs per egg mass to the size of the available cap stone?

Females of several trichopteran families (Glossosomatidae, Hydropsychidae, Polycentropidae, Psychomyiidae) oviposit on submerged substrata. Many exhibit modified leg morphology, which can influence swimming behavior (Deutsch, 1985). In *Agapetus fuscipes*, the length : width ratios of the tibiae and first tarsi were 6.5 and 6.1, respectively, for males ($n = 10$) and 5.0 and 3.4 for females ($n = 10$) and are significantly different ($p < 0.001$, Becker, 1991). Clear sexual dimorphism was therefore apparent in this species, as in other glossosomatids (Deutsch, 1985).

7.8.2.9 Egg Number and Egg Masses per Female

Newly emerged imagines were used to determine the number of eggs per female and body wet weight. Females were raised in the laboratory from pupae collected in the upper stretch of the Breitenbach (Becker, 1991). The number of eggs per female varied between 92 and 235, with a mean of 147 ± 44 and the mean number of eggs per egg mass was 13 ± 8 (Table 7.27), each female producing an average of 11 egg masses. The number of egg masses oviposited in a single diving period is unknown, but females of *Hydropsyche contubernalis* were able to produce three egg masses from their stock of mature eggs in one diving period (Becker, 1987a, 1987b).

7.8.2.10 Distribution and Abundance of Egg Masses in the Breitenbach

Nine perpendicular sampling profiles across the upper part of the Breitenbach revealed egg masses across the entire stream width, although usually with fewer nearer the banks (Figure 7.133). The highest egg mass abundance, 2925 m^{-2}, was found at a mean current velocity of 0.12 $m\,s^{-1}$ and mean water depth of less than 5 cm. The lowest abundance, 125 egg masses m^{-2}, was observed at a current velocity of 0.29 $m\,s^{-1}$ and water depth of 14 cm (Becker, 1991). These correspond to mean egg numbers of between 1625 and 38 025 m^{-2}, respectively. Egg mass density was negatively correlated with current velocity ($r^2 = -0.62$, $P < 0.05$) and water depth ($r^2 = -0.67$, $P < 0.05$).

Had the water level fallen, egg masses in shallow water near the banks would have been killed by desiccation. However, hydropsychid females in larger rivers behave differently, avoiding shallow water near the bank and ovipositing in deeper areas, down to several metres (Fremling, 1960; Deutsch, 1984; Becker, 1987a, 1987b). This protects the egg masses against drying up.

Table 7.27 Summary statistics for egg numbers, cap stone weights, and female wet weights.

Number of eggs per female	147	44	92	235	45
Number of eggs per egg mass	13	8	1	54	170
Weight of cap stones (mg)	13.6	20.5	0.7	213.3	209.0
Wet weight of females (mg)	2.97	1.08	1.48	5.12	27

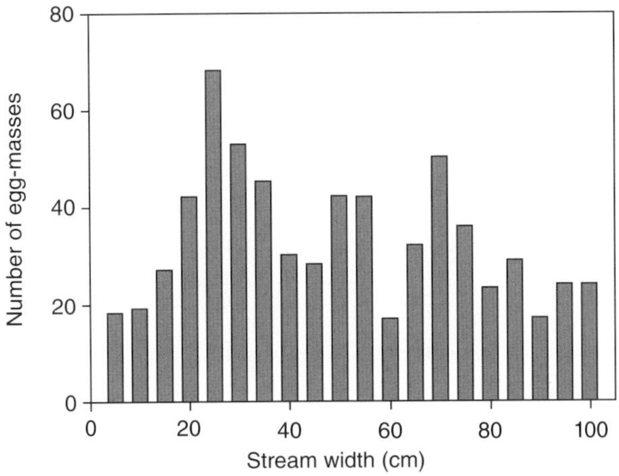

Figure 7.133 Egg mass frequency distribution in profiles across the stream ($n = 9$). All profiles were transformed to a standard width of 1 m.

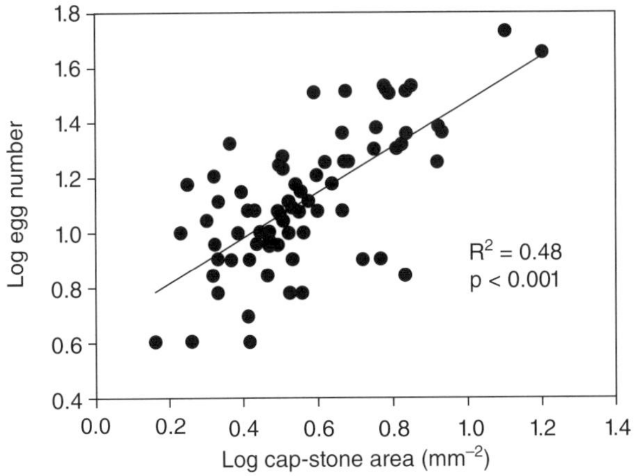

Figure 7.134 Correlation between egg masses and cap stone planar surface areas covering eggs in the field.

7.8.2.11 **Correlation between Egg Number and Cap Stone Size**

The size of the substratum used as base stones by the females varied greatly in the field. Females generally chose stones with a diameter of less than 1 cm, although sometimes larger stones, <10 cm, were used. The number of eggs and the cap stone planar surface area and mass were determined for 70 egg masses from the upper stretch of the Breitenbach. Significant relationships between egg number per egg mass and cap stone planar surface area ($r^2 = 0.48$, $P < 0.001$, Figure 7.134) and cap stone mass ($r^2 = 0.37$, $P < 0.001$) were found.

Laboratory experiments to determine the preferred cap stone size involved offering ovipositing females five size classes of cap stones. Stones between 1.6 and 3.15 mm in diameter were preferred (76%), and, as in the field there was significant correlation between egg number and cap stone mass ($r^2 = 0.16$, $P < 0.001$). The laboratory experiments confirmed that females always covered the eggs completely with a cap stone. Therefore the possibility that absence of eggs protruding from beneath cap stones in the field might be due to grazing can be excluded. Comparing mean female wet weight (Table 7.27) with mean cap stone weight shows that females were able to utilize stones many times their own wet weight (mean 4.8 times, maximum 72 times).

The results suggest that females were able to adjust the number of eggs per egg mass according to the size of available cap stone, especially as they appeared to look for a cap stone first and oviposit afterwards. The ecological significance of this egg-laying behavior lies with the ability of the cap stones to protect the underlying eggs against grazers and desiccation, particularly those at the edge of the egg mass. The interesting question as to how they estimate the cap stone size remains unsettled.

7.8.2.12 Ovipositing Behavior of Glossosomatidae

Little is known about the egg masses of Glossosomatidae. The oviposition strategy of covering egg masses with cap stones has been described for other European glossosomatid species, for example, *Synagapetus indipennis* and *Synagapetus moselyi* (Fischer, 2003), but cap stones are not used by all the species of this family. Anderson and Bourne (1974) reported that eggs of *Agapetus bifidus* are not covered, but are instead deposited in small crevices in stone surfaces. The ovipositing strategies of semi-terrestrial insects in general merit further investigation because they seem to have a decisive influence on species distribution in different types and zones of running waters.

7.8.3 Importance of Case Construction for Crawling Behavior and the Coexistence of Three Trichopteran Scrapers in the Upper Reach of the Breitenbach

7.8.3.1 Functions of Different Case Constructions for Trichopteran Species

Different ways of using silk lead to diverse case and retreat construction, niche partitioning and a high diversity of Trichoptera (Ross, 1967; Mackay and Wiggins, 1979; Weaver and Morse, 1986). Cases, tubes, and other types of retreat are constructed by aquatic animals of different taxonomical groups (Dudgeon, 1990). However, larval Trichoptera produce a greater diversity of retreats and cases than any other order of insects (Wiggins, 1996). This section discusses the different functions of trichopteran cases.

Larval cases may protect against predators, camouflaging the larva against the substratum and protecting it from damage (e.g., Otto, 1982; Nislow and Molles, 1993; Johansson and Nilsson, 1992; Boyero, Rincon and Bosch, 2006), providing resistance to shifting mineral substrata (Mackay and Wiggins, 1979;

Williams, Read and Moore, 1983; Statzner, Merigoux and Leichtfried, 2005), improving respiration (e.g., Williams, Tavares and Bryant, 1987), and also acting as ballast in lotic habitats, reducing the risk of being accidentally displaced (e.g., Waringer, 1989a, 1989b, 1993; Otto and Johansson, 1995). For the larvae, cases represent important investments in time and energy, and sometimes contested resources (e.g., Merrill, 1969; Englund and Otto, 1991). Case constructions reveal information about the ecological niche of trichopteran larvae. Molles (1984) suggested that case construction is related to resource partitioning among coexisting species, and may be important for the outcome of interspecific interactions (Otto, 1976). However, the relationship between feeding, building, and ontogenetic changes in the selection of particles for case construction is unclear (Dudgeon, 1990), and it is not easy to understand the adaptive values or functional advances of particular case architecture. The trade-offs between particle size choice and energy expenditure during case construction and transportation are an interesting research field. For example, detailed analyzes of the relationship between case mass (due to case construction) and biomass of the larval constructor are rare (Otto, 1987).

The adaptive importance of species-specific case construction for crawling behavior, foraging, resource partitioning, and coexistence of three trichopteran scrapers from the upper stretch of the Breitenbach, and the mechanistical coupling between these aspects were studied (Becker, 2001).

7.8.3.2 Case Construction with Respect to Size, Shape, Composition, and Mass

All instars of the coexisting trichopteran scrapers, *Agapetus fuscipes* (Glossosomatidae), *Apatania fimbriata* and *Drusus annulatus* (Limnephilidae), build portable cases from silk and sand. The case-building behavior differs between the Limnephilidae and the Glossosomatidae. Limnephilid larvae widen and lengthen the case as they grow, and continuously remove the posterior end. Glossosomatid larvae build a new case after each molt and the larval case cannot be altered within an instar (Hansell, 1968a; Anderson and Bourne, 1974; Bohle and Fischer, 1983).

Case construction showed significant differences between the three species (Figure 7.135). *Drusus annulatus* had the longest cases, followed by *A. fuscipes* and *A. fimbriata* (Table 7.28). The species-specific shape of the cases also differed markedly. Both limnephilid larvae build gradually tapered and curved cases, the case of *A. fimbriata* being more conical. *Agapetus fuscipes* has a hemispherical case with alternately used openings on the ventral surface. Cases of *D. annulatus* and *A. fimbriata* were widest at their anterior ends, those of *A. fuscipes* in the center (Table 7.28).

The dry mass of *A. fuscipes* cases was significantly higher than that of *D. annulatus* and *A. fimbriata* cases, whose dry mass also differed significantly. The ash-free dry mass (AFDM) also differed between species (Table 7.28), AFDM highest in *D. annulatus*, followed by *A. fimbriata* and *A. fuscipes*. The ratio of mineral to organic case mass also differed significantly between *A. fuscipes* and the two limnephilid species, but not between the latter.

a)

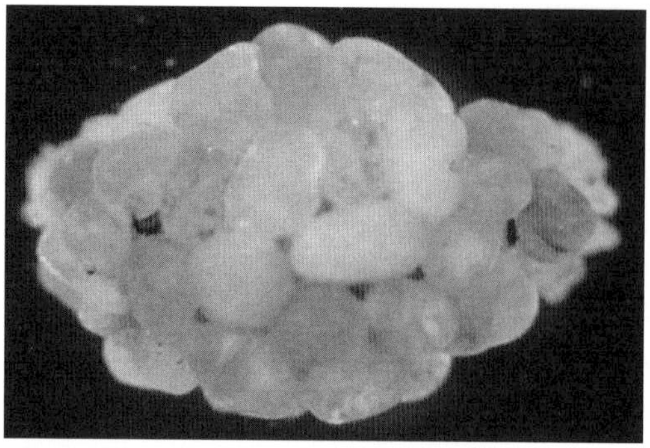

b)

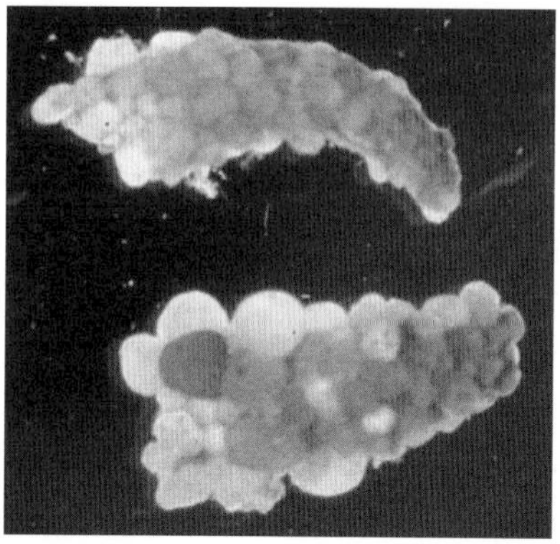

c)

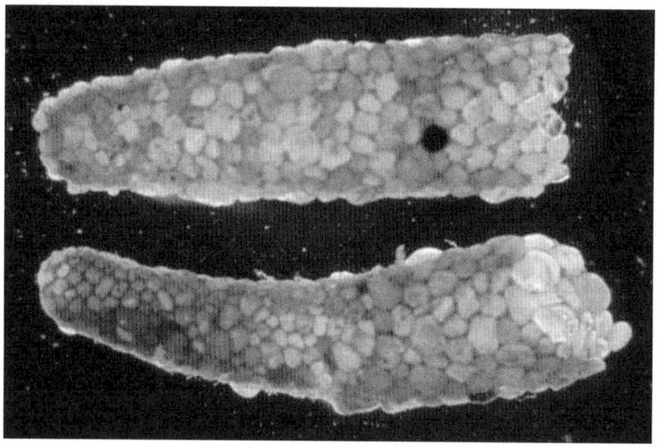

Figure 7.135 Cases of last instar larvae of: (a) *A. fuscipes*, dorsal, (b) *A. fimbriata*, and (c) *D. annulatus*, dorsal and lateral view.

Table 7.28 Case length, width, DM, AFDM, and the ratio of mineral to organic case mass of the final larval instars of the scrapers *A. fuscipes*, *A. fimbriata*, and *D. annulatus*, from the upper reach of the Breitenbach.

	n	*A. fuscipes*	*A. fimbriata*	*D. annulatus*
Case length (mm)	18	7.2 ± 0.5	6.6 ± 1.2	11.6 ± 0.8
Case width (mm) centrum or anterior end	18	4.2 ± 0.2	2.5 ± 0.6	3.2 ± 0.2
Case DM (mg)	23	53.5 ± 9.7	29.4 ± 6.2	43.8 ± 4.8
Case AFDM (mg)	23	0.21 ± 0.13	0.35 ± 0.13	0.68 ± 0.59
Ratio of mineral to organic case mass	23	255	84	64

7.8.3.3 Number and Size of Sand Grains

To determine why *A. fuscipes* cases had a significantly higher mass than the two limnephilid cases, the species-specific size frequencies of sand grains from five larval cases of each species were analyzed using image analysis (Lucia, Nikon; Becker, 2001). The composition of the cases was significantly different (Kolmogorov–Smirnov two-sample test, $P < 0.001$). *Drusus annulatus* (219 ± 24, mean ± SD) used significantly more sand grains than *A. fimbriata* (173 ± 23) and *A. fuscipes* (120 ± 8), and the size range of grains selected by *A. fuscipes* larvae was greater than by the limnephilids. *Agapetus fuscipes* used a high percentage of small grains, with a peak in the 0.3 mm mean-size class. However, they also contained larger grains (>1.3 mm), which were not used by the limnephilid larvae. The most frequent grain size category in the limnephilid cases was 0.5 mm. Although the number of grains in the cases of *A. fuscipes* was lower than in *A. fimbriata* and *D. annulatus*, the case mass of *A. fuscipes* was significantly higher. *Agapetus fuscipes* used larger grains to construct the dome than the underside of the case, and many small grains were used around the ventral openings. Only 14% of grains in the cases of *A. fuscipes* were larger than 1.3 mm, but these grains comprised 67% of the case mass.

The link between habitat and use of particles for building cases is apparent from the higher proportion of species that build mineral cases in lotic than in lentic waters (Webster and Webster, 1943; Otto, 1982). Some limnephilid species exhibit case polymorphism, using both mineral and organic material. Larvae of the caddisfly, *Potamophylax latipennis* (Limnephilidae), can use chemical cues to discriminate between types of predator and modify their case choice behavior according to predation risk (Boyero *et al.*, 2006). Caddis larvae can manipulate the size and shape of organic particles, but they must select the most appropriate particles when they build mineral cases (e.g., Merrill, 1965; Bohle, 1972a, 1972b; Hansell, 1974). *Agapetus fuscipes* (Hansell, 1968b) and other species use size rather than the mass of sand grains as a criterion (e.g., Teyrovsky, 1939; Hanna, 1961;

Nepomnyashchikh, 1993). Optimal sized mineral grains can be a limited resource for hydropsychid larvae when they build their pupal cases at preferred riffle sites, with high flow and strong erosion (Statzner, Merigoux and Leichtfried, 2005).

7.8.3.4 Use of Silk, Case Construction Costs, and Case: Larval Mass Quotient

The organic portion of the *A. fuscipes* cases was significantly lower than in the limnephilid cases. *Agapetus fuscipes* has many small respiration gaps in its dome and, in this part of the case, the grains are only joined by spots of silk. In contrast, the sand grains of both limnephilid cases are cemented together so that the cases only have openings at the ends. In addition, only the floor of the *A. fuscipes* case is thickly lined with silk, whereas the entire interior surface is thickly lined in the limnephilid cases (e.g., Bohle and Fischer, 1983; Cianficconi, Bicchierai and Moretti, 1993). Dudgeon (1990) considered the trade-off between construction costs and the range of particles employed. Mineral cases involve higher construction costs than most organic cases because the larger number of particles in mineral cases requires a higher volume of silk to cement them together (Smart, 1976) and the search intensity is higher (Otto and Svensson, 1980). Silk is a proteinaceous material (Wallace *et al.*, 1990) and is relatively costly to synthesize (Craig *et al.*, 1999). Otto (1974) showed that silk used in building a mineral case contributed 12% of energy expenditure in the final instar larvae of *Potamophylax cingulatus* (Limnephilidae). The high energy expenditure of case building was demonstrated by experiments on two caddis fly species (Stevens *et al.*, 1999, 2000). Rebuilding a new case by fifth instar larvae caused a significant increase in silk production and a decrease in mass and size of body parts of the resultant adults.

Larval dry mass of *A. fuscipes* (1.18 mg) was significantly lower than that of the limnephilid species (Mann–Whitney U-test, $P < 0.001$, Figure 7.136a). *Apatania fimbriata* (1.80 mg) and *D. annulatus* (4.67 mg) also differed significantly ($P < 0.001$). The relationship between case mass (DM) and larval mass (DM) showed species-specific differences (Figure 7.136b): about 44 for *A. fuscipes*, 16 for *A. fimbriata*, and 9 for *D. annulatus*. Thus, the small, light weight larvae of *A. fuscipes* drag relatively heavy cases, and the relationship between case mass and larval mass was 44 times higher than in species that construct cases from plant material (compare Otto, 1987).

The relationship between larval mass (AFDM) and the organic component of the cases (mostly silk, Table 7.28) did not differ significantly between species (Kruskal–Wallis test, $P < 0.2$). This relationship was 4.35 ± 1.79 (mean ± SD) for *D. annulatus*, 4.19 ± 1.34 for *A. fuscipes*, and 3.59 ± 1.9 for *A. fimbriata*. That means that, with respect to the proportion of silk used for case construction, the costs were similar for the three species (Becker, 2001).

7.8.3.5 Crawling Velocity of Larvae

To gain an insight into locomotory responses to substratum texture, the crawling velocity of the larvae was analyzed on artificial substrata lacking biofilms. A mosaic of nine clay tiles was used (Becker, 2001). The tiles were placed in a laboratory channel (length 3 m, width 0.26 m, height 0.40 m) connected to a 250 l

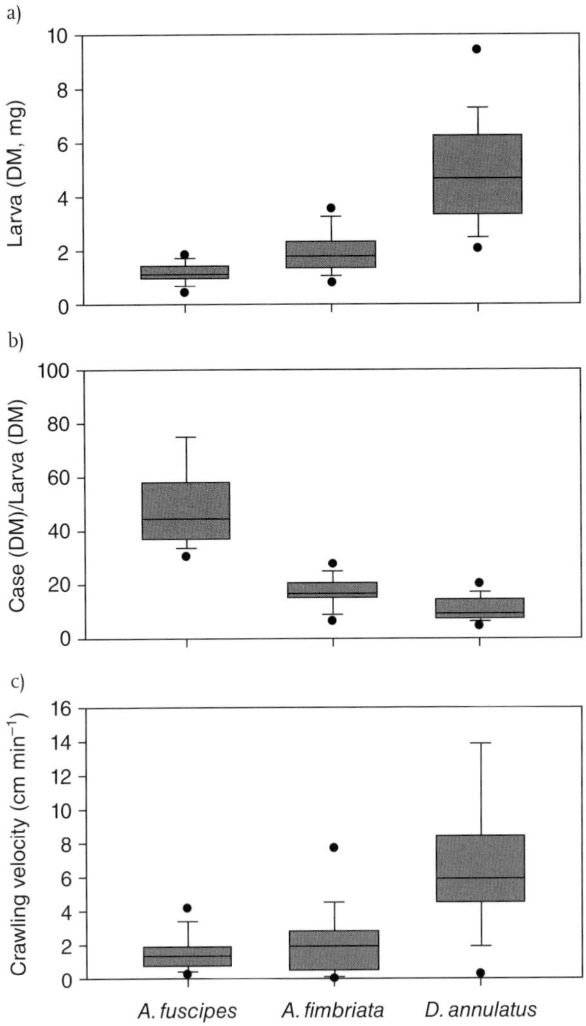

Figure 7.136 (a) Larval dry mass, (b) relationship between case and larval dry mass, and (c) crawling velocity of the last instar larvae of *A. fuscipes*, *A. fimbriata*, and *D. annulatus* on a smooth substratum surface without any biofilm. Box plots show median, 10th, 25th, 75th, and 90th percentiles. The closed circles show the 5th and 95th percentiles.

reservoir of stream water. Because the upstream crawling velocity of trichopteran larvae is negatively correlated to the current velocity (Bournaud, 1972), relatively low current velocities of 5–8 cm s^{-1} were chosen for the experiments. These low current velocities enabled the larvae to move across the tiles without being displaced by the current, even when the case had a lateral or downstream orientation to the current. Current velocity was measured with a vane anemometer (Schildknecht, mini-water, sensor diameter 15 mm). The water temperature was 14 ± 1 °C and the light/dark cycle in the laboratory was adjusted to ambient field conditions. Individual larvae (final instar) were placed at the center of the nine-tile mosaic. Larval movements were then filmed for 30 min using a black and white video camera (Panasonic, WV-CD22) connected to a video recorder (Sony VO-9600P, Umatic-SP). The path of each individual was traced on a digitizing tablet and analyzed with an interactive image analysis system (VIDS IV, AI Tektron GmbH). Crawling time, crawling distance, and crawling velocity were recorded for each larva.

The three species showed significant differences in their mobility (Kruskal–Wallis test, $P < 0.001$, Figure 7.136c). The species-specific crawling velocity of last instar larvae on smooth tiles, which were mostly traversed quickly along a straight path, was 6.2 cm min^{-1} for *D. annulatus*, followed by *A. fimbriata* (2.1 cm min^{-1}) and *A. fuscipes* (1.4 cm min^{-1}). It was positively correlated to larval mass and negatively to the case : larva mass ratio. The large, fast-moving larvae of *D. annulatus*, which have relatively light and close-fitting cases, can cover large distances in short periods of time (Menzler, 1995). These larvae can even drag their cases above the water surface, which was not observed in *A. fuscipes* or *A. fimbriata* (personal observation). *Agapetus fuscipes* cases have a large volume and a high mass relative to the small larval size and low larval mass. Glossosomatid larvae have been observed leaving their heavy cases behind, to drift during unfavorable conditions (Merrill, 1969; see Section 7.8.2.3).

7.8.3.6 **Advantages of Different Case Constructions for Epilithic Scrapers**

The ballast effect of mineral cases reduces the risk of displacement by the current (Waringer, 1989a, 1989b, 1993; Delgado and Carbonell, 1997). Waringer (1989a, 1989b) pointed out that a heavy case reduces the cost of staying in place, especially in low currents, where the adhesive friction force is higher than the drag force of the current. This is because the energy needed to attach the case to the substratum is almost entirely provided by the adhesive friction of the case. Thus, the heavy case of *A. fuscipes* may be advantageous for scraping in areas exposed to low currents. In fact, *A. fuscipes* prefers low current speeds in the Breitenbach, which contrasts with the limnephilid species (Becker, 1990; Figure 7.155). Dragging a heavy mineral case has high energetic costs (Otto and Svensson, 1980), especially on a vertical surface where the case has no contact with the surface (Bournaud, 1974a). Bournaud (1974b) found that, in *Micropterna testacea* (Limnephilidae), most of the energy used for locomotion was needed to drag the mineral case. The locomotory costs should therefore be higher in *A. fuscipes* than in *A. fimbriata* and *D. annulatus*. Thus the selection of case material is of great bionomic importance for the Trichoptera. Constructing and dragging a mineral case requires more energy than an organic case, and these costs seem to affect the generation time, final larval length, and pupal weight in trichopteran species (Otto and Svensson, 1980; Otto, 1982; Ito, 1987).

High crawling velocity is advantageous to scrapers as *D. annulatus* which move between algal patches and resting areas (Menzler, 1995). However, fast movements can be disadvantageous because predators, such as fish, that detect prey visually, focus their attention on such mobile prey (e.g., Kohler and McPeek, 1989; Johansson, 1991; Johansson and Englund, 1995). Thus the fast-moving larvae of *D. annulatus* would be at higher risk of attack by brown trout in the Breitenbach than the slow-moving larvae of *A. fimbriata* and, especially, *A. fuscipes*.

7.8.3.7 **Effect of Substratum Roughness on Larval Crawling Velocity**

In the Breitenbach, trichopteran larvae scrape biofilms covering rough sandstone surfaces, but the effects of substratum roughness on scraper foraging is poorly understood. Arens (1990) showed that mouthparts are worn more in insect larvae that scrape rough substrata than those that scrape smooth substrata. Nevertheless, scrapers still occur on various types of substratum. Clifford *et al.* (1989, 1992) found that nymphs of *Epeorus* spp. and *Cingymula* spp. (Ephemeroptera, Heptageniidae) favored smooth tiles, whereas nymphs of *Baetis* favored rough tiles. This suggests that substratum texture may affect the foraging behavior of scrapers.

Tiles were classified as smooth (unglazed), intermediate (unglazed) and rough (glazed). A friction factor was determined to quantify the roughness of each type (Waringer, 1989a, 1989b). A piece of Bunter Sandstone from the Breitenbach was weighed, tied to a spring balance and pulled horizontally across the tiles, which had been submerged in still water. When the stone started to move the reading on the spring balance was noted. The friction factor, which was calculated by dividing the mean spring balance reading by the mass of the sandstone, differed significantly between the three types of tile (Kruskal–Wallis test, $P < 0.001$; Becker, 2001). On average, the lowest friction factor was 0.37 for the smooth tiles, followed by the intermediate tiles (0.51) and the rough tiles (0.68).

The larvae of all three species showed an immediate trial and error response in their search for food, repeatedly testing the tile surface with their mouthparts as they moved forward and tried to scrape the surface. In the absence of biofilms, the larvae vacated the tiles in a relatively straight path after a few minutes, obeying the "giving-up rule" of the optimal foraging theory (e.g., Charnov, 1976; Pyke *et al.*, 1977). However, larvae showed species-specific responses to substratum roughness.

Agapetus fuscipes showed significant differences in the time spent crawling (Kruskal–Wallis test, $P < 0.001$), and in crawling velocity ($P < 0.001$), on the three substrata (Figure 7.137). Crawling time was shortest on the smooth tiles, on which a relatively large distance was covered at a relatively high crawling velocity. Larvae remained longer on the intermediate surface and covered significantly shorter distances at a lower velocity. They remained much longer, with the lowest velocity, on the rough tiles, but covered similar distances on the rough and smooth tiles. A similar pattern was observed with *A. fimbriata*, although the crawling times were not significantly different on different substrata. However the larvae crawled significantly faster on smooth tiles than on the intermediate and rough tiles.

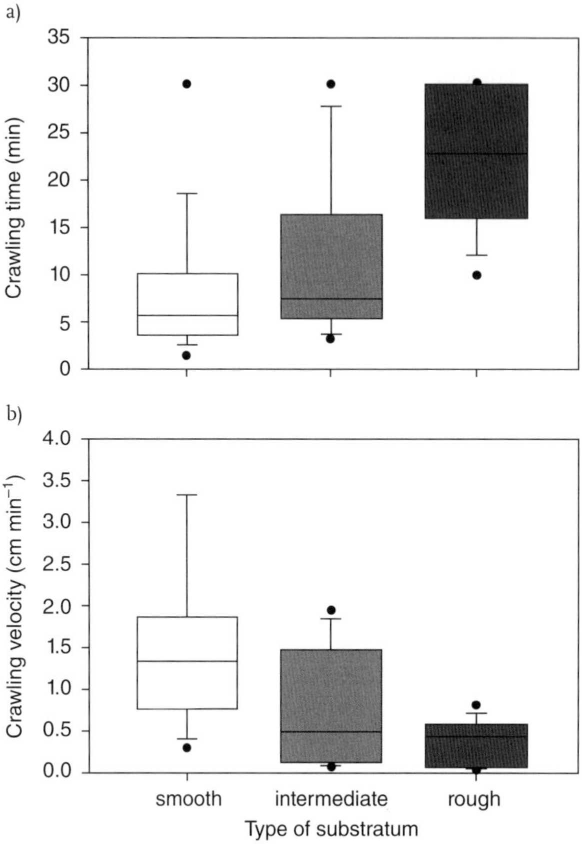

Figure 7.137 (a) Crawling time on substratum and (b) crawling velocity of sixth and seventh instar larvae of *A. fuscipes* on artificial substrata of different roughness. Box plots show median, 10th, 25th, 75th, and 90th percentiles. The closed circles show the 5th and 95th percentiles.

Substratum roughness had less influence on *D. annulatus* than on the other two species (Figure 7.138). Crawling time was longest on rough tiles, followed by smooth and intermediate tiles. Crawling velocity was higher on smooth and intermediate tiles than on rough tiles (Kruskal–Wallis test, $P < 0.01$; Becker, 2001).

But why should larval crawling behavior vary with substratum roughness? Roughness is important for the effectiveness of larval attachment to the substratum (Waringer, 1989a, 1989b), and effective attachment is a precursor to case-towing. When they move over a horizontal surface, larvae lift the front end of their cases and drag the back end over the substratum. The adhesive force of a mineral case on the substratum is related to the roughness of the substratum (Waringer, 1989a, 1989b). Thus, the efficiency of larval attachment and the passive adhesive friction of the mineral cases should be greatest on rough tiles, and lowest on

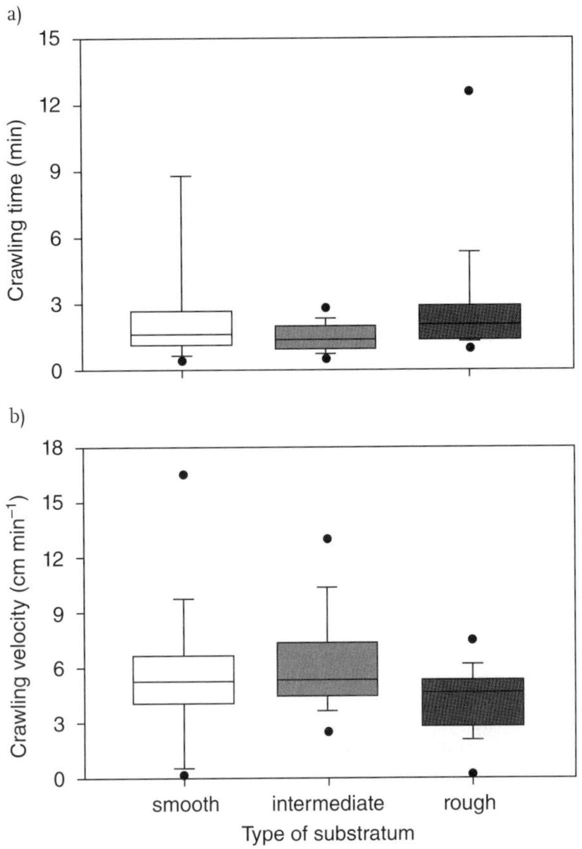

Figure 7.138 (a) Crawling time on substratum and (b) crawling velocity of fourth and fifth instar larvae of *D. annulatus* on artificial substrata of different roughness. Box plots show median, 10th, 25th, 75th, and 90th percentiles. The closed circles show the 5th and 95th percentiles.

smooth tiles. However, the cost of dragging a case over a rough substratum should be much higher than over a smooth substratum due to the greater adhesive friction factor on a rough substratum. The shape of the case also seems to be important. The tapered and curved cases of the limnephilid species may result in a smaller contact area between the case and the substratum surface, than for *A. fuscipes*. Because the different adhesive friction of mineral cases on hard substrata depends on substratum roughness, case mass and case shape, case construction may affect larval crawling velocity as well as larval size. This would explain why the relatively large larvae of *D. annulatus*, with relatively light cases, show smaller differences in their behavior on substrata of different roughness than the smaller larvae of *A. fuscipes*, which have heavier cases.

The larvae of *A. fuscipes* and *A. fimbriata* searched more intensively for food on intermediate and rough tiles than on smooth tiles. The positive response of larvae to rough substrata seems to be appropriate because rough substrata, which have a greater surface area and surface heterogeneity, provide a more suitable surface for microphytobenthic colonization (Hamilton and Duthie, 1984; Antoine and Benson-Evans, 1985). Laboratory experiments and field studies in the Breitenbach showed that *A. fuscipes* colonized biofilm patches that had a significantly lower chlorophyll-*a* content than sites where *D. annulatus* scraped intensively (Becker and Holfeld, 2001). It is assumed that larvae that scrape thin biofilms in the Breitenbach are influenced by the relatively rough surface texture of the Bunter Sandstone substratum. However, the species-specific responses may not be distinct if a thick biofilm is present, because substratum surface texture can be masked by the polysaccharide layers of biofilms (Blinn *et al.*, 1980; Ogilvie, 1988). Finally, it is difficult to decide whether the friction force between case and substratum, or the more intensive searching behavior significantly reduces the crawling velocity of *A. fuscipes* and *A. fimbriata* on rough substrata.

Considering the particle selection and case architecture of these trichopteran species of the same functional feeding group, species-specific differences in case building strategies are obvious. These result from significant differences in the relationship between case and larval mass, and crawling velocity. There are significant functional advantages to differing case constructions. The species-specific differences in physiological and behavioral adaptations, with respect to case construction and crawling behavior, seem to be important for resource partitioning and habitat use in these scrapers, enabling them to coexist in small uphill streams, such as the Breitenbach.

7.8.4 Interactions between Scrapers and Periphyton

7.8.4.1 Introduction

Research into trophic relationships is an important aspect of ecosystem analysis. Interactions between herbivores and autotrophic organisms are clearly processes that control ecosystem structure and functioning. Several studies have shown the complex interactions between bottom-up effects (nutrients, light intensity) and top-down effects (consumers) on autotrophic biomass and composition (e.g., Lamberti *et al.*, 1989; Leibold, 1997; Hillebrand *et al.*, 2002; Hillebrand, 2005; Darcy-Hall and Hall, 2008). Most studies on the interactions between herbivores and their food resources in diverse aquatic habitats showed that mean autotrophic biomass was reduced by grazer presence, and that high densities of grazers altered the spatial distribution of autotrophs (Steinman, 1996; Hillebrand, 2008). Many investigations have focussed on interactions between microphytobenthos and scrapers in benthic zones of running waters (reviewed by Feminella and Hawkins, 1995; Lamberti, 1996; Steinman, 1996) and have revealed the influence of periphyton abundance and patchiness on the distribution, growth and development of grazers (e.g., Feminella and Resh, 1990; Hill, 1992; Hill *et al.*, 1992), as well as the effects of grazer type and density on standing crop, turnover and species composition of periphyton communities (e.g., Lamberti *et al.*, 1987a; Hill and Knight, 1988; Coletti *et al.*, 1987; Steinman *et al.*, 1987b; Stief and Becker, 2005). Many of these studies were conducted in unshaded headwater streams of temperate regions (e.g., McAuliffe, 1984a; Feminella and Resh, 1991; Kohler, 1992), with distinctly higher primary production rates than forested headwater streams (e.g., Fisher and Likens, 1973; Webster and Meyer, 1997d; McNeely and Power, 2007; McNeely *et al.*, 2007).

Macroinvertebrate grazers can affect both the biomass and assemblage structure of benthic microalgae, as well as functional components, such as productivity (reviewed by Steinman, 1996). However, the outcome of grazer-periphyton interactions largely depends on grazer type, its body size, mobility, density, mouthpart morphology, grazing rate and efficiency, and metabolic requirements (Lamberti *et al.*, 1987a, 1987b; DeNicola *et al.*, 1990; Stief and Becker, 2005). Several studies showed that snails and caddisfly larvae are more effective than mayflies at removing algae because their radular and blade-like mouthparts, respectively, allow them to feed on low-lying forms in algal biofilms (e.g., Lowe and Hunter, 1988; Tuchman and Stevenson, 1991; Holomuzki and Biggs, 2006), which are unavailable to mayfies with sweeping brush-like mouthparts (Hill and Knight, 1988; Holomuzki *et al.*, 2006). The latter are better suited to feeding on loosely attached algae. On the other hand, *Baetis* uses its mouthparts to scrape and gouge the epilithon from the substratum (Arens, 1989), and might be able to feed on tightly attached, adnate algae (Wellnitz and Poff, 2006). Large grazing caddisflies can be more effective than snails at removing periphyton, due to their higher foraging activity and grazing rates (Lamberti *et al.*, 1987a; Lamberti *et al.*, 1987b; Hill and Knight, 1988; Hill *et al.*, 1992; Holomuzki and Biggs, 2006). When ungrazed algal patches experience intensive grazing, the community changes from erect diatoms and filamentous chlorophytes, to adnate diatoms and basal cells of chlorophytes (e.g., Lamberti *et al.*, 1987a, 1987b, 1989 Steinman *et al.*, 1987a, 1987b Holomuzki and Biggs, 2006; Holomuzki *et al.*, 2006; Connelly *et al.*, 2008).

Grazers reduce producer biomass, but, in many cases, grazing results in younger algal assemblages with higher biomass-specific rates of primary production (e.g., Gregory, 1983; Lamberti and Resh, 1983), because grazers alter the availability of limiting resource, such as light and nutrients. By reducing the overstorey layer, detritus and mineral sediments, when they thin the periphyton patch they decrease shading and increase the rate of light penetration and nutrient transport to lower photoautotrophic strata because they create gaps in the algal matrix (e.g., Hart, 1985b; Power *et al.*, 1988; Lamberti *et al.*, 1989; McCormick and Stevenson, 1991; Mulholland *et al.*, 1991; Hillebrand, 2005). Furthermore, when they serve as nutrient pumps and remineralize nutrients, grazers may help regulate the availability of nutrients to stream algae, which may stimulate production and reduce the turnover time of algal cells (e.g., McCormick and Stevenson, 1989; McCormick, 1990; Hill *et al.*, 1992; Evans-White and Lamberti, 2005; Stief and Becker, 2005).

High grazer efficiency keeps algal communities in an early successional stage. Early algal assemblages are composed mostly of diatoms and unicellular green algae, whereas late communities are dominated by senescent, filamentous, green and blue-green algae (e.g., Lamberti *et al.*, 1989). Senescent filamentous algae may be less nutritious for grazers than vigorously growing diatoms (Lamberti and Moore, 1984; Steinman *et al.*, 1987a). Mature algal filaments are often considered grazer resistant. Intense grazing on early developmental, edible stages of filamentous algae prevents the establishment of inedible, old stages of filamentous algae in algal patches (DeNicola *et al.*, 1990; Dudley and D'Antonio, 1991; Steinman *et al.*, 1992; Darcy-Hall and Hall, 2008).

The productive capacity of autotrophs is a major determinant of the distribution, growth and development of consumers in stream ecosystems (e.g., Lamberti *et al.*, 1989). Standing biomass of periphyton may only represent a small fraction of total microphytobenthic production in streams (Gregory, 1983). Because of high turnover rates, streams may have a small amount of autotrophs, but high herbivore biomass (e.g., Gregory, 1983; Lamberti and Resh, 1983; McAuliffe, 1983; Lamberti and Moore, 1984). However, when consumer requirements greatly exceed productivity capacity, grazers may compete for limited food resources (Hart, 1987; Lamberti *et al.*, 1987b)

But benthic algae are not only eaten by herbivores. Grazers increase the export of algal cells from high standing crops by dislodgement as they feed and move in the algal patches (e.g., Lamberti *et al.*, 1989; Barnese and Lowe, 1992). The algal patches then drift downstream and build a nutritious food resource for filter feeders (Benke and Wallace, 1980; Doi *et al.*, 2007) and detritivores (Lamberti and Moore, 1984).

Ten grazer species regularly occur on Bunter sandstone substrata in the Breitenbach, six trichopterans, three ephemeropterans, and the gastropod, *Ancylus fluviatilis*. The larvae of *Agapetus fuscipes, Apatania fimbriata, Drusus annulatus,* and *Silo pallipes* (all with mineral cases) and of *Micrasema longulum* (with silken cases) build mobile retreats, while the larvae of *Tinodes rostocki* live in tubular retreats and forage by grazing on the biofilms of their retreats (Stief and Becker, 2005). *Micrasema longulum* has been found on stones as well as submerged mosses. Larvae of the case-building species use silken safety lines to avoid slipping off the stone surface in the current (Menzler, 1995; Becker, personal observations). These trichopteran species have similar life cycles in the Breitenbach. They are univoltine, begin emerging in the spring (Sandrock, 1978; Section 7.6.4.5) and do not show marked differences in their temporal adaptations. The larvae of the mayflies *Baetis vernus* and *B. rhodani* are mobile grazers/gatherers and have been found in high densities on the sandstone substrata. The mayfly *Ecdyonurus* sp. and the snail *A. fluviatilis* are typical epilithic grazers. Long-term emergence studies have shown that both baetids and *A. fuscipes* are dominant species in the Breitenbach (Illies, 1978, 1982; Sandrock, 1978; Sections 7.6.4.2 and 7.6.4.5). However, *T. rostocki, A. fimbriata,* and *D. annulatus* and *S. pallipes* also develop large populations. In contrast, *M. longulum, Ecdyonurus* sp., and *A. fluviatilis* are generally less prevalent.

Long-term emergence studies demonstrated the high production of insects with grazing larvae in the Breitenbach (Illies, 1978, 1982; Sandrock, 1978). In their capacity as primary consumers, the high larval density of the grazers would appear to have an important influence on stream metabolism.

The main goal of the following studies was to investigate the ecological segregation of these grazers, focusing on the trichopteran species, and to understand the feeding strategies and coexistence of these species along the Breitenbach. Most attention was given to the foraging behavior, examining the quantity and quality of the epilithic periphyton and the composition of larval gut contents along the stream, the spatial and temporal development of grazers and periphyton in two sections of the stream, the response to different food conditions: (i) natural epilithic biofilms grown in the Breitenbach, and (ii) unialgal layers cultured in laboratory flumes. Intra- and interspecific competition for food resources between grazers, and the strength of the top-down control of the epilithon by these consumers, were investigated. Furthermore metabolic demand, as well as food choice, growth and development at a food supply of diverse qualities, were analyzed. In addition, the diurnal foraging activity of two species was examined.

7.8.4.2 **Comparison of the Dietary Composition of Epilithic Trichopteran Species**

An investigation of periphyton on stony substrata in the Breitenbach (Cox, 1990b) revealed a high degree of uniformity in the diatom flora along the stream. Smaller species, such as *Navicula* were dominant. *Pinnularia* and *Eunotia* spp. were dominant on submerged mosses. *Chamaesiphon incrustans* and filamentous *Anabaena* and *Oscillatoria* spp. were prominent on the stony substrata. The filamentous green algae *Microspora amoena* (Chlorophyceae) and *Tribonema viride* (Xantophyceae) were also found, as well as various colonial and coccoid forms (but see Section 5.5.1).

The aim of the field study, conducted between April 1986 and October 1987, was to reveal variability in the larval diets of six trichopteran species through the year, and to investigate whether qualitative differences in food availability along the stream could influence the longitudinal distribution of the larvae. Later-staged larvae were sampled from eight sites with stony substrata in the lower, permanently flowing 2 km of the stream on 11 occasions. At this time, *A. fuscipes, A. fimbriata* and *D. annulatus* colonized the entire stream length, but were most abundant in the upper reaches. In contrast, *T. rostocki, M. longulum*, and *S. pallipes* were only found in the middle and lower reaches (Becker, 1990).

Gut contents obtained from 1321 larvae scraping stony substrata, were extracted, dispersed using brief sonication, drawn down onto a filter membrane (Cummins, 1973; Reidelbach, 1982), classified into nine components and quantified using interactive, color-image analysis (VIDS IV, AI Tektron GmbH). The relative areas of nine gut content components were quantified; three microphytobenthos fractions (diatoms, cyanobacteria, green algae), three detritus fractions (POM $< 5\,\mu m$, POM $> 5\,\mu m$, leaf fragments), fungi, bryophytes, and inorganic material. Periphyton samples were also taken by brushing three stones at each sampling site. A total of 280 membrane filters were analyzed (gut contents, $n = 220$, periphyton, $n = 60$). At least 20, randomly chosen fields per filter were analyzed, examining at least $8000\,\mu m^2$ gut content material (Becker, 1990).

7.8.4.3 Gut Content and Periphyton Analyses

There was a high degree of variability in the gut contents of each species. To obtain an overview, an ANOVA was used to reveal the importance of species, sampling month and site on gut content composition (Becker, 1990). The results showed significant differences between species for seven of the dietary components (diatoms, cyanobacteria, POM < 5 μm, POM > 5 μm, bryophytes, green algae, inorganic material). There were no significant differences for leaf fragments or fungi. Significant differences between sampling months were also found for seven dietary components (diatoms, cyanobacteria, POM < 5 μm, POM > 5 μm, leaf fragments, fungi, inorganic material). Only green algae, which were rarely consumed, and bryophytes showed no significant differences between sampling month. Significant differences between sampling sites were only found for diatoms.

7.8.4.4 Comparison between Species

The guts of *M. longulum*, *T. rostocki*, and *D. annulatus* contained above-average proportions of diatoms, differing significantly from those of the other species (Figure 7.139a; Becker, 1990). The highest proportions of diatoms occurred in *M.*

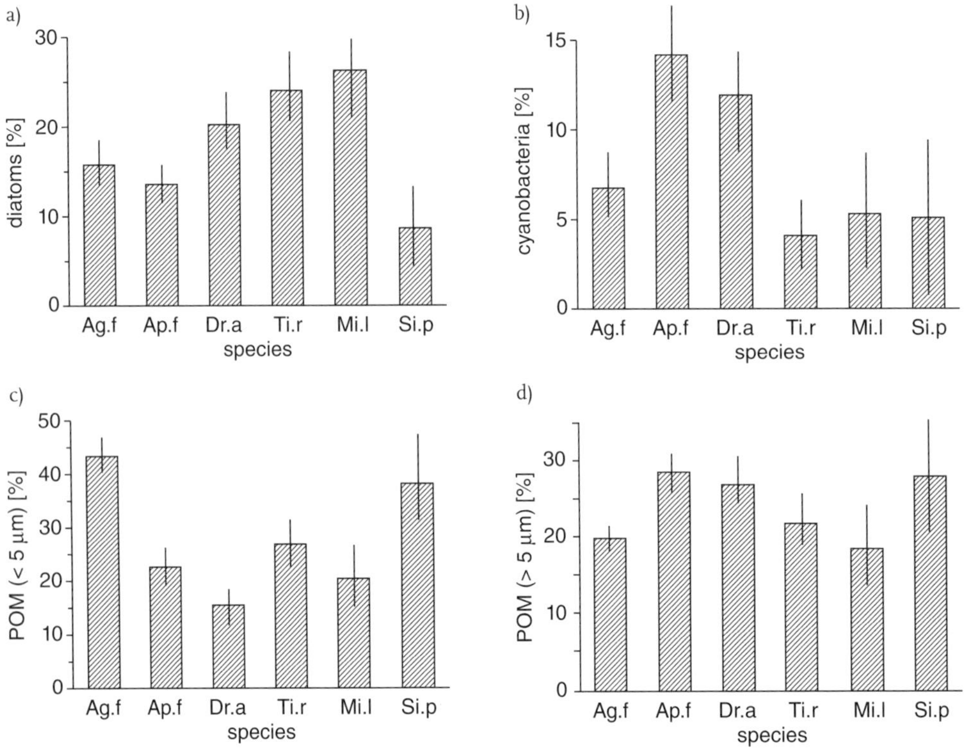

Figure 7.139 Mean percentage areas of: (a) diatoms, (b) cyanobacteria, (c) POM < 5 μm, and (d) POM > 5 μm of the gut contents of the different species, for all samples. Ag (*A. fuscipes*), Ap (*A. fimbriata*), Dr (*D. annulatus*), Ti (*T. rostocki*), Mi (*M. longulum*), Si (*S. pallipes*).

longulum and *T. rostocki*, where they represented up to 60% of the gut-contents in spring. *Apatania fimbriata* and *D. annulatus* consumed relatively high proportions of cyanobacteria (Figure 7.139b), significantly higher than in *T. rostocki*, *A. fuscipes*, and *M. longulum*. In the field, *A. fimbriata* and *D. annulatus* larvae were often observed on cyanobacterial mats and lichens, in the upper and middle reaches of the stream. The larvae of *A. fuscipes* were rarely found on these mats and contained high proportions of POM (<5 μm), differing significantly from those in *A. fimbriata*, *D. annulatus*, *M. longulum*, and *T. rostocki* (Figure 7.139c). High proportions of POM (<5 μm) were also found in *S. pallipes*, which differed significantly from *A. fimbriata*, *D. annulatus*, and *M. longulum*. Above average proportions of POM (>5 μm) were observed in *A. fimbriata*, *D. annulatus*, and *S. pallipes* (Figure 7.139d), but low proportions in *A. fuscipes* and *M. longulum*. *Micrasema longulum* consumed relatively high proportions of moss (mean 13%, max. 46%). In contrast, moss was rarely found in *A. fimbriata* or *D. annulatus*, and was totally absent from the other species. Fragments of green algae were rarely found in larval guts and so appeared to be less important for the nutrition of these species. On the whole, the other food types were quantitatively unimportant. The lowest proportions of green algae were found in *A. fuscipes* and *S. pallipes*, the highest in *A. fimbriata* and *M. longulum*. Mean proportions of fungi ranged between 1 and 3%. In autumn, there were up to 29.5% of leaf fragments in *D. annulatus* and 35.3% in *A. fimbriata*, however, the mean proportions ranged between 1.5% and 4.0%.

The results show that there are significant differences with respect to the utilization of various food components by the six trichopteran scrapers. This was despite high variability in their gut content composition, partly caused by seasonal changes in periphyton composition, as well as the distinct overlap in their food resources and microhabitats (Becker, 1990). The differing gut content compositions may reflect differences in the feeding strategies of these species. This is indicated by differences in mandible morphology. *Agapetus fuscipes*, *A. fimbriata*, *D. annulatus*, and *S. pallipes* have scoop-shaped mandibles, typical of trichopteran scrapers (Lamberti and Moore, 1984; Arens, 1989), while the mandibles of *T. rostocki* shave fine-toothed structures, and resemble those of filter-feeding caddisflies (Hydropsychidae), and the mandibles of *M. longulum* resemble those of shredders (Hickin, 1967; Wiggins, 1996).

7.8.4.5 **Gut Content and Periphyton Composition throughout the Year**

Seasonal variability in periphyton composition was partly reflected in the larval gut contents of all species (Figures 7.140 and 7.141; Becker, 1990). A spring diatom maximum in both the periphyton (Figure 7.140a) and the larval gut contents (Figure 7.140b) was followed by a summer minimum. There was a second periphyton maximum in autumn, but only a slight increase in the gut contents. Patterns were similar through the year, although the gut content maxima and minima lagged behind those of the periphyton. The opposite pattern was observed for cyanobacteria. In periphyton (Figure 7.141a) and gut contents (Figure 7.141b) there were low proportions of cyanobacteria in spring, followed by a summer maximum and an autumnal decrease. The gut content maximum also lagged

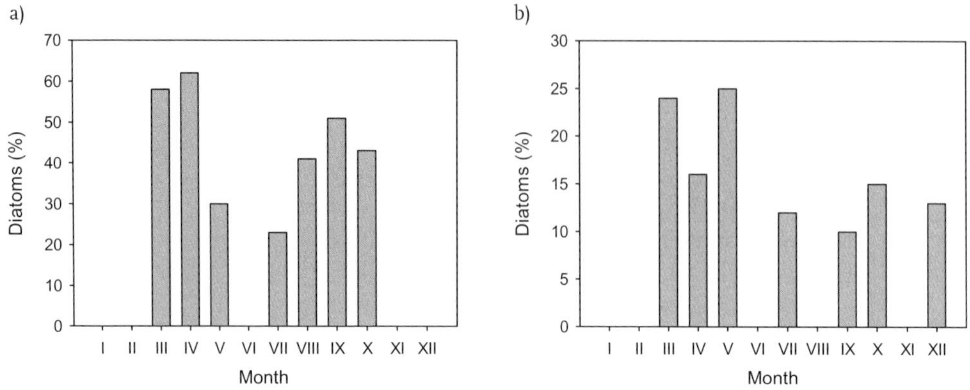

Figure 7.140 Comparison of mean percentage areas of diatoms in: (a) periphyton and (b) gut contents of all species (all samples).

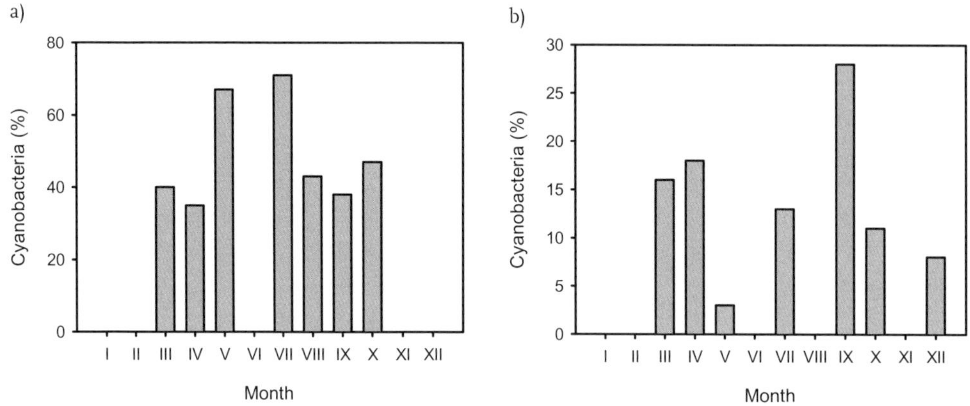

Figure 7.141 Comparison of mean percentage areas of cyanobacteria in: (a) periphyton and (b) gut contents of all species (all samples).

behind that of the periphyton. An increase in the proportion of green algae in the periphyton from August to October, up to an average of 6%, was not reflected in the gut contents.

The consumption of diverse dietary components differed significantly between individual larval species across the year (Becker, 1990). The guts of all species contained relatively high proportions of diatoms in the spring, but there were distinct differences in their diatom consumption in the autumn (Figure 7.142). The proportion of diatoms consumed by *A. fuscipes* remained at the spring level through the summer, but decreased in autumn (Figure 142a). After relatively high proportion in spring, a decrease through the year was observed in *A. fimbriata* (Figure 142b). Both species contained low proportions of diatoms in December.

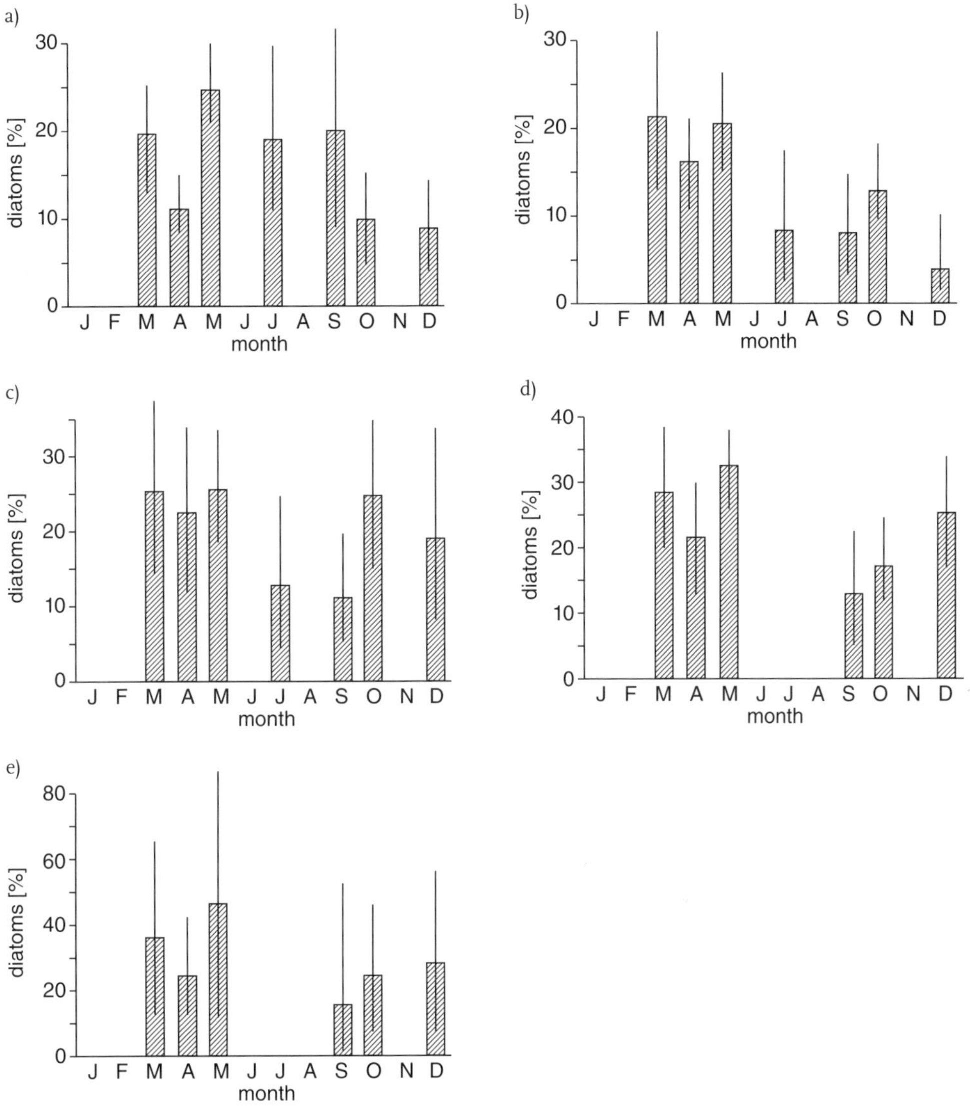

Figure 7.142 Mean percentage areas of diatoms in gut contents of *A. fuscipes* (a), *A. fimbriata* (b), *D. annulatus* (c), *T. rostocki* (d), and *M. longulum* (e) throughout the year (all sampling sites).

In contrast, there were similar patterns in the diatom proportions in the guts of *D. annulatus*, *T. rostocki* and *M. longulum* through the year, albeit at different levels (Figure 142c–e). The highest proportions were observed in spring, the lowest in September. The diatom proportions returned to their spring levels in October and December. All species showed a decrease in diatom consumption in April. The gut contents of *A. fimbriata* and *D. annulatus* reflected the summer increase in

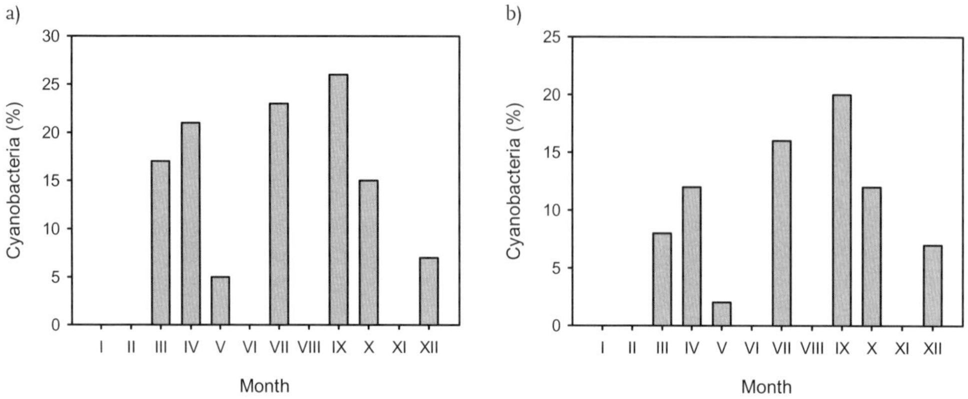

Figure 7.143 Mean percentage areas of cyanobacteria in gut contents of *A. fimbriata* (a) and *D. annulatus* (b) throughout the year (all sampling sites).

cyanobacteria (Figure 7.143a,b). Larvae of *A. fuscipes* consumed almost the same proportion of POM (<5 μm) through the year, while there were differences between the warm and cold seasons in *T. rostocki*. On average, twice as much was consumed in April, May and September than in October, December and March. There were also significant differences in the POM (>5 μm) fractions consumed in five species, apart from *D. annulatus*. A marked increase in this gut-content fraction at the end of the year (October and December) coincided with macrophyte die-back. There was less variation in *A. fuscipes* than the other species, such as *M. longulum*. In the latter species, the proportion consumed increased between September and December, while consumption of POM (>5 μm) in October and December differed significantly from that in the spring.

The low proportions of diatoms in *A. fimbriata* compared with other species, may have been due to differences in hibernation strategies. Information on hibernation is available for three of the Breitenbach species. In the mild winter of 1987/1988 winter dormancy was observed in *A. fimbriata* (Aurich, 1989), while larvae of *T. rostocki* (Becker, 1993) and *A. fuscipes* (Becker, 2005) developed throughout the entire winter. In this study most winter-found *A. fimbriata* larvae were attached to the undersides of stones. The gut contents of larvae sampled feeding on the upper stone surfaces contained an average of 60% POM (>5 μm). There was a clear shift between a microphytobenthos-rich diet in spring, to a detritus-rich diet in autumn and winter in this species (Becker, 1990).

Another influence on larval gut content composition could be stream discharge. The relatively low proportions of diatoms observed in larval guts in April (Figures 7.140b and 7.142) appear to have been due to the spring spates, between the end of March and the beginning of April (Becker, 1990). The proportions of diatoms in periphyton were obviously high at this time (Figure 7.140a). Observations made directly after flooding revealed that substrata in the center of the stream, which were otherwise densely populated throughout the year, had been vacated by most of the larvae, especially by *A. fuscipes*. However, high densities of these larvae were found near the bank at this time. The high proportions of POM (<5 μm) in their diet indicate that the larvae were feeding in low velocity areas, where detrital sedimentation was high. In addition, the relatively high proportions of detritus and inorganic material in gut-contents in the summer probably reflect higher sedimentation rates on stone surfaces, as a result of lower stream discharge. This would also have been accompanied by a decrease in microphytobenthos abundance at this time, typical of other streams too (Marker, 1976; Tett *et al.*, 1978; Robinson and Minshall, 1986).

7.8.4.6 **Variability in Gut Content and Periphyton Composition along the Stream**

Combining all sampling occasions for microphytobenthos composition along the stream, demonstrated that the only notable differences between sites were in the proportion of periphytic diatoms in the Georgsborn, as well as the proportions of cyanobacteria in the Georgsborn and the spring-fed tributary (Becker, 1990). The generally low proportions of periphytic green algae were not significantly different between sites. When the gut contents of all the species were considered together, the only significant differences occurred between the high proportions of diatoms in the guts of individuals sampled in the middle reaches of the stream, and the low proportions in animals from the spring-fed tributary. The lack of significant differences between the gut contents of any individual species sampled along the stream, as well as the lack of significant correlations between sampling sites and the proportions of single dietary components consumed, suggests that the longitudinal distribution of these larval species was not significantly influenced by food availability along the stream. Daily and annual amplitudes in water temperature may have been more important factors (Sections 7.9, 8.3, and 8.4).

Although no clear increase or decrease of the proportion of diatoms in the periphyton and larval gut contents along the Breitenbach, diatoms in the epilithic periphyton increased in mean cell length at the eight sampling sites, from the source to the mouth of the stream, with respect to changing species composition along the stream (Figure 7.144a). Distinct differences in mean diatom cell lengths were found between the larval scraper guts. Summing the data from all sampling sites showed that the highest amount of small diatom cells was found in the guts of *A. fuscipes* (88% of cells < 20 μm) and *D. annulatus* (76% of cells < 20 μm), two species with a high abundance in the upper part of the stream. The gut contents of *T. rostocki* (67% of cells < 20 μm) and *M. longulum* (61% of cells < 20 μm), two species that colonized the middle and lower sections of the stream, showed a higher proportion of larger diatom cells. The mean diatom cell length pattern in *Drusus* guts along the stream was well correlated with the cell length pattern in the periphyton (Figure 7.140c). In contrast, mean diatom cell length in *Agapetus* guts increased only slightly along the stream (Figure 7.140b). Thus, *D. annulatus* larvae consumed an average sample of available periphytic diatoms, while *A. fuscipes* larvae fed at specific sites with small diatom cells (Becker, unpublished data).

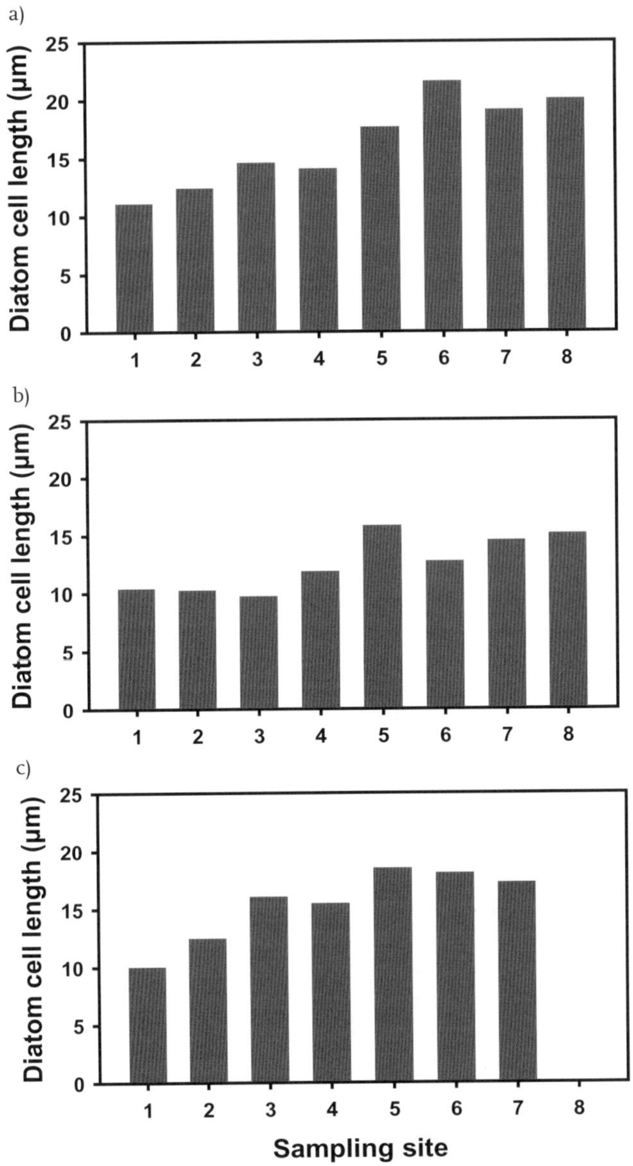

Figure 7.144 Diatom cell length of periphyton (a) and in the guts of *A. fuscipes* (b) and *D. annulatus* (c) larvae along the Breitenbach.

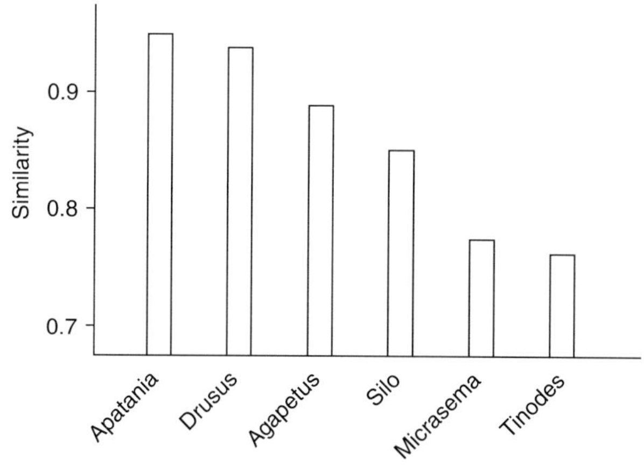

Figure 7.145 Similarity of the microphytobenthic composition in periphyton and larval gut contents.

7.8.4.7 Primary Producers, Food Quality, and Resource Overlap

A comparison of the microphytobenthic composition of gut contents with the local periphyton was used to reveal whether any particular species consumed microphytobenthos proportionally to the natural periphytic occurrence (Becker, 1990). The closest correspondence was found for *A. fimbriata*, closely followed by *D. annulatus* (Figure 7.145). Both species consumed high proportions of cyanobacteria. The greatest deviation from the periphytic microphytobenthic composition was observed for *T. rostocki* and *M. longulum*, which were characteristically diatom-specific consumers. For *T. rostocki* this is explained by the specific feeding strategy of the larvae, which graze on the retreat walls, which contain a high abundance of diatoms (Hasselrot *et al.*, 1996; but see Section 7.8.1). In *A. fuscipes* and *S. pallipes*, microphytobenthic composition was dominated by diatoms and cyanobacteria, respectively, although they consumed a low proportion of primary producers.

Figure 7.146 shows the dietary components as three groups: high quality components (microphytobenthos), lower quality components (detritus), and the remainder (Becker, 1990). Distinct differences were observed between the larval species. The highest proportions of microphytobenthos were found in *D. annulatus* and *M. longulum*, followed by *A. fimbriata* and *T. rostocki*. In contrast, *A. fuscipes* and *S. pallipes* consumed small proportions of microphytobenthos, but high proportions of detritus. The remaining dietary components appeared to be less important.

Taking all the dietary components into consideration, a calculation of niche overlap gave values of 0.67 to 0.89. This indicates a clear overlap in food resources between the various larval species. A cluster analysis revealed three groupings (Figure 7.147). The most distinct resource overlap occurred between *A. fuscipes* and *S. pallipes*, followed by *A. fimbriata* and *D. annulatus*, which were more

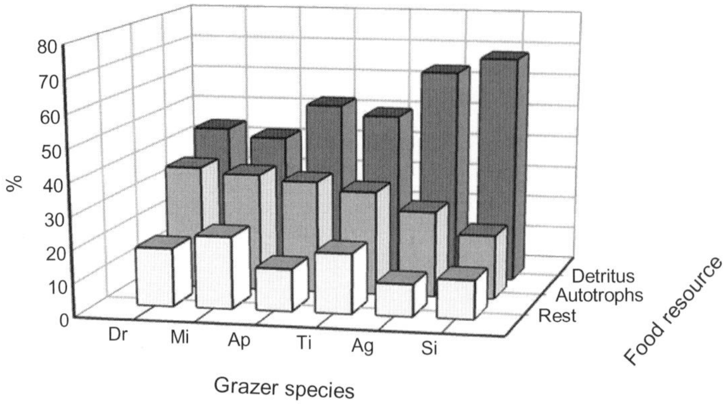

Figure 7.146 Comparison of the mean proportions of microphytobenthos, detritus and the remaining food components between larval species. Ag (*A. fuscipes*), Ap (*A. fimbriata*), Dr (*D. annulatus*), Ti (*T. rostocki*), Mi (*M. longulum*), Si (*S. pallipes*).

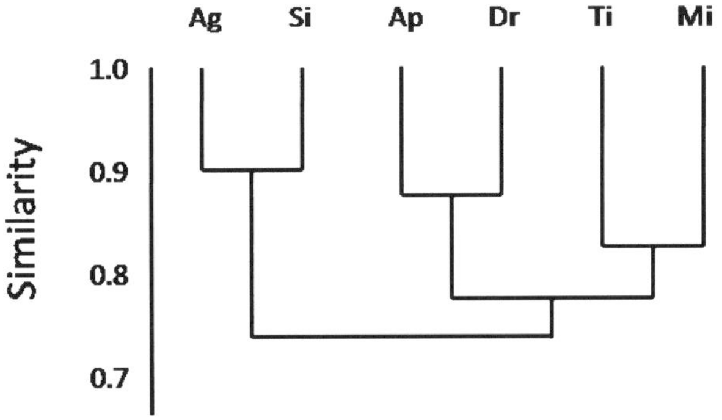

Figure 7.147 Resource overlap dendrogram (all food components included). Ag (*A. fuscipes*), Ap (*A. fimbriata*), Dr (*D. annulatus*), Ti (*T. rostocki*), Mi (*M. longulum*), Si (*S. pallipes*).

abundant in the upper part of the Breitenbach. *Tinodes rostocki* and *M. longulum* grouped together at a lower level. These species, with apparently diverse feeding strategies were more abundant in the middle and lower sections of the Breitenbach (Becker, 1990).

The full significance of species-specific differences in the consumption of qualitatively differing food components is revealed when the assimilation efficiency of the food components is considered. Benke and Wallace (1980) reported a 30% assimilation efficiency for microphytobenthos by Hydropsychidae, but only 10% efficiency for FPOM and CPOM. The assimilation efficiency for diatoms was 55% in Simuliidae and 49% in Hydropsychidae (McCullough *et al.*, 1979). Schweder (1985) reported a 69% assimilation efficiency for diatoms in *Ecdyonurus venosus*. If the trichopteran species considered here have comparable assimilation efficiencies, the interspecific differences in food quality is much greater than is indicated by the proportional distribution of food components in the guts. Diatoms appear to be particularly important to aquatic scrapers. The species investigated here concluded their larval development during the diatom spring maximum in the Breitenbach (Becker, 2005). For the Trichoptera, up to 80% total larval weight gain occurs in the final larval stage (Becker, 1987a, 1987b), during which significant lipid reserves are built up. In addition to proteins, the relatively high levels of lipids in diatoms (Opute, 1974) would also appear to be important for larval development at this stage. 75% of glossosomatid production (Mayer and Likens, 1987) and 84% of scraper production (Huryn and Wallace, 1988) has been attributed to the consumption of diatoms. Steinman *et al.* (1987a) suggested that the high level of polyunsaturated lipids in diatoms may be crucial to the nutrition of aquatic insects, which are mostly unable to synthesize their own polyunsaturated fatty acids (Downer, 1978) or sterols (Robbins *et al.*, 1971). Most diatoms found in larval guts had been digested with their cell walls intact. Thus it appears that the larvae can digest the diatom cell contents through pores in the cell walls, as also observed by other authors (Lamberti and Moore, 1984; Klonowska, 1986).

Green algae, especially filamentous species, appear to be unimportant in the nutrition of these trichopteran species (but see Section 7.8.4.9). Mature filaments are often considered grazer-resistant (Dudley and D'Antonio, 1991, Steinman *et al.*, 1992), and cellulose cell walls can cause problems for their digestion (Neumann, 1961; Lamberti and Moore, 1984; Schweder, 1985; Hoffmann, 1986). Cyanobacteria cell walls, which can be broken down by proteases, appear easier to digest, although this may be partly dependent on cell wall thickness (Hoffmann, 1986). Avoidance of filamentous microphytobenthos by scraping trichopteran larvae has been previously observed (Jacoby, 1987). *Leucotrichia pictipes* larvae removed filamentous blue-green algae from their foraging areas, but did not ingest them (McAuliffe, 1984b; Hart, 1985b).

The main nutritional value of detritus (POM < 5 μm, POM > 5 μm, leaf fragments), which is rich in cellulose and lignin, and of inorganic material, is probably derived from the attached bacteria. Bacterial protein is likely to be an important source of nitrogen (White, 1978). Marxsen (1988a, 1988b) reported that the highest bacterial activity on fine sediments in the Breitenbach occurred in the summer. The potential importance of protozoa and rotifers to larval nutrition has yet to be assessed. Although probably abundant (Packroff and Zwick, 1998), they were not found in filtered gut and periphyton samples.

Although there was a clear overlap in food resources for the trichopteran grazers and a lack of any distinct temporal segregation of the caddisfly species, there are significant differences in the larval diets. It is suggested that these differences are mainly the result of species-specific foraging strategies (Becker, 1990).

7.8.4.8 Interspecific Competition between Scrapers and Top-Down Control of Biofilms

It is presumed that high larval abundances of trichopteran scrapers caused intra- and interspecific competition for high quality food components, such as diatoms (e.g., Hart, 1981, 1987; Hill and Knight, 1988). In October 1988, *A. fuscipes* attained abundances of 17 229 ± 7880 young and middle-stage larvae m^{-2} (mean ± SD, $n = 4$) under favorable conditions in the upper stretch of the Breitenbach, near trap T1; *A. fimbriata* 4728 ± 1858, and *D. annulatus* 4851 ± 3490 (Becker, unpublished data). Territorial behavior, known for *Leucotrichia pictipes* (McAuliffe, 1984b; Hart, 1985b), has not been observed for these species.

In autumn a field experiment was carried out in the upper part of the Breitenbach to investigate interspecific competition between the trichopteran scrapers (Becker and Menzler, unpublished data). 12 batches, each with 12 smooth unglazed clay tiles (5 × 5 cm) were deposited in the stream for several weeks at sites with a low current velocity, which are preferred by *A. fuscipes* larvae. Larval immigration onto the artificial substrata and migration was not restricted. After some weeks, the substrata were colonized by high densities of *A. fuscipes* larvae, and an equilibrium between larval abundance and biofilm standing crop was established. At this time we only found a few individual larvae of *D. annulatus* and baetids on the substrata; *A. fimbriata* larvae were not active, because of their winter dormancy (Aurich, 1989).

At this stage the field experiment was initiated and *A. fuscipes* larvae were removed daily from one part of the substratum, for a period of five weeks. Larvae of *D. annulatus* were not removed. On the other part of the substratum (the control), the abundance of *A. fuscipes* remained unmanipulated. The density of all larval scrapers was recorded daily on all tiles. The development of the microphytobenthos on the manipulated substrata and controls was analyzed with an *in vivo* fluorescence method (Becker *et al.*, 1997). To do this, tiles were removed from the stream, the fluorescence of the biofilms was measured in the laboratory, and the tiles were returned to their specific positions in the stream. At the beginning and at tend of the experiment (after 5 weeks), the chlorophyll-a content was analyzed using a spectrophotometer, and the AFDM of the biofilm was measured (for details, see Becker *et al.*, 1997).

Agapetes fuscipes always colonized the controls throughout the experimental period in high densities. The result of the intensive grazing activity on the controls was a thin biofilm (AFDW < 40 $\mu g\,cm^{-2}$). When *A. fuscipes* larvae were removed daily from the substrata, the biofilm grew up markedly. After five weeks the AFDM of these biofilms had increaed more than 20 times than of the controls (Figure 7.148). Individuals of other scrapers, such as *D. annulatus* and *Baetis* spp. were feeding unhindered, although apparently at low densities, they were unable to reduce the standing crop of the biofilm significantly. Along with the food biomass, the abundance of *D. annulatus* increased significantly on the manipulated substrata compared to the controls (Figure 7.148).

These results show the significant top-down effect of *A. fuscipes* on biofilms at sites with a low current velocity and demonstrate asymmetric competition for food between these trichopteran scrapers with different, species-specific, foraging strategies. If high densities of *A. fuscipes* larvae are feeding intensively under low current velocity, they reduce the standing crop to such an extent that the biofilms are unattractive to competitors such as *D. annulatus*. The much larger *D. annulatus*

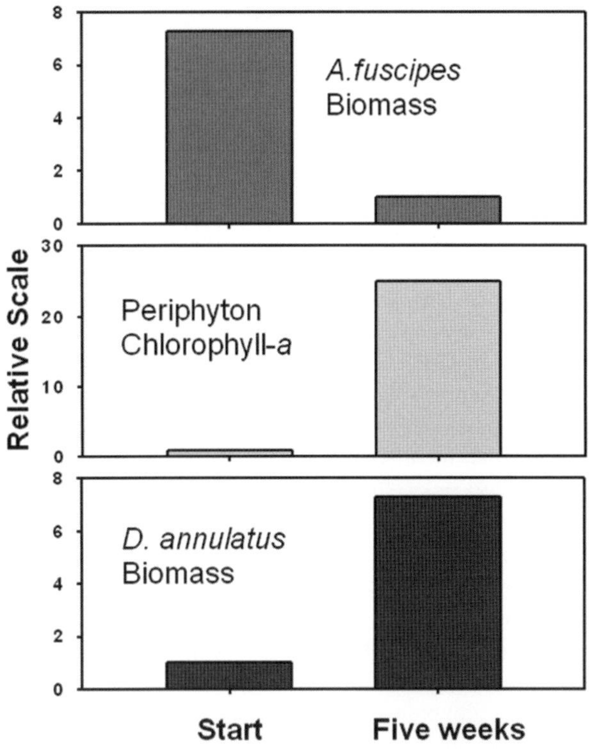

Figure 7.148 Top-down effect of *A. fuscipes* on the standing crop of periphyton, and exploitative competition between *A. fuscipes and D. annulatus* for periphyton.

larvae leave these areas and switch to thicker biofilms on surfaces under higher current velocity. Thus, *D. annulatus* is more tolerant of high current velocity than *A. fuscipes*. However, food quantity has a decisive influence on the microdistribution of *D. annulatus*, because *D. annulatus* also fed on surfaces under low current velocity, when competing *A. fuscipes* was removed (Becker and Menzler, unpublished data).

However, some interesting results from Menzler (1995) showed that, in the upper section of the stream (near trap T1), the relationships between the three abundant trichopteran scrapers are more complex than just described (see above). Throughout a spring spate with high current velocities, a relatively thick biofilm was found on a sandstone transect, but without any scrapers. As water level decreased, larvae of *D. annulatus* were found to colonize the stone surfaces from the downstream side, followed by *Apatania* larvae, and started to graze the thick biofilm layers. Over the next days, the biofilm layer was markedly reduced over parts of the stone surface, and most of the larvae (of both species) were feeding intensively on the thick biofilm patches at the border between the thick and reduced biofilms. When the water level fell further, *A. fuscipes* larvae moved onto the stones, also on the downstream side. However, the *Agapetus* larvae aggregated (at high densities) on the reduced biofilm patches and were clearly not feeding on the edge of the thick biofilm patch. Some days later no thick biofilm patches could be detected on the stones, most of *Drusus* and *Apatania* larvae had left the stones, but *Agapetus* larvae were still feeding intensively at high densities on the thin biofilm layers. The biomass, and presumably also the heterogeneity, of algal mats on these stones changed markedly over a period of about two weeks, from an ungrazed, thick layer to a highly diverse, thin algal layer. Whereas *Drusus* and *Apatania* larvae grazed intensively in thick algal patches, *Agapetus* larvae simultaneously grazed thin algal patches on the same stone, leading to a relatively even, thin, and strongly reduced algal layer, on which only *Agapetus* larvae remained, dramatically controlling the algal community (Menzler, 1995). This example shows that three grazers of different body size, with presumably different nutritional reqirements, can regulate algal patchiness, and demonstrates the importance of the temporal aspect with respect to the role of grazers, increasing or decreasing autotrophic spatial heterogenity (e.g., Alvarez and Peckarsky, 2005; Hillebrand, 2008).

The conclusion of these findings is that *D. annulatus* and *A. fimbriata* preferentially feed on thick biofilms, in contrast to *A. fuscipes* larvae, which prefer thin biofilms and are unable to use thick periphyton mats. However, when limnephilid species reduce thick biofilms to thin layers, they prepare feeding areas offering optimum conditions for *A. fuscipes* larvae. If the current velocity is relatively low, *A. fuscipes* larvae colonize these areas at high densities, and reduce the standing crop to such an extent that the biofilm biomass may be below the minimum metabolic demand of the coexisting limnephilids. At these sites intense herbivory exerts a profound top-down control on the biomass and species composition of the autotrophic community. Hart (1987) found a higher food quality (lower C : N ratio) in periphyton that was intensively grazed by *Glossosoma nigrior* larvae compared to ungrazed periphyton patches. The small, slow-moving *A. fuscipes* larvae may have low energetic requirements and be more likely to remain permanently in thin periphyton patches, and hence exert greater overall impact on the microalgal community than the larger, more mobile, *Drusus* and *Apatania* larvae, with higher energetic requirements.

However, current velocity also plays a crucial role in the exploitation of algal patches by *A. fuscipes*. If current velocity at a feeding area is high, *A. fuscipes* larvae are unable to colonize the sites at high densities and the biofilm standing crop will grow up with high renewal rates to a level that is attractive to *D. annulatus* and *A. fimbriata* larvae. These findings are in agreement with the study of Hoffman *et al.* (2006), who found that *Agapetus boulderensis* also preferred smooth surfaces with thin, diatom-dominated layers under slow current velocities. Several other studies have demonstrated a strong reduction in periphyton by glossosomatid larvae (e.g., Hart, 1987; Kohler, 1992; Dudley and D'Antonio, 1991; Poff and Ward, 1995; McNeely and Power, 2007).

In years with relatively high discharge and, as a result, a large range of current velocities, sufficient feeding areas with diverse current conditions should be available for coexisting trichopteran larvae. However, in years with low discharge, few feeding areas with high current velocity would be available in the upper part of the Breitenbach. Under such conditions the large population of *A. fuscipes* larvae will be able to colonize nearly all stone surfaces in the upper section of the stream. It is assumed that *D. annulatus* and *A. fimbriata* larvae are then outcompeted and forced to leave this section of the stream.

7.8.4.9 Spatial and Temporal Development of Trichopteran and Ephemeropteran Grazers and Periphyton along the Breitenbach

The distribution of benthic organisms along stream courses still raises many questions. Long-term emergence studies of insects show that many species differ in their distribution along the Breitenbach (e.g., Illies, 1982), however, causal relationships between species-specific along-stream distribution, and the biotic and abiotic factors are difficult to determine (but see Sections 7.9 and 8.3). Becker (1990) only found modest differences in the food composition of individual trichopteran grazers along the Breitenbach. Werneke (1997) could not find strong correlation between the growth and development of baetids and the microphytobenthos standing crop in the middle part of the Breitenbach.

Interactions between grazers, and between grazers and their food resources, epilithic biofilms, were investigated between May and October 2003, along the Breitenbach (Gergs, 2004). The development of grazers and their food resources have seldom been analyzed simultaneously on natural unmanipulated, stream substrata (but see Katano *et al.*, 2005, 2007). The aim of this field research was to analyze growth, development and the seasonal distribution of the dominant grazers and the amount of periphyton at different sites in the Breitenbach, to determine the spatial and temporal segregation of different grazers and their influence on periphyton quantity and composition.

On nine occasions from May until October 2003, quantitative samples of grazers and periphyton were taken at two sites, one in the upper reaches near the source (T1) and another in the lower stretch, near the confluence with the River Fulda (upstream T6). Grazers were sampled using a surber-sampler and biomass (AFDM) was determined. Periphyton was scraped from the stone surfaces using ultrasound (Werneke, 1997), and chlorophyll-*a* concentration and biomass (AFDM) were determined. The photosynthetic pigments were analyzed by HPLC (see also Section 7.8.1). The relative amounts of the different pigments indicate the microphytobenthos composition (Lucas and Holligan, 1999).

Ten grazer species were regularly found in the Breitenbach, six Trichoptera, three Ephemeroptera, and the gastropod *Ancylus fluviatilis* (see Sections 7.5 and 7.6). Both baetids belong to the dominant macroinvertebrates of the Breitenbach (Section 7.6.4.2) and may be the most important competitors for epilithic biofilms with trichopteran grazers. Some of the species showed significant differences in their spatial distribution along the stream and in their temporal development (Gergs, 2004). *Agapetus fuscipes* and *B. vernus* were the dominant species in the upper part, *T. rostocki* and *B. rhodani* in the lower part of the stream. The Trichoptera and Ephemeroptera showed distinct differences in their life cycles. The caddisflies mostly completed their larval development in spring, while baetids showed the strongest increase in biomass in early summer. In the upper stream section the coexisting Trichoptera, *A. fuscipes* and *D. annulatus,* dominated the grazer biocenosis in spring and late summer, while *B. vernus* showed a biomass maximum in the middle of July, after most Trichoptera individuals had emerged. In the lower stream section, the caddisfly *T. rostocki* dominated early in the year and in late summer, while the maximum biomass of *B. rhodani* occurred in early July, when density of *T. rostocki* larvae was at its lowest. The maximum growth rates of the two baetid species, with a distinct biomass increase in the older larvae, also showed a temporal pattern. *Baetis vernus* completed its larval development earlier than *B. rhodani* in both stretches of the stream. It is assumed that these different spatial and temporal patterns reduce interspecific competition for food among the grazer species (Gergs, 2004).

Periphytic abundance showed clear differences between the two stream sections. In the upper reaches a decrease in chlorophyll-a concentration, from a mean of 5.0 to 2.5 $\mu g\,cm^{-2}$ was observed in the middle of August, while mean chlorophyll-*a* concentration (5.0 $\mu g\,cm^{-2}$) in the lower section of the stream did not change markedly throughout the research period. A significant negative correlation between *A. fuscipes* biomass and the periphytic chlorophyll-*a* concentration showed that low, the late summer–autumn, periphyton abundance in the upper creek section was probably the result of intensive grazing activity, particularly of younger glossosomatid larvae (Gergs, 2004). A distinct top-down effect on periphyton has also been observed in other species of this genus (e.g., Hart, 1987; Kohler, 1992; Dudley and D'Antonio, 1991; Poff and Ward, 1995; McNeely and Power, 2007). However, in the lower section, the influence of grazers on the periphyton was less clear than in the upper reaches. However, due to the high density of *T. rostocki*, it is likely that, due to their retreats and specific foraging behavior, this species had a clear influence on the structure, taxonomic composition and primary production of the periphyton (Becker, 1993; Stief and Becker, 2005; see also Section 7.8.1). There were significant differences between the upper and lower part of the Breitenbach with respect to photopigment composition of the microphytobenthos communities and food quality. Relatively high fucoxanthin and zeaxanthin content and low chlorophyll-*b* content indicated that diatoms and cyanobacteria were dominant in the upper part of the stream, while a relatively high chlorophyll-*b* content indicated a higher abundance of Chlorophyta in the lower part of the Breitenbach.

The results provide an insight into the complex interactions between grazers and periphyton, and the spatial and temporal segregation of grazers in a first-order stream. Several factors may be responsible for the different distribution of grazers along the Breitenbach. The low diurnal and yearly amplitude of water temperature in the upper part, compared with its high amplitude in the lower part, changing food composition and quality with respect to diatom cell size and the contrasting proportions of diatoms, cyanophytes and green algae based on photopigment ratios along the stream, and interspecific competition among the grazers must all be taken into consideration.

7.8.4.10 Significance of Phototrophic Microorganisms for the Mortality, Growth, and Development of Trichopteran Scrapers

Laboratory experiments showed the influence of food quality on larval survival, growth and development of the trichopteran scrapers, *A. fuscipes*, *A. fimbriata*, and *D. annulatus*. Different larval instars were collected in spring, summer, and autumn. Larvae were cultured for a period of four weeks at 14 °C in small laboratory stream systems and provided with food of distinctly different nutritional quality. Periphyton and detritus from the Breitenbach and conditioned nettle broth were offered in unlimited amounts. Detritus was maintained in aerated tap water in the dark for a minimum of 14 days to suppress algal growth.

Agapetus fuscipes and *D. annulatus* showed markedly higher mortality when larvae were fed exclusively on detritus and nettle broth, compared to larvae fed on periphyton. The result was different for *A. fimbriata* and varied with season. When larvae sampled in August and February were fed on detritus and nettle broth, mortality was higher compared to larvae fed on periphyton, but this did not apply to larvae collected in April and September. Growth rates of larvae fed on periphyton were distinctly higher than for detritus-fed larvae of all three species. The growth rates also varied with the larval collection date. The highest growth rates for *Drusus* larvae were found in spring, for *Apatania* larvae in August. Larval development was analyzed in detail for *A. fimbriata* and *D. annulatus*. In earlier and later instars development was handicapped when larvae were fed exclusively on detritus and nettle broth. The molting and pupation rates were significantly reduced (Becker, unpublished data).

The results show the importance of microphytobenthos as an essential food component for survival, growth and development of these trichopteran scrapers. The larvae were not able to compensate for the absence of high quality food, such as microphytobenthos, by increased ingestion rates of low quality food, such as detritus. The differences in larval mortality between the collecting dates were probably caused by variation in larval physiology through the year, or by significant seasonal changes in periphyton and detritus quality in the Breitenbach.

7.8.4.11 The Diurnal Foraging Behavior of *A. fuscipes* and *D. annulatus*

Diurnal foraging activity has been examined for two trichopteran scrapers. Video studies in laboratory flumes, using a black and white camera for daytime and a night vision camera for nighttime examination, demonstrated that late larval instar individuals of the glossosomatid *A. fuscipes* remained on the surface of their stony substrata for several days under favorable food conditions and fed during the day and at night (Becker, unpublished data).

In contrast, the larvae of *D. annulatus* undertake diurnal microhabitat movements, between epilithic grazing patches and resting areas within the sediment, as shown by individual labeling of larval cases (Menzler, 1995). The grazing activity of *D. annulatus* exhibits a circadian rhythm, which differs significantly from other patterns of day, night and crepuscular activity in insects (Saunders, 1976), particularly of trichopteran larvae (e.g., Elliott, 1969, 1970; Lehmann, 1972; Wagner, 1991;

Bergey and Resh, 1994). The activity maximum, as reflected by the highest larval densities on exposed stone surfaces, occurred at least one hour before sunset, while the activity minimum occurred after dawn. Thus in winter *D. annulatus* grazed most actively between noon and midnight. This behavior has been observed in the upper and middle reaches of the Breitenbach and also in laboratory stream systems (Menzler, 1995). In contrast to the findings of Elliott (1970), who described night activity in this species, *D. annulatus* fed in the afternoon and at night on stone surfaces in the Breitenbach (Menzler, 1995). This rhythm does not therefore seem to be a strategy to avoid attacks by visual predators such as salmonids, which are present in the Breitenbach. The larval activity pattern was obviously well correlated with the diurnal temperature pattern of early morning minimum and late afternoon maximum. Further experiments are necessary to determine whether this rhythm is a specific adaptation of *D. annulatus* to changing food quality during the day, for example, with respect to the accumulation of photosynthetic products in the algae during the light period.

7.8.4.12 **Food Choice in Laboratory Experiments**

The feeding selectivity of five epilithic, case-building, trichopteran species from the Breitenbach was examined using laboratory experiments (Becker, 1994). The aim of this study was to investigate larval behavioral responses to variation in food quality and to correlate these results with the dietary composition of the larvae in the field (Becker, 1990; Section 7.8.4.4).

Final instar larvae had the choice of three food types: (i) periphyton, freshly scraped from sandstone in the Breitenbach, (ii) detritus, washed from sediments in a lentic zone of the Breitenbach and maintained in aerated tap water in the dark for a minimum of 14 days to suppress algal growth, and (iii) conditioned nettle broth as an "artificial" food source (Becker, 1987a, 1987b). The three food types, on nine glass fiber filters positioned at random in an aerated container (30 × 30 cm, Figure 7.149) were offered to the larvae. Each species was tested individually to prevent possible interspecific competition. The number of larvae associated with each filter was noted at 15 or 30 min intervals. Experiments lasted between 3 and 4 h. With each food type, visible inspection suggested an equal distribution and comparable layer (ca. 1 mm) of material on the filters. Thus, the experimental design assumed random, low degree of variability in the quantity of food available on the filters. Since three filters of each food type were used in each experiment, it was unlikely that any single food type was consistently available in higher quantities because experiments were replicated with fresh sets of filters. It can therefore be assumed that the larvae would have reacted to food quality rather than quantity (Becker, 1994).

The first experiment, with three species, showed that larvae vacated clean filters within a few minutes of being placed at their centres. *Drusus annulatus* left the filters within 2 min ($n = 15$), *A. fimbriata* within 3.5 min ($n = 10$), and *A. fuscipes* within 10 min ($n = 10$; but see Section 7.8.4.13).

Significant differences between larval densities associated with the various food types, and the particularly low density of larvae associated with the nettle broth

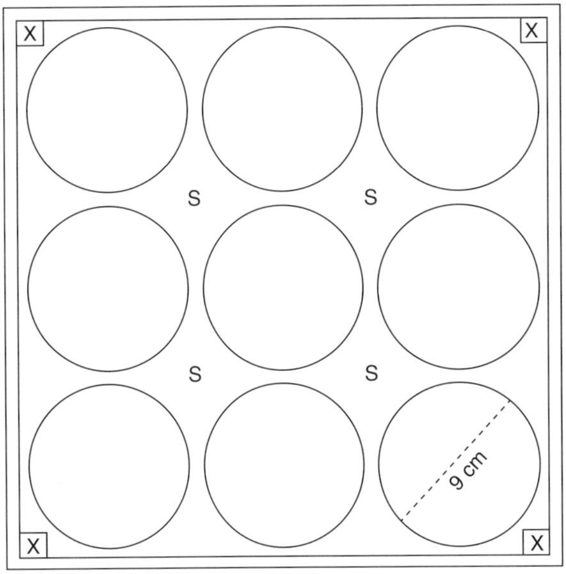

Figure 7.149 Plan of the experimental setup. X = aerator, S = starting point.

(max. 25%), indicated that all five trichopteran species had a basic ability of distinguishing between different food types (Figure 7.150). Except for *A. fimbriata*, numbers of larvae on nettle broth filters were lower than for the other food types. The results reveal distinct, species-specific differences in the choice between detritus and periphyton. *Drusus annulatus* (72%), *M. longulum* (56%), and *A. fimbriata* (50%) showed clear preferences for periphyton. In contrast, *A. fuscipes* and *S. pallipes* showed no preference for periphyton over detritus, and *A. fuscipes* showed a preference for detritus (43%). Although all are scrapers, larval responses to the various food types differed between species. For example, the larvae of a typical trichopteran scraper, such as *A. fuscipes*, did not necessarily show a distinct preference for periphyton rather than detritus.

Foraging behavior complexity became apparent when pieces of moss (*Chiloscyphus polyanthos*: Hepaticae) were added to the periphyton filters for *M. longulum* (Becker, 1994): 60% of the larvae were found on filters containing moss (Figure 7.151). Larvae of *M. longulum*, which were observed feeding on submerged bryophytes (*Chiloscyphus polyanthos* and *Scapania undulata* (Hepaticae) as well as stone surfaces in the stream (Becker, personal observations), showed a distinct preference for the moss-supplemented filters over those with periphyton or detritus alone. It is difficult to interpret this result because epiphytic periphyton cannot be separted from the moss. On the one hand, larvae could use the moss as a substratum, while on the other hand epiphytic algae could be an additional food resource. In contrast to other scrapers, pieces of moss were found in the guts of *M. longulum*, but they appeared to remain undigested (Becker, 1990). The question therefore

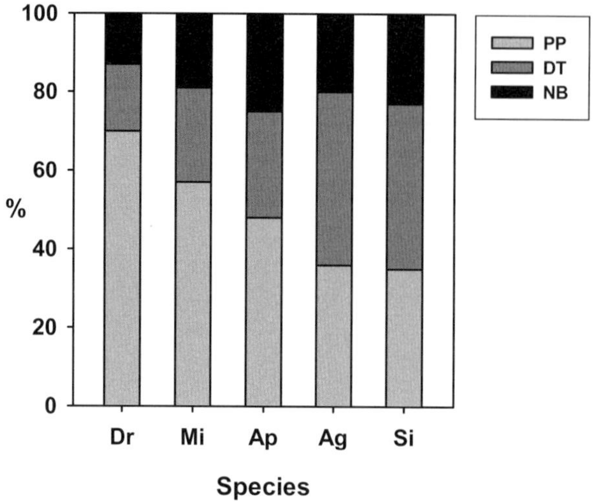

Figure 7.150 Proportions of larvae on filters offering different food types. NB = nettle broth, DT = detritus, PP = periphyton, Dr (*D. annulatus*), Mi (*M. longulum*), Ap (*A. fimbriata*), Ag (*A. fuscipes*), Si (*S. pallipes*).

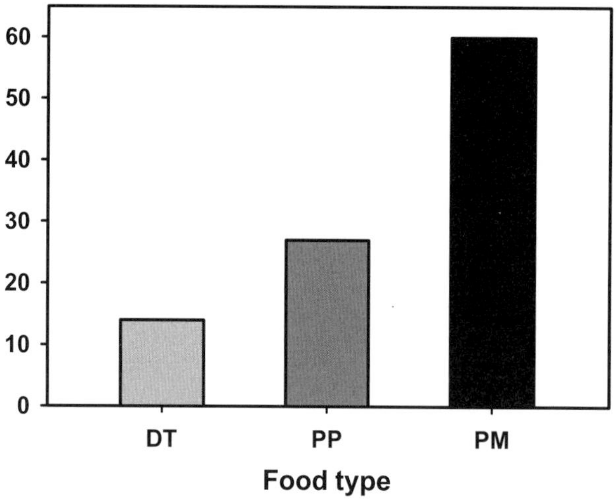

Figure 7.151 Proportions of *M. longulum* larvae on filters offering different food types. DT = detritus, PP = periphyton, PM = periphyton with added moss.

arises as to whether the larvae were attracted to the moss because of the epiphytic algae, or whether the structure of the moss itself was the attractant.

Species-specific adaptations in foraging behavior were reflected in preferences for particular food types as well as the gut-content composition (Becker, 1990, 1994). Figure 7.152a shows the relationship between larval frequencies on peri-

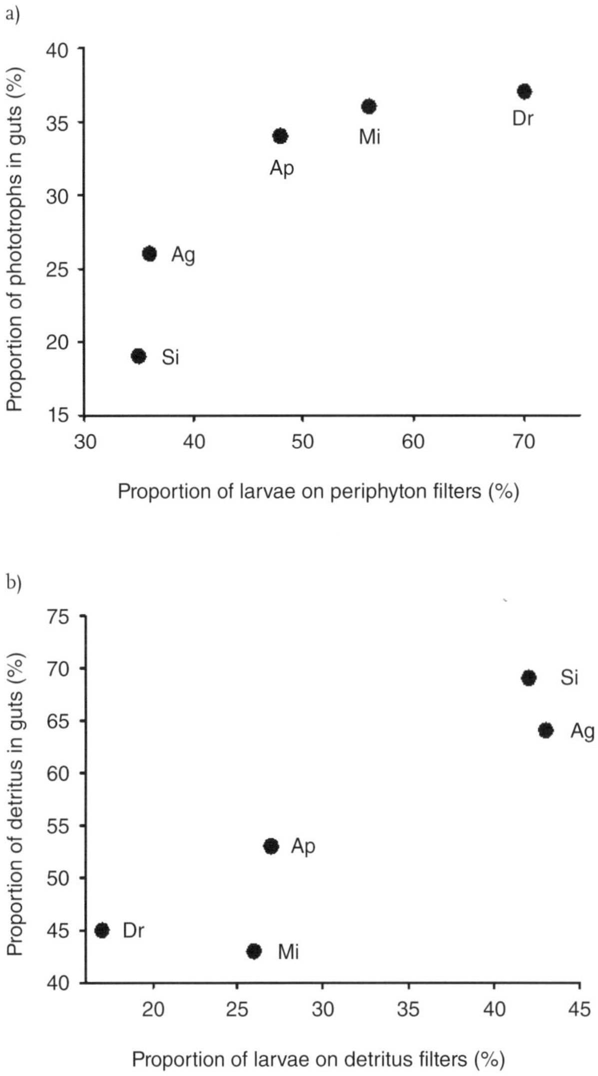

Figure 7.152 Correlations between species-specific preferences for periphyton-covered filters and percentage areas of autotrophs in larval guts (a) and preferences for detritus-covered filters and percentage areas of detritus in larval guts (b).

phyton filters in laboratory experiments and mean percentage areas of algae in the gut-contents of larvae sampled in the Breitenbach (Section 7.8.4.7). Species that preferentially selected periphyton in the laboratory experiments (*D. annulatus, M. longulum, A. fimbriata*) also consumed the highest proportions of algae in the field. Figure 7.152b shows the relationship between larval frequencies on detritus filters in laboratory experiments and mean percentage areas of detritus in the gut contents of field-sampled larvae. *Agapetus fuscipes* and *S. pallipes*, which selected detritus more frequently, had markedly lower proportions of algae, but higher proportions of detritus in their guts.

Species-specific microdistributions of grazers in the field can be influenced by a number of environmental factors. In addition to periphyton abundance (e.g., Streit, 1976; Bohle, 1978; Kohler, 1984; Schweder, 1985; Lodge, 1986; Smock and Roeding, 1986; Vaughn, 1986; Hershey *et al.*, 1988; Richards and Minshall, 1988; Willoughby and Mappin, 1988), current velocity and substratum structure can also be important (e.g., Scott, 1958; Kovalak, 1979; McElhone and Davies, 1983; Teague, Knight and Teague, 1985; Williams and Moore, 1985; Barmuta, 1990; Poff and Ward, 1995; Hart and Finelli, 1999; Finelli, Hart and Merz, 2002; Poff, Wellnitz and Monroe, 2003; Hoffman *et al.*, 2006). Together, these factors can induce complex behavioral patterns in the larvae. In our laboratory experiments, it was possible to exclude the influence of factors such as current, substratum and interspecific interactions, because larvae could select food under similar environmental conditions without any competition from other species. It is assumed that any potential differences in food quantity between filters were unimportant since each experiment began with a freshly prepared set of filters, each with a comparable layer of food. The results of this study showed that food quality has an important influence on the microdistribution of these species (Becker, 1994).

Schweder (1985), Lodge (1986), and Rosillon (1988) also described the influence of food quality on food selection by grazers. However, Vaughn (1986) found that food quality was less important than abundance to the trichopteran scraper, *Helicopsyche borealis* (Helicopsychidae). Feeding preferences have also been found among other trichopteran scrapers (Oemke, 1984; Hart, 1985a; Vaughn, 1986; Peterson, 1987; Hill and Knight, 1988), but the stimuli for this selectivity are unclear.

Compared with microalgae, fine detritus is regarded as a poor food resource. Only the attached microorganisms, which only form a small proportion of the total weight of detritus (Paerl, 1977 data, in Lamberti and Moore, 1984), have high nutritional values (e.g., Benke and Wallace, 1980; Mattson, 1980; Lamberti and Moore, 1984; Smock and Roeding, 1986). A greater number of adults of *Helicopsyche borealis* emerged from cultures containing green algae or diatoms than those containing bacteria alone (Vaughn, 1986). The high frequencies of *A. fuscipes* and *S. pallipes* on detritus filters, and the high proportions of detritus in their guts (Becker, 1990), suggest that detritus, or possibly the associated microbial flora, is important for the nutrition of these species. However, laboratory experiments showed that microphyotobenthos is also an essential food resource for *A. fuscipes* (see Section 7.8.4.10).

At the Breitenbach, *S. pallipes* larvae can be found on stone surfaces as well as fine sediments (Becker, personal observations), an observation also reported by Ormerod (1988). *Agapetus fuscipes* preferred stone surfaces under low currents (0.05–0.2 $m\,s^{-1}$), where the rate of detrital sedimentation was relatively high. Benedetto Castro (1975) also found high proportions of detritus (82%) in the guts of *A. fuscipes* larvae from the Breitenbach. In contrast, *D. annulatus* and *M. longulum* larvae were more rheophilous. At such sites in the Breitenbach the periphyton contained high proportions of epiphytic algae, and low proportions of detritus and inorganic material (Becker, personal observations).

Species-specific foraging behavior was reflected in the gut contents of larvae taken from the field (Becker, 1990). Species that searched actively for epiphytic algae in laboratory experiments had higher proportions of algae in their guts. Dudgeon and Lam (1985) reported a similar trend for gastropods; the fastest (*Physella acuta*) had a specialized algal diet. However, *A. fuscipes* and *S. pallipes*, which showed no preference for periphyton, were clearly able to make qualitative distinctions between food types, since they were not randomly distributed on the filters and avoided the nettle broth. However, they spent a lot of time on the detritus filters, a nutrient-poor food resource. In relation to their foraging for epiphytic algae, *A. fuscipes* and *S. pallipes* did not appear to follow the "area-restricted searching" and "giving up" rules of the optimal foraging theory (as reviewed by Wiley and Kohler, 1984). This contrasted with the other species, which aggregated on filters providing periphyton, a food resource of a higher quality.

7.8.4.13 **Response of Trichopteran Grazers to Biofilms of Diverse Quantity and Quality**

The analysis of the influence of food quality and food quantity on growth and development of trichoperan grazers is complicated because most species in temperate streams have an univoltine life history and grow relatively slowly. Thus, it is not possible to offer biofilms of almost the same composition, quality, and quantity over a long period, which would be necessary if the dependence of final larval biomass and adult fitness of the adults on food conditions is to be analyzed. Therefore, it was necessary to find another way to determine the nutritional requirements of different grazer species.

Foraging behavior that involves gathering and processing of food resources is a regulatory behavior that increases the control an individual can exert over its own metabolic status. Detailed knowledge of grazing behavior is therefore important to understanding the factors that affect the foraging behavior of scrapers in a stream. Hart and Resh (1980) mentioned that two-thirds of the total time–activity budget of aquatic larval insects is dedicated to feeding, and these authors assumed that the patterns of movement largely reflect food acquisition activities. Mobility of stream insects increases when food availability is reduced experimentally (e.g., Gallepp, 1977; Hart and Resh, 1980) and grazers respond to spatial patchiness by concentrating their activity within patches with high food levels (e.g., Kohler, 1984; Hart *et al.*, 1991; Alvarez and Peckarsky, 2005). The rate of movement across small-scale food patches is negatively correlated with the amount of periphyton present per patch (Hart and Resh, 1980).

The short-term experiments had two major aims. The first was to determine whether it is possible to obtain detailed information about the nutritional recruitment of three epilithic trichopteran grazer species from the upper section of the Breitenbach to different natural food conditions using short experiments. The second was to test whether larvae showed species-specific behavioral responses to a range of diverse biofilm quantities and qualities (Becker, unpublished data).

Plexiglass plates with a mosaic comprising nine unglazed, clay tiles (each 5×5 cm), were placed in channels above the Breitenbach, which were supplied with stream water. These substrata were incubated in the Breitenbach for different time periods between February and April, between one week and 3.5 months, to obtain biofilms of different ages, biomass, and composition. The artificial substrata were then placed in a laboratory channel connected to a reservoir containing about 250 l of stream water. The crawling velocity of trichopteran larvae is strongly correlated with current speed (Bournaud, 1972; Poff and Ward, 1992; Poff and Ward 1995; Hoffman *et al.*, 2006), therefore a relatively low current velocity, of between 5 and 8 $cm\,s^{-1}$ was chosen, which enabled all the larvae to move in all directions on the tile surfaces without being forced off by the current. The movements of *A. fuscipes*, *A. fimbriata*, and *D. annulatus* larvae were filmed for a period of 20 min, using a black and white video camera and analyzed by an interactive image analysis setup (VIDS IV, AI Tektron GmbH). The experiments provided information on larval search behavior within food patches, and on crawling distance, time, and velocity of each larva. A total of 1621 trichopteran larval pathways with 12 different food conditions were tested and the chlorophyll-*a* content and the AFDM of the biofilms analyzed.

All species showed an immediate response in their food search. Area-restricted searching and vacating of food patches were observed (Pyke *et al.*, 1977; Kohler, 1984). Controls showed that *A. fuscipes*, *A. fimbriata*, and *D. annulatus* larvae vacated tiles lacking periphyton within 10 min. The crawling velocity of *D. annulatus* larvae decreased strongly with increasing chlorophyll-*a* content. At higher chlorophyll-*a* content, larvae moved with distinctly reduced velocities while searching for food and feeding intensively (Figure 7.153). The residence time increased significantly with chlorophyll-*a* content of the biofilms. Compared to *D. annulatus*, the crawling velocity of *A. fimbriata* larvae decreased more strongly with increasing chlorophyll-*a* content. Larvae needed more time to leave the control tiles without biofilm, because they moved slowly compared to *D. annulatus* (but see Section 6.3). However, even at a chlorophyll-*a* content of 1.1 $\mu g\,cm^{-2}$, the larvae remained on tiles for a long time. A similar reduction in crawling velocity and positive correlation between residence time and relatively low chlorophyll-*a* content (1.1 $\mu g\,cm^{-2}$) has been observed in *A. fuscipes* larvae (Figure 7.154). However, the correlation was not as strong as for the other two species, due to increasing crawling velocity and decreasing residence time with high chlorophyll-*a* content (6.3 $\mu g\,cm^{-2}$).

Distinct differences have been found between the three species. A relatively low chlorophyll-*a* concentration was necessary to reduce crawling velocity and increase residence time of *Agapetus* and *Apatania* larvae compared to *D. annulatus*, and *Agapetus* larvae left algal patches with a higher chlorophyll-*a* content at an higher crawling velocity. Thus, larvae showed distinct and species-specific responses to different natural Breitenbach biofilms in these short-term experiments. This method provides information about the food preferences and nutritional requirements of trichopteran grazers (Becker, unpublished data).

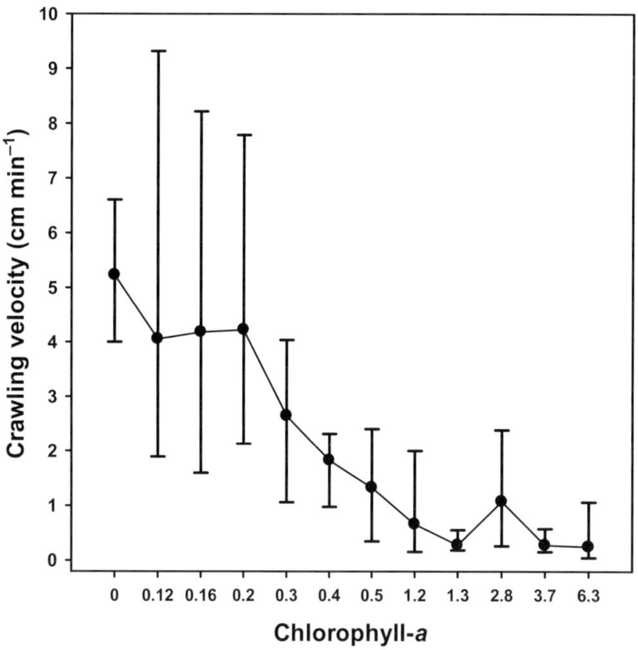

Figure 7.153 Response of *D. annulatus* larvae to the chlorophyll-*a* content of periphyton on artificial substrata.

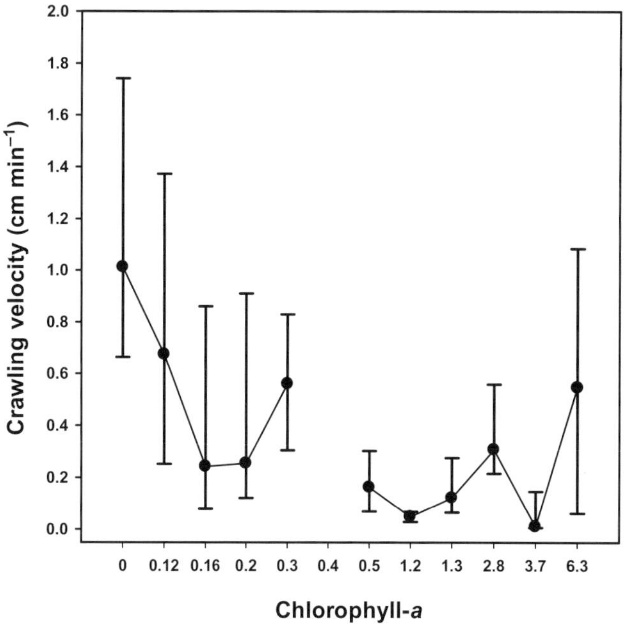

Figure 7.154 Response of *A. fuscipes* larvae to the chlorophyll-*a* content of periphyton on artificial substrata.

In a continuation of this work, larval responses to unialgal layers of diverse algal species and densities (cultured by H. Holfeld) were tested in short term experiments to analyze in more detail the response of larval to different food quantity and quality. Two diatom species with different growth forms, and one cyanophyte species, were cultured in laboratory flumes: *Diatoma mesodon* which forms filaments; *Achnanthes lanceolata*, which occurs as single cells in a relatively even layer; and *Oscillatoria* sp., a filamentous cyanophyte. A mosaic comprising 12 unglazed clay tiles (see above), supporting microphythobenthic biomasses with low variance in chlorophyll-*a* content and AFDW, were transferrred from the flowing culture to an experimental flume and behavioral experiments were conducted on the same day. Chlorophyll-*a* content, cell numbers, and AFDW of the biofilm were determined at the end. A new video technique was developed to analyze larval behavior. Cases of trichopteran larvae were labeled with colored dots of high contrast compared with the background, and larval positions were recorded by tracking eight dots simultaneously, using a video camera and a dotfinder (Jander Videometric Products) for a 10 min test period. The total substratum area, that is, the algal patch, was automatically subdivided into 625 fields by the dotfinder software, and the regularly registered counts per field allowed calculation of the path length, crawling velocity and residence time of the larvae on various types of biofilm layers.

The results show that larval residence times are strongly influenced by microphythobenthic quantity and species. At low densities of *Achnanthes lanceolata* (349 cells mm^{-2}, Chl.a: 0.1 $\mu g\,cm^{-2}$), 45% *Agapetus* larvae, 65% *Apatania* larvae, and 100% *Drusus* larvae vacated the mosaic within 10 min. At higher densities of *Achnanthes lanceolata* (7770 cells mm^{-2}, Chl.a: 2.3 $\mu g\,cm^{-2}$) only 6% *Agapetus* larvae, 25% *Drusus* larvae, but 48% *Apatania* larvae vacated the mosaic within 10 min. Densities of *Diatoma mesodon* up to 1748 cells mm^{-2} (Chl.a: 0.7 $\mu g\,cm^{-2}$) were apparently unattractive to *Apatania* (80% vacation rate) and *Drusus* larvae (92% vacation rate). However, 65% *Agapetus* larvae remained on the mosaic under these conditions. Different results were found with the *Oscillatoria* layers. At low cell densities (trichome length: 1.5 mm mm^{-2}, Chl.a: 0.09 $\mu g\,cm^{-2}$) 87% *Agapetus* larvae, 56% *Apatania* larvae, but only 4% *Drusus* larvae remained on the mosaic. However, at higher algal densities (trichome length: 25 mm mm^{-2}, Chl.a: 1.1 $\mu g\,cm^{-2}$) only 24% *Agapetus* larvae, but 52% *Apatania* larvae and 50% *Drusus* larvae remained on the mosaic (Becker and Holfeld, unpublished data).

These data reflect the complex relationships between scrapers and the periphyton, and show species-specific adaptations of the three scrapers to different food patches. Not only is the quantity of algal cells important for the larval response, but also the quality of the different autotrophs. *Agapetus fuscipes* showed a distinctly positive response to very thin autotrophic layers, 55% larvae remained on patches of *A. lanceolata* (0.1 $\mu g\,cm^{-2}$ Chl-a), 65% on *D. mesodon* patches (0.7 $\mu g\,cm^{-2}$ Chl-a), and 87% on *Oscillatoria* patches (0.09 $\mu g\,cm^{-2}$ Chl-a). 94% *Agapetus* larvae remained in a patch with a higher density of *A. lanceolata* (2.3 $\mu g\,cm^{-2}$ Chl-a), but most of the larvae (75%) avoided a higher density of *Oscillatoria* cells. *Drusus annulatus* showed a distinctly negative response to thin autotrophic layers ($<0.8\,\mu g\,cm^{-2}$

Chl-a) of all the autotrophs (vacation rates 80–100%), a positive response to the higher density of *A. lanceolata* (residence rate 75%) and an intermediate response to a higher density of *Oscillatoria* cells (residence rate 50%). The response of *Apatania* larvae was unexpected, because the larvae did not demonstrate a clear response to any of the tested autotrophic layers. Some 65–80% *A. fimbriata* larvae vacated the thin diatom layers and, in contrast to the larvae of the other species, only 52% larvae remained in patches of a higher density of *A. lanceolata* cells and 56 and 52% remained on lower and higher densities of *Oscillatoria* cells, respectively.

In addition to variability in the quality and quantity of microphytobenthic species, the physical conditions of the biofilm also changed, depending on the algal density and growth form. The physical conditions of thick periphyton layers, especially of thick layers of filamentous algal species, may influence larval crawling behavior and make it more difficult for the larvae to fix their silken safety lines to the substratum to prevent being swept off by the current. The results are in general agreement with field observations in the Breitenbach. *Agapetus fuscipes* colonized patches with low-biomass periphyton and avoided thick biofilms, whereas *D. annulatus* preferred higher periphyton biomass.

7.8.4.14 **Feeding Strategies and Ecological Segregation of Trichopteran Grazers in the Breitenbach**

Along with a lack of clear temporal segregation of the trichopteran larval species, high numbers of scrapers in the Breitenbach appear to result in the limitation of high quality food, such as epilithic microphytobenthos. Comparing field and laboratory results, different species-specific physiological, morphological, and behavioral adaptations, such as differences in body size, nutritional requirements, life cycles (number of larval stages, winter dormancy), foraging behavior, diurnal activity, competitive behavior (interference, exploitative competition), and retreat-building have been observed in the trichopteran grazers. At least some of the grazer species live in competitive coexistence, as shown in the upper section of the Breitenbach in the field experiment. It is suggested that interspecific competition between trichopteran and ephemeropteran grazers is reduced by temporal segregation, because baetids finish their development in the upper and lower sections of the stream after most of the trichopteran grazers had emerged. It is assumed that these interspecific life history differences and the different distributions along the stream facilitate coexistence in a small stream such as the Breitenbach.

The following section summarizes the results of Sections 7.8.1 to 7.8.4, which are relevant to grazer–periphyton interactions, species-specific foraging strategies, and the ecological segregation of the most abundant trichopteran grazers of the Breitenbach.

Agapetus fuscipes is the trichopteran species with the highest abundance in the Breitenbach. Most of the larvae colonize the upper and middle sections of the stream. *Agapetus fuscipes* has seven larval instars and, in the upper section of the stream, most larvae pupate between May and July. However, if egg and adult stages

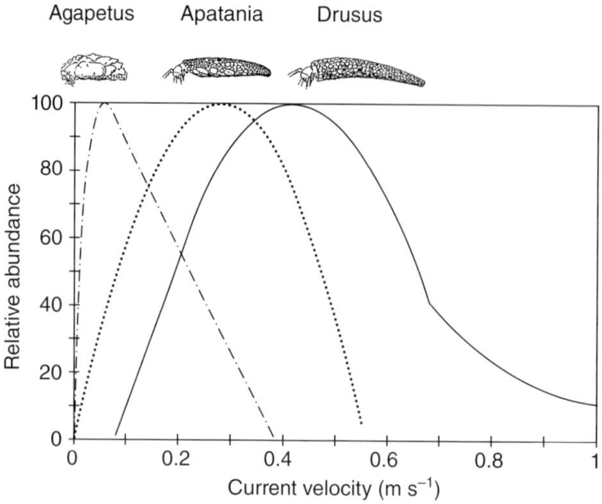

Figure 7.155 Relationship between the current velocity and abundance of the three grazers, *A. fuscipes, A. fimbriata*, and *D. annulatus*, in the upper reach of the Breitenbach.

are included, at least five different developmental stages coexist at any given time of the year (Becker, 2005). The small larvae are slow movers, carry heavy cases with respect to their lightweight larval biomass (Becker, 2001), and prefer microhabitats with a low current velocity (Figure 7.155). The larvae do not travel far and avoid thick epilithic algal layers. Field and laboratory investigations showed a positive larval response to thin autotrophic layers, and a negative response to thick periphyton mats. When the current velocity is low, larvae remain in restricted feeding areas at high densities for long periods, feed at day and night, and do not show diurnal migration between feeding areas and resting sites. It is assumed that the metabolic requirements of *A. fuscipes* larvae are lower compared to the larger larvae of *D. annulatus* and *A. fimbriata. Agapetus* larvae use both ventral case openings to feed, with high efficiency on low-biomass periphyton and demonstrate an efficient top-down control of the epilithic microbial communities, down to thin layers (Gergs, 2004), as well as exploitative competition with coexisting grazer species, such as *D. annulatus, A. fimbriata*, and baetids. It is known that intensive grazing favors small, fast-growing diatom species (Hart, 1985b; Coletti *et al.*, 1987; Steinman *et al.*, 1987b; Lowe and Hunter, 1988), because these autotrophs find growing conditions when the competition for light and nutrients is low favorable (Lamberti and Resh, 1983). Also any settling detritus, which could hinder microphytobentic growth (Marker, 1976; Tett *et al.*, 1978) would be constantly removed. These findings explain the high proportion of small-sized diatom cells (88% < 20 μm) and the high proportion of detritus in the gut contents of *A. fuscipes* larvae in the Breitenbach (Becker, 1990, 1994). It is assumed that the standing crop, taxonomic composition and productivity of the epilithic microorganisms, as well as the distribution of coexisting grazers in the upper and middle section of the Breitenbach, are significantly affected by the highly efficient grazing activity of *A. fuscipes* larvae.

The highest densities of ***D. annulatus*** larvae were also found in the upper and middle sections of the stream. The relatively large larvae, compared to *A. fuscipes*, are highly mobile and were sometimes found on periphyton patches above the water surface. The case to larval mass quotient is relatively low compared to *A. fuscipes* (Becker, 2001). The high locomotive activity of the larvae enables them to move fairly rapidly between food patches, even involving upstream movements over short distances at rates of up to $1\,m\,h^{-1}$ (Menzler, 1995). This can partially compensate for downstream movement due to drifting. *Drusus annulatus* larvae tolerate a wide range of stream currents up to $1\,m\,s^{-1}$ (Figure 7.155). The larvae undertake diurnal movements between feeding and resting areas, burrowing into fine sediments in the field and laboratory cultures, and the grazing activity shows a circadian rhythm, with a maximum 1 h before sunset and a minimum after dawn (Menzler, 1995). This correlates well with variation in water temperature. The larvae obviously select patches with thick periphyton mats, rapidly vacating substrata with low food concentrations in the field, and show a negative response to thin biofilms but a positive response to thick biofilms in laboratory experiments. Of all the species examined, *D. annulatus* larvae consumed the highest proportions of autotrophs, both diatoms and cyanobacteria. The relatively low proportion of detritus in the gut contents may reflect the high current velovity at their feeding sites, which may have prevented detrital sedimentation (Becker, 1990, 1994). It is suggested that the highly mobile larvae have a relatively high metabolic demand. Feeding areas at low current velocities, which are mostly exploited by *A. fuscipes* larvae, are unattractive for the large larvae of *D. annulatus*, which are forced to switch to a high current velocity feeding areas and thick biofilms, which are not colonized by the competing *Agapetus* larvae.

Larvae of ***A. fimbriata*** colonize the upper and middle sections of the Breitenbach. The case to larval mass quotient is relatively low compared to *A. fuscipes* (Becker, 2001). The larvae, which were more mobile than *A. fuscipes* and *S. pallipes*, but slow compared to *D. annulatus*, avoided strong currents (Figure 7.155). In contrast to the other trichopteran grazers, *A. fimbriata* larvae pass through winter dormancy (Aurich, 1989). Cyanobacteria appear to be an important food resource for *A. fimbriata*. The larvae consumed cyanobacteria in significantly higher proportions in summer than in spring and autumn (Becker, 1990). In the field, aggregates of these larvae were often found grazing on black plaques of periphyton containing colonies of cyanobacteria and lichens, sometimes together with larvae of *D. annulatus*. *Apatania fimbriata* larvae responded more to thin natural autotrophic patches in laboratory experiments compared to *D. annulatus* and remained in thicker algal patches for a long time (Menzler, 1995). However, the larvae did not show a clear response to unialgal layers of diatoms of different growth form, or to *Oscillatoria* sp.

Tinodes rostocki is a sedentary species, with a tubular retreat attached to the stone surface, belonging to the dominant macroinvertebrate species in the middle and lower reaches of the Breitenbach, where substrata are densely covered by their retreats (Becker, 1993). As a consequence of crowding, the high larval densities induce intraspecific competition for space. The larvae show territorial behavior and defend their retreats intensively. Field studies in the Breitenbach showed that large larvae, rather than the younger instar larvae, were forced to leave their overcrowded substrata (Becker, 1993). The retreat functions as a protection against predators, desiccation and currents, and act as a favorable substratum for microphytobenthos. The older part of the retreat has a higher photosynthetic activity and microphytobenthic biomass than the nearby stone biofilm, and the composition of the autotrophic community of the retreats is dominated by diatom, in contrast to the stone biofilm (Stief and Becker, 2005). The high proportion of diatoms and low proportion of cyanobacteria in the guts of this species suggest that the larvae did not feed by scraping nearby stones (Becker, 1990). Larvae elongate their retreat at one end by adding new mineral particles and break down the other end in order to feed on the organic matter that is attached to the particles. The complex larval behavior has a significant effect on the quality of their food resources, the biofilms of their retreats, and demonstrates "gardening" by *Tinodes* larvae throughout their development (Kahlert and Baunsgaard, 1999; Stief and Becker, 2005). Field investigations showed the crucial role of retreat size and the attached organic material for the growth of late instar larvae in spring. When larvae feed intensively on the attached organic material of their retreats, larval biomass increases significantly and is negatively correlated with the decrease in retreat size. It is suggested that the microbial community of epilithic biofilms in the middle and lower sections of the Breitenbach are significantly affected by the structuring impact of *T. rostocki* with respect to the larva-induced changes in taxonomic composition, spatial organization, and metabolic activity of microorganisms.

Larvae of ***S. pallipes*** mainly colonize the middle and lower sections of the Breitenbach. They are slightly more mobile than *A. fuscipes* larvae, but the low proportion of microphytobenthos in their gut contents suggests that they do not actively search for sites of high microphytobenthic abundance (Becker, 1990). The larvae were found mostly on stone surfaces at sites with a reduced current velocity, as well as on fine sediments, as also observed by Ormerod (1988). This agrees with the result of the laboratory food choice experiment, in which larvae showed no preference for periphyton over detritus (Becker, 1994). Silo diets showed the clearest overlap with the diet of *A. fuscipes* (Becker, 1990).

Larvae of ***M. longulum*** colonize the middle and lower sections of the Breitenbach at low densities, and appear to have an affinity for higher currents. They were mostly found on submerged mosses in the center of the stream, or, as was generally the case in the spring, on stone surfaces in the main current. The larvae are highly mobile, carry a lightweight silken case, and appear to consume specifically diatoms (Becker, 1990). Cyanobacteria are rarely consumed. In general, it can be concluded that the most obviously mobile species, namely *D. annulatus* and *M. longulum*, consumed the highest proportions of phototrophic microorganims (Becker, 1990). Dudgeon and Lam (1985) reported a similar trend for gastropods. Mobile species searched actively for sites of high microphytobenthic abundance. In laboratory experiments, *Micrasema* larvae preferred filters with moss fragments and a high proportion of moss fragments was found in their gut contents (Becker, 1990, 1994). The larvae have coarse-toothed mandibles resembling those of shredders (Hickin, 1967; Wiggins, 1996). An examination of their feces revealed that most of the moss cells remained undigested when they passed through the gut (Becker, 1990). Thus it is likely that the moss was consumed as a substratum for high quality food, such as diatoms, and it was mainly the diatoms that were digested. A similar observation was made for the mayfly *Ephemerella ignita* (Ephemeridae) by Willoughby and Mappin (1988).

7.8.4.15 **Conclusions**

On the basis of several grazer populations with high larval densities, and the relatively low standing stock of periphyton throughout the year along the stream (chlorophyll-*a* content about $5\,\mu g\,cm^{-2}$), it is assumed that the microphytic community of the Breitenbach is strongly controlled by grazing insects. The high productive capacity of the relatively small standing crop of the epilithon may support the large number of grazers growing and developing in this small stream. The significant increase in periphyton biomass over the course of the exclusion experiment in autumn demonstrated the high productive capacity of the epilithic autotrophic community in the upper section of this stream, and the strong reduction of periphyton by *A. fuscipes*. The continuous biomass accumulation over the five weeks of the field experiment indicated that the autotrophs are not limited by nutrient concentrations. Due to high turnover rates, streams may have a small quantity of algae but high herbivore biomass (e.g., Lamberti and Resh, 1983; McAuliffe, 1983), resulting in an inverted trophic pyramid (Gregory, 1983; Lamberti and Moore, 1984). However, when consumptive demand greatly exceeds productivity capacity, grazers may compete for limited food resources (Hart, 1987; Hill and Knight, 1987; Lamberti *et al.*, 1987b). The interspecific exploitative competition between *A. fuscipes* and *D. annulatus* and other coexisting grazers in the upper section for epilithon, and the interference (territorial behavior) by *T. rostocki* in the middle and lower sections demonstrate that periphyton is an intensively competed-for resource in the Breitenbach. Food limitation and density-dependent growth of grazers in streams have been reported several times (e.g., Hart, 1987; Lamberti *et al.*, 1989; Hill, Weber and Stewart, 1992; McNeely *et al.*, 2007).

High grazing efficiency maintains autotrophic epilithic communities in an early successional stage, and mostly adnate diatoms and basal cells of chlorophytes dominate the epilithic community. Mature filaments are often considered grazer-resistant (Dudley and D'Antonio, 1991; Steinman *et al.*, 1992). When grazers consume the vulnerable juvenile forms of filamentous algae in heavily grazed environments they prevent the establishment of inedible adult forms (DeNicola *et al.*, 1990; Dudley and D'Antonio, 1991). Thick epilithic mats, with apparently senescent and dead algal cells that have lost resistance to shear stress and been lifted as a large patch from the substatum by the current have never been observed in the Breitenbach. The permanently strong control by grazers of the autotrophic community of the Breitenbach was obvious in spring 1986. An accident with the insecticide cypermithrin killed the macroinvertebrate community in the lower, about 2 km, section of of the stream. In the following weeks in the absence of strong grazing pressure long filaments of the chlorophyte *Microspora* (length up to 100 cm), occurred along the stream. They had never been observed before, nor have they been since, in the years after this accident.

7.9
Effects of Water Temperature along the Breitenbach

Rüdiger Wagner

The impact of temperature on the development of aquatic invertebrates is well known (e.g., Roux, 1975; Svensson, 1977; Sweeney, 1978; Butler, 1984; Sweeney and Vannote, 1986; Sweeney *et al.*, 1986; Giberson and Rosenberg, 1992). Competing hypotheses consider water temperature as the main variable, explaining the success of species throughout their distribution range and trying to identify "thermally optimal sites" and to predict temperature effects at those sites. Success is defined as female size and weight (reproductive fitness) or as population success (density, abundance, biomass).

The "thermal equilibrium hypothesis" (Vannote and Sweeney, 1980) considers that water temperature is the main variable explaining size (fore wing length or body weight) throughout a species distributional range. A thermal regime was considered optimal at sites where female body weight and fecundity were maximized; this should also coincide with the greatest population success. Based on mayfly data from two Texas streams, Stanley and Short (1988) stated that population-level characteristics (density, biomass) were more appropriate for determining a thermal site optimum than body size. Hogue and Hawkins (1991, with further references) argued that the interaction of growth timing and water temperature was responsible for size variation in some aquatic insect adults, but genetic effects must also be considered (Wool, 1979).

7.9.1
Water Temperature as an Important Variable for Explaining Body Size

Compared with Ephemeroptera and Plecoptera, Trichoptera have a limited number of instars that affect the final size and weight of adults, in particular females. Data from various life cycle studies demonstrate that last instar larvae are "weight effective" accumulating approximately 80% (from 60 to >90%) of the body mass of the subsequent pupae and adults (Figure 7.86); that is, the better the general growth conditions during this period (temperature, food), the fitter the subsequent adult.

Because members of various functional feeding groups vary in size, weight, and abundance, their preconditions and successful seasonal strategies may differ. Limiting variables (e.g., temperature, food availability and quality, population density, competition) must be taken into account separately for each group, or even species.

These assumptions provoke questions: Are populations, and the weight of adult caddisflies limited by one or several of the above-mentioned variables? Are there differences between shredders and grazers? Do temperature and food affect size and weight of adults?

Size and weight data for adult caddisflies are available for the period 1983–1991 for sites T1, T2, T4 and T6. The effects of the above mentioned variables on adults were tested using combined laboratory studies and field experiments with the caddisfly, *Chaetopteryx villosa*, and the crustacean, *Gammarus fossarum*. In addition, effects of annual water temperature patterns at individual sites were examined.

7.9.2
Functional Feeding Groups

At the Breitenbach, caddisfly grazers, shredders, and predators may be distinguished by individual weight and abundance. The abundance of caddisfly grazers is high, but individual weight is low; shredders are low in abundance but individuals are heavy; predators are intermediate (Figure 7.156).

Significant negative correlations of female dry weight with total number of individuals per site and with dry weight of standing stock at individual sites over all years were found for the shredder, *C. villosa* (Figure 7.157), and the scraper, *A. fuscipes* (Figure 7.160). This suggested that both were resource limited (food) in the field. The nutrient supply for large shredders (CPOM, FPOM) appeared spatially and temporally unlimited (Chapter 4). Larvae of caddisfly shredders do not solely depend on the autumnal input of leaves, but may also feed on fresh, green

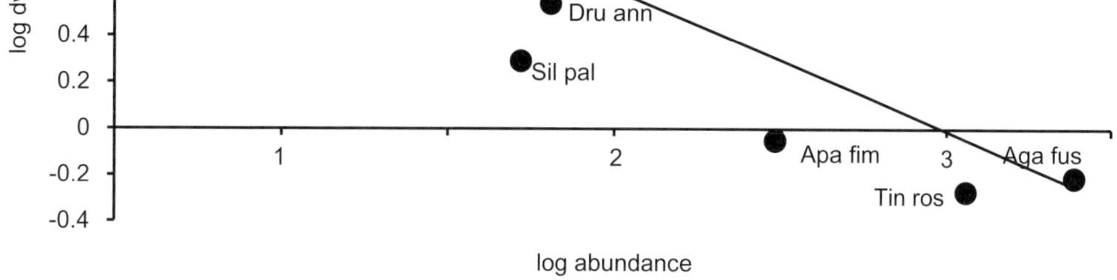

Figure 7.156 Linear regression of log-log data of caddisfly abundance on mean female DW (data pooled for nine years and four emergence traps): log$y$ = 1.807 – 0.604 log$x$; $r^2 = 0.69$, $p < 0.002$.

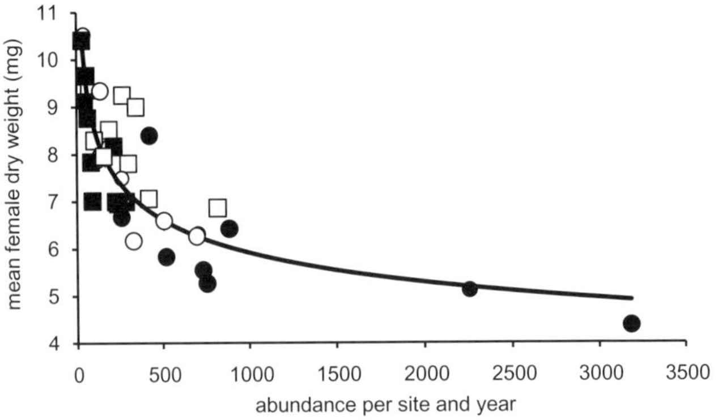

Figure 7.157 Regression of *C. villosa* female mean dry weight to specimen numbers at four sites along the Breitenbach (1983–1991): $y = a \times b$; $a = 18.0$; $b = -0.16$; $r^2 = 0.72$.

bank vegetation hanging in the stream. On the other hand, grazers presumably depend on the seasonally variable biofilm, whose availability is lowest in winter. Two maxima of algal abundance, in spring and in autumn, have been reported (Benedetto Castro, 1975; Cox, 1990b).

7.9.3 Experiments with *Chaetopteryx villosa*

7.9.3.1 Life Cycle and Adult Size at Different Stream Sites

Chaetopteryx villosa was the most abundant shredding caddisfly, females are significantly larger than males. The sex ratio was remarkably constant, two males to one female (Table 7.19). Analysis of life cycle duration and adult weight at different stream sites appear contradictory. Although the species is univoltine at all sites, the period during which larvae occur decreases with increasing distance from the source, ranging from approximately 9.5 months (T1) to only 8.0 months (T4) (Figure 7.158), but specimens are larger and heavier at T4 than at T1.

The reason for this is the annual stream temperature pattern. Embryogenesis is faster (6 weeks) in the upstream section (T1) than at T3 to T4 (10–12 weeks), due to decreasing water temperature downstream in winter. In summer, the comparatively low temperature at T1 keeps daily growth rates of the weight-decisive last larval instar down, prolonging larval and pupal development. Higher summer water temperature at the downstream sites more than compensates for the temporal disadvantage (Wagner, 1986, 1990a, 1990b; Figure 7.158).

7.9.3.2 Between-Sexes Size Difference

Extensive studies on all life cycle stages elucidated the effects of temperature and food on females and males at sites along the stream. Size differences between sexes depend on temporal life cycle differences. Larvae of future females rapidly

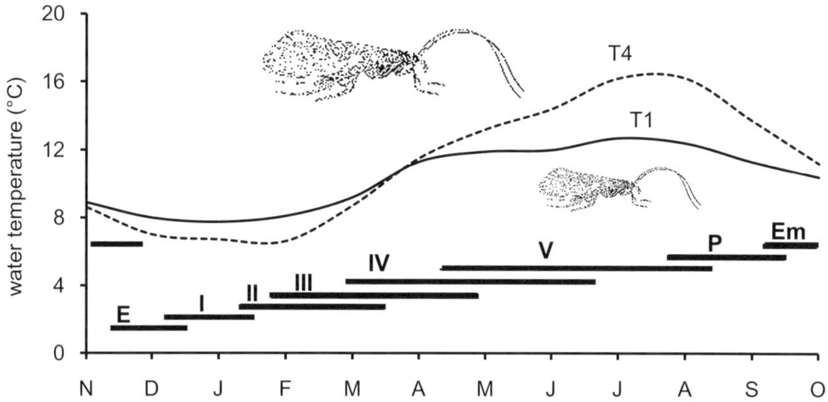

Figure 7.158 Long-term (1983–1993) maximum monthly water temperature at sites T1 and T4 along the Breitenbach, and site-specific life history patterns of *C. villosa*. E = embryogenesis, I to V = larval instars, P = pupa, Em = emergence.

Table 7.29 Significance of Pearson correlation coefficients between individual dry weights of female/male caddisflies with variables (trap = T1, T2, T4, T6); year = 1983–1991 (1–9); pattern (1–3); *n*/trap = abundance of same species per trap; *n*/patt = abundance of same species per pattern; dw/y = DW of all specimens per year; dwFFG = DW of all specimens in the respective functional feeding group). Positive correlations are underlined.

	Trap	Year	Pattern	*n*/trap	*n*/patt	dw/y	dwFFG
Apa_fim	–	–	–	–	0.018	–	–
Aga_fus	0.035/–	–	–	0.0001	–	0.005/0.032	0.006/0.03
Dru_ann	0.005/0.027	–	–	–	–	0.036/0.035	–
Sil_pal	0.0006/0.03	–	–	–	–	–	–
Tin_ros	–	0.0001/0.014	–	–	–	–	–
Cha_vil	0.012/0.002	–	–	0.0001	0.042	0.0001	0.002/0.0001
Pot_cin	0.003/–	–	–	–	–	0.048/–	–
Pot_luc	0.03/0.002	–	–	–/0.02	–	0.006/0.003	0.004/0.005
Ser_per	0.0001	–	–	0.036/–	–	–	0.04/–
Ple_con	–	–	–	–	0.0001	–	–
Rhy_fas	–	–	–	–	–	0.024/–	–/0.048

pass through instars I to III, reaching instar IV after about 6 weeks. Over the whole life cycle, this means that prospective females feed at their highest ingestion rates (instar IV and V) for a longer period (Wagner, 1990a. At the beginning of their larval life females invest in body size ("space" for developing eggs) and later fill up this space. Males spent longer as instars I to III, but shorter as instars IV and V, remaining smaller and lighter. Higher water temperature in summer, particularly in section T2 and downstream, increased ingestion and growth rates, resulting in the observed size and abundance pattern of *C. villosa* (Figure 7.158, Table 7.29).

7.9.3.3 Field Experiment – Effects of Food and Temperature

Food quality and water temperature effects, and the source of specimens (upstream, downstream) were tested with multifactorial enclosure experiments at sites T1 (water temperature almost constant) and T4 (pronounced diurnal and seasonal temperature fluctuations; compare Richardson, 1991).

At the start, weights of instar IV or early instar V larvae did not differ between sites. For any individual instar, no two groups of larvae could be interpreted as prospective males or females in terms of dry weight or fat content. The fat content of larvae from individual sites ranged between 22 and 35% dw (standard error 1–2%).

Fifty larvae from the T1 and T4 reaches were added at approximately natural densities (450 individuals m^{-2}) to each of 16 enclosures at the same sites (two replicates × FPOM or CPOM × two origins of larvae). At weekly intervals, live larvae were counted and returned to the channels together with new food. Pupated specimens were transferred to the laboratory and kept individually until emergence to determine sex, forewing length, and fat content.

Successful pupation was 52–96% in the CPOM treatments, but only 2–52% in the FPOM treatments. Although permanently available in high proportions the quality of FPOM is comparatively low characterized by high lignin particle content, and C:N ratios below 1:13 (Neumann, 1995). Death rates during pupation were high if specimen weight was low, indicating that food quality affected successful pupation and emergence. If FPOM was the only food source, this resulted in high numbers of unpupated specimens and few emergent adults. The effect of low food quality was even more obvious at site T4 with higher water temperature.

Pupation started in mid-August and 74–98% of larvae survived in the CPOM treatments, 66–98% in the FPOM enclosures. At the end of the experiment (December), the proportion of unpupated specimens was 4–44% in the FPOM group and 0–10% in the CPOM group.

Successful adult emergence was 34–76% for the CPOM group and 2–34% for the FPOM treatments. The increased percentage of pupated specimens and successfully emerged adults in the CPOM treatments was attributed to better food quality. In the stream, access of *C. villosa* larvae to CPOM was not limited, because during periods of low allochthonous input larvae feed on almost all aquatic and semi-aquatic macrophytes on the banks, and even leave the water to feed on grass (Jabobsen, 1993; Newman, 1991; Sand-Jensen and Madsen, 1989; Suren and Lake, 1989; Wagner, 1992, 2002c).

Synchronization of emergence was better in the CPOM-fed specimens. Pupation and emergence periods of FPOM-fed individuals were longer, similar to the emergence patterns observed in the field (Figure 7.92). At T1, with the smallest specimens, emergence occurred from September to December and in October and November at the other sites.

Dry weight and absolute fat content differed between sexes, but no sex-dependent difference was found in relative fat content. Significant differences between sites and food quality occurred in females but not in males. Forewing length, dry weight, and fat content were linearly related. Sex, food quality, and experimental site affected the size, dry weight, and fat content of emergent adults. These factors accounted for almost 95% of the observed weight variation (ANOVA). The origin of specimens had no significant effect, probably because 80% of the final larval weight is gained during the final instar. This appears to counteract any other environmental effects to which the earlier larval instars had been exposed. MANOVA indicated the same significant main and between-subject effects on males and females, but the combination "food × site" disproportionally benefited females.

Low adult weight at T1 is therefore not the consequence of high population density (food limitation) but an effect of low temperature during the development of the "weight decisive" last instar. From laboratory studies it is known that mean relative growth rate (RGR) is reduced from 0.0137 to 0.0064 mg^{-1} day^{-1} between 14 and 10°°C (Wagner, 1990a, 1990b). It is obvious that simply analyzing the density dry weight relationship will lead to a conclusion of food shortage, but this "masks" the relevant temperature effect that affects growth efficiency of individuals in the same way as food quality or quantity – it is another limiting resource.

7.9.3.4 ***Chaetopteryx villosa*** **– Growth Models**

The life cycle data obtained from laboratory and field experiments (egg development, growth rates and duration of individual instars (♂ and ♀); Wagner, 1986, 1990a, 1990b, 2002a, 2002b, 2002c, 2002d) were combined with stream temperature data (Schmidt, personal communication) from sites T1 and T4 in 1989 to test whether, in combination, both were sufficient to calculate larval growth and adult weight at both field sites (Figure 7.159).

The admittedly rough models reveal temperature sum differences of about 440 degree days between both sites over approximately 250 days. Weight differences between females and males at every site and between sites for each sex were predicted correctly. However, compared with the data of emerged specimens in the respective year, the model overestimates weights by an order of 10 or 60%; lowest deviation (9 or 39%) with females at site T1, highest (60%) for males at site T4.

7.9.4 *Gammarus fossarum* – Another Example

A similarly designed experiment was carried out with *G. fossarum* (Winkler, 1992). The effects of food quality (FPOM or CPOM), water temperature (T1, T4) and "origin of specimens" (transferred between experimental sites) were tested. Results were compared with reference specimens from the stream. At the start of the experiment it was evident that the stream populations at both sites had developed quite differently in winter before the experiment. At T1 mating couples were observed in winter, and sufficient specimen numbers were available as early as April, but equal numbers were not found until June at T4.

Results of the (food quality × exchange × site) experiment demonstrated that specimens at T4 grew faster than at T1 during summer. Examination of individual variables showed significant effects of food quality ($P < 0.0001$) and water

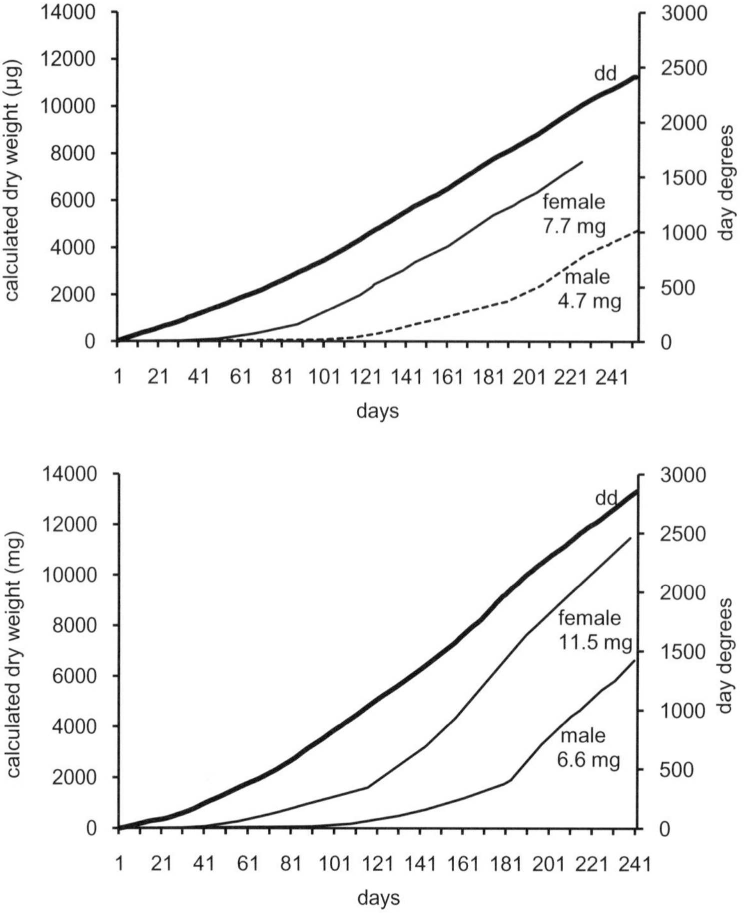

Figure 7.159 Growth model of males and females of *C. villosa* combined from laboratory data (daily growth rates per instar at given temperature) and field temperature data (daily mean, 1989) of sites T1 (left) and T4 (right).

temperature ($P < 0.002$; ANOVA). However, the pairwise interactios revealed only one significant effect, "water temperature × origin of specimens" ($P < 0.0001$). "Origin of specimens" included the site specific growth conditions (food, water temperature) before the experiment.

The results fit information that *Gammarus* spp. can survive and reproduce as well with high- as low-quality food (e.g., Koch, 1991). The reproductive period was clearly limited by water temperature at site T4 (no reproduction after mid-November; Meijering and Teichmann, 1978; Teichmann, 1982) but was not limited at site T1, even during winter. As for *C. villosa*, water temperature was again the decisive variable affecting the growth of *G. fossarum*.

7.9.5 Water Temperature – A Limiting Resource

The initial hypothesis of food limitation in *C. villosa* was rejected by field and laboratory experiments (Wagner, 1986, 1990a, 2002a, 2002c). Decreasing abundance and increasing individual weight downstream were seemingly unaffected by each other. Reduced growth rates indicated limiting effects due not to food but to water temperature. With *C. villosa* the negative abundance versus DW relationship was striking, because no similar relationship existed for other caddisfly shredders, such as *S. personatum* or *Potamophylax* species. Life cycles (Section 7.7) and physiological abilities may explain this difference.

Comparing the results of the feeding and exchange experiments of *G. fossarum* and *C. villosa* it was evident that water temperature is a particularly important variable. It acts like any other resource – limitation causes deficiency, with harsh conditions, and has effects on populations (density) or individuals (size, weight, reproductive fitness).

7.9.6 Scrapers – Indication of Site-Specific Intra- and Interspecific Competition Based on Emerged Adults

Competition occurs when a number of specimens of the same or different species utilize common resources, the supply of which is short. If resources are not in short supply, competition occurs when the organisms seeking that resource harm each other in the process (Birch, 1957). This definition includes the idea of exploitative competition, that assumes depletion of resources such that another individual is disadvantaged, and interference competition as the direct interaction of usually aggressive nature (Allan, 1995). Hypotheses of competition include an assumption that this occurs at any time. Resource partitioning assumes many possible competitors, but may be the starting point of competition if marginal conditions change with time. Testing the Breitenbach caddisfly data leads one to conclude that, although competition is an ever-present phenomenon, its effects (size, weight, fitness) are difficult to prove, at least based on emerging insects.

Scraping caddisflies are faced with the problems that: (i) the development of their main food source (biofilm, algae) is seasonally variable (e.g., Benedetto Castro, 1975; Cox, 1990b) and (ii) their population density (which is affected by the flow regime) may vary dramatically between years, with probable nutrient limitation due to intra- and interspecific competition. The following data are from emergence traps over several years, based on adults that grew in the stream; they complement Becker's stream experiments (Section 7.8).

7.9.6.1 Caddisfly Scrapers – Size Differences along the Breitenbach

Different feeding modes by scrapers along the stream lead to complicated interactions with gastropods, mayflies and other caddisflies (Becker, 1991, 1994, 2001; Becker and Menzler, 1998). At site T1, where *A. fuscipes* was by far the dominant scraper, relationships between individual dry weight (DW) and abundance were significant. At low densities, female DW was twice to three times higher than at high densities. The relationship of *A. fuscipes* female DW to total scraper biomass was also high. Dependency of female and male DW on the abundance of its own species was best described by linear regression: $y = a + b^{\exp}x$ ($a = 0.35$, $b = 0.00045$; $r^2 = 0.74$, $P < 0.003$ for females/$a = 0.67$, $b = 0.00039$; $r^{-2} = 0.69$, $P < 0.006$ for males). At site T1, abundance variability from 1983 to 1991 was so large that a model for all stream sites was still significant, even though no significant correlation was detected for individual sites, T2, T4 and T6.

Comparison of female DW and species abundance between individual and all traps again provided exclusively site-specific results. Abundance of *A. fuscipes* and DW of female *D. annulatus* were only negatively correlated at T1. DW of *T. rostocki, D. annulatus*, and *A. fimbriata* were positively correlated.

At T2 and T4 abundance of *A. fuscipes* and DW of female *A. fimbriata* and *D. annulatus* were negatively correlated, as were specimen numbers of *A. fimbriata* and female DW of that species and of *D. annulatus*. Female DW of *A. fimbriata, D. annulatus*, and *T. rostocki* were positively correlated, as were specimen numbers of *A. fuscipes, A. fimbriata*, and *T. rostocki*. At T4, correlations of *A. fimbriata* female DW with *T. rostocki, D. annulatus*, and *S. pallipes* were positive. There was no significant relationship at site T6 where most caddisfly scrapers were rare. *Baetis* (mayfly) species and the mollusc, *Ancylus fluviatilis*, were dominant grazers there (Section 7.8).

7.9.6.2 **Within- and Between-Species Resource Limitation**

Resource limitation is a major precondition for competition whose effects are evident in the weight and size of individuals in the scraper functional feeding group. Competition within or between grazer species occurred irregularly (Section 7.8). The main cause was an increased abundance of *A. fuscipes* in years with low discharge – a very local and temporally "unusual" situation. This simultaneously affected other species, intensifying within and between species competition. Only at high densities, when larval cases almost totally covered all stone surfaces, that is, ~1200 to 4000 adults m^{-2} were collected in the nearby trap T1, did competition effects within *A. fuscipes*, or between it and other scrapers, become evident (Figure 7.160); the more specimens collected, the lighter the mean weight. High *A. fuscipes* densities also affected the weight of the competing species, *D. annulatus* and *A. fimbriata* (Figure 7.160); the higher the density of *A. fuscipes*, the lighter the females of both competitors.

The feeding behavior of *A. fuscipes* larvae and their occupation and defense of a feeding area (Becker, 1991, 2001) lead one to presume that all specimens of a population use the same biofilm as well as feeding on the algae growing in and on their own cases (Cox and Wagner, 1989). This was sufficient to keep many specimens alive, but each with lower weight.

In a series of experiments, Becker and Menzler (1998) demonstrated asymmetric and exploitative competition between *A. fuscipes* and *D. annulatus* larvae. High densities of *A. fuscipes* forced the mobile *D. annulatus* larvae to sub-optimal feeding

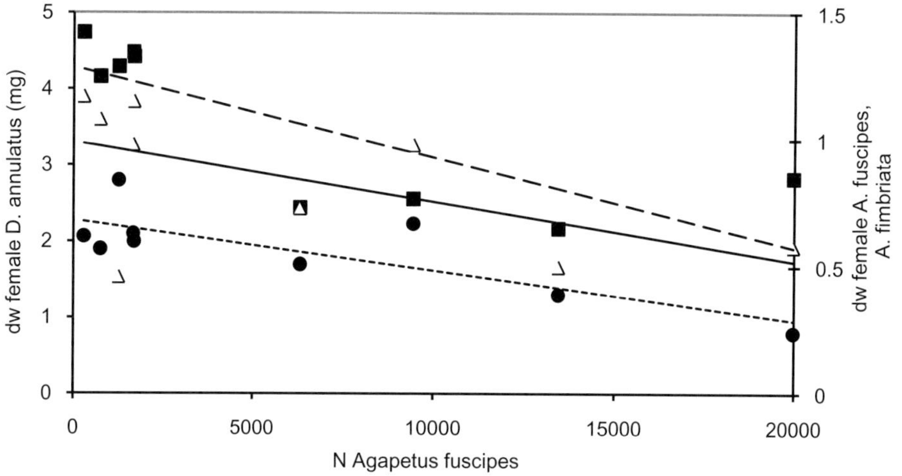

Figure 7.160 Dependence of the female DW of *Agapetus fuscipes* (full circles, dashed line; $y = 0.68 - 0.00002x$; $P < 0.01$, $r^2 = 0.64$), of *Apatania fimbriata* (open triangles, full line; $y = 0.991 - 0.0002x$; $P < 0.11$, $r^2 = 0.33$), and of *Drusus annulatus* (full squares, broken line; $y = 4.29 - 0.0001x$, $P < 0.0001$, $r^2 = 0.63$) on the abundance of *Agapetus fuscipes*.

sites (Becker, 1991, 2001; Becker and Menzler, 1998). It is therefore plausible that competition during larval life at high densities will affect the size and weight of *A. fuscipes* and its competitors.

The negative correlation with *T. rostocki* at site T1 was probably not the result of competition but depended on the different distribution centres of both species along the stream, *A. fuscipes* preferring the upper, *T. rostocki* the middle and lower sections. This was supported by the results at T4, where both species were quite abundant, but lack of correlation indicated no interaction due to the different habitats and feeding strategies of the species (Section 7.8).

Comparisons of *C. villosa* and *A. fuscipes* dry weights versus species abundance (Figures 7.158 and 7.160) both suggest density dependence effects of nutrient limitation. However, carefully planned experiments can reveal these effects. The along-stream size difference in *C. villosa* was clearly affected by stream temperature at unlimited food supply. For *A. fuscipes,* only increased densities near the spring, as a result of low discharge periods and a particular site revealed the effects of competition. Nutrient limitation affected scrapers, water temperature limited the growth of shredders.

8
Ecosystem Breitenbach

8.1
The Sandy Streambed Areas of the Breitenbach

Rüdiger Wagner, Jürgen Marxsen, and Hans-Heinrich Schmidt

8.1.1
Importance of Sediments in Streams

The stream bottom, including the stream bed sediment ("hyporheic habitat"), has received increasing interest in ecological studies since the 1960s, beginning mainly with the studies of Orghidan (1959) and Schwoerbel (1961), who defined the hyporheic habitat to be the saturated interstitial spaces in fluvial consolidated rock below and adjacent to running waters as a border zone between streams and rivers vs. groundwater. It is a most important compartment of the stream in its valley (Hynes, 1975), which is highly significant for the stream communities as well as for stream metabolism (Jones and Mulholland, 2000). The simple, popular definition regards the hyporheic zone as saturated interstitial spaces below the stream bank that contain some proportion of channel water (White, 1993). However, more developed concepts describe this environment as a dynamic ecotone which fluctuates in response to sediment characteristics and hydrological exchange and which bounds the stream and the groundwater (Gibert *et al.*, 1990; Vervier *et al.*, 1992; for a detailed review on conceptual developments, see Boulton *et al.*, 2010). The basic questions related to this compartment concern the extension of its open space below the stream bottom, its temporal variation, its contribution to stream metabolism, and its utilization as habitat and food resource by benthic invertebrates (e.g., Bretschko, 1981, 1984; Bretschko and Klemens, 1986; Findlay and Sobczak, 2000; Godbout and Hynes, 1982; Hakenkamp and Palmer, 2000; Hynes *et al.*, 1976; Kaplan and Newbold, 2000; Wallis, 1981; Williams, 1989; Williams and Hynes, 1974).

The Breitenbach sediment was the subject of several research attempts with very different foci, in particular the dependence of its structure on hydrology (Wagner *et al.*, 1993; see Section 3.2), the chemistry of its interstitial waters (Wagner *et al.*,

Central European Stream Ecosystems: The Long Term Study of the Breitenbach, First Edition. Edited by Rüdiger Wagner, Jürgen Marxsen, Peter Zwick, Eileen J. Cox.

1993; see Sections 3.3 and 4.3), the flux of carbon (Fiebig, 1995), microbial structure and activity (Beier *et al.*, 2008; Marxsen, 2001; Marxsen and Fiebig, 1993; see Chapter 6), and its colonization by micro- and macroinvertebrates including its faunal biomass production (see Chapter 7). This chapter integrates the main findings to provide a comprehensive overview of the importance of the sandy streambed areas of the Breitenbach, in particular for the composition of the stream communities as well as for its metabolism.

8.1.2
The Breitenbach Streambed

During weathering of Bunter Sandstone bedrock small sand particles and cement are released. Sand particles and cement generate sediments and clay along the stream. Thus, clay layers are deposited over large areas below the Breitenbach (Kupfahl, 1965). However, these clay layers do not completely separate the stream from the surrounding groundwater, but only to an unknown extent. In most parts of the stream, through measurements using piezometers that had been inserted into the streambed sediment in its middle and lower courses, Fiebig (1995) observed more or less continuous inflow of groundwater throughout the year. Particularly in the areas around the spring horizons above T1 and T4, the stream receives large amounts of groundwater. The groundwater that enters the Breitenbach via diffuse perfusion through the streambed over large areas of the Breitenbach especially contributes large amounts of DOM, which is immobilized in the streambed during groundwater perfusion and utilized by the streambed community (Fiebig, 1995). This occurs mainly via the large, very active microbial communities (Marxsen and Fiebig, 1993; Marxsen, 2001; see Chapter 6) and increases the food resources of the stream community considerably (Marxsen, 2006; see Section 8.7).

The streambed sediment usually consists of stones, coarse and fine gravel, sand, and silt deposits (Figure 8.1) mixed with varying amounts of organic matter. Grain size structure and POM content of the sediment are dependent on the discharge in the antecedent period, as demonstrated by observations on the relationship between discharge and grain size structure (Figure 8.2), and between discharge and sediment POM content (Figure 8.3) in the middle reach of the Breitenbach (Wagner *et al.*, 1993). Sediment POM content was inversely related to discharge. During periods of high discharge organic content was below 1%, but increased within a few weeks during periods of decreasing discharge, to about 7% (Figure 8.3). The relationship between POM content (POM) and discharge (Disch) is best described by the equation $\text{POM} = 7.94 - 2.19\,(\ln \text{Disch})$ ($r^2 = 0.49$, $P < 0.05$). Thus about 50% POM content fluctuations in the sandy sediments are explained by discharge. This is generally the case in the Breitenbach, as POM content and the proportion of the small grain size fractions decrease with increasing discharge (Wagner *et al.*, 1993). Typical POM concentrations for lotic sites in the Breitenbach are between 0.3 and 1.0%, but may increase up to about 15% in lentic areas (Thomas, 1990; Neumann, 1995; Marxsen, unpublished data).

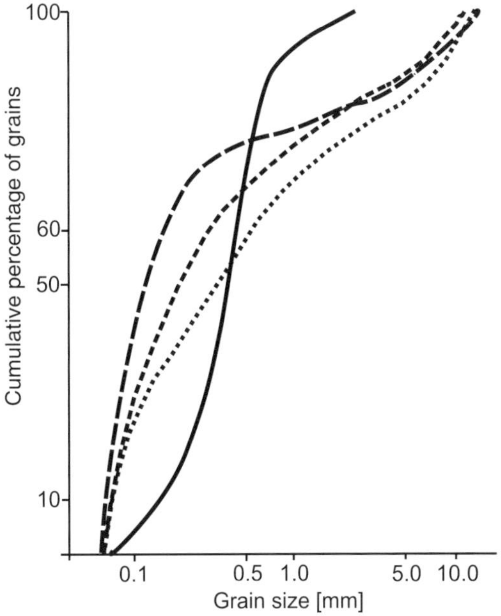

Figure 8.1 Grain size distribution by dry weight in an artificially sanded stream section (full line) and in various "natural" sections in the area between T2 and T4 (dashed lines).

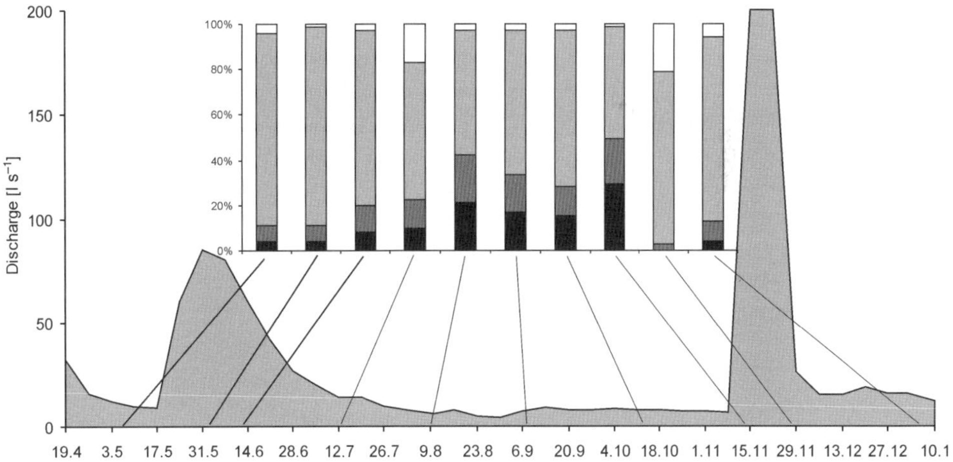

Figure 8.2 Discharge of the Breitenbach (weekly maxima at GT6). Insert shows proportion of the grain size fractions <0.063 mm (black), 0.063–0.1 mm (dark gray), 0.1–0.63 mm (light gray), and >0.63 mm (white) below trap T5 from April 1984 to January 1985, demonstrating the dependence of grain size structure on discharge in the antecedent period (after Wagner *et al.*, 1993).

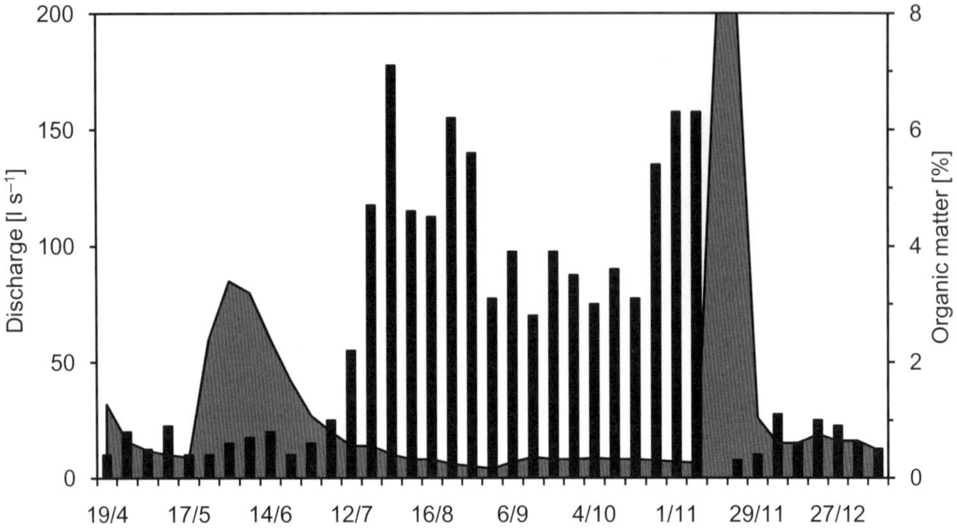

Figure 8.3 Discharge of the Breitenbach (weekly maxima at GT6, gray area) and percentage of organic material in the sandy sediment (at trap T6, columns) from April 1984 to January 1985.

The extent of the stream bottom area fluctuates very little according to discharge (for details, see Section 3.1.5). Only at very low water or during floods does the area decrease or increase significantly. An example of an increase in streambed area in the lower Breitenbach resulted from a flood event in 1994 (Figure 8.4). However, the proportions of different substrata covering the stream-bed fluctuate according to discharge history (Figures 8.4 and 8.5). In the example from 1994, early in the year after extended periods of higher discharge, the percentage ratio between fine-grained and coarse substrata was 38:62. But in September, after several months of low discharge (at gauge GT2 decreasing from less than $20 l s^{-1}$ more or less continuously to less than $10 l s^{-1}$ without any intervening flood events, Figure 8.5; cf. Figure 6.32 for discharge), fine-grained sediment dominated (86:14). However, the coverage of the stream bottom by different types of sediments can be drastically different between different years, significantly affecting the structure of the macrozoobenthic community (see Section 8.2).

8.1.3
Interstitial Water: Chemical Characteristics

The vertical extension of aerobic zones in the sediment was roughly estimated using iron rods driven vertically into the stream bed from March to June 1984. After removal, the rods appeared black in reducing layers, but brownish at layers with oxidizing conditions (Wagner *et al.*, 1993). The extent of oxic conditions in the sediment depended on grain sizes deposited by the current over the preceding

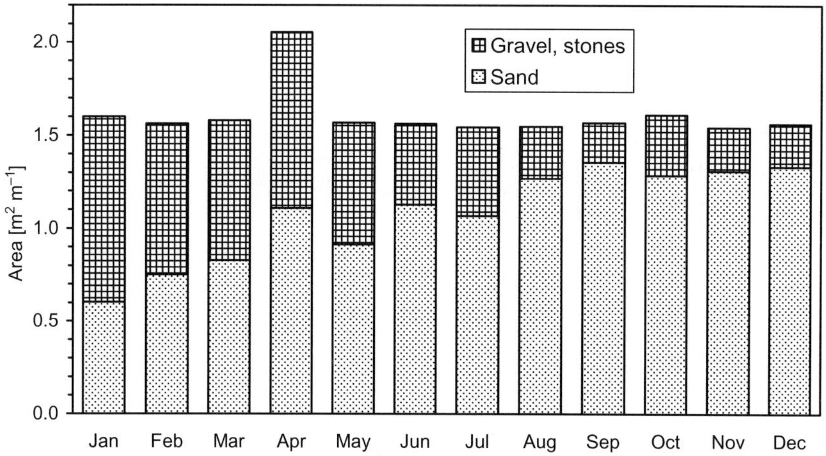

Figure 8.4 Distribution of fine-grained sandy sediments and coarse substrata at the transition between the middle and lower reach of the Breitenbach during 1994 (Marxsen unpublished data). It is visible that the stream width is somewhat independent of discharge, except when extreme floods occur (cf. Chapter 3), as happened in April 1994 (compare Figure 6.32 for discharge data).

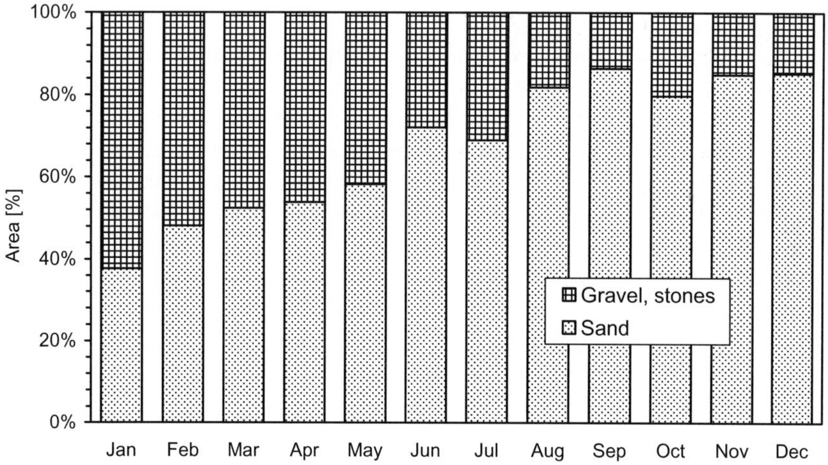

Figure 8.5 Relative distribution of fine-grained sandy sediments and coarse substrata at the transition between the middle and lower reach of the Breitenbach during 1994 (Marxsen, unpublished data). The percentage of fine-grained sediments increases almost continuously throughout the whole year, reaching 80% after several months of low discharge (compare Figure 6.32 for discharge data).

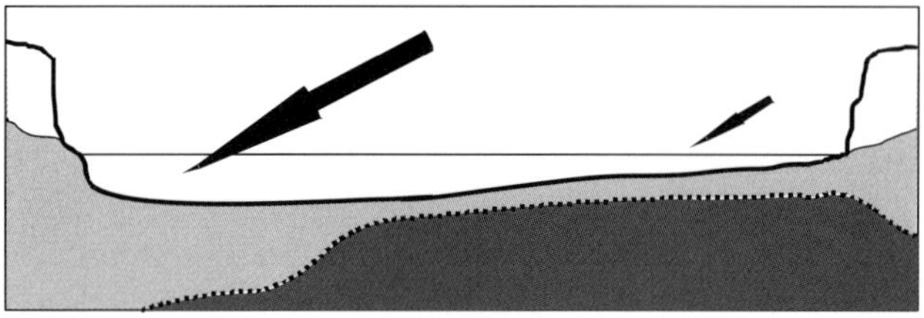

Figure 8.6 Simplified cross-section of the Breitenbach stream bed and adjacent soil. The boundary between sediment/soil and water/air is indicated by bold, the water surface by thin lines. Areas with interstitial spaces containing oxic water, anoxic water, or air are light gray, dark gray, and white, respectively. Arrows indicate different flow velocities.

period. At sites with high current velocity and large grain size, oxygen-containing layers extended deeper (down to about 20 cm) than in areas with low current velocity and sandy sediments, where reducing conditions could sometimes occur only a few mm below the surface (Figure 8.6). During periods of permanent high discharge (>60 l s^{-1}), conditions seemed to be temporally stable and spatially similar. At lower flow rates (<20 l s^{-1}) deposition of fine-grained sediments and FPOM was initiated, with pronounced fluctuations in POM content over time and space. Grain size generally decreases towards the banks and with low current velocities. Thus, combined with high FPOM deposition rates these sediments often exhibit anaerobic conditions only a few millimetres below the surface (Wagner *et al.*, 1993; Thomas, 1990; Neumann, 1995; Marxsen, unpublished data; cf. Figures 8.6 and Figure 6.33).

The oxygen gradients within the interstitial spaces were often monitored by pumping water out of steel gauze cages buried in the sediment, left in place for repeated sampling over weeks or months (cf. Chapter 3). This approach was also used to determine chemical variables, for example, during a sampling campaign from 1986 (Wagner *et al.*, 1993), when two steel gauze cages were buried 20 and 40 cm deep below trap T6. Throughout the year no oxygen was detected at these depths.

Stream water pH is typically close to 7 (see Chapter 3), fluctuating only slightly, except after snow melt or rainfall induced floods. With increasing sediment depth pH decreases, to about 6 in both layers, with small temporal variation (Figure 8.7). No differences were found in conductivity between stream and interstitial water apart from slight concurrent variations, between approximately 120 and 140 $\mu S_{25}\,cm^{-1}$.

Nitrate concentration in the stream water was between 600 and 800 $\mu g\,N\,l^{-1}$, but no nitrate was detected in interstitial water. Stream water ammonium concentrations were below 100 $\mu g\,N\,l^{-1}$, but reached 350–500 $\mu g\,N\,l^{-1}$ within the sediment. Temporal variation in NH_4^+ and NO_3^- was low. However in April and May a strong increase in NO_3^- occurred with the spring flood, although other measured compounds were unaffected.

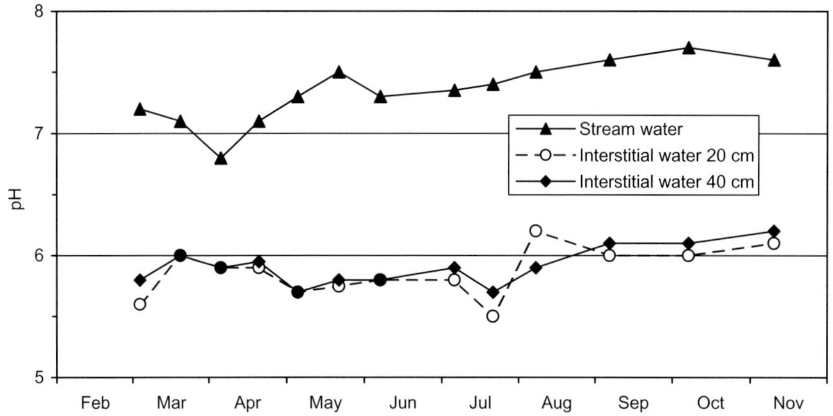

Figure 8.7 pH in stream water and in interstitial water from different depths in the Breitenbach lower course below trap T5, February–November 1986.

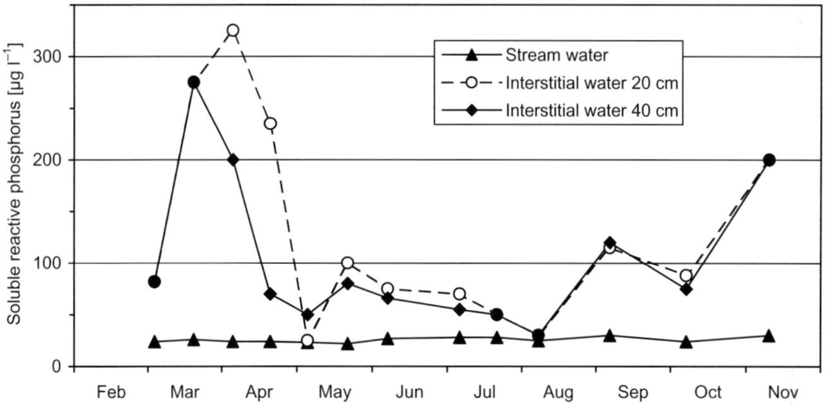

Figure 8.8 Concentrations of soluble reactive phosphorus in stream water and in interstitial water from different depths in the Breitenbach lower course below trap T5, March–November 1986.

Soluble reactive phosphate (SRP) concentrations in the stream were typically low (<40 µg P l^{-1}), but rose to about 300 µg P l^{-1} in the sediment. Concentrations in the stream water remained almost constant over the study period, but fluctuations were marked in the sediment (Figure 8.8). During the spring flood, when carbon and nitrate in the stream water increased strongly, almost no change in SRP concentration was detected in the stream water, although there were remarkable increases at both sediment depths.

During discharge of about $20 \, l \, s^{-1}$ the amount of dissolved organic carbon (DOC) in the surface water was low (about 1.5–3 $mg \, C \, l^{-1}$), compared with water in the sediment (about 3–5 $mg \, C \, l^{-1}$), as is usual for such situations (Figure 8.9; cf. Section 4.3). In spring DOC increased with discharge in the stream water, up to nearly 10 $mg \, C \, l^{-1}$ (Figure 8.10), which is also generally observed with floods. This large, short-term increase in DOC in stream water can be linked to the high amounts

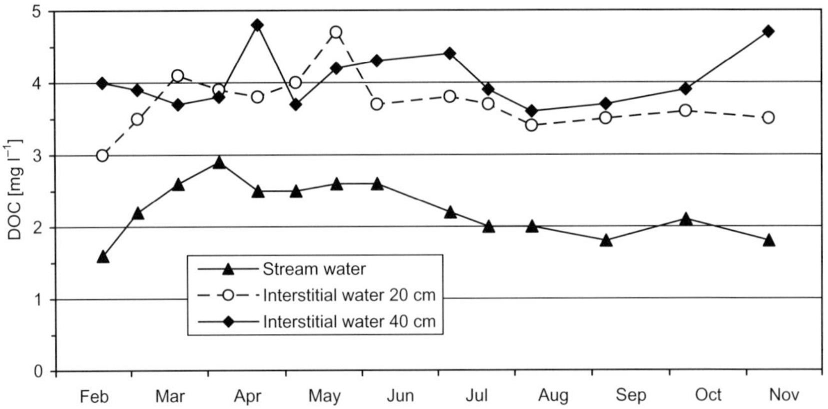

Figure 8.9 Concentrations of dissolved organic carbon (DOC) in stream water and in interstitial water from different depths in the Breitenbach lower course below trap T5, February–November 1986.

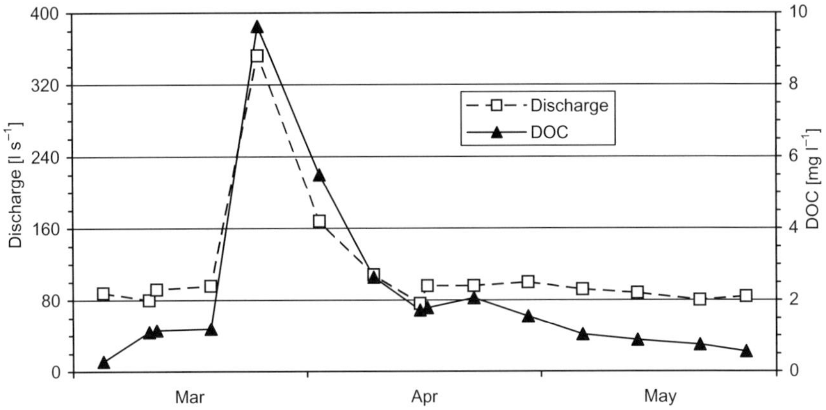

Figure 8.10 Concentrations of dissolved organic carbon (DOC) in stream water in the Breitenbach lower course below trap T5, March–May 1986. Water samples were taken at shorter intervals than samples from interstitial water (Figure 8.9), showing large short-term fluctuations in DOC concentrations. However, because samples from interstitial water are missing from situations with high DOC concentrations in stream water, the increase in interstitial water DOC values found by Fiebig (1995) during high water could not be observed.

of groundwater with strongly elevated organic matter content which enter the stream during floods (Fiebig, 1995). A considerable amount of these materials is immobilized on the sediment particle surfaces during the perfusion process. The microbial communities are able to utilize these fixed substrates over the following weeks and months, thus making it available to higher trophic levels (Fiebig and Marxsen, 1992; Fiebig, 1997).

8.1.4
Breitenbach Sandy Sediments as Habitat

The chemical and biological conditions in the Breitenbach differ from those in many streams in other regions because a permanent and continuous, deep, porous, oxygen-containing hyporheic habitat does not exist due to the geological conditions. The sandstone substratum exerts significant effects on the structure of the interstitial spaces, affecting the sediment porosity, and leading to often steep, fluctuating chemical gradients. However, there is intense upwelling and downwelling of water (Fiebig, 1995), but at very different intensities and extents during and between years, depending mainly on the specific hydrological circumstances.

Studies in streams from geological strata other than Bunter Sandstone (e.g., Schwoerbel, 1967; Tilzer, 1968) demonstrated the continuous availability of the hyporheic habitat for macroinvertebrate communities. In contrast, the availability of this habitat to macroinvertebrates is restricted in large stretches along the Breitenbach. During low flow periods the small interstitial spaces in the large sandy streambed areas are filled up with silt and fine particulate organic material. Distinct chemical discontinuities with large oxygen-deficient zones, which can rapidly develop, inhibit the exchange of mobile organisms between surface and deeper sediment layers. Clogged interstitial spaces often restrict even the exchange of water itself. The fine-grained sandy deposits, usually with high fine particulate organic matter content, can become anoxic only a few millimeters below the surface (Figure 8.6). This agrees well with results of Simon (1985) and Whitman and Clark (1982), who studied fine-grained sediments in streams and demonstrated the absence of oxygen at similarly short distances from the surface. Because sufficient oxygen for macrofaunal life is absent over large parts of the sandy areas in the Breitenbach, these areas are unavailable as a permanent habitat for macroinvertebrates and cannot function as stable shelter during floods.

Exceptionally, the mayfly *Baetis vernus* remains almost unaffected by floods because, for the greater part of the year, it remains attached to the underside of stones as a diapausing egg. Other organisms that can use the oxygen-deficient zones of the Breitenbach are adapted to anaerobic conditions or are able to receive oxygen from the overlying stream water or even from the atmosphere. Some specialists, predominantly larvae of the Ptychopteridae, Tipulidae, and Limoniidae (Diptera) live in this type of sediment (e.g., Mattingly, 1987). They stick vertically in the sediment, with spiracles at the water surface receiving atmospheric oxygen.

Sandy surfaces are too unstable for permanent and abundant colonization because they are susceptible to discharge variation. The high FPOM content of the sand sediments provides sufficient energy sources to be used by deposit-feeders (Mattingly, 1987, 1988); *Ptychoptera townesi* consumed FPOM, less than 0.5 μm diameter, from the sediment and compacted it to considerably larger fecal pellets. In the Breitenbach, larvae of the false crane fly *Ptychoptera paludosa* prefer the sandy sediments (Wolf and Zwick, 2001).

The temporal stability of sandy sediments and the occurrence of FPOM deposits depend mainly on discharge. During times of low flow sandy sediments remain stable over longer periods and are often colonised by benthic algae that create a smooth surface, increasing the critical drag force required to initiate sand grain movement (e.g., Raudkivi, 1982).

Alongside the oxygen gradients, steep chemoclines occur, which potentially affect the extension of the colonisable interstitial space in the Breitenbach sediment. Williams (1989) tried to define the hyporheic zone in terms of several, predominantly chemical, discontinuities (break lines) from the river margin down under the banks. However, forepart from with oxygen, most species showed no correlation with the chemistry of the interstitial water. More importantly for the faunal gradient, Brunke and Gonser (1999) identified that the ratio of particulate organic carbon to total fine particles (POC/TFP) explained more than 50% of the variation of total animal abundance within the sediment of the Töss, a third-order calcareous gravel bed river in Switzerland. This ratio can be taken as an indicator of POM availability for detritivorous sediment feeders and particle feeders.

Although the extension of the habitable hyporheic for macrobenthos is quite limited, and the abundance of typical subterranean taxa is low (Husmann, 1971), the areas without macroinvertebrates are nevertheless colonized by abundant, highly active communities that are of great importance for the stream system. These communities are composed of bacteria and fungi, protozoa and microinvertebrates (meiofauna; all described in detail in previous chapters of this book), which establish highly productive, complex food chains (Marxsen, 2006).

8.2
Aquatic Insects Emerging over an Artificially Changed Stream Bottom

Rüdiger Wagner

Several attempts have been made to classify running water courses, for example, classification as a whole, based on production, communities or individual taxa (e.g., Brinck, 1949; Naumann, 1932; Thienemann, 1936), classification by zonation, based on community (e.g., Illies and Botosaneanu, 1963; Vannote *et al.*, 1980; Thienemann, 1925), or classification by substratum (e.g., Thienemann, 1912; Marlier, 1951).

From the 1960s to the 1980s in particular, many investigations were carried out into the problems of the interaction between environmental factors/parameters

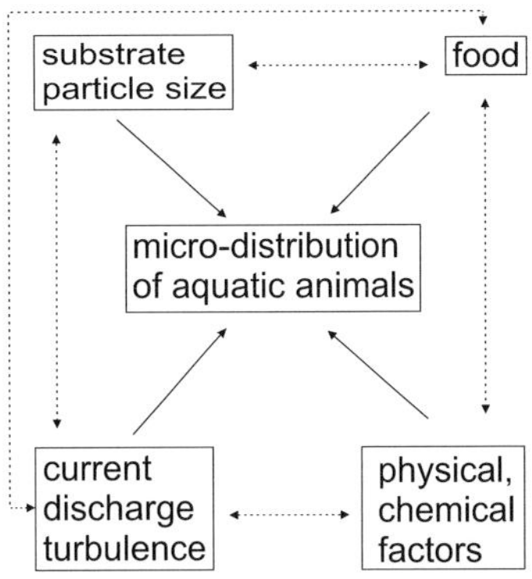

Figure 8.11 Dependence of aquatic invertebrate macro- and microdistribution and major environmental variables. Arrows with full lines indicate effects on microdistribution, arrows with broken lines dependencies among variables; arrowheads on both ends indicate mutual effects (modified from Cummins and Lauf, 1969).

and the microdistribution of benthic stream macroinvertebrates on their substratum. It was evident that species composition differed markedly between substratum types, and numerous studies tried either empirically (field investigation) or experimentally (in laboratory or field studies) to relate species distribution to substratum composition (see references in Cummins, 1966; Thorup, 1966; Hynes, 1970; Tolkamp, 1980). Four major classes of environmental variables were taken into account by Cummins and Lauf (1969): current, temperature and chemical variables that limit the macrodistribution (i.e., habitat tolerance), whereas substratum particle size and food could determine the microdistribution of stream invertebrates (Figure 8.11).

Other physical and biological factors that were investigated, that is, flow regime, drought, spates, suspended solids, competition, predation, drift, (e.g., Hynes, 1970a, 1970b; Macan, 1961, 1962) turned out to be more or less closely related to the above-mentioned "net" of environmental variables.

Empirical and experimental studies each have their own problems. Empirical field studies are time-consuming and it is often difficult to distinguish clearly between the effects of nutrient and substratum. Naturally developed habitats are disturbed by each sampling event, and field sampling requires many parallel collections (statistical treatment).

Laboratory experiments are often used to support the results of empirical studies, but are usually run at constant current speeds, which must be slow enough to ensure that the smallest particle size under investigation is undisturbed. Under similar conditions in the field, spaces between larger particles are filled with sand and organic matter. In the field, a mosaic of constantly changing habitats are formed by wide amplitudes of current speed, reversed current near the banks and increasing speeds towards the middle of the stream.

Figure 8.12 Interior view of the sand trap on the Breitenbach.

8.2.1 The "Silted" Emergence Trap

Because a small-scale mosaic of substrata are usually found distributed over the stream bottom in running waters, we tried to deposit one substratum type over a large area of the stream bottom under an emergence trap and a few metres upstream. The simplest way to achieve this was be building a barrage just downstream from an emergence trap, reducing the current speed so that even the finest sediment size classes remained in place (Figure 8.12).

In this way, from 1981 to 1983, the stream bottom under one emergence trap (T3, S) was covered with approximately 10–20 cm sand (siltation), allowing an *in situ* experiment. Emerging insects were collected as usual. Specimen numbers from the "sand trap" were compared with those from an adjacent trap with an "unmodified" substratum. Abundances from the periods before and during the siltation period were also obtained. The opposite condition was provided by a trap under which the stream bottom was covered almost exclusively by stones, partly covered by aquatic mosses. Observations were collected for a single year only (1983).

Each week sediment samples (100 cm^3, three replicates) were collected from the upper 3 cm of stream bed just below the trap. Comparison of the sum curves of percent contribution of size classes between "normal" and "sandy" sediments clearly shows that for sand, the maximum grain size was about 2 mm, and more than 90% (weight) was smaller than 0.5 mm (Figure 8.1). The mean weekly discharge over that period was calculated as the arithmetic mean of daily discharge (one reading per day from a nearby V-shaped weir). This was compared with proportions of sediment size classes (Figure 8.2) and a clear relationship between AFDW, the percentage of smallest sand size classes and mean weekly discharge was found.

8.2.2
Laboratory Experiments

In the laboratory tests were performed with individual species. Eight sediment size classes (diameter: 0.1–0.2 mm, <0.4 mm, <0.8 mm, <1.6 mm, <3.15 mm, <5.0 mm, <10 mm, >10 mm) were provided for the larvae at different current speeds. Most specimens rapidly left smaller substratum size classes (<0.8 mm diameter) by actively entering the drift. Only larvae of the mayfly *Ephemera danica*, the stonefly *Leuctra nigra*, and the caddisfly *Sericostoma personatum* burrowed into the ground. With respect to larger particle size substrata, most specimens rapidly withdrew from the sediment surface into the open spaces between the particles. However, shredding caddisfly larvae were substratum nonspecific. They were attracted a substratum size class only if their preferred food (alder leaves) was available/offered. For this reason, results from the laboratory experiments were either evaluated extremely cautiously or even rejected.

8.2.3
Emerging Aquatic Insects

From the field experimental stream section, a total of 283 taxa were identified. From 1981 to 1983, 37 species (~13%) displayed significant abundance differences between the manipulated and the "natural" traps; numbers of 27 species (~9.5% of total) decreased, and 10 (~3.5% of total) increased (Table 8.1). Larvae of species that increased in abundance under the sand trap could be characterized as generalists or collectors; those that decreased in abundance were predominantly predators, grazers or those that preferred habitats in plant stands and moss-carpets on stones (Table 8.2).

The most substantial change occurred in the dominant Ephemeroptera; all except *Ephemerella* spp. were markedly affected by the modified bottom substratum. *Baetis vernus* abundance dropped to 16–27%, while *B. rhodani* was only 1–20% of the control. In contrast, numbers of *C. luteolum*, *P. submarginata* and *H. fusca* significantly increased. However, the loss of numbers and biomass of the *Baetis* spp. was far from compensated by the increase of the above three species (Table 8.2).

About every third Plecopteran species showed significant inter-site changes. Abundance of *Leuctra nigra* increased two- to fivefold in the sandy trap, that of the rare *Nemuralla picteti* (not treated statistically) three- to 11-fold. For *Protonemoura auberti* and *P. intricata*, abundance in the sand trap was only 7–30% that in the reference trap, for *P. meyeri* 25–50%, and for *Isoperla goertzi*, 10–14%.

The abundance of four Trichoptera was lower in the sand trap: *Rhyacophila fasciata* (22–33%), *Agapetus fuscipes* (7–11%), *Drusus annulatus* (20–64%), *Tinodes rostocki* (32–128%; Table 8.2). Larvae of the last three species are scrapers, which prefer stone surfaces where they compete for the biofilm. *R. fasciata* larvae are caseless predators that require large stones for pupation; none were expected to occur over the sandy area.

Table 8.1 Total species number (N), species number of aquatic insect orders and Diptera families with significantly changed abundance (NC) and declined (DE) or increased (IN) abundance over the "sand trap" from 1981 to 1983.

	N	NC	DE	IN
Ephemeroptera	7	5	2	3
Plecoptera	16	5	4	1
Trichoptera	33	4	4	0
Neuroptera	1	0	0	0
Megaloptera	2	0	0	0
Coleoptera	2	1	1	0
Diptera	222	22	16	6
Tipulidae	30	2	2	0
Limoniidae	65	10	7	3
Dixidae	4	0	0	0
Ptychopteridae	2	1	0	1
Psychodidae	25	4	3	1
Culicidae	2	0	0	0
Thaumaleidae	3	0	0	0
Stratiomyidae	5	1	1	0
Dolichopodidae	50	1	1	0
Empididae	21	1	1	0
Tabanidae	8	1	0	1
Syrphidae	5	0	0	0
Rhagionidae	2	1	1	0
Total	283	37	27	10
%		13	9.5	3.5

The abundances of about 10% of the Diptera (22 species) was distinctly different between sites. Sixteen species declined in numbers; some are aquatic predators (Pediciinae and Limnophilinae–Limoniidae *sensu lato*, Rhagionidae, Empididae, Dolichopodidae), others are moss-dwelling Psychodidae and phytosaprophageous Stratiomyidae and Limoniidae. Increasing abundance in the sand trap was noted for two phytosaprophagous and one carnivorous Limoniidae and one species each of the Tabanidae and Psychodidae. The greatest increase in abundance was noted for *Ptychoptera paludosa*. Larvae are collectors (gatherers) that plunge vertically into the soft sediments where they are able to benefit from large amounts of FPOM in oxygen-poor areas. Other species cannot compete because the *Ptychoptera* larvae are larger and possess an elongate, rectractable sipho that enables them to feed with their head deep in the anoxic substratum while obtaining oxygen from the air via the terminal stigmata.

Chironomidae were only analyzed for one year due to the array of specimens and species (Becker and Wagner, 2004). About 156 000 Chironomidae (110 taxa) were recorded from the "sand" and "stone" traps. Species numbers were higher from the sand trap (99) than the "stone-trap" (85), as were specimen numbers (88 510–sand; 67 340–stone). Proportions of Prodiamesinae and Chironomini

Table 8.2 Under- or overrepresentation (%) of aquatic insect orders, Dipteran families and species in the "sand trap" compared to the "natural" trap (=100) from 1981 to 1983.

Species	1981	1982	1983	Species	1981	1982	1983	Species	1981	1982	1983
Ephemeroptera	22	10	35	Diptera				Diptera			
Bae. vernus	21	15	27	Tipulidae	61	84	24	Psychodidae	102	36	47
Bae. rhodani	20	1	15	*Tip. maxima*	75	147	48	*Ber. illiesi*	36	12	68
Cen. luteolum	0	1825	7600	*Tip. pagana*	43	7	21	*Ber. unispinosa*	33	12	29
Par. submarginata	350	1400	900	*Tip. rufina*	0	0	10	*Ber. manicata*	0	0	8
Hab. fusca	0	0	700					*Sat. trivialis*	270	137	131
				Limoniidae	89	67	95				
Plecoptera	124	69	47	*Dic. pavida*	32	10	33	Stratiomyidae	0	15	10
Leu. nigra	454	190	218	*Rap. claripennis*	9	0	20	*Ber. clavipes*	0	15	13
Pro. auberti	7	3	3	*Elo. maculata*	633	202	493				
Pro. intricata	20	19	8	*Neo. nemoralis*	28	45	35	Dolichopodidae	82	170	30
Pro. meyeri	6	0	43	*Eri. griseipennis*	21	36	70	*Her. cupreus*	50	22	13
Iso. goertzi	10	11	13	*Eri. lutea*	2200	4800	169				
				Orm. depilata	53	35	46	Empididae	126	124	104
Trichoptera	47	44	26	*Mol. ater*	13	7	24	*Che. trapezina*	36	15	17
Rhy. fasciata	22	33	31	*Mol. bifidus*	7	22	45				
Aga. fucipes	10	11	7	*Dic. modesta*	1120	736	49	Tabanidae	209	160	94
Tin. rostocki	32	128	38					*Chr. caecutiens*	1200	0	1100
Dru. annulatus	40	21	64	Ptychopteridae	279	1883	620				
Cha. villosa	105	46	100	*Pty. paludosa*	279	1883	620	Rhagionidae	4	25	20
								Chs. auratus	4	18	20
Coleoptera											
Hel. minuta	35	41	82								

Table 8.3 Total species numbers of Chironomidae subfamilies and tribes at T2 and two different stream bottom types at T3.

Subfamily/tribe	T2	T3 sand	T3 stone
Tanypodinae	5–11	10	7
Diamesinae	4–6	6	5
Prodiamesinae	2	2	1
Orthocladiini	24–38	37	36
Metriocnemini	13–32	21	16
Chironomini	7–16	12	11
Tanytarsini	4–10	11	10
Σ	59–108	99	86

increased in the sand trap, while the abundance of Tanytarsini increased by almost the same amount in the stone trap (Figure 7.114, Table 8.3).

The ten most abundant species accounted for 82.8% of specimens in the stone trap, but only 62.6% in the sand trap. Species numbers in the stone trap were 13% less than in the sand trap, and total specimen numbers were 24% less. Biomass was 37% lower. Similarity indices between traps were less than 50%. Total production (stone ~1.9 $mg\,m^{-2}\,year^{-1}$; sand ~2.7 $mg\,m^{-2}\,year^{-1}$) was low and characteristic of oligotrophic habitats.

Thienemann's rule that communities from increasingly monotonous, uniform habitats were less diverse than those from diverse habitats was not validated. This is partly due to a methodological weakness of emergence traps: reduced current speed beneath a trap improves the chance of successful emergence from drifting pupae. Furthermore, an 8 m stretch along the stream was probably too short to reveal the effects of sand sedimentation for an entire stream.

Nevertheless, several abundant species whose larvae avoid sandy sediments occurred in the "sand trap". This allowed a rough calculation of the distance over which specimens are displaced by the current during the emergence process.

In general, the stream community over the sandy stretch was altered so that it was dominated by aquatic Diptera, both in specimens numbers and biomass (Figure 8.13). Larvae of the stonefly *Leuctra nigra*, the mayflies *Paraleptophlebia submarginata* and *Centroptilum luteolum*, and the caddisfly *Sericostoma personatum* also prefer this sort of sediment. The greatest losses occurred in the Ephemeroptera, Plecoptera and Trichoptera. However, the loss of specimens and biomass of sand-avoiding species (mainly predators, scrapers and species with larval habitats in plant stands and moss carpets) was far from compensated by the sand-preferring species. The biomass of emerging insects over the sand trap was markedly lower (48–68%) than at the unaltered reference condition.

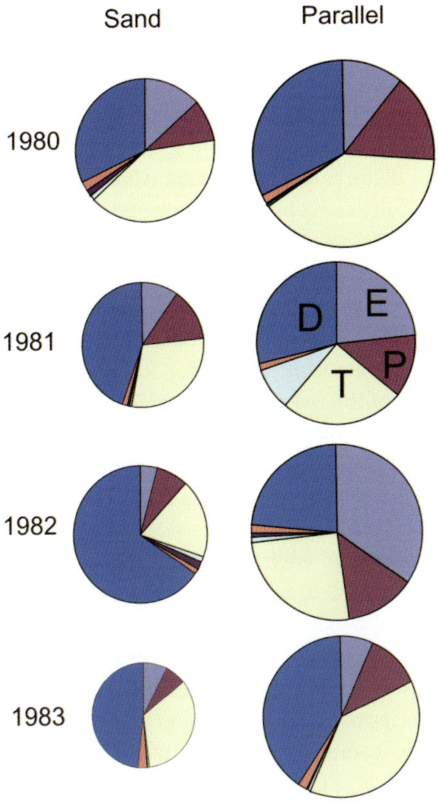

Figure 8.13 Biomass of aquatic insect orders in paired traps over the artificially silted stream bottom and an unaffected, parallel stream area, 1980 both unaffected. (E = Ephemeroptera, P = Plecoptera, T = Trichoptera, D = Diptera; light green = Neuroptera, orange Helodes spp., Diptera without Chironomidae).

8.3
Discharge Patterns Largely Determine Species Abundance and Community Diversity

Rüdiger Wagner and Hans-Heinrich Schmidt

Many environmental variables affect stream communities simultaneously, of which discharge is most often discussed. The amount of water running downstream in different seasons shapes the stream channel and the bottom substratum (Chapters 3 and 8.1). The overall effect of environmental heterogeneity on patterns and processes in streams has been repeatedly debated (e.g., Cooper *et al.*, 1997; Palmer and Poff, 1997; Palmer *et al.*, 1997; Downes *et al.*, 1997) and discharge variability is most frequently cited as the reason for changes in benthic population density (e.g., Wlosinski and Minshall, 1983; Chesson, 1986; Hildrew and Townsend, 1987; Death and Winterbourn, 1994; Wood *et al.*, 2000). Variation in "water depth", as well as short- and long-term floods, have contrasting effects on species and communities (Kajak, 1959; Thorup, 1970; Hoopes, 1974; Molles, 1985). Phenomenological studies have revealed the rates and directions of benthic macroinvertebrate succession after spates (e.g., Boulton *et al.*, 1992). These and many other studies are pieces of a jigsaw puzzle which have promoted the development of ecological theories such as the habitat templet concept (Southwood 1977, 1988; Minshall, 1988), the intermediate disturbance hypothesis (Connell, 1978; Ward and Stanford, 1983), the patch dynamics concept (e.g., Townsend, 1989) and the *harsh–benign concept* (Peckarsky, 1983).

Over the past decades attempts have focussed on using long-term ecological data to test ecological theory in releation to freshwater systems (e.g., Johnson *et al.*, 1994; Poff and Ward, 1990; Statzner, 1987; Statzner and Resh, 1993; Statzner *et al.*, 1994a, 1994b; Poff and Allen, 1995). Data from the Upper Rhône (France) did not confirm the theoretical predictions of the river habitat templet (Townsend and Hildrew, 1994) and questions were raised over whether the underlying theory was wrong, or the tests were fair (Resh *et al.*, 1994). However, Townsend *et al.* (1997) found some agreement between species traits (Stearns, 1976) and disturbance intensity and frequency in New Zealand streams.

Long-term research data on aquatic insects and environmental variables in the Breitenbach also allow ecological theory to be tested, as well as to answer the question why there should be extreme changes in abundance of aquatic insects in different years (Illies, 1974, 1978). In a recent paper we (Wagner and Schmidt, 2004) analyzed the EPT emergence data from 1969 to 1993 (site T2) for parallel or inverse temporal development among species pairs, and in relation to the variables describing the environment over the period leading up to emergence (most species are univoltine). We assumed that the number of emerging individuals was determined primarily by habitat conditions on the stream bottom during larval and pupal development. We expected that, at least for univoltine species, the number of emerging adults did not only depend on one variable, or an individual disturbance, but on the entire "annual pattern" of environmental variation.

Temperature and particularly precipitation and discharge affected the community, while other variables were insignificant. Water temperature was of minor importance, probably due to the fact that only the data from site T2 were used. However, temperature regimes of streams are important in the evolutionary ecology of aquatic organisms (Ward, 1985). Increasing temperature effects would be expected if more sites with greater seasonal temperature variability were analyzed (Chapter 8.4).

Deviations from the long-term seasonal trend for variables, particularly discharge in a given year, can generate "harsh" conditions for one part of the community but "benign" for another. It was obvious that, in a given year, specimen numbers and community assemblage were to a certain extent predictable (Wagner *et al.*, 2006). Our assumption that the number of emerging aquatic insects, that is, the success of a population in the stream in a particular year, can be deduced from "annual patterns" of key variables during larval development was proved. Discharge had a primary effect on community assemblage, even when precipitation and water temperature were taken into account.

Correlations indicated that species with similar habitat preferences have essentially parallel long-term abundance changes, while species with contrasting habitat requirements were negatively correlated. No predator–prey relationship was detected.

Here we use the same methods to test whether the results from 25 years (Wagner and Schmidt, 2004) concur with those from a 37-year data set. For the period 1991–2006, only two of the formerly distinguished (four) discharge patterns remained, probably indicating climatic change. Thus in some cases data from three periods are compared: (i) total period (1969–2006), (ii) pre-1990 (1969–1990), and (iii) post-1990 (1991–2006), that is, a series of "dry" and "nonseasonal discharge" years.

8.3.1
Community and Variables Data Treatment

Although the distribution of many species and environmental variables was unimodal (SPSS, 2007), nonparametric tests were preferentially used to ensure comparability with other analyses. Data were usually untransformed, but some programs automatically standardize variables to zero mean and unit variance.

Community data comprised the annual specimen numbers of 43 EPT taxa with known life cycles and larval feeding modes (Table 8.4). Annual specimen numbers

Table 8.4 Basic life cycle characteristics of 42 EPT taxa in the Breitenbach. Abbreviations: G = generations per year, Ld = larval development, Pd = pupal development, E = emergence period, Ffg = functional feeding group (g = grazer, s = shredder, p = predator, f = collector/fine particle feeder). Numbers indicate months (e.g., 1 = January; 12 = December), when a life stage is present in the stream (11–8 = between November and August). Superscripts: [a]Long egg diapause to January next year; [b]short diapause; [+] larval dormancy from November to February next year. Relevant references are given at the foot of the table.

Species	G	Ld	Pd	E	Ffg
Ephemeroptera					
Baetis rhodani Pictet	2.0	1–12	–	4–10	g
B. vernus Curtis	1.0	1–5[a]	–	6–8	g
Ephemerella mucronata Bengtsson	1.0	6–4	–	5–7	f
E. ignita Poda	1.0	8–5	–	7–9	f
Paraleptophlebia submarginata Stephens	1.0	7–4	–	5–8	f
Ephemera danica Müller	0.5	1–12	–	6–7	f
Ecdyonurus venosus Fabricius	1.0	7–5	–	6–7	g
Centroptilum luteolum Müller	2.0	1–12	–	6–10	f
Plecoptera					
Leuctra prima Kempný	1.0	4–12	–	1–4	f
L. nigra (Olivier)	0.5–1.0	1–12	–	4–8	f, s
L. digitata Kempný	1.0	11–8	–	9–10	f

Continued

Table 8.4 *Continued*

Species	G	Ld	Pd	E	Ffg
Nemoura marginata Pictet	1.0	6–3	–	4–5	f
N. flexuosa Aubert	1.0	6–3	–	3–5	f
N. cambrica Stephens	1.0	6–3	–	4–5	f
N. cinerea Retzius	1.0?	8–4	–	5–7	f
Nemurella pictetii Klapálek	2.0–3.0	1–12	–	4–9	f, s
Protonemura auberti Illies	1.0	10–5	–	6–9	f
P. intricata Ris	1.0	8–4[b]	–	5–9	f
P. meyeri Pictet	1.0	5–2	–	3–4	f
P. nitida Pictet	1.0	10–8	–	9–10	f
Amphinemura standfussi Ris	1.0	7–4	–	5–8	f
Isoperla goertzi Illies	0.5	9–4	–	5–7	p
Siphonoperla torrentium Pictet	1.0	7–5	–	5–7	p
Brachyptera risi Morton	1.0	6–3	–	3–6	f
Trichoptera					
Apatania fimbriata (Pictet)	1.0	7–4[+]	3–6	4–8	g
Rhyacophila fasciata Hagen	0.6–1.0	1–12	3–9	5–10	p
Drusus annulatus Stephens	0.6–1.0	1–12	4–9	5–10	g
Tinodes rostocki MacLachlan	1.0	6–3	4–7	5–8	g
Sericostoma personatum Kirby and Spence	0.3–1.0	1–12	4–7	5–8	s
Silo pallipes Fabricius	1.0	6–3	4–6	5–7	g
Agapetus fuscipes Curtis	1.0	6–3	4–9	5–9	g
Hydropsyche instabilis Curtis	1.0	8–5	6–7	7	f, p
H. saxonica MacLachlan	1.0	6–6	4–9	5–9	f, p
Plectrocnemia conspersa Curtis	1.0	7–4	5–8	6–9	p
Adicella reducta MacLachlan	1.0	7–4	5–6	6–7	s
Potamophylax cingulatus Stephens	1.0	11–8	8–9	8–10	s
Potamophylax luctuosus Piller and Mitterb.	1.0	7–3	4–5	5–6	s
Chaetopteryx villosa Fabricius	1.0	12–9	9–10	10–11	s
Halesus digitatus Schrank	1.0	10–7	8	9	s
Micrasema longulum MacLachlan	1.0	8–3	4–6	5–7	g, s
Wormaldia occipitalis (Pictet)	1.0	1–12	2–10	3–12	c
Odontocerum albicorne Scopoli	1.0	8–5	6–7	7–8	p

References: Aurich (1992), Becker (1990, 1993, 1994), Benedetto (1975), Bohle (1969), Cliffort (1982), Elliott (1987), Enders and Wagner (1996), Knöfel (1983), Marten (1990), Marten and Zwick (1989), Saltveit (1979), Schmidt (1984), Schwarz (1970), Wagner (1986, 1990a, 1990b, 1991, unpublished data), Wernecke and Zwick (1992), Zwick (1981, 1986, 1991, personal communication).

were variable, and the abundance of some taxa was low in some years; these were omitted from a few analyses.

For EPT, coefficients of species stability (SS = s.d./$\bar{x}$, annual numbers for 37 years) were computed. Populations were sorted empirically into three groups, distinguished by stability: that is, unstable populations (SS ≤ 0.8), populations with intermediate stability (0.8 < SS < 1.2), and stable populations (SS ≥ 1.2). The synchrony of abundance variation of the 32 most abundant taxa ($\bar{x} > 20$ specimens per year) over time was tested by pairwise correlation (SPSS, 2007).

Precipitation, discharge, and water temperature data were treated by discharge year, 1st October one year to 30th September next year (Chapter 3). Few missing data were replaced by long-term means. For canonical analyses 3-month periods were sufficient (avoiding collinearities with monthly data). Other variables (air temperature, number of rainy days or combinations of these with precipitation) that might affect the reproductive success of adults during their terrestrial lives (Wernecke and Zwick, 1992; Enders and Wagner, 1996) had no significant effects on species abundance.

Variables were examined for predictability (P), constancy (C), and contingency (M; Colwell, 1974; Stearns, 1981). P measures variation over successive periods, it is maximized if the same monthly pattern of a variable is repeated every year. C describes a variable's characteristic to remain in a certain state throughout the year, it is maximized if the variable is the same for all months. Contingency (M) is maximized if the variable is different in all months but its fluctuations are predictable ($P = C + M$; P_{max}, C_{max}, M_{max} are all 1). The C/P and M/P ratios express constancy or contingency relative to predictability ($C/P + M/P = 1$). Data matrices consisted of 12 months over 38 years. Precipitation data were organized into five classes (≤40, ≤80, ≤120, ≤160, >160 $l\,m^{-2}\,month^{-1}$), water temperature in eleven classes of two-degree steps (≤2, ≤4, ≤6 . . . ≤20, >20 °C), and discharge in seven classes (≤10, ≤20, ≤40, ≤80, ≤160, ≤320, . . . >320 $l\,s^{-1}$; Chapter 3).

Classification of years by four discharge patterns was based on three measures: monthly mean ($\bar{x}$), standard deviation (s.d.), and discharge variability (CV = $\bar{x}$/s.d.). In addition, for pattern contours and distinction the annual mean of the monthly maxima, the annual sums of the monthly SD and CV were calculated (Table 8.4). We assumed that lowest variability within each discharge pattern group exists if the annual sum of monthly SD, and the sum of the monthly CV are minimal.

Relationships of species abundance to discharge patterns were checked with the Kruskal–Wallis (KW) test, with "discharge pattern" as the grouping variable. If the KW test was significant, U-tests were performed for pairwise differences between discharge patterns.

Species/year matrices were related using Correspondence Analysis (CA) via CANOCO (ter Braak, 1988, 1990; Jongman, ter Braak and van Tongeren, 1987; Palmer, 1993; Šmilauer, 1992; ter Braak and Šmilauer, 1998 and ter Braak and Verdonschot 1995). A detrended correspondence analysis (DCA) revealed gradients well below 4 s.d., indicating that unimodal-based methods were appropriate. Year groups discriminated on discharge pattern are visualized by the "envelope" option of CANODRAW. Species richness, Shannon–Wiener and N2 diversity measures were read from CANODRAW, and regressions on canonical axes 1 and 2 were performed with Excel™ and SPSS™. Finally, a discriminant analysis with "*a priori*" selection of "discharge pattern" as the grouping variable was performed (SPSS, 2007).

Table 8.5 Predictability (P), constancy (C), contingency (M), and percentage shares C/P and M/P of precipitation, discharge, and water temperature to predictability for the periods 1969–2006, 1969–1990 and 1991–2006 (all values significant, $P = 0.01$, χ^2).

	Precipitation			Discharge			Water temperature		
	1969–2006	1969–1990	1991–2006	1969–2006	1969–1990	1991–2006	1969–2006	1969–1990	1991–2006
P	1.00	0.99	1.00	0.76	0.76	0.79	0.81	0.81	0.85
C	0.62	0.63	0.62	0.73	0.73	0.74	0.68	0.67	0.71
M	0.38	0.36	0.45	0.03	0.05	0.05	0.14	0.14	0.15
C/P	62	64	58	97	95	94	83	83	83
M/P	38	36	42	3	5	6	17	17	17

8.3.2
Environmental Variables

Predictability (P), constancy (C), and contingency (M) of precipitation, discharge, and water temperature were significant, but C and M contributed contrastingly to P (Table 8.5).

P, C, M, and shares (C/P, M/P) for water temperature remained almost the same for the three periods (total, before and after 1990). This is due to the comparatively low temperature increase over the period 1969 to 2005 (Chapter 3), compared with the 2 °C steps used for grouping. The limited effect of water temperature on the ordination was also due to a single analyzed site.

P of precipitation was high for every period; C remained about the same, but M increased from 1991 to 2006. M/P rose from 36 to 42%, but C/P decreased from 64 to 58% over the same period. Correlation of monthly precipitation and maximum discharge for the entire period (1969–2006) was weak but significant (0.236, $P < 0.0001$, Spearman); it was better (0.284, $P < 0.0001$) for 1969 to 1990 than for 1991 to 2006 (0.172, $P = 0.021$).

P of discharge remained almost constant for the three periods (Table 8.5). It was highest from 1991 to 2006, a period with predominantly low flow and some summer spates. C remained almost the same, and M was slightly higher in the periods before and after 1990 than over the entire period. This resulted in decreased C/P and increased M/P for the same periods, indicating a minor increase in seasonal variability since 1991.

Comparison of discharge patterns (Table 8.6) indicated that predictability was similar between groups B and C. It was highest in the "seasonal" groups (D, E, F), particularly group E (spring spates). Contingency increased within group B and was highest again in group E. C/P was highest in the "spring spates" group (15%), and less marked in years with nonseasonal flow. It remained low in the other groups.

Table 8.6 Predictability (P), constancy (C), and contingency (M) and percentage shares C/P and M/P of discharge patterns B, C, D, E, and F in the Breitenbach (compare Figure 8.1; all values significant, $P = 0.01$, χ^2).

	B	C	D	E	F
P	0.77	0.79	0.80	0.86	0.83
C	0.74	0.73	0.75	0.72	0.79
M	0.03	0.06	0.05	0.15	0.04
C/P	96	92	93	83	95
M/P	4	8	7	17	5

Classification into year groups with similar discharge patterns was empirical (Figure 8.14, Table 8.6). They were arranged hierarchically below the long-term annual discharge pattern (group A). Firstly, group C (nonseasonal flow, including years with spates at "unusual" times in a year) was distinguished from group B (seasonal flow, with seasonally predictable runoff). In a second step, group B was divided into three subgroups: D (wet years, with permanent good flow without marked spates or low flow periods), E (spring spates, with winter or spring spates and decreasing discharge in summer and autumn), and F (dry years, long-lasting low discharge with only slight increase in winter).

The annual mean of monthly mean discharge was lowest in group F, about twice as high in group D, and three times as high in groups C and E. High (>1.0) or low (<0.5) CV in different months occurred in most groups. Groups A and C had low CV in October, but no other group had CV <0.7 in any month. High variability in October in group A depended almost exclusively on the data for group C (Table 8.7).

Group A (all years) had high monthly variability (CV) in February, March, May, and July, group C (nonseasonal discharge (in the same months, plus August and September; variability was extraordinarily high (>2.0) in July. In group B (seasonal flow) variability was high in October, February, May, and June. In subgroups D (wet years), E (spring spates), and F (dry years) variability increased in 8 or 9 months. Very high values (>2) were only found for group E (March, May) during periods of high flow due to snowmelt or spring rains. Compared to mean flow, the particular patterns are characterized by high variability at intermediate flow (D), high variability at high flow (E), or low variability at low flow (F). Discharge variability is an interesting measure but is only meaningful in relation to the long-term mean. The same variability measure at low mean flow refers to different absolute discharge values and thus has a different effect on stream bottom sediments and the community, than at intermediate or high flow.

The most striking fact was that since 1991 discharge patterns for all years only belong to either "dry years" or "nonseasonal flow" groups. This can be taken as an indication of climatic change. It is obvious that the proportion of snow in the annual precipitation sum decreases noticeably after 1990 (Chapter 3). If ground

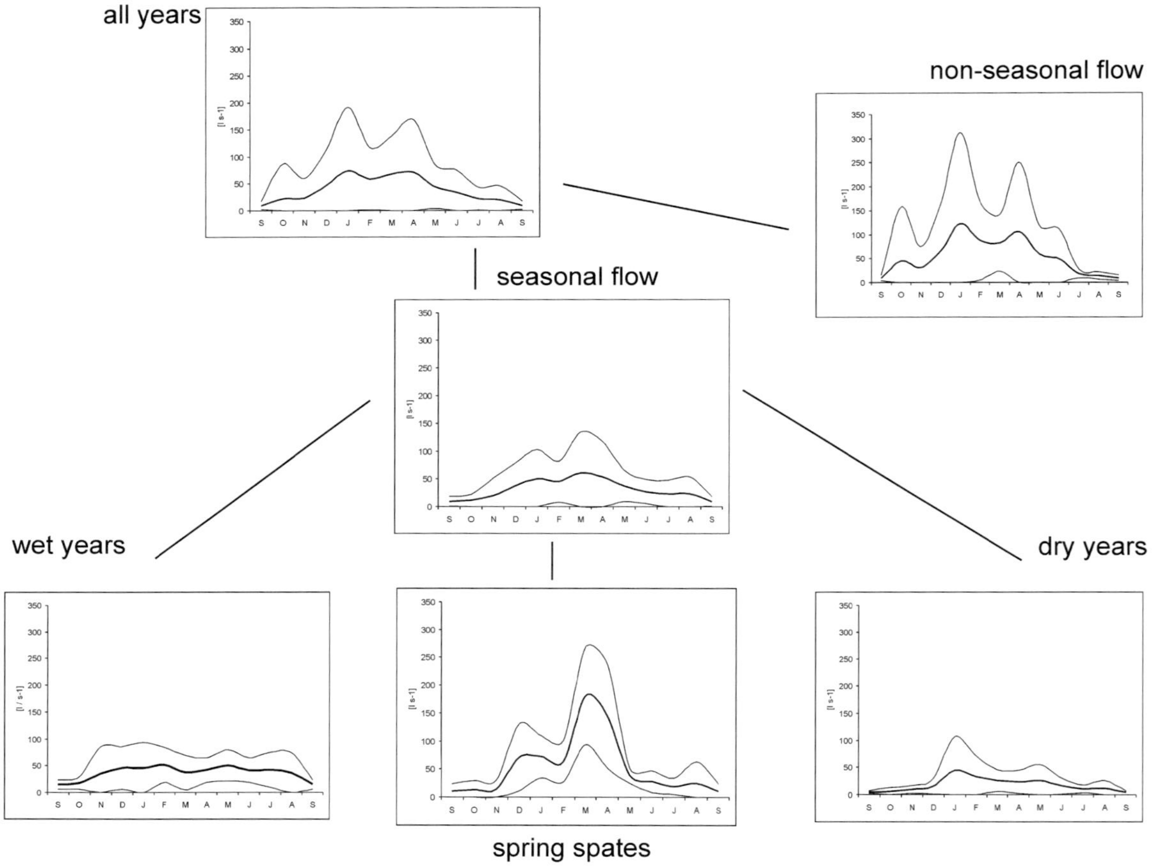

Figure 8.14 Discharge patterns of the Breitenbach, within group monthly mean (±1 s.d.), 37-year discharge (A), years with seasonal discharge (B), years with nonseasonal discharge (C), good discharge in wet years (D), spring spates after snowy winters (E), long-term low flow in dry years (F).

Table 8.7 Within group monthly mean discharge, standard deviation, and monthly discharge variability (mean/s.d.; mean and s.d. are l/s); right column: the within group annual mean of monthly mean discharge and the sum of within groups monthly s.d., for the six discharge patterns A–F (compare Figure 8.1).

	Oct	Nov	Dec	Jan	Feb	Mar	Apr	May	Jun	Jul	Aug	Sep	Annual
All years (A)													
Mean	21.9	23.0	46.5	73.6	58.9	67.8	71.0	44.4	34.4	21.9	20.1	9.6	41.09
s.d.	64.3	36.8	65.4	118.5	57.8	69.8	97.8	41.4	40.5	21.1	25.7	7.7	646.92
Variability	0.3	0.6	0.7	0.6	1.0	1.0	0.7	1.1	0.8	1.0	0.8	1.2	
Seasonal (B)													
Mean	11.6	19.5	37.2	50.4	45.3	60.8	54.0	37.2	27.5	23.6	23.0	9.7	33.30
s.d.	11.2	32.5	41.0	52.7	37.2	74.3	62.7	27.8	21.5	25.0	30.6	8.6	424.97
Variability	1.0	0.6	0.9	1.0	1.2	0.8	0.9	1.3	1.3	0.9	0.7	1.1	
Nonseasonal (C)													
Mean	44.9	30.4	65.7	122.1	87.2	82.5	106.5	59.4	48.8	18.3	14.2	9.5	57.45
s.d.	113.1	45.1	98.6	190.1	81.5	59.6	143.8	59.7	63.5	8.7	8.2	5.8	877.70
Variability	0.4	0.7	0.7	0.6	1.1	1.4	0.7	1.0	0.8	2.1	1.7	1.6	
Wet years (D)													
Mean	17.1	34.9	45.6	45.1	51.2	37.0	41.6	50.1	41.4	42.0	36.0	15.0	38.09
s.d.	11.7	50.6	39.9	49.1	32.7	31.9	22.4	29.6	22.6	32.7	38.1	8.6	369.72
Variability	1.5	0.7	1.1	0.9	1.6	1.2	1.9	1.7	1.8	1.3	0.9	1.7	
Spring spate (E)													
Mean	13.2	14.2	70.3	72.0	63.4	180.4	143.9	38.2	27.6	19.1	24.0	10.7	56.42
s.d.	15.5	16.2	59.5	38.4	37.0	87.2	92.0	13.8	19.5	14.5	39.0	12.3	444.97
Variability	0.9	0.9	1.2	1.9	1.7	2.1	1.6	2.8	1.4	1.3	0.6	0.9	
Dry years (F)													
Mean	6.4	9.3	15.3	44.9	32.2	25.8	23.2	26.2	15.9	10.5	11.8	4.9	18.88
s.d.	6.0	6.6	14.0	62.0	39.1	20.1	20.7	28.2	14.9	7.6	13.7	2.5	235.56
Variability	1.1	1.4	1.1	0.7	0.8	1.3	1.1	0.9	1.1	1.4	0.9	1.9	

water is predominantly restored during periods of snow cover, and the amount of groundwater is a prerequisite for discharge, the long-term water balance of the Breitenbach drainage area has been changed by a decrease in the proportion of snow to total precipitation since the beginning of the 1990s (Chapter 3). This will produce longer low flow periods, and summer thunderstorms with heavy rains will produce nonseasonal floods, probably with low or even no positive effects on groundwater replenishment.

Long low flow periods (extending over several years), with floods in normally flood-free periods negatively affect most species in the community. Low flow periods are depositional periods when the amount of fine sediment and FPOM increaseon the stream bottom. Stones are covered by fine material, and so these areas generally become unavailable to many invertebrates that prefer solid surfaces overgrown with biofilm.

8.3.3
Community

Community characteristics over the entire study period reflect the dominating effect of discharge. Species stability (SS) of 42 EPT species (Table 8.8) over thirty-seven years was interpreted as an effect of discharge-mediated substratum dynamics on populations. Large differences between minimum and maximum abundance were common for most species.

Seven species (four caddis flies, two stoneflies, one mayfly) had "stable" populations (SS > 1). Remarkable life cycle traits of the mayfly, *B. vernus* (up to eight months egg diapause; Schmidt, 1984), sustain the population through most discharge-mediated disturbances, and high female fecundity (<2500 eggs; Wernecke and Zwick, 1992) supports the long-term stability of the population. For the stonefly, *I. goertzi,* semivoltinism may be a successful strategy (Chapter 7); *L. prima* emerges early in the year and females oviposit some weeks later, thus avoiding the "regular" flood period in late winter or early spring.

Stable caddisfly populations included species with large, heavy, mineral larval cases (*C. villosa, P. cingulatus,* and *S. personatum*) providing protection against predation and acting as ballast (Dobson *et al.*, 1997; Otto, 1976; Otto and Svensson, 1980; Otto and Johannsson, 1995; Wiggins, 1977). However, Waringer (1989) demonstrated that the adhesive friction imposed by case weight was less than the drag induced by the increased surface area of the case. A semi-voltine life cycle, as in *R. fasciata* and *S. personatum,* may be interpreted as risk spreading; *S. personatum* larvae further dig into the stream bottom to avoid floods (Wagner, 1991).

Species with an intermediate SS included small caddisfly grazers with stone cases (*D. annulatus, S. pallipes, T. rostocki,* and the generalist *P. conspersa*), and some of the most abundant stoneflies. Intermediate stability of *B. rhodani* is probably related to bivoltinism, and flexible larvae that may swim to food patches and can handle FPOM as food. *Baetis rhodani* usually has a scarce spring generation preceding a large autumn generation (Chapter 7). High loss rates in winter (larvae of the future generation suffer from low temperature, low algal production, and

Table 8.8 Forty EPT species divided into groups with high (SS ≥ 1.2), intermediate (0.8 < SS < 1.2), and low (SS ≤ 0.8) index of species stability (SS, 1969–2005). Values from the 25-year analysis for species that subsequently changed groups are given in parentheses.

High (SS ≥ 1.2)		Intermediate (0.8 < SS < 1.2)		Low (SS ≤ 0.8)	
Ephemeroptera					
B. vernus	1.37	*B. rhodani*	0.81	*E. mucronata*	0.54
				E. ignita	0.62 (0.83)
				P. submarg.	0.38
				Ep. danica	0.18
				C. luteolum	0.39
				Ec. subalpinus	0.40
Plecoptera					
L. prima	1.28	*P. auberti*	1.05	*S. torrentium*	0.74
I. goertzi	1.18	*P. intricata*	0.99	*N. cambrica*	0.68
		N. marginata	0.99 (0.79)	*N. pictetii*	0.55
		L. digitata	0.97	*P. nitida*	0.41
		L. nigra	0.92	*B. risi*	0.49
				A. standfussi	0.48
				P. meyeri	0.43
				N. cinerea	0.33
				N. flexuosa	0.31
Trichoptera					
R. fasciata	1.73	*T. rostocki*	1.11 (1.26)	*H. digitatus*	0.75 (0.92)
S. personat.	1.65	*P. conspersa*	1.02 (0.71)	*A. reducta*	0.75 (0.83)
P. cingulatus	1.32	*D. annulatus*	0.95 (1.38)	*A. fimbriata*	0.71 (0.91)
C. villosa	1.27	*S. pallipes*	0.89 (1.41)	*A. fuscipes*	0.70 (0.89)
				P. luctuosus	0.61 (1.02)
				M. longulum	0.50
				H. saxonica	0.45
				H. instabilis	0.44
				W. occipitalis	0.50
				O. albicorne	0.34

high discharge) were compensated for by the summer generation of the preceding year. However, in recent years specimen numbers remained low. In principle, the Breitenbach population could be supplemented by specimens from the nearby River Fulda (Chapter 7), but this is unlikely based on decreasing specimen numbers in recent years, indicating a more general process of abundance decline, also including *B. rhodani* in the River Fulda.

Mayflies and stoneflies (except *A. standfussi* and *S. torrentium*) in the low stability group mainly include rare species and species related to fine sediments or macrophyte stands. Larvae live in habitats with low temporal stability, such as sand (Parry, 1981), and often feed on FPOM. As the stability of small inorganic and organic particles on the stream bottom always remains low, this creates a permanent risk for species related to sand sediments.

Rare caddisflies also occur in this group, but there are three remarkable exceptions: *A. fuscipes* and *Ap. fimbriata* were for many years among the most abundant species (before 1990), while *W. occipitalis* became the most abundant caddisfly after 1990. If abundance measures are included we discover that species showing the greatest specimen numbers may increase or decrease.

Comparison of the 37-year with the 25-year data indicates that 26% of the species changed between groups (Table 8.8): *P. conspersa* from low to intermediate stability, the remaining species from high to intermediate, or from intermediate to low stability. Only one (of eight) Ephemeroptera, one (of 16) Plecoptera, but nine (of 18) Trichoptera changed groups.

Results of pairwise correlation of the 32 most abundant species over 37 years corroborate results from species stability scores (Table 8.9). Of the 496 possible correlations, 181 (36.5%) were significant: 139 (76.8%) positively, and 42 (23.2%) negatively. Species pairs with larvae living in similar microhabitats showed parallel abundance trends (increase or decrease), and species in habitats affected by discriminative variability (stone versus sand) showed inverse trends. Positive correlations were found in scrapers (*B. rhodani, A. fuscipes, A. fimbriata, T. rostocki*) and in species from depositional areas (*P. submarginata, C. luteolum, L. nigra, N. picteti, S. personatum*). Negative correlations were found in species with larvae preferring contrasting habitats, that is, *L. nigra – B. rhodani* (sand – stone), *W. occipitalis – A. fuscipes, A. fimbriata* (gravel, stone) or differing nutritional properties (collector, shredder – grazer), or for pairs from either up- or downstream sections (*A. fuscipes – P. luctuosus*). Not a single negative correlation was attributed to biologically competing species or predator–prey interactions. It is remarkable that the most abundant *B. vernus* showed only one (negative) correlation, with *T. rostocki*. Again this can be attributed to its unique life cycle. In summary, the results indicate that environmental variability affects species and community more than biological interactions.

8.3.4
Abundance and Discharge Pattern

There were about 40 most abundant EPT species but no new species successfully colonised the Breitenbach during the study period (1969–2005). Four different community assemblages were discriminated. These “community states” were related to annual discharge patterns that provide different ranges and qualities of food and habitats, which can be successfully used by some species but not by others. Most species favoured one pattern over others explained by significant abundance differences (Table 8.10). There were 79 significant abundance differences between the discharge patterns (51 with $P < 0.05$, and 28 with $P < 0.01$); 36 occurred between the “nonseasonal” and “seasonal” patterns. There were 18 differences between years with permanently high or permanently low discharge (patterns D/F), and 10 or 15 between the remaining pairs. This indicates the relevance of precipitation–discharge interactions and the spatial and temporal

Table 8.9 Pairwise correlation of the 32 most abundant EPT species over 37 years (1969–2006) at site T2, sorted by decreasing correlations (Spearman $*P < 0.05$, $**P < 0.01$; ( ) = negative correlation).

	Wor_occ	Bra_ris	Ple_con	Par_sub	Bae_rho	Pot_luc	Eph_ign	Cen_lut	Nrl_pic	Aga_fus	Leu_dig	Iso_goe	Nem_cam	Apa_fim	Ser_per	Cha_vil	Nem_mar	Amp_sta	Dru_ann	Pro_aub	Pro_mey	Sip_tor	Leu_pri	Leu_nig	Nem_fle	Pro_int	Sil_pal	Eph_muc	Nem_WW	Rhy_fas	Tin_ros	Bae_ver
Wor_occ																																
Bra_ris	**																															
Ple_con	**	**																														
Par_sub	**	**	**																													
Bae_rho	(**)	(*)	(**)	(**)																												
Pot_luc	**	**	**	**	(**)																											
Eph_ign	(**)	(**)	(*)		**	(**)																										
Cen_lut	**	**	**	**	(*)	**																										
Nrl_pic	*	**	*	**	(*)	*		*																								
Aga_fus	(**)	(**)	(**)	(**)	**	(**)	**	(*)	(**)																							
Leu_dig		**	*	**	(*)			**																								
Iso_goe	**	**	**	*		*	(*)		*																							
Nem_cam	(*)						*		*		*																					
Apa_fim	(**)	(*)		(**)	**	(**)	**	(*)		**		*																				
Ser_per	*	*	*	**	(**)	*	(**)																									
Cha_vil	(**)	(**)	(*)	(**)	**	(*)	*	(**)	(*)	**				**																		
Nem_mar	*	*	*					**				**																				
Amp_sta		**						*	**				**																			
Dru_ann	**	**	*			**	(**)			(*)							**															
Pro_aub	*	**	*							*		**					**	**	*													
Pro_mey							*				**		**	*				*														
Sip_tor									**		*	*	**					*			*											
Leu_pri	*	*	**	**				**				*					**															
Leu_nig				**	(**)				**	(*)	*		*		*							**										
Nem_fle							*				**		**	*				*			**	*										
Pro_int	(*)										**		*					**			**	**										
Sil_pal	*	*	**	*		**								(*)	*				**													
Eph_muc				**		*					*		*		**											*						
Nem_WW											**		**				*				**		**		**			*				
Rhy_fas		**			*							*					*	*		**						*						
Tin_ros	(*)			*	**		**																				**					
Bae_ver																															(*)	

Table 8.10 Kruskal–Wallis test (3 d.f.) for significant abundance differences between four discharge patterns, with pairwise test for comparison between discharge patterns (U-test, 2-tailed asymptotic significance), and preference of 43 taxa for certain discharge patterns. C = Red = years with nonseasonal discharge; D = green = wet summer; E = blue = snowy winter; F = orange = dry summer.

Species	Chi^2	Sign	Patterns	C–D	C–E	C–F	D–E	D–F	E–F
Bae_rho	24.66	0.00	D.E > F,C	0.00	0.00	–	–	0.00	0.00
Bae_ver	14.67	0.00	C > D,E,F	0.01	0.03	0.00	–	–	–
Eph_muc	13.07	0.00	D < E,F,G	0.00	–	–	0.00	0.02	–
Eph_ign	17.30	0.00	D,E > F,C	0.00	0.00	–	–	0.02	0.01
Par_sub	14.91	0.00	D < E,F,C	0.00	–	–	0.01	0.00	–
Epe_dan	2.20	n.s.		–	–	–	–	–	–
Ecd_sub	5.60	n.s.		–	–	–	–	0.02	–
Cen_lut	6.16	n.s.	D < E,F,C	0.05	–	–	0.02	–	–
Leu_pri	8.13	0.04	D < F	–	0.01	–	0.05	–	–
Leu_nig	10.55	0.01	F > D,E,C	–	–	0.01	–	0.01	0.03
Leu_dig	5.54	n.s.		–	–	–	–	0.05	–
Nem_mar	1.74	n.s.		–	–	–	–	–	–
Nem_fle	8.03	0.05	E > D,F,C	–	0.01	–	0.00	–	–
Nem_cam	7.07	n.s.	E > D,F,C	–	0.01	–	0.01	–	–
Nem_WW	8.85	0.03	E > D,F,C	–	0.01	–	0.01	–	–
Nem_cin	1.26	n.s.		–	–	–	–	–	–
Nrl_pic	10.38	0.02	F > C	–	–	0.05	0.04	0.01	–
Pro_aub	0.39	n.s.		–	–	–	–	–	–
Pro_int	6.33	n.s.		–	–	–	–	–	–
Pro_mey	5.61	n.s.		–	0.02	–	–	–	–
Pro_nit	14.34	0.00	D > E,F,C	0.00	–	–	0.01	0.01	–
Amp_sta	1.12	n.s.		–	–	–	–	–	–
Iso_goe	3.31	n.s.		–	–	–	–	–	–
Sip_tor	11.41	0.01	E,F > D.C	–	0.04	0.00	–	–	–
Bra_ris	9.53	0.02	D < E,F,C	0.01	–	–	0.00	0.00	–
Apa_fim	23.44	0.00	D,E > F,C	0.00	0.00	–	–	0.00	0.00
Rhy_fas	4.74	n.s.		–	–	–	–	–	0.05
Dru_ann	4.51	n.s.		–	–	–	–	–	–
Tin_ros	10.34	0.02	D,E > F,C	0.04	0.02	–	–	0.02	0.02
Ser_per	10.57	0.01	D,E < F,C	0.05	–	–	–	0.00	0.04
Sil_pal	0.69	n.s.		–	–	–	–	–	–
Aga_fus	18.79	0.00	D > E > F,C	0.00	–	–	0.04	0.00	0.04
Hyd_sax	5.77	n.s.		–	0.03	–	–	–	–
Hyd_ins	16.05	0.00	E > D,F,C	–	0.00	–	0.02	–	0.01
Ple_con	9.78	0.02	D,C < E,F	–	–	0.05	0.02	0.01	–
Adi_red	6.33	n.s.		–	0.05	–	0.01	–	–
Pot_luc	19.25	0.00	D < E,F,C	0.00	–	–	0.00	0.00	–
Pot_cin	1.52	n.s.		–	–	–	–	–	–
Cha_vil	8.92	0.03	D > F,C	0.01	–	–	–	0.01	–
Hal_dig	5.82	n.s.		–	–	–	–	–	–
Mic_lon	8.34	0.04	D,E > F,C	0.03	0.01	–	–	–	–
Odo_alb	2.30	n.s.		–	–	–	–	–	–
Wor_occ	13.07	0.00	D < F,C	0.01	–	–	–	0.00	0.03

transitions in habitat and nutrition to the success of aquatic insects (collected in the emergence traps).

The 15 most abundant species per discharge pattern are shown in Table 8.11. In terms of relative or absolute population size increase, permanent good flow (D) provided favorable conditions for *A. fuscipes*, *B. rhodani*, *B. vernus*, *T. rostocki*, and others. Winter and spring floods (E) were positive for *P. intricata*, *L. digitata*, both *Baetis* species, with the highest relative increases for *E. ignita*, *N. flexuosa*, and *S. torrentium*. With pattern F (long-term low flow) *L. nigra* had the highest absolute and relative increases, and the increased percentage shares of rarer species (*P. submarginata*, *C. luteolum*, *N. picteti*, and *S. personatum*) were even higher. Unpredictable floods (C) were advantageous only for *B. vernus* as a result of its unique life cycle; the absolute and relative abundance of all other species decreased. Species without marked preferences for any pattern had either low abundances (e.g., *E. danica*, *M. longulum*, *O. albicorne*), or are widespread in small streams (e.g., *P. auberti*, *A. standfussi*, *I. goertzi*, *R. fasciata*, *D. annulatus*, *P. cingulatus*, *C. villosa*).

8.3.5 Ordination

Axes 1 and 2 of a correspondence analysis (CA) described two strong gradients (Figure 8.15; eigenvalues (e) of axes 1–4: 0.30, 0.18, 0.11, 0.10) with cumulative percentage variance of species data: 29.6%, 47.7%, 57.9%, 67.4%. Based on community composition, year groups affected by the three seasonal patterns D, E, and F, were arranged along the first axis, with increasing numbers of months with low discharge from the right to the left. Years with discharge pattern C were arranged along the second axis. The year 1986 (insecticide input) lies in the "spring spates" group, a correct assignment with respect to flow pattern, but surprising if specimen numbers are compared. Arrangement in year groups based on variables other than discharge pattern was not meaningful. Similar discharge patterns resulted in a similar community composition (Figure 8.15, Table 8.11). The respective year groups are characterized as "community states", that is, a distinct, community assemblage is characteristic of a discharge pattern.

8.3.6 Discriminant Analysis

An a priori discriminant analysis with "discharge pattern" as a grouping variable resulted in significant differences between all groups (Figure 8.16). Discriminant functions 1 and 2 explained more than 95% of the total variation (Table 8.12), that is, "discharge pattern" is sufficient to discriminate between the different "community states". Year points of groups C, E, and F were relatively close to their group centroid, whereas parts of group (D) were rather scattered. Results were similar to indirect gradient analysis, nonseasonal flow years are some distance from years with seasonal flow. The sequence within the seasonal group

Table 8.11 Mean annual specimen numbers of the 15 most abundant EPT species in emergence trap T2 in all years (white), and in years with nonseasonal discharge (red), in wet years (green), in spring spates (blue), and in dry years (orange). Last line shows sums of specimen numbers for respective years.

Species	All years	Species	Nonseasonal	Species	Wet year	Species	Spring spates	Species	Dry year
Bae_ver	3021	*Bae_ver*	4946	*Aga_fus*	4495	*Pro_int*	1991	*Leu_nig*	3099
Aga_fus	1575	*Leu_nig*	1044	*Bae_rho*	2306	*Bae_ver*	1975	*Bae_ver*	1798
Leu_nig	1546	*Pro_int*	926	*Bae_ver*	2183	*Bae_rho*	1620	*Pro_int*	1209
Pro_int	1102	*Aga_fus*	694	*Tin_ros*	1355	*Leu_dig*	1321	*Leu_dig*	1039
Bae_rho	942	*Leu_dig*	656	*Leu_nig*	844	*Tin_ros*	1257	*Sip_tor*	527
Tin_ros	870	*Tin_ros*	653	*Pro_int*	743	*Aga_fus*	1041	*Tin_ros*	522
Leu_dig	785	*Pro_aub*	388	*Apa_fim*	546	*Leu_nig*	1008	*Wor_occ*	491
Pro_aub	343	*Bae_rho*	329	*Cha_vil*	414	*Leu_pri*	555	*Nrl_pic*	392
Leu_pri	306	*Leu_pri*	206	*Leu_dig*	391	*Nem_cam*	461	*Leu_pri*	367
Sip_tor	283	*Cha_vil*	196	*Pro_aub*	271	*Pro_aub*	449	*Amp_sta*	362
Cha_vil	244	*Amp_sta*	158	*Leu_pri*	246	*Sip_tor*	426	*Aga_fus*	360
Amp_sta	202	*Wor_occ*	126	*Sip_tor*	185	*Apa_fim*	288	*Pro_aub*	294
Apa_fim	201	*Iso_goe*	118	*Eph_ign*	87	*Amp_sta*	238	*Pro_mey*	233
Wor_occ	179	*Ser_per*	110	*Nem_cam*	78	*Eph_muc*	205	*Par_sub*	207
Nem_cam	163	*Sip_tor*	108	*Iso_goe*	67	*Cha_vil*	201	*Nem_cam*	195
Sum	11 762		10 659		14 211		13 036		11 095

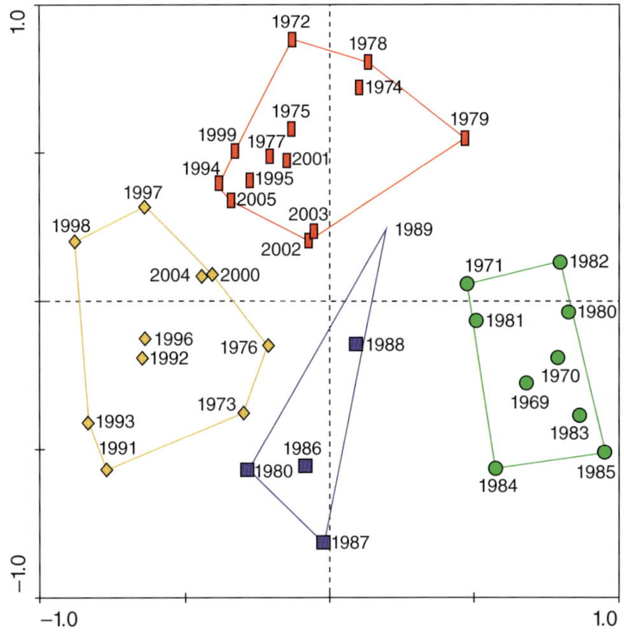

Figure 8.15 Correspondence analysis of years based on EPT emergence (42 species) in the Breitenbach 1969–2005. Envelopes distinguish groups of years with similar discharge pattern in the year prior to emergence, compare Figure 8.1. Red = Years with nonseasonal discharge; green = wet year; blue = spring spates; orange = dry year.

differs in that the group with winter and spring floods, and the wet-year group changed position.

8.3.7
Community Diversity

Individual "community states" were characterized by total abundance and diversity measures (Table 8.13). Diversity scores in group C years (nonseasonal flow) were lower than in seasonal flow years (B). Within the seasonal patterns, diversity in dry years (F) and years with spring spates (E) was higher than in wet years (D). Linear regressions of Shannon–Wiener diversity on CA axes 1 and 2 were significant (Figure 8.17). Site points of group C years were distinguished from the seasonal groups on axis 2 by a gap at the right-hand end of the regression line, and they were mainly some distance below the regression line on axis 1. Site points of groups D, E and F with seasonal discharge patterns were arranged along the regression line on axis 1 (Figure 8.17, left), with dry years on the left and wet years on the right. These groups were mixed along axis 2 (Figure 8.17, right) and distinguished from the site points of "nonseasonal flow" on the right-hand end. The effect of these group C years on the accuracy of the regression

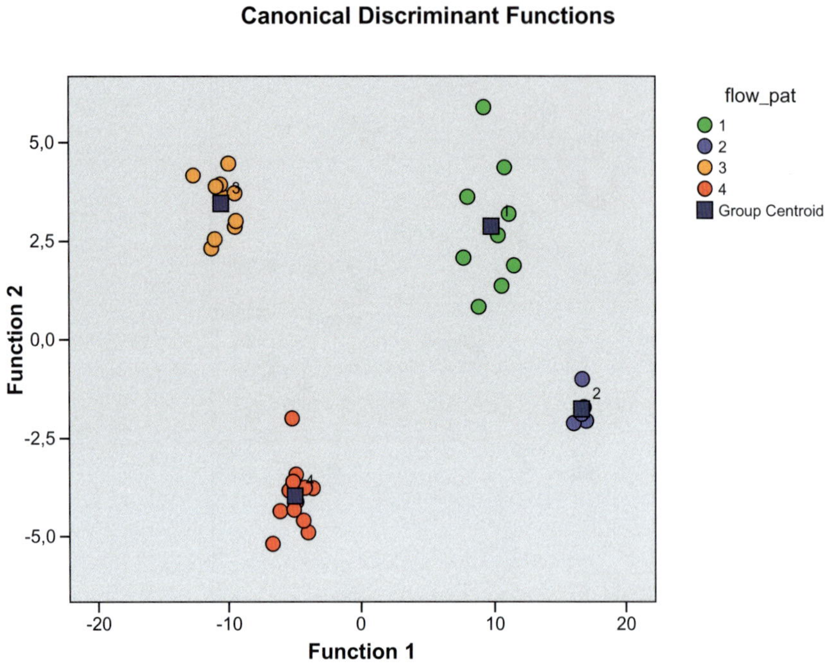

Figure 8.16 Canonical discriminant analysis of 42 EPT species in trap T2 under four pre-selected discharge patterns for the period 1969–2006. Red = Years with nonseasonal discharge; green = wet summer; blue = snowy winter; orange = dry summer.

Table 8.12 Test statistics (Eigenvalues, Wilks' lambda) for the first three discriminant functions for the EPT community at T2 at four discharge patterns. Pre-selected as the grouping variable.

Function	Eigenvalue	Variance (%)	Cumulative %	Canonical correlation
Eigenvalues				
1	111 162	85.6	85.6	0.996
2	12 544	9.7	95.3	0.962
3	6090	4.7	100.0	0.927
Test of function(s)	Wilks' lambda	Chi-square	df	Sig.
Wilks' lambda				
1 through 3	0.000	167.123	96	0.000
2 through 3	0.010	82.164	62	0.044
3	0.141	35.257	30	0.233

Table 8.13 Shannon–Wiener diversity. N2 diversity and annual abundance of 42 EPT species (mean and range) for all years (A), seasonal discharge (B), nonseasonal discharge (C), wet years (D), dry years (F), and years with spring spates (E). Cf. Table 8.7.

	A	B	C	D	E	F
SW diversity						
Mean + s.d.	2.60	2.60	2.45	2.41	2.79	2.60
Mean – s.d.	1.93	2.40	1.67	2.07	2.39	2.40
Mean	2.27	2.50	2.06	2.24	2.59	2.50
N2 diversity						
Mean + s.d.	8.84	9.26	7.27	7.18	11.39	9.03
Mean – s.d.	3.95	5.25	2.34	5.22	7.03	6.85
Mean	6.39	7.25	4.81	6.20	9.21	7.94
Number of specimens						
Mean	13 067	13 924	11 484	14 719	16 883	12 806
Min	6119	6119	6156	9897	11 264	7132
Max	27 285	27 285	19 551	19 441	19 646	27 285

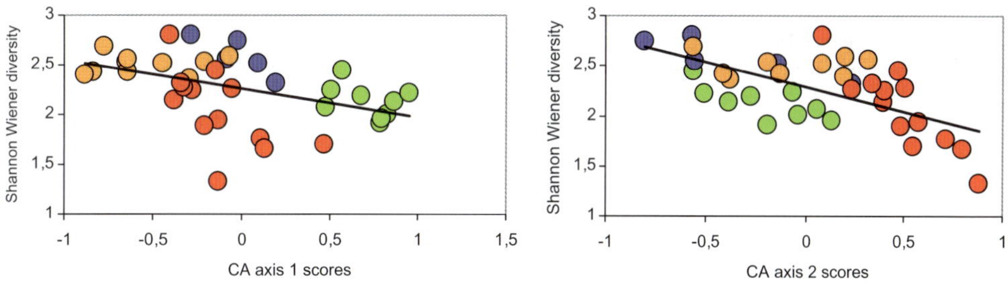

Figure 8.17 Linear regressions of Shannon Wiener diversity scores of discharge years against CA axis 1 (a, left; $y = -0.284 + 2.260x$; $r^2 = 0.19$; $P < 0.016$) and CA axis 2 (b, right; $y = -0.492 + 2.291x$; $r^2 = 0.40$; $P < 0.001$). Red = Years with nonseasonal discharge; green = wet summer; blue = snowy winter; orange = dry summer.

line becomes clear if they were omitted. The regression on the first axis is then $y = -0.286x + 2.39$ ($r^2 = 0.51$), and on the second axis $y = -0.257x + 2.33$ ($r^2 = 0.11$). On the first axis the nonseasonal years conflict with the "better" regression coefficient, and on the second axis they improve the regression coefficient. Diversity along the first CA axis increases with increasing number of months with low discharge (patterns D–F). Diversity along the second axis increased with seasonal discharge predictability. Mean, minimum, and maximum specimen numbers differed between groups, highest or lowest values did not correspond to any one group (Table 8.13).

Table 8.14 Percentage increase or decrease of four species with increased or decreased specimen numbers after 1990 (for absolute numbers, compare Table 8.11).

Species	%
Wor. occ.	12 173
Pot. luc.	793
Cen. lut.	663
Nrl. pic.	522
Eph. ign.	–85.6
Apa. fim.	–86.0
Bae. rho.	–87.3
Aga. fus.	–89.9

8.3.8
Periods before and after 1990

If we compare species with the greatest percentage abundance changes before and after 1990 there are clear differences (Table 8.14). The highest relative increase after 1990 mainly occurred in species with filter-feeding and collecting larvae, whereas species with the greatest losses were predominantly grazers. But there are a lot of clearly flexible species in which other effects of the changing environment may play a role in relation to life cycle and specific habitat requirements. Comparison of the "community states" (e.g., Table 8.11) and discharge patterns indicates the same basic trend of decreasing numbers of scrapers and increased abundance of shredders, filtering and gathering collectors. These changes are further indications of climatic change.

8.3.9
The EPT Community of the Breitenbach and Environmental Variation

The EPT community of the Breitenbach is typical of many small Central European streams. Although abundance fluctuations of all species in subsequent years were large, none became extinct between 1969 and 2005. Even accidental insecticide input in 1986, which dramatically reduced specimen numbers (Zwick, 1992; see Chapter 7), did not eliminate any species, and specimen numbers of aquatic Empididae remained almost unaffected (Wagner and Gathmann, 1996). The community recovery after the insecticide input indicates the resilience of the ecosystem, due at least partly to the presence of unaffected up- or downstream areas, and to the River Fulda, from which rapid re-colonization occurred (Chapter 7). An attempt to establish stone flies that were previously uncommon in the stream was unsuccessful (Chapter 7). Disregarding this human-induced disturbance, we have evidence that local environmental, and large-scale climatic, variability change aquatic insect populations and the stream community (e.g., Bradley and Ormerod, 2001; Wood *et al.*, 2000).

Basically, it is necessary to determine whether environmental variability relates to life cycles and species traits, and whether it is stochastic or regular (Townsend *et al.*, 1997b). We observed that EPT emergence was more or less regular every year, and thus we focussed our analysis on the idea that the long-term, recurrent, seasonal sequence and the life cycles of species are linked. We concluded that years with similar environmental variability will yield similar species abundances, and thus similar population measures. Results from species models (Wagner *et al.*, 2006; see Section 8.6) provided further evidence.

Effects on the community of precipitation and discharge were more important than water temperature. Compared with precipitation, enhanced discharge seasonality can be linked to climatic preconditions, and the sponge-like characteristic of the catchment. Precipitation is stored for longer periods as snow (winter) or as soil humidity, or is rapidly released via evapo-transpiration (summer), and constant baseflow (all year; see Chapter 3). Snowmelt in rainy periods with smooth icy surfaces may cause floods, as can heavy thunderstorms in summer. However, under "normal" climatic conditions, flash floods, and droughts are infrequent and seasonally unpredictable at the Breitenbach. But precisely for this reason floods and droughts have a great effect on species, populations, and community composition.

Effects of the small increase in precipitation and discharge after 1990 were interpreted as an indication of climatic change. Since then, only dry or nonseasonal flow years have been recorded. This affects the extent and availability of habitats on the stream bottom, not only at a particular moment but over the entire period of larval and pupal development over several years.

Every discharge variation changes substratum composition and simultaneously provides harsh conditions for one part of the community, but is benign for another (Peckarsky, 1983), and may leave others unaffected. Sediment texture is primarily a translation of discharge to stream bottom habitat composition, affecting specimens, populations, and the community. Thus, it was not unexpected that many differences of individual species abundances were related to discharge, and that four discharge patterns discriminated four well distinguishable "community states". These effects were so clear that the community assemblage could even discriminate streams (and disturbance regimes) from separate geographical areas (e.g., Resh *et al.*, 1988; Poff *et al.*, 2006).

As there are fundamental interactions between habitats and species traits (Stearns, 1976), linkage between a species' life cycle and a predictable environment is beneficial, whereas stochasticity is detrimental. Species traits for disturbance avoidance include emergence (Lytle, 2001, 2002; see Chapter 7), egg-diapause (Schmidt, 1984), larval dormancy, (Aurich, 1992), larvae anchoring to the substratum (Aurich, 1992; Bacher and Waringer, 1996; Bohle and Fischer, 1983; Menzler, 1995), and risk spreading (Wagner 1990b, 1991; see Chapter 7). These strategies are successful responses to restricted environmental predictability.

Thus, nonseasonal floods negatively affected all species except *B. vernus*, which for the greater part of the year is an "inactive" community member.

The fundamental influence of discharge-mediated substratum stability and fast community recovery at sites with coarse substrata and high slope were important findings of a long-term study in southeast England (Armitage and Gunn, 1996); and they provide additional insight into disturbance effects (Reice, 1985; Matthei *et al.*, 1997; Lancaster, 1999). In the Breitenbach, floods at "inappropriate" times had detrimental effects on most species. This indicates that invertebrates have become adapted to, or "selected" by, limits set by the long-term discharge pattern on the one hand, and population plasticity on the other.

With respect to dislodgement, the availability of flow refuges is also crucial for the maintenance of the population (Lancaster and Belyea, 1997; Winterbottom *et al.*, 1997; Townsend *et al.*, 1997a). Similarly, if seasonally predictable floods last too long, or are extremely high, a greater proportion of the refuges will be destroyed, with increasingly detrimental effects on most species.

If similar discharge patterns generate similar "community states", "community states" can be representative of discharge patterns. This interrelationship was confirmed by all statistical analyses, and was interpretable from species ecology and empirical knowledge. These results support Armitage (1995), who stated that the crucial factor in maintaining stream integrity, including the community, is the timing and nature of disturbance.

The relationship between life cycles and the environment is not precise in a mathematical sense. Increasing of the linkage precision between life cycle and environmental variability increases the risk that species becomes extinct if an "expected" event does not occur at a certain time, or if "unexpected" events occur. Within the limits of the long-term environmental variability flexible life cycles can maintain species in streams even under unfavorable conditions, albeit at low abundance. High female fecundity and favorable environment are then the preconditions for rapid population recovery in a changing environment, while other populations decline (e.g., *W. occipitalis* and *A. fuscipes*). In any case, if annual flow patterns predominantly affect the community assemblage, they can also be used as a predictive tool, for example, for the effects of environmental change. The example of *W. occipitalis* indicates that rare species should not be excluded from ecosystem analyses.

8.3.10 Multiple Stable Community States in Running Waters?

A recent study on the effects of drought after a series of dry years on fish and invertebrate communities in Californian streams showed no recovery of invertebrate community composition, but invertebrate taxon richness and abundance recovered quickly (Bêche *et al.*, 2009). High spatial synchrony was observed in the temporal changes at the four study sites, and three periods, based on the time in a year when the stream fell dry, were distinguished. Precipitation and flow permanence were identified as the main variables influencing community composition, essentially similar dominant factors to the present study. These variables are part of a cascade of environmental effects, including climate and local weather, which, in conjunction with the attributes of the drainage area, affect the respective annual flow pattern. They determine the temporal availability of instream habitats which are prerequisites for the developmental success of any species. Different temporal preconditions result in different community assemblages or “community states”.

In the Californian study, 11 of the 24 indicator taxa identified, were characteristic for the pre-drought, eight for the drought, and five for the post-drought period. These three periods are comparable to the “discharge patterns” we distinguished for the Breitenbach. They describe annual flow characteristics that affect temporal and spatial habitat availability, thus determining species success and the community measures in turn.

The high spatial synchrony of the temporal community changes in the Californian streams is also similar to our study. Comparison of the Breitenbach and another stream in the area (see Section 8.5) indicated high synchrony of environmental variability, resulting in parallel increases and decreases in macroinvertebrate abundance. In addition, similar synchronous abundance changes were observed over an 11-year period at four sites along the Breitenbach (see Section 8.4). High spatial synchrony of temporal variability depends on the predictability of the climatic, geological, and other site-specific conditions.

Results from both studies suggest that, following different environmental impacts (whether manmade or natural), different “community states” can be interpreted using the ecological concept of “multiple stable states” (references in Gunderson, 2000; Folke *et al.*, 2004). The idea of “multiple stable states” has already been demonstrated for several ecosystem types, such as temperate and tropical lakes, wetlands, estuaries, coastal seas, coral reefs, kelp forests, savannas, and more (Folke *et al.*, 2004). It is mainly applied in an ecosystem management context and has developed from the hypothesis of “a stable and infinitely resilient environment”, but must recognize that “the self-repairing capacity of ecosystems should no longer be taken for granted’ (Folke *et al.*, 2004).

Terms like resilience, stability, stable state and others are defined differently by different authors, and remain interpretable (Gunderson, 2000). As a result of our study of an almost pristine ecosystem, predominantly applied studies on running waters should be reviewed in the light of the “multiple stable states concept”. If systems have different stable community states the idea of “reference situations” becomes debateable. If a system is not in the “desired” reference situation when samples are taken, expensive restoration may be required. If we define resilience as “the capacity of a system to absorb disturbance and reorganize while undergoing change so as to retain essentially the same function, structure identity and feedbacks’ (see Folke *et al.*, 2004), it supports the engineers’ view that ecosystems must “serve” man. To manage ecosytems to serve man requires a profound understanding of their “natural” variability. Long-term studies are then particularly required to learn how, over time, ecosystems react to a naturally variable environment and to human intervention.

8.4
Eleven Years Emergence at Four Sites along the Breitenbach and the Effects of Discharge Patterns on the EPTD Community

Rüdiger Wagner and Hans-Heinrich Schmidt

Running waters are "open" ecosystems with a through-flow of inorganic and organic matter and of energy. Chemical and physical water attributes change gradually from spring to mouth due to the temporally (time of exposure) and spatially (catchment size) increasing effects from the terrestrial environment. Environmental variables affect the community, although their influence is often modified by water attributes. Variables measured at successive sites along a stream develop gradients that cause predictable changes with area, distance from source, or time. Such gradients are preferred "instruments" along which ecological manipulation experiments are planned, or they are helpful for interpreting empirical environmental data. A good example is provided by the temperature pattern of unaltered stream–river systems. With increasing distance from the spring, differences between the summer maximum and the winter minimum temperature increase, which should markedly affect species and communities in running waters (e.g., Illies and Botosaneanu, 1963; Ward, 1985).

Searching for a basic pattern for the multitude of running water systems, Illies and Botosaneanu (1963) summarized the recorded attempts and published the Rhithron–Potamon concept (RPC), which discriminated streams and rivers based on community characteristics (i.e., morphological adaptation to either turbulent flow in streams or laminar flow in rivers), flow characteristics, annual temperature patterns, and morphological species "traits". One measure that distinguished streams from rivers was temperature. In streams the greatest difference between maximum and minimum monthly mean temperature should be less than 20 °C, in rivers more than 20 °C. However, from measuring the along stream and river gradients it became evident that field conditions were more complicated. The biological consequences of the thermal regime of streams are manifold; it affects the activity, development, and growth of specimens, or limits the distribution of species and the productivity of the community (Allan, 1995). Adaptations of the body shape of aquatic invertebrates to the variable hydraulic conditions were found to be quite similar around the world (Statzner, 1981, 1987; Statzner and Higler, 1986; Statzner, Gore and Resh, 1988). Particular adaptations were tested using various methods by, among others, Ambühl (1959), Statzner and Holm (1989), and König and Waringer (2008).

In contrast, the River Continuum Concept (RCC; Vannote *et al.*, 1980) postulated that gradual changes in stream communities resulted from the impact of a changing gradient of environmental variables per site, along which the community should change predictably. It included the increasing influence of catchment size on the community, for example, nutritional conditions, temperature, discharge. The RCC was a bold attempt to describe the functions of lotic ecosystems and to accommodate variation between sites that results from differences in their terrestrial environment, but several questions remained unanswered (e.g., Statzner and Higler, 1985).

Between 1983 and 1993, emerging insects were collected from four sites (T1, T2, T4, and T6) along the Breitenbach using identical traps and collecting methods (see Chapters 2 and 7). The site arrangement made it possible simultaneously to analyze variability of the distribution of individual taxa and the community along this small, short stream (spatial variation) over subsequent years (temporal variation).

The objectives were:

- Are species evenly distributed along the stream or centered on individual sites?
- Is there a succession of species of individual genera along the stream?
- Do community measures regularly develop along the Breitenbach?
- Is the abundance variation of an individual species similar at all sites over time?
- Do sites show a reasonable order based on community data?
- Do variables provide an ecologically sound and interpretable order among sites?

8.4.1
Variables, Community, and Statistics

Variables considered against species data over the 37-year study were precipitation, discharge, and water temperature, because they described the greatest share of environmental variability (see Section 8.3). Monthly maxima were usually used, but three months data were sufficient (to avoid numerous collinearities) for some tests. The four collecting sites and the three discharge patterns (see Section 8.3) span a spatial and temporal templet for the analysis of population and community variability in the Breitenbach. Discharge patterns were grouped according to Wagner and Schmidt (2004; see Section 8.3); discharge years run from October one year to September the next. The remaining variables were treated for the same periods. Some 83 EPTD species were treated as the community (Table 8.17).

Environmental variables were tested for predictability (P), constancy (C), and contingency (M) according to Colwell (1974) and Stearns (1981). Methods and size classes of variables were the same as in Section 8.3.

As several data sets were not normally distributed (Kolmogorov–Smirnov test), nonparametric tests were applied. Species data were tested for spatial and temporal variation, with "sites", and "discharge pattern" as grouping variables (Kruskal–Wallis test). If the Kruskal–Wallis test indicated significance, Mann–Whitney U-tests were performed to test for differences between pairs of sites or discharge patterns (SPSS, 2007).

Species/year and year/environmental variables matrices were related by indirect (correspondence analysis; CA) and direct (canonical correspondence analysis; CCA) gradient analysis using CANOCO for Windows 4.5 (ter Braak 1988, 1990; Jongman *et al.*, 1987; Palmer, 1993; ter Braak and Verdonschot, 1995).

Discriminated site groups (spatial, temporal, spatio-temporal) were visualized by "envelopes" (CANODRAW, ter Braak and Šmilauer, 1998).

For CCA, variables were maximum and minimum water temperature (Ta, Ti – per site), precipitation sum (P) and discharge maximum (D) for up to six three-month periods before emergence (the period of larval development) – and both were presumed to be the same for all sites. The periods in the year preceding emergence were January–March, April–June, July–September, October–December; precipitation July to September of the year before emergence was designated as P 7–9*x*). Data remained untransformed. Inter-set correlation of environmental variables and species-derived sample scores were read from a solution file. Site groups with the relevant sets of the temperature, precipitation and discharge variables were visualized in CANODRAW. Significance of the first and all canonical axes was tested with Monte Carlo permutation tests, as implemented in CANOCO (ter Braak, 1988; ter Braak and Šmilauer, 1998). Diversity, evenness, and species richness were read from CANODRAW and exported to other programs for further analysis.

Finally, "*a priori*" discriminant analyses (SPSS, 2007) was performed to test whether the spatial and temporal "summary variables" (discharge pattern – temporal; "site along the stream" – spatial) are sufficient to classify the site scores into interpretable groups.

8.4.2 Environmental Variables

8.4.2.1 Precipitation

Precipitation effects were considered identical at all sites along the stream. Maximum annual precipitation for the period 1983–1993 was recorded in 1984 ($780\,l\,m^{-2}$), the minimum in 1991 ($490\,l\,m^{-2}$). Mean precipitation per three months ranged between 140 and $180\,l\,m^{-2}$, minimum ($50.6\,l\,m^{-2}$) or maximum ($277\,l\,m^{-2}$) values were well beyond these limits. Predictability (P) was moderate on a monthly time scale. Seasonal constancy (C, 87%) contributed relatively more to predictability than seasonal variability (M, 13%; Table 8.15).

8.4.2.2 Discharge

Discharge was determined at the source and at site T2. Predictability was highest at the source (BS), mainly due to its relative constancy (93%). At T2 predictability decreased, 86% due to constancy, and 14% to contingency; the contribution of seasonal aspects (M) was insignificant (Table 8.15). Based on the annual patterns, discharge-years were attributed to three groups: permanent high discharge, winter and spring spates, and long-term low flow (patterns D, E, F; Wagner and Schmidt, 2004; see Section 8.3); the nonseasonal pattern did not occur in this period. Within-group P and C were higher than they were for the entire period, indicating increased similarity within a "pattern group" and reasonably discriminated patterns. Although less reliable because of the fewer data points within groups, with an increase in predictable fluctuation in the period 1986 to 1990 values were similar among groups.

Table 8.15 Predictability (P), constancy (C), contingency (M), percentage constancy (C/P) and contingency (M/P) of precipitation, discharge, and water temperature (monthly maximum) at five stream sites (BS, A, B, E, G), maximum and minimum air temperature and number of days with precipitation for the Breitenbach, 1982–1993 (HQ since July 1983); $*P < 0.05$, χ^2.

	Precip	Discharge		Water temperature					Air temp.		prec.
	MR	BS	T2	HQ	T1	T2	T4	T6	Max	Min	Days
P	0.64*	0.85*	0.74*	0.89*	0.87*	0.80*	0.81*	0.85*	0.84*	0.92*	0.77*
C	0.56*	0.79*	0.63*	0.86*	0.70*	0.61*	0.59*	0.58*	0.68*	0.81*	0.75*
M	0.09	0.06	0.11	0.03	0.17	0.19	0.22	0.27*	0.16*	0.11*	0.02
C/P	87	93	86	96	81	76	73	68	81	88	97
M/P	13	7	14	4	19	24	27	32	19	12	3

8.4.2.3 **Water Temperature**

At four stream sites water temperature formed an almost linear gradient from spring to mouth. Highest predictability (P) occurred at the source, with 96% constancy and 4% contingency. P remained above 0.8 at all sites, but P and C decreased with increasing distance from the spring. The synchronous increase in predictable fluctuation was evident throughout, but only significant at T6. From T1 to T6, percent constancy (C/P) decreased from 81 to 68%, whereas contingency (M/P) increased from 19 to 32% (Table 8.15).

8.4.2.4 **Air Temperature and Number of Rainy Days**

Minimum and maximum air temperature are highly predictable, and C (the majority of P) and M are significant. Finally, the number of rainy days was also highly predictable, based on constancy.

8.4.3 Community

8.4.3.1 **Distribution of Species along the Breitenbach**

Sixty of the 83 EPTD species (72%) showed significant differences in abundance between traps, 31 (37%) between discharge patterns, and for 27 (32%), abundance varied simultaneously with "site" (space), and "discharge pattern" (time). For summaries of the Kruskal–Wallis and Mann–Whitney tests, and for information on life cycles and ecological aspects, see Table 8.17 and Chapter 7 in this book.

The abundance of all Ephemeroptera, 13 of 15 Plecoptera, and 17 of 18 Trichoptera differed significantly between at least one pair of traps. All mayflies were rare at site T1, and abundance increased downstream. Eleven-year sums for *B. vernus* and *P. submaculata* were highest at T2, for *B. rhodani, E. mucronata, C. luteolum, E. submontanus*, and *E. sylvicola* at T4, and for *E. ignita* and *E. danica* at T6.

Of the stoneflies, three, *L. prima*, *N. marginata*, and *A. standfussi*, had their highest abundance at T1, eight at T2, and three at T6, but none at T4. Specimen numbers increased downstream for *L. nigra*, *L. digitata*, and *N. flexuosa*. Within genus sequences of species along the stream occurred in *Leuctra*, with the maximum abundance of *L. prima* upstream (T1), *L. nigra* in the mid region (T2), and *L. digitata* downstream (T6), and in *Protonemura*, with *P. auberti* (T1), *P. intricata* and *P. meyeri* (T2), and *P. nitida* (T6). Variations in pairs of species (upstream versus downstream) occurred between *Nemoura marginata*–*N. flexuosa*, and *N. cambrica*–*N. cinerea* (see Chapter 7), *Nemurella pictetii* was most abundant at T1, the remaining species at T2. Six were more or less evenly distributed along the stream; for four species results were close to significance, treated as a "tendency" for the preference of a certain stream section. However, the exact physiological adaptations (e.g., feeding, growth efficiency) of many species at different temperatures remain unknown.

For caddisflies, along-stream succession occurred only in *Potamophylax*, with *P. nigricornis* at the source and its run-off, *P. cingulatus* in the upstream area around T1, and *P. luctuosus* at T4 and T6. The emergence periods of *P. luctuosus* (May) and *P. cingulatus* (August) are seasonally discriminated. For the genus *Rhycophila*, *R. fasciata* is the dominant species, its close relative *R. nubila* was rarely collected in the downstream area (T6). Of the other caddisfly species, *A. fuscipes*, *D. annulatus*, *S. personatum*, *C. villosa*, *P. cingulatus*, and *W. occipitalis* were most abundant at T1, *P. conspersa* and *A. fimbriata* at T2. With the exception of the generalist *P. conspersa*, these taxa are crenophilous or prefer low order streams. The six species with the highest abundance at T4 (*R. fasciata*, *T. rostocki*, *S. pallipes*, *H. instabilis*, *P. luctuosus*, and *M. longulum*) are typical of mid-sized streams, whereas the four dominant species at T6 (*A. reducta*, *H. digitatus*, *H. instabilis*, and *O. albicorne*) prefer large streams or, like *H. instabilis* in particular, belong to the fauna of small rivers like the nearby River Fulda.

The abundance of most Diptera was higher in the downstream (T4, T6) than upstream sections. *Ptychoptera paludosa* increased downstream, with a maximum at site T6, while *Dixa nebulosa* and *D. puberula* and the Empididae, *Chelifera flavella*, *C. precabunda*, and *C. precatoria*, *Phyllodromia melanocephala*, *Hemerodromia unilineata*, *Clinocera fontinalis*, and *D. irrorata*, also had their highest abundance at T6. The Psychodids with highest abundance at T6, *Berdeniella manicata*, *Satchelliella trivialis*, *Bazarella subneglecta*, prefer larger streams, are bi-voltine (spring, autumn generation) or polyvoltine, for example, *Ps. phalaenoides*.

Six Empididae, *Clinocera wesmaeli*, *Wiedemannia lamellata*, *W. zetterstedti*, *Dolichocephala irrorata*, and *D. oblongoguttata*, were the only species with maximum abundance at T4. The abundance of *D. irrorata* decreased upstream and *D. oblongoguttata* decreased downstream. *Tipula maxima*, *Chelifera trapezina*, *Wiedemannia bohemani*, *D. oblongoguttata*, *Berdeniella illiesi*, *B. unispinosa*, *Pericoma blandula*, and *Psychoda parthenogenetica* had their highest abundances at T2. *Dixa submaculata*, *Chelifera diversicauda*, *C. pyrenaica*, and *Clinocera stagnalis* favored the spring region.

8.4.3.2 Species Affected by Discharge Pattern

For less than half the species there were significant effects of discharge patterns on abundance (Tables 8.16 and 8.17). Most Diptera had their highest specimen numbers in dry years (pattern F), although none of the Empididae was involved (small larvae, living in mosses and sediments, oxygen from water, closed spiracles), and most were Psychodidae (larvae small, in mosses, open spiracles). All Dixidae were affected by the discharge pattern, their larvae and pupae live exposed to the current at the water meniscus in the hygropetric zone. Surprisingly no effect was detected for *Ptychoptera paludosa*; its larvae usually prefer the abundant, stable sandy substrata deposited during low flow. Along-stream differences may be masked by large temporal differences in abundance.

Caddisfly species with larvae that construct collecting nets or live in galleries, that is, *Plectrocnemia conspersa*, *Wormaldia occipitalis*, and *Tinodes rostocki*, were mainly affected by the changing discharge pattern. *Wormaldia occipitalis* and *P. conspersa* were only successful under low-flow conditions with little variation, *T. rostocki* favored good flow with little variation.

The stoneflies *Leuctra nigra*, *Nemurella picteti*, and *Siphonoperla torrentium* preferred low-flow conditions, while *L. digitata* was successful in summer periods with good discharge.

Three mayflies (*Epeorus assimilis*, *Ephemera danica*, *Centroptilum luteolum*), all usually quite rare, preferred low discharge in summer. *Baetis rhodani* required adequate summer discharge for the development of the autumn generation; *B. vernus* and the other species remained unaffected.

In general, the small number of taxa affected significantly by discharge patterns can be explained by the absence of "nonseasonal floods" (which have most drastic effects on the sediments and hence on the community) from 1983 to 1993.

Table 8.16 Species number per taxon (N), shared species numbers that vary between traps (n-trap), between discharge patterns (n-pattern), or between both (n-both).

Taxonomic groups	N	n-trap	n-pattern	n-both
Ephemeroptera	9	9 (100%)	4 (44%)	2 (22%)
Plecoptera	15	9 (60%)	5 (33%)	2 (13%)
Trichoptera	18	17 (94%)	6 (33%)	5 (27%)
Diptera	41	19 (46%)	9 (22%)	4 (10%)
Tipulidae	1	–	–	–
Ptychopteridae	1	1	–	–
Dixidae	3	1	2	–
Psychodidae	9	5 (55%)	6 (66%)	3 (33%)
Brachystomatidae	1	–	–	–
Empididae	26	12 (46%)	1 (4%)	1 (4%)
Σ	83	54 (65%)	24 (29%)	13 (16%)

Table 8.17 Life cycle characteristics of 83 EPTD taxa in the Breitenbach. ld = Larval development; E = emergence period; ffg = functional feeding group. Numbers indicate months (from 1 = January to 12 = December), when a life stage is present in the stream (11–8 = November to August of the following year). g = Grazer; s = shredder; p = predator; f = collector/fineparticle feeder (E = 9; P = 15; T = 18; D = 41). Kruskal–Wallis (KW) test statistics, and summary of Mann–Whitney U-test (MWU) for species abundance comparisons between traps and discharge patterns.

	Ld	E	ffg	Trap		Pattern	
				KW	MWU	KW	MWU
Ephemeroptera							
Baetis rhodani	1–12	4–10	g	0.015	1 < 2,4,6	0.000	D > E > F
Baetis vernus	1–5	6–8	g	0.048	1 < 6	n.s.	(D < F)
Ephemerella mucronata	6–4	5–7	f	0.000	1 < 2,4,6-2 < 4	n.s.	
Ephemerella ignita	8–5	7–9	f	0.000	1 < 2,4,6-2 < 4	n.s.	
Paraleptophlebia submarginata	7–4	5–8	f	0.000	1 < 2,4,6	n.s.	
Ephemera danica	1–12	6–7	f	0.016	1 < 6	n.s.	
Ecdyonurus venosus	7–5	6–7	g	0.001	1 < 4,6-2 < 6	0.009	D,E < F
Centroptilum luteolum	1–12	6–10	f	0.002	1 < 2,4,6	0.019	D,E < F
Epeorus sylvicola	7–5	5–6	g	0.031	1 < 6	0.001	D,E < F
Plecoptera							
Leuctra prima	4–12	1–4	f	0.005	1,2 > 4,6	n.s.	(D < E < F)
Leuctra nigra	1–12	4–8	f	n.s.	(1 < 6)	0.000	D,E < F
Leuctra digitata	11–8	9–10	f	0.000	1 < 2,4,6	0.017	D,E > F
Nemoura marginata	6–3	4–5	f	0.000	1,2 > 4,64	n.s.	–
Nemoura flexuosa	6–3	3–5	f	0.000	1 < 2,4 < 6	n.s.	–
Nemoura cambrica	6–3	4–5	f	n.s.	(1 < 2)	n.s.	–
Nemoura cinerea	8–4	5–7	f	0.032	1,4 < 6	n.s.	(E < F)
Nemurella pictetii	1–12	4–9	f	0.008	1,2,4 > 6	0.001	D,E < F
Protonemura auberti	10–5	6–9	f	0.000	1 < 2,4,6-2 > 4,6	n.s.	–
Protonemura intricata	8–4	5–9	f	0.020	1 < 6-4 < 2,6	n.s.	–
Protonemura meyeri	5–2	3–4	f	n.s.	–	n.s.	–
Protonemura nitida	10–80	9–10	f	n.s.	–	0.000	D > E,F
Amphinemura standfussi	7–4	5–8	f	0.004	2 > 4,6	n.s.	–
Isoperla goertzi	9–4	5–7	p	n.s.	(2 > 6)	n.s.	–
Siphonoperla torrentium	7–5	5–7	p	n.s.	(1 < 2)	0.002	D,E < F
Trichoptera							
Apatania fimbriata	7–4	4–8	g	0.000	1,2,4 > 6	0.024	D > E,F
Rhyacophila fasciata	1–12	5–10	p	0.000	1 < 2,4,6-2 < 4	n.s.	(E < F)
Drusus annulatus	1–12	5–10	g	0.000	1 > 2,4,6-2,4 > 6	n.s.	–
Tinodes rostocki	6–3	5–8	g	0.001	1 < 2,4,6	0.002	D,E > F3
Sericostoma personatum	1–12	5–8	s	0.020	1 < 4,6	n.s.	–
Silo pallipes	6–3	5–7	g	0.000	1 < 2,4,6-2 < 4 > 6	n.s.	–
Agapetus fuscipes	6–3	5–9	f	0.000	1 > 2,4 > 6	n.s.	–
Hydropscyhe saxonica	6–4	5–9	f,p	0.011	1 < 4,6	n.s.	(E < F)
Hydropsyche instabilis	8–5	7	f,p	0.000	1 < 2,4 < 6	n.s.	–
Plectrocnemia conspersa	7–4	6–9	p,f	n.s.	–	0.002	D,E < F
Adicella reducta	7–4	6–7	s	0.000	1,2 < 4,6	n.s.	–
Potamophylax luctuosus	7–3	5–6	s	0.001	1,2,4 < 6	n.s.	–
Potamophylax cingulatus	11–8	8–10	s	0.000	1 > 2,4,6	n.s.	–

Table 8.17 *Continued*

	Ld	E	ffg	Trap		Pattern	
				KW	MWU	KW	MWU
Chaetopteryx villosa	12–9	10–11	s	0.000	1 > 2,4 > 6	n.s.	(D > E)
Halesus digitatus	10–7	9	s	0.000	1,2,4 > 6	0.033	D > F
Micrasema longulum	8–3	5–7	g,s	0.005	1 < 4,6	0.023	D,E > F
Odontocerum albicorne	8–5	7–8	p	0.000	1,2,4 < 6	n.s.	–
Wormaldia occipitalis	11–9	9–11	ff	0.000	1,2 > 4,6	0.018	D,E < F
Diptera							
Tipula maxima	7–4	5–7	s,f	n.s.	–	n.s.	–
Ptychoptera paludosa	7–4	5–7	f	0.000	1 < 2,4,6-2 < 4,6	n.s.	–
Dixa nebulosa	1–12	2–12	f	0.000	1,2 < 4,6	n.s.	(D < F)
Dixa puberula	1–12	2–12	f	n.s.	–	0.000	E < D,F
Dixa submaculata	1–12	2–12	f	n.s.	(1,2 > 6)	0.007	D < F
Berdeniella unispinosa	6–3	5–7	f	n.s.	(2 > 6)	n.s.	(D < E)
Berdeniella illiesi	6–3	5–7	f	n.s.	(2 > 4)	0.015	D < E,F
Berdeniella manicata	6–5	5–9	f	0.000	1 < 2,4 < 64	0.046	(D > E > F)
Berdeniella females	1–12	5–9	f	n.s.	(1 < 2)	0.044	D > E
Satchelliella trivialis	5–3	4–9	f	0.005	1,2,4 < 6	n.s.	–
Bazarella submaculata	5–4	5–9	f	0.000	1,2,4 < 6	n.s.	–
Pericoma blandula	5–4	4–10	f	0.002	1 < 2,4,6	0.015	D > E
Psychoda parthenogenetica	1–12	1–12	f	n.s.	–	0.000	D < E < F
Psychoda phalaenoides	1–12	1–12	f	0.041	1,2,4 < 6	0.048	D > E,F
Chelifera astigma	6–5	6	p	n.s.	–	n.s.	–
Chelifera concinnicauda	6–4	5–9	p	0.020	–	n.s.	–
Chelifera diversicauda	5–3	5–11	p	0.000	1,2 > 4,6	n.s.	–
Chelifera flavella	7–5	7	p	n.s.	–	n.s.	–
Chelifera precabunda	5–3	5–10	p	0.000	1 < 2,4,6	n.s.	–
Chelifera precatoria	5–3	5–10	p	0.000	1 < 2,4,6	n.s.	–
Chelifera pyrenaica	5–3	5–11	p	0.000	1 < 2,4,6	n.s.	–
Chelifera subangusta	6–3	5	p	n.s.	–	n.s.	–
Chelifera trapezina	5–3	5–11	p	0.013	2 > 1,4,6	n.s.	–
Phyllodromia melanocephala	7–4	6–9	p	n.s.	–	n.s.	–
Hemerodromia adulatoria	8–6	8	p	n.s.	–	n.s.	–
Hemerodromia oratoria	7–5	7	p	n.s.	–	n.s.	–
Hemerodromia unilineata	8–6	7–9	p	n.s.	–	n.s.	–
Heleodromia immaculata	9–6	8–9	p	n.s.	–	n.s.	–
Hydrodromia fontinalis	9–6	8	p	0.047	–	n.s.	–
Hydrodromia stagnalis	1–12	1–12	p	n.s.	–	n.s.	–
Hydrodromia wesmaeli	1–12	1–7	p	n.s.	–	n.s.	–
Kowarzia bipunctata	6–5	7	p	n.s.	–	n.s.	–
Kowarzia madicola	4, 10	6, 11	p	n.s.	(1 < 4,6)	n.s.	–
Kowarzia plectrum	9–8	10–12	p	n.s.	–	n.s.	–
Wiedemannia bistigma	6–4	6	p	n.s.	–	n.s.	–
Wiedemannia bohemani	5–3	5–10	p	0.017	1 < 2,4,6	n.s.	–
Wiedemannia lamellata	5–3	5–11	p	0.004	1,2,4 < 6	n.s.	–
Wiedemannia zetterstedti	6–5	7–9	p	0.011	4 > 1,2,6	n.s.	–
Dolichocephala irrorata	5–4	5–10	p	0.000	1 < 4,6-2 < 4,6	n.s.	–
Dolichocephala oblongoguttata	7–5	6–10	p	0.000	–	n.s.	–
Dolichocephala thomasi	7–5	7	p	n.s.	–	n.s.	–

8.4.3.3 **Species with Simultaneous Spatial and Temporal Variability**

Simultaneous effects of spatial and temporal variability only appeared for a restricted number of species: *Leuctra digitata* and *Nemurella picteti* (Plecoptera); *A. fimbriata, T. rostocki,* and *W. occipitalis* (Trichoptera); *B. rhodani, C. luteolum,* and to some extent, *B. vernus* (Ephemeroptera). For the Diptera, the abundance of the Psychodidae (*B. subneglecta, P. blandula,* and *Ps. phalaenoides*) and the empidid *D. irrorata* was affected by both spatial and temporal variability (Tables 8.16 and 8.17). Particular ecological life cycle features could not be deduced for these species.

8.4.3.4 **Correspondence Analysis**

Correspondence analysis indicated that site scores (based on the abundance of 83 species) were best arranged along the first two canonical axes. Eigenvalues of axis 1 (0.364) and axis 2 (0.212) were quite high, and the cumulative variation explained by both was 41.4%. Eigenvalues of axes 3 and 4 were smaller, and the cumulative percentage variance of axes 1–4 was 60.9%. Classifying the resulting scores into groups representing sites along the Breitenbach, discharge patterns, or both using the envelope option of CANODRAW revealed the effect of these determinants more clearly.

Site score groups were well distinguished from one another (Figure 8.18). Although there is some overlap between T2 and T4, a clear trend is evident along axis 1, with the site closest to the spring on the right hand end and the most downstream site on the left hand side. If the site scores are grouped based on "discharge patterns", the "dry years" are distinguished from "wet years" (Figure 8.19), while the remaining years with heavy snow cover in winter and "spring spates" have some overlap with the "wet years". If scores were simultaneously grouped after "site" and "discharge pattern", the overlap between the resulting smaller groups is low (Figure 8.20). The "dry years", with sites ordered along the

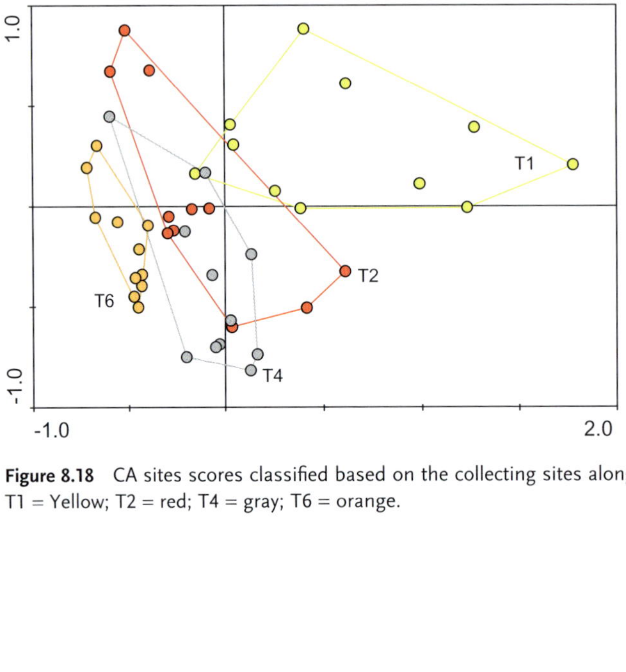

Figure 8.18 CA sites scores classified based on the collecting sites along the Breitenbach. T1 = Yellow; T2 = red; T4 = gray; T6 = orange.

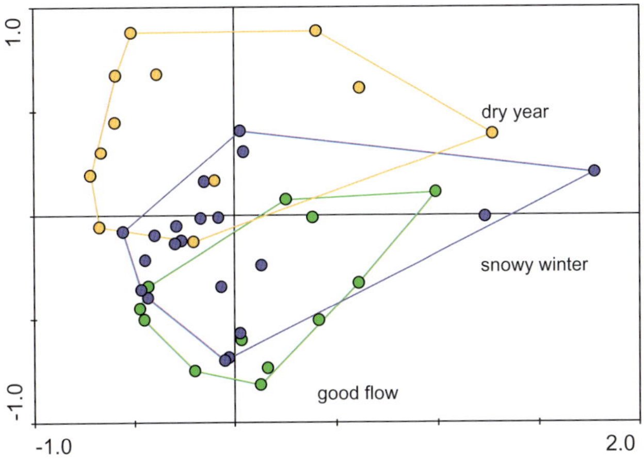

Figure 8.19 CA sites scores classified based on the discharge patterns. Wet years = Green lines, spring spates = blue lines, dry years = orange lines.

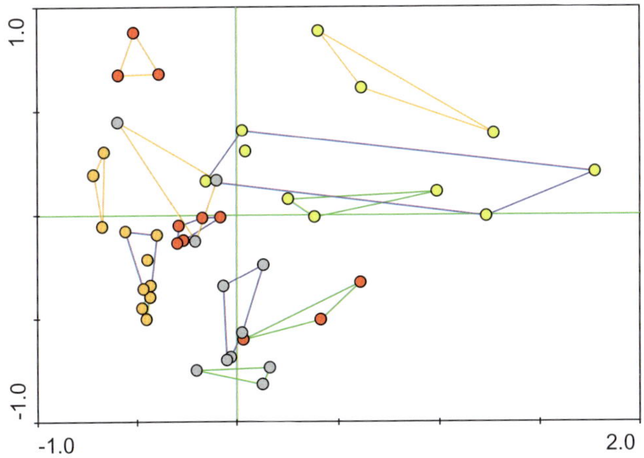

Figure 8.20 CA sites scores classified by collecting sites along the Breitenbach (T1 = yellow dots, T2 = red dots, T4 = gray dots, T6 = orange dots), and discharge pattern (wet years = green lines, spring spates = blue lines, dry years = orange lines).

stream, lie in the upper part of the diagram; the wet years, with the same order of sites, in the lower part. Years with "spring spates" after snowy winters lie inbetween. It is surprising that some site scores of the "spring spate" pattern at T2 overlap with the dry year scores of T4, although Figure 8.19 predicts large overlap between "wet year" and "spring spates".

Results of direct gradient analysis (CCA) are presented in Figures 8.21–8.23, showing "envelopes" of sites along the stream based on CA. The trace (sum of all eigenvalues) was significant (trace = 0.987, $F = 3.028$, $P < 0.002$), that is, species

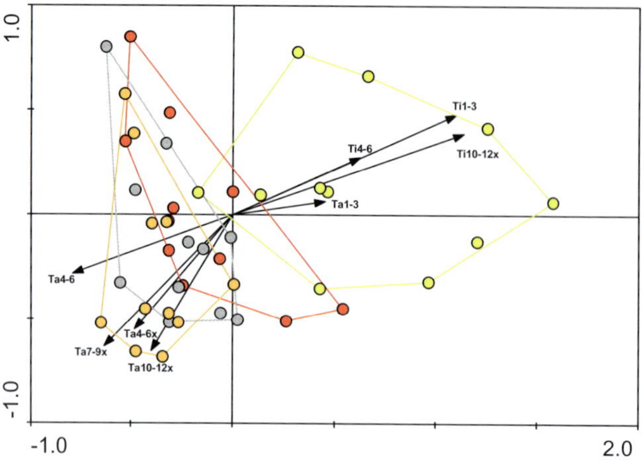

Figure 8.21 CCA scores grouped by collecting sites along the Breitenbach (T1 = yellow, T2 = red, T4 = gray, T6 = orange), and maximum (Ta) and minimum (Ti) water temperature (three-month periods in the year of, and the year before, emergence = x).

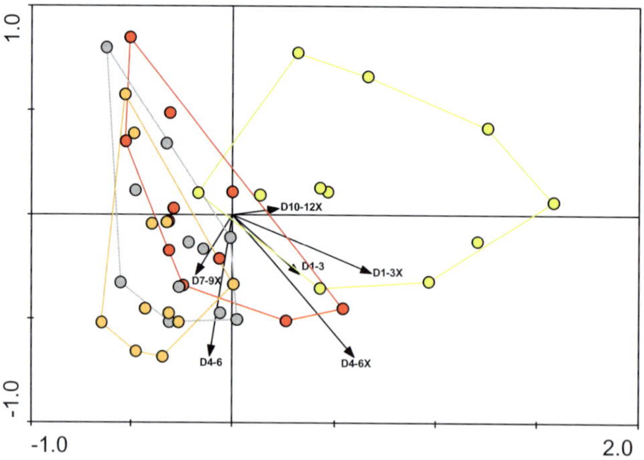

Figure 8.22 CCA scores grouped by collecting sites along the Breitenbach (T1 = yellow, T2 = red, T4 = gray, T6 = orange), and discharge (three-month periods in the year of, and the year before, emergence = x).

data were significantly related to the set of environmental variables. Canonical axis 1 was likewise significant ($e = 0.34$, F-ratio = 7.658, $P < 0.002$; Monte-Carlo test, 499 permutations). The CCA distinction between "sites along the stream" and "discharge patterns" was similar to that with CA. Depicting the total set of variables in one figure was confusing. However, the temperature data (Figure 8.21) show a strong gradient at an angle to the horizontal axis. Low winter temperature (Ti) lies

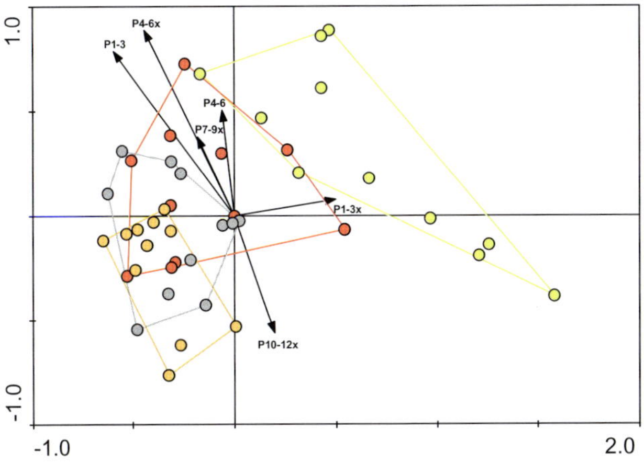

Figure 8.23 CCA scores, axes 1 and 3, grouped by collecting sites along the Breitenbach (T1 = yellow, T2 = red, T4 = gray, T6 = orange), and precipitation (three-month periods in the year of, and the year before, emergence = x).

to the upper right hand end and high summer temperatures (Ta) to the lower left hand end of the vertical axis.

Discharge data are less clear. Effects were greatest in spring (D 4–6*x*, D 4–6). Discharge arrowheads generally run almost perpendicular to the temperature arrows, and confirm the "group characteristics" from correspondence analysis (Figure 8.22). Precipitation data are best visualized with axes 1 and 3. Together they indicate a gradient along axis 3 (Figure 8.23); high precipitation in autumn is negatively related to axis 3, all other measures are positively related. Precipitation in the winter preceding emergence had no effect.

Water temperature, discharge and precipitation from January to March of the year of emergence, and discharge from April to June of the years of and before emergence (and reproduction of most species) were best related to axis 1, 2 or 3 (Table 8.18). Most temperature variables were related to axis 2, precipitation to axis 3.

8.4.3.5 **Community Measures**

N2 diversity (inverse of Simpson index), evenness and variation in specimen number developed almost in parallel at all sites along the stream. In general the lowest mean values occurred at T1 and the highest at T6. At T2 and T4, values were approximately the same and intermediate. Species number increased downstream with the highest increase between T1 and T2 (Table 8.19). Highest species number occurred at T6, particularly in periods with high mean flow and low variation (1983–1985). Between 1986 and 1990 all values decreased at the downstream sites, T4 and T6, compared with the upstream sites, probably due to variable effects of winter and spring floods on the two stream sections. Species number remained quite similar between discharge patterns, but was highest at T4 and T6. Regression of N2 diversity on the CA-axis 1 was highly significant (Figure 8.24).

Table 8.18 Significant (* $P < 0.05$) inter-set correlations of environmental variables with CCA axes 1–3.

Axis 1		Axis 2		Axis3	
Ti 1–3	0.64*	Ta 4–6 V	0.42*	Ta 4–6 V	–0.55*
Ti 10–12 V	0.66*	Ta 7–9 V	0.49*	Ta 4–	–0.32*
Ta 4–6	–0.64*	Ta 10–12 V	0.50*		
		Ta 4–6	0.22		
D 1–3 V	0.26	D 4–6 V	–0.34*	D 1–3 V	–0.41*
D 4–6 V	0.23	D 4–6	–0.34*	D 4–6	0.36*
D 1–3	0.12				
P 1–3	–0.23	P 4–6	–0.45*	P 4–6 V	0.55*
				P 10–12 V	–0.35*
				P 1–3	0.49*
				P 4–6	0.32

Table 8.19 Species and specimen numbers, Shannon–Wiener diversity, evenness and N2 diversity at four sites along the Breitenbach, under three discharge patterns (1983–1985 = normal flow; 1986–1990 = spring spates; 1991–1993 = permanent low flow).

	T1	T2	T4	T6	Mean along
Number of species					
1983–1985	55	58	63	67	61
1986–1990	52	58	62	62	58
1991–1993	50	62	63	63	59
Mean – time	52	59	62	64	59
Number of specimens					
1983–1985	11977	14401	14436	14198	13753
1986–1990	15593	17680	13242	9960	14044
1991–1993	16707	26309	17396	14780	18798
Mean – time	14911	19139	14700	12294	15532
SW diversity					
1983–1985	2.56	2.56	2.59	2.84	2.64
1986–1990	2.38	2.88	2.76	3.05	2.77
1991–1993	2.31	2.15	2.90	2.97	2.76
Mean – time	2.41	2.79	2.75	2.97	2.72
Evenness					
1983–1985	0.64	0.63	0.62	0.67	0.64
1986–1990	0.60	0.71	0.67	0.74	0.68
1991–1993	0.59	0.69	0.70	0.72	0.68
Mean – time	0.61	0.68	0.67	0.72	0.67
N2 diversity					
1983–1985	26.6	28.8	31.1	32.6	29.8
1986–1990	25.9	30.0	31.0	30.7	29.4
1991–1993	25.3	31.5	32.3	30.3	29.8
Mean – time	26.0	30.1	31.5	31.2	29.7

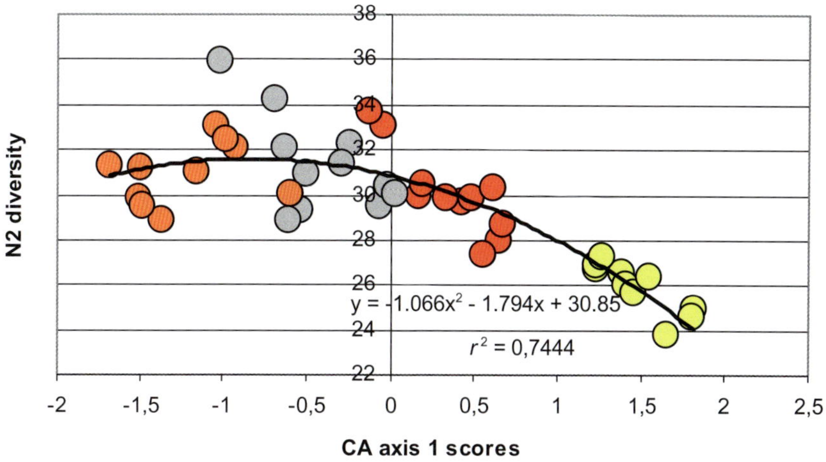

Figure 8.24 N2 diversity indices at four sites along the Breitenbach on CA axis-1 scores read from a correspondence analysis (CA) of 83 EPTD taxa, with the quadratic regression line $y = -1.066x^2 - 1.794x + 30.85$; $P < 0.0001$. T1 = Yellow; T2 = red; T4 = gray; T6 = orange.

Total specimen numbers, diversity and evenness increased with decreasing mean flow and discharge variability. From 1986 to 1990 at T4 and T6, specimen numbers and diversity decreased due to the above effects. However, taxa persisted in the stream, both spatially and temporally. No species became extinct during the study period.

8.4.3.6 **Discriminant Analysis**

Discrimination analysis based on four sites and three discharge patterns showed highly significant relationships. The variables, "site" and "discharge pattern", represent fundamental spatial and temporal dimensions. With the help of sites the distinction was even clearer than for discharge patterns, probably a result of the predictability of the respective variables at individual sites in space or in time. In both cases the first axis accounts for the greatest share (94%–site, 90%–pattern) of the total variance, the second axis for the remaining 5 or 10% (Tables 8.20 and 8.21; Figures 8.25 and 8.26).

Discrimination due to "sites" scores were spaced along axis 1, with T2, T4 and T6 along axis 2. However, T1 broke rank and was shifted along axis 2. "Discharge pattern" discriminated sites between "dry" and "wet" years'. Along the second axis, years with high mean flow and large variation were distinguished from years with high flow and low variation. (Figure 8.9, Table 8.21).

8.4.4
Discussion

Although the length of the Breitenbach is very short, the aquatic insect community (EPTD) showed differences both between sites along the stream and between discharge patterns. Even along this short flow course the community appears to

Table 8.20 Results of canonical discriminant functions of 44 site points (83 species) under three discharge patterns at the Breitenbach 1983–1993 (compare Figure 8.25).

Function	Eigenvalue	Variance (%)	Cumulative (%)	Canonical correlation
Eigenvalues				
1	349 640[a)]	89.8	89.8	0.999
2	39 799[a)]	10.2	100.0	0.988
Test of function(s)	Wilks' lambda	Chi-square	d.f.	Significance
Wilks' lambda				
1 through 2	0.000	200 937	82	0.000
2	0.025	77 882	40	0.000

a) First two canonical discriminant functions were used in the analysis.

Table 8.21 Results of canonical discriminant functions of 44 site points (83 species) on four collecting sites at the Breitenbach 1983–1993 (compare Figure 8.26).

Function	Eigenvalue	Variance (%)	Cumulative (%)	Canonical correlation
Eigenvalues				
1	1 394 984[a)]	94.1	94.1	1.000
2	70 255[a)]	4.7	98.	0.993
3	17 394[a)]	1.2	100.0	0.972
Test of function(s)	Wilks' lambda	Chi-square	d.f.	Significance
Wilks' lambda				
1 through 3	0.000	302 812	120	0.000
2 through 3	0.001	150 744	78	0.000
3	0.054	61 152	38	0.010

a) First three canonical discriminant functions were used in the analysis.

be well organized. Precipitation–discharge and temperature were the ruling factors, providing a sound and interpretable order in space and in time. Sites were ordered in the analyses, either along the stream (spatially) or by the discharge pattern (temporally). Predictability in an environment is advantageous for individual species and communities. It is a crucial precondition for the proper sequence of life cycles, including all physiological dependencies from egg development, larval feeding efficiency to adult fitness. Temperature is the spatial variable, its seasonal variation being highly predictable. Seasonal maximum and minimum temperatures span a reproducible gradient along the stream.

Ordered by significance, discharge has large short-term effects on the community, mainly as a result of its lower seasonal predictability. It depends on local or regional precipitation, reflected in the direct gradient analysis, where precipitation

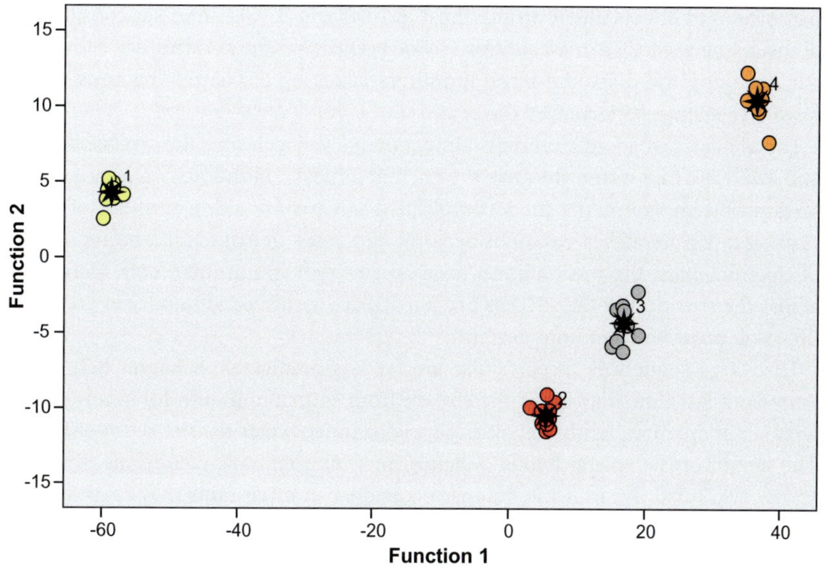

Figure 8.25 Graphic presentation of a canonical discriminant analysis of 44 site points (83 species) at four collecting sites. T1 = Yellow; T2 = red; T4 = gray; T6 = orange; stars denominate group means.

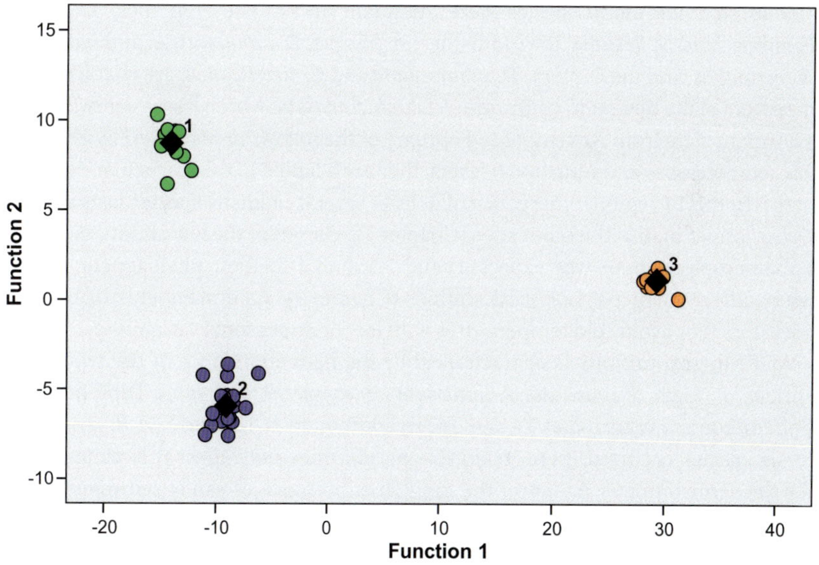

Figure 8.26 Graphic presentation of a canonical discriminant analysis of 44 site points (83 species) under three discharge patterns (wet years = green; spring spates = blue; dry years = orange; diamonds denominate group means).

had profound effects albeit on the third ordinal axis. Lower inter-set correlations of discharge with CCA indicate less obvious effects on the community. However, summarizing the disparate three month variables by discharge "patterns", they form a repeating pattern over the year.

Larval stages of aquatic animals almost exclusively colonize the stream bottom and avoid the free water. Because the current almost continuously rearranges the stream bottom sediments (in a complicated interaction with a number of other variables) it generates permanent dynamic processes of extinction and recreation of microhabitats, the precondition for a species rich community, both along and across the stream. For the success of a particular species, conditions over an entire life cycle must be taken into account.

Discharge sequences in particular are far less predictable (Chapter 8.3). Long periods of low flow may affect the stream from source to mouth for several years with sedimentation, while wet summers with much water restrict sedimentation. The height of the spring floods depends on winter climatic conditions, whether snow cover exists or not, while intense occasional summer rains may cause floods. Different instars of the same species are variously affected by such events, for example, summer floods may particularly affect "defenseless" eggs or young larvae, whereas older larvae or protected eggs of species like *Baetis vernus* are less affected.

As expected, most species were not evenly distributed along the stream but preferred certain stream sections. This applied to almost all species, as well as some community measures, producing rather "typical" spatial community patterns. Evenly distributed species are euryecious and have large distributions, such as *L. nigra, S. torrentium, P. conspersa, T. maxima*, and some aquatic Empididae.

Cold stenothermous species were abundant at T1. The stoneflies, *L. prima, P. auberti*, and *N. flexuosa*, the caddisflies, *A. fuscipes, D. annulatus, A. fimbriata*, and *W. occipitalis*, and the Diptera, *D. submaculata* and *C. diversicauda*, are characteristic members of the upstream community. Like *A. fimbriata*, whose larvae overwinter in a dormant state from November to February of the following year, thereby avoiding low temperatures and nutritional levels, they are adapted to this particular environment. In the T1 region, emergence of at least several caddisfly species lasts several weeks longer than at the other sites (Chapter 7). Therefore the low abundance of all Ephemeroptera there was expected, and attributed to their phylogenetic origin which affects their physiological abilities (Chapter 7). As descendents of tropical ancestors they avoid cold temperatures with low or no seasonal variability.

At T2 the community is characterized by the high abundance of the two *Baetis* species, *L. nigra, P. intricata, A. fimbriata, P. conspersa*, and some Diptera. Most Ephemeroptera occurred at T4 and T6. In addition to *B. rhodani* and *B. vernus*, up to six species occurred there. Life cycle peculiarities may affect this pattern, but for the semivoltinous *E. danica* the long-term presence of sandy sediments, into which larvae burrow, seems to be more relevant. Particularly high abundances of *Baetis vernus* were not detected along the stream. These only occurred in years with a "nonseasonal" discharge pattern (Chapter 8.3) and such years did not occur in the study period 1983–1993. *Leuctra nigra, L. digitata, T. rostocki*, and some Plecoptera, Trichoptera, and several Diptera were abundant downstream.

The ecological requirements of individual species seem to be the major reasons for the observed patterns along the stream and between the years. Species and individuals must respond to the environmental conditions. Water temperature is particularly seasonally predictable, its main effects occurring during limited periods in winter and in summer. Species life cycles and the seasonal sequence of these quite predictable variables seem to be closely correlated (or species of communities at particular sites were "selected"/screened by the environment). As a result, species can more or less successfully "handle" the environmental preconditions through their inherited attributes.

In contrast to temperature, discharge is temporally less predictable and its variation affects the variability of the stream bottom and banks, and in turn, the success of populations in the stream. The unpredictability, and often rapid and far-reaching effects, of flash floods are immediate and eye-catching. However temperature changes occur over a longer period and their effects on life cycles are more subtle (see Chapter 3 and Section 8.8).

Community measures do not show a regular pattern along the Breitenbach. The upstream region (T1) differs from the other sites mainly in lower species number, diversity, and evenness. A few species, mainly *A. fuscipes,* can attain extremely high abundance in some years while mayflies are rare, hence the lower diversity. The three points on the right hand side of Figure 8.24 indicate the years 1989–1991, a transition period from "spring spates" to "dry years", that is, a change in the water content of the stream over a short period. The remaining dry years, 1992 and 1993, are also on the right hand end of CA axis 1, but with increased diversity. Remaining values for the other sites are similar. The diversity increase along the stream was only marked between T1 and T2. The position of the site points follows the distribution along the stream (CA axis 1).

Variation in abundance over time was not synchronous for most species at all four sites, that is, approximately similar increases in the same years at all four sites are the exception, but if several years are considered, parallel increases and decreases are evident. However, the fundamental temporal abundance trends, parallel increase or decrease over several years, occur in all species. The increase or decrease depends on the particular site (stream section). Abundance variation was greatest at sites where specimen numbers were usually high. In conclusion, these results indicate an additional scale at which environmental variability acts. In particular, discharge affects individual stream sections differently depending on the particular local conditions, environmental as well as community. Such causal relationships may be a major reason for the ecological flexibility or resilience of stream communities and ecosystems.

8.5
The Erlenbach – a Comparison with the Breitenbach

Rüdiger Wagner

One recurrent question is how two similar streams and their communities may behave under almost identical environmental conditions without any direct interaction between the streams. How similar do the environmental variables that provide the template for the community behave? Based on these preconditions, do populations of the same species behave similarly in each stream in terms of start, duration and end of the emergence period, and patterns of variation between annual abundances? Are there similar interactions between environmental variables and species, or do stochastic processes prevail?

We therefore looked for another stream with as similar as possible a community to that of the Breitenbach, but some distance from its drainage area, so that mutual impact was as slight as possible, if not negligible. Although the entire landscape around Schlitz appears near-natural, we became aware that most small streams have been more or less intensively used by man in different ways. Today hobby fish farming is important, but in earlier decades or centuries, the meadowland was watered with stream water (irrigation rights). In the end, one small stream, the "Erlenbach", between the villages of Grebenau and Udenhausen appeared a suitable match.

The "Erlenbach" (a common name for streams in Germany) is a small tributary of the Jossa, which is itself a tributary of the River Fulda, entering the latter about 15 km north of the Breitenbach entry (Figure 2.2). The Jossa enters the Fulda from the west, the Breitenbach from the east. Both the Breitenbach and Erlenbach, flow in a northwesterly direction. The sampling site on the Erlenbach is approximately 12.5 km northwest of the Breitenbach site T2. This distance seems sufficient to prevent permanent, active faunal exchange between the two streams, at least with respect to the dominant taxa.

The Erlenbach drainage area covers approximately 2.3 km^2. The stream flows in a north-westerly direction from a source area at about 320 m a.s.l. (9°29'05″E, 50°43'09″N) for about 1750 m through a narrow valley with extensively used meadowland. The mountainsides are covered with mixed forests, dominated by beech (*Fagus sylvatica*) intermixed with some spruce (*Picea abies*) and larch (*Larix decidua*). An artificial pond (about 100 m long, surface area ~4400 m^2) is situated about 950 m below the source, that is, approximately 150 m downstream of the collecting site.

A small emergence trap was installed at the collecting site (9°28'48″E, 50°43'23″N). The collecting trough was filled with a 4% formalin solution plus drops of detergent and was emptied two to three times per week in summer and weekly in winter. Sorting, determination, and further analyses were identical to the methods described for the Breitenbach.

Air temperature was read from a maximum–minimum thermometer (accuracy ±1 °C); water temperature was determined by data loggers at 15 min intervals (accuracy ± 0.2 °C). Data were transferred to a computer every 4–6 weeks and compared to hand measurements in the meantime. On every collecting occasion, water samples were taken back to the laboratory to determine pH, conductivity, oxygen content, and other variables (same methods as for the Breitenbach). Discharge was determined at a pipe using the Manning–Strickler formula as recommended by Naudascher (1992).

We asked the following questions: How similar are the important environmental variables and their seasonal changes? How similar are the communities in the two

Table 8.22 Mean, maximum and minimum pH and conductivity of the Erlenbach from January 1998 to November 2005.

	pH	Cond
Mean	6.8	162.3
Max	7.2	197.2
Min	6.5	121.0

streams? How comparable are changes in abundance of the most abundant species in different years?

8.5.1 Environmental Variables

8.5.1.1 Conductivity and pH

Conductivity and pH were approximately constant (Table 8.22). The water was slightly acid and minimum and maximum pH values did not differ markedly over the seven-year period. Conductivity was also quite constant, ranging between 120 and about 200 μS cm^{-1}. Correlations of pH and conductivity with discharge were significant, pH decreasing, and conductivity increasing with high flow. Monthly maximum pH values showed a significantly negative relationship (−0.584, $P < 0.0001$; Spearman) with maximum discharge, whereas conductivity and discharge were positively correlated (+0.353, $P < 0.0001$; Spearman).

8.5.1.2 Discharge

The seasonal pattern of discharge was almost always parallel between both streams (Figure 8.1). The main difference between the Erlenbach and Breitenbach is the amount of water in the streams at any one time. As the drainage area of the Breitenbach is about four times that of the Erlenbach, discharge was also about four times higher, based on almost identical geological conditions. Comparison of the monthly discharge maxima using non-parametric methods indicated a highly significant relationship (Spearman 0.79; $P < 0.0001$). This is an important precondition for hypothesizing that changes in taxon abundance should be similar in both streams in different years.

8.5.1.3 Air and Water Temperature

Air temperature varied seasonally between more than 30 °C (maximum) and approximately −10 °C (minimum). Mean daily water temperature varied between 7.0 and 9.2 °C. Maximum water temperature in summer increased up to 19.0 °C, and in winter minimum water temperature dropped to 1.6 °C (Table 8.23, Figure 8.27). Temperature variation follows the regional and seasonal conditions. Comparison of the monthly water temperature maxima of the Breitenbach and Erlenbach indicated high temporal similarity (Figure 8.28; Spearman 0.903; $P < 0.0001$).

Table 8.23 Mean, maximum and minimum monthly mean, maximum and minimum water temperature, and maximum and minimum air temperature at the Erlenbach collecting site from January 1998 to November 2005.

	Water (°C)			Air (°C)	
	Mean	Max	Min	Max	Min
Mean	8.0	9.2	7.0	19.0	2.4
Max	15.5	19.0	13.7	36.2	12.4
Min	2.0	2.3	1.6	0.3	–10.2

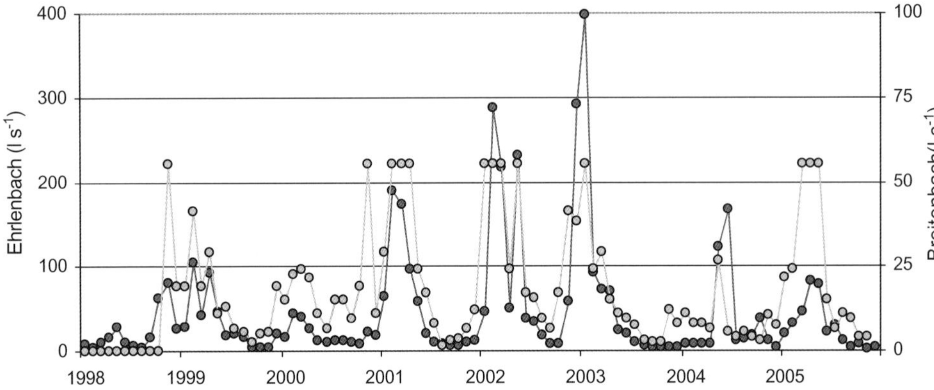

Figure 8.27 Monthly maximum discharge (l s^{-1}) in the Erlenbach (gray line with dots, left ordinate) and Breitenbach (black line, right ordinate) between January 1999 and June 2005.

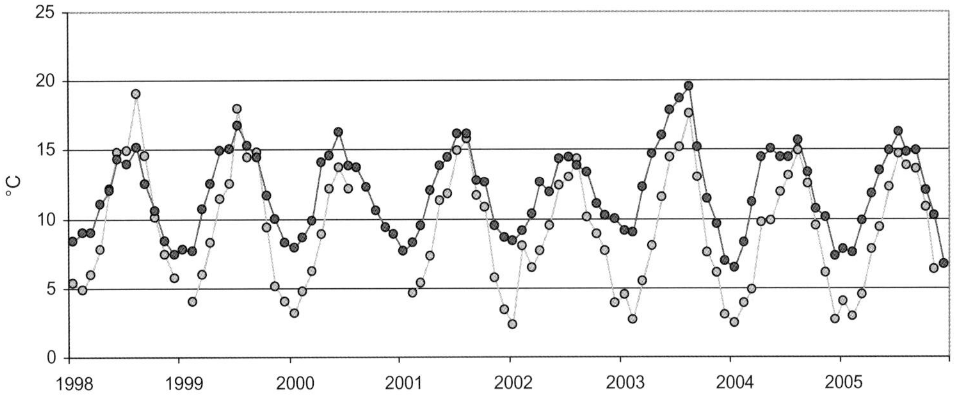

Figure 8.28 Variation in maximum monthly water temperature in the Erlenbach (gray) and the Breitenbach (black) in 1998–2005. Data for the Erlenbach incomplete in 2000.

Water temperature and discharge were negatively correlated (Spearman –0.462; $P < 0.0001$) for the Erlenbach, but the coefficient was not significant for the Breitenbach over the same period, 1998–2005.

In summary, the basic variables, discharge and water temperature, indicate great similarity in the environmental conditions in both streams. This should affect the communities in several ways. In the majority of cases the same species occur in both streams, and emergence periods and abundance variation are similar in particular years.

8.5.2
Community

The community of the Erlenbach is very similar to that of the Breitenbach, Table 8.24 showing the major species of the EPT and Diptera communities. Ten mayfly species were collected on the EB, two (both Heptageniidae) only occur there and not in the Breitenbach. *Baetis rhodani* and *B. vernus* were the dominant mayflies in both streams. Two of the 15 stoneflies and 10 of a total of 36 caddisflies were only found in the EB. With the exception of *Nemoura sciurus* and *Chaetopteryx major*, the species restricted to the Erlenbach were quite rare and only collected occasionally in some years. The Sørensen similarity index for the EPT communities was high (QS = 89.7), indicating that the EPT communities of both streams are almost 90% similar.

The appearances of aquatic Neuropteroidea, and aquatic or water-dependent Diptera in both streams, are also shown in Table 8.24. It is evident that the greater number of species occurs in both streams, however there are some differences in abundance. Including the Diptera and Neuropteroidea, the Sørensen similarity index dropped to 73.1, indicating that the difference between the two stream communities was largely due to the Diptera. However, the similarity of the communities would presumably increase with increased collecting from the Erlenbach, because Diptera in particular often occur in low numbers and only in certain years.

Based on the above information, emergence periods and abundance variation can be compared from 1998 to 2005 for a number of species that are frequent on both streams.

Emergence periods were almost identical in any one year, but it was more interesting to determine whether the well documented inter-annual abundance changes were the same in both streams or were accidental. Thirty-three species of Ephemeroptera, Plecoptera, Trichoptera, Diptera, Neuroptera, and Megaloptera, which had sufficiently high abundances in both streams for nine years, were tested with the Wilcoxon rank test. For 22 species the comparisons were significantly similar, but for 11 taxa they were not (Table 8.25). Abundance data for four species in Figure 8.29 show how well pairs of lines, that is, yearly abundances, can match.

Generally the abundances of most species in both streams run in parallel; there were no examples of abundances running in opposite directions. Several comparisons of species abundances changes (Table 8.25) that were significantly similar still show similar annual "increases" and "decreases". In particular, the "increases"

Table 8.24 Ephemeroptera, Plecoptera, Trichoptera, Neuropteroidea, and Diptera species of the Erlenbach community.

Ephemeroptera	**BTB**	**EB**	**Neuropteroidea**	**BTB**	**EB**	**Diptera**	**BTB**	**EB**
Baetidae leach			**Neuroptera, Megaloptera**			**Thaumaleidae**		
Baetis vernus	×	×	*Osmylus fulvicephalus*	×	×	*Thaumalea bezzii*	×	×
B. rhodani	×	×	*Sialis fuliginosa*	×	×	*T. cebennica*	×	×
Centroptilum luteolum	×	×	*S. lutaria*	×	×	*T. testacea*	×	×
Heptageniidae			*Sysira fuscata*		×			
Rhithrogena picteti		×						
Electrogena ujhelyii		×	Diptera			Dixidae		
E. rivuscellana		×	Ptychopteridae			*Dixa nebulosa*	×	×
Ephemerellidae			*Ptychoptera albimana*	×	×	*D. nubilipennis*	×	×
Ephemerella mucronata	×	×	*P. longicauda*		×	*D. puberula*	×	×
E. ignita	×	×	*P. paludosa*	×	×	*D. submaculata*	×	×
Leptophlebiidae			*P. contaminata*		×	*Dixella aestivalis*	×	×
Paraleptophlebia submarginata	×	×	*P. lacustris* Meigen, 1830		×	*Dix. autumnalis*		×
Ephemeridae						*Dix. obscura*	×	×
Ephemera danica	×	×	Psychodidae			*Dix. serotina*	×	
			Sycorax silacea	×	×			
Plecoptera			*Bazarella subneglecta*	×	×	Brachystomatidae		
Taeniopterygidae			*Berdeniella globulifera*		×	Subfamily trichopezinae		
Brachyptera risi	×	×	*B. illiesi*	×		*Heleodromia immaculata*	×	
Leuctridae			*B. manicata*	×		*Trichopeza longicornis*	×	×
Leuctra digitata	×	×	*B. unipsinosa*	×				
L. fusca	×	×	*Pericoma blandula*	×	×	Empididae		
L. nigra	×	×	*P. pseudoexquisita*	×		Subfamily hemerodromiinae		
Leuctra prima	×	×	*Satchelliella canescens*		×	*C. astigma*	×	
Nemouridae			*S. mutua*	×	×	*C. concinnicauda*	×	
Amphinemura standfussi	×	×	*S. nubila*	×		*C. diversicauda*	×	×
N. cambrica	×	×	*S. palustris*	×	×	*C. flavella*	×	
N. cinerea	×	×	*S. pilularia*		×	*C. precabunda*	×	×
N. flexuosa	×	×	*S. stammeri*	×	×	*C. precatoria*	×	×

N. marginata	×	×	*S. trivialis*	×	×	*C. pyrenaica*	×	×
N. sciurus		×	*Clytocerus ocellaris*	×	×	*C. stigmatica*	×	
N. picteti	×	×	*C. tetracorniculatus*		×	*C. subangusta*	×	
Protonemura auberti	×	×	*Tonnoiriella nigricauda*	×		*C. trapezina*	×	×
P. intricata	×	×	*T. pulchra*	×		*Chelifera* sp.	×	
Perlodidae			*Ulomyia cognata*	×	×	*Hemerodromia adulatoria*	×	
Isoperla grammatica		×	*U. fuliginosa*	×	×	*H. oratoria*	×	
I. goertzi	×		*Jungiella danica*	×	×	*H. unilineata*	×	×
			J. hassiaca	×	×	*Chelipoda melanocephala*	×	×
Trichopetra			*J. interna*	×		*C. vocatoria*	×	×
Rhyacophilidae			*J. soleata*	×		*C. albiseta*		×
Rhyacophila fasciata	×	×	*Panimerus albifacies*	×				
Glossosomatidae			*Paramormia polyascoidea*	×		Subfamily Clinocerinae		
Agapetus fuscipes	×	×	*Par. ustulata*	×		*Clinocera fontinalis*	×	×
Philopotamidae			*Philosepedon humeralis*	×	×	*C. stagnalis*	×	×
Wormaldia occipitalis	×	×	*Mormia albicornis*	×		*C. wesmaeli*	×	×
Psychomyiidae			*M. caliginosa*	×		*Kowarzia bipunctata*	×	
Lype reducta	×	×	*M. eatoni*	×		*K. plectrum*	×	
Polycentropodidae			*M. revisenda*	×		*K. madicola*	×	×
Plectrocnemia conspersa	×	×	*M. vaillanti*	×		*Wiedemannia bistigma*	×	
Polycentropus flavomaculatus		×	*Telmatoscopus longicornis*		×	*W. bohemani*	×	×
Hydropsychidae			*T. schlitzensis*	×	×	*W. lamellata*	×	
Hydropsyche siltalai	×	×	*T. ellisi*	×		*W. zetterstedti*	×	×
H. guttata		×	*T. havelkai*	×		*Dolichocephala guttata*	×	
Lepidostomatidae			*T. labeculosus*	×		*D. irrorata*	×	×
Lepidostoma basale	×	×	*T. pseudolongicornis*	×		*D. oblongoguttata*	×	×
Limnephilidae			*T. rothschildi*	×		*D. ocellata*	×	
Drusus annulatus	×	×	*Peripsychoda auriculata*	×	×	*D. thomasi*	×	
Ironoquia dubia		×	*Threticus lucifugus*	×				
Anabolia nervosa	×	×	*Psychoda albipennis*	×	×	Rhagionidae		
Glyphotaelius pellucidus	×	×	*P. brevicornis*	×	×	*Chrysopilus aff. flaveolus*	×	
Limnephilus auricular	×	×	*P. cinerea*	×	×	*C. erythrophthalmus*	×	×
L. centralis	×	×	*P. crassipennis*	×		*C. cristatus*	×	
L. fuscicornis		×	*P. erminea*	×	×	*C. asiliformis*	×	

Continued

Table 8.24 *Continued*

Ephemeroptera	BTB	EB	Neuropteroidea	BTB	EB	Diptera	BTB	EB
Baetidae leach			**Neuroptera, Megaloptera**			**Thaumaleidae**		
L. ignavus		×	*P. gemina*	×	×	*C. auratus*		×
L. sparsus	×	×	*P. grisescens*	×	×			
Enoicyla pusilla	×	×	*P. lobata*	×	×	Stratiomyidae		
Halesus digitatus	×	×	*P. minuta*	×	×	*Beris chalybata*	×	×
Micropterna lateralis	×	×	*P. mycophila*	×		*B. clavipes*	×	×
M. sequax	×	×	*P. parthenogenetica*	×	×	*B. strobli*	×	×
Potamophylax cingulatus	×	×	*P. phalaenoides*	×	×	*B. vallata*	×	x
P. latipennis		×	*P. pusilla*	×		*Oxycera pardalina*	×	
P. luctuosus	×	×	*P. setigera*	×	×			
P. nigricorni	×	×	*P. surcoufi*	×		Tabanidae		
Hydatophylax infumatus		×	*P. trinodulosa*	×	×	*Chrysops caecutiens*	x	x
Annitella obscurata	×	×	*Tinearia alternate*		×	*C. relictus*	x	
Chaetopteryx major		×	*Tinearia lativentris*	×		*C. viduatus*	x	
Chaetopteryx villosa	×	×				*Haematopota pluvialis*	×	×
Goeridae						*Hybomitra mühlfeldi*	×	×
Silo pallipes	×	×				*Tabanus bromius*	×	×
Leptoceridae						*T. sudeticus*	×	×
Adicella reducta	×	×						
Oecetis testacea		×						
Sericostomatidae								
Sericostoma personatum	×	×						
Beraeidae								
Beraea pullata	×	×						
Beraeodes minutus		×						

Table 8.25 Wilcoxon Signed rank test for parallel abundance variation in EPTD species in emergence traps on the Erlenbach and Breitenbach (1998–2005).

EPTD species	Z	$
Baetis vernus	−2.67	0.01
B. rhodani	−2.67	0.01
Paraleptophlebia submarginata	−2.67	0.01
Centroptilum luteolum	−2.52	0.01
Leuctra prima	−2.55	0.01
L. nigra	−2.67	0.01
Nemoura marginata	−2.55	0.01
N. cambrica	−1.72	0.09
Protonemura auberti	−2.67	0.01
Amphinemura standfussi	−0.47	0.64
Brachyptera risi	−2.67	0.01
Sericostoma personatum	−2.67	0.01
Agapetus fuscipes	−1.24	0.21
Plectrocnemia conspersa	−2.67	0.01
Potamophylax cingulatus	−1.95	0.05
P. luctuosus	−2.67	0.01
Chaetopteryx villosa	−2.67	0.01
Ptychoptera paludosa	−1.36	0.17
Tipula maxima	−1.13	0.26
Osmylus fulvicephalus	−0.77	0.44
Sialis fuliginosa	−2.08	0.04
Satchelliella trivialis	−2.55	0.01
Jungiella hassiaca	−1.13	0.26
Clytocersu ocellaris	−1.84	0.07
Dixa submaculata	−1.95	0.05
Anepsiomyia flaviventris	−2.55	0.01
Campsicnemus curvipes	−0.12	0.91
Dolichopus nigricornis	−1.24	0.21
D. popularis	−2.67	0.01
D. ungulatus	−2.19	0.03
Hercostomus brevicornis	−2.55	0.01
Hypophyllus obscurellus	−2.67	0.01
Rhaphium riparium	−0.06	0.95

are just abundance overshoots at one site, which probably depend on very local (small scale) conditions (accidental). In no case do such very local and temporal events seem to modify the longer-term abundance trends.

In summary, the variability of the main environmental variables is parallel by variability in the major members of both communities. Similar environmental and geological conditions predetermine the communities of both streams in a similar but not identical manner.

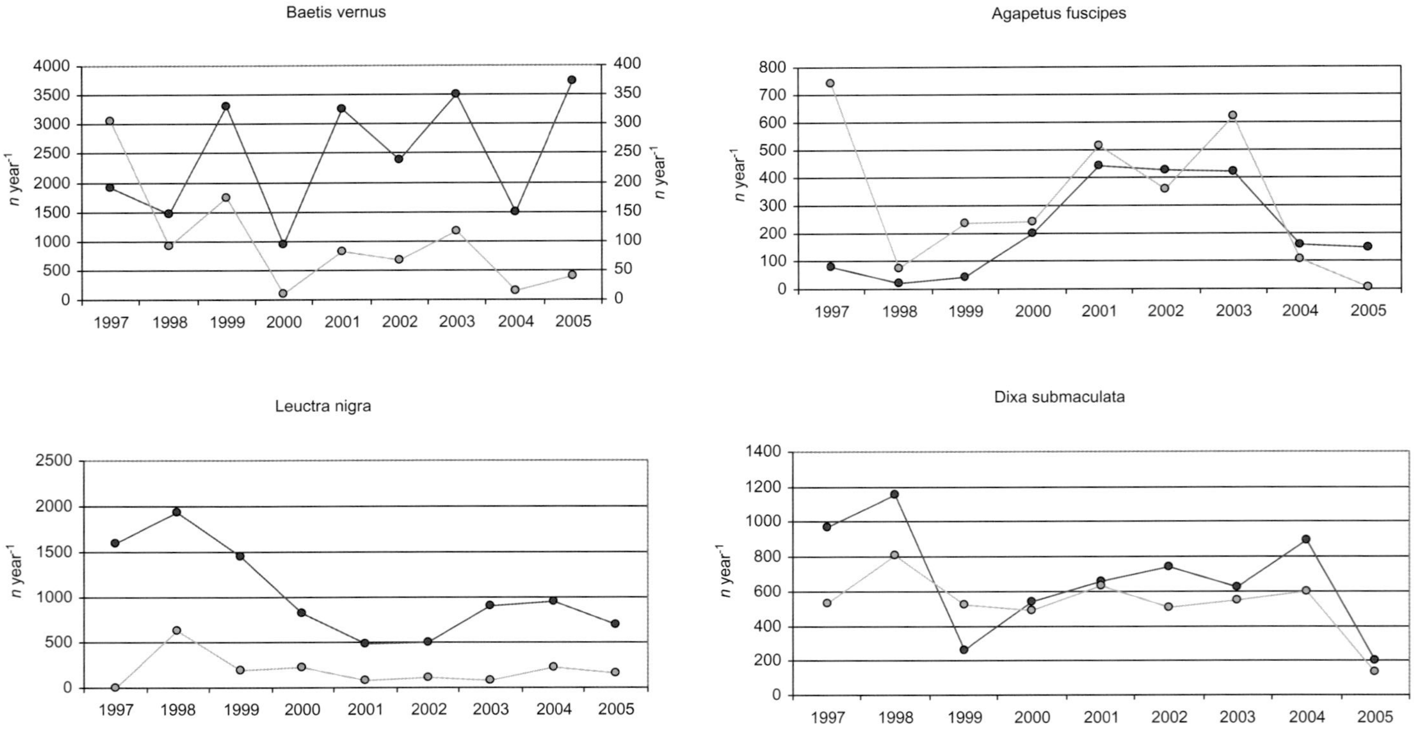

Figure 8.29 Abundance (specimens per year) of four EPTD species in the Erlenbach and Breitenbach (1997–2005).

8.5.3 Conclusion

Since the start of collections of emerging insects by the Limnologische Flussstation (Illies, 1971), repeated attempts have been made to compare different streams or springs, not only around Schlitz (e.g., Gümbel, 1976; Sandrock, 1978), but also in Austria (e.g., Malicky 1976, 1989). At the same time several studies dealing with individual taxa were published (e.g., Jacob, 1986; Malicky, 1976; Mey, 1992), and the suitability of emergence traps for the study of along stream variability was also discussed (Illies, 1978, 1982). However, most of these studies failed to meet their expectations because of weaknesses in the collecting and recording methods and because the choice of collecting sites was often pragmatic (e.g., accessibility of sites, taxonomical preferences). Finally, because these studies were based on production biology theory, and environmental variability dependencies of the communities on the environment were neither fully considered nor fully understood, an unsorted variety of individual results resulted. Malicky (1989) concluded that the emergence method cannot provide "quantitative area-related data", however its value for the study of community structure and dynamics and other ecological studies remained unaffected.

Thus, for the present study, the selection of a comparative stream to the Breitenbach was primarily based on the similarity of the taxonomic community, and of the environmental conditions. The dependency of the communities on the environment should then be very similar in both streams. Regional proximity, similar geology, climate, and land use (extensively used meadowland) were additional prerequisites. The comparable temporal characteristics of the environmental variables and community parameters resulted in the selection of the Erlenbach for comparison with the Breitenbach.

Our expectations that all the major factors (environmental variables, community, seasonal variation in factors and species, and abundance variation over time) act in a similar way were fulfilled; they were quite similar, but not identical. Seasonal changes in the major variables affecting the community, discharge and water temperature, were highly comparable. The substantial similarity in seasonal variation, as well as changes over several years, is essential to the development of similar communities, life cycles constraints and emergence periods of aquatic insects, and for parallel abundance trends of at least the dominant taxa in particular years.

The high similarity of communities in both streams is clear, but the Erlenbach was chosen to have a community that was as similar as possible to the Breitenbach. Although taxon abundance was not identical, trends in abundance variation were the same over the nine-year period. The statistical tests used to quantify temporal trends did not take into account the fact that abundance overshoots may be related to very local circumstances. However, the latter do not change the temporal trends, that is, turn an "increase" into a "decrease".

The Breitenbach data revealed water temperature and discharge to be the main factors affecting the stream and, in turn, its individuals, species and community.

They act in the same way and with similar results in the Erlenbach. With the experience from the long-term study on the Breitenbach we can conclude that the study of the structure, function, and dynamics of both communities and ecosystems requires long-term investigation, where long-term means decades, not simply years.

8.6 Modeling Discharge and Insect Abundance with Artificial Neural Networks (ANNs)

Michael Obach and Rüdiger Wagner

The use of empirical statistical and analytical methods for ecosystem analysis and prediction are often limited by spatially complex and temporally dynamic processes. Not only the physical and chemical properties of running waters and their effects on communities are driven by numerous environmental variables, but the underlying interactions and dependencies are only partially understood. Because, for example, the knowledge of species–habitat interrelations is still inadequate, an integrative and predictive assessment of ecosystem properties remains difficult, and typically data on the interrelations between variables and species do not follow normal distributions (e.g., Statzner *et al.*, 1994; Townsend and Hildrew, 1994; Townsend, 1989; Vannote *et al.*, 1980).

Mathematical methods, such as artificial neural networks (ANNs), provide attractive alternative tools to analyze and model ecological data due to their specific features, such as nonlinearity, adaptivity, the ability to generalize, and model independence (no *a priori* model needed). Because ANNs learn from examples, representative (for a given problem, site or study period) and compatible data are the main requirement. We repeatedly applied ANNs to various fields of aquatic sciences and engineering, such as modeling water quality and relating community characteristics to environmental variables, and to the long-term Breitenbach data to elucidate the effect of water temperature, precipitation and discharge on the annual abundances of emerging Ephemeroptera, Plecoptera, and Trichoptera (Obach, 2003; Obach *et al.*, 2001; Schleiter *et al.*, 1999, 2003; Wagner and Schmidt, 2004; Werner and Obach, 2001).

Here we provide examples of precipitation–discharge relationships, and the prediction (from a set of environmental variables) of monthly and annual abundances of emerging aquatic insects (EPT). In streams, water temperature, precipitation, and discharge are generally the most relevant variables affecting species abundance and community composition. Water temperature effects are often discussed in the context of environmental change, but they are less eye-catching and persist for longer periods than the often dramatic, short-term effects of floods that drastically change (disturb) the physical environment (stream bottom sediments).

Model quality was described by the root mean squared error (RMSE) of [0, 1] scaled test data, which were not used for building the models or to adapt its parameters

$$E_{RMS} = \sqrt{\frac{1}{n}\sum_{i=1}^{n}[y_i - \hat{y}(x_i)]^2}$$

and the determination coefficient

$$D = 1 - \frac{\sum_{i=1}^{n}[y_i - \hat{y}(x_i)]^2}{\sum_{i=1}^{n}(y_i - \bar{y})^2},$$

with observed values y_i, predictions $\hat{y}(x_i)$ and the mean value $\bar{y}$.

8.6.1
Precipitation – Discharge

Precipitation–discharge models for the Breitenbach and its drainage area are important for a better understanding of the dependencies of discharge on weather and climate, for numerical and qualitative simulations that use these models, for estimating missing values of runoff data, and finally for protecting valuable measuring equipment adjacent to, or within the stream from the risk of inundation.

The remainder of this subsection is based on the results of Obach (2003). ANNs were successfully applied for runoff predictions by Zhu *et al.* (1994) and others. The models prepared for the Breitenbach used daily precipitation data from the weather observation stations (Michelsrombach, Schlitz) close to the drainage area (Chapter 2). Average daily air temperature and discharge were measured at Breitenbach T2. Discharge measures of more than $120\,\mathrm{l\,s^{-1}}$ were mapped onto this maximum value, to reduce the weight given to extreme flood events. All values were linearly scaled onto the interval [0,1]. Learning data were from 1 January 1990 to 31 May 1998, and measurements from 1 June 1998 to 31 December 2000 were used to test the model.

The most important of the 20 predictors were precipitation sums over days 0–7, 0–14, 14–28 (Schlitz), day 0–28 (Michelsrombach), and air temperature (T2) over days 0–28, 0–80 and 80–160 before the target day (Obach, 2003). The first prediction model for the daily mean discharge was independent of previous discharge and consisted of a combination of a radial basis function (RBF) network and a self organizing map (Obach, 2003). Cross validations on the training data with general regression neural networks (see, e.g., Specht, 1991) found seven relevant predictors. Some outputs of this network, with two varying variables and the remaining five kept constant, are shown in Figure 8.30.

A comparison of predicted and measured daily discharge values is presented in Figure 8.31. Root mean squared error based on [0,1] scaled test data was $E = 0.1057$. The determination coefficient was $D = 0.3439$.

In addition to the above predictors, a second ANN-based model to predict daily discharge of the Breitenbach included discharge values of previous days and weeks. Its determination coefficient for the test data was $D = 0.92$, much higher

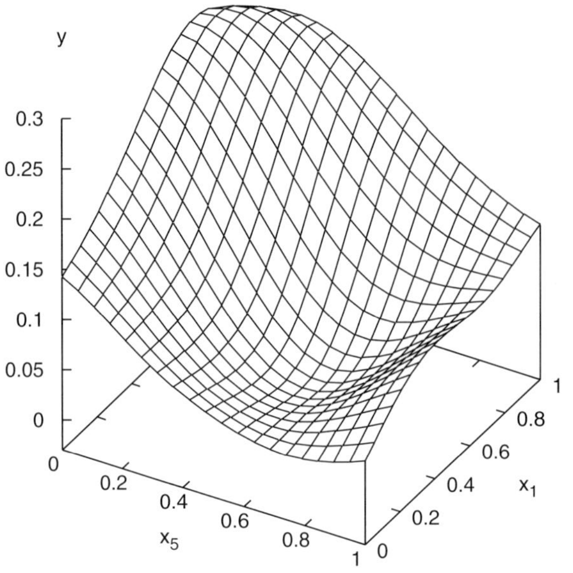

Figure 8.30 Three-dimensional illustration of the outputs y of an RBFSOM network predicting daily mean discharge values of the Breitenbach based on varying precipitation over the preceding week ([0,1] scaled values, x_1) and mean air temperature over the preceding four weeks (x_5); the other five input variables were set constant to their values on 15 September 1999.

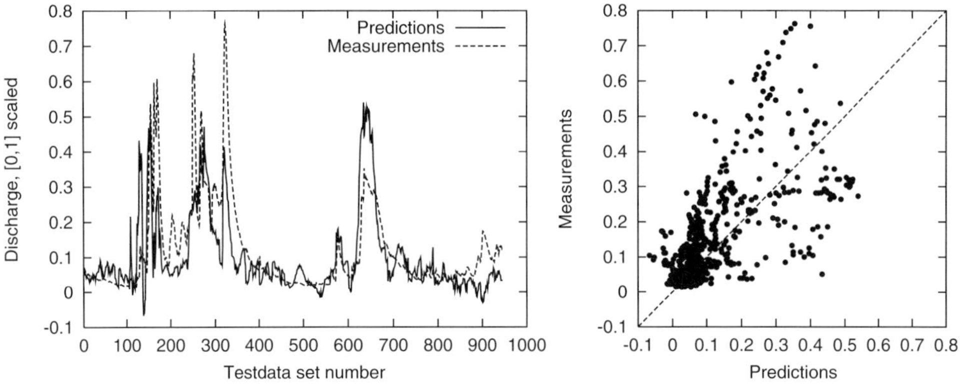

Figure 8.31 Comparison of measured and predicted values of the mean daily discharge of the Breitenbach based on seven predictors, excluding discharge of the preceding day, using an RBFSOM network with linear cells for the predictions.

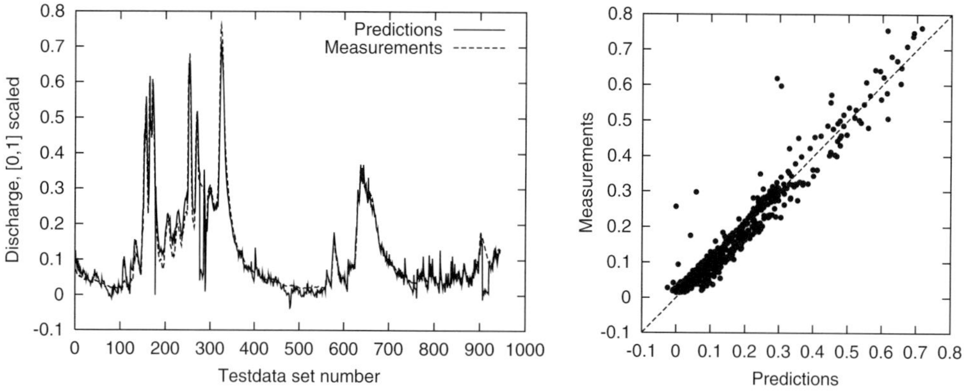

Figure 8.32 Comparison of measured and predicted values of the mean daily discharge of the Breitenbach based on 20 predictors, including discharge of the preceding day, using an RBFSOM network with linear cells for the predictions.

than the above model. Figure 8.32 shows a comparison of the measured and predicted values for the test data. This shows that, in principle, discharge can be adequately predicted from weather and climatic data alone, but the inclusion of available past discharge data significantly increased the quality of the model.

In addition to increasing our understanding of discharge dependencies on climate and weather, runoff models provide an opportunity roughly to estimate the composition of stream bottom substrata. So it is at least possible to distinguish sedimentation periods from erosion periods. This information, particularly their time and duration is important for understanding why, based on ecological requirements, some species are successful and become abundant in some periods but not in others. Without doubt, a combination of local weather and runoff can also be useful for forecasting the possible effects of climatic changes on local and regional water resources.

8.6.2
Prognosis of Yearly Abundance of 17 EPT Species

To predict the annual abundance of EPT species, monthly species abundance, precipitation, maximum temperature and discharge data, from April 1969 to December 1992, were arranged as 273 52-dimensional vectors to apply the sliding window technique (month after month). The modeling process comprised five steps: (i) dimension reduction during pre-processing, (ii) data scaling [0,1], (iii) testing the generalization ability and relevance of the selected parameters, (iv) prediction, and (v) testing the predictions. Three methods for dimension reduction (pre-selection of the five most relevant predictors by regression, neural sensitivity analysis, stepwise linear regression) or the total information of the 52-dimensional vectors were used (for further details, see Dapper, 1998; Wagner *et al.*, 2006). Of these data, 80% were used for ANN training, 20% for verification. Dimension

Table 8.26 Overview of predictions of the monthly abundances of aquatic insects in emergence traps with sliding windows, best determination coefficient (bold). All = all information of environmental variables and abundance of the parent generation included; regres = five "most relevant" predictors selected by regression analysis; sensi = five "most relevant" predictors selected by sensitivity analysis; linear = linear regression. TG = Taxonomic group: E = Ephemeroptera, P = Plecoptera, T = Trichoptera.

Species	TG	Determination coefficient			
		All	Regres	Sensi	Linear
Baetis rhodani	E	**0.56**	0.40	0.55	0.40
B. vernus	E	0.54	0.63	**0.71**	0.54
Leuctra prima	P	0.65	0.50	**0.74**	0.44
L. nigra	P	**0.94**	0.65	0.32	0.51
L. digitata	P	0.67	**0.91**	0.89	0.48
Protonemura auberti	P	0.32	0.38	0.43	**0.48**
P. intricata	P	0.21	0.45	0.30	**0.48**
P. meyeri	P	0.87	**0.91**	**0.91**	0.34
Amphinemura standfussi	P	0.86	0.91	**0.93**	0.45
Isoperla goertzi	P	0.54	**0.75**	0.59	0.53
Siphonoperla torrentium	P	0.34	**0.57**	0.05	0.38
Apatania fimbriata	T	0.63	0.86	**0.93**	0.86
Chaetopteryx villosa	T	0.35	**0.66**	0.64	**0.66**
Drusus annulatus	T	0.24	0.49	**0.56**	0.08
Rhycophila fasciata	T	0.58	**0.70**	0.67	0.39
Tinodes rostocki	T	0.45	**0.57**	0.31	0.54
Agapetus fuscipes	T	**0.56**	0.37	0.44	0.34

reduction was tested to determine how reduction of data quantity affects model quality.

As expected, the quality of abundance prediction varied among species, among ANN types, and between ANNs and the linear regression model (Table 8.26). The determination coefficient *D* of the best abundance prediction models was generally high. For five species it was >0.9, for nine >0.7 and for 15 species >0.5. Only for the stoneflies, *P. intricata* and *P. auberti,* was *D* just below 0.5. In general, linear regression models had the lowest *D*. The difference between linear regression and the best ANN was below 0.2 for most species; it was less than 0.05 in *P. intricata, C. villosa,* and *T. rostocki.* Exceptionally, the linear models for *P. auberti* and *P. intricata* were better than the ANN models. Various variable scaling options did not as a rule improve the model. Dimension reduction, to reduce computation effort without reduction of model quality, was in most cases appropriate, particularly during ANN training. The resulting models were easy to analyze, ecologically interpretable, and allowed recognition of the relevant predictors (Dapper, 1998; Wagner *et al.*, 2006).

Pre-selection of the best five variables often improved *D* for the ANNs. Models with dimension reduction by regression or sensitivity analysis gave similar quality results (i.e., difference of $D \leq 0.05$). Clear differences occurred for *L. prima* and *P. auberti* (pre-selection by sensitivity analysis), and for *I. goertzi, P. intricata, S. torrentium*, and *T. rostocki* (pre-selection by regression). For three species, all available information was required to compute a model of reasonable quality. For the stonefly, *L. nigra*, the use of all 51 predictors dramatically increased D (by 0.29 or even 0.62) compared with pre-selection of five variables. For *B. rhodani* and *A. fuscipes* improvement in model quality was less dramatic (approx. 0.1).

The need for extensive data for *L. nigra* is related to its ecological demands, its particular life cycle attributes and sediment requirements of the developing larvae. The latter live on and in the temporally least stable, stream bottom substratum, that is, sandy sediments. They are disturbed by almost every variation in discharge throughout the year, because the small sand particles are quickly displaced or deposited by even slight changes in discharge. Furthermore, the population falls into two cohorts, with a one- or two-year life cycle, and larvae are therefore present in the stream at all times of the year (Chapter 7). Enhanced ecological flexibility of a population in a temporally variable and permanently susceptible microhabitat is probably the chief reason that more extensive information is needed. Similar considerations apply to *B. rhodani*, a bivoltine mayfly with a permanent larval population presence in the stream, and to the caddisfly, *A. fuscipes*. However, larvae of both these species prefer stony substrata.

8.6.3
Seasonal Effects of Variables on Species in Models

Because the life cycles of the 17 species start in different months, the same number for a variable denotes different months in different models. To determine whether species models are affected by variables at the same time of the year or not, the variable numbers were synchronized so that individual months were unambiguously identified. February, March, April, May, and October were the months with the highest entries of occurrences in the models (Figure 8.33).

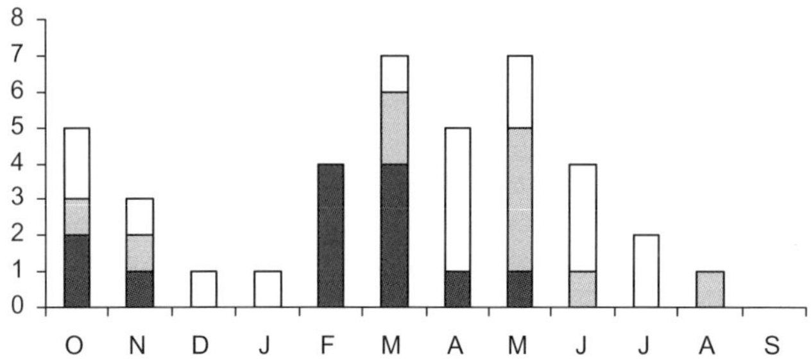

Figure 8.33 Monthly numbers of species models affected by environmental variables: discharge (black), precipitation (gray), or temperature (white) in 17 EPT species in the Breitenbach (T2 data) for the period 1969–1993.

Abundance was a predictor 47 times. Abundance of the parent generation (months 1–3) was relevant for every species ($n = 21$), but that of the month preceding the target month ($n = 23$) was not discussed, because it just describes a "slope" towards abundance in the target month.

Environmental variables were identified as relevant predictors 40 times, most in March–May, least in August, September, and November–January (Figure 8.33). Temperature was relevant in all months, except February, August, and September. Precipitation and discharge were predominantly relevant in February and March, but also in April, May, October, and November. They had little effect in December, January, and the period June–September.

Precipitation or discharge affected the models for all Plecoptera (exception *L. nigra*), *Baetis vernus*, and *Tinodes rostocki*. The effects on *Baetis vernus* (April) and *I. goertzi* (May) occurred just before the start of emergence. No discharge effects were detected on *Baetis rhodani*, *Leuctra nigra*, *Drusus annulatus*, and *Agapetus fuscipes*.

From July to September temperature or precipitation were abundance predictors for *D. annulatus* and *C. villosa* (Trichoptera) which emerge until, or from, September onwards. The *A. fimbriata* model is affected by October temperature and discharge, readily interpreted by the onset of larval winter-dormancy underneath stones. The model that emerges for *L. prima* from January onwards shows discharge effect in October and a temperature effect in December. For *R. fasciata* the model indicates temperature effects in October and discharge effects in November. *Baetis rhodani* and *L. nigra* abundances were affected by temperature in November and January respectively, but no temperature effect was seen in the models for all *Protonemura* species, *I. goertzi* and *B. vernus*.

8.6.4
Effects of Discharge Parameters on Predictions of Monthly Abundances of *Baetis rhodani*

Because discharge was found to be the best abundance predictor for *B. rhodani*, we compared models including the monthly maximum or minimum discharge, the long-term mean pattern, and deviance from the long-term pattern. The long-term mean was the best predictor. Deviation of discharge from the long-term mean was the second best model (Table 8.27). Using maximum discharge data produced the poorest model, while the minimum discharge model only increased the determination coefficient slightly.

It was surprising that *D* was lowest for models based on minimum or maximum discharge;, maximum discharge should have been an appropriate measure of disturbance. The second best model, deviation of actual from long-term discharge, is however interpreted similarly. Additional variables (discharge, water temperature, precipitation) increased the model quality (*D*). This indicates a temporal link between the recurrent sequence of seasonal variables and the life cycle of *B. rhodani*, with negative effects on populations due to deviations from the long-term pattern. It also shows that, rather than single events, a number of variables

Table 8.27 Determination coefficient (D) of models to predict *B. rhodani* abundance with different discharge parameters; ltm = long-term mean.

Year	1972	1974	1976	1978	1980	1982	1984	1986	1988	1990	1992
D	0.65	0.97	0.00	0.96	0.30	0.51	0.64	–	0.95	0.88	0.81

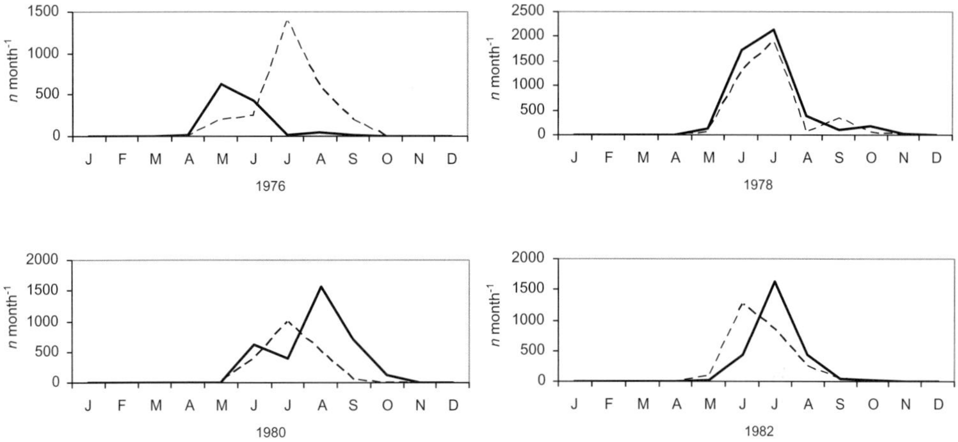

Figure 8.34 Good and bad quality of prognoses of yearly abundance patterns (12 consecutive months) for *Baetis vernus* in trap T2 (model–broken line; actual emergence–full line); for respective D compare Table 8.3.

and their temporal sequence during the growth period of a population "determine" the success of aquatic insect populations in streams.

8.6.5 Predictions of Yearly Abundance Patterns

Using *Baetis vernus* as an example, the quality of the abundance pattern predictions varied, with some good and some poor results in particular years (13 years training data, 11 years for verification; Table 8.27). A poor prediction occurred if the magnitude of the abundance was predicted inaccurately and if the maximum abundance month was missed for more than one month (Figure 8.34: 1976, 1980). In the years with best results for abundance magnitude the monthly abundance course was accurately predicted. For all years, mean *D* was 0.52; in five out of ten years *D* was greater than 0.8, and below 0.5 in only two years. Best predictions with highest *D* occurred in years that, based on "discharge patterns", were "nonseasonal" (1972, 1974, 1978) or with "spring floods" (1988, 1990). Predictions were worst in dry years (1976) and in years with low variation in flow (1980, 1982, 1984) (Table 8.28). These results fit the known characteristics of the species that, based on its long egg

Table 8.28 Determination coefficient (D) for the prognosis of monthly emergence patterns for the mayfly *B. vernus* over 11 years (no coefficient calculated for 1986 when the community was disturbed by insecticide input).

***B. rhodani* abundance**	***D***
Maximum discharge	0.11
Minimum discharge	0.18
ltm discharge	0.37
Difference from ltm discharge	0.28

diapause (Schmidt, 1984), is usually very successful, and more abundant in years with a "nonseasonal" flow pattern. Eggs are stuck to the underside of large stones and thus are less susceptible to disturbance in streams than the mobile larvae of most other species. As a result, a model that pre-selects the five most relevant variables provided sufficient information for this particular species.

8.6.6 Abundance Predictions for the Caddisfly *Apatania fimbriata* at Four Sites along the Stream

For further tests of ANNs generalization ability, data on the caddisfly *Apatania fimbriata* were chosen because, over 25 years, they had the best predictions at T2 (Table 8.26). For the period 1983–1993, respective models were generated for T1, T2, T4, T6 and later applied to data from the other sites.

Best generalization was achieved for T1 with high abundance. At sites with "intermediate" abundances (T2, T4) *D* was lower, abundance variation between years was highest at T4, and accompanied by the lowest *D*. The high *D* at T6 is related to the low abundance there, in fact a simple presence–absence prediction.

As data from T2 were part of the 25-year data set, both 11-year and 25-year models were compared. Narrower variability of the data over the shorter period resulted in increased *D* in the 11-year data ($D = 0.93$ versus $D = 0.71$). With all parameters the determination coefficient was 0.63, with five regressors 0.86 and, for the selection of five sensitive parameters, 0.93. This is because over the 11-year period there were no years with nonseasonal flow patterns, which usually increase the among-years abundance variability and so decrease the determination coefficient *D*.

With the same procedures results for another caddis fly, *C. villosa*, were similar; highest *D* was detected at sites with high or low abundance (presence–absence). Compared with the 25-year data the best *D* increased from 0.66 to 0.88 for the 11-year period for the same reasons as for *A. fimbriata*.

In both cases, when models developed for one site were transferred to another site (e.g., model developed for T2 used with data from T1) the generalization ability of the ANN was limited. In every case, *D* decreased drastically. The inclusion of additional data (spatial and seasonal water temperature per site) corrupted the models, because the additional spatial, "along stream" dimension was not adequately trained. Models were therefore extrapolations (outside the trained space) rather than generalizations.

8.6.7 General Aspects

Several ecological theories deal with the effects of environmental variables, especially flooding, on individual species or stream communities. The "intermediate disturbance hypothesis" (e.g., Connell, 1978; Ward and Stanford, 1983) predicts highest species richness and diversity at intermediate disturbance intensity and frequency, with predictable changes in species and species traits. The "habitat templet concept" (e.g., Southwood, 1977; Minshall, 1988) expects highest diversity and species richness with increasingly random and unpredicatable discharge because communities consist mainly of taxa that are adapted to extreme environmental conditions. The "patch dynamics concept" (e.g., Townsend, 1989) describes the effects of discharge on sediment texture and macroinvertebrates that may resettle newly developed ecological niches, and the "harsh–benign concept" (Peckarsky, 1983) emphasizes both large environmental variation and variation in the relative importance of abiotic and biotic interactions. Low flow conditions do not attract much interest compared to eye-catching floods, because changes in sediment composition are slow and seem less dramatic. However, sedimentation of fine material may cover gravel and stones, and, over months or even years, this will affect habitat and food availability, at for least grazers, that is, it acts as a perpetual disturbance. No single concept adequately describes all communities (Townsend, 1989), but all try to understand and predict communities and their internal structure. In addition, theoretical predictions were not supported by analyzed data, leading to debate over whether the underlying theory is wrong, or the tests fair (Resh *et al.*, 1994).

Of the mentioned hypotheses the "patch dynamics concept" (Townsend, 1989) and the "harsh–benign concept" (Peckarsky, 1983) seem to be more appropriate for the interpretation of our data and results. Both focus on the interdependency of populations and environmental variables, that is, the relationships of species to a patchy and temporally variable environment, particularly in streams. Our models at least indicate that the "major" environmental variables (precipitation–discharge, temperature) substantially affect these interdependencies, and are the main reasons for the observed inter-annual changes in aquatic insect community in the Breitenbach.

We can state that ANNs can be successfully applied to the analysis of cause–effect relationships (e.g., species/species traits with habitat), including the identification and assessment of complex impact factors, and for modeling long-term population dynamics. They have particular advantages if the relationships are unknown, complex or nonlinear. Combined with specific procedures for the selection of the most important impact variables, they account reasonably for the reduction of model complexity. Generalization performance and model simplification increase and permit better understanding of the underlying relations.

We expect that, with the aid of neural models and specific dimension-reducing, pre-processing methods, bioindication, ecological prediction, and the analysis of cause–effect relations can be substantially improved. Generally, modeling complex nonlinear relationships can be handled much better with ANNs and supplementary modeling techniques than with classical methods. The best results may be obtained with specific combinations of linear and nonlinear techniques.

8.7 Food Chains and Carbon Flow

Jürgen Marxsen

8.7.1 The Organic Matter Budget of the Breitenbach

Organic matter enters the Breitenbach from multiple external and internal sources. Photoautotrophic and chemoautotrophic primary production are internal sources, although the latter is quantitatively unimportant in streams like the Breitenbach (Wuhrmann, 1974; Marxsen, 2006). External sources are dissolved and particulate organic matter (Table 8.29). Once within the stream's boundaries, organic matter is transformed from one state to another, respired, transported downstream to the River Fulda, or lost to the terrestrial environment. These dynamics are strongly pulsed, mainly depending on the growing season and hydrological events (Allan,

Table 8.29 Inputs of organic carbon to the Breitenbach, including primary production and allochthonous material (from Marxsen, 2006). Most organic carbon inputs to the Breitenbach are from dissolved organic substances via groundwater and spring water. However, autochthonous primary production and particulate organic carbon inputs from the terrestrial environment are also considerable.

Source	Value ($kg\,C\,m^{-2}\,year^{-1}$)	References
Gross primary production (algae and macrophytes)	0.40	Marxsen (1980a), Marxsen *et al.* (1997), Horvath (unpublished data)
Particulate organic carbon	0.42	Marxsen (1980a), Marxsen *et al.* (1997)
DOC by groundwater and springs	1.02	Fiebig (1995), Marxsen *et al.* (1997)
DOC from interflow and overland flow	?	
DOC from macrophytes	?	
Total	1.84	

Table 8.30 Outputs of organic carbon from the Breitenbach by transport, respiration and emergence (from Marxsen, 2006). The sum of heterotrophic respiration is 0.60 kg C m^{-2} $year^{-1}$, whereas total respiration is 0.80 kg C m^{-2} $year^{-1}$. Thus bacterial respiration accounts for 78% of heterotrophic and 59% of total respiration.

Source	Value (kg C m^{-2} $year^{-1}$)	References and notes
Particulate transport of organic carbon	0.34	Gerhardt (1986), Marxsen *et al.* (1997), Marxsen (2006)
Dissolved transport of organic carbon	0.74	Fiebig (1995), Marxsen *et al.* (1997), Marxsen (2006)
Autotrophic respiration	0.20	Marxsen *et al.* (1997)
Heterotrophic respiration: bacteria	0.47	After Marxsen (2001)
Heterotrophic respiration: fungi	0.04	Estimation based on Suberkropp (1991)
Heterotrophic respiration: protozoa	0.05	Estimation based on Nagata (1988) and Laybourn-Parry and Parry (2000)
Heterotrophic respiration: metazoa	0.04	Estimation based on Benke and Wallace (1980)
Emergence	0.002	Illies (1982)
Total output	1.88	

1995). The time scale of the different processes varies from minutes to months, or even longer. A substantial amount of organic matter may be buried in the sediments for tens or hundreds of years. However, in a small stream like the Breitenbach, with major storms in most years, such long-term storage is less probable. Hence on a time scale of a few decades, which is used for the average annual organic matter budget provided in this synopsis, the balance should be more or less even. Loss of organic matter from the Breitenbach mainly occurs via downstream transport to the River Fulda (dissolved and dead particulate organic matter as well as organisms) and as respiration (Table 8.30, Figure 8.35). Another ecologically significant path, although quantitatively not important, is the emergence of insects. The immigration of organisms from the River Fulda into the Breitenbach is quantitatively unimportant.

Based on all available input and output data determined between 1973 and 1998, the organic matter budget for the Breitenbach appears fairly balanced (Tables 8.29 and 8.30, Figure 8.35, Marxsen, 2006). The inputs of gross primary production, particulate organic matter and dissolved organic matter from groundwater amount to 1840 g C m^{-2} $year^{-1}$, whereas the total outputs from autotrophic and heterotrophic respiration and transport of dissolved and particulate organic materials to the River Fulda is 1880 g C m^{-2} $year^{-1}$. Thus, the OM output surpasses input by only 40 g C m^{-2} $year^{-1}$. However, there are some input fractions which were not measured. These

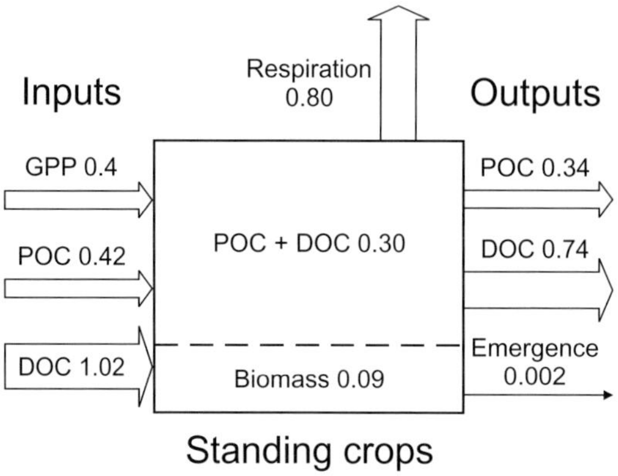

Figure 8.35 Annual carbon budget for the Breitenbach (from Marxsen, 2006, with permission of Wiley-Blackwell). The data are given in kg C m^{-2} or kg C m^{-2} $year^{-1}$, respectively. DOC = dissolved organic carbon, GPP = gross primary production, POC = particulate organic carbon.

are the liberation of DOC from macrophytes and algae, which can make up a considerable part of primary production (Kaplan and Bott, 1989; Allan, 1995), and DOC input from inter- and overland flow. This last source can contribute significant amounts of organic matter to streams (Thurman, 1985; Allan, 1995). The small difference determined for the Breitenbach organic matter budget can be regarded as more or less negligible because of the imprecision of many of the measurements and estimations, and the inaccuracies associated with the many measurements performed in different years. Compared to studies on organic matter budgets for other streams and rivers (Webster and Meyer, 1997a; see Table 8 in Marxsen, 2006), the budget is already markedly balanced, even without any adjustments. Although small differences between input and output measurements sometimes occur in such studies, much larger deviations from balance than obtained for the Breitenbach were often observed (Webster and Meyer, 1997c). Also, well balanced budgets often are the result of assumptions that undetermined fractions of the budget are equal to the missing amounts of matter.

The P/R ratio has been used as an important characteristic of lotic systems since the time of H.T. Odum (1957b). An ecosystem that respires the same amount of energy as it fixes by primary producers has a P/R ratio of 1. But in most streams P/R ratios were below 1 (see Table 8 in Marxsen, 2006). Thus the stream communities require the import of energy from the terrestrial environment to maintain their metabolism. From the budget data for the period 1973–1998 a P/R ratio of 0.50 was determined for the Breitenbach (Marxsen, 2006). However, from the community metabolism studies between 1997 and 2004, a lower value of 0.27 was obtained (Schmidt, unpublished data; see Section 5.3). Thus, at least half of the energy for the stream's metabolism must be acquired from terrestrial sources. This is in agreement with the input measurements, which show that about 70% of the determined organic matter inputs for the Breitenbach are from allochthonous sources. This is well within the range reported for other streams (Marxsen,

2006). Webster and Meyer (1997c) correlated the P/R ratios from many streams with mean annual discharge and stream order. The data for the Breitenbach are very close to their regression lines between discharge and P/R ratio, as well as between stream order and P/R ratio. This was also found for correlations between stream order or discharge and the ratio of NPP/total carbon input (NPP + POC + DOC), demonstrating that, with respect to carbon flow, the Breitenbach exhibits the typical characteristics of a first-order stream.

8.7.2 Bacteria in the Food Web System of the Breitenbach

The structure of the food web system in the Breitenbach was investigated in detail, mainly with respect to the potential role of bacteria as a food resource for higher trophic levels (Marxsen, 1988a). Enough evidence was available to consider switching from the traditional concept of bacteria as only mineralisers of nutrients fixed by primary producers to the additional concept of bacteria as a source of biomass for protozoan–metazoan food webs (Kemp, 1990; Wetzel, 2001; Marxsen, 2006). Most bacteria in streams, as in other benthic systems, are associated with particles (detritus, sediment grains). This allows not only protozoa and small metazoa, but also larger metazoa to consume bacteria (Hall and Meyer, 1998). There are therefore fewer trophic transfers between bacteria and top consumers in a stream system compared to lentic systems. Hence, the microbial food chain is a more important link here than in pelagic environments (Meyer, 1994).

Bacterial carbon production in the Breitenbach was determined to be about $200\,g\,C\,m^{-2}\,year^{-1}$ (Marxsen, 2001; Table 8.31). Assuming 30% growth efficiency

Table 8.31 Production by the different groups of heterotrophic organisms in the Breitenbach. Data from Marxsen (2006).

Source	Value ($kg\,C\,m^{-2}\,year^{-1}$)	References and notes
Bacteria	0.20	Marxsen (2001)
Fungi	0.02	Estimation based on Sinsabaugh and Findlay (1995) and Baldy *et al.* (2002)
Protozoa	0.02	Ciliates Packroff (1994) and Packroff and Zwick (1998); others: estimates based on Arndt *et al.* (2000)
Metazoa	0.04	For insects, estimation based on data from Benedetto (1975), Illies (1982), Wolf *et al.* (1997), Wolf and Zwick (2001); for amphipods, Pieper (1978); for microinvertebrates, estimates based on Borchardt and Bott (1995); for nematodes Christl (2008, personal communication)
Total	0.28	

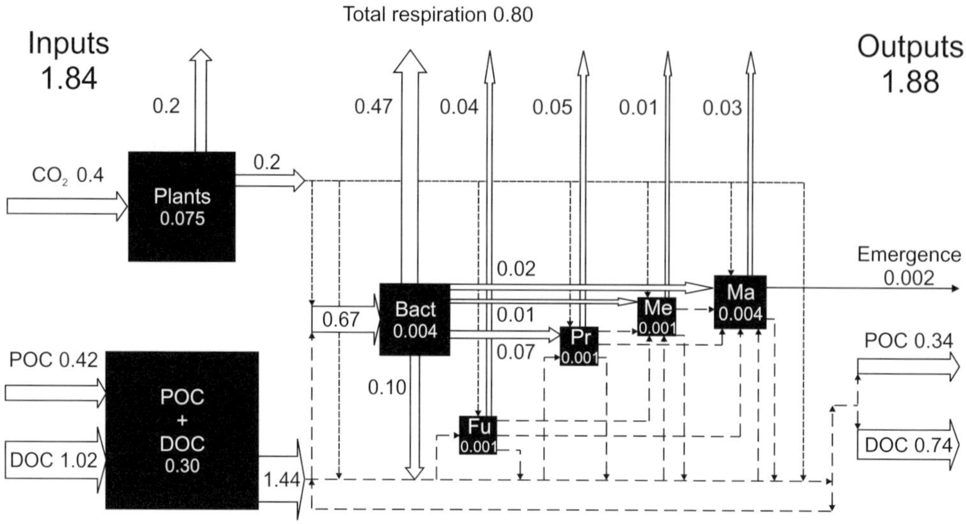

Figure 8.36 Diagram of organic carbon flow in the Breitenbach, focused on the role of bacteria (from Marxsen, 2006, with permission of Wiley-Blackwell). The data are calculated on an annual basis in $kg\,C\,m^{-2}$ or $kg\,C\,m^{-2}\,year^{-1}$, respectively. Black boxes symbolize the standing stocks, arrows the fluxes. When data for fluxes are available, solid arrows are used, if not the arrows are dashed. Unquantified fluxes originating from autochthonous primary production are symbolized by arrows with short dashed lines. The reverse arrow at the bottom of the figure indicates that excretion from organisms in the stream and dead organisms are considered within the POC/DOC flux. Bact = bacteria, Fu = fungi, Pr = protozoa, Me = meiofauna (microinvertebrates), Ma = macroinvertebrates, POC = particulate organic carbon, DOC = dissolved organic carbon.

for bacteria (del Giorgio and Cole, 1998), $670\,g\,C\,m^{-2}\,year^{-1}$ is required to sustain this amount of BCP (Figure 8.36). This is a little more than one-third (37%) of the total organic matter entering the stream (Table 8.29) and 80% of the amount that is not lost from the system by transport (Table 8.30).

Protozoan biomass production in the Breitenbach was estimated to be up to $20\,g\,C\,m^{-2}\,year^{-1}$, but perhaps lower (Marxsen, 2006; Table 8.31, Figure 8.36). If the losses from respiration are included (Table 8.30), about $70\,g\,C\,m^{-2}\,year^{-1}$ are needed to sustain these organisms. This means that up to one-third of the BCP might be consumed by protozoa (Figure 8.36). The question whether protozoa are significant "top down" controllers of bacterial abundance in the Breitenbach cannot be answered. However, from the current state of knowledge this only seems possible if metazoans are included. But there is no doubt that protozoa are the main consumers of bacteria. The bacteria–protozoa path is also confirmed by investigations in the Breitenbach to be an important trophic link between metazoans and carbon sources otherwise unavailable to these organisms (Schönborn, 1981, 1984; Carlough and Meyer, 1989; Bott and Kaplan, 1990; Meyer, 1994; Marxsen, 2006).

Meiofaunal (metazoa < 0.5 mm) biomass production for the Breitenbach was estimated to be about $10\,g\,C\,m^{-2}\,year^{-1}$ (Marxsen, 2006; Table 8.31, Figure 8.36). Twice that amount of carbon is necessary to sustain these organisms if respiration is included. The role of these organisms in stream food chains is remarkably understudied (Borchardt and Bott, 1995), although they undoubtedly play a significant role in the trophic link between bacteria and the macrobenthos (Traunspurger, 2000; Stead *et al.*, 2003). They utilize a wide range of food resources, including not only bacteria, but also algae, living animals, and detritus (Borchardt and Bott, 1995; Schmid and Schmid-Araya, 2002). Thus, on average less than three trophic transfers occur, if the complete food chain from DOC and POC via bacteria and protozoa is traveled. If less than $20\,g\,C\,m^{-2}\,year^{-1}$ is consumed by meiofauna in the Breitenbach, probably not more than 10 g bacterial C $m^{-2}\,year^{-1}$ is utilized via this pathway (Figure 8.36).

Most of the bacteria in streams are associated with particles. Thus macroinvertebrates are also able to ingest them in considerable amounts (Meyer, 1994). The very detailed studies on the Coweeta streams (GA, USA; e.g., Findlay *et al.*, 1986; Edwards and Meyer, 1987, 1990; Hall and Meyer, 1998; Hall *et al.*, 2000) were compared to data from the Breitenbach (Marxsen, 2006). Based on total macrofaunal production, including respiration of $60\,g\,C\,m^{-2}\,year^{-1}$ (Table 8.31) and 30% bacterial carbon as the organic matter resource (with the rest being derived from leaf tissue, amorphous detritus, primary production and animals; Hall and Meyer, 1998; Hall *et al.*, 2000), about 20 g bacterial $C\,m^{-2}\,year^{-1}$ are assimilated (Figure 8.36). This is about 10% of the total annual bacterial production. Although this is a considerable amount, for the macrofaunal community as a whole their carbon demand is met to a larger extent by dead POC than by bacterial biomass. However, there are still species that feed mainly on bacteria (Edwards and Meyer, 1987, 1990), and juvenile forms in particular might meet their carbon demand to a much greater extent from bacteria.

The animal community (protozoa and metazoa together) assimilates half the bacterial production of the Breitenbach (~$100\,g\,C\,m^{-2}\,year^{-1}$ from 200 g BCP $m^{-2}\,year^{-1}$; 70, 10, $20\,g\,C\,m^{-2}\,year^{-1}$ by protozoa, meiofauna, macrofauna, respectively). The fate of the other 50% is not yet known, although the main paths are obvious (Marxsen, 2006):

- A proportion is damaged through animal feeding activities without being assimilated, thus re-entering the pools of POC and DOC as dead biomass.
- A considerable amount will probably be lethally infected by viruses. No data are available on the role of viruses in streams, but it is reasonable to assume that they play no more minor a role in streams than they do in other aquatic environments (Rohwer *et al.*, 2009).
- Bacteria are also thought to undergo autolysis.
- Another important fate is export to the River Fulda. Such export is in the range 10–$50\,g\,C\,m^{-2}\,year^{-1}$, counterbalanced by a much lower input from terrestrial sources (Marxsen, 2006).

There are only few investigations that suggest faunal assemblages are able to control bacterial abundance in aquatic sediments (for reviews, see Kemp, 1990, mainly containing studies on marine systems; Perlmutter and Meyer, 1991 for stream environments). Although about half of BCP is utilized by animals in the Breitenbach, this loss is probably too low to exert a general control on bacteria, except perhaps under exceptional conditions, temporally, and spatially limited.

8.8
Global Change and Timing of Insect Emergence at the Breitenbach

Peter Zwick and Rüdiger Wagner

8.8.1
Introduction

Continental drift, mid-oceanic rift valleys, the rise of the Himalayas and the Andes, fossil forests on Antarctica, and a subtropical Eocene swamp at Messel near Frankfurt are all indicators of continuous global change, whose ultimate cause is probably the fluid dynamics of the hot interior of the earth and its impact on the cool, rigid continental crust.

In the northern hemisphere, there is much evidence of natural post-pleistocene change over a few thousand years. In Europe, the rise of the Scandinavian land mass after the Pleistocene ice cap melted continues today, and many central European landscapes were shaped during the glacial periods. Palynology, Baltic Sea and lake sediment studies, stone age human art in caves, and other lines of evidence document the drastic changes in the Central European biota over the past few thousand years. Natural climate change was accompanied by shifts in fauna and flora. For example, beech (*Fagus sylvatica*), the presently dominant, central European forest tree, only spread into the area about 2500 years ago (Freitag, 1962). Zoogeographers, especially de Lattin (1967), identified Pleistocene faunal refugia as centres of long-range postglacial dispersal. More recently, phylogeography and its molecular methods can reveal details of the historical ranges and distribution spread of the species concerned.

Superimposed on natural postglacial change is the recent, much more rapid change induced by human activities during the industrial era. Increasing amounts of CO_2, produced during the combustion of fossil fuels, end up in the atmosphere and, together with H_2O, CH_4, and others, act as so-called greenhouse gases, increasing the heat-trapping ability of the earth. This leads to the phenomenon known as "global warming" whose impacts differ between continents, and even between regions on a given continent.

Effects on aquatic ecosystems potentially differ from those on terrestrial ones because of the contrasting physical properties of water and air. In aquatic ecosystems the consequences of increasing temperature also differ between running and standing waters. The uptake of energy from the environment essentially occurs at the water surface. In deep, stratified, standing waters, temperature effects on the hypolimnion may be dampened or delayed. Running waters typically lack vertical stratification; the effects of increasing atmospheric temperature therefore affect running waters immediately.

Warming of the earth since about 1850 is well documented (Schönwiese, 2002), and several recent temperature extremes have been recorded (http://www.energie-fakten.de/pdf/klima.pdf). In Central Europe, the vegetation growth period has become longer and phenological change in the flora has been observed (Streitfert *et al.*, 2008), with often complex consequences for animals (e.g., Frommer, 2010). Several spectacular, warm-loving European insects are spreading north (see, e.g., Drees, 1997; Frommer, 2006). Pix (2005) introduced the highly descriptive term "thermische Faunendrift".

These changes do not appear to be severe, as long as they do not exceed physiological tolerances. However, there is widespread concern about indirect effects and the ecological consequences of global warming (IPPC, 2007). This is also reflected in recent fundings of related research (http://www.bmbf.de/de/917.php), revealing that the interaction of rising temperatures with other environmental factors may affect competition between taxa and contribute to changes in community structure (e.g., Woodward *et al.*, 2002). Different organismal reactions may lead to the uncoupling of vital interactions or functions. Some examples involving popular birds have also raised interest in the general press (e.g., Dawson, 2005; http://www.sueddeutsche.de/wissen/der-kuckuck-stirbt-aus-dem-die-stunde-schlaegt-1.938370).

Ultimately, warming will affect the availability of water. The upper courses of small European streams are already affected today. Severe drought and water shortness, as in areas with a Mediterranean climate may become more widespread and more frequent. Regularly intermittent streams have specifically adapted fauna (e.g., Aubert, 1963; Dieterich and Anderson, 1995; Sommerhäuser, 2000; Stewart and Anderson, 2010; Williams, 2006), but unpredictable droughts affect the fauna severely and in different ways. Bêche *et al.* (2009) describe some examples and review recent literature on the subject.

Nevertheless detailed information on the effects of warming on stream fauna is scanty, and not all suggested examples withstand critical examination. Ongoing studies at some research institutes aim experimentally to document accelerated growth and precocious emergence of stream insects. However, there is some evidence for species replacement along stream continua, for example in the River Fulda (Zwick and Zwick, 2010a).

For the Breitenbach, a small, temperate, foothill stream it seems that indirect effects *via* the catchment are most important. Rising temperatures alter local air circulation systems, weather, amounts and seasonal patterns of precipitation, increasing water demand by vegetation *via* raised evapotranspiration, and so on. Together, these effects impact the hydrological regime and stream discharge, which in turn determine sediment structure and induce a sequence of in-stream changes (see Section 8.3).

However, there is also circumstantial evidence of direct temperature effects on the growth and development of several Breitenbach insects leading to changes in timing and duration of emergence over the many years of study which we present here.

8.8.2
Methods

The present section investigates changes in emergence phenology of mayflies, stoneflies and caddisflies in Trap 2 of the Breitenbach, for the years 1969–2005. However, to be suitable for the intended analyses, insect populations must meet the requirements listed below. The accuracy of the original emergence data also requires attention. To avoid errors due to outliers, cumulative percentage emergence was analyzed, recording dates of 5%, 50% or mean emergence, and of 95% emergence.

8.8.2.1 Selection Criteria

Low specimen numbers often result in discontinuous emergence and staggered cumulative curves, which lead to apparent delays in emergence. Therefore, total annual emergence of fewer than 20 specimens was not considered. All remaining data were examined for continuity of emergence. Brief phases of staggered emergence also occurred, if emergence began hesitantly or (mainly) when late specimens extended the asymptotic top end of the curve. As a result, the 95% cumulative emergence level varied more irregularly than the 5% and 50% emergence dates.

For the Plecoptera, post-emergence arrival of old, ovipositing females was documented for several species (Chapter 7.6.4.3). Immigration of old females cannot be routinely distinguished from late female emergence and therefore only males were analyzed for the Plecoptera.

In cases of plurimodal emergence, cumulative emergence over the entire year was only used to distinguish plateaus separating successive cohorts or generations. Data were split accordingly, and cumulative emergence calculated separately for each generation. The following notes on individual species also apply.

Baetis rhodani and *B. vernus*: The difficult taxonomic separation of these two species lead to uncertainty over the identity of single or few specimens emerging outside the main emergence periods. Date of mean (50%) emergence always fell within the main emergence pulse and was not noticeably affected by the taxonomic problem. However, dates of 5% and 95% emergence cannot be determined with confidence.

Nemurella pictetii: In cumulative emergence curves Julian day 200 marks a clear breakpoint between the two generations, which were then analyzed separately.

Drusus annulatus and *Rhyacophilca fasciata* both exhibit bimodal emergence. Although specimens of subsequent peaks probably do not represent successive generations (see Section 7.6.4.5), they were analyzed as distinct cohorts. Julian day 200 was the approximate breakpoint between the cohorts for both species.

The following taxa exhibited extended, or occasionally bimodal, emergence but either cohort separation failed, or bimodality occurred too rarely, to permit separate analysis: *Centroptilum luteolum, Protonemura auberti, Isoperla goertzi, Leuctra nigra.*

8.8.2.2 **Accuracy of Emergence Data and Possible Sources of Error**

Differences in emergence time that were possibly caused by global warming are currently presumed to be of the order of days or weeks. The temporal accuracy of the original data must be considered, because it determines the reliability with which possible change can be estimated.

Sampling. For most of the study period, samples were removed from the preservation trays every Monday, Wednesday, and Friday, but the 2-2-3-day pattern disappears when data are pooled by weeks. However, weekly intervals may be close to the phenological time difference that we are seeking. Therefore, we used the original sampling data, expressed as Julian days. Although samples are assigned to the date of collection, actual emergence may have occurred up to two days earlier for Monday samples. Inevitably therefore there is a potential error of one to two days in the original data.

Lost year. On 26 May 1986, an insecticide accident affected all taxa in Reaches A–C and also many species further downstream. For taxa whose emergence ended abruptly, an unrealistically early, false, date of mean cumulative emergence resulted. For taxa resuming emergence after some delay (survivors, or specimens drifting in from sites upstream of the point-source of the poison), the resulting date is unnaturally late. Useful data were only obtained for species that had already completed their annual emergence before Julian day 147, that is, the first generation of *Baetis rhodani*, and most Plecoptera. Other species were excluded from the analysis.

Seventeen mayfly and stonefly species, and 12 caddisflies satisfied the requirements listed above. Because four species occurred in separate generations, the number of operational taxa is 33, but the number of cases amenable to analysis differs between taxa and is indicated in Tables 8.32 and 8.33.

8.8.2.3 **Shift in Emergence Dates**

The first approach investigated whether date of emergence of individual species had shifted over the 35 years of study. Thus, Julian days at which 5%, 50%, and 95% emergence occurred were calculated for each species in each year, and linear regressions calculated for calendar years. Shift of emergence time was defined as the difference between emergence dates calculated from these regressions for 1969 and 2005 (Ephemeroptera, Plecoptera) (Table 8.32), or 2006 for Trichoptera (Table 8.33).

8.8.2.4 **Emergence Dates and Temperature Regimes**

The second approach compared between year differences in thermal regime to timing of emergence, over 35 study years. Several temperature data (based on daily means recorded at Measuring Station B) were calculated:

- Yearly cumulative degree-days above freezing point.
- Cumulative degree-days from the middle of one year to the middle of the next year. For most taxa, this period matched the length of the entire life cycle.
- Cumulative degree-days from the beginning to the middle of each year.
- Cumulative degree-days from the start of the year to Week 33.
- Cumulative degree-days from Week 48 of one year to Week 4 of the next year (caddisflies only).

Temperatures in individual years were expressed as percentages of long-term averages of all other years in the study period. Values <100% identified relatively cool years and values >100% relatively warm years.

Table 8.32 Number of days by which the dates of 5%, 50%, and 95% cumulative emergence of mayflies and stoneflies shifted in the Breitenbach between 1969 and 2005. Estimates are from regressions between calendar years and actual dates in individual years. n = Number of observations. Levels of significance: gray, n.s.; blue, $P < 0.05$; yellow, $P < 0.01$; green, $P < 0.001$.

	5%	50%	95%	*n*
Amphinemura standfussi	−35	−45	−32	24
Ephemerella ignita	−17	−26	−29	16
Baetis rhodani, second generation	−8	−26	−29	24
Protonemura meyeri	−19	−23	−29	21
Siphonoperla torrentium	−18	−20	−15	37
Leuctra nigra	−12	−18	−24	37
Nemoura flexuosa	−29	−11	−15	5
N. marginata	−14	−10	1	23
Ephemerella mucronata	−18	−10	−20	22
Nemurella pictetii, second generation	−14	−10	−11	18
Baetis vernus	−27	−10	−37	34
Nemurella pictetii, first generation	−14	−8	1	22
Protonemura intricata	−8	−7	0	37
Nemoura cambrica	−5	−6	−3	35
Leuctra digitata	−9	−4	−8	37
Isoperla goertzi	6	7	59	28
Leuctra prima	1	10	10	35
Protonemura auberti	59	40	22	34
Baetis rhodani, first generation	19	46	60	18
	n.s.	*	**	***

Table 8.33 Number of days by which date of mean (=50% cumulative) emergence of caddisflies shifted in the Breitenbach between 1969 and 2005. Negative values indicate earlier, positive ones delayed emergence. Estimates are from regressions between calendar years and actual dates in individual years. n = number of observations; significance levels: gray not significant; blue, $P < 0.05$; yellow, $P < 0.01$; green, $P < 0.001$.

Species	5%	50%	95%	n
Rhyacophila fasciata, second generation	−11	−17	2	36
Silo pallipes	−22	−17	−18	36
Wormaldia occipitalis	−54	−15	−24	26
Agapetus fuscipes	−4	−12	−27	37
Apatania fimbriata	−10	−11	−16	34
Chaetopteryx villosa	0	−3	−12	37
Potamophylax luctuosus	3	−1	−4	29
Plectrocnemia conspersa	13	2	−10	37
Rhyacophila fasciata, first generation	−6	3	−6	37
Tinodes rostocki	6	7	−4	37
Sericostoma personatum	3	9	15	37
Drusus annulatus, first generation	5	10	13	37
Potamophylax cingultaus	13	15	2	33
Drusus annulatus, second generation	7	21	36	37
	n.s.	*	**	***

Mayflies, stoneflies, and most caddisflies showed congruent temporal patterns of emergence, that is, almost all were relatively early or late, in a given year. For each species or cohort, dates of emergence (Julian days), that is, the number of days required to attain a certain cumulative emergence, were converted to percentages of long-term averages of all other years during the study period. Values <100% identified relatively early, values >100% relatively late, emergence. Conversion to percentages also allowed direct comparisons between species emerging in different seasons. Therefore, the overall mean percentage for all taxa together expressed early or late emergence of all species in a single figure.

8.8.3 Results

8.8.3.1 Shift in Emergence Time of Individual Ephemeroptera and Plecoptera Species

Table 8.32 shows the shift in the estimated dates of 5%, 50%, and 95% cumulative emergence. Negative values indicate earlier, positive values later, emergence in recent years compared to 1969. There is a general tendency for earlier emergence today, but there also seem to be two cases of delayed emergence.

Six species (top section of Table 8.32) now emerge distinctly earlier than in the past. This is most obvious for the 50% emergence dates, but is also true of the 5% and 95% emergence dates. However, how many days earlier taxa emerge cannot be stated with precision. The regressions from which shifts of emergence time were estimated had low coefficients of determination (R^2 ranging from 0.12 in *Nemoura cambrica* to 0.56 in the first generation of *Nemurella pictetii*). Standard errors of slopes and intercepts of regressions for 50% emergence dates averaged about 30%. Numbers of days by which emergence shifted have similarly high potential errors.

For several other species (center of Table 8.32) estimated shifts of 50% emergence are nonsignificant, and for several, the forward shift of mean emergence is less important than that of 5% and 95% emergence, which is occasionally significant.

In contrast, the emergence of *Protonemura auberti* and of the very early, first generation of *Baetis rhodani* (bottom of Table 8.32), appear to be significantly delayed.

8.8.3.2 **Shift in Emergence Time of Trichoptera Species**

Table 8.33 shows the shift in estimated dates of 5%, 50% and 95% emerged specimens for the holometabolous Trichoptera, for the period 1969 to 2006. There was no general tendency for earlier emergence in 2006 in caddisflies as a whole, but five species emerged earlier, two later and, for seven species, there was no significant tendency for change in either direction.

As for the Ephemeroptera and Plecoptera, determination coefficients of regressions from which emergence shifts were estimated were mostly low (R^2 between 0.16 in *C. villosa* to 0.31 in *A. fuscipes*) but significant; *P. cingulatus* (R^2 = 0.81) was a notable exception. It was surprising that for caddisflies the 95% values (end of emergence) had quite high determination coefficients, between 0.4 and 0.56. Particularly for *Agapetus fuscipes* (~4 weeks), *Apatania fimbriata* (~2 weeks) and autumn-emerging *C. villosa* (~2 weeks) emergence ended earlier. However emergence of *D. annulatus* (second generation) was delayed by approximately five weeks, and prolonged for about four weeks. Comparison of the 5% and 95% emergence data indicated that *Wormaldia occipitalis* prolonged its emergence period by 78 days (~2.5 months), starting earlier and ending later. This formerly autumnal species is now on the wing from April to November at T1. For most of the remaining species overall emergence was either shifted earlier or later. However, these results are only valid for the area around T2.

In most other Trichoptera (Table 8.33 center) emergence shifts were not significant although emergence of *Plectrocnemia conspersa* started two weeks later and ended ten days earlier. The emergence of two "late summer" species, the second generation of *D. annulatus* and *P. cingulatus* started one or two weeks earlier, and the 50% value was two or three weeks later.

8.8.3.3 Emergence Dates and Thermal Regime

The inter-annual temperature pattern varied irregularly, but the four different thermal regime estimates varied similarly in given years. Degree-days above freezing accumulated during the first 33 weeks of each year are shown. They seem to rise from about 1980, but fell again in 2004–2005. The rise over the study period is significant at the 5% level, but apparently only because of extreme values in 1999–2003 (Figure 8.35).

Dates of emergence also varied irregularly from one year to the next, although 5%, 50% and 95% cumulative emergence trends were consistent in given years. There was a statistically nonsignificant trend of increasingly early 50% emergence of the pooled taxa.

The temporal pattern of the thermal regime curve is a mirror image of the changes in emergence date curve, which strongly suggests a causal relationship. For all temperature parameters, deviations from long-term means were negatively correlated with relative dates of mean cumulative emergence. The correlation was closest for cumulated degree days over the first 33 weeks of each year (Table 8.34, Figures 8.37 and 8.38), and weakest for thermal sums between the solstices of successive years. Plots of thermal sums against the corresponding mean

Table 8.34 Correlation coefficients between deviations from long-term means for four different parameters of temperature regime and cumulative 50% emergence of 19 operational taxa (Ephemeroptera and Plecoptera).

Cumulative degree-days, weeks 1–33	−0.6647
Cumulative degree-days, weeks 1–26	−0.6385
Annual cumulative degree-days	−0.5877
Cumulative degree-days, Solstitium I – Solstitium II	−0.3543

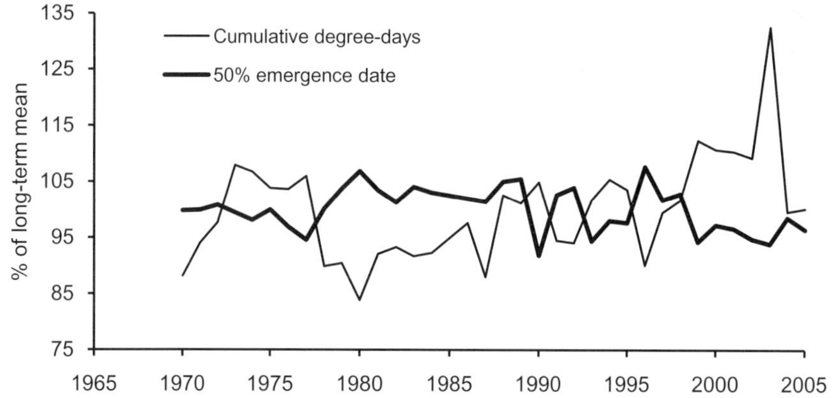

Figure 8.37 Cumulative degree days during the first 33 weeks of each year and 50% emergence dates of 19 operational taxa over the study period. Data are expressed as percentages of long-term means (ordinate). Data points connected by auxiliary lines.

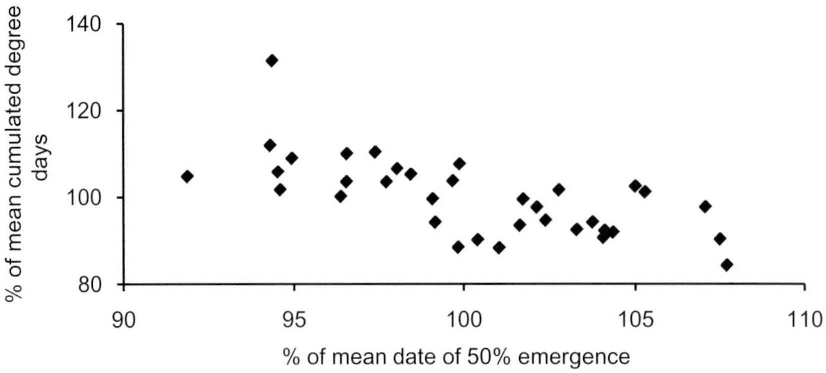

Figure 8.38 Cumulative degree days during calendar weeks 1–33 of given years plotted against the corresponding date of mean emergence of the 19 operational taxa of Ephemeroptera and Plecoptera. Values are expressed as percentages of long-term averages (1969–2005).

emergence dates clearly show the negative correlation: the warmer the year, the earlier the emergence (Figure 8.38).

In general the Trichoptera results are similar to those for Ephemeroptera and Plecoptera. However, correlation coefficients between 50% emergence and long-term temperature measures were only –0.51 ($P < 0.01$) for solstice to solstice and for cumulative degree days for weeks 1–26. For annual cumulative degree days correlation was even lower although significant (–0.36; $P < 0.05$). The "early" species, with emergence before Julian day 200 (*A. fuscipes, A. fimbriata, Silo pallipes, T. rostocki*, first cohorts of *R. fasciata* and *D. annulatus*–most of them scrapers) particularly tended to emerge earlier the warmer the year (Figure 8.39 and left part of Figure 8.40). No significant trend occurred if all or only the "late" species were pooled. However, two late summer and autumn emerging species (*P. cingulatus*, second cohort of *D. annulatus*) (Table 8.33 and Figure 8.40, right hand side) emerged later.

When the data were analyzed at the species level it was found that the yearly temperature sum (middle of one year to middle of the next year) affected two species of *Potamophylax* in different ways. Emergence of the spring species, *Potamophylax luctuosus*, was a few days earlier, but emergence of the late summer species, *P. cingulatus*, was delayed by two weeks (Figure 8.40). Emergence of *A. fimbriata*, with larval dormancy in winter, was also negatively correlated with temperature sums of weeks 48 one year to week 4 the next year (Spearman: –0.368; $P = 0.032$).

8.8.4
Discussion

Climate change may have different effects. Changes in community composition and reduced abundance, with the eventual disappearance of sensitive taxa, were

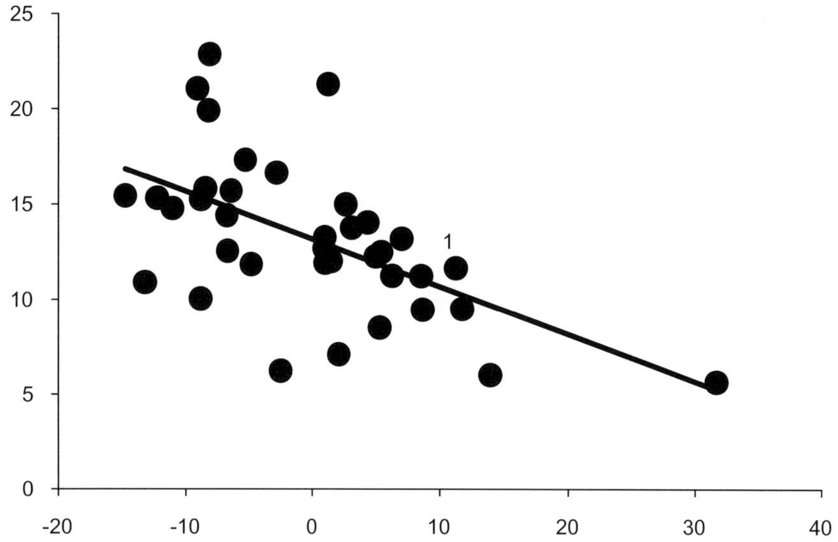

Figure 8.39 Deviation (%) of emergence of early emerging caddisflies (abscissa) and deviation from long-term mean temperature (sum days 1–203, ordinate) in individual years (1969–2006; regression line: $y = -0.25x + 13.16$; $R^2 = 0.32$; $P < 0.05$).

documented or postulated for a 25-year study of British headwater streams (Durance and Ormerod, 2007). At present, there is no evidence that any taxa have been lost from the Breitenbach. On the contrary, *Leuctra braueri,* which does not occur anywhere else near Schlitz, has recently established itself in the Breitenbach (R. Wagner, personal communication). Changes in abundance (see Section 8.3) may be climate-related, but cannot directly be ascribed to changing temperature.

A tendency to emerge earlier today than in the past is evident for several species. However, the absolute numbers of days by which emergence has shifted (Tables 8.32 and 8.33) must be viewed with great caution because coefficients of determination of the respective regressions were low and standard errors high. The 1- to 2-day error inherent in the original data was not even considered in these estimates.

Some species emerge significantly earlier than before (Tables 8.32 and 8.33, top sections). For *Ephemerella ignita* this is probably a result of earlier termination of egg diapause due to warmth (Bohle, 1972).

Factors that terminate egg diapause in *Amphinemura standfussi* are not known. 50% cumulative emergence is usually attained around Julian day 180, and peak oviposition must follow shortly. From the recorded duration of diapause (Zwick, 1993), larval hatching probably occurs before midwinter, and therefore, larval growth occurs at essentially the same time as in other spring or summer species.

The same is true of *Protonemura intricata,* whose variable egg diapause during summer leads to relatively synchronous autumnal hatching.

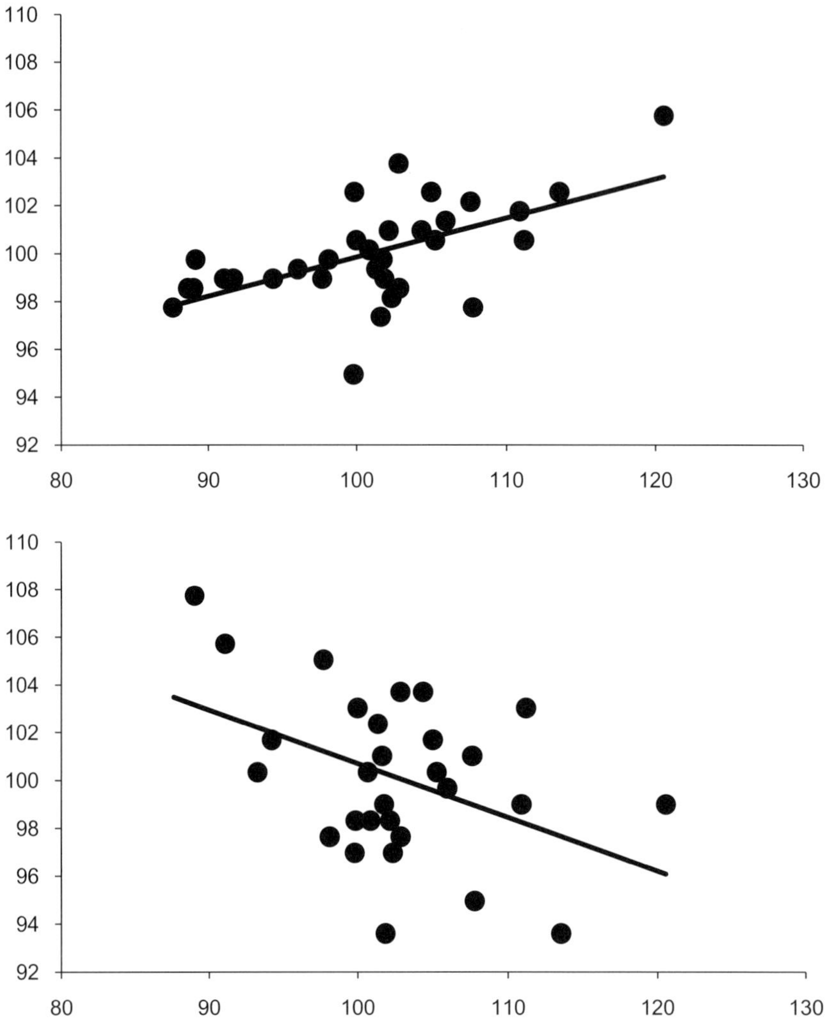

Figure 8.40 Deviation (%; abscissa) of emergence of the caddisflies *Potamophylax luctuosus* (left) and *P. cingulatus* (right) against deviation from long-term mean temperature (sum days 187 one year to 186 next year, ordinate) in individual years (1969–2006). Regression lines: *P. luctuosus* $y = -0.225x + 123.16$; $R^2 = 0.19$, $P < 0.05$; *P. cingulatus* $y = 0.163x + 83.55$; $R^2 = 0.35$; $P < 0.01$).

Protonemura meyeri probably emerges earlier because the threshold temperature for emergence (Figure 7.44) is attained earlier. This probably also applies to the first generation of *Nemurella*.

Larval *Siphonoperla torrentium* are predators, mainly of small nematoceran larvae, but we do not know whether duration of availability or abundance of the latter have changed. Adult *Siphonoperla* depend on conifer pollen for food. Boyer

(1973) showed the dependence of pollination day of an American *Pinus* sp. on cumulative degree-days in spring. Even if European *Pinus* and *Picea* spp. near the Breitenbach now pollinate earlier, an adaptive shift of emergence phenology in response to conifer pollination appears improbable.

Leuctra nigra is mostly univoltine, but part of the population may take more than a year to develop and is thus semivoltine (Figure 7.39). Suppression of the semivoltine component may explain its earlier emergence, but this cannot be proved.

The earlier emergence of the, usually large, second generation of *Baetis rhodani* (Table 8.32, top section) conflicts with the apparently delayed emergence of its first generation (Table 8.32, bottom section), which may be an artifact of the paucity of data points. Only occasionally were there enough specimens to be considered in calculations. In fact, for several years the near absence of the first generation raises doubts over whether the summer generation is descended from indigenous Breitenbach females or from ovipositing females immigrating from the River Fulda (Chapter 7.2).

Leuctra prima emerges while stream temperatures are still minimal (Figure 7.44), at times when emergence may be depressed by low air temperature (Figure 7.51). It emerges no earlier than before, possibly even a little later.

Protonemura auberti is the only species which now seems to emerge distinctly later than 35 years ago. Emergence is often bimodal (Figure 7.48) but consistent separation of two cohorts (and separate calculation) was not possible. Figure 8.41 shows that timing of the invariably autumnal emergence peak changes little between years. The apparent delay in emergence probably simply results from the recent insignificance, or absence, of the vernal emergence peak. How this early peak forms is unclear. It may be that, in some years, specimens drift in from the almost nonseasonal subpopulation in the Georgsborn and thereby modify the pattern. It is also possible that the population in the main course of the Breitenbach follows an alternative developmental pathway, as do *Leuctra nigra* or *Nemurella pictetii*, and that certain temperature regimes suppress the early emerging group in some way.

The emergence of *Wormaldia occipitalis* now starts about two months earlier. In the past the species was very rare, but its abundance has increased in the spring area since the 1990s. In parallel to its increased abundance the emergence period has been extended from September–November to May–November. Thus earlier emergence not only depends on increased temperature, but also seems to be affected by an increase in abundance in this species, initially in the spring region and subsequently in the middle part of the stream.

It is remarkable that three of the remaining four Trichoptera are scrapers emerging in spring. We have no observational data, but possibly earlier growth of diatoms in the biofilm, their preferred food, may accelerate larval development, leading to earlier pupal development and emergence. Because adult weight is determined during the final instar, earlier and better food supply with slightly increased temperature may be effective. However, scraping mayflies did not show similar effects.

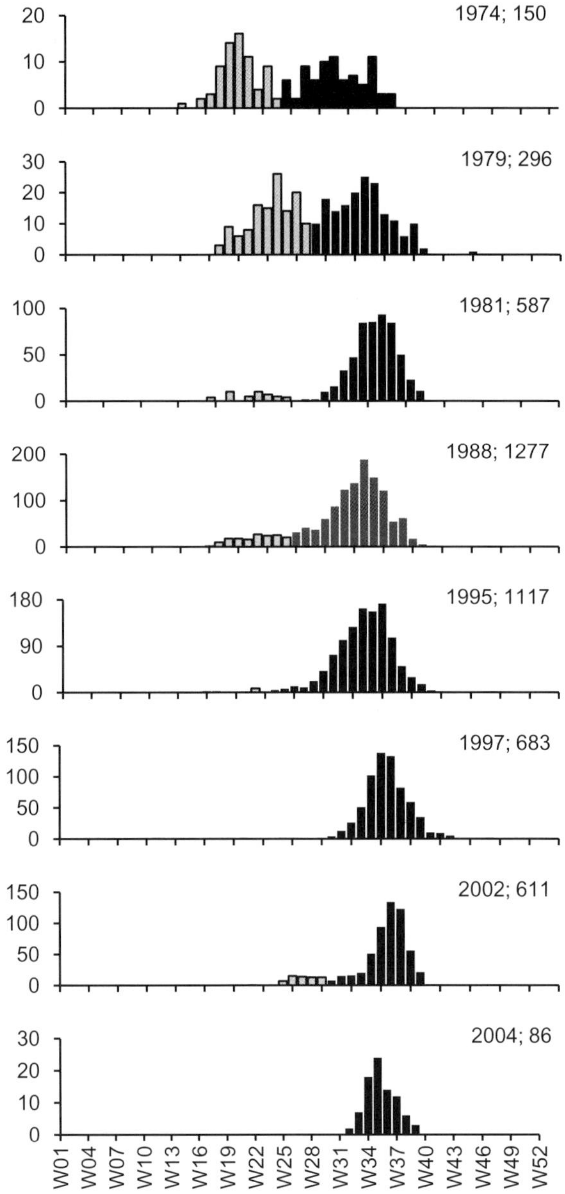

Figure 8.41 Numbers of *Protonemura auberti* collected per week in Trap 2 in selected years. Shades distinguishing two groups were assigned to show that the second group maintains its date of emergence almost unchanged over the study period.

The earlier emergence of the usually larger, second cohort of *R. fasciata* differs markedly from the unaltered emergence of the first cohort but we cannot suggest any reasons for this.

Emergence periods of two caddisflies are shorter. The emergence of *Plectronemia conspersa* begins later and ends earlier, with an unaltered emergence mean. *Chaetopteryx villosa* is similar, with an unaltered start to emergence but a tendency to terminate two weeks earlier. *Potamophylax cingulatus* emergence starts about two weeks earlier but termination has not changed. Temperature may be responsible for the more synchronous emergence of these species.

Sericostoma personatum shows a tendency to emerge later. Emergence of the first cohort of *Drusus annulatus* is now terminated about two weeks later, while the mean and end of emergence of the second cohort is delayed for three or even five weeks.

The negative correlation between cumulative degree days and length of time required to emergence (Figures 8.37 and 8.38, Table 8.34) suggests that the shifts are related to overall warming, even if the latter cannot be directly documented for the Breitenbach (see Chapter 3). Whenever emergence dates deviated by 5% or more from long-term averages, other possible causes were considered, especially hydrological regime or extreme abundance of the taxon, but no alternative explanations can be offered.

Interestingly, temperatures had the greatest apparent influence on timing of emergence during the first 33 (26 in caddis) weeks of the year, thermal sums accumulated over the entire life cycle of the species were less closely correlated with time of emergence. For most taxa, midwinter temperature affects small larvae that accumulate very little biomass. After the winter solstice, daylength increases and temperature rises (e.g., Figures 7.34, 7.41, 7.43) and rapid growth and significant biomass accrual begin. Warm spells in spring that accelerate biochemical and physiological processes (van't Hoff's rule) coincide with periods of maximum biological activity and are therefore probably most effective.

9
Summary

Georg Becker, Eileen J. Cox, Jürgen Marxsen, Hans-Heinrich Schmidt, Rüdiger Wagner, and Peter Zwick

The Breitenbach is an unpolluted, first-order stream flowing through Bunter Sandstone foothills in Hesse, Central Germany (50°39′N, 9°38′E), between 430 and 220 m a.s.l. It is typical of streams in the larger area. Studying this stream was the most intense and long-term research activity of the Limnologische Fluss-Station at Schlitz (a section of the Max-Planck-Institute of Limnology at Plön, Germany), operating from 1951 to 2006.

The earliest studies (until 1965) focused on organismal drift (periodicity and its causes), while attempts to estimate secondary production from numbers of adult insects emerging from the stream, and autecological studies of *Gammarus* spp., dominated from 1969 to 1982.

The present book concentrates on work carried out since 1983 when the scientific remit was broadened. From that year, all functions, processes and organisms in the stream itself, or related to it, were considered. Both descriptive and experimental approaches were adopted, in the field and in the laboratory.

9.1
Climate, Environment

Extensive climatic and environmental data were collected using several measuring stations distributed over the entire drainage area. In addition, data from official weather stations in the region were considered for comparison and modeling. The aquatic milieu was characterized with both physical and chemical methods, seasonally, monthly, daily, or at even shorter intervals.

Alongside the geological characteristics of the catchment, irradiance, precipitation, wind and air temperature determine the prevailing conditions and influence the subsequent development of the Breitenbach and its community. The strength of the external influences affects the environment and the communities living there.

Central European Stream Ecosystems: The Long Term Study of the Breitenbach, First Edition. Edited by Rüdiger Wagner, Jürgen Marxsen, Peter Zwick, Eileen J. Cox.

In general, on the plateau the wind most often blows from the north-west (with rain) but the situation in the valley is different. Throughout the year in the valley, the wind regularly blows from the south-east, creating a downflow, but only at night. During the day, another wind system crossing the valley develops, starting at 0800 in summer and 1100 in winter.

With respect to climate change, only variation in air temperature indicates changes consistent with global warming. Other trends were either insignificant or the observation period was too short; a minimum observation period of forty years is required to address this issue.

The range of abiotic factors remained largely the same over the monitoring period, with predictable changes along the stream. Only the analysis of the variability of meteorological, physical and hydrological factors provided insights into causal effects.

With respect to the stream community, two causal relationships were observed:

1) The hydrological chain, which affects habitat availability via stream bed sediment morphology.
2) The thermal chain, which affects the growth and development of organisms via water temperature (an ecophysiological parameter).

9.1.1 Hydrological Chain

All climatic factors (irradiation, air temperature, precipitation, wind) affect the catchment's ability to absorb, retain, and release water. Based on geological preconditions a proportion of the rainfall re-emerges as stream discharge. The temporal relationship of precipitation (cause) and discharge (effect) may be immediate, but depending on the season, it may also be postponed by several months. The spatial and temporal discharge characteristics affect the basic morphological characteristics (habitats) of the stream bottom and even the physical and chemical nature of the stream and sediment water. Macrophyte stands may modify the shaping effect of water. Depending on the hydraulic effect, directly on specimens or indirectly via habitat generation and availability, individual organisms as well as populations may or may not have adequate habitats.

9.1.2 Thermal Chain

Water temperature is essentially controlled by irradiation, air, and ground temperature. Particularly in low flow periods, peaks in discharge may severely interfere with this relationship. During the growing season, macrophyte stands reduce exchange processes at the water–air interface. For this reason, unlike air temperature, water temperature has changed only moderately over the study period.

Hydrology and temperature are the dominant variables in the abiotic environment. Other variables indicate the community's response to the dynamic variability of the meteorological and physical factors. This is particularly shown by the diurnal and seasonal oxygen content variation. Hardly any variable that is directly or indirectly related to photoperiod fails to fluctuate. Where this does not apply, effects of the discharge regime come to the fore.

9.1.3
Water Chemistry

Acid-binding capacity (ABC) is mainly affected by discharge, with pH values following ABC. pH fluctuations are also governed by the photoperiod. Low ABC values occur with high discharge, and high values with low flow. Bicarbonate varies similarly, but the relative amount of bicarbonate to ABC increases with increasing distance from the spring. A long-term trend to lower ABC was not significant. pH is usually more or less neutral, but low pH occurs under high flow and high pH under low flow.

Oxygen content was always around saturation point. Correlation between irradiation, water temperature and oxygen concentration differed from the "classical" scheme – the %-saturation curve in the stream water followed the irradiance intensity immediately. The difference is attributed to the lower primary production of the Breitenbach compared with streams where early diurnal oxygen curves were determined by Odum in the 1950s. Of the chemical components, nitrate and ammonium vary in step with the photoperiod, with distinct diurnal and seasonal variation. However, there was low temporal variation in phosphate. Sulfate and chloride remained approximately constant; there was no reliable data on silicon.

9.2
Detrital Energy Sources

Detritus (dead particulate and dissolved organic matter) provides important sources of matter and energy to the heterotrophic organisms of the Breitenbach, as is typical for such lotic environments. This includes not only organic matter from primary producers within the stream, but also mainly from allochthonous origin. Annual inputs (long-term averages) of particulate organic matter are $0.42\,kg\,C\,m^{-2}$, and dissolved organic matter $1.02\,kg\,C\,m^{-2}$, thus clearly surpassing the instream primary production of $0.2\,kg\,C\,m^{-2}\,year^{-1}$. Annual outputs of organic matter to the River Fulda (via transport) were 0.34 and $0.74\,kg\,C\,m^{-2}$ for particulate and dissolved organic matter, respectively. Annual average standing stocks were $0.30\,kg\,C\,m^{-2}$ for particulate and $0.8\,g\,C\,m^{-2}$ for dissolved organic matter. However, because of the high flow rates through the sediments, the availability of DOM to the microbial communities is probably even higher.

The composition, amount, and seasonal distribution of allochthonous CPOM were characteristic for individual sites. Leaf shedding in autumn and early winter provides about 70% of the total annual input of CPOM. In forested regions input was highest, up to about 2700 g OM m^{-2} year^{-1}, lower in managed grassland, and lowest in fallow areas, where it was less than 100 g OM m^{-2} year^{-1}.

Concentrations of dissolved organic carbon in the stream water are typical of unpolluted low order streams, fluctuating between 1 mg C l^{-1} at low water and 10 mg C l^{-1} at exceptionally high spates, 1.5–3.5 mg C l^{-1} for most of the year. No trends were observed with temperature but there were pronounced discharge effects. Elevated discharge coincided with enhanced dissolved organic carbon concentrations. DOC concentrations in interstitial water slightly exceeded those in stream water, but they were distinctly higher (occasionally more than 10-fold) in groundwater adjacent to the stream.

Dissolved carbohydrates provide a significant source of carbon and energy, especially for heterotrophic bacteria. Six individual free dissolved carbohydrates were regularly, or at least usually, detected in stream water: saccharose, glucose, ribose, fructose, 2-deoxy-glucose, and arabinose, at concentrations up to 100 nmol l^{-1}. In interstitial water concentrations were on average 10-fold higher, and more carbohydrates were regularly detected. The remarkably elevated level of ribose in the interstitial water was attributed to the high degradation activity of the microbial communities in this habitat. Elevated concentrations of galactose and arabinose are indicative of a nonwoody angiosperm origin of organic matter, which is in agreement with the particulate organic matter sources for the Breitenbach.

9.3 Primary Producers

The predominant primary producers in the Breitenbach are macrophytes (angiosperms, bryophytes) and photoautotrophic algae. Angiosperms are the dominant macrophytes while the algae comprise a diverse range of microscopic eukaryotes, dominated by diatoms, but also containing representatives of almost all freshwater algal groups, especially chlorophytes. The algal assemblages are primarily benthic, on sediments, stones and macrophytes; as in most lotic systems, suspended photoautotrophs are less important. Cyanobacteria also make a significant contribution under particular conditions.

Total primary production was estimated to about 400 g C m^{-2} year^{-1} (gross) and 200 g C m^{-2} year^{-1} (net primary production), from1973 to 1998, but calculations on the basis of daily oxygen curves (1997–2004) suggest lower rates. Total community respiration appeared at about 800 g C m^{-2} year^{-1}. The average production/respiration (P/R) ratios of 0.50 for the earlier and 0.27 for the more recent years of investigation reveal that at least half the energy for the stream's metabolism is acquired from terrestrial sources, confirming the importance of allochthonous organic matter for the stream community.

Algal biomass was estimated using chlorophyll-a, revealing much greater algal biomass (standing crop) on sediments (from 15 to >40 μg cm^{-2}) than stones (<7 μg cm^{-2}), although the timing of peaks differed between sites and substrata.

Short-term studies revealed the potential for rapid variation in biomass over short time periods (days). Estimates of diatom cell densities were usually around 10^4 cells cm^{-2} sediment, although densities were lower in January–February, and also fell after spates. Algal biomass is also controlled by grazers (or their absence). *In situ* grazing experiments revealed the impact of caddisfly larvae on algal standing crop (reducing it), while algal biomass increased strikingly after the insecticide accident.

Detailed investigations on algae in the Breitenbach focussed on diatoms, which grow on a variety of substrata, showing marked fluctuations in species composition and biomass over the course of a year. Epipelic assemblages in the Georgsborn showed the greatest stability, but in the Breitenbach they fluctuated seasonally, as well as showing an increase in species richness in a downstream direction. Epilithic and epiphytic assemblages were less diverse and more uniform throughout the stream. The significance of microhabitat conditions for assemblage composition is clear, while inter-annual variation in assemblage composition could be linked to discharge patterns.

The role of discharge in modifying the stream substratum was clearly reflected in the diatom assemblages, with some taxa being consistently found on coarser sediments, others particularly on finer sediments, for example, *Navicula* spp. Variation in species occurrences with microhabitat in the stream indicated that care must be taken when using diatom species to infer water quality. While epilithic assemblages probably show a clearer response to water chemistry, epipelic and epiphytic assemblages (including on bryophytes) are probably growing in a modified environment, as reflected by the more acidophilic taxa in bryophytes and more eutrophic taxa on finer sediments. Using simple elution experiments, finer sediments contained higher concentrations of phosphorus than coarser sediments.

The use of live counts for this study ensured that the autochthonous taxa were documented, whereas relying on cleaned material (as is usually the case for diatoms) does not allow discrimination between autochthonous and allochthonous material with the potential for skewing ecological interpretations, Similarly the discrepancies between percentage abundances and density measures raise questions about the use of the former for site comparisons and the inference of species optima. The importance of long-term studies spanning many years was also supported by the results of the seasonal studies; if the algae in this system had only been investigated for two years it would have been more difficult to understand the causes of some of the assemblage variation.

9.4 Bacteria

Knowledge of the general role of bacteria in streams was significantly advanced by the decades of research on the Breitenbach. Bacteria were found to be the main decomposers, accounting for 78% of heterotrophic and 59% of total community respiration. This confirmed the classical understanding of the ecological function of bacteria. However, bacteria were also found to constitute a large part (36%) of the total heterotrophic biomass, and to account for most of the heterotrophic production (71%), contrasting with the classical interpretation of their role. Bacterial production is of a similar magnitude to photoautotrophic net primary production: both about 0.20 kg C m^{-2} $year^{-1}$ in the Breitenbach.

Bacterial biomass therefore forms an important food resource for higher trophic levels. About 50% bacterial carbon production enters the food chain: about 35% BCP via protozoa and 15% directly to the metazoa. About 10% BCP (20 g C m^{-2} $year^{-1}$) is assimilated by the macrofauna. However, the total carbon demand of the macrofaunal community is met to a larger extent by dead particulate organic matter than by bacterial biomass.

Thus the role of bacteria is not restricted to the traditional concept as mineralisers of organic matter and nutrients fixed by primary producers, but must include their important food resource contribution to protozoan–metazoan food webs.

Molecular biological investigations on the bacterial community structure in the Breitenbach were just developing when the institute was closed. Thus many open questions remain. Among the final results was the observation that distinct bacterial communities occur in different stream habitats. In particular there were striking differences between stream water and sediment communities. It also seems that the classical assumption that the very small size of bacteria in environments such as the Breitenbach is due to the presence of mainly starved cells must be questioned. In recent years large numbers of very small bacteria (ultramicrobacteria <0.1 μm^3 in size) have been observed in freshwaters (e.g., *Polynucleobacter*-like organisms). Among the bacterial 16S rRNA gene sequences detected in the Breitenbach and other nearby streams, those similar to, or even identical with, *Polynucleobacter* sequences were most frequent.

The role of bacteria remains far from fully understood. For example, research into the fate of bacterial biomass is missing (e.g., how much is lysed by viruses?), as are studies of bacterial production in different stream habitats (e.g., in biofilms on macrophyte surfaces or in CPOM accumulations). The potentially important role of fungi in the Breitenbach is also remarkably understudied

9.5
Organic Matter Balance and Carbon Flow

For the period 1973–1998, the average organic matter balance of the Breitenbach was fairly even, with total inputs of 1.88 kg C m^{-2} $year^{-1}$ and total outputs of 1.84 kg C m^{-2} $year^{-1}$. It shows that, even in an open grassland stream like the Breitenbach, allochthonous organic matter (particulate as well as dissolved) is the main carbon and energy source for the stream community, three or four times more than gross primary production.

Our synthesis of the role of bacteria and organic matter dynamics in the Breitenbach was based on a considerable amount of data generated over more than 30 years. Although it still contains many assumptions, estimates, and sometimes even speculation, it presents a consistent picture of the general role of bacteria in the carbon flux of the Breitenbach, serving as an example of the general importance of bacteria in streams.

9.6 Fauna

The faunal inventory contains more than 2000 species, several new to science. The size spectrum of the fauna peaked at very small species. This may be an adaptation to the smallness of hyporheic refugia and/or to the strong predominance of FPOM and bacterial biomass in the organic matter pool. However, two largely bacterivorous groups produced only small amounts of biomass (Nematoda, *circa* $0.6\,g\,C\,m^{-2}$ $year^{-1}$), or were estimated to consume less than 1% bacterial biomass (Ciliophora).

General biological information and accounts of Breitenbach representatives have been provided for all groups, but studies focused on the pterygote insects, more than 820 species. The primarily terrestrial insects met the main challenges of life in water, respiration and osmoregulation, in many different ways.

The Breitenbach fauna was not unusual, most species inhabit numerous other rhithral streams in the area. However, montane taxa, species requiring wide open interstices in bottom sediments, or those that live in stable galleries in fine deposits were absent. Titanophilous species were also absent, or rare. The Ciliophora included species known from streams and standing waters, from very clean to polluted.

9.6.1 The Dominant Pterygota

Information on the life history, developmental biology, trophic relations of adults and larva, and other aspects of the ecology of the dominant orders (Ephemeroptera, Plecoptera, Trichoptera, Diptera, Coleoptera) is presented for many years. Adults were caught with emergence traps, aquatic stages mainly with a Surber sampler.

Emergence Traps

One trap, operated continuously from 1969 to 2006, provided information on abundance variability affected by environmental factors over time. A multitrap arrangement (1983–1993) also revealed differences in along stream distribution or timing of emergence, as well as upstream oviposition flights along the Breitenbach. Over 500 terrestrial species of Coleoptera, plus other terrestrial insects collected in the traps, proved that emergence traps also acted as activity traps.

The fauna was remarkably stable and the dominant taxa remained almost the same over the 37 year study period. However, numbers emerging varied markedly from year to year, sometimes by orders of magnitude.

Combined approaches suggested that the emergence of adult insects was controlled in several ways:

- The strength of year classes of individual taxa was largely determined by the hydrological regime and related environmental variables.
- The timing of emergence in different years depended on photoperiod and degree-days during development. It therefore depended on the thermal regime in individual years and was susceptible to global change.
- Adverse weather conditions temporarily suppressed adult emergence.

Surber Samples

Nine parallel samples per month showed good agreement on the size structure of populations, providing detailed information on growth and development. Specimen numbers between parallel samples varied too much for reliable estimates of secondary benthic production. More parallel samples would have been destructive to the small stream and impossible to handle.

9.6.1.1 Ephemeroptera

There was considerable overlap with the fauna of the nearby River Fulda. Of the 16 Breitenbach species, only some were regularly abundant in the entire stream. Faunal exchange was suggested by the sudden appearance of several species in the Breitenbach after efficient waste water treatment had largely restored the River Fulda. The same species were absent from similar streams near Schlitz, which did not drain into the River Fulda. Grazers (two *Baetis* spp., several Heptageniidae) and grazer–gatherers (two *Ephemerella* spp.) dominated, while detritivores constituted a minor faunal component.

Irregular emergence patterns and variable voltinism of the dominant *Baetis vernus* and *B. rhodani* seemed to result from faunal exchange interfering with population dynamics. Larval drift and cohort-splitting also played a role. In most other species, known unimodal standard emergence patterns were observed. Abundance of individual taxa differed along the stream. In the sand-dwelling *Centroptilum luteolum* emergence patterns also differed between sites and years. Seasonal size decline occurred in several species. Limited knowledge of structural change during metamorphosis did not support early emergence at a small size as predator avoidance strategy of as has been suggested in the literature.

In the univoltine *Ephemerella* species, temporal separation of emergent adults was maintained during larval life. The *Baetis* spp. were at best partially temporally separate.

Mating and oviposition occurred in flight, but *Baetis* females crept under water to oviposit on emergent hard substrata. The availability of oviposition sites varied with discharge. Experimental manipulation of oviposition sites permitted estimation of loss and survival rates, which were found to be of the same order of magnitude during the several months of larval development and the few days of adult life.

Numerous larval moults and stepwise increase of biomass in mayflies, stoneflies, and many beetles permitted finer resolution of their life cycles than for taxa with few instars and important intrastadial weight gain, for example the Trichoptera.

9.6.1.2 Plecoptera

Stoneflies emerged during the entire frost-free period, beginning in January, but with most in late spring to early summer. Emergence was synchronous and unimodal except where cohort-splitting and partial semivoltinism (*I. goertzi, L. nigra*) or partial plurivoltinism (*N. pictetii*) occurred.

All stoneflies occurred regularly in the Breitenbach; none was shared with the River Fulda. Single errant adults of potamal species were occasionally caught in the trap closest to the River Fulda, but their larvae were never found in the Breitenbach. Both *Brachyptera* species preferred upstream sites, *Nemoura flexuosa* was commonest in the downstream section. They were only recorded when several traps were operated along the entire length of the stream.

The scraping *Brachyptera* spp. mainly grazed on epilithic algae. Periphyton was extensively consumed by *Nemurella pictetii*, a mainly opportunistic detritivore, as were most other Breitenbach Plecoptera. *Isoperla goertzi* and *S. torrentium* were among the few predators in the stream.

Earlier published claims of temporal segregation of congeners *via* very short specific flight periods were disproved. Emergence periods of three closely related *Nemoura* spp. overlapped considerably. Development was synchronous, without evidence of competition or competition-avoidance. Partial temporal and spatial separation was observed in *Protonemura* spp. The three distantly related *Leuctra* species emerged and oviposited at different times. Nevertheless, larval size spectra overlapped through cohort splitting in *L. nigra*. After an insecticide spill the empty stream was re-stocked with more than 120 000 hatchlings of *Leuctra inermis* from the Rhön Mountains, which developed together with hatchlings from fresh oviposition by the other Plecoptera. These species disappeared again after a few years.

Emergence cues. As soon as air temperature rose above freezing point, emergence began even when stream water temperatures were lowest. There was no evidence of threshold water temperatures. In *Nemurella pictetii*, thermal regime determined how many cohorts emerged per year. There was evidence of photoperiod acting as an emergence cue.

Reproduction, development, and seasonal size decline. In *Leuctra* spp. mating occurred upon emergence, after adult feeding in other Nemouroidea and often far from the stream. Except for *Isoperla*, egg biomass was generated by terrestrial adult feeding leading to distinct weight gains. Most adults consumed fine detritus or grazed on epiphytic algae on tree trunks, but *Siphonoperla* gathered large amounts of conifer pollen. Eggs were laid into water in flight and dispersed; in *Nemurella*, batches were laid on substrata in moist littoral sites.

Eggs of most species developed directly. Incubation periods depended on temperature but were also characteristic of individual families. A few species passed through facultative or obligatory diapause. Hatch success was generally close to 100%. Most species exhibited no pause in growth during winter, but biomass accumulation was accelerated when daylength increased.

Stoneflies passed through a variable number of instars. Their fixed three-step-mode of wing development pre-determined adult size months before emergence, in the antepenultimate instar, possibly even earlier. In *Nemurella pictetii* the antepenultimate instar exhibited a size decline with time. Directional change of photoperiod and daylength have been postulated as increasingly intense developmental triggers, inducing metamorphosis in successively smaller individuals over time. The reversed autumnal size trend of several late-flying species was in line with this interpretation.

9.6.1.3 Trichoptera

Seventy-four caddis flies were collected. The identification of two taxa remains uncertain and requires molecular methods. The taxocenose contains several widespread euryecious species and is affected by the nearby River Fulda and by small ponds in the valley, but rheophilic, cold stenothermous stream species are dominant. Larvae construct fixed shelters, filter nets or transportable cases. Adults are on the wing from March to December, emergence lasting three to five, or even more, months. In the upstream area, life cycles of several species appear desynchronized with extended emergence periods. In general, a life cycle comprises egg, five larval instars, pupa, and adult, but *Agapetus fuscipes* had seven larval instars. Instar number remains unclear for *Sericostoma personatum*. Most species were univoltine, but larval populations of *Drusus annulatus* and *Rhyacophila fasciata* were more complex and exhibited bimodal emergence patterns, suggesting cohort-splitting and partial semivoltinism. Development of *S. personatum* may last up to three years or more.

Life cycles of several species were examined in detail in the field and in the laboratory. The most abundant species produced 60–65% males. Adult life lasts two to three days in female *A. fimbriata* and up to four weeks in *C. villosa*. For the Breitenbach species, a colonization cycle with upstream flight of mated females was not found. Pheromones were identified from several species and successfully tested with *A. fimbriata* in the field. Females deposit single egg masses of different size and shape on specific sites. *Agapetes fuscipes* females produce several egg masses, crawling under water to oviposit and carrying a cap stone which is placed over the eggs, probably to protect against mechanical disturbance, enemies and drying. Egg development usually starts after oviposition, but in *S. personatum* some egg masses entered a diapause. Embryogenesis was temperature-dependent and described by power functions. An upper temperature limit for successful development (18–20 °C) was detected for *A. fimbriata* and *D. annulatus*. Diurnal temperature fluctuation (laboratory) reduced the developmental period by up to 40% in *C. villosa* and *Anabolia nervosa*.

Up to 80% or more of adult weight is gained during last larval instar. Field experiments with shredding *C. villosa* larvae (and *Gammarus* spp.) indicated that the seasonal course of water temperature limits growth rather than food availability. Only strict experimental exclusion from CPOM showed significant effects of food supply.

Apatania fimbriata larvae exhibit anti-predatory behavior. They use fatty acids against predators (*Rhyacophila* larva) invading the case, but these are not effective against larger predators (e.g., fish). *Apatania fimbriata* larvae also pass through winter dormancy. *Sericostoma personatum* larvae have a diurnal feeding cycle; they feed at night on CPOM on the stream bottom and rest in the sediment by day, excreting FPOM. Hatchlings of *P. conspersa* live for some days in a collective net, probably thereby increasing their survival rate.

Seasonal size decline was observed in caddis flies. In years when the grazer *A. fuscipes* became very abundant the weight of individuals decreased (intraspecific competition). The weight of the competing *A. fimbriata* and *D. annulatus* (interspecific competition) was also affected. By late summer, after most specimens had emerged, weights of specimens again increased (decreased competition).

9.6.1.4 **Diptera**

Twenty-two families of aquatic Diptera, with over 800 species, were determined from the emergence traps. Most families were not studied continuously due to large specimen and species numbers, and the absence of taxonomic specialists. The species lists indicate the enormous biodiversity in the stream.

Chironomidae were analyzed for five subsequent years, and for one year of a field experiment with an artificially modified stream bottom. Simuliidae were studied intensively for several years; eight species regularly populate the stream, four of which are abundant. *Simulium cryophilum* dominates in the crenic region, *S. vernum* in the middle reach, and *S. ornatum* downstream; *P. tomosvaryi* is distributed evenly along the stream. Life cycle studies indicate that *P. tomosvaryi* is absent from the stream in summer, larvae develop synchronously over the winter. The life cycles of *S. cryophilum* and *S. vernum* are similar and highly desynchronized, due to extended flight and oviposition periods, and possibly retarded embryonic development. Population dynamics of *S. ornatum* were complex, partly due to effects of the River Fulda population. Females invaded the Breitenbach valley in large numbers from the River Fulda and oviposited on stream vegetation. Conspicuous numbers of other species that undoubtedly emerged from the River Fulda were even captured in the uppermost section of the Breitenbach, about 3.5 km from the River Fulda.

Larvae of *Ptychoptera paludosa*, an abundant phantom cranefly, feed several centimeters deep in soft substrata, but deposit the FPOM-rich feces on the sediment surface, making the organic material available to other detrital feeders. More information on other families is beyond the scope of this text and will be published elsewhere.

9.6.1.5 **Coleoptera**

The many aquatic beetles included some very rare species. However, most were typical inhabitants of standing waters and good flyers. Most Dytiscidae and all Haliplidae caught in the Breitenbach were probably in transit, having drifted in from ponds. *Limnius perrisi*, *Elmis aenea* and *E. maugetii* (Elmidae) formed the core of the Breitenbach fauna coexisting everywhere in the stream. An insecticide poisoning left a long-lasting mark in population structures. From this, a longer generation time than reported in literature was estimated for the Breitenbach population of *Limnius perrisi*. Many *Oulimnius tuberculatus*, plus additional taxa, drifted into the study area after a long winter drought but their larvae were not, or only exceptionally, found. *Oulimnius* remained rare after the insecticide accident.

Odeles marginata (semivoltine) and *Elodes minuta* were abundant Scirtidae. Larvae pass through many more instars than are reported in the literature. *Elodes* exhibits cohort splitting, some larvae requiring more than one year until they pupate on land. Studies at the Breitenbach stimulated additional research into previously unknown aspects of scirtid biology and morphology.

9.6.2
Scrapers of the Breitenbach Community

The functional feeding group "scrapers" is fascinating because the interactions of photoautotrophs and herbivores are very important in unshaded streams. They provide ideal systems to explore "bottom up" and "top down" effects, or to test for intra- and interspecific competition and its consequences on the structure and function of the phototrophic and grazer community.

The microphytobenthic community of the biofilm is strongly controlled by scrapers. The photoautotrophs are not limited by nutrient concentrations, but when grazing pressure on the microphytobenthos greatly exceeds productivity, grazers must compete for a limited food resource.

High feeding efficiency reduces microphytobenthos biomass and thereby keeps the epilithic community in an earlier (younger) successional stage, which affects its production rates positively. In turn, the productive capacity of the autotrophs affects growth and development of the consumers positively. Specimens of different species do not use the biofilm uniformly, but specific feeding, behavioral adaptations, and strategies permit the coexistence of species.

Ten grazers were sufficiently abundant to be used in field studies and laboratory experiments: six Trichoptera, three Ephemeroptera, and the mollusc *Ancylus fluviatilis*. However, there was special emphasis on caddisflies. In general their larvae live in cases or retreats, providing physical protection, camouflage, and acting as ballast and a respiration device. In the upper and middle part of the stream three abundant caddisflies and several instars per species coexist at any one time.

Architecture and particle size selection for case construction differed between species. *Drusus annulatus, Apatania fimbriata, Silo pallipes,* and *Micrasema longulum* build portable cases of silk and small sand grains, which are widened and lengthened as the larvae grow. *Agapetus fuscipes* constructs dome-shaped cases of larger sand grains and builds a completely new, larger case after each molt. Exceptionally, this caddisfly has seven (instead of the usual five) instars, of which up to five coexist at any one time. The two additional molts may be an adaptive response to increased mouthpart abrasion on the rough Bunter sandstone. Finally, *Tinodes rostocki* larvae spin silken "galleries" on stones, which are enlarged as the larvae grow and incorporate inorganic and organic particles into the retreat walls.

Among other factors, the proportion of case to larval weight differed between species. It was highest in *A. fuscipes* (44), intermediate in *A. fimbriata* (16), and low in *D. annulatus* (9). Case weight affects locomotive ability. *Agapetus fuscipes* larvae are least mobile. They occupy small feeding areas, which they defend against interspecific competitors for the biofilm standing crop. Mobility is higher in *A. fimbriata* and particularly *D. annulatus*, but comparatively low in *T. rostocki*.

Although there is high within-species variability, dietary composition differs between species. Laboratory experiments showed that larvae are able to distinguish certain food items and identify sites that fulfill their specific requirements. *Drusus annulatus*, *A. fimbriata*, and *M. longulum* preferred periphyton over detritus, *S. pallipes* and *A. fuscipes* had no preference. *Agapetus fuscipes* avoided thick biofilms while *D. annulatus* preferred high periphyton biomass. Of the larvae collected in the field, *A. fuscipes* and *S. pallipes* had also high proportions of detrital POM in their guts. In summer, *A. fimbriata* and *D. annulatus* preferred cyanobacteria, while *T. rostocki*, *D. annulatus*, and *M. longulum* selected high proportions of diatoms throughout the year, and *M. longulum* also fed on mosses.

Seasonal variability of the gut contents of larvae collected in the field mirrored the periphyton composition of the stone surfaces with a certain time lag. High amounts of detritus in the guts reflected periods of sedimentation. Distribution of scrapers along the stream may not be affected by food quality or availability.

Agapetus fuscipes larvae were the only taxon able to reduce the standing crop of the biofilm significantly, the other species could not. It fed at sites with thin biofilm layers and small diatoms, while *D. annulatus* consumed all available periphyton cells at sites with thicker biofilms. *Drusus annulatus* larvae were outcompeted at sites with high *A. fuscipes* densities. Due to their higher mobility and lower case/larval mass proportion, they can quite rapidly switch to food supplies under higher current velocity, out of reach for *A. fuscipes*. After feeding *D. annulatus* returns to low current velocity areas where they obviously rest.

Crowding problems and feeding strategies were studied in *Tinodes rostocki*. Larvae build silken retreats on stones in which phototrophs, detritus and mineral particles are incorporated. *Tinodes* seem to control the microenvironment within the retreats. In the light, oxygen concentration (>150%) and pH (up to 10) increased compared to the external environment. Larvae feed exclusively on the phototrophs, predominantly diatoms, growing on the retreat silk, and on the trapped detritus. This "gardening" activity affects the composition of the biofilm on the retreats (=food quality). *Tinodes rostocki* larvae consume the older parts of the retreats which are richer in nutrients. The retreat wall biofilm community differs from that on the surrounding stones. Manipulation experiments showed that large larvae in particular leave positions where there is increased intraspecific competition.

Resource and niche overlap between the caddisfly species, and the absence of temporal segregation in development, lead to increasing within- and between-species competition. The multitude of specific physiological, morphological, behavioral, and nutritional requirements, together with body size, life cycle, and case construction, affect within- and between-species competition.

9.7
Ecosystems

9.7.1
The Stream Bottom

The stream bottom is the most important compartment of the stream with respect to metabolism and organismal abundance. Weathering of Bunter sandstone generates stones, gravel, sand, silt, and large clay deposits below the stream. Grain size structure and POM content of the upper sediment layers depend on the discharge in antecedent periods. After periods of low flow, fine-grained sediments dominate on the stream bottom. The extent of the oxygenated hyporheic habitat that can be utilized by the macrobenthos is highly variable between seasons and years. The abundance of typical subterranean taxa was low. The extent of the oxygen-saturated zone depends on discharge. At high discharge and in coarse substrata, it extends to at least 20 cm, at low discharge and in fine-grained sediments with high FPOM content, it is often only a few millimeters. Complex food webs, of high importance for the stream system, also occur in poorly oxygenated and anoxic hyporheic habitats. Large amounts of groundwater with high DOM load enter the stream via diffuse perfusion through the streambed. The DOM is, to a large extent, immobilized on the surfaces of sediment particles and utilized later by the microbial community, thus entering the food chain and being made available to higher trophic levels.

In a field experiment the stream bottom under an emergence trap was artificially covered with sand. Compared with an untreated trap the abundance of 13% of the aquatic insects changed, particularly the abundance of the grazing mayflies. *Baetis vernus* and *B. rhodani* decreased by more than 95%. Abundance of *Centroptilum luteolum, Leuctra nigra* and others increased. Losses in sand-avoiding species were not compensated by sand-preferring species, so that the biomass over the sand trap was lower than over the unaltered reference trap.

9.7.2
Ordination

Precipitation, temperature, and discharge were the dominant factors explaining the major portion of the annual variability of the insect community (EPT). Long-term variation in temperature was seasonally predictable, precipitation (monthly scale) almost constant. Neither explained the observed variation in the community; however, the discharge year did (October one year to September the next year). Seasonal discharge variability could be categorized as nonseasonal, wet year, spring spate, and dry year. Indirect and direct ordination methods successfully related the respective "community state" with one of these discharge patterns. Species abundance changes in relation to discharge pattern were plausible particularly in relation to their ecological requirements and life cycles. Since the early 1990s only nonseasonal and low flow patterns have occurred – probably a harbinger of global change. This correlates with the hydrological pattern.

From 1983 to 1993, four emergence traps were operated simultaneously, at about equal distances from the source. Results agreed with the long-term data above. In addition to the temporal variation in the community, conspicuous spatial organization along the stream was evident. Ordination arranged sites along two axes, the first in relation to discharge, the second with annual temperature maxima and minima at sites. Community states at individual sites were characteristic for discharge pattern and temperature. "*A priori*" discriminant analyses with spatial and temporal "summary variables" (four sites, three discharge patterns) classified the sites into significantly distinguished and readily interpretable groups. In both cases the first axis accounted for at least 90% of the total variance.

9.7.3 Modeling

Artificial neural nets (ANN) were used to predict interrelationships between environmental variables and between variables and species. Precipitation-discharge models, based on temperature, long- and short-term precipitation, had quite low determination coefficients ($D = 0.34$). If previous days' discharge was included, D increased to 0.92. The quality of prognoses of annual EPT species abundance based on monthly precipitation, discharge, water temperature data, and abundance of the parent generation varied among species and between models. Effects were greatest from February to May, precipitation and discharge particularly in February and March. Determination coefficients for individual species models were generally adequate. Prognoses for monthly emergence patterns of *B. vernus* were usually high ($D > 0.8$; five of ten cases); D was below 0.3 in two cases. Models developed for a particular site are difficult to transfer to other situations. With the aid of neural models and specific dimension-reducing, pre-processing methods, bioindication, ecological prediction, and the analysis of cause–effect relationships can be substantially improved.

9.7.4 Erlenbach

The question whether individual streams are unique was invesigated by comparing the Erlenbach and the Breitenbach. The communities and the land use in the respective environments were almost identical, discharge of the Erlenbach was about 10% of the Breitenbach. It was expected that all the major environmental variables, community, seasonal variation in factors and species, and variation in abundance over time would act in a similar way. From 1998 to 2005 abundance varied in parallel for almost all EPT species and Diptera. The high similarity of both communities is a significant finding. Although the taxon abundances were not identical the trends were the same over almost nine years.

Thus, we concluded that the streams were similar but not identical. Geological setting and climate were also similar to many other rhithral streams in Bunter sandstone highlands. The fauna also largely agrees between these streams. The closer comparison and agreement with the Erlenbach suggest a general similarity between headwater streams in the area. Findings on one particular stream, the Breitenbach, therefore seem valid and applicable to numerous other streams in Central European highlands.

9.7.5
Global Change

The dates of mean cumulative emergence (50%) of the common Ephemeroptera, Plecoptera, and Trichoptera were expressed as Julian days and compared to overall mean emergence dates over the study years. Earlier/later than average emergence occurred synchronously in most species and was correlated with degree-days during development. Regressions between calendar year and mean emergence date were negative in most species, indicating a trend for increasingly earlier emergence, but were significant in only some species. Emergence of the estival–autumnal *Protonemura auberti* and the very early first generation of *Baetis rhodani* was significantly delayed. For Trichoptera, species emerging before Julian day 200 (*A. fuscipes, A. fimbriata, S. pallipes, T. rostocki,* first cohorts of *R. fasciata* and *D. annulatus*–mainly scrapers) tended to emerge earlier the warmer the year. Whether this is correlated with earlier development of the biofilm/algae remains unknown. At the species level, *Potamophylax cingulatus,* and the second cohort of *D. annulatus* emerged later. For the genus *Potamophylax,* the vernal species, *P. luctuosus,* emerged earlier, the autumnal *P. cingulatus* later.

References

Adis, J., and Kramer, E. (1975) Formaldehyd-Lösung attrahiert *Carabus problematicus* (Coleoptera – Carabidae). *Entomol. Ger.*, **2**, 121–125.

Adler, P.H., and Crosskey, R.W. (2009) World Blackflies (Diptera: Simuliidae): a comprehensive revision of the taxonomic and geographical inventory. http://entweb.clemson.edu/biomia/pdfs/blackflyinventory.pdf.

Ainsworth, A.M., and Goulder, R. (2000a) Epilithic and planktonic leucine aminopeptidase activity and leucine assimilation along the River Tweed, Scottish Borders. *Sci. Total Environ.*, **251/252**, 83–93.

Ainsworth, A.M., and Goulder, R. (2000b) Downstream change in leucine aminopeptidase activity and leucine assimilation by epilithic microbiota along the River Swale, northern England. *Sci. Total Environ.*, **251/252**, 191–204.

Alderson, R. (1969) Studies of the larval biology of caddisflies on the family Psychomyiidae. Ph.D. thesis, University of Wales, Cardiff.

Alecke, C. (1998) Ökologie und Habitatbindung von *Tinodes unicolor* und *Tinodes pallidulus* (Trichoptera; Insecta). Ein Beitrag zur Typisierung und Bewertung von Tieflandbächen. Ph.D. thesis, University of Münster, Münster.

Alecke, C., Spänhoff, B., and Meyer, E.I. (2005) Avoidance behaviour in the gallery construction of *Tinodes unicolor* (Psychomyiidae, Trichoptera) to prevent intraspecific encounters. *Arch. Hydrobiol.*, **164**, 335–343.

Alf, A. (1991) Neu- und wiedergefundene Arten des Makrozoobenthon im Neckar. *Lauterbornia*, **8**, 71–76.

Allan, J.D. (1981) Determinants of diet of brook trout (*Salvelinus fontinalis*) in a moutain stream. *Can. J. Fish. Aquat. Sci.*, **38**, 184–192.

Allan, J.D. (1995) *Stream Ecology. Structure and Function of Running Waters.* Chapman and Hall, London, 388 pp.

Altschul, S.F., Madden, T.L., Schäffer, A.A., Zhang, J., Zhang, Z., Miller, W., and Lipman, D.J. (1997) Gapped BLAST and PSI-BLAST: a new generation of protein database search programs. *Nucleic Acids Res.*, **25**, 3389–3402.

Alvarez, M., and Pardo, I. (2005) Life history and production of *Agapetus quadratus* (Trichoptera: Glossosomatidae) in a temporary, spring-fed stream. *Freshwater Biol.*, **50**, 930–943.

Alvarez, M., and Peckarsky, B.L. (2005) How do grazers affect periphyton heterogeneity in streams? *Oecologia*, **142**, 576–587.

Amann, R., and Fuchs, B.M. (2008) Single-cell identification in microbial communities by improved fluorescence in situ hybridization techniques. *Nat. Rev. Microbiol.*, **6**, 339–348.

Amann, R., Binder, B.J., Olson, R.J., Chisholm, S.W., Devereux, R., and Stahl, D.A. (1990) Combination of 16S rRNA-targeted oligonucleotide probes with flow cytometry for analyzing mixed bacterial populations. *Appl. Environ. Microbiol.*, **56**, 1919–1925.

Amann, R.I. (1995) Fluorescently labeled, rRNA-targeted oligonucleotide probes in

Central European Stream Ecosystems: The Long Term Study of the Breitenbach, First Edition. Edited by Rüdiger Wagner, Jürgen Marxsen, Peter Zwick, Eileen J. Cox.

the study of microbial ecology. *Mol. Ecol.*, **4**, 543–554.

Amann, R.I., Ludwig, W., and Schleifer, K.-H. (1995) Phylogenetic identification and in situ detection of individual microbial cells without cultivation. *Microbiol. Rev.*, **59**, 143–169.

Amann, R.I., Fuchs, B.M., and Behrens, S. (2001) The identification of micro-organisms by fluorescence in situ hybridisation. *Curr. Opin. Biotechnol.*, **12**, 231–236.

Ambühl, H. (1959) Die Bedeutung der Strömung als ökologischer Faktor. *Schweiz. Z. Hydrol.*, **21**, 133–264.

Anderson, N.H. (1973) The eggs and oviposition behaviour of *Agapetus fuscipes* Curtis (Trich., Glossosomatidae). *Entomol. Mon. Mag.*, **109**, 128–131.

Anderson, N.H. (1976) Carnivory by an aquatic detritivore *Clistoronia magnifica* (Trichoptera Limnephilidae). *Ecology*, **57**, 1081–1085.

Anderson, N.H., and Bourne, J.B. (1974) Bionomics of three species of glossosomatid caddis flies (Trichoptera: Glossosomatidae) in Oregon. *Can. J. Zool.*, **52**, 405–411.

Angus, R. (1992) Insecta, Coleoptera, Hydrophilidae, Helophorinae, in *Süßwasserfauna von Mitteleuropa 20* (eds J. Schwoerbel and P. Zwick), Gustav Fischer, Heidelberg, 144 pp.

Antoine, S.E., and Benson-Evans, K. (1985) Colonisation rates of benthic algae on four different rock substrates in the River Ithon, Mid. Wales, U.K. *Limnologica*, **16**, 307–313.

Antoine, S.E., and Benson-Evans, K. (1986) The epipelic algal flora of the River Wye system, Wales, U.K. Part 2. Algal phyla and species population dynamics. *Int. Rev. Ges. Hydrobiol.*, **71**, 813–839.

Arens, W. (1989) Comparative functional morphology of the mouthparts of stream animals feeding on epilithic algae. *Arch. Hydrobiol. Suppl.*, **83**, 253–354.

Arens, W. (1990) Wear and tear of mouth-parts: a critical problem in stream animals feeding on epilithic algae. *Can. J. Zool.*, **68**, 1896–1914.

Armitage, P.D. (1995) Faunal community change in response to flow manipulation. in *The ecological basis for river management.* (eds D. Harper and A. Ferguson), J. Wiley and Sons Ltd, Chichester; pp. 59–78.

Armitage, P.D., and Gunn, R.J.M. (1996) Different response of benthos to natural and anthropogenic disturbance in 3 lowland streams. *Int. Revue Ges. Hydrobiol.*, **81**, 161–181.

Arndt, H., Dietrich, D., Auer, B., Cleven, E.J., Gräfenhan, T., Weitere, M., and Mylnikow, A.P. (2000) Functional diversity of heterotrophic flagellates in aquatic ecosystems, in *The Flagellates: Unity, Diversity and Evolution* (eds B.S.C. Leadbeater and J.C. Green), Taylor and Francis, Basingstoke, pp. 240–268.

Atkinson, D., and Sibly, R.M. (1997) Why are organisms usually bigger in colder environments? Making sense of a life history puzzle. *Trends Ecol. Evol.*, **12**, 235–239.

Aubert, J. (1946) Les Plécoptères de la Suisse Romande. *Mitt. Schweiz. Entomol. Ges.*, **20**, 8–128.

Aubert, J. (1963) Les Plécoptères des cours d'eau temporaires de la péninsule ibérique. *Mitt. Schweiz. Entomol. Ges.*, **35**, 301–315.

Aurich, M. (1989) Der Lebenszyklus von *Apatania fimbriata* Pictet (Trichoptera) im Breitenbach. Diploma thesis, University of Bremen, Bremen, 82+9 pp.

Aurich, M. (1992) The life-cycle of *Apatania fimbriata* Pictet in the Breitenbach. *Hydrobiologia*, **239**, 65–78.

Aurich, M., Wagner, R., Reder, E., and Veith, R. (1991) Defensive behaviour of the larva of *Apatania fimbriata* (Pictet). *Verh. Int. Ver. Theor. Angew. Limnol.*, **24**, 2901.

Avigad, G. (1968) A modified procedure for the colorimetric, ultramicro determination of reducing sugars with the alkaline ferricyanide reagent. *Carbohydr. Res.*, **7**, 94–97.

Aykulu, G. (1982) The epipelic algal flora of the River Avon. *Br. Phycol. J.*, **17**, 27–38.

Azam, F., Fenchel, T., Field, J.G., Gray, J.S., Meyer-Reil, L.A., and Thingstad, F. (1983) The ecological role of water-column microbes in the sea. *Mar. Ecol. Prog. Ser.*, **10**, 257–263.

Bach, H.J., Hartmann, A., Schloter, M., and Munch, J.C. (2001) PCR primers and functional probes for amplification and detection of bacterial genes for extracellular peptidases in single strains and in soil. *J. Microbiol. Methods*, **44**, 173–182.

Bacher, I., and Waringer, J.A. (1996) Hydraulic microdistribution of cased caddis larvae in an Austrian mountain brook. *Int. Revue Ges. Hydrobiol.*, **81**, 541–554.

Bährmann, R., and Bellstedt, R. (1988) Beobachtungen und Untersuchungen zum Vorkommen der Lonchopteriden auf dem Gebiet der DDR, mit einer Bestimmungstabelle der Arten. *Dtsch Entomol. Z. N.F.*, **35**, 265–279.

Baldy, V., Chauvet, E., Charcosset, J.Y., and Gessner, M.O. (2002) Microbial dynamics associated with leaves decomposing in the mainstem and floodplain pond of a large river. *Aquat. Microb. Ecol.*, **28**, 25–36.

Baltes, B. (2000) Einfluss der Gewässerversauerung auf aquatische Insekten. *Mitt. Dtsch Ges. Allg. Angew. Entomol.*, **12**, 231–235.

Barmuta, L.A. (1990) Interaction between the effects of substratum, velocity and location on stream benthos: an experiment. *Aust. J. Mar. Freshwater Res.*, **41**, 557–573.

Barnese, L.E., and Lowe, R.L. (1992) Effects of substrate, light, and benthic invertebrates on algal drift in small streams. *J. N. Am. Benthol. Soc.*, **11**, 49–59.

Barr, W.B. (1982) Attachment and silk of larvae of *Simulium vittatum* Zetterstedt (Diptera: Simuliidae). MS Thesis, Edmonton, Alberta.

Battin, T.J. (2000) Hydrodynamics is a major determinant of streambed biofilm activity: from the sediment to the reach scale. *Limnol. Oceanogr.*, **45**, 1308–1319.

Battin, T.J., Wille, A., Sattler, B., and Psenner, R. (2001) Phylogenetic and functional heterogeneity of sediment biofilms along environmental gradients in a glacial stream. *Appl. Environ. Microbiol.*, **67**, 799–807.

Baud, F. (1973) Biologie et cytologie de cinq espèces du genre *Lonchoptera* Meig. (Dipt.) don't l'une est parthénogénétique et les autres bisexuées avec quelques remarques d'ordre taxonomique. *Rev. Suisse Zool.*, **80**, 473–515.

Bauernfeind, E., and Humpesch, U. (2001) *Die Eintagsfliegen Zentraleuropas (Insecta: Ephemeroptera): Bestimmung und Ökologie*, Verlag des Naturhistorischen Museums, Vienna, 240 pp.

Bêche, L.A., Connors, P.G., Resh, V.H., and Merenlender, A.M. (2009) Resilience of fishes and invertebrates to prolonged drought in two California streams. *Ecography*, **32**, 778–788.

Becker, C., and Wagner, R. (2004) Chironomidae (Diptera) Emergence from two adjacent stream reaches with contrasting bottom substrata of a brook in Hesse/Germany in 1983. *Entomol. Gen.*, **27**, 105–122.

Becker, G., (1987a) Lebenszyklus, Reproduktion und ökophysiologische Anpassungen von *Hydropsyche contubernalis*, einer Köcherfliege mit Massenvorkommen im Rhein. Ph.D. thesis, University of Cologne, Cologne, 108 pp.

Becker, G., (1987b) Net-building behaviour, tolerance and development of two caddisfly species from the river Rhine (*Hydropsyche contubernalis* and *H. pellucidula*) in relation to the oxygen content. *Oecologia*, **73**, 242–250.

Becker, G. (1990) Comparison of the dietary composition of epilithic trichopteran species in a first order stream. *Arch. Hydrobiol.*, **120**, 13–40.

Becker, G. (1991) Ovipositing behaviour of *Agapetus fuscipes* (Trichoptera: Glossosomatidae) in a small Central European upland stream, in *Proc. Sixth Int. Symp. Trichoptera, 1989* (ed. C. Tomaszewski), Adam Mickiewicz University Press, Poznan, pp. 143–147.

Becker, G. (1993) Age structure and colonization of natural substrata by the epilithic caddisfly *Tinodes rostocki* (Trichoptera: Psychomyiidae). *Arch. Hydrobiol.*, **127**, 423–436.

Becker, G. (1994) Food preference by five trichopteran scrapers. *Hydrobiologia*, **273**, 171–178.

Becker, G. (2001) Larval size, case construction, and crawling velocity at different substratum roughness in three scraping caddis larvae. *Arch. Hydrobiol.*, **151**, 317–334.

Becker, G. (2005) Life cycle of *Agapetus fuscipes* (Trichoptera, Glossosomatidae) in a first-order upland stream in central Germany. *Limnologica*, **35**, 52–60.

Becker, G., and Holfeld, H. (2001) Response of trichopteran grazers to unialgal layers. *Verh. Int. Ver. Theor. Angew. Limnol.*, **27**, 2396–2397.

Becker, G., and Menzler, D.A. (1998) Einfluß der Weidetätigkeit von *Agapetus fuscipes* (Trichoptera) auf Biofilme und Nahrungskonkurrenten. *DGL Tagungsbericht 1997 (Frankfurt/M.), Krefeld*, 1, pp. 477–479.

Becker, G., Holfeld, H., Hasselrot, A.T., Fiebig, D.M., and Menzler, D.A. (1997) Use of a microscope photometer to analyze in vivo fluorescence intensity of epilithic microalgae grown on artificial substrata. *Appl. Environ. Microbiol.*, **63**, 1318–1325.

Beer-Stiller, A., and Zwick, P. (1995) Biometric studies of some stoneflies and a mayfly (Plecoptera and Ephemeroptera). *Hydrobiologia*, **299**, 169–178.

Béguin, P. (1990) Molecular biology of cellulose degradation. *Annu. Rev. Microbiol.*, **44**, 219–248.

Béguin, P., and Aubert, J.P. (1994) The biological degradation of cellulose. *FEMS Microbiol. Rev.*, **13**, 25–58.

Beier, M. (1948) Zur Kenntnis von Körperbau und Lebensweise der Helminen (Elminthidae, Dryopoidea). *EOS Rev. Esp. Entomol.*, **24**, 123–211.

Beier, M. (1949) Körperbau und Lebensweise der Larve von *Helodes hausmanni* Gredler (Col. Helodidae). *EOS Rev. Esp. Entomol.*, **25**, 49–100.

Beier, M. (1952) Bau und Funktion der Mundwerkzeuge bei den Helodiden-Larven (Col.). *Trans. Ninth Int. Congr. Entomol.*, **1**, 135–138.

Beier, S. (2003) Analyse der Bakteriengemeinschaften in Fließgewässern mittels TGGE und Sequenzanalyse. Diploma thesis, Technical University of Munich, Munich, 94 pp.

Beier, S., Witzel, K.P., and Marxsen, J. (2008) Bacterial community composition in Central European running waters examined by temperature gradient gel electrophoresis and sequence analysis of 16S rRNA genes. *Appl. Environ. Microbiol.*, **74**, 188–199.

Bellstedt, R., Stark, A., and Meyer, H. (1999) Dolichopodidae, in *Checkliste der Dipteren Deutschlands* (eds H. Schumann, R. Bährmann, and A. Stark), Studia Dipterologica *Supplement*, 2, Ampyx-Verlag, Halle-Saale, pp. 92–99.

Bendix, J. (2004) *Geländeklimatologie*, Gebrüder Bornträger, Verlagsbuchhandlung.

Benech, V. (1972a) La fécondité de *Baetis rhodani* Pictet. *Freshwater Biol.*, **2**, 337–354.

Benech, V. (1972b) Le polyvoltinisme chez *Baetis rhodani* Pictet (Insecta, Ephemeroptera) dans un ruisseau à truites des Pyrénées-Atlantiques, le Lissuraga. *Ann. Hydrobiol.*, **3**, 141–171.

Benech, V., and Vignes, J.V. (1972) Étude expérimentale de l'incubation des œufs de *Baetis rhodani* Pictet. *Freshwater Biol.*, **2**, 243–252.

Benedetto Castro, L. (1975) Ökologie und Produktionsbiologie von *Agapetus fuscipes* Curt. im Breitenbach 1971–1972. Schlitzer produktionsbiologische Studien (11). *Arch. Hydrobiol. Suppl.*, **45**, 305–375.

Bengtsson, J. (1973) Vækst og livscyklus hos *Baëtis rhodani* (Pict.) (Ephemeroptera). *Flora Fauna*, **79**, 32–34.

Bengtsson, J. (1988) Autecological studies on *Baëtis rhodani* (Pict.) (Ephemeroptera). *Arch. Hydrobiol.*, **111**, 615–623.

Benick, L. (1924) Zur Biologie der Käferfamilie Helodidae. *Mitt. Geogr. Ges. Lübeck (2)*, **29**, 47–75.

Benick, L. (1925) Beiträge zur Kenntnis der Tierwelt norddeutscher Quellgebiete. *Entomol. Blätt. Biol. Syst. Käfer*, **21**, 164–172.

Benke, A.C. (1984) Secondary production of aquatic insects, in *The Ecology of Aquatic Insects* (eds V.H. Resh and D.M. Rosenberg), Praeger, New York, pp. 289–322.

Benke, A.C., and Wallace, J.B. (1980) Trophic basis of production among net-spinning caddisflies in a southern Appalachian stream. *Ecology*, **61**, 108–118.

Benke, A.C., Huryn, A.D., Smock, L.A., and Wallace, J.B. (1999) Length–mass relationships for freshwater macroinvertebrates in North America with particular reference to the southeastern United States. *J. N. Am. Benthol. Soc.*, **18**, 308–343.

Benner, R., Lay, J., K'nees, E., and Hodson, R.E. (1988) Carbon conversion efficiency for bacterial growth on lignocellulose: implications for detritus-based food webs. *Limnol. Oceanogr.*, **33**, 1514–1526.

Benton, M.J. (1989) Energy budgets and reproductive ecologies of mayflies occupying disparate thermal environments. *Can. J. Zool.*, **67**, 2782–2791.

Benton, M.J., and Pritchard, G. (1988) New methods of mayfly instar number determination and growth curve estimation. *J. Freshwater Ecol.*, **4**, 361–367.

Berger, H., Foissner, W., and Kohmann, F. (1997) *Bestimmung und Ökologie der Mikrosaprobien nach DIN 38410*, Gustav Fischer Verlag, Stuttgart, Jena, I–X, 291 pp.

Bergey, E.A. (1995) Local effects of a sedentary grazer on stream algae. *Freshwater Biol.*, **33**, 401–409.

Bergey, E.A., and Resh, V.H. (1994) Effect of burrowing by a stream caddisfly on case-associated algae. *J. N. Am. Benthol. Soc.*, **13**, 379–390.

Bergmann, J., Löfstedt, C., Ivanov, V.D., and Francke, W. (2002) Electrophysiologically active compounds identified from six species of caddisflies (Trichoptera). In Proc. 10th Int. Symp. Trichoptera – Nov. Suppl. Entomol. (ed. W. Mey), vol. 15, pp. 37–46.

Bergquist, P.L., Gibbs, M.D., Morris, D.D., Te'o, V.S.J., Saul, D.J., and Morgan, H.W. (1999) Molecular diversity of thermophilic cellulolytic and hemicellulolytic bacteria. *FEMS Microbiol. Ecol.*, **28**, 99–110.

Berthélemy, C. (1979) Elmidae de la région paléarctique occidentale: systématique et répartition (Coleoptera Dryopoidea). *Ann. Limnol.*, **15**, 1–102.

Berthélemy, C., and de Riols, J. (1965) Les larves d'*Elmis* du groupe d' *E. maugetii* [Coléoptères Dryopoidea]. *Ann. Limnol.*, **1**, 21–38.

Berthélemy, C., and Ductor, M. (1965) Taxonomie larvaire et cycle biologique de six espèces d'*Esolus* et d'*Oulimnius* européens (Coleoptera Dryopoidea). *Ann. Limnol.*, **1**, 257–276.

Biermann, J. (1990) Abhängigkeit der Selbstreinigung von der Naturnähe der Gewässer. *Mitteilungen des Deutschen Verbandes* für Wasserwirtschaft und Kulturbau e.V (DVWK), Band 21, Bonn.

Biggs, B.J.F., and Close, M.E. (1989) Periphyton biomass dynamics in gravel bed rivers: the relative effects of flows and nutrients. *Freshwater Biol.*, **22**, 209–231.

Biggs, B.J.F., and Smith, R.A. (2002) Taxonomic richness of stream benthic algae: effects of flood disturbance and nutrients. *Limnol. Oceanogr.*, **47**, 1175–1186.

Bikandi, J., Millan, R.S., Rementeria, A., and Garaizar, J. (2004) In silico analysis of complete bacterial genomes: PCR, AFLP-PCR, and endonuclease restriction. *Bioinformatics*, **22**, 798–799.

Birch, L.C. (1957) The meaning of competition. *Am. Nat.*, **91**, 5–18.

Bishop, J.E., and Hynes, H.B.N. (1969) Downstream drift of the invertebrate fauna in a stream ecosystem. *Arch. Hydrobiol.*, **66**, 56–90.

Bisswanger, H. (1994) *Enzymkinetik. Theorie und Methoden*, 2nd Aufl., VCH Verlagsgesellschaft, Weinheim, 310 pp.

Blanckenhorn, W.U. (1997) Effects of temperature on growth, development, and diapause in the yellow dung fly – against all rules? *Oecologia*, **111**, 318–324.

Blinn, D.W., Fredericksen, A., and Korte, V., (1980) Colonization rates and community structure of diatoms on three different rock substrata in a lotic system. *Br. Phycol. J.*, **15**, 303–310.

Blomquist, G.J., Borgeson, C.E., and Vundla, M. (1991) Polyunsaturated fatty acids and eicosanoids in insects. *Insect Biochem.*, **21**, 99–106.

Bohle, H.W. (1969) Untersuchungen über die Embryonalentwicklung und die embryonale Diapause bei *Baëtis vernus* Curtis und *Baëtis rhodani* (Pictet) (Baëtidae, Ephemeroptera). *Zool. Jahrb. Abt. Anat. Ontog. Tiere*, **53**, 493–575.

Bohle, H.W. (1972a) Die Temperaturabhängigkeit der Embryogenese und der embryonalen Diapause von *Ephemerella ignita* (Poda) (Insecta, Ephemeroptera). *Oecologia (Berl.)*, **10**, 253–268.

Bohle, H.W. (1972b) Vergleichende Untersuchungen über den frühlarvalen Köcherbau der Brachycentridenarten *Micrasema minimum* MacLachlan und *Brachycentrus montanus* Klapalek (Trichoptera, Insecta). *Zool. Jahrb. Abt. Syst. Ökol. Geog. Tiere*, **99**, 507–544.

Bohle, H.W. (1974) Der frühlarvale Köcherbau von *Micrasema longulum* McL. (Trichoptera, Brachycentridae). *Int. Rev. Ges. Hydrobiol.*, **59**, 403–420.

Bohle, H.W. (1978) Beziehungen zwischen dem Nahrungsangebot, der Drift und der räumlichen Verteilung bei Larven von *Baetis rhodani* (Pictet) (Ephemeroptera: Baetidae). *Arch. Hydrobiol.*, **84**, 500–525.

Bohle, H.W. (2004) Larval and pupal case building in Trichoptera: a comment to articles of G.B. Wiggins and H. Malicky in Braueria 27 and 28. *Braueria*, **31**, 9–10.

Bohle, H.W., and Fischer, M. (1983) Struktur und Entstehung der Larven- und Puppengehäuse einiger Glossosomatidae und Rhyacophilidae, insbesondere bei *Synagapetus iridipennis* (Trichoptera: Rhyacophiloidea). *Entomol. Gen.*, **9**, 17–34.

Bohn, U. (1996) *Vegetationskarte der Bundesrepublik Deutschland 1:200 000 – Potentielle natürliche Vegetation – Blatt CC 5518 Fulda*, Schr. Reihe Veg. 15/2, erweiterte Auflage, BDR, Bonn-Bad Godesberg, 364 pp.

Boon, P.I. (1989) Organic matter degradation and nutrient regeneration in Australian freshwaters: I. Methods for exoenzyme assays in turbid aquatic environments. *Arch. Hydrobiol.*, **115**, 339–359.

Boon, P.I. (1990) Organic matter degradation and nutrient regeneration in Australian freshwaters: II. Spatial and temporal variation, and relation with environmental conditions. *Arch. Hydrobiol.*, **117**, 405–436.

Boon, P.I. (1991a) Enzyme activities in billabongs of southeastern Australia, in *Microbial Enzymes in Aquatic Environments* (ed. R. Chróst), Brock/Springer Series in Contemporary Bioscience, Springer, New York, pp. 286–297.

Boon, P.I. (1991b) Organic-matter degradation and nutrient regeneration in Australian fresh waters. 3. Size fractionation of phosphatase activity. *Arch. Hydrobiol.*, **126**, 339–360.

Boorman, J. (1997) Diptera Ceratopogonidae, in *Contributions to a Manual of Palaeartic Diptera Vol. 2* (eds L. Papp and P. Darvas), Science Herald, Budapest, pp. 349–368.

Borchardt, M.A., and Bott, T.L. (1995) Meiofaunal grazing of bacteria and algae in a Piedmont stream. *J. N. Am. Benthol. Soc.*, **14**, 278–298.

Bormann, F.H., Likens, G.E., and Eaton, J.S. (1969) Biotic regulation and solution losses from a forest ecosystem. *Bioscience*, **19**, 600–610.

Boschker, H.T.S., and Cappenberg, T.M. (1998) Patterns of extracellular enzyme activities in littoral sediments of Lake Gooimeer, The Netherlands. *FEMS Microbiol. Ecol.*, **25**, 79–86.

Botosaneanu, L., and Malicky, H. (1978) Trichoptera, in *Limnofauna Europaea* (ed. J. Illies), G. Fischer, Stuttgart, pp. 333–359.

Bott, T.L. (1996) Primary productivity and community respiration, in *Methods in Stream Ecology* (eds F.R. Hauer and G.A. Lamberti), Academic Press, Toronto, pp. 533–556.

Bott, T.L., and Kaplan, L.A. (1990) Potential for protozoan grazing of bacteria in streambed sediments. *J. N. Am. Benthol. Soc.*, **9**, 336–345.

Böttger, K. (1970) Die Ernährungsweise der Wassermilben (Hydrachnellae, Acari). *Int. Rev. Ges. Hydrobiol.*, **55**, 895–912.

Böttger, K. (1976) Types of parasitism by larvae of water mites (Acari: Hydrachnellae). *Freshwater Biol.*, **6**, 497–500.

Böttger, K. (1977) The general life cycle of fresh water mites (Hydrachnellae, Acari). *Acarologia*, **18**, 496–502.

Boulton, A.J., and Boon, P.I. (1991) A review of methodology used to measure leaf litter decomposition in lotic environments: time to turn over an old leaf? *Aust. J. Mar. Freshwater Res.*, **42**, 1–43.

Boulton, A.J., Peterson, C.G., Grimm, N.B., and Fisher, S.G. (1992) Stability of an aquatic macroinvertebrate community in a multiyear hydrologic disturbance regime. *Ecology*, **73**, 2192–2207.

Boulton, A.J., Thibault, D., Kasahara, T., Mutz, M., and Stanford, J.A. (2010)

Ecology and management of the hyporheic zone: stream–groundwater interactions of running waters and their floodplains. *J. N. Am. Benthol. Soc.*, **29**, 26–40.

Bournaud, M., (1972) Influence de la vitesse du courant sur l'activité de locomotion des larves de *Micropterna testacea* (Gmel.) (Trichoptera, Limnephilidae). *Ann. Limnol.*, **8**, 141–216.

Bournaud, M., (1974a) Rôle d'une force dans l'activité locomotrice des larves de *Micropterna testacea* (Gmel.) (Trichoptera, Limnephilidae): Comparaison avec l'activité dans le courant. *Arch. Hydrobiol.*, **73**, 417–463.

Bournaud, M. (1974b) A progress report on the locomotion behaviour of a larva of Limnephilidae (*Micropterna testacea*) in water currents, in *Proc. 1st Int. Symp. Trichoptera, 1974* (ed. H. Malicky), Junk, The Hague, pp. 203–204. (Beachten 1974a und 1974b)

Boyer, W.D. (1973) Air temperature, heat sums, and pollen shedding phenology of Longleaf Pine. *Ecology*, **54**(2), 420–426.

Boyero, L., Rincon, P.A., and Bosch, J. (2006) Case selection by a limnephilid caddisfly [*Potamophylax latipennis* (Curtis)] in response to different predators. *Behav. Ecol. Sociobiol.*, **59**, 364–372.

Bradley, D.C., and Ormerod, S.J. (2001) Community persistence among stream invertebrates tracks the North Atlantic Oscillation. *J. Anim. Ecol.*, **70**, 987–996.

Bratbak, G., (1993) Microscope methods for measuring bacterial biovolume: epifluorescence microscopy, scanning electron microscopy, and transmission electron microscopy, in *Handbook of Methods in Aquatic Microbial Ecology* (eds P.F. Kemp, B.F. Sherr, E.B. Sherr, and J.J. Cole), Lewis Publishers, Boca Raton, pp. 309–317.

Brehm, J., and Meijering, M.P.D. (1996) *Fließgewässerkunde: Einführung in die Ökologie der Quellen, Bäche und Flüsse*, 3., überarb. Aufl., Quelle und Meyer, Wiesbaden, 302 pp.

Bretschko, G. (1981) Vertical distribution of zoobenthos in an alpine brook of the Ritrodat-Lunz study area. *Verh. Int. Verein. Limnol.*, **21**, 873–876.

Bretschko, G. (1984) Quantitative sampling of the fauna of gravel streams (Project Ritrodat-Lunz). *Verh. Int. Verein. Limnol.*, **22**, 2049–2052.

Bretschko, G., and Klemens, W.E. (1986) Quantitative methods and aspects in the study of the interstitial fauna of running waters. *Stygologia*, **2**, 297–316.

Bretschko, G., and Leichtfried, M. (1987) The determination of organic matter in river sediments. *Arch. Hydrobiol. Suppl.*, **68**, 403–417.

Bretschko, G., and Moser, H. (1993) Transport and retention of matter in riparian ecotones. *Hydrobiologia*, **251**, 95–101.

Briers, R.A., Cariss, H.M., and Gee, J.H.R. (2002) Dispersal of adult stoneflies (Plecoptera) from upland streams draining catchments with contrasting land-use. *Arch. Hydrobiol.*, **155**, 627–644.

Briers, R.A., Cariss, H.M., Geoghegan, R., and Gee, J.H.R. (2005) The lateral extent of the subsidy from an upland stream to riparian lycosid spiders. *Ecography*, **28**, 165–170.

Brinck, P. (1949) Studies on Swedish stoneflies (Plecoptera). *Opusc. Entomol. Suppl.*, **11**, 1–250.

Brinkmann, R. (1991) Zur Habitatpräferenz und Phänologie der Limoniidae, Tipulidae und Cylindrotomidae (Dipteral) im Bereich eines norddeutschen Tieflandbaches, *Faun.-Ökol. Mitt. Suppl.*, **11**, 156 pp.

Brittain, J.E. (1978) Semivoltinism in mountain populations of *Nemurella pictetii* (Plecoptera). *Oikos*, **30**, 1–6.

Brittain, J.E. (1982) Biology of Mayflies. *Annu. Rev. Entomol.*, **27**, 119–147.

Brittain, J.E. (1990) Life history strategies in Ephemeroptera and Plecoptera, in *Mayflies and Stoneflies: Life Histories and Biology* (ed. I.C. Campbell), Kluwer Academic Publishers, Dordrecht, pp. 1–12.

Brittain, J.E., and Lillehammer, A. (1987) Variability in the rate of egg development of the stonefly, *Nemoura cinerea* (Plecoptera). *Freshwater Biol.*, **17**, 565–568.

Brittain, J.E., and Sartori, M. (2003) Ephemeroptera (Mayflies), in *Encyclopedia of Insects* (eds V.H. Resh and R.T. Cardé),

Academic Press, Amsterdam, pp. 373–380.

Brittain, J.E., Lillehammer, A., and Saltveit, S.J. (1984) The effect of temperature on intraspecific variation in egg biology and nymphal size in the stonefly *Capnia atra* (Plecoptera). *J. Anim. Ecol.*, **53**, 161–169.

Brock, V. (1985) Vergleichende Untersuchungen zur Entwicklung der Steinfliegen (Insecta, Plecoptera). Ph.D. thesis, University of Hamburg, Hamburg, VIII + 152 pp.

Brown, H.P. (1987) Biology of riffle beetles. *Annu. Rev. Entomol.*, **32**, 252–273.

Brunke, M., and Gonser, T. (1997) The ecological significance of exchange processes between rivers and groundwater. *Freshwater Biol.*, **37**, 1–33.

Brunke, M., and Gonser, T. (1999) Hyporheic invertebrates – the clinal nature of interstitial communities structured by hydrological exchange and environmental gradients. *J. N. Am. Benthol. Soc.*, **18**, 344–362.

Budde, H. (1928) Die Algenflora des Sauerländischen Gebirgsbaches. *Arch. Hydrobiol.*, **19**, 433–520.

Burgherr, P., and Meyer, E.I. (1997) Regression analysis of linear body dimensions vs. dry mass in stream macroinvertebrates. *Arch. Hydrobiol.*, **139**, 101–112.

Bürgis, H. (1993) Die Wasserschlupfwespe *Agriotypus armatus*, ein Köcherfliegen-parasit. *Nat. Mus.*, **125**, 140–148.

Burkart, H., and Zwick, P. (2006) Life cycle and seasonal habitat change in the rove beetle, *Lesteva longelytrata* (Goeze) (Coleoptera: Staphylinidae: Omalinae). *Entomol. Bl.*, **101**, 87–105.

Burkholder, J.M., and Sheath, R.G. (1984) The seasonal distribution, abundance and diversity of desmids (Chlorophyta) in a softwater, north temperate stream. *J. Phycol.*, **20**, 159–172.

Burns, A., and Ryder, D.S. (2001) Response of bacterial extracellular enzymes to inundation of floodplain sediments. *Freshwater Biol.*, **46**, 1299–1307.

Butcher, R.W. (1940) Studies in the ecology of rivers. IV. Observations on the growth and distribution of the sessile algae in the River Hull, Yorkshire. *J. Ecol.*, **28**, 210–223.

Butler, M.G. (1984) Life histories of aquatic insects, in *The Ecology of Aquatic Insects* (eds V.H. Resh and D.M. Rosenberg), Academic Press, New York, pp. 24–55.

Butz, L., and Schuwerk, T. (1993) Beobachtungen zum Massenflug von *Ephoron virgo* (Olivier 1791) an der Donau Ostbayerns (Insecta, Ephemeroptera). *Lauterbornia*, **13**, 25–26

Calow, P. (1973) The food of *Ancylus fluviatilis* (Müll.), a littoral stone-dwelling herbivore. *Oecologia*, **13**, 113–133.

Campbell, I.C. (1991) Size allometry in some Australian mayfly nymphs (Insecta: Ephemeroptera). *Aquat. Insects*, **13**, 79–86.

Carlough, L.A., and Meyer, J.L. (1989) Protozoans in two southeastern blackwater rivers and their importance to trophic transfer. *Limnol. Oceanogr.*, **34**, 163–177.

Caspers, N. (1983) Chironomiden-Emergenz zweier Lunzer Bäche, 1972. *Arch. Hydrobiol. Suppl.*, **65**, 484–549.

Caspers, N., and Wagner, R. (1982) Emergenz-Untersuchungen an einem Mittelgebirgsbach bei Bonn. VII. Empididen- und Dolichopodiden-Emergenz 1976 (Insecta, Diptera, Brachycera). Schlitzer produktions-biologische Studien (46). *Arch. Hydrobiol.*, **93**, 209–237.

Chappell, K.R., and Goulder, R. (1992) Epilithic extracellular enzyme-activity in acid and calcareous headstreams. *Arch. Hydrobiol.*, **125**, 129–148.

Chappell, K.R., and Goulder, R. (1994) Seasonal variation of epilithic extracellular enzyme activity in three diverse headstreams. *Arch. Hydrobiol.*, **130**, 195–214.

Chappell, K.R., and Goulder, R. (1995) A between-river comparison of extracellular-enzyme activity. *Microb. Ecol.*, **29**, 1–17.

Charnov, E.L. (1976) Optimal foraging, the marginal value theorem. *Theor. Popul. Biol.*, **9**, 129–136.

Cheng, L. (2003) Marine insects, in *Encyclopedia of Insects* (eds V.H. Resh and R.T. Cardé), Academic Press/Elsevier Science, New York, pp. 679–682.

Chesson, P.L. 1986: Environmental variation and the coexistance of species, in *Community ecology* (eds J.M. Diamond and

T.J. Case), Harper and Row, New York; pp. 240–256.

Chmielewski, C.M., and Hall, R.J. (1993) Changes in the emergence of blackflies (Diptera: Simuliidae) over 50 years from Algonquin Park streams: is acidification the cause? *Can. J. Fish. Aquat. Sci.*, **50**, 1517–1529.

Christl, H. (1991) Die Simuliiden des Breitenbaches – eine Emergenzstudie. Diploma Thesis, University of Marburg, 141 pp.

Christl, H. 2008. Die Nematoden des Breitenbachs – Struktur, Dynamik und mögliche Funktion im Ökosystem eines Mittelgebirgsbachs. Ph.D. Thesis, University of Marburg, Marburg, 401 pp.

Chróst, R.J. (1989) Chracterization and significance of β-glucosidase activity in lake water. *Limnol. Oceanogr.*, **34**, 660–672.

Chróst, R.J. (1990) Microbial ectoenzymes in aquatic environments, in *Aquatic Microbial Ecology. Biochemical and Molecular Approaches* (eds J. Overbeck and R.J. Chróst), Brock/Springer Series in Contemporary Bioscience, Springer, New York, pp. 47–78.

Chróst, R.J. (1991) Environmental control of the synthesis and activity of aquatic microbial ectoenzymes, in *Microbial Enzymes in Aquatic Environments* (ed. R.J. Chróst), Brock/Springer Series in Contemporary Bioscience, Springer, New York, pp. 29–59.

Chróst, R.J., and Overbeck, J. (1987) Kinetics of alkaline phosphatase activity and phosphorus availability for phytoplankton and bacterioplankton in Lake Plußsee (north German eutrophic lake). *Microb. Ecol.*, **13**, 229–242.

Chróst, R.J., Weislo, R., and Halemejko, G.Z. (1986) Enzymatic decomposition of organic matter by bacteria in an eutrophic lake. *Arch. Hydrobiol.*, **107**, 145–165.

Chutter, F.M. (1972) A reappraisal of Needham and Usinger's data on the variability of a stream fauna when sampled with a Surber sampler. *Limnol. Oceanogr.*, **17**, 139–141.

Cianficconi, F., Bicchierai, M.C., and Moretti, G. (1993) Silk glands and silk weave in trichopteran larvae, in *Proc. 7th Int. Symp. Trichoptera, 1992* (ed. C. Otto), Bachhuys, Leiden, pp. 33–38.

Ciborowski, J.J.H. (1987) Dynamics of drift and microdistribution of two mayfly populations: a predictive model. *Can. J. Fish. Aquat. Sci.*, **44**, 832–845.

Clayton, R.A., Sutton, G., Hinkle, Jr., P.S., Bult, C., and Fields, C. (1995) Intraspecific variation in small-subunit rRNA sequences in GenBank: why single sequences may not adequately represent prokaryotic taxa. *Int. J. Syst. Bacteriol.*, **45**, 595–599.

Clifford, H.F. (1970) Analysis of a northern mayfly (Ephemeroptera) population, with special reference to allometry of size. *Can. J. Zool.*, **48**, 305–316.

Clifford, H.F. (1972) Comparison of samples of stream bottom fauna collected during the day and at night. *Limnol. Oceanogr.*, **17**, 479–481.

Clifford, H.F. (1982) Life cycles of mayflies (Ephemeroptera), with special reference to voltinism. *Quaest. Entomol.*, **18**, 15–90.

Clifford, H.F., Gotceitas, V., and Casey, R.J. (1989) Roughness and color of artificial substratum particles as possible factors in colonization of stream invertebrates. *Hydrobiologia*, **175**, 89–95.

Clifford, H.F., Casey, R.J., and Saffran, K.A. (1992) Short-term colonization of rough and smooth tiles by benthic macroinvertebrates and algae (chlorophyll a) in two streams. *J. N. Am. Benthol. Soc.*, **11**, 304–315.

Cohen-Kupiec, R., and Chet, I. (1998) The molecular biology of chitin digestion. *Curr. Opin. Biotechnol.*, **9**, 270–277.

Cole, J.J., Findlay, S., and Pace, M.L. (1988) Bacterial production in fresh and saltwater: a cross-system overview. *Mar. Ecol. Prog. Ser.*, **43**, 1–10.

Cole, J.R., Chai, B., Farris, R.J., Wang, Q., Kulam-Syed-Mohideen, A.S., McGarrell, D.M., Bandela, A.M., Cardenas, E., Garrity, G.M., and Tiedje, J.M. (2007) The ribosomal database project (RDP-II): introducing myRDP space and quality controlled public data. *Nucleic Acids Res.*, **35** (Database issue), D169–D172, doi: 10.1093/nar/gkl889.

Cole, J.R., Wang, Q., Cardenas, E., Fish, J., Chai, B., Farris, R.J., Kulam-Syed-Mohideen, A.S., McGarrell, D.M., Marsh, T., Garrity, G.M., and Tiedje, J.M. (2009)

The Ribosomal Database Project: improved alignments and new tools for rRNA analysis. *Nucleic Acids Res.*, **37** (Database issue), D141–D145; doi: 10.1093/nar/gkn879.

Coletti, P.J., Blinn, D.W., Pickart, A., and Wagner, V.T. (1987) Influence of different densities of the mayfly grazer *Heptagenia criddlei* on lotic diatom communities. *J. N. Am. Benthol. Soc.*, **6**, 270–280.

Collier, K.J., and Smith, B.J. (1995) Sticky trapping of adult mayflies, stoneflies and caddisflies alongside three contrasting streams near Hamilton, New Zealand. *N. Z. Nat. Sci.*, **22**, 1–9.

Collier, K.J., Bury, S., and Gibbs, M. (2002) A stable isotope study of linkages between stream and terrestrial food webs through spider predation. *Freshwater Biol.*, **47**, 1651–1659.

Colwell, R.K. (1974) Predictability, constancy, and contingency of periodic phenomena. *Ecology*, **55**, 1148–1153.

Connell, J.A. (1978) Diversity in tropical rainforests and coral reefs. *Science*, **199**, 1302–1310.

Connelly, S., Pringle, C.M., Bixby, R.J., Brenes, R., Whiles, M.R., Lips, K.R., Kilham, S., and Huryn, A.D. (2008) Changes in stream primary producer communities resulting from large-scale catastrophic amphibian declines: can small-scale experiments predict effects of tadpole loss? *Ecosystems*, **11**, 1262–1276.

Consalvey, M., Paterson, D.M., and Underwood, G.J.C. (2004) The ups and downs of life in a benthic biofilm: migration of benthic diatoms. *Diatom Res.*, **19**, 181–202.

Cooper, S.D., Barmuta, L., Sarnelle, O., Kratz, K., and Diehl, S. (1997) Quantifying spatial heterogeneity in streams. *J. North Am. Benthol. Soc.*, **16**, 174–188.

Corbet, P.S. (1999) *Dragonflies: Behaviour and Ecology of Odonata*, Harley Books, Martins, 830 pp.

Corkum, L.D., Ciborowski, J.J.H., and Poulin, R.G. (1997) Effects of emergence date and maternal size on egg development and sizes of eggs and first-instar nymphs of a semelparous aquatic insect. *Oecologia*, **111**, 69–75.

Cotner, J.B., and Biddanda, B.A. (2002) Small players, large role: microbial influence on biogeochemical processes in pelagic aquatic ecosystems. *Ecosystems*, **5**, 105–121.

Cousin, S., Brambilla, E., Yang, J., and Stackebrandt, E. (2008) Culturable aerobic bacteria from the upstream region of a karst water rivulet. *Int. Microbiol.*, **11**, 91–100.

Cowie, G.L., and Hedges, J.L. (1984) Carbohydrate sources in a coastal marine environment. *Geochim. Cosmochim. Acta*, **48**, 2075–2087.

Cox, E.J. (1984) Observations on some benthic diatoms from North German lakes. The effects of substratum and light regime. *Verh. Int. Verein. Limnol.*, **22**, 924–928.

Cox, E.J. (1987) Studies on the diatom genus *Navicula* Bory. VI. The identity, structure and ecology of some freshwaters species. *Diatom Res.*, **2**, 159–174.

Cox, E.J. (1988a) Has the role of the substratum been under-estimated for algal distribution patterns in freshwater ecosystems? *Biofouling*, **1**, 49–63.

Cox, E.J. (1988b) Variation within the genus *Pinnularia* Ehrenb.: further evidence for the use of live material in diatom systematics? Proc. 9th Int. Diatom Symp., Bristol, 1986 (ed. M. Round), pp. 437–447.

Cox, E.J. (1990a) Studies on the algae of a small softwater stream. I. Occurrence and distribution with particular reference to the diatoms. *Arch. Hydrobiol. Suppl. (Monogr. Beitr.)*, **83**, 525–552.

Cox, E.J. (1990b) Studies on the algae of a small softwater stream. II. Algal standing crop (measured by chlorophyll-a) on soft and hard substrata. *Arch. Hydrobiol. Suppl. (Monogr. Beitr.)*, **83**, 553–566.

Cox, E.J. (1990c) Studies on the algae of a small softwater stream. III. *Interaction* between discharge, sediment composition and diatom flora. *Arch. Hydrobiol. Suppl. (Monogr. Beitr.)*, **83**, 567–584.

Cox, E.J. (1990d) Microdistributional patterns of freshwater diatoms in relation to their use as bioindicators. In Proc. 10th Diatom Symp., Joesun, 1988 (ed. H. Simola), pp. 521–528.

Cox, E.J. (1991) What is the basis for using diatoms as monitors of river quality? in *The Use of Algae for Monitoring Rivers* (eds

B.A. Whitton, E. Rott, and G. Friedrich), Halle, Innsbruck, pp. 33–40.
Cox, E.J. (1993) Freshwater diatom ecology: developing an experimental approach as an aid to ecological interpretation. *Hydrobiologia*, **269/270**, 447–452.
Cox, E.J. (1994) Ecological tolerances and optima – real or imaginary? *Verh. Int. Verein. Limnol.*, **25**, 2238–2241.
Cox, E.J., and Wagner, R. (1989) Does *Agapetus fuscipes* cultivate algae in its case? *Hydrobiologia*, **175**, 117–120.
Craig, C.L., Hsu, M., Kaplan, D., and Pierce, N.E. (1999) A comparison of the composition of silk proteins by spiders and insects. *Int. J. Biol. Macromol.*, **24**, 109–118.
Crosby, T.K. (1973) Dyar's rule predated by Brooks' rule. *N. Z. Entomol.*, **5**, 175–176.
Crosskey, R.W. (1990) *The Natural History of Blackflies*, John Wiley & Sons, Ltd, Chichester, 711 pp.
Cummins, K.W. (1973) Trophic relations of aquatic insects. *Annu. Rev. Entomol.*, **18**, 183–206.
Cummins, K.W. (1974) Structure and function of stream ecosystems. *BioScience*, **24**, 631–641.
Cummins, K.W., and Klug, M.J. (1979) Feeding ecology of stream invertebrates. *Annu. Rev. Ecol. Syst.*, **10**, 147–172.
Cummins, K.W., and Lauf, G.H. (1969) The influence of substrate particle size on the microdistribution of stream macrobenthos. *Hydrobiologia*, **34**, 145–181.
Cummins, K.W., Coffman, W.P., and Roff. P.A. (1966) Trophic relationships in a small woodland stream. *Verh. Int. Verein. Limol.*, **16**, 627–638.
Dall, P.C., Heegaard, H., and Fullerton, A.F. (1984) Life-history strategies and production of *Tinodes waeneri* (L.) (Trichoptera) in Lake Esrom, Denmark. *Hydrobiologia*, **112**, 93–104.
Danecker, E. (1961) Studien zur hygropetrischen Fauna. Biologie und Ökologie von *Stactobia* and *Tinodes* (Insecta, Trichoptera). *Int. Rev. Ges. Hydrobiol.*, **46**, 214–254.
Danks, H.V. (1987) *Insect Dormancy: An Ecological Perspective*, National Museum of Natural Sciences, Ottawa, 439 pp.
Dapper, T. (1998) Dimensionsreduzierende Vorverarbeitungen für Neuronale Netze mit Anwendungen in der Umwelt-modellierung; Verlag Shaker, Aachen, 243 pp.
Darcy-Hall, T.L., and Hall, S.R. (2008) Linking limitation to species composition: importance of inter- and intra-specific variation in grazing resistance. *Oecologia*, **155**, 797–808.
Daubner, I. (1969) The effect of some ecological factors on bacteria in water. *Verh. Int. Verein. Limnol.*, **17**, 731–743.
Daubner, I. (1972) *Mikrobiologie des Wassers*, BLV Verlagsgesellschaft, Munich, 440 pp.
Daubner, I., and Trzilova, B. (1973) Ergebnisse der mikrobiologischen Donauuntersuchung in einem Punktprofil (1969–1971). *Biológia (Bratislava)*, **28**, 775–781.
Dawson, A. (2005) The effect of temperature on photoperiodically regulated gonadal maturation, regression and moult in starlings – potential consequences of climate change. *Funct. Ecol.*, **19**, 995–1000.
De Beer, D., Schramm, A., Santegoeds, C.M., and Kühl, M. (1997) A nitrite microsensor for profiling environmental biofilms. *Appl. Environ. Microbiol.*, **63**, 973–977.
De Jong, H., Oosterbroek, P., Gelhaus, J., Reusch, H., and Young, C. (2008) Global diversity of craneflies (Insecta, Diptera: Tipulidea or Tipulidae sensu lato) in Freshwater, in *Developments in Hydrobiology 198, Freshwater Animal Diversity Assessment* (eds E. Balian, C. Lévêque, H. Segers, and K. Martens), Springer, pp. 457–467.
de Lattin, G. (1967) *Grundriss der Zoogeographie*, G. Fischer, Jena, 602 pp.
De Moor, F.C., and Ivanov, V.D. (2008) Global diversity of caddisflies (Trichoptera: Insecta) in freshwater. *Hydrobiologia*, **595**, 393–407.
Dean, J.C., and Cartwright, D.I. (1987) Trichoptera of a Victorian forest stream: species composition and life histories. *Aust. J. Mar. Freshwater Res.*, **38**, 845–860.
Death, R.G., and Winterbourn, M.J. (1994) Environmental stability and community persistence, a multivariate perspective. *J. North Am. Benthol. Soc.*, **13**, 125–139.
Degrange, C. (1954) Deux cas de parthénogenèse chez les Ephéméroptères: *Siphlonurus aestivalis* Eat, et *Centroptilum*

luteolum Müll. *C. R. Acad. Sci., D, Sci. Nat.*, **239**, 1082–1083.

Degrange, C. (1958) Premiers cas de parthénogenèse chez les Plécoptères. *C. R. Hebd. Seances Acad. Sci.*, **246**, 1100–1102.

Degrange, C. (1960) Recherche sur la reproduction des Ephéméroptères. *Trav. Labor. Hydrob. Pisc. Univ. Grenoble*, **50/51**, 7–193.

Del Giorgio, P.A., and Cole, J.J. (1998) Bacterial growth efficiency in natural aquatic systems. *Annu. Rev. Ecol. Syst.*, **29**, 503–541

Delgado, J.A., and Carbonell, R. (1997) Case features of caddisfly larvae (*Sericostoma selysi*) as related to water velocity and potential to drift. *J. Freshwater Ecol.*, **12**, 193–197.

Den Boer, P.J. (1968) Spreading of risk and stabilization of animal numbers. *Acta Biotherm.*, **18**, 165–194.

DeNicola, D.M., McIntire, C.D., Lamberti, G.A., Gregory, S.V., and Ashkenas, L.R. (1990) Temporal patterns of grazer–periphyton interactions in laboratory streams. *Freshwater Biol.*, **23**, 475–489.

Denis, C. (1981) Données sur la biologie et les stades du développment de *Sericostoma galeatum* Rambur (Trichoptera, Sericostomatidae). *Riv. Idrobiol.*, **20**, 425–439.

Deniseger, J., Austin, A., and Lucey, W.P. (1986) Periphyton communities in a pristine mountain stream above and below heavy metal mining operations. *Freshwater Biol.*, **16**, 209–218.

Descy, J.-P. (1984) Écologie et distribution de diatomées benthiques dans le basin belge de la Meuse. Documents du travail, *Inst. Roy. Sci. Nat. Belg.*, **18**, 25.

Descy, J.-P., and Mouvet, C. (1984) Impact of the Tihange nuclear power plant on the periphyton and phytoplankton of the Meuse River (Belgium). *Hydrobiologia*, **119**, 119–128.

Desertová, B. (1981) Relationship between chlorophyll-a-concentration and phytoplankton biomass in several reservoirs in Czechoslovakia. *Int. Rev. Ges. Hydrobiol.*, **66**, 153–169.

Deufel, J. (1972) Die Bakterien- und Keimzahlen im Oberlauf der Donau bis Ulm. *Arch. Hydrobiol. Suppl.*, **44**, 1–9.

Deutsch, W.G., (1984) Oviposition of Hydropsychidae (Trichoptera) in a large river. *Can. J. Zool.*, **62**, 1988–1994.

Deutsch, W.G. (1985) Swimming modifications of adult female Hydropsychidae compared with other Trichoptera. *Freshwater Invertebr. Biol.*, **4**, 35–40.

Dieterich, M., and Anderson, N.H. (1995) Life cycles and food habits of mayflies and stoneflies from temporary streams in western Oregon. *Freshwater Biol.*, **34**, 47–60.

Disney, R.H.L. (1999) *British Dixidae (Meniscus midges) and Thaumaleidae (Trickle midges: Key with Ecological Notes)*, Freshwater Biological Association, Scientific Publication 56. Titus Wilson and Son, Kendal, p. 129.

Dobson, M., Cariss, H., and Murray, B. (1997) Effects of a flash flood on channel morphology and some insects of a small stony stream in the Peak district, Derbyshire. *Freshwater Forum*, **9**, 2–13.

Doi, H., Takemon, Y., Ohta, T., Ishida, Y., and Kikuchi, E. (2007) Effects of reach-scale canopy cover on trophic pathways of caddisfly larvae in a Japanese mountain stream. *Mar. Freshwater Res.*, **58**, 811–817.

Donath, H. (1989) Verbreitung und Ökologie der Zweigestreiften Quelljungfer, *Cordulegaster boltoni* (Donovan, 1907) in der DDR (Insecta, Odonata : Cordulegasteridae). *Fauni. Abh. Staatl. Mus. Tierk. Dresden*, **16**, 97–106.

Dorfer, W. (2008) Erstnachweis von *Isoptena serricornis* (Plecoptera, Chloroperlidae) in Bayern/First record of *Isoptena serricornis* (Plecoptera, Chloroperlidae) in Bavaria. *Lauterbornia*, **62**, 53–55.

Douglas, B. (1958) The ecology of the attached diatoms and other algae in a small stony stream. *J. Ecol.*, **46**, 295–322.

Downer, R.G.H. (1978) Functional role of lipids in insects, in *Biochemistry of Insects* (ed. M. Rockstein), Academic Press, New York, pp. 57–92.

Downes, J.A., and Wirth, W.W. (1981) Ceratopogonidae, in *Manual of Nearctic Diptera Vol. 1* (eds J.F. McAlpine, B.V. Peterson, G.E. Shewell, H.J. Teskey, J.R. Vockeroth, and D.M. Wood), USDA,

Research Branch Agriculture Monograph, No. 27, pp. 393–421.

Downes, B.J., Glaister, A., and Lake, P.S. (1997) Spatial variation in the force requirement to initiate rock movement in 4 upland streams: implications for estimating disturbance frequencies. *J. North Am. Benthol. Soc.*, **16**, 203–220.

Drees, M. (1997) Beobachtungen zur Ausbreitung thermophiler Wanzen (Heteroptera) im Raum Hagen. *Decheniana*, **150**, 233–234.

Dudgeon, D. (1983) An investigation of the drift of aquatic insects in Tai Po Kau Forest Stream, New Territories, Hong Kong. *Arch. Hydrobiol.*, **96**, 434–447.

Dudgeon, D. (1990) Functional significance of selection of particles and their use by aquatic animals in the construction of external structures, in *The Biology of Particles in Aquatic Systems* (ed. R.S. Wotton), CRC Press, Boca Raton, pp. 263–288.

Dudgeon, D., and Lam, P.K.S. (1985) Freshwater gastropod foraging strategies: intraspecific comparisons, in *The Malakofauna of Hong Kong and Southern China, Vol. 2* (eds B. Morton and D. Dudgeon), *Proc. 2nd. Int. Workshop on the Malakofauna of Hong Kong and Southern China*, Hong Kong University Press, Hong Kong, pp. 591–600.

Dudley, T.L., and D'Antonio, C.M. (1991) The effects of substrate texture, grazing, and disturbance on macroalgal establishment in streams. *Ecology*, **72**, 297–309.

Dunstan, G.A., Volkman, J.K., Barrett, S.M., Leroi, J.M., and Jeffrey, S.W. (1994) Essential polyunsaturated fatty-acids from 14 species of diatoms (Bacillariophyceae). *Phytochemistry*, **35**, 155–161.

Durance, I., and Ormerod, S.J. (2007) Climate change effects upon upland stream macroinvertebrates over a 25-year period. *Global Change Biol.*, **13**, 942–957. doi: 10.1111/j-1365-2486.2007.01340.x

Dyar, H.G. (1890) The number of moults of Lepidoptera larvae. *Psyche*, **5**, 420–422.

Eckhardt, K. (2008) A comparison of baseflow indices, which were calculated with seven different baseflow separation methods. *J. Hydrol.*, **352**, 168–173.

Edwards, R.T., and Meyer, J.L. (1987) Metabolism of a sub-tropical low gradient blackwater river. *Freshwater Biol.*, **17**, 251–263.

Edwards, R.T., and Meyer, J.L. (1990) Bacterivory by deposit-feeding mayfly larvae (*Stenonema* spp.). *Freshwater Biol.*, **24**, 453–462.

Edwards, R.W., and Owens, M. (1962) The effects of plants on river conditions. IV. The oxygen balance of a chalk stream. *J. Ecol.*, **50**, 207–220.

Elliott, J.M. (1968) The life histories and drifting of Trichoptera in a Dartmoor stream. *J. Anim. Ecol.*, **37**, 615–625.

Elliott, J.M. (1969) Life history and biology of *Sericostoma personatum* Spence (Trichoptera). *Oikos*, **20**, 110–118.

Elliott, J.M. (1970) The diel activity patterns of caddis larvae (Trichoptera). *J. Zool.*, **160**, 279–290.

Elliott, J.M. (1972) Effect of temperature on the time of hatching in *Baetis rhodani* (Ephemeroptera: Baëtidae). *Oecologia*, **9**, 47–51.

Elliott, J.M. (1978) Effect of temperature on the hatching time of eggs of *Ephemerella ignita* (Poda) (Ephemeroptera: Ephemerellidae). *Freshwater Biol.*, **8**, 51–58.

Elliott, J.M. (1984) Hatching time and growth of *Nemurella pictetii* (Plecoptera: Nemouridae) in the laboratory and a Lake District stream. *Freshwater Biol.*, **14**, 491–499.

Elliott, J.M. (1987a) Egg hatching and resource partitioning in stoneflies: the six British *Leuctra* spp. (Plecoptera). *J. Anim. Ecol.*, **56**, 415–426.

Elliott, J.M. (1987b) Temperature-induced changes in the life cycle of *Leuctra nigra* (Plecoptera: Leuctridae) from a Lake District stream. *Freshwater Biol.*, **18**, 177–184.

Elliott, J.M. (1987c) Life cycle and growth of *Leuctra geniculata* (Stephens) (Plecoptera: Leuctridae) in the river Leven. *Entomol. Gaz.*, **38**, 129–134.

Elliott, J.M. (1988a) Interspecific and intraspecific variations in egg hatching for British populations of *Taeniopteryx nebulosa* and *Brachyptera risi* (Plecoptera: Taeniopterygidae). *Hol. Ecol.*, **11**, 55–59.

Elliott, J.M. (1988b) Egg hatching and resource partitioning in stoneflies (Plecoptera) – Ten British species in the family Nemouridae. *J. Anim. Ecol.*, **57**, 201–216.

Elliott, J.M. (1991a) The effect of temperature on egg hatching for three populations of *Isoperla grammatica* and one population of *Isogenus nubecula* (Plecoptera: Perlodidae). *Entomol Gaz.*, **42**, 61–65.

Elliott, J.M. (1991b) The effect of temperature on egg hatching for three populations of *Perla bipunctata* (Plecoptera: Perlidae). *Entomol. Gaz.*, **42**, 99–103.

Elliott, J.M. (1995) The effect of temperature on egg hatching for three populations of *Sialis lutaria* (L.) and two populations of *Sialis fuliginosa* Pictet (Megaloptera: Sialidae). *Entomol. Gaz.*, **46**, 155–159.

Elliott, J.M. (2009) Inter- and intraspecific differences in the number of larval instars in British populations of 24 species of stoneflies (Plecoptera). *Freshwater Biol.*, **54**, 1271–1284.

Elliott, J.M., and Minshall, G.W. (1968) The invertebrate drift of the River Duddon, English Lake District. *Oikos*, **19**, 39–52.

Elliott, J.M., and Tullett, P.A. (1977) The downstream drifting of larvae of *Dixa* (Diptera: Dixidae) in two stony streams. *Freshwater Biol.*, **7**, 403–407.

Elpers, C., and Tomka, I. (1992) Struktur der Mundwerkzeuge und Nahrungsaufnahme bei den Larven von *Oligoneuriella rhenana* Imhoff (Ephemeroptera: Oligoneuriidae). *Mitt. Schweiz. Entomol. Ges.*, **65**, 119–139.

Elwood, J.W., and Turner, R.R. (1989) Streams: Water Chemistry and Ecology *Analysis of Biogeochemical Cycling Processes in Walker Branch Watershed* (eds D.W. Johnson and R.I. Van Hook), Springer Verlag, pp. 301–350.

Enders, G. (1994) Eiablage, Embryonal- und frühe Larvalentwicklung von *Apatania fimbriata* und *Drusus annulatus* (Insecta: Trichoptera) aus dem Breitenbach. Diploma thesis, University of Würzburg, Würzburg, 106 pp.

Enders, G., and Wagner, R. (1996) Mortality of *Apatania fimbriata* (Insecta, Trichoptera) during embryonic, larval and adult life stages. *Freshwater Biol.*, **36**, 93–104.

Engblom, E. (1996) Ephemeroptera, Mayflies, in *Aquatic Insects of North Europe – A Taxonomic Handbook* (ed. A.N. Nilsson), Apollo Books, Stenstrup, pp. 13–53, pls. 1–12.

Englund, G. (1992) Effects of density and food availability on habitat selection in a net-spinning caddis larva, *Hydropsyche siltalai*. Paper IV, in *Competition in caddis larvae*. Dissertation, University of Umeå, Umeå.

Englund, G., and Olsson, T.I. (1990) Fighting and assessment in the net-spinning caddis larva *Arctopsyche ladogensis*: a test of the sequential assessment game. *Anim. Behav.*, **39**, 55–62.

Englund, G., and Otto, C. (1991) Effects of ownership status, weight asymmetry, and case fit on the outcome of case contests in two populations of *Agrypnia pagetana* (Trichoptera: Phryganeidae) larvae. *Behav. Ecol. Sociobiol.*, **29**, 113–120.

Erlandsson, A., Malmqvist, B., Andersson, K.G., Herrmann, J., and Sjoestroem, P. (1988) Field observations on the activities of a groupliving semiaquatic bug, *Velia caprai. Arch. Hydrobiol.*, **112**, 411–419.

Erwin, T.L. (1982) Tropical forests: their richness in Coleoptera and other arthropod species. *Coleopt. Bull.*, **36**, 74–75.

Evans, M.R., and Owens, J.D. (1973) Soil-bacteria in land-drainage water. *Water Res.* **7**, 1295–1300

Evans-White, M.A., and Lamberti, G.A. (2005) Grazer species effects on epilithon nutrient composition. *Freshwater Biol.*, **50**, 1853–1863.

Fazi, S., Amalfitano, S., Pernthaler, J., and Puddu, A. (2005) Bacterial communities associated with benthic organic matter in headwater stream microhabitats. *Environ. Microbiol.*, **7**, 1633–1640.

Feminella, J.W., and Hawkins, C.P. (1995) Interactions between stream herbivores and periphyton – a quantitative-analysis of past experiments. *J. N. Am. Benthol. Soc.*, **14**, 465–509.

Feminella, J.W., and Resh, V.H. (1990) Hydrologic influences, disturbance and intraspecific competition in a stream caddisfly population. *Ecology*, **71**, 2083–2094.

Feminella, J.W., and Resh, V.H. (1991) Herbivorous caddisflies, macroalgae, and epilithic microalgae: dynamic interactions in a stream grazing system. *Oecologia*, **87**, 247–256.

Ferris, M.J., and Ward, D.M. (1997) Seasonal distributions of dominant 16S rRNA-defined populations in a hot spring microbial mat examined by denaturing gradient gel electrophoresis. *Appl. Environ. Microbiol.*, **63**, 1375–1381.

Fiałkowski, W. (1986) Life cycles and microdistribution of *Nemoura cinerea* (Retz.) and *Nemurella picteti* Klap. (Plecoptera: Nemouridae) from two small lowland streams in Southern Poland. *Acta Hydrobiol.*, **28**, 199–213.

Fiebig, D.M. (1992) Fates of dissolved free amino acids in groundwater discharged through stream bed sediments. *Hydrobiologia*, **235/236**, 311–319.

Fiebig, D.M. (1995) Groundwater discharge and its contribution of dissolved organic carbon to an upland stream. *Arch. Hydrobiol.*, **134**, 129–155.

Fiebig, D.M. (1997) Microbiological turnover of amino acids immobilized from groundwater discharged through hyporheic sediments. *Limnol. Oceanogr.*, **42**, 763–768.

Fiebig, D.M., and Marxsen, J. (1992) Immobilization and mineralization of dissolved free amino acids by stream-bed biofilms. *Freshwater Biol.*, **28**, 129–140.

Fiebig, D.M., Lock, M.A., and Neal, C. (1990) Soil water in the riparian zone as a source of carbon for a headwater stream. *J. Hydrol.*, **116**, 217–237.

Findlay, S. (1995) Importance of surface–subsurface exchange in stream ecosystems: the hyporheic zone. *Limnol. Oceanogr.*, **40**, 159–164.

Findlay, S., and Sobczak, W.V. (2000) Microbial communities in hyporheic sediments, in *Streams and Ground Waters* (eds J.B. Jones and P. Mulholland), Academic Press, San Diego, pp. 287–306.

Findlay, S., Meyer, J.L., and Risley, R. (1986) Benthic bacterial biomass and production in two blackwater rivers. *Can. J. Fish. Aquat. Sci.*, **43**, 1271–1276.

Findlay, S., Hickey, C.W., and Quinn, J.M. (1997) Microbial enzymatic response to catchment-scale variations in supply of dissolved organic carbon. *N. Z. J. Mar. Freshwater Res.*, **31**, 701–706.

Findlay, S., Sinsabaugh, R.L., Fischer, D.T., and Franchini, P. (1998) Sources of dissolved organic carbon supporting planktonic bacterial production in the tidal freshwater Hudson River. *Ecosystems*, **1**, 227–239.

Findlay, S., Tank, J., Dye, S., Valett, H.M., Mulholland, P.M., McDowell, W.H., Johnson, S.L., Hamilton, S.K., Edmonds, J., Dodds, W.K., and Bowden, W.B. (2002) A cross-system comparison of bacterial and fungal biomass in detritus pools of headwater streams. *Microb. Ecol.*, **43**, 55–66.

Findlay, S.E.G., Meyer, J.L., and Edwards, R.T. (1984) Measuring bacterial production via rate of incorporation of [^{3}H]thymidine into DNA. *J. Microbiol. Methods*, **2**, 57–72.

Finelli, C.M., Hart, D.D., and Merz, R.A. (2002) Stream insects as passive suspension feeders: effects of velocity and food concentration on feeding performance. *Oecologia*, **131**, 145–153.

Fink, T.J. (1980) A comparison of mayfly (Ephemeroptera) instar determination methods, in *Advances in Ephemeroptera Biology* (eds J.F. Flannigan and K.E. Marshall), Plenum, New York, pp. 367–380.

Fink, T.J. (2008) Morphology, Ultrastructure and Function of the Palmén Body and Contact Mallet. *International Advances in the Ecology, Zoogeography, and Systematics of Mayflies and Stoneflies* (eds F.R. Hauer, J.A. Stanford, and R.L. Newell), University of California Press, Berkeley, pp. 369–393.

Fink, T.J., Soldán, T., Peters, J.G., and Peters, W.L. (1991) The reproductive life history of the predacious, sand-burrowing mayfly *Dolania americana* (Ephemeroptera: Behningiidae) and comparisons with other mayflies. *Can. J. Zool.*, **69**, 1083–1093.

Fischer, C. (1984) The swarming behaviour of stream inhabiting mayflies (Ephemeroptera). in Proc. 4th Int. Conf. Ephemeroptera (eds V. Landa, T. Soldán, and M. Tonner), České Budějovice, pp. 281–282.

Fischer, C. (1990) Fuldische Straße oder Fischerpfad. *Stud. Schlitzerländer Gesch.*, **1**, 66–118.

Fischer, F. (2003) Das Nischenkonzept und seine Bedeutung für die Erklärung regionaler Verbreitungsmuster am Beispiel dreier Glossosomatidenarten (Trichoptera, Glossosomatidae). [The role of Hutchinson's (1957) niche concept in explaining regional patterns of distribution using the example of three glossosomatid species (Trichoptera, Glossosioma)]. Ph.D. thesis, Philipps-University Marburg. URL: http://archiv.ub.uni-marburg.de/diss/z2004/0073/ (accessed 12 December 2010). URN: urn:nbn:de:hebis:04-z2004-00733

Fischer, H., and Pusch, M. (2001) Comparison of bacterial production in sediments, epiphyton and the pelagic zone of a lowland river. *Freshwater Biol.*, **46**, 1335–1348.

Fischer, H., Pusch, M., and Schwoerbel, J. (1996) Spatial distribution and respiration of bacteria in stream-bed sediments. *Arch. Hydrobiol.*, **137**, 281–300.

Fischer, H., Wanner, S.C., and Pusch, M. (2002) Bacterial abundance and production in river sediments as related to the biochemical composition of particulate organic matter (POM). *Biogeochemistry*, **61**, 37–55.

Fischer, J., Fischer, F., Schnabel, S., Wagner, R., and Bohle, H.W. (1998) Spring fauna of the Hessian Mittelgebirge: population structure, adaptative strategies, and relations to habitats of the macroinvertebrates, as exemplified by springs in the Rhenisch metamorphic shield and in the East-Hessian sandstone plate, in *Studies in Crenobiology: The Biology of Springs and Springbrooks* (ed. L. Botosaneanu), Fischer, Leiden, pp. 182–199.

Fisher, S.G., and Carpenter, S.R. (1976) Ecosystem and macrophyte primary production of the Fort River, Massachusetts. *Hydrobiologia*, **49**, 175–187.

Fisher, S.G., and Likens, G.E. (1973) Energy flow in Bear Brook, New Hampshire: An integrative approach to stream ecosystem metabolism. *Ecol. Monogr.*, **43**, 421–439.

Fittkau, E.J. (1964) Remarks on limnology of central-Amazon rain forest streams. *Verh. Int. Verein. Limnol.*, **15**, 1092–1096.

Fittkau, E.J. (1992) Vierzig Jahre Limnologische Flußstation Schlitz – ein Kapitel der Geschichte deutscher Limnologie. in: *Vierzigjähriges Jubiläum der Limnologischen Flußstation Schlitz des Max-Planck-Instituts für Limnologie am 4. Juni 1992. – Festvorträge.* (ed. Limnologische Flußstation); self published, Limnologische Flußstation Schlitz, pp. 1–17.

Fittkau, E.J. (2001) 50 Jahre Fuldastation Schlitz. in: *Feierstunde anlässlich des 50-jährigen Bestehens der Limnologischen Fluss-Station Schlitz des Max-Planck-Instituts für Limnologie, Schlitz, am 13. Juni 2001.* (ed. Limnologische Flußstation); self published, Limnologische Flußstation Schlitz, pp. 4–6.

Fittkau, J., and Reiss, F. (1978) Chironomidae, in *Limnofauna Europaea (2nd Edn)* (ed. J. Illies), Gustav Fischer Verlag, Stuttgart, pp.404–440.

Fjellheim, A., and Raddum, G.G. (1998) Life cycle and drift of *Glossosoma intermedia* (Trichoptera: Glossosomatidae) in western Norway. *Verh. Int. Verein. Limnol.*, **26**, 2048–2052.

Flecker, A.S., and Allan, J.D. (1988) Flight direction in some rocky mountain mayflies (Ephemeroptera), with observations of parasitism. *Aquat. Insects*, **10**, 33–42.

Flecker, A.S., Allan, J.D., and McClintock, N.L. (1988) Male body size and mating success in swarms of the mayfly *Epeorus longimanus*. *Hol. Ecol. (now Ecography)*, **11**, 280–285.

Foissner, W. (1997) Faunistic and taxonomic studies on ciliates (Protozoa, Ciliophora) from clean rivers in Bavaria (Germany), with descriptions of new species and ecological notes. *Limnologica*, **27**, 179–238.

Foissner, W., Blatterer, H., Berger, H., and Kohmann, F. (1991) Taxonomische und ökologische Revision der Ciliaten des Saprobiensystems. Band I: Cyrtophorida, Oligotrichida, Hypotrichia, Colpodea. *Informationsberichte des Bayerischen Landesamtes für Wasserwirtschaft*, 1/91, 478 pp.

Foissner, W., Berger, H., Blatterer, H., and Kohmann, F. (1995) Taxonomische und ökologische Revision der Ciliaten des Saprobiensystems. Band IV: Gymnostomatea, Loxodes, Suctoria. Informationsberichte des Bayerischen Landesamtes für Wasserwirtschaft, (ed Bayerisches Landesamt für Wasserwirtschaft) 1/95, 540 pp.

Folke, C., Carpenter, S., Walker, B., Scheffer, M., Elmqvist, T., Gunderson, L., and Holling, C.S. (2004) Regime shifts, resilience, and biodiversity in ecosystem management. *Annu. Rev. Ecol. Evol. Syst.*, **33**, 557–581.

Freeman, C., Lock, M.A., Marxsen, J., and Jones, S. (1990) Inhibitory effects of high molecular weight dissolved organic matter upon metabolic processes in biofilms from contrasted rivers and streams. *Freshwater Biol.*, **24**, 159–166.

Freeman, C., Lock, M.A., and Marxsen, J. (1993) Poly-beta-hydroxy alkanoate and the support of river biofilm metabolism following radical changes in environmental conditions. *Hydrobiologia*, **271**, 159–164.

Freitag, H. (1962) *Einführung in die Biogeographie von Mitteleuropa unter besonderer Berücksichtigung von Deutschland*, Gustav Fischer Verlag, Stuttgart, XIV + 214 pp.

Fremling, C.R. (1960) Biology and possible control of nuisance caddisflies of the upper Mississippi River. *Iowa State Univ. Agric. Home Econ. Exp. S. Res. Bull.*, **483**, 356–379.

Friederichs, K. (1921) Untersuchungen über Simuliiden (Teil II.). *Z. Angew. Entomol.*, **8**, 31–92, Berlin.

Fritsche, W. (1985) *Umwelt-Mikrobiologie. Mikrobiologie des Umweltschutzes und der Umweltgestaltung*, Akademie-Verlag, Berlin, 195 pp.

Frommer, U. (2006) Die Große Holzbiene *Xylocopa violacea* (Linnaeus 1758) im Lahngebiet (Hymenoptera, Apidae). *Hess. Faun. Briefe*, **25**, 25–32.

Frommer, U. (2010) Beobachtungen zum Ausbreitungsmodus der Efeu-Seidenbiene *Colletes hederae* Schmidt & Westrich, 1993 (Hymenoptera, Apidae) in Hessen und die Bedeutung des blühenden Efeus (*Hedera helix* L.). *Hess. Faun. Briefe*, **29**, 1–20.

Frutiger, A. (1987) Investigations on the life-history of the stonefly *Dinocras cephalotes* Curt. (Plecoptera: Perlidae). *Aquat. Insects*, **9**, 51–63.

Fuhrman, J.A., and Azam, F. (1980) Bacterioplankton secondary production estimates for coastal waters of British Columbia, Antarctica and California. *Appl. Environ. Microbiol.*, **39**, 1085–1095.

Fuss, C.L., and Smock, L.A. (1996) Spatial and temporal variation of microbial respiration rates in a blackwater stream. *Freshwater Biol.*, **36**, 339–349.

Gallepp, G.W. (1977) Responces of caddisfly larvae (*Brachycentrus* spp.) to temperature, food availability, and current velocity. *Am. Midl. Nat.*, **98**, 59–84.

Gao, X., Olapade, O.A., and Leff, L.G. (2005) Comparison of benthic bacterial community composition in nine streams. *Aquat. Microb. Ecol.*, **40**, 51–60.

Garcia Criado, F., and Fernandez Alaez, M. (1995) Aquatic Coleoptera (Hydraenidae and Elmidae) as indicators of the chemical characteristics of water in the Orbigo River basin (N-W Spain). *Ann. Limnol.*, **31**, 185–199.

Garms, R., and Walsh, J.F. (1988) The migration and dispersal of blackflies: *Simulium damnosum* s.l., the main vector of human onchocerciasis, in *Blackflies. Ecology, Population Management, and Annotated World List* (eds K.C. Kim and R.W. Merrit), Pennsylvania State University, University Park, pp. 201–214.

Geesey, G.G., Mutch, R., Costerton, J.W., and Green, R.B. (1978) Sessile bacteria: an important component of the microbial population in small mountain streams. *Limnol. Oceanogr.*, **23**, 1214–1223.

Gerecke, R. (1994) Süßwassermilben (Hydrachnellae). Ein Bestimmungsschlüssel für die aus der Westpaläarktis bekannten Gattungen der Hydrachnellae mit einer einführenden Übersicht über die im Wasser vorkommenden Milben. *Lauterbornia*, **18**, 1–84.

Gergs, R. (2004) Quantitative Analyse der saisonalen Entwicklung von Weidegängern und Periphyton im Längsverlauf eines Mittelgebirgsbaches. Diploma thesis, University of Freiburg, Freiburg, 100 pp.

Gerhardt, A. (1986) Qualitative und quantitative Untersuchungen zum Detritusgehalt des Breitenbaches (Osthessen). Diploma thesis, University of Freiburg, Freiburg, 110 pp.

Gessner, M.O., and Chauvet, E. (1993) Ergosterol-to-biomass conversion factors for aquatic hyphomycetes. *Appl. Environ. Microbiol.*, **59**, 502–507.

Giberson, D.J., and Rosenberg, D.M. (1992) Effects of temperature, food quantity, and nymphal rearing density on life-history traits of a northern population of *Hexagenia* (Ephemeroptera: Ephemeridae). *J. N. Amer. Benthol. Soc.*, **11**, 181–193.

Gibert, J., Dole-Olivier, M.J., Marmonier, P., and Vervier, P. (1990) Surface water-groundwater ecotones, in *The Ecology and Management of Aquatic-Terrestrial Ecotones* (eds R.J. Naiman and H. Décamps), United Nations Educational, Scientific, and Cultural Organization, Paris and Parthenon Publishers, Carnforth, pp. 199–226.

Gilbert, H.J., and Hazlewood, G.P. (1993) Bacterial cellulases and xylanases. *J. Gen. Microbiol.*, **139**, 187–194.

Gilbert, N., and Raworth, D.A. (1996) Insects and temperature – a general theory. *Can. Entomol.*, **128**, 1–13.

Gilkes, N.R., Henrissat, B., Kilburn, D.G., Miller, Jr. R.C., and Warren, R.A.J. (1991) Domains in microbial β-1,4-glycanases: sequence conservation, function, and enzyme families. *Microbiol. Rev.*, **55**, 303–315.

Gillies, M.T., and Knowles, R.J. (1990) Colonization of a parthenogenetic mayfly (Caenidae: Ephemeroptera) from Central Africa, in *Mayflies and Stoneflies*, (ed. I.C. Campbell), Kluwer Academic Publishers, Dordrecht, pp. 341–345

Gillooly, J.F., and Dodson, S.I. (2000) The relationship of egg size and incubation temperature on embryonic development time in univoltine and multivoltine aquatic insects. *Freshwater Biol.*, **44**, 595–604.

Gittenberger, E., and Janssen, A.W. (eds) (1998) De Nederlandse zoetwassermollusken. Recente en fossile weekdieren uit zoet en brak water in *Nederlandse fauna 2*, Nationaal Naturrhistorisch Museum Naturalis KNNV Uitgiverij, European Invertebrate Survey – Nederland, 288 pp.

Glöckner, F.O., Amann, R., Alfreider, A., Pernthaler, J., Psenner, R., Trebesius, K., and Schleifer, K.-H. (1996) An in situ hybridization protocol for detection and identification of planktonic bacteria. *System. Appl. Microbiol.*, **19**, 403–406.

Glöckner, F.O., Zaichikov, E., Belkova, N., Denissova, L., Pernthaler, J., Pernthaler, A., and Amann, R. (2000) Comparative 16S rRNA analysis of lake bacterioplankton reveals globally distributed phylogenetic clusters including an abundant group of Actinobacteria. *Appl. Environ. Microbiol.*, **66**, 5053–5065.

Glöer, P. (2002) Die Süßwassergastropoden Nord- und Mitteleuropas. Bestimmungsschlüssel, Lebensweise, Verbreitung, in *Die Tierwelt Deutschlands*, Teil 73, ConchBooks, Hackenheim, 328 pp.

Godbout, L., and Hynes, H.B.N. (1982) The three dimensional distribution of the fauna in a single riffle in a stream in Ontario. *Hydrobiologia*, **97**, 87–96.

Golini, V.I., and Davies, D.M. (1988) Oviposition of black flies, in *Blackflies. Ecology, Population Management and Annotated World List* (eds K.C. Kim and R.W. Merrit), Pennsylvania State University, University Park (1987, publ. 1988), pp. 261–275.

Gooday, G.W. (1990) The ecology of chitin degradation. *Adv. Microb. Ecol.*, **11**, 387–430.

Gordon, N.D., McMahon, T.A., and Finlayson, B.L. (2004) *Stream Hydrology: An Introduction for Ecologists*, John Wiley & Sons, Ltd, Chichester, 444 pp.

Goulder, R. (1988) Epilithic bacteria in an acid and a calcareous headstream. *Freshwater Biol.*, **19**, 405–416.

Graf, W., Murphy, J., Dahl, J., and Zamora-Munoz, C. (2008) Trichoptera, in *Distribution and Ecological Preferences of European Freshwater Organisms*, vol. 1 (eds A. Schmidt-Kloiber and D. Hering), Pensoft, Sofia/Moscow, 388 pp.

Grandi, M. (1941) Contributi allo studio degli Efemerotteri italiani. III. Cloëon dipterum L. *Boll. Ist. Entomol. Univ. Bologna*, **13**, 29–71.

Grant, J., Bathmann, U.V., and Mills, E.L. (1986) The interaction between benthic diatom films and sediment transport. *Estuar. Coast. Shelf Sci.*, **23**, 225–238.

Grasser, U. (1994) The Elmidae (Coleoptera) communities of the crystalline Bohemian Mountains in northern Austria in relation to the distribution patterns indicating environmental changes. *Verh. Int. Verein. Limnol.*, **25**, 1661–1666.

Gray, T.R.G., and Williams, S.T. (1971) Microbial productivity in soil, in *Microbes und Biological Productivity* (eds D.E. Hughes and A.H. Rose), Society for General Microbiology, Cambridge, pp. 255–286.

Gregor, T. (1992) Flora und Vegetation des Schlitzerlandes. Dissertation, Technical University, Berlin, 471 pp.

Gregor, T., and Wolf, B. (2001) Wasser- und Ufermoose des Breitenbaches. *Philippia*, **101**, 39–47.

Gregory, S.V. (1983) Plant–herbivore interactions in stream systems, in *Stream Ecology*(eds J.R. Barnes and G.W. Minshall), Premium Press, New York, pp. 157–189.

Gremm, T.J., and Kaplan, L.A. (1997) Dissolved carbohydrates in streamwater determined by HPLC and pulsed amperometric detection. *Limnol. Oceanogr.*, **42**, 385–393.

Guerold, F., Vein, D., and Jacquemin, G. (1991) Les peuplements d'ephéméroptères, de plécoptères et de trichoptères des ruisseaux acides et non acides du massif vosgien: Première approche. *Rev. Sci. Eau*, **4**, 299–314.

Gümbel, D. (1976) Emergenz-Vergleich zweier Mittelgebirgsquellen 1973. *Arch. Hydrobiol. Suppl.*, **50**, 1–53.

Gunderson, L.H. (2000) Ecological resilience – in theory and application. *Annu. Rev. Ecol. Syst.*, **31**, 425–439.

Gupta, S., Gupta, A., and Meyer-Rochow, V.B. (2000) Post-embryonic development of the lateral eye of *Cloeon* sp. (Ephemeroptera: Baetidae) as revealed by scanning electron microscopy. *Entomol. Fenn.*, **11**, 89–96.

Hahn, M.W. (2003) Isolation of strains belonging to the cosmopolitan *Polynucleobacter necessarius* cluster from freshwater habitats located in three climatic zones. *Appl. Environ. Microbiol.*, **69**, 5248–5254.

Hahn, M.W. (2006) The microbial diversity of inland waters. *Curr. Opin. Biotechnol.*, **17**, 256–261.

Hahn, M.W., Pöckl, M., and Wu, Q.L. (2003) Low intraspecific diversity in a *Polynucleobacter* subcluster population numerically dominating bacterioplankton of a freshwater pond. *Appl. Environ. Microbiol.*, **71**, 4539–4547.

Hakenkamp, C.C., and Palmer, M.A. (2000) The ecology of hyporheic meiofauna, in *Streams and Ground Waters* (eds J.B. Jones and P. Mulholland), Academic Press, San Diego, pp. 307–336.

Hall, R.O., and Meyer, J.L. (1998) The trophic significance of bacteria in a detritus-based stream food web. *Ecology*, **79**, 1995–2012.

Hall, R.O., Wallace, J.B., and Eggert, S.L. (2000) Organic matter flow in stream food webs with reduced detrital resource base. *Ecology*, **81**, 3445–3463.

Hamilton, P.B., and Duthie, H.C. (1984) Periphyton colonization of rock surfaces in a boreal forest stream studied by scanning electron microscopy and track autoradiography. *J. Phycol.*, **20**, 525–532.

Hanna, H.M. (1961) Selection of materials for case building by larvae of caddis flies (Trichoptera). *Proc. R. Entomol. Soc. Lond. (A)*, **36**, 37–47.

Hannappel, U. (1991) Die Larven von *Helodes tournieri* Kiesw. und *Helodes gredleri* Kiesw. und ihre Stellung im phylogenetischen System (Coleoptera, Helodidae). *Z. Arb. Österr. Entomol.*, **43**, 89–100.

Hannappel, U., and Paulus, H.F. (1994) 29. Familie: Scirtidae, in *Die Larven der Käfer Mitteleuropas 2. Band Myxophaga Polyphaga Teil 1* (ed. B. Klausnitzer), Goecke and Evers, Krefeld, pp. 74–87.

Hansell, M.H. (1968a) The selection of house building materials by the caddis-fly larva, *Agapetus fuscipes* Curtis. *Rev. Comp. Anim.*, **2**, 91–102.

Hansell, M.H. (1968b) The house building behaviour of the caddis-fly larva *Silo pallipes* Fabricius: I. The structure of the house and method of house extension. II. Description and analysis of the selection of small particles. III. The selection of large particles. *Anim. Behav.*, **16**, 558–584.

Hansell, M.H. (1974) Regulation of building unit size in the house building of the caddis larva *Lepidostoma hirtum*. *Anim. Behav.*, **22**, 133–143.

Happey-Wood, C.M., and Priddle, J. (1984) The ecology of epipelic algae of five Welsh lakes, with special reference to Volvocalean green flagellates (Chlorophyceae). *J. Phycol.*, **20**, 109–124.

Hargrove, W.W., and Pickering, J. (1992) Pseudoreplication: a sine qua non for regional ecology. *Landsc. Ecol.*, **6**, 251–258.

Harker, J.E. (1992) Swarm behaviour and mate competition in mayflies (Ephemeroptera). *J. Zool. (Lond.)*, **228**, 571–587.

Harker, J.E. (1997) The role of parthenogenesis in the biology of two species of mayfly (Ephemeroptera). *Freshwater Biol.*, **37**, 287–297.

Harper, P.P., and Hynes, H.B.N. (1972) Life-histories of Capniidae and Taeniopterygidae (Plecoptera) in Southern Ontario. *Arch. Hydrobiol. Suppl.*, **40**, 274–314.

Harrison, S.S.C., and Hildrew, A.G. (1998) Patterns in the epilithic community of a lake littoral. *Freshwater Biol.*, **39**, 477–492.

Harrison, S.S.C., and Hildrew, A.G. (2001) Epilithic communities and habitat heterogeneity in a lake littoral. *J. Anim. Ecol.*, **70**, 692–707.

Hart, D.D. (1981) Foraging and resource patchiness: field experiments with a grazing stream insect. *Oikos*, **37**, 46–52.

Hart, D.D. (1983) The importance of competitive interactions within stream populations and communities, in *Stream Ecology* (eds J.R. Barnes and G.W. Minshall), Plenum Press, New York, pp. 99–136.

Hart, D.D. (1985a) Causes and consequences of territoriality in a grazing stream insect. *Ecology*, **66**, 404–414.

Hart, D.D. (1985b) Grazing insects mediate algal interactions in a stream benthic community. *Oikos*, **44**, 40–46.

Hart, D.D. (1987) Experimental studies of exploitative competition in a grazing stream insect. *Oecologia*, **73**, 41–47.

Hart, D.D., and Finelli, C.M. (1999) Physical-biological coupling in streams: The pervasive effects of flow on benthic organisms. *Annu. Rev. Ecol. Syst.*, **30**, 363–395.

Hart, D.D., and Resh, V.H. (1980) Movement patterns and foraging ecology of a stream caddisfly larva. *Can. J. Zool.*, **58**, 1174–1185.

Hart, D.D., and Robinson, C.T. (1990) Resource limitation in a stream community: phosphorus enrichment effects on periphyton and grazers. *Ecology*, **71**, 1494–1502.

Hart, D.D., Kohler, S.L., and Carlton, R.G. (1991) Harvesting of benthic algae by territorial grazers: the potential for prudent predation. *Oikos*, **60**, 329–335.

Hartley, J.C. (1958) The root-piercing spiracles of the larva of *Chrysogaster hirtella* Loew (Diptera: Syrphidae). *Proc. R. Entomol. Soc. Lond. (A)*, **33**, 81–87.

Hartley, J.C. (1961) A taxonomic account of the larvae of some British Syrphidae. *Proc. Zool. Soc. Lond.*, **136**, 505–573.

Harz, K. (1966) Eine Paarungsgesellschaft von Steinfliegen. *Mitt. Dtsch Entomol. Ges.*, **25**, 75–76.

Hasselrot, A.T. (1993) Insight into a Psychomyiid life. Ph.D. Thesis, University of Uppsala, Uppsala.

Hasselrot, A.T., Becker, G., Holfeld, H., Menzler, D.A., and Marxsen, J. (1996) Strong indication for larval caddisfly gardening inside the retreat. *Bull. N. Am. Benthol. Soc.*, **13**, 133.

Havelka, P. (1975) Ceratopogoniden in der Emergenz am Breitenbach und Rohrwiesenbach (1971–1972). *Arch. Hydrobiol. Suppl.*, **50**, 54–95.

Head, I.M., Hiorns, W.D., Embley, T.M., McCarthy, A.J., and Saunders, J.R. (1993) The phylogeny of autotrophic ammonia-oxidizing bacteria as determined by analysis of 16S ribosomal RNA gene sequences. *Gen. Microbiol.*, **139**, 1147–1152.

Hecht, M. (1992) Ernährungsökologische Untersuchungen an Ciliaten unter besonderer Berücksichtigung der Bakterien-Konsumption. Diploma thesis, University Marburg, Marburg, 121 pp.

Helioclim (2010). Helioclim, www.helioclim.net (accessed 12 December 2010).

Hendel, B. (1999) Der mikrobielle Abbau von Holz und Laub im Breitenbach unter besonderer Berücksichtigung der Bedeutung extrazellulärer Enzyme. Dissertation, University of Gießen, Gießen, 126 pp.

Hendel, B., and Marxsen, J. (1997) Measurement of low level extracellular enzyme activity in natural waters using fluorigenic model substrates. *Acta Hydrochim. Hydrobiol.*, **25**, 253–258.

Hendel, B., and Marxsen, J. (2000) Extracellular enzyme activity associated with degradation of beech wood in a Central European stream. *Int. Rev. Hydrobiol.*, **85**, 95–105.

Hendel, B., and Marxsen, J. (2005a) Viscosimetric determination of endocellulase activity, in *Methods to Study Litter Decomposition: A Practical Guide* (eds M.A.S. Graça, F. Bärlocher, and M.O. Gessner), Springer, Berlin, pp. 255–260.

Hendel, B., and Marxsen, J. (2005b) Fluorometric determination of the activity of β-glucosidase and other extracellular hydrolytic enzymes, in *Methods to Study Litter Decomposition: A Practical Guide* (eds M.A.S. Graça, F. Bärlocher, and M.O. Gessner), Springer, Berlin, pp. 261–266.

Hendel, B., Sinsabaugh, R.L., and Marxsen, J. (2005) Spectrophotometric determination of lignin-degrading enzyme activities (phenoloxidase and peroxidase), in *Methods to Study Litter Decomposition: A Practical Guide* (eds M.A.S. Graça, F. Bärlocher, and M.O. Gessner), Springer, Berlin, pp. 273–277.

Hendricks, S.P. (1993) Microbial ecology of the hyporheic zone: a perspective integrating hydrology and biology. *J. N. Am. Benthol. Soc.*, **12**, 70–78.

Hendricks, S.P. (1996) Bacterial biomass, activity, and production within the hyporheic zone of a north-temperate stream. *Arch. Hydrobiol.*, **136**, 467–487.

Hershey, A.E., Hiltner, A.L., Hullar, M.A.J., Miller, M.C., Vestal, J.R., Lock, M.A., Rundle, S., and Peterson, B.J. (1988) Nutrient influence on a stream grazer: *Orthocladius* microcommunities respond to nutrient input. *Ecology*, **69**, 1383–1392.

Hessische Landesanstalt für Umwelt und Geologie. (1964). Hessische Landesanstalt für Umwelt und Geologie, www.hlug.de, http://www.hlug.de/medien/luft/recherché.

Hessischer Minister für Landwirtschaft und Forsten (ed.) (1964) *Wasserwirtschaftlicher Rahmenplan FULDA*, FULDA, Wiesbaden.

Hetrick, N.D., Morse, J.C., and West, J.L. (1998) Descriptions and phylogeny of four limnephiloid caddisflies (Trichoptera) based on first instars. *Ann. Entomol. Soc. Am.*, **91**, 497–514.

Hickin, N.E. (1967) *Caddis Larvae. Larvae of the British Trichoptera*. Hutchinson, London, XI + 476 pp.

Hildrew, A.G. and Ormerod, S.J. (1995) Acidification–causes, consequences and solutions, in *Ecological Basis for River Management* (ed. A.G. Hildrew), John Wiley & Sons Ltd, London, pp. 147–160.

Hildrew, A.G., and Townsend, C.R. (1987) Organization of freshwater benthic communities, in *Organization of communities* (eds J.H.R. Gee and P.S. Giller), The 27th symposium of the British Ecological Society, Aberystwyth, 1986; pp. 347–371

Hildrew, A.G., and Wagner, R. (1992) The briefly colonial life of hatchlings of the net-spinning caddisfly *Plectrocnemia conspersa*. *J. N. Am. Benthol. Soc.*, **11**, 60–68.

Hildrew, A.G., Townsend, C.R., and Henderson, J. (1980) Interactions between larval size, microdistribution, and substrate in the stoneflies of an iron-rich stream. *Oikos*, **35**, 387–396.

Hill, W.R. (1992) Food limitation and interspecific competition in snail-dominated streams. *Can. J. Fish. Aquat. Sci.*, **49**, 1257–1267.

Hill, W.R., and Knight, A.W. (1987) Experimental analysis of the grazing interaction between a mayfly and stream algae. *Ecology*, **68**, 1955–1965.

Hill, W.R., and Knight, A.W. (1988) Concurrent grazing effects of two stream insects on periphyton. *Limnol. Oceanogr.*, **33**, 15–26.

Hill, W.R., Weber, S.C., and Stewart, A.J. (1992) Food limitation of two lotic grazers: quantity, quality, and size-specificity. *J. N. Am. Benthol. Soc.*, **11**, 420–432.

Hillebrand, H. (2005) Light regime and consumer control of autotrophic biomass. *J. Ecol.*, **93**, 758–769.

Hillebrand, H. (2008) Grazing regulates the spatial variability of periphyton biomass. *Ecology*, **89**, 165–173.

Hillebrand, H., Kahlert, M., Haglund, A.L., Berninger, U.G., Nagel, S., and Wickham, S. (2002) Control of microbenthic communities by grazing and nutrient supply. *Ecology*, **83**, 2205–2219.

Hisdal, H., Tallaksen, L.M., Clausen, B., Peters, E., and Gustard, A. (2004) Hydrological drought characteristics, in *Hydrological Drought Processes and Estimation Methods for Streamflow and Groundwater. Developments in Water Sciences vol 48* (eds L.M. Tallaksen and H.A.J. van Lanen), Elsevier Science Publisher, Rotterdam, pp. 139–198.

Hoffman, A.L., Olden, J.D., Monroe, J.B., Poff, N.L., Wellnitz, T., and Wiens, J.A. (2006) Current velocity and habitat patchiness shape stream herbivore movement. *Oikos*, **115**, 358–368.

Hoffmann, H.J. (2008) Zur Verbreitung der Grundwanze *Aphelocheirus aestivalis* (Fabricius, 1794) in Deutschland, nebst Angaben zur Morphologie, Biologie, Fortpflanzung und Ökologie der Art und zum Fund eines makropteren Exemplars (Heteroptera). *Entomol. Nachr. Ber.*, **52**, 149–180.

Hoffmann, J. (1986) Ernährungsökologie und Populationsdynamik von *Valvata piscinalis* Müll. (Gastropoda, Prosobranchia). Diplomarbeit, University Köln, Köln, 86 pp.

Hogue, J.N., and Hawkins, C.P. (1991) Morphological variation in adult aquatic insects: associations with developmental temperature and seasonal growth patterns. *J. N. Am. Benthol. Soc.*, **10**, 309–321.

Holcik, J. (ed.) (1986) *The Freshwater Fishes of Europe*, Aula Verlag, Berlin, 1/I: Petromyzontiformes, 313 pp.

Holder-Franklin, M.A. (1981) Methods of studying population shifts in aquatic bacteria in response to environmental change. Scientific Series, 124. Inland Waters Directorate, Ottawa.

Holland, D.G. (1972) *A Key to the Larvae, Pupae, and Adults of the British Species of Elminthidae*, Freshwater Biological Association, London, *Scientific Publication* 26, 58 pp.

Holomuzki, J.R., and Biggs, B.J.F. (2006) Food limitation affects algivory and grazer performance for New Zealand stream macroinvertebrates. *Hydrobiologia*, **561**, 83–94.

Holomuzki, J.R., Lowe, R.L., and Ress, J.A. (2006) Comparing herbivory effects of stream macroinvertebrates on microalgal patch structure and recovery. *N. Z. J. Mar. Freshwater Res.*, **40**, 357–367.

Hölzel, H. (2002) Insecta: Megaloptera, in *Süßwasserfauna von Mitteleuropa 15* (eds J. Schwoerbel and P. Zwick), Spektrum Akademischer Verlag, Berlin, pp. 1–30.

Hölzel, H., and Weißmair, W. (2002) Insecta: Neuroptera, in *Süßwasserfauna von Mitteleuropa 16* (eds J. Schwoerbel and P. Zwick), Spektrum Akademischer Verlag, Berlin, pp. 31–68.

Honek, A., and Kocourek, F. (1990) Temperature and development time in insects: a general relationship between thermal constants. *Zool. Jahrb. Abt. Syst. Okol. Geog. Tiere*, **117**, 401–439

Hoof, M. (2001) Autökologische Untersuchungen an Scirtiden temporärer Tümpel unter besonderer Berücksichtigung der Art *Microcara testacea* Linnaeus, 1767. Diploma thesis, University Marburg, Marburg, 80 pp.

Hoopes, R.L. (1974) Flooding, as the results of Hurricane Agnes, and its effect on the macrobenthic community in an infertile headwater stream in Central Pennsylvania. *Limnol. Oceanogr.*, **19**, 853–857.

Hoppe, H.G. (1983) Significance of exoenzymatic activities in the ecology of brackish water: measurements by means of methylumbelliferyl-substrates. *Mar. Ecol. Prog. Ser.*, **11**, 299–308.

Hoppe, H.G. (1993) Use of fluorogenic model substrates for extracellular enzyme activity (EEA) measurement of bacteria, in *Handbook of Methods in Aquatic Microbial Ecology* (eds P.F. Kemp, B.F. Sherr, E.B. Sherr, and J.J. Cole), Lewis Publishers, Boca Raton, pp. 422–431.

Hoppe, H.G., Arnosti, C., and Herndl, G.F. (2002) Ecological significance of bacterial enzymes in the marine environment, in *Enzymes in the Environment. Activity, Ecology, and Applications* (eds R.G. Burns and R.P. Dick), Marcel Dekker, New York, pp. 73–107.

Höppner, J. (1999) Struktur und Funktion nitrifizierender Bakterien in

Fließgewässern. Diploma thesis, University Gießen, Gießen, 83 pp.
Horion, A. (1955) Faunistik der mitteleuropäischen Käfer IV. Kommissionsverlag Dr. Ewald Reitter, München, XXIV+ 280 pp.
Horvath, T.G. (2004) Retention of particulate matter by macrophytes in a first-order stream. *Aquat. Bot.*, **78**, 27–36.
Hossell, J.S., and Baker, J.H. (1976a) A note on the enumeration of epiphytic bacteria by microscopic methods with particular reference to two freshwater plants. *J. Appl. Bacteriol.*, **46**, 87–92.
Hossell, J.S., and Baker, J.H. (1976b) Estimation of the growth of epiphytic bacteria and *Lemna minor* in a river. *Freshwater Biol.*, **9**, 319–327.
Hossell, J.S., and Baker, J.H. (1976c) Epiphytic bacteria of the freshwater plant *Ranunculus penicillatus*: enumeration, distribution and identification. *Arch. Hydrobiol.*, **86**, 322–337.
Houghton, D.C., and Stewart, K.W. (1998) Life-history and case-building behavior of *Culoptila cantha* (Trichoptera, Glossosomatidae) in the Brazos River, Texas. *Ann. Entomol. Soc. Am.*, **91**, 59–70.
Hudson, J.J., Roff, J.C., and Burnison, B.K. (1992) Bacterial productivity in forested and open streams in southern Ontario. *Can. J. Fish. Aquat. Sci.*, **49**, 2412–2122.
Hullar, M.A.J., Kaplan, L.A., and Stahl, D.A. (2006) Recurring seasonal dynamics of microbial communities in stream habitats. *Appl. Environ. Microbiol.*, **72**, 713–722.
Humpesch, U.H. (1980) Effect of temperature on the hatching time of parthenogenetic eggs of five *Ecdyonurus* spp. and two *Rhithrogena* spp. (Ephemeroptera) from Austrian streams and English rivers and lakes. *J. Anim. Ecol.*, **49**, 927–937.
Hurlbert, S.H. (1984) Pseudoreplication and the design of ecological field experiments. *Ecol. Monogr.*, **54**, 187–211.
Huryn, A.D., and Wallace, J.B. (1988) Community structure of Trichoptera in a mountain stream: spatial patterns of production and functional organization. *Freshwater Biol.*, **20**, 141–155.
Husmann, S. (1971) Ecological Studies on Freshwater Meiobenthon in Layers of Sand and Gravel. *Smithson. Contr. Zool.*, **76**, 161–169.
Hynes, H.B.N. (1942) A study of the feeding of adult stoneflies (Plecoptera). *Proc. R. Entomol. Soc. Lond. Ser. A, Gen. Entomol.*, **A17**, 81–82.
Hynes, H.B.N. (1963) Imported organic matter and secondary productivity in streams. *Int. Congr. Zool.*, **16**, 324–329.
Hynes, H.B.N. (1970a) The ecology of stream insects. *Annu. Rev. Entomol.*, **15**, 25–42.
Hynes, H.B.N. (1970b) *The Ecology of Running Waters*. Liverpool University Press, Liverpool, 555 pp.
Hynes, H.B.N. (1971) 1.4 Benthos of flowing water, in *A Manual on Methodes for the Assessment of Secondary Productivity in Fresh Waters* (eds W.T. Edmondson and G.G. Winberg), IBP Handbook 17, Blackwell Scientific Publications, Oxford, pp. 66–74.
Hynes, H.B.N. (1975) The stream and its valley. *Verh. Internat. Verein. Limnol.*, **19**, 1–19.
Hynes, H.B.N., and MacPhee, F.M. (1970) The maxillae of the British species of *Isoperla* (Plecoptera: Perlodidae). *Proc. R. Entomol. Soc. Lond. Ser. A, Gen. Entomol. A*, **45** (7/9), 123–124.
Hynes, H.B.N., Williams, D.D., and Williams, N.E. (1976) Distribution of the benthos within the substratum of a Welsh mountain stream. *Oikos*, **27**, 307–310.
ICZN (International Commission for Zoological Nomenclature) (1995) Opinion 1812. Elmidae Curtis, 1839 (Insecta, Coleoptera): conserved as the correct spelling, and the gender of *Elmis* Latreille, 1802 ruled to be feminine. *Bull. Zool. Nomencl.*, **52**, 217–218.
Ide, F.P. (1935) Post embryonal development of Ephemeroptera (Mayflies). External characters only. *Can. J. Res.*, **12**, 433–478.
Illies, J. (1952a) Die Plecopteren und das Monard'sche Prinzip. *Ber. Limnol. Flussstn. Freudenthal*, **3**, 53–69.
Illies, J. (1952b) Die europäischen Arten der Plecopterengattung *Isoperla* Banks (=*Chloroperla* Pictet). *Beitr. Entomol.*, **2**, 369–424.
Illies, J. (1953a) Die Besiedlung der Fulda (insbes. das Benthos der Salmonidenregion) nach dem jetzigen

Stande der Untersuchung. *Ber. Limnol. Flussstn Freudenthal*, **5**, 1–28.

Illies, J. (1953b) Die deutschen Arten der Dryopidengattung *Lathelmis* Rttr. *Entomol. Bl. Biol. Syst. Käfer*, **49**, 173–184.

Illies, J. (1955) Steinfliegen oder Plecoptera, *Tierwelt Deutschlands 43*, VEB Gustav Fischer Verlag, Jena, pp. IV + 150.

Illies, J. (1961) Versuch einer allgemeinen biozönotischen Gliederung der Fließgewässer. *Int. Revue ges. Hydrobiol.*, **46**, 205–213.

Illies, J. (1968) Ephemeroptera (Eintagsfliegen), *Handb. Zool.*, **4** (2), 2–5, 63 pp.

Illies, J. (1971) Emergenz 1969 im Breitenbach. Schlitzer produktions-biologische Studien (1). *Arch. Hydrobiol.*, **69**, 14–59.

Illies, J. (1972) Emergenzmessung als neue Methode zur produktionsbiologischen Untersuchung von Fließgewässern. Schlitzer produktionsbiologische Studien (2). *Verh. Dt. Zool. Ges.*, 65–68.

Illies, J. (1974) Emergenzschwankungen – ein produktionsbiologisches Problem. Schlitzer produktions-biologische Studien Nr. 7. *Verh. Ges. Ökol.*, **3**, Junk, The Hague, pp. 131–142.

Illies, J. (1978) Vergleichende Emergenzmessung im Breitenbach 1969–1976 (Ins.: Ephemeroptera, Trichoptera, Plecoptera). Schlitzer produktionsbiologische Studien (25). *Arch. Hydrobiol.*, **82**, 432–448.

Illies, J. (1983) Ökosystemforschung an einem Mittelgebirgsbach (Emergenzanalyse). Schlitzer produktions-biologische Studien (49). *Verh. Ges. Ökol.*, **11**, 247–253.

Illies, J., (1982) Längsprofil des Breitenbachs im Spiegel der Emergenz (Ins.: Ephemeroptera, Plecoptera, Trichoptera). *Arch. Hydrobiol.*, **95**, 157–168.

Illies, J., and Botoşaneanu, L. (1963) Problèmes et méthodes de la classification et de la zonation écologique des eaux courantes, considerées surtout du point de vue faunistique. *Mitt. int. Ver. Limnol.*, **12**, 1–57.

Illies, J., and Masteller, E.C. (1977) A possible explanation of emergence patterns of *Baetis vernus* Curtis (Ins.: Ephemeroptera) on the Breitenbach. Schlitz studies on productivity, No. 22. *Int. Rev. Ges. Hydrobiol.*, **62**, 315–321.

IPCC Intergovernmental Panel on Climate Change IPCC (2007) Klimaänderung 2007. http://www.bmbf.de/pub/IPCC2007.pdf (accessed 12 December 2010).

Irons, J.G. (1988) Life history patterns and trophic ecology of Trichoptera in two Alaskan (U.S.A.) subarctic streams. *Can. J. Zool.*, **66**, 1258–1265.

Iserentant, R., and Blancke, D. (1986) A transplantation experiment in running water to measure the response rate of diatoms to changes in water quality. In Proc. 8th Int. Diatom Symp., Bristol (ed. F.E. Round), pp. 347–354.

Ito, T. (1987) A review of bionomics of the Japanese Lepidostomatidae (Trichoptera), with particular reference to the change of case materials and its ecological significance, in *Proc. 5th Int. Symp. Trichoptera, 1986 Lyon* (eds M. Bournaud and H. Tachet), Dr. W. Junk Publishers, Dordrecht, pp. 269–273.

Iversen, T.M. (1973) Life cycle and growth of *Sericostoma personatum* Spence (Trichoptera, Sericostomatidae) in a Danish spring. *Entomol. Scand.*, **4**, 323–327.

Iversen, T.M. (1976) Life cycle and growth of Trichoptera in a Danish spring. *Arch. Hydrobiol.*, **78**, 482–493.

Iversen, T.M. (1978) Life cycle and growth of three species of Plecoptera in a Danish spring. *Entmol. Meddr.*, **46**, 57–62.

Jäch, M. (1992) 42. Familie: Dryopidae, in *Die Käfer Mitteleuropas 13* (eds G.A. Lohse and W.H. Lucht), Goecke and Evers, Krefeld, [2. Supplementband mit Katalogteil], pp. 67–82.

Jäch, M.A. (1994) Ein Massenanflug von *Helophorus (Atracthelophorus) brevipalpis* Bedel (Helophoridae, Col.). *Entomologisches Nachrichtenblatt.*, **1**(3/4), 7–10.

Jäch, M.A. (1997) Daytime swarming of rheophilic water beetles in Austria (Coleoptera: Elmidae, Hydraenidae, Haliplidae). *Latissimus*, **9**, 10–11.

Jäch, M. (1998) Annotated check list of aquatic and riparian/littoral beetle families of the world (Coleoptera), in *Water Beetles of China, II* (eds M. Jäch and L. Ji), Halle, Vienna, pp. 25–42.

Jäch, M.A. (2003) *Ancyronyx* Erichson: new faunistic records, and description of a new species from Sulawesi (Indonesia) (Coleoptera: Elmidae). *Koleopterol. Rundsch.*, **73**, 255–260.

Jacob, U. (1986) Analyse der Ephemeroptera-Jahresemergenz des Breitenbaches bei Schlitz/Hessen (Bundesrepublik Deutschland). *Arch. Hydrobiol.*, **107**, 215–248.

Jacobsen, D. (1993) Trichopteran larvae as consumers of submerged angiosperms in running waters. *Oikos*, **67**, 379–383.

Jacoby, J.M. (1987) Alterations in periphyton characteristics due to grazing in a Cascade foothill stream. *Freshwater Biol.*, **18**, 495–508.

Jahnke, R.S., and Craven, D.B. (1997) Quantifying the role of heterotrophic bacteria in the carbon cycle: a need for respiration rate measurements. *Limnol. Oceanogr.*, **40**, 436–441.

Jannasch, H.W. (1955) Zur Ökologie der zymogenen planktischen Bakterienflora natürlicher Gewässer. *Arch. Mikrobiol.*, **23**, 146–180.

Jansson, M., Olsson, H., and Petterson, K. (1988) Phosphatases; origin, characteristics and function in lakes. *Hydrobiologia*, **170**, 157–175.

Jazdzewska, T. (1973) Notes on the biology and ecology of the mayfly *Ametropus eatoni* Brodskij (Ephemeroptera). *Pol. Pismo Entomol.*, **43**, 469–477.

Jedlička, L. (1978) Distribution of *Odagmia ornata* (Meigen, 1818) and *Odagmia spinosa* (Doby et Deblock, 1957) in Slovakia (Diptera, Simuliidae). *Acta F. R. N. Univ. Comen. Zoologia*, **23**, 81–89.

Jenkins, M.F. (1960) On the method by which *Stenus* and *Dianous* (Coleoptera: Staphylinidae) return to the banks of a pool. *Trans. R. Entomol. Soc. Lond.*, **112**, 1–14.

Jenkins, R.A. (1977) Notes on the distribution of psychomyiid larvae (Trichoptera) in south-west Wales. *Entomol. Rec. J. Var.*, **89**, 57–61.

Jervis, M.A., Boggs, C.L., and Ferns, P.N. (2005) Egg maturation strategy and its associated trade-offs: a synthesis focusing on Lepidoptera. *Ecol. Entomol.*, **30**, 359–375.

Jin, H.S., and Ward, G.M. (2007) Life history and secondary production of *Glossosoma nigrior* Banks (Trichoptera: Glossosomatidae) in two Alabama streams with different geology. *Hydrobiologia*, **575**, 245–258.

Johansson, A. (1991) Caddis larvae cases (Trichoptera, Limnephilidae) as antipredatory devices against brown trout and sculpin. *Hydrobiologia*, **211**, 185–194.

Johansson, A., and Englund, G. (1995) A predator–prey game between bullheads and case-making caddis larvae. *Anim. Behav.*, **50**, 785–792.

Johansson, A., and Nilsson, A.N. (1992) *Dytiscus latissimus* and *D. circumcinctus* (Coleoptera, Dytiscidae) larvae as predators on three case-making caddis larvae. *Hydrobiologia*, **248**, 201–213.

Johnson, P.D., Brown, K.M., and Colwell Jr, C.V. (1994) A comparison of the macroinvertebrate assemblage in Doe Run Creek, Kentucky, 1960 and 1990. *J. North Am. Benthol. Soc.*, **13**, 496–510.

Johnson, R., Richards, T., and Blinn, D.W. (1975) Investigation of diatom populations in rhithron and potamon communities in Oak Creek, Arizona. *Southwest. Nat.*, **20**, 197–204.

Jones, J.B., and Mulholland, P. (eds) (2000) *Streams and Ground Waters*. Academic Press, San Diego, 425 pp.

Jones, S.E., and Lock, M.A. (1993) Seasonal determination of extracellular hydrolytic activities in heterotrophic and mixed heterotrophic/autotrophic biofilms from two contrasting rivers. *Hydrobiologia*, **257**, 1–16.

Jongman, R.H.G., ter Braak, C.J.F., and van Tongeren, O.F.R. (1987) Data analysis in community and landscape ecology. Pudoc, Wageningen, 299 pp.

Joost, W. (1980) *Hemerodromia illiesi* sp.n. aus Georgien (UdSSR) (Diptera, Empididae). *Reichenbachia*, **18**, 81–84.

Jørgensen, N.O.G., and Jensen, R.E. (1994) Microbial fluxes of free monosaccharides and total carbohydrates in freshwater determined by PAD-HPLC. *FEMS Microbiol. Ecol.*, **14**, 79–94.

Juday, C. (1940) The annual energy budget of an inland lake. *Ecology*, **21**, 438–450.

Jüttner, I., Rothfritz, H., and Ormerod, S.J. (1996) Diatoms as indicators of river quality in the Nepalese Middle Hills with consideration of the effects of habitat-specific sampling. *Freshwater Biol.*, **36**, 475–486.

Jüttner, I., Sharma, S., Dahal, B.M., Ormerod, S.J., Chimonides, P.J., and Cox, E.J. (2003) Diatoms as indicators of stream quality in the Kathmandu Valley and Middle Hills of Nepal and India. *Freshwater Biol.*, **48**, 2065–2084.

Kahlert, M., and Baunsgaard, M.T. (1999) Nutrient recycling – a strategy of a grazer community to overcome nutrient limitation. *J. N. Am. Benthol. Soc.*, **18**, 363–369.

Kaiser, H., and Schwank, P. (1985) Mermithidae (Nematoda) aus oberhessischen Fließgewässern. *Arch. Hydrobiol.*, **103**, 347–369.

Kajak, Z. (1959) Role of rise in water level in the removal and introduction of benthic fauna in environments connected with a river. *Ekol. Pol., Ser. B*, **5**, 47–54.

Kaplan, L.A., and Bott, T.M. (1989) Diel fluctuations in bacterial activity on streambed substrata during vernal algal blooms: effects of temperature, water chemistry, and habitat. *Limnol. Oceanogr.*, **34**, 718–733.

Kaplan, L.A., and Newbold, J.D. (2000) Surface and subsurface dissolved organic carbon, in *Streams and Ground Waters* (eds J.B. Jones and P. Mulholland), Academic Press, San Diego, pp. 237–258.

Katano, I., Doi, H., Houki, A., Isobe, Y., and Oishi, T. (2007) Changes in periphyton abundance and community structure with the dispersal of a caddisfly grazer, *Micrasema quadriloba*. *Limnology*, **8**, 219–226.

Katano, I., Mitsuhashi, H., Isobe, Y., Sato, H., and Oishi, T. (2005) Reach-scale distribution dynamics of a grazing stream insect, *Micrasema quadriloba* Martynov (Brachycentridae, Trichoptera), in relation to current velocity and periphyton abundance. *Zool. Sci.*, **22**, 853–860.

Katzner, A. (2011) Vorkommen von gelösten Kohlenhydraten in einem naturnahen mitteleuropäischen Bach (Breitenbach, Osthessen). Ph.D. thesis, University of Gießen, Gießen.

Kaushik, N.K., and Hynes, H.B.N. (1971) The fate of dead leaves that fall into streams. *Arch. Hydrobiol.*, **68**, 465–515.

Kawase, T., Saito, A., Sato, T., Kanai, R., Fujii, T., Nikaidou, N., Miyashita, K. and Watanabe, T. (2004) Distribution and phylogenetic analysis of family 19 chitinases in Actinobacteria. *Appl. Environ. Microbiol.*, **70**, 1135–1144.

Kawecka, B. (1986) Sessile algae of the River Nida (Southern Poland) in the area of experimentally fertilized bankside soils. *Acta Hydrobiol. (Cracow)*, **28**, 371–378.

Kazimírová, M. (1981) Sezonna dynamika liahnutia a produkcia muskovitych (Simuliidae, Diptera) y alpskych potokoch. Saisondynamik des Schlüpfens und die Produktion der Kriebelmücken (Diptera, Simuliidae) in Alpenbächen. MS Thesis, University of Bratislava, Bratislava, 178 pp.

Kelly, M.J., and Whitton, B.A. (1995) The trophic diatom index: a new index for monitoring eutrophication in rivers. *J. Appl. Phycol.*, **7**, 433–444.

Kemp, P.F. (1990) The fate of benthic bacterial production. *Aquat. Sci.*, **2**, 109–124.

Kerney, M.P., Cameron, R.A.D., and Jungbluth, J.H. (1983) *Die Landschnecken Nord- und Mitteleuropas*, Paul Parey, Hamburg, 384 pp.

Kerst, C.D., and Anderson, N.H. (1974) Emergence patterns of Plecoptera in a stream in Oregon, USA. *Freshwater Biol.*, **4**, 205–212.

Khoo, S.G. (1964) Studies on the biology of *Capnia bifrons* (Newman) and notes on the diapause in the nymphs of this species. *Gewäss. Abwäss. Düsseldorf*, **34/35**, 23–30.

Khoo, S.G. (1968) Experimental studies on diapause in stoneflies. III. Eggs of *Brachyptera risi* (Morton). *Proceedings of the Royal Entomological Society London, Series A General Entomol*, **43**, 10/12, 141–146.

Kiel, E. (1996) Effects of Aufwuchs on colonization by simuliids (Simuliidae, Diptera). *Int. Rev. Ges. Hydrobiol.*, **81**, 565–576.

Kiel, E., and Röder, T. (2002) Gelelectrophoretic studies on labial gland secretions of immature blackflies

(Simuliidae, Diptera). *Limnologica*, **32**, 201–205.

Kiel, E., Reidelbach, J., Rühm, W., and Rupp, K. (1989) Verhaltensbiologische Studie an Simuliiden (Simuliidae, Diptera): Landen und Ansiedeln auf einem Substrat. *Z. Angew. Zool.*, **76**, 385–401.

Kiel, E., Dickmann, K., and Rühm, W. (1998) Effects of frequent drift events on blackfly (Simuliidae, Diptera) development: a laboratory study. *Limnologica*, **28**, 301–305.

Kimmins, D.E. (1954) *A revised key to the adults of the British species of Ephemeroptera*. Freshwater Biological Association, *Scientific Publications*, 15, 72 pp.

Kirchman, D.L., K'nees, E., and Hodson, R.E. (1985) Leucine incorporation and its potential as a measure of protein synthesis by bacteria in natural aquatic systems. *Appl. Environ. Microbiol.*, **49**, 599–607.

Kirschfeld, D., Wenk, E. (1976) The dorsal compound eye of Simuliid flies. *Z. Naturforsch.*, **31c**, 764–765.

Kjeldsen, L. (1991) Flying pattern of stoneflies (Plecoptera) at culverts and concrete pipes. *Nat. Jutl.*, **23**, 45–56.

Kjer, K.M., Blahnik, R.J., and Holzenthal, R.W. (2002) Phylogeny of caddisflies (Insecta, Trichoptera). *Zool. Scr.*, **31**, 83–91.

Klausing, O. (1974) Die Naturräume Hessens. Hessische Landesanstalt für Umwelt, Wiesbaden, 86 pp.

Klausnitzer, B. (1971) Zur Ökologie und Verbreitung tiergeographisch bemerkenswerter Arten der Gattung *Cyphon* Payk. in der Oberlausitz (Col., Helodidae). *Abh. Naturkundmus. Görlitz*, **46**, 1–5.

Klausnitzer, B. (1996) *Käfer im und am Wasser (2. überarbeitete Auflage), Neue Brehm Bücherei 567*, Westarp Wissenschaften Magdeburg, Spektrum Akademischer Verlag, Berlin, 200 pp.

Klausnitzer, B. (2009) Insecta: Coleoptera: Scirtidae, in *Süßwasserfauna von Mitteleuropa. Bd. 20/17* (eds J. Schwoerbel and P. Zwick), Gustav Fischer, Berlin. XIV + 326 pp.

Klausnitzer, B., and Richoux, P. (1996) *40. Familie: Elmidae.* in *Die Larven der Käfer Mitteuropas 3* (ed. B. Klausnitzer), Gustav Fischer Verlag, Jena, pp. 110–143.

Klemp, H. (1979) Die Copepoden (Crustacea) des Breitenbaches (Schlitz/ Oberhessen). Diploma thesis, University of Darmstadt, Darmstadt, 87 pp.

Klima. (1950) Deutscher Wetterdienst in der US-Zone, Klima-Atlas von Hessen, Bad Kissingen.

Klingenberg, C.P., and Spence, J.R. (1997) On the role of body size for life-history evolution. *Ecol. Entomol.*, **22**, 55–68.

Klonowska, M. (1986) The food of some mayfly (Ephemeroptera) nymphs from the streams of the Krakow-Czestochowa Jura (Southern-Poland). *Acta Hydrobiol. (Cracow)*, **28**, 181–197.

Knie, J. (1977) Ökologische Untersuchungen der Käferfauna von ausgewählten Fließgewässern des Rheinischen Schiefergebirges (Insecta: Coleoptera). *Decheniana*, **130**, 151–221.

Kniepert, F.W. (2000) Tabanidae, in *Süßwasserfauna von Mitteleuropa Insecta. Diptera. 21/18,19: Stratiomyidae und Tabanidae* (eds J. Schwoerbel and P. Zwick), Gustav Fischer, Berlin, pp. 113–204.

Knöfel, B. (1983) Beiträge zur Biologie und Ernährungsbeziehungen einer carnivoren Insektenlarve, *Rhyacophila faciata* Hagen 1859 (Trichoptera, Rhyacophilidae). *Beitr. Naturk. Osthessen*, **19**, 91–134.

Koch, K.D. (1991) Ernährungsökologische Untersuchungen an *Gammarus pulex* (L.) und *Gammarus fossarum* KOCH, 1835 (Crustacea, Amphipoda) in einem Wiesenbach und einem Waldbach. *Beitr. Naturkde. Osthessen*, **26**, 3–126.

Kodada, J., and Jäch, M.A. (2005) 18.2. Elmidae Curtis, 1830, in *Handbook of Zoology, Volume IV (Part 38), Coleoptera, Beetles, Volume 1: Morphology and Systematics (Archostemata, Adephaga, Myxophaga, Polyphaga partim)* (eds R.G. Beutel and R.A.B. Leschen), Walter de Gruyter, XI+567, Berlin, pp. 471–496.

Kohler, S.L. (1984) Search mechanism of a stream grazer in patchy environments: the role of food abundance. *Oecologia*, **62**, 209–218.

Kohler, S.L. (1992) Competition and the structure of a benthic stream community. *Ecol. Monogr.*, **62**, 165–188.

Kohler, S.L., and McPeek, M.A. (1989) Predation risk and foraging behavior of competing stream insects. *Ecology*, **70**, 1811–1825.

König, E., and Waringer, J. (2008) The ballast stones in *Silo nigricornis* cases (Insecta: Trichoptera): drift resistance and ecological benefits, investigated by acoustic Doppler velocimetry. *Arch. Hydrobiol./Suppl. Beil. Large Rivers*, **18**, 311–328.

Kopelke, J.P., and Müller-Liebenau, I. (1981) Eistrukturen bei Ephemeroptera und deren Bedeutung für die Aufstellung von Artengruppen am Beispiel der europäischen Arten der Gattung *Baetis* Leach, 1815. Teil II: rhodani-, vernus- und fuscatus-Gruppe. *Spixiana*, **4**, 39–54.

Koss, R.W. (1968) Morphology and taxonomic use of Ephemeroptera eggs. *Ann. Entomol. Soc. Am.*, **61**, 696–721.

Kovalak, W.P. (1979) Day–night changes in stream benthos density in relation to current velocity. *Arch. Hydrobiol.*, **87**, 1–18.

Kracht, V. (1983) Die Bakterienflora in der fließenden Welle eines oligosaproben Fließwassersystems im jahreszeitlichen Wechsel. Ph.D. dissertation, University of Tübingen, Tübingen, 161 pp.

Kriska, G., Horvath, G., and Andricovics, S. (1998) Why do mayflies lay their eggs en-masse on dry asphalt roads – water-imitating polarized-light reflected from asphalt attracts Ephemeroptera. *J. Exp. Biol.*, **201**, 2273–2286.

Kuehn, K.A., Lemke, M.J., Suberkropp, K., and Wetzel, R.G. (2000) Microbial biomass, and production associated with decaying leaf litter of the emergent macrophyte *Juncus effusus*. *Limnol. Oceanogr.*, **45**, 862–870.

Kuhbier, S., Lorch, H.J., and Ottow, J.C.G. (1999) Bakterielle Abundanz, und Enzymaktivitäten im Sediment unterschiedlich belasteter Zonen eines Fließgewässers'(Horloff/Vogelsberg). *Verh. Ges. Ökol.*, **29**, 473–479.

Kuhbier, S., Lorch, H.J., and Ottow, J.C.G. (2002) Enzymatic response to pollution in sediments and aquatic systems, in *Enzymes in the Environment. Activity, Ecology, and Applications* (eds R.G. Burns and R.P. Dick), Marcel Dekker, New York, pp. 433–446.

Kühtreiber, J. (1934) Die Plekopterenfauna Nordtirols. Ber. Naturwiss. Med. Ver. Innsbruck, *43/44*, I–VII, 1–219, Tafel 1–6 + 1 Karte.

Kupfahl, H.G. (1964) *Geologische Karte von Hessen 1:25 000 Blatt Nr. 5323 Schlitz*, Hessisches Landesamt für Bodenforschung, Wiesbaden.

Kupfahl, H.G. (1965) *Erläuterungen zur Geologischen Karte von Hessen 1:25 000 Blatt Nr. 5323 Schlitz*. Hessisches Landesamt für Bodenforschung, Wiesbaden, 257 pp.

Kureck, A. (1992) Das Massenschwärmen der Eintagsfliegen am Rhein. Zur Rückkehr von *Ephoron virgo* (Olivier 1791). *Nat. Landsch.*, **67**, 407–409.

Kusnetsov, S.I., and Romanenko, V.J. (1966) Produktion der Biomasse heterotropher Bakterien und die Geschwindigkeit ihrer Vermehrung im Rybinsk-Stausee. *Verh. Int. Ver. Limnol.*, **16**, 1493–1500.

Kuusela, K., and Huusko, A. (1996) Post-emergence migration of stoneflies (Plecoptera) into the nearby forest. *Ecol. Entomol.*, **21**, 171–177.

Kyrieleis, H. (2000) Ökologische Untersuchungen an Faltenmücken (Diptera, Ptychopteridae). Diploma thesis, University of Kiel, Kiel, 70 pp.

Lacoursière, J.O., and Craig, D.A. (1993) Fluid transmission and filtration efficiency of the labral fans of black fly larvae (Diptera: Simuliidae) – hydrodynamic, morphological, and behavioral aspects. *Can. J. Zool.*, **71**, 148–162.

Ladle, M., and Radke, R. (1990) Burrowing and feeding behaviour of the larva of *Ephemera danica* Müller Ephemeroptera: Ephemeridae). *Entomol. Gaz.*, **41**, 113–118.

Lake, P.S., Barmuta, L.A., Boulton, A.J., Campell, I.C., and St Clair, R.M. (1985) Australian streams and Northern Hemisphere stream ecology: comparisons and problems. *Proc. Ecol. Soc. Aust.*, **14**, 61–82.

Lamberti, G.A. (1996) The role of periphyton in benthic food webs, in *Algal Ecology: Freshwater Benthic Ecosystems* (eds R.J. Stevenson, M.L. Bothwell, and R.L. Lowe), Academic Press, San Diego, pp. 534–572.

Lamberti, G.A., and Moore, J.W. (1984) Aquatic insects as primary consumers, in

The Ecology of Aquatic Insects (eds V.H. Resh and D.M. Rosenberg), Praeger Publishers, New York, pp. 164–195.

Lamberti, G.A., and Resh, V.H. (1983) Stream periphyton and insect herbivores: an experimental study of grazing by a caddisfly population. *Ecology*, **64**, 1124–1135.

Lamberti, G.A., Ashkenas, L.R., Gregory, S.V., and Steinman, A.D. (1987a) Effects of three herbivores on periphyton communities in laboratory streams. *J. N. Am. Benthol. Soc.*, **6**, 92–104.

Lamberti, G.A., Feminella, J.W., and Resh, V.H. (1987b) Herbivory and intraspecific competition in a stream caddisfly population. *Oecologia*, **73**, 75–81.

Lamberti, G.A., Gregory, S.V., Ashkenas, L.R., Steinman, A.D., and McIntire, C.D. (1989) Productive capacity of periphyton as a determinant of plant–herbivore interactions in streams. *Ecology*, **70**, 1840–1856.

Lancaster, J. (1999): Small scale movements of lotic macroinvertebrates with variation in flow. *Freshw. Biol.*, **41**, 605–619.

Lancaster, J., and Belyea, L.R. (1997) Nested hierarchies and scale-dependence of mechanisms of flow refugium use. *J. North Am. Benthol. Soc.*, **16**, 221–236.

Lawrence, J.F., and Brittain, E.B. (1991) Coleoptera (Beetles), in *The insects of Australia. A textbook for students and research workers, vol. 2, 2nd edn.* (ed. CSIRO (Div. of Entomology)), Melbourne University Press, pp. 543–683.

Laybourn-Parry, J., and Parry, J. (2000) Flagellates and the microbial loop, in *The Flagellates: Unity, Diversity and Evolution* (eds B.S.C. Leadbeater and J.C. Green), Taylor and Francis, Basingstoke, pp. 216–239.

Ledger, M.E., and Hildrew, A.G. (2000) Resource depression by a trophic generalist in an acid stream. *Oikos*, **90**, 271–278.

Ledger, M.E., and Hildrew, A.G. (2001) Growth of an acid-tolerant stonefly on epilithic biofilms from streams of contrasting pH. *Freshwater Biol.*, **46**, 1457–1470.

Leese, F. (2004) Molecular genetic, chemotaxonomic, and autecological investigations of European Sericostomatidae (Insecta: Trichoptera). Diploma thesis, Ruhr-University Bochum, Bochum, 136 pp.

Leese, F., and Wagner, R. (2005) The "Sericostoma–problem" – molecular genetic, chemotaxonomic and autecological approaches (Trichoptera: Sericostomatidae). *Lauterbornia*, **54**, 161–163.

Lehmann, J. (1971) Chironomiden der Fulda. *Arch. Hydrobiol. Suppl.*, **37**, 466–555.

Lehmann, U. (1972) Tagesperiodisches Verhalten und Habitatwechsel der Larven von *Potamophylax luctuosus* (Trichoptera). *Oecologia*, **9**, 265–278.

Leibold, M.A. (1997) Do nutrient-competition models predict nutrient availabilities in limnetic ecosystems. *Oecologia*, **110**, 132–142.

LeSage, L., and Harper, P.P. (1976) Cycles biologiques d'Elmidae (Coléoptères) de ruisseaux des Laurentides, Québec. *Ann. Limnol.*, **12**, 139–174.

Liesack, W., Weyland, H., and Stackebrandt, E. (1991) Potential risks of gene amplification by PCR as determined by 16S rDNA analysis of a mixed-culture of strict barophilic bacteria. *Microb. Ecol.*, **21**, 191–198.

Lieske, R. (2005) Autökologische Studien an der Steinfliege *Nemurella pictetii* (Plecoptera: Nemouridae). Ph.D thesis, Philipps-University Marburg, Marburg.

Lieske, R., and Zwick, P. (2007) Food preference, growth and maturation of *Nemurella pictetii* (Plecoptera: Nemouridae). *Freshwater Biol.*, **52**, 1187–1197.

Lillehammer, A. (1975) Norwegian stoneflies. III. Field studies on ecological factors influencing distribution. *Norw. J. Entomol.*, **22**, 71–80.

Lillehammer, A. (1985) Temperature influence on egg incubation period and nymphal growth of the stoneflies *Leuctra digitata* and *L. fusca* (Plecoptera: Leuctridae). *Entomol. Gen.*, **11**, 59–68.

Lillehammer, A. (1987) Intraspecific variation in the biology of eggs and nymphs of Norwegian populations of *Leuctra hippopus* (Plecoptera). *J. Nat. Hist.*, **21**, 29–41.

Lillehammer, A., Saltveit, S.J., and Brusven, M. (1991) The influence of variable temperatures on the incubation period of stonefly eggs (Plecoptera), in *Overview and Strategies of Ephemeroptera and Plecoptera* (eds J. Alba-Tercedor and A. Sanchez-Ortega), Sandhill Crane Press, Gainesville, pp. 377–381.

Lillehammer, A., Brittain, J.E., Saltveit, S.J., and Nielsen, P.S. (1989) Egg development, nymphal growth and life cycle strategies in Plecoptera. *Hol. Ecol.*, **12**, 173–186.

Lindegaard, C. (1997) Chironomidae, in *Aquatic Insects of North Europe – A Taxonomic Handbook, Vol. 2,* (ed. A.N. Nilsson), Apollo Books, Stenstrup, pp. 133–144.

Lindeman, R.L. (1942) The trophic–dynamic aspect of ecology. *Ecology*, **23**, 399–417.

Linsley, R.K., Kohler, M.A., and Paulhus, J.L.H. (1975) *Hydrology for Engineers*, 2nd edn, McGraw-Hill, New York, 340 pp.

Löbl, I., and Smetana, A. (2003) Archostemata – Myxophaga – Adephaga, in *Catalogue of Palaearctic Coleoptera 1* (ed. Smetana, A.), Apollo Books, Stenstrup, 820 pp.

Lock, M.A., and Hynes, H.B.N. (1975) The disappearance of four leaf leachates in a hard and soft water stream in south Western Ontario, Canada. *Int. Rev. Ges. Hydrobiol.*, **60**, 847–856.

Lodge, D.M. (1986) Selective grazing on periphyton: a determinant of freshwater gastropod microdistributions. *Freshwater Biol.*, **16**, 831–841.

Loferer-Krößbacher, M., Klima, J., and Psenner, R. (1998) Determination of bacterial cell dry mass by transmission electron microscopy and densitometric image analysis. *Appl. Environ. Microbiol.*, **64**, 688–694.

Löhr, P.W. (1988) Schwebfliegen (Diptera: Syrphidae) aus den Emergenzfallen der Limnologischen Flußstation in Schlitz am Breitenbach (Schlitzerland). *Beitr. Naturk. Osthessen*, **23**, 81–93. (1987).

Löhr, P.W. (1991) Beobachtungen und Versuche zum Territorialverhalten der Männchen von *Xylota florum* Fabricius, 1805 (Diptera, Syrphidae). *Beitr. Naturk. Osthessen*, **26**, 135–140.

Löhr, P.W. (2004a) Schwebfliegenbeifänge (Diptera: Syrphidae) aus Emergenzfallen des Hermelin-Quellsystems vom Breitenbach im Vogelsberg. *Hess. Faun. Briefe*, **23**, 1–8.

Löhr, P.W. (2004b) Schwebfliegen (Diptera: Syrphidae) von Erlenbach und Breitenbach im Vogelsberg. Beifänge aus Emergenz-Untersuchungen. *Beitr. Naturk. Osthessen*, **39**, 31–43.

Lonsdale, D.J., and Levinton, J.S. (1985) Latitudinal differentiation in copepod growth: an adaptation to temperature. *Ecology*, **66**, 1397–1407.

Lowe, R.L., and Hunter, R.D. (1988) Effect of grazing by *Physa integra* on periphyton community structure. *J. N. Am. Benthol. Soc.*, **7**, 29–36.

Lowe, R.L., Golladay, S.W., and Webster, J.R. (1986) Periphyton response to nutrient manipulation in streams draining clear-cut and forested watersheds. *J. N. Am. Benthol. Soc.*, **5**, 221–229.

Lozupone, C.A., and Knight, R. (2007) Global patterns in bacterial diversity. *Proc. Natl Acad. Sci. USA*, **104**, 11436–11440.

Lucas, C.H., and Holligan, P.M. (1999) Nature and ecological implications of algal pigment diversity on the Molenplaat tidal flat (Westerschelde estuary, SW Netherlands). *Mar. Ecol. Prog. Ser.*, **180**, 51–64.

Ludwig, W., Strunk, O., Westram, R., Richter, L., Meier, H., Yadhukumar, J., Buchner, A., Lai, T., Steppi, S., Jobb, G., Förster, W., Brettske, I., Gerber, S., Ginhart, A.W., Gross, O., Grumann, S., Hermann, S., Jost, R., König, A., Liss, T., Lüßmann, R., May, M., Nonhoff, B., Reichel, B., Strehlow, R., Stamatakis, A., Stuckmann, N., Vilbig, A., Lenke, M., Ludwig, T., Bode, A., and Schleifer, K.-H. (2004) ARB: a software environment for sequence data. *Nucleic Acids Res.*, **32**, 1363–1371.

Lytle, D.A. (2001) Disturbance regimes and life-history evolution. *Am. Nat.*, **157**, 525–536.

Lytle, D.A. (2002) Flash floods and aquatic insect life-history evolution: evaluation of multiple responses. *Ecology*, **83**, 370–385.

Macan, T.T. (1961) Factors that limit the range of freshwater animals. *Biol. Rev.*, **36**, 151–198.

Macan, T.T. (1962) The ecology of aquatic insects. *Annu. Rev. Entomol.*, **62**, 261–288.

MacArthur, R., and Wilson, E.O. (1967) The Theory of Island Biogeography, in *Monographs in Population Biology 1* (ed. R. MacArthur), Princeton University Press, Princeton, 204 pp.

Mackay, R.J., and Wiggins, G.B. (1979) Ecological diversity in Trichoptera. *Annu. Rev. Entomol.*, **24**, 185–208.

Macneale, K.H., Peckarsky, B.L., and Likens, G.E. (2005) Stable isotopes identify dispersal patterns of stonefly populations living along stream corridors. *Freshw. Biol.*, **50**, 1117–1130.

Mahmoud, H.M.A., Goulder, R., and Carvalho, G.R. (2005) The response of epilithic bacteria to different metals regime in two upland streams: assessed by conventional microbiological methods and PCR-DGGE. *Arch. Hydrobiol.*, **163**, 405–427.

Maitland, P.S. (2003) *Ecology of the River, Brook and Sea Lamprey. Conserving Natura 2000 Rivers*, Ecology Series No. 5, English Nature, Peterborough, 54 pp.

Majecki, J., Schot, J., Verdonschot, P.F.M., and Higler, L.W.G. (1997) Influence of sand cover on mortality and behavior of *A. fuscipes* larvae (Trichoptera: Glossosomatidae). In Proc. 8th Symp. Trichoptera, Minneapolis and Lake Itasca, USA, 1995 (eds J.C. Morse and R.W. Holzenthal), Ohio Biological Survey, Columbus, pp. 283–288.

Makarieva, A.M., Gorshkov, V.G., and Li, B.-L. (2005) Temperature associated upper limits to body size in terrestrial poikilotherms. *Oikos*, **111**, 425–436.

Malicky, H. (1976) Trichopteren-Emergenz in zwei Lunzer Bächen. *Arch. Hydrobiol.*, **77**, 51–65.

Malicky, H. (1982) Adaptations of two stonefly species to intermittent streams in Crete. *Ann. Mus. Goulandris*, **5**, 179–182.

Malicky, H. (1989) Fünfzehn Jahre Emergenzuntersuchungen in Lunz, Niederösterreich. In Verh. IX SIEEC, Gotha 1986, Dresden, 1989, pp. 311–316.

Malicky, H. (2000) Which caddis larvae construct a new case for pupation? *Braueria*, **27**, 19–20.

Malicky, H. (2002a) A quantitative field comparison of different types of emergence traps in a stream: general, Trichoptera, Diptera (Limoniidae and Empididae). *Ann. Limnol.*, **38**, 133–149.

Malicky, H. (2002b) A quantitative field comparison of emergence traps with open and covered bottoms in a stream: general and Trichoptera. *Ann. Limnol.*, **38**, 241–246.

Malicky, H. (2004) *Atlas der europäischen Köcherfliegen*, 2nd edn, Springer, Doordrecht, 359 pp.

Malmqvist, B., and Sackmann, G. (1996) Changing risk of predation for a filter feeding insect along a current velocity gradient. *Oecologia*, **108**, 450–458.

Mann, K.H. (1975) Patterns of energy flow, in *River Ecology. Studies in Ecology 2* (ed. B.A. Whitton), University of California Press, Berkeley, pp. 248–263.

Manz, W., Amann, R., Ludwig, W., Wagner, M., and Schleifer, K.H. (1992) Phylogenetic oligodesoxynucleotide probes for the major subclasses of Proteobacteria: problems and solutions. *System. Appl. Microbiol.*, **15**, 593–600.

Manz, W., Amann, R., Ludwig, W., Vancanneyt, M., and Schleifer, K.H. (1996) Application of a suite of 16S rRNA-specific oligonucleotide probes designed to investigate bacteria of the phylum Cytophaga-Flavobacter-Bacteroides in the natural environment. *Microbiology*, **142**, 1097–1106.

Marchant, R., and Hehir, G. (1999) Growth, production and mortality of two species of *Agapetus* (Trichoptera: Glossosomatidae) in the Acheron River, south-east Australia. *Freshwater Biol.*, **42**, 655–671.

Marchler-Bauer, A., and Bryant, S.H. (2004) CD-Search: protein domain annotations on the fly. *Nucleic Acids Res.*, **32**, W327.

Marker, A.F.H. (1976) The benthic algae of some streams in Southern England. I. Biomass of the epilithon in some small streams. *J. Ecol.*, **64**, 343–358.

Marker, A.F.H., and Casey, H. (1982) The population and production dynamics of benthic algae in an artificial recirculating hard-water stream. *Philos. Trans. R. Soc. Lond., B, Biol. Sci.*, **298**, 265–308.

Marlier, C. (1951) La biologie d'un ruisseau de plaine, le Smohain. *Inst. Roy. Sci. Natur. Belg. Mem.*, **114**, 1–98.

Marten, M. (1983) Die Ephemeropteren, Plecopteren, Heteropteren und

Coleopteren der Fulda. Diploma thesis, Freie University, Berlin, 148 pp.

Marten, M. (1987) Labor- und Freilanduntersuchungen zur Autökologie von *Protonemura*-Arten (Plecoptera: Insecta)unter besonderer Berücksichtigung des Einflusses der Temperatur auf die Embryonal- und Larvalentwicklung. Ph.D. thesis, University of Kassel, Kassel, 162 pp.

Marten, M. (1990) Interspecific variation in temperature dependence of egg development of five congeneric stonefly species (*Protonemura* Kempny, 1898, Nemouridae, Plecoptera). *Hydrobiologia*, **199**, 157–172.

Marten, M. (1991) The effect of temperature on the egg incubation period of *Perlodes microcephalus, Perlodes dispar* (Plecoptera: Perlodidae) and *Perla burmeisteriana* (Plecoptera: Perlidae), *Overview and Strategies of Ephemeroptera and Plecoptera* (eds J. Alba-Tercedor and A. Sanchez-Ortega), Sandhill Crane Press, Gainesville, pp. 387–401.

Marten, M. and Zwick, P. (1989) The temperature dependence of embryonic and larval development in *Protonemura intricata* (Plecoptera: Nemouridae). *Freshwater Biol.*, **22**, 1–14.

Marten, M., and Zwick, P. (1999) 3.10 Emergenz, in *Methoden der Biologischen Gewässeruntersuchung 2* (eds W. von Tümpling and G. Friedrich), Gustav Fischer, Jena, pp. 227–238.

Marxsen, J. (1976) Untersuchungen zu Struktur und Funktion der Bakterienpopulationen der fließenden Welle von Mittelgebirgsbächen. Ph.D. thesis, University of Kiel, Kiel, 306 pp.

Marxsen, J. (1980a) Untersuchungen zur Ökologie der Bakterien in der fließenden Welle von Bächen. I. Chemismus, Primärproduktion, CO_2-Dunkelfixierung und Eintrag von partikulärem organischen Material. *Arch. Hydrobiol. Suppl.*, **57**, 461–533.

Marxsen, J. (1980b) Untersuchungen zur Ökologie der Bakterien in der fließenden Welle von Bächen. II. Die Zahl der Bakterien im Jahreslauf. *Arch. Hydrobiol. Suppl.*, **58**, 26–55.

Marxsen, J. (1980c) Untersuchungen zur Ökologie der Bakterien in der fließenden Welle von Bächen. III. Aufnahme gelöster organischer Substanzen. *Arch. Hydrobiol. Suppl.*, **58**, 207–272.

Marxsen, J. (1982a) Ein neues Verfahren zur Untersuchung der bakteriellen Besiedlung grundwasserführender sandiger Sedimente. *Arch. Hydrobiol.*, **95**, 221–233.

Marxsen, J. (1982b) Zur Anwendbarkeit zweier bakteriologischer Verfahren bei der Untersuchung limnischer sandiger Sedimente. *Beitr. Naturk. Osthessen*, **18**, 3–11.

Marxsen, J. (1988a) Evaluation of the importance of bacteria in the carbon flow of a small open grassland stream, the Breitenbach. *Arch. Hydrobiol.*, **111**, 339–350.

Marxsen, J. (1988b) Investigations into the number of respiring bacteria in groundwater from sandy and gravelly deposits. *Microb. Ecol.*, **16**, 65–72.

Marxsen, J. (1989) Untersuchungen zur pH-Toleranz von Bakterien aus quellnahen Bächen mit unterschiedlichen pH-Werten. *DVWK-Mitteilungen*, **17**, 383–391.

Marxsen, J. (1996) Measurement of bacterial production in stream-bed sediments via leucine incorporation. *FEMS Microbiol. Ecol.*, **21**, 313–325.

Marxsen, J. (1999) Importance of bacterial production in the carbon flow of an upland stream, the Breitenbach. *Arch. Hydrobiol., Spec. Issues Advanc. Limnol.*, **54**, 135–145.

Marxsen, J. (2001) Bacterial production in different streambed habitats of an upland stream: sandy versus coarse gravelly sediments. *Arch. Hydrobiol.*, **152**, 543–565.

Marxsen, J. (2006) Bacterial production in the carbon flow of a Central European stream, the Breitenbach. *Freshw. Biol.*, **51**, 1838–1861.

Marxsen, J., and Fiebig, D.M. (1993) Use of perfused cores for evaluating extracellular enzyme activity in stream-bed sediments. *FEMS Microbiol. Ecol.*, **13**, 1–11.

Marxsen, J., and Moaledj, K. (1988) On the structure of bacterial communities in a Central European open grassland stream, the Breitenbach. *Verh. Int. Ver. Limnol.*, **23**, 1855–1860.

Marxsen, J., and Schmidt, H.-H. (1993) Extracellular phosphatase activity in sediments of the Breitenbach, a Central European mountain stream. *Hydrobiologia*, **253**, 207–216.

Marxsen, J., and Witzel, K.P. (1990) Measurement of exoenzymatic activity in streambed sediments using methylumbelliferyl-substrates. *Arch. Hydrobiol. Beih. Ergebn. Limnol.*, **34**, 21–28.

Marxsen, J., and Witzel, K.P. (1991) Significance of extracellular enzymes for organic matter degradation and nutrient regeneration in small streams, in *Microbial Enzymes in Aquatic Environments* (ed. R. Chróst), Brock/Springer Series in Contemporary Bioscience, Springer, New York, pp. 270–285.

Marxsen, J., and Witzel, K.P. (2005) Nachweis von Chitinase-Genen in Fließgewässern – ein Beispiel für die Analyse von Genen polymerabbauender Enzyme. *DGL Tagungsbericht 2004* (Potsdam), Berlin 2005, *Weißensee Verlag*, pp. 578–582.

Marxsen, J., Schmidt, H.-H., and Fiebig, D.M. (1997) Organic matter dynamics in the Breitenbach, Germany. *J. N. Am. Benthol. Soc.*, **16**, 28–32.

Marxsen, J., Tippmann, P., Heininger, H., Preuß, G., and Remde, A. (1998) Enzymaktivität, in *Mikrobiologische Charakterisierung aquatischer Sedimente – Methodensammlung* (ed. Vereinigung für Allgemeine und Angewandte Mikrobiologie), Oldenbourg Verlag, München, pp. 87–114.

Marxsen, J., Zoppini, A., and Wilczek, S. (2010) Microbial communities in streambed sediments recovering from desiccation. *FEMS Microbiol. Ecol.*, **71**, 374–386. DOI: 10.1111/j.1574-6941.2009.00819.x.

Matthei, C.D., Uehlinger, U., and Frutiger, A. (1997) Response of benthic macroinvertebrates to natural versus experimental disturbance in a Swiss prealpine river. *Freshw. Biol.*, **37**, 61–77.

Mattingly, R.L. (1987) Resource utilization by the freshwater deposit feeder *Ptychoptera townesi* (Diptera: Ptychopteridae). *Freshwater Biol.*, **18**, 241–253.

Mattingly, R.L. (1988) Pelletization of sediment by deposit-feeding invertebrate, *Ptychoptera townesi* (Diptera: Ptychopteridae). *Arch. Hydrobiol.*, **112**, 421–431.

Mattson, W.J. (1980) Herbivory in relation to plant nitrogen content. *Annu. Rev. Ecol. Syst.*, **11**, 119–161.

Mayer, M.S., and Likens, G.E. (1987) The importance of algae in a shaded headwater stream as food for an abundant caddisfly (Trichoptera). *J. N. Am. Benthol. Soc.*, **6**, 262–269.

McAuliffe, J.R. (1983) Competition, colonization patterns, and disturbance in stream benthic communities, in *Stream Ecology – Application and Testing of General Ecology Theory* (eds J.R. Barnes and G.W. Minshall), Plenum Press, New York, pp. 137–156.

McAuliffe, J.R. (1984a) Resource depression by a stream herbivore: effects on distributions and abundances of other grazers. *Oikos*, **42**, 327–333.

McAuliffe, J.R. (1984b) Competition for space, disturbance, and the structure of a benthic stream community. *Ecology*, **65**, 894–908.

McCormick, P.V. (1990) Direct and indirect effects of consumers on benthic algae in isolated pools of an ephemeral stream. *Can. J. Fish. Aquat. Sci.*, **47**, 2057–2065.

McCormick, P.V., and Stevenson, R.J. (1989) Effects of snail grazing on benthic algal community structure in different nutrient environments. *J. N. Am. Benthol. Soc.*, **8**, 162–172.

McCormick, P.V., and Stevenson, R.J. (1991) Grazer control of nutrient availability in the periphyton. *Oecologia*, **86**, 287–291.

McCreadie, J.W., Adler, P.H., and Masteller, E.C. (1994) Long-term emergence patterns of black flies (Diptera: Simuliidae) in northwestern Pennsylvania. *Hydrobiologia*, **288**, 39–46.

McCullough, D.A., Minshall, G.W., and Cushing, C.E. (1979) Bioenergetics of lotic filter-feeding insects *Simulium* spp. (Diptera) and *Hydropsyche occidentalis* (Trichoptera) and their function in controlling organic transport in streams. *Ecology*, **60**, 585–596.

McDowell, W.H., and Wood, T. (1984) Podsolization: Soil processes control dissolved organic carbon concentrations in stream water. *Soil Sci.*, **137**, 23–32.

McElhone, M.J., and Davies, R.W. (1983) The influence of rock surface area on the microdistribution and sampling of attached riffle dwelling Trichoptera in Hartley Creek, Alberta. *Can. J. Zool.*, **61**, 2300–2304.

McIntire, C.D. (1968) Structural characteristics of benthic algal communities in a tributary system of Lake Ontario. *Can. J. Bot.*, **50**, 1663–1647.

McIntosh, A.R., Peckarsky, B.L., and Taylor, B.W. (2002) The influence of predatory fish on mayfly drift: Extrapolating from experiments to nature. *Freshwater Biol.*, **47**, 1497–1513.

McNeely, C., and Power, M.E. (2007) Spatial variation in caddisfly grazing regimes within a Northern California watershed. *Ecology*, **88**, 2609–2619.

McNeely, C., Finlay, J.C., and Power, M.E. (2007) Grazer traits, competition, and carbon sources to a headwater-stream food web. *Ecology*, **88**, 391–401.

McShaffrey, D., and McCafferty, W.P. (1990) Feeding behavior and related functional morphology of the mayfly *Ephemerella needhami* (Ephemeroptera: Ephemerellidae). *J. Insect Behav.*, **3**, 673–688.

McShaffrey, D., and McCafferty, W.P. (1991) Ecological association of the mayfly *Ephemerella needhami* (Ephemeroptera: Ephemerellidae) and the green alga *Cladophora* (Chlorophyta: Cladophoraceae). *J. Freshwater Ecol.*, **6**, 383–394.

Meier, H., Amann, R., Ludwig, W., and Schleifer, K.H. (1999) Specific oligonucleotide probes for in-situ detection of a major group of gram-positive bacteria with low DNA G+C content. *Syst. Appl. Microbiol.*, **22**, 186–196.

Meijering, M.P.D., and Teichmann, W. (1978) Zeitpläne limnischer Gammariden unter naturnahen Bedingungen. *Verh. Ges. Ökol., Kiel*, **1977**, 191–199.

Mendl, H., and Müller, K. (1970) Zur Laufaktivität von *Capnia atra* Morton (Plecoptera). *Oikos Suppl.*, **13**, 75–79.

Menzler, D. (1995) Mikrohabitatwahl und circadiane Aktivitätsmuster von *Drusus annulatus* (Tichoptera: Limnephilidae), einem dominanten Weidegänger des Breitenbachs (Hessen). Diploma thesis, University of Göttingen, Göttingen, 86 pp.

Merrill, D. (1965) The stimulus of case building activity in caddis-worms (Trichoptera). *J. Exp. Zool.*, **158**, 123–132.

Merrill, D. (1969) The distribution of case recognition behaviour in ten families of caddis larvae (Trichoptera). *Anim. Behav.*, **17**, 486–493.

Messner, B. (1965) Bemerkungen zur Biologie von *Agriotypus armatus* Walk. (Hymenoptera, Agriotypidae). *Zool. Anz.*, **174**, 354–362.

Methvin, B.R., and Suberkropp, K. (2003) Annual production of leaf-decaying fungi in two streams. *J. N. Am. Benthol. Soc.*, **22**, 554–564.

Mey, W. (1992) Die Trichoptera-Emergenz im Mittellauf eines Bergbaches des Thüringer Waldes (Insecta, Trichoptera). *Int. Revue ges. Hydrobiol.*, **77**, 265–289.

Meyer, E. (1989) The relationship between body length parameters and dry mass in running water invertebrates. *Arch. Hydrobiol.*, **117**, 191–203.

Meyer, H., and Heydemann, B. (1990) Faunistisch-ökologische Untersuchungen an Dolichopodiden und Empididen (Diptera: Dolichopodidae u. Empididae, Hybotidae) in Küsten- und Binnenlandbiotopen Schleswig-Holsteins. *Faun.-Ökol. Mitt.*, **6**, 147–172.

Meyer, J.L. (1988) Benthic bacterial biomass and production in a black-water river. *Verh. Int. Verein. Limnol.*, **23**, 1832–1838.

Meyer, J.L. (1994) The microbial loop in flowing waters. *Microb. Ecol.*, **28**, 195–199.

Meyer, J.L., McDowell, W.H., Bott, T.L., Elwood, J.W., Ishizaki, C., Melack, J.M., Peckarski, B.L., Peterson, B.J., and Rublee, P.A. (1988) Elemental dynamics in streams. *J. N. Am. Benthol. Soc.*, **7**, 410–432.

Michiels, N.K. (1989) Morphology of male and female genitalia in *Sympetrum danae* (Sulzer), with special reference to the mechanism of sperm removal during copulation (Anisoptera: Libellulidae). *Odonatologica*, **18**, 21–31.

Miller, P.L. (1995) Sperm competition and penis structure in some libelluloid dragonflies (Anisoptera). *Odonatologica*, **24**, 63–72.

Minshall, G.W. (1988) Stream ecosystem theory, a global perspective. *J. North Am. Benthol. Soc.*, **7**, 263–288.

Moaledj, K. (1984) Über schnelle miniaturisierte Bestimmungsverfahren für Bakterienpopulationen in aquatischen Ökosystemen. *Arch. Hydrobiol.*, **100**, 99–121.

Moaledj, K. (1986) Comparison of Gram-staining and alternate methods, KOH test and aminopeptidase activity in aquatic bacteria: their application to numerical taxonomy. *J. Microbiol. Methods*, **5**, 303–310.

Møller-Andersen, N. (1982) The Semiaquatic Bugs (Hemiptera, Gerromorpha) Phylogeny, Adaptations, Biogeography and Classification. *Entomonograph*, 3, E. J. Brill Publishers, Leiden, 455 pp.

Molles, Jr., M.C. (1984) Patterns of case building by caddisflies in the southern Rocky Mountains, in *Proc. 4th Int. Symp. Trichoptera, 1983* (ed. J.C. Morse), Junk, The Hague, pp. 235–242.

Molles, M.C. (1985) Recovery of a stream invertebrate community from a flash flood in Tesuque Creek, New Mexico. *Swest. Nat.*, **30**, 279–287.

Moog, O., and Jäch, M.A. (1995) Elmidae, in *Fauna Aquatica Austriaca, edition May 1975* (ed. O. Moog), Wasserwirtschaftskataster, Bundesministerium für Land- und Forstwirtschaft, Vienna, 6 pp.

Moog, O., Humpesch, U.H., and Konar, M. (1995) The distribution of benthic invertebrates along the Austrian stretch of the River Danube and its relevance as an indicator of zoogeographical and water quality patterns – part 1. *Arch. Hydrobiol. Suppl.*, **101**, 121–213.

Moore, J.W. (1974) The benthic algae of Southern Baffin Island. I. Epipelic communities in rivers. *J. Phycol.*, **10**, 50–57.

Moore, J.W. (1976) Seasonal succession of algae in rivers. I. Examples from the Avon, a large slow-flowing river. *J. Phycol.*, **12**, 342–349.

Moore, J.W. (1977) Seasonal succession of algae in a eutrophic stream in southern England. *Hydrobiologia*, **53**, 181–192.

Moore, J.W. (1981) Seasonal cycles of rare species of epipelic diatoms in a hardwater temperate-zone stream. *Nova Hedwigia*, **34**, 247–258.

Mopper, K. (1977) Sugars and uronic acids in sediment and water from the black sea and north sea with emphasis on analytical techniques. *Mar. Chem.*, **5**, 585–603.

Mopper, K., Schultz, C.A., Chevolot, M., Germain, C., Revuelta, R., and Dawson, R. (1992) Determination of sugars in unconcentrated seawater and other natural waters by liquid-chromatography and pulsed amperometric detection. *Environ. Sci. Technol.*, **26**, 133–138.

Morin, A., Mousseau, T.A., and Roff, D.A. (1987) Accuracy and precision of secondary production estimates. *Limnol. Oceanogr.*, **32**, 1342–1352.

Motzka, R. (1968a) *Geologische Karte von Hessen 1:25 000 Blatt Nr. 5324 Hünfeld*, 2nd Aufl., Hessisches Landesamt für Bodenforschung, Wiesbaden.

Motzka, R. (1968b) *Erläuterungen zur Geologischen Karte von Hessen 1:25 000 Blatt Nr. 5324 Hünfeld*, 2nd Aufl., Hessisches Landesamt für Bodenforschung, Wiesbaden, 156 pp.

Mucha, V. (1967) Die Mikrobiologie der Donau, in *Limnologie der Donau, Lieferung 3*, Hessisches Landesamt für Bodenforschung, Wiesbaden, pp. 132–157.

Mulholland, P.J., Steinman, A.D., Palumbo, A.V., and Elwood, J.W. (1991) Role of nutrient cycling and herbivory in regulating periphyton communities in laboratory streams. *Ecology*, **72**, 966–982.

Müller, K. (1970) Tages- und Jahresperiodik der Drift in Fließgewässern in verschiedenen geographischen Breiten. *Oikos Suppl.*, **13**, 31–44.

Müller, K. (1973) Life cycles of stream insects. *Aquilo Ser. Zool.*, **14**, 105–112.

Müller, K. (1982) The colonization cycle of freshwater insects. *Oecologia*, **52**, 202–207.

Müller-Haeckel, A. (1966) Diatomeendrift in Fliessgewässern. *Hydrobiologia*, **28**, 73–87.

Münster, U. (1984) Distribution, dynamic and structure of free dissolved carbohydrates in the Plußsee, a North German eutrophic lake. *Verh. Int. Verein. Limnol.*, **22**, 107–120.

Münster, U. (1991) Extracellular enzyme activity in eutrophic and polyhumic lakes, in *Microbial enzymes in aquatic environments. Brock/Springer Series in Contemporary Bioscience* (ed. R. Chróst), Springer, New York, pp. 96–122.

Münster, U., Einiö, P., and Nurminen, J. (1989) Evaluation of the measurements of extracellular enzyme activities in a polyhumic lake by means of studies with 4-methylumbelliferyl-substrates. *Arch. Hydrobiol.*, **115**, 321–337.

Muyzer, G. (1999) DGGE/TGGE a method for identifying genes from natural ecosystems. *Curr. Opin. Microbiol.*, **2**, 317–322.

Muyzer, G., and Smalla, K. (1998) Application of denaturing gradient gel electrophoresis (DGGE) and temperature gradient gel electrophoresis (TGGE) in microbial ecology. *Antonie Van Leeuwenhoek Int. J. Gen. Mol. Microbiol.*, **73**, 127–141.

Muyzer, G., De Waal, E.C., and Uitierlinden, A.G. (1993) Profiling of complex microbial populations by denaturing gradient gel electrophoresis analysis of polymerase chain reaction-amplified genes coding for 16S rRNA. *Appl. Environ. Microbiol.*, **59**, 695–700.

Myklestad, S.M., Skanoy, E., and Hestmann, S. (1997) A sensitive and rapid method for analysis of dissolved mono- and polysaccharides in seawater. *Mar. Chem.*, **56**, 279–286

Naegeli, M.W., and Uehlinger, U. (1997) Contribution of the hyporheic zone to ecosystem metabolism in a prealpine gravel-bed river. *J. N. Am. Benthol. Soc.*, **16**, 794–804.

Nagata, T. (1988) The microflagellate–picoplankton food linkage in the water column of Lake Biwa. *Limnol. Oceanogr.*, **33**, 504–517.

Naiman, R.J. (1983) A geomorphic approach for examining the role of periphyton in large watersheds, in *Periphyton of Freshwater Ecosystems* (ed. R.G. Wetzel), Dr. W. Junk, The Hague, pp. 191–198.

Naudascher, E. (1992) *Hydraulik der Gerinne und Gerinnebauwerke.* Springer Verlag, Wien/New York.

Naumann, E. (1932) Grundzüge der regionalen Limnologie. Die Binnengewässer 11; E. Schweitzerbart science publishers Stuttgart, 176 pp.

Needham, P.R., and Usinger, R.L. (1956) Variability of the macrofauna of a single riffle in Prosser Creek, California, as indicated by the Surber sampler. *Hilgardia*, **24**, 383–409.

Neef, A. (1997) Anwendung der in situ-Einzelzell-Identifizierung von Bakterien zur Populationsanalyse in komplexen mikrobiellen Biozönosen. Ph.D. thesis, University of München, München, 141 pp.

Negrobov, O.P. (1991) Dolichopodidae, *Catalogue of palearctic Diptera Vol. 7, Dolichopodidae – Platypezidae* (eds Á. Sóos and L. Papp), Dr. W. Junk, The Hague, pp. 11–142.

Nelson, D.J., and Scott, D.C. (1962) Role of detritus in the productivity of a rock outcrop community in a Piedmont stream. *Limnol. Oceanogr.*, **7**, 369–413.

Nepomnyashchikh, V.A. (1993) The control of selection of particles for case building by larvae of *Chaetopteryx villosa* Fabr. (Trichoptera: Limmnephilidae), in *Proc. 7th Int. Symp. Trichoptera, 1992,* (ed. C. Otto), Backhuys, Leiden, pp. 263–265.

Nesterovitch, A., and Zwick, P. (2003) The development of *Nemurella pictetii* Klapálek (Plecoptera: Nemouridae) in two springstreams in central Europe. *Limnologica*, **33**, 231–243.

Neumann, D. (1961) Ernährungsbiologie einer rhipidoglossen Kiemenschnecke. *Hydrobiologia*, **17**, 133–151.

Neumann, P. (1989) Biozönotische Untersuchungen am Breitenbach (Hessen), unter besonderer Berücksichtigung der Eibiologie bei der Gattung *Brachyptera* (Plecoptera, Insecta). Diploma thesis, University of Kiel, Kiel, 146 pp.

Neumann, P. (1995) Untersuchungen zur Nahrungsqualität von benthischem feinpartikulärem Detritus für Feinpartikelsammler, unter dem Aspekt seiner biochemischen Zusammensetzung im Breitenbach. Ph.D, thesis University Marburg, Marburg, 135 pp.

Neveu, A. (1973) Estimation de la production de populations larvaires du genre *Simulium* (Diptera, Nematocera). *Ann. Hydrobiol.*, **4**, 183–199.

Newell, S.Y., Moran, M.A., Wicks, R., and Hodson, R.E. (1995) Productivities of microbial decomposers during early stages of decomposition of leaves of a freshwater sedge. *Freshwater Biol.*, **34**, 135–148.

Newman, R.M. (1991) Herbivory and detritivory on freshwater macrophytes by invertebrates, a review. *J. N. Am. Benthol. Soc.*, **10**, 89–114.

Newport, G. (1851) On the anatomy and affinities of *Pteronarcys regalis* Newm.: with a postscript, containing descriptions of some American Perlidae, together with notes on their habits. *Trans. Linn. Soc. Lond. Zool.*, **20**, 425–452.

Nielsen, A. (1942) Über die Entwicklung und Biologie der Trichopteren mit besonderer Berücksichtigung der Quelltrichopteren Himmerlands. *Arch. Hydrobiol. Suppl.*, **17**, 255–631.

Nijboer, R. (2004) The ecological requirements of *Agapetus fuscipes* Curtis (Glossosomatidae), a characteristic species in unimpacted streams. *Limnologica*, **34**, 213–223.

Nislow, K.H., and Molles, Jr., M.C. (1993) The influence of larval case design on vulnerability of *Limnephilus frijole* (Trichoptera) to predation. *Freshwater Biol.*, **29**, 411–417.

Noble, R.T., and Fuhrman, J.A. (1998) Use of SYBR Green I for rapid epifluorescence counts of marine viruses and bacteria. *Aquat. Microb. Ecol.*, **14**, 113–118.

Nolte, U. (1990) Chironomid communities of lotic mosses. *Acta Biol. Debr. Oecol. Hung.*, **3**, 249–255. (1989).

Nolte, U. (1991) Seasonal dynamics of moss-dwelling chironomid communities. *Hydrobiologia*, **222**, 197–211.

Nolte, U., and Hoffmann, T. (1992a) Life cycle of *Pseudodiamesa branickii* (Chironomidae) in a small upland stream. *Neth. J. Aquat. Ecol.*, **26**, 309–314.

Nolte, U., and Hoffmann, T. (1992b) Fast life in cold water: *Diamesa incallida* (Chironomidae). *Ecography*, **15**, 25–30.

Norland, S. (1993) The relationship between biomass and volume of bacteria, in *Handbook of Methods in Aquatic Microbial Ecology* (eds P.F. Kemp, B.F. Sherr, E.B. Sherr, and J.J. Cole), Lewis Publishers, Boca Raton, pp. 303–307.

Nowak, B. (ed.) (1990) Beiträge zur Kenntnis hessischer Pflanzengesellschaften. Ergebnisse der pflanzensoziologischen Sonntagsexkursionen der Hessischen Botanischen Arbeitsgemeinschaft. *Bot. Natursch. Hessen, Beih.*, **2**, 207 pp.

Nyholm, T. (1969) Über Bau und Funktion der Kopulationsorgane bei den Cyphones (Col. Helodidae) Studien über die Familie Helodidae. X. *Entomol. Tidsskr.*, **90**, 233–271.

Nyholm, T. (1972) Die nordeuropäischen Arten der Gattung *Cyphon* Paykull (Coleoptera) Taxonomie, Biologie, Ökologie und Verbreitung. *Entomol. Scand. Suppl.*, **3**, 100 pp.

O'Neill, R.V. (2001) Is it time to bury the ecosystem concept? (With full military honors of course!) *Ecology*, **82**, 3275–3284.

Obach, M. (2003) Zuverlässigkeit der Prognosen von hybriden Neuronalen Netzwerken und ihre Visualisierung mit Anwendungen in der Limnologie. Dissertation.de – Verlag im Internet; ISBN 3-89825-660-X.

Obach, M., Wagner, R., Werner, H., and Schmidt, H.H. (2001) Modelling population dynamics of aquatic insects with artificial neural networks. *Ecol. Model.*, **146**, 207–217.

Obere Naturschutzbehörde Kassel (2004) Umsetzung der Flora-Fauna-Habitat-(FFH) Richtlinie in Hessen. Gebietsvorschlag im Regierungsbezirk Kassel. 5323-301 Breitenbachtal bei Michelsrombach. URL: http://141.90.2.24/static/themen/naturschutz/ffh/5323301.pdf (accessed 20 July 2009). (Archived by WebCite® at http://www.webcitation.org/5iPA8CbCS)

Oberndorfer, R.Y., and Stewart, K.W. (1977) The life cycle of *Hydroperla crosbyi* (Plecoptera: Perlodidae). *Great Basin Nat.*, **37**, 260–273.

Odum, E.P. (1962) Relationships between structure and function in the ecosystem. *Jpn J. Ecol.*, **12**, 108–119.

Odum, H.T. (1956a) Primary production in flowing waters. *Limnol. Oceanogr.*, **27**, 102–117.

Odum, H.T. (1956b) Efficiencies, size of organisms, and community structure. *Ecology*, **37**, 592–597.

Odum, H.T. (1957a) Trophic structure and productivity of Silver Springs, Florida. *Ecol. Monogr.*, **27**, 55–112.

Odum, H.T. (1957b) Primary production measurements in eleven Florida springs and a marine turtle-grass community. *Limnol. Oceanogr.*, **2**, 85–97.

Oemke, M.P. (1984) Diatom feeding preferences of *Glossosoma nigrior* (Banks) larvae, stream grazers in two Southern Michigan streams, in *Proc. 4th Int. Symp. Trichoptera*, (ed. J. Morse), Dr. W. Junk, The Hague, pp. 285–290.

Oemke, P.M., and Burton, T.M. (1986) Diatom colonization dynamics in a lotic system. *Hydrobiologia*, **139**, 153–166.

Ogilvie, G. (1988) The effects of periphyton manipulations on the micro-distribution of grazing macroinvertebrates. *Verh. Int. Ver. Limnol.*, **23**, 1101–1106.

Ohle, W. (1972) Gelöste organische Stoffe, Aufnahme und Abgabe durch Planktonorganismen im See. *Gewässerschutz – Wasser – Abwasser (GWA)*, **8**, 1–56.

Ohmura, A. (2006) Observed long-term variations of solar irradiance at the earth surface. *Space Sci. Rev.*, **125**, 111–128.

Oliveira, M.A., and Goulder, R. (2006) The effects of sewage-treatment-works effluent on epilithic bacterial and algal communities of three streams in Northern England. *Hydrobiologia*, **568**, 29–42.

Oliver, D.R. (1981) Chironomidae, in *Manual of Nearctic Diptera Vol. 1* (eds J.F. McAlpine, B.V. Peterson, G.E. Shewell, H.J. Teskey, J.R. Vockeroth, and D.M. Wood), USDA, Research Branch Agriculture Monograph, No. 27, pp. 423–458.

Olmi, M. (1976) *Fauna d`Italia, Coleoptera Dryopidae, Elminthidae.* Edizioni Calderini, Bologna, I–X, pp. 1–280.

Oosterbroek, P. (1978) The western Palaearctic species of *Nephrotoma,* Meigen 1803, Diptera: Tipulidae) Part I. *Beaufortia*, **27**, 1–138.

Oosterbroek, P. (1979a) The western Palaearctic species of *Nephrotoma,* Meigen 1803, Diptera: Tipulidae) Part III. *Beaufortia*, **28**, 157–204.

Oosterbroek, P. (1979b) The western Palaearctic species of *Nephrotoma,* Meigen 1803, Diptera: Tipulidae) Part IV. *Beaufortia*, **29**, 129–198.

Oosterbroek, P., and Theowald, B. (1992) Family Tipulidae, in *Catalogue of the Palaearctic Diptera, Vol. 1. (Trichoceridae – Nymphomyiidae)* (eds Á. Soós, L. Papp, and P. Oosterbroek), Dr. Junk, pp. 56–178.

Opute, F.I. (1974) Lipid and fatty-acid composition of diatoms. *J. Exp. Bot.*, **25**, 823–835.

Orghidan, T. (1959) Ein neuer Lebensraum des unterirdischen Wassers: Der hyporheische Biotop. *Arch. Hydrobiol.*, **55**, 392–414.

Ormerod, S.J. (1988) The micro-distribution of aquatic macroinvertebrates in the Wye river system: the result of abiotic or biotic factors? *Freshwater Biol.*, **20**, 241–247.

Oswald, D., Kureck, A., and Neumann, D. (1991) Populationsdynamik, Temperaturtoleranz und Ernährung der Quellschnecke *Bythlnella dunkeri* (Frauenfeld 1856). *Zool. Jahrb. Syst. Abt. Syst. Ökol. Geog. Tiere*, **118**, 65–78.

Otter, H.H. (1989) Larvalentwicklung und Larvalbiologie von Glossosomatiden (Trichoptera) eines kleinen Waldbaches. Diploma thesis, University of Marburg, Marburg.

Otto, A., and Westermann, F. (2003) Erstnachweis von *Isoptena serricornis* (Pictet 1841) (Plecoptera, Chloroperlidae) in Rheinland-Pfalz / First record of *Isoptena serricornis* (Pictet 1841) (Plecoptera, Chloroperlidae) in Rhineland-Palatinate. *Lauterbornia*, **46**, 103–106.

Otto, C. (1974) Growth and energetics in a larval population of *Potamophylax cingulatus* (Steph.)(Trichoptera) in a South Swedish stream. *J. Anim. Ecol.*, **43**, 339–361.

Otto, C. (1976) Habitat relationships in the larvae of three Trichoptera species. *Arch. Hydrobiol.*, **77**, 505–517.

Otto, C. (1982) Habitat, size and distribution of Scandinavian limnephilid caddisflies. *Oikos*, **38**, 355–360.

Otto, C. (1987) Behavioural adaptions by *Agrypnia pagetana* (Trichoptera) larvae to cases of different value. *Oikos*, **50**, 191–196.

Otto, C., and Johansson, A. (1995) Why do some caddis larvae in running waters construct heavy, bulky cases. *Anim. Behav.*, **49**, 473–478.

Otto, C., and Svensson, B.S. (1980) The significance of case material selection for the survival of caddis larvae. *J. Anim. Ecol.*, **49**, 855–865.

Overbeck, J. (1965) Primärproduktion und Gewässerbakterien. *Naturwissenschaften*, **51**, 145.

Overbeck, J. (1972) Bakterien in der Nahrungskette im See. *Umschau*, **72**, 358.

Overbeck, J. (1974) Microbiology and biochemistry. *Mitt. Int. Ver. Limnol.*, **20**, 198–228.

Overbeck, J. and Chróst, R.J. (eds.) (1990) *Aquatic Microbial Ecology. Biochemical and Molecular Approaches.* Brock/Springer Series in Contemporary Bioscience, Springer, New York, 190 pp.

Owens, M., and Edwards, R.W. (1961) The effects of plants on river conditions. II. Further crop studies and estimates of net productivity of macrophytes in a chalk stream. *J. Ecol.*, **49**, 119–126.

Packroff, G. (1994) Die Ciliaten des Breitenbachs; faunistisch-ökologische Untersuchungen an einem Mittelgebirgsbach. Ph.D. thesis, University of Marburg, Marburg, 178 pp.

Packroff, G. and Zwick, P. (1996) The ciliate fauna of an unpolluted foothill stream, the Breitenbach, 1: qualitative aspects. *Limnologica*, **26**, 255–262.

Packroff, G., and Zwick, P. (1998) The ciliate fauna of an unpolluted German foothill stream, the Breitenbach, 2: quantitative aspects of the ciliates (Ciliophora; Protozoa) in fine sediments. *Eur. J. Protistol.*, **34**, 436–445.

Paerl, H.W. (1977) Bacterial sediment formation in lakes: trophic implications, in: *Interactions between sediments and freshwater* (ed. H.L. Golterman), Dr. W. Junk Publishers, The Hague; pp. 40–47.

Pahlich, E. (1984) *Prinzipien der Geschwindigkeitskontrolle von Stoffwechselabläufen.* Friedr. Vieweg and Sohn, Braunschweig/Wiesbaden, 148 pp.

Palissa, A. (2000) Insecta: Collembola, in *Süßwasserfauna von Mitteleuropa 10* (eds J. Schwoerbel and P. Zwick), Spektrum Akademischer Verlag, Gustav Fischer, Heidelberg, Berlin, 166 pp.

Palmer, M.A., and Poff, N.L. (1997) The influence of environmental heterogeneity on patterns and processes in streams. *J. North Am. Benthol. Soc.*, **16**, 169–173.

Palmer, M.A., Hakenkamp, C.C., and Nelson-Baker, K. (1997) Ecological heterogeneity in streams: why variance matters. *J. North Am. Benthol. Soc.*, **16**, 189–202

Palmer, M.W. (1993) Putting things in even better order, the advantages of canonical correspondence analysis. *Ecology*, **74**, 2215–2230.

Paradise, C.J., and Kuhn, K.L. (1999) Interactive effects of pH and leaf litter on a shredder, the scirtid beetle, *Helodes pulchella*, inhabiting treeholes. *Freshwater Biol.*, **41**, 43–49.

Parry, G.D. (1981) The meanings of r- and K-selection. *Oecologia*, **48**, 260–264.

Patrick, R. (1949) A proposed biological measure of stream conditions based on a survey of Conestoga Basin, Lancaster County, Pennsylvania. *Proc. Acad. Nat. Sci. Philadelphia*, **101**, 277–431.

Patrick, R. (1961) A study of the numbers and kinds of species found in rivers in eastern United States. *Proc. Acad. Nat. Sci. Philadelphia*, **113**, 215–258.

Patrick, R., Cairns, Jr., J., and Roback, S.S. (1966) An ecosystematic study of the fauna and flora of the Savannah River. *Proc. Acad. Nat. Sci. Philadelphia*, **118**, 109–407.

Pauls, S.U., Graf, W., Haase, P., Lumbsch, H.T., and Waringer, J. (2008) Grazers, shredders and filtering carnivores – The evolution of feeding ecology in Drusinae (Trichoptera: Limnephilidae): insights from a molecular phylogeny. *Mol. Phylogenet. Evol.*, **46**, 776–791.

Peach, W.J., and Fowler, J.A. (1986) Life cycle and laboratory culture of *Dixella autumnalis* Meigen (Dipt., Dixidae). *Entomol. Mon. Mag.*, **122**, 59–62.

Peckarsky, B.L. (1983) Biotic interactions or abiotic limitations? A model of lotic community structure, in *Dynamics of Lotic Ecosystems* (eds T.D. Fontaine III and S.M. Bartell). Ann Arbor Science Publications, Ann Arbor, Michigan, pp. 303–323.

Peckarsky, B.L., Taylor, B.W., and Caudill, C.C. (2000) Hydrologic and behavioral constraints on oviposition of stream insects: implications for adult dispersal. *Oecologia*, **125**, 186–200.

Peckarsky, B.L., Taylor, B.W., McIntosh, A.R., McPeek, M.A., and Lytle, D.A. (2000) Variation in mayfly size at metamorphosis as a developmental response to risk of predation. *Ecology*, **82**, 740–757.

Perlmutter, D.G., and Meyer, J.L. (1991) The impact of a stream-dwelling harpacticoid copepod upon detritally associated bacteria. *Ecology*, **72**, 2170–2180.

Pernthaler, A., Pernthaler, J., and Amann, R. (2004) Sensitive multi-color fluorescence in situ hybridization for identification of environmental microorganisms, in *Molecular Microbial Ecology Manual, 2nd Edition, 3.11* (eds G.A. Kowalchuk, F.J. Bruijn, I.M. Head, A.D. Akkermans, and J.D. van Elsas), Springer, Dordrecht, pp. 711–726.

Peters, J.G., Peters, W.L., and Fink, T.J. (1986) Seasonal synchronization of emergence in *Dolania americana* (Ephemeroptera: Behningiidae). *Can. J. Zool.*, **65**, 3177–3185.

Petersen, I., Winterbottom, J.H., Orton, S., Friberg, N., Hildrew, A.G., Spiers, D.C., and Gurney, W.S.C. (1999) Emergence and lateral dispersal of adult Plecoptera and Trichoptera from Broadstone Stream, U.K. *Freshwater Biol.*, **42**, 401–416.

Petersen, L.B.M. (1985) Food preference in three species of Hydropsyche (Trichoptera). *Verh. Int. Ver. Limnol.*, **22**, 3270–3274.

Petersen, R.C. and Cummins, K.W. (1974) Leaf processing in a woodland stream. *Freshwater Biol.*, **4**, 343–368

Peterson, C.G. (1987) Gut passage and insect grazer selectivity of lotic diatoms. *Freshwater Biol.*, **18**, 455–460.

Peterson, C.J., and Stevenson, R.J. (1989) Substratum conditioning and diatom colonisation in different current regimes. *J. Phycol.*, **25**, 790–793.

Petersson, E., and Hasselrot, A.T. (1994) Mating and nectar feeding in the psychomyiid caddis fly *Tİnodes waeneri*. *Aquat. Insects*, **16**, 177–187.

Piechocki, A. (1985) Die Mollusken-Fauna des Breitenbachs und des Rohrwiesenbachs (Schlitzerland/Hessen). *Beitr. Naturk. Osthessen*, **20**, 84–95. (1984).

Pieper, H.G. (1978) Ökophysiologische und produktionsbiologische Untersuchungen an Jugendstadien von *Gammarus fossarum* Koch, 1835. *Arch. Hydrobiol. Suppl.*, **54**, 237–327.

Pitsch, T. (1993) Zur Larvaltaxonomie, Faunistik und Ökologie mitteleuropäischer Fließwasser-Köcherfliegen (Insecta: Trichoptera). Ph.D. thesis, Freie University, Berlin, 189 pp.

Pix, A. (2005) Die Libellen der Ballertasche – mit Gedanken zur thermischen Faunendrift. *Gött. Naturk. Schrift.*, **6**, 41–54.

Plaganyi, E.E., and Branch, G.M. (2000) Does the limpet *Patella cochlear* fertilize its own algal garden? *Mar. Ecol. Prog. Ser.*, **194**, 113–122.

Poff, N.L., and Allan, J.D. (1995) Functional organization of stream fish assemblages in relation to hydrological variability. *Ecology*, **76**, 606–627.

Poff, N.L., and Ward, J.V. (1990) Physical habitat template of lotic systems: recovery in the context of historical pattern of spatiotemporal heterogeneity. *Environ. Manage.*, **14**, 629–645.

Poff, N.L., and Ward, J.V. (1992) Heterogeneous currents and algal resources mediate in situ foraging activity of a mobile stream grazer. *Oikos*, **65**, 465–478.

Poff, N.L., and Ward, J.V. (1995) Herbivory under different flow regimes: a field experiment and test a model with benthic stream insects. *Oikos*, **72**, 179–188.

Poff, N.L., Wellnitz, T., and Monroe, J.B. (2003) Redundancy among three herbivorous insects across an experimental current velocity gradient. *Oecologia*, **134**, 262–269.

Poff, N.L., Bledsoe, B.P., and Cuhaciyan, C.O. (2006) Hydrologic variation with land use across the contiguous United States: Geomorphic and ecological consequences for stream ecosystems. *Geomorphology*, **79**, 264–285.

Pohlon, E., Marxsen, J., and Küsel, K. (2010) Pioneering bacterial and algal communities and extracellular enzyme activities of stream biofilms. *FEMS Microbiol. Ecol.*, 364–373. DOI: 10.1111/j.1574-6941.2009.00817.x.

Pollet, M. (2000) A documented Red List of the dolichopodid flies (Diptera: Dolichopodidae) of Flanders [in Dutch with English summary]. *Communications*

of the Institute of Nature Conservation, 8, *Brussels*, 190 pp.
Pomeroy, L.R. (1974) The ocean's food web, a changing paradigm. *BioScience*, **24**, 499–504.
Pomeroy, L.R. (1994) The microbial food web. *Oceanus*, **35**, 28–35.
Porter, K.G., and Feig, S.Y. (1980) The use of DAPI for identifying and counting aquatic microflora. *Limnol. Oceanogr.*, **25**, 943–948.
Power, M.E., Stewart, A.J., and Matthews, J. (1988) Grazer control of algae in an Ozark Mountain stream: effects of short-term exclusion. *Ecology*, **69**, 1894–1898.
Preisig, E., and Leumann, A. (2008) *Palingenia longicauda* (Olivier, 1791) in einem seiner letzten Refugien Europas. Video.
Pringle, C.M. (1985) Effects of chironomid (Insecta: Diptera) tube-building activities on stream diatom communities. *J. Phycol.*, **21**, 185–194.
Pringle, C.M., and Bowers, J.A. (1984) An in situ substratum fertilization technique: Diatom colonization on nutrient enriched sand substrata. *Can. J. Fish. Aquat. Sci.*, **41**, 1247–1251.
Pritchard, G., Harder, L.D., and Mutch, R.A. (1996) Development of aquatic insect eggs in relation to temperature and strategies for dealing with different thermal environments. *Biol. J. Linn. Soc. Lond.*, **58**, 221–224.
Prügel, M. (1988) Temperatur-Abhängigkeit der Puppen-Entwicklung und -Mortalität bei zwei Kriebelmücken-Arten (Diptera: Simuliidae). *Entomol. Gen.*, **13**, 29–45.
Prygiel, J. (1991) Use of benthic diatoms in surveillance of the Artois-Picardie basin hydrobiological quality, in *The Use of Algae for Monitoring Rivers* (eds B.A. Whitton, E. Rott, and G. Friedrich), Springer, Berlin, pp. 89–96.
Pusch, M., Andersen, H.E., Bäthe, J., Behrendt, H., Fischer, H., Friberg, N., Gancarczyk, A., Hoffmann, C.C., Hachol, J., Kronvang, B., Nowacki, F., Pedersen, M.L., Sandin, L., Schöll, F., Scholten, M., Stendera, S., Svendsen, L.M., Wnuk-Glawdel, E., and Wolter, C. (2009) Rivers of the Central European Highlands and Plains, in *Rivers of Europe* (eds K. Tockner, U. Uehlinger, and C.T. Robinson), Academic Press, London, pp. 525–576.
Pyke, G.H., Pulliam, H.R., and Charnov, E.L. (1977) Optimal foraging: a selective review of theory and tests. *Q. Rev. Biol.*, **52**, 137–154.
Rasmussen, K. (1978) Livscyklus og habitat praeference hos *Helodes marginata* Fabr. og *Helodes minuta* L. (Coleoptera), og de to artes udbredelse i Jylland. *Flora Fauna*, **82**, 80–84.
Ratte, H.T. (1985) Temperature and insect development, in *Environmental Physiology and Biochemistry of Insects* (ed. K.H. Hoffman), Springer, Berlin, pp. 33–66.
Raudkivi, A.J. (1982) *Grundlagen des Sedimenttransports*, Springer, Berlin, 255 pp.
Raušer, J. (1956) K poznání československých larev rodu *Protonemura* [Zur Kenntnis der tschechoslowakischen *Protonemura*-Larven.]. *Pr. Acta Acad. Sci. Csl (Brno)*, **28**, 449–498.
Raušer, J. (1980) Rád Pošvatky – Plecoptera, in *Klic Vodnich Larev Hmyzu* (ed. R. Rozkosny), Cesk. Akad. Ved. Cesk. Akad. Ved., Prague, pp. 86–132.
Ravizza, C. (1974) Ricerche ecologico-faunistiche sui Plecotteri della Val Staffora (Appennino Lombardo) (Plecoptera). *Boll. Mus. Civ. Stor. Nat. Verona*, **1**, 195–248.
Razumov, A.S. (1932) *Mikrobiologiya*, **1**, 131–146. [cited by Sorokin, Y.I., and Kadota, H. (1972) Techniques for the assessment of microbial production and decomposition in fresh waters].
Razumov, A.S. (1947) Methods of microbiological studies of water. Moscow. Vodgeo (cited by Sorokin, Y.I., and Kadota, H. (1972): Techniques for the assessment of microbial production and decomposition in fresh waters).
Recasens, L., and Murillo, L. (1986) Life cycle of *Agapetus fuscipes* Curtis, 1834 and *Hydropsyche siltalai* Döhler, 1963 (Trichoptera) in Vallvidrera streams (Catalunya, N. E. Spain). *Jorn. Asoc. Espan. Entomol.*, **8**, 1224–1232.
Reice, S.R. (1985) Experimental disturbance and the maintainance of diversity in a stream community. *Oecologia*, **67**, 90–97.
Reichenbach, H. (1999) The order Cytophagales, in *The Prokaryotes, 3rd Edition, Release 3.0*, (eds M. Dworkin, S.

Falkow, E. Rosenberg, K.-H. Schleifer, and E. Stackebrandt), Springer, New York, pp. 549–591.
Reidelbach, J. (1982) Methoden zur Untersuchung der Nahrungsökologie von Simuliiden-Larven (Diptera, Nematocera) aus Freiland-Populationen. Diploma thesis, University of Konstanz, Konstanz, 174 pp.
Reidelbach, J. (1994) Untersuchungen zur Populationsdynamik der Kriebelmücken (Diptera: Simuliidae) des Breitenbachs. Ph.D. thesis, University of Marburg, Marburg, 327 pp.
Reidelbach, J. (2002) A new type of artificial substratum for sampling aquatic stages of blackflies (Diptera: Simuliidae) in fast-flowing streams. *Stud. Dipterol.*, **8**, 577–595.
Reidelbach, J., and Christl, H. (2002) A quantitative investigation into the temporal and spatial variations in the emergence of adult blackflies (Diptera: Simuliidae) from the Breitenbach, a small upland stream in Germany. *Limnologica*, **32**, 206–235.
Reidelbach, J., and Kiel, E. (1990) Observations on the behavioural sequences of looping and drifting by blackfly larvae (Diptera: Simuliidae). *Aquat. Insects*, **12**, 49–60.
Rennerich, J., and Schuhmacher, H. (1984) Some aspects of the case-building behaviour of the burrowing caddis-larvae of *Sericostoma personatum* Kirby and Spence 1826, in *Proceeding of the 4th international Symposium on Trichoptera, Clemson, South Carolina, 1983* (ed. J.M. Morse), Dr. W. Junk, The Hague, pp. 301–305.
Resh, V.H., Flynn, T.S., Lamberti, G.A., McElravy, E.P., Sorg, K.L., and Wood, J.R. (1981) Responses of the sericostomatid caddisfly *Gumaga nigricula* (McL.) to environmental disruption, in *Proc. 3rd Int. Symp. Trichoptera, 1980* (ed. G.P. Moretti), Junk, The Hague, pp. 311–318.
Resh, V.H., Brown, A.V., Covich, A.P., Gurtz, M.E., Li, H.W., Minshall, G.W., Reice, S.S., Sheldon, A.L., Wallace, J.B., and Wissmar, R.C. (1988) The role of disturbance in stream ecology. *J. North Am. Benthol. Soc.*, **7**, 433–455.
Resh, V.H., Hildrew, A.G., Statzner, B., and Townsend, C.R. (1994) Theoretical habitat templets, species traits, and species richness, a synthesis of long-term ecological research on the Upper Rhône River in the context of concurrently developed ecological theory. *Freshw. Biol.*, **31**, 539–554.
Reusch, H., and Oosterbroek, P. (1997) Diptera Limoniidae and Pediciidae, short-palped crane flies, in *Aquatic Insects of North Europe – A Taxonomic Handbook. Vol. 2* (ed. A.N. Nilsson), Apollo, Stenstrup, pp. 105–132.
Reusch, H. and Schrankel, I. (2006) Schnakenartige (Tipulomorpha) in *Quellen im Nationalpark Berchtesgaden. Lebensgemeinschaften als Indikatoren des Klimawandels* (ed. Nationalparkverwaltung Berchtesgaden), Forschungsbericht 51, Berchtesgaden, pp. 177–182.
Reuter, K. (1974) *Elmis coiffaiti* Berth. and Clav. neu für Deutschland (Elminthidae). *Ent. Bl. Biol. Syst. Käfer*, **70**, 61.
Revsbech, N.P. (1989) An oxygen microsensor with a guard cathode. *Limnol. Oceanogr.*, **34**, 474–478.
Rheinheimer, G. (1977) Mikrobiologische Untersuchungen in Flüssen. I. Fluoreszenzmikroskopische Analyse der Bakterienflora einiger norddeutscher Flüsse. *Arch. Hydrobiol.*, **81**, 106–118.
Rheinheimer, G. (1985) *Mikrobiologie der Gewässer*. 4. Aufl., Gustav Fischer Verlag, Stuttgart, 262 pp.
Riaño, P., Basaguren, A., and Pozo, J. (1997) Diet variations of *Ephemerella ignita* (Poda) (Ephemeroptera: Ephemerellidae) in relation to the developmental stage in Ephemeroptera and Plecoptera: Biology – Ecology – Systematics, (eds P. Landolt and M. Sartori), MTL, Fribourg, pp. 60–54.
Richards, C., and Minshall, G.W. (1988) The influence of periphyton abundance on *Baetis bicaudatus* distribution and colonization in a small stream. *J. N. Am. Benthol. Soc.*, **7**, 77–86.
Richardson, J.S. (1991) Seasonal food limitation of detritivores in a montane stream: an experimental test. *Ecology*, **72**, 873–887.
Riemann, B., and Bell, R.T. (1990) Advances in estimating bacterial biomass and

growth in aquatic systems. *Arch. Hydrobiol.*, **118**, 385–402.
Riemann, B., Simonsen, P., and Stensgaard, L. (1989) The carbon and chlorophyll content of phytoplankton from various nutrient regimes. *J. Plankt. Res.*, **11**, 1037–1045.
Rimes, C.A., and Goulder, R. (1986a) Quantitative observations on the ability of epiphytic bacteria to contribute to the populations of suspended bacteria in two dissimilar headstreams. *Freshwater Biol.*, **16**, 301–311.
Rimes, C.A., and Goulder, R. (1986b) Temporal variation in density of epiphytic bacteria on submerged vegetation in a calcareous stream. *Lett. Appl. Microbiol.*, **3**, 17–21.
Ringe, F. (1974) Chironomiden-Emergenz 1970 in Breitenbach und Rohrwiesenbach. Schlitzer produktionsbiologische Studien (10). *Arch. Hydrobiol. Suppl.*, **45**, 212–304.
Ringe, F. (1976) *Heleniella serratosioi* n. sp., eine neue Orthocladiine (Dipt., Chir.) aus der Emergenz von Rohrwiesenbach und Kalkbach. *Arch. Hydrobiol.*, **77**, 254–266.
Ritter, H. (1985) Die Ephemeropteren des Stocktalbaches (Kühtai, Tirol). Ph.D. thesis, University of Innsbruck, Innsbruck, 153 pp.
Ritter, H. (1990) Ephemeroptera emergence from a high mountain stream in Tyrol, Austria, in *Mayflies and Stoneflies: Life Histories and Biology*. (ed. I.C. Campbell), Kluwer Academic Publishers, Dordrecht, pp. 563–569
Rivosecchi, L., (1978) Simuliidae in *Fauna d'Italia XIII*, Edizioni Calderini, Bologna, 353 pp.
Robbins, W.E., Kaplanis, J.M., Svoboda, J.A., and Thompson, M.J. (1971) Sterol metabolism in insects. *Annu. Rev. Entomol.*, **16**, 53–72.
Robert, B. (2001) Verzeichnis der Köcherfliegen (Trichoptera) Deutschlands. In *Die Köcherfliegen Fauna Deutschlands: Ein kommentiertes Verzeichnis mit Verbreitungsangaben. Entomofauna Germanica* (ed. B. Klausnitzer), 5, pp. 107–151. (Beiheft 6, Entomologische Nachrichten.und Berichte).
Robinson, C.T., and Minshall, G.W. (1986) Effects of disturbance frequency on stream benthic community structure in relation to canopy cover and season. *J. N. Am. Benthol. Soc.*, **5**, 237–248.
Rodina, R., (1961) Microbiological methods in application to hydrobiology. *Verh. Int. Ver. Limnol.*, **14**, 831–834.
Roff, D.A. (1992) *The Evolution of Life Histories: Theory and Analysis.* Chapman and Hall, New York, pp. XII + 535.
Rohwer, F., Prangishvili, D., and Lindell, D. (2009) Roles of viruses in the environment. *Environ. Microbiol.*, **11**, 2771–2774. DOI: 10.1111/j.1462-2920.2009.02101.x.
Romaní, A.M., and Marxsen, J. (2002) Extracellular enzymatic activities in epilithic biofilms of the Breitenbach: microhabitat differences. *Arch. Hydrobiol.*, **155**, 541–555.
Romaní, A.M., and Sabater, S. (1998) A stromatolitic cyanobacterial crust in a Mediterranean stream optimizes organic matter use. *Aquat. Microb. Ecol.*, **16**, 131–141.
Romaní, A.M., and Sabater, S. (1999) Epilithic ectoenzyme activity in a nutrient-rich Mediterranean river. *Aquat. Sci.*, **61**, 122–132.
Romaní, A.M., and Sabater, S. (2000) Influence of algal biomass on extracellular enzyme activity in river biofilms. *Microb. Ecol.*, **40**, 16–24.
Romaní, A.M., and Sabater, S. (2001) Structure and activity of rock and sand biofilms in a Mediterranean stream. *Ecology*, **82**, 3232–3245.
Romaní, A.M., Guasch, H., Muñoz, I., Ruana, J., Vilalta, E., Schwartz, T., Emtiazi, F., and Sabater, S. (2004) Biofilm Structure and Function and Possible Implications for Riverine DOC Dynamics. *Microb. Ecol*, **47**, 316–328.
Rosciszewska, E. (1996) Egg capsule structure of the stonefly *Protonemura intricata* (Ris 1902) (Plecoptera: Nemouridae). *Acta Biol. Cracoviensa, Ser. Zool.*, **38**, 41–48, plate 7.
Rosillon, D. (1988) Food preference and relative influence of temperature and food quality on life history characteristics of a grazing mayfly, *Ephemerella ignita* (Poda). *Can. J. Zool.*, **66**, 1474–1481.
Roskošný, R. (1973) The Stratiomyioidea (Diptera) of Fennoscandia and Denmark, in *Fauna Entomologica Scandinavica,*

Volume 1 (ed. R. Roskošný), Scandinavian Science Press, Gadstrup, 151 pp.

Roskošný, R. (2000) Insecta: Diptera: Stratiomyidae, in *Süßwasserfauna von Mitteleuropa 21/18,19* (eds J. Schwoerbel and P. Zwick), Spektrum Akademischer Verlag, Gustav Fischer, Heidelberg, Berlin, pp. 1–110.

Ross, D.H., and Craig, A.D. (1980) Mechanisms of fine particle capture by larval black flies (Diptera: Simuliidae). *Can. J. Zool.*, **58**, 1186–1192.

Ross, H.H. (1967) The evolution and past dispersal of the Trichoptera. *Annu. Rev. Entomol.*, **12**, 169–206.

Rosselló-Mora, R., and Amann, R. (2001) The species concept for prokaryotes. *FEMS Microbiol. Rev.*, **25**, 39–67.

Rotheray, G.E., and Gilbert, F. (1999) Phylogeny of Palaearctic Syrphidae (Diptera): evidence from larval stages. *Zool. J. Linn. Soc.*, **127**, 1–112.

Rothfritz, H., Jüttner, I., Suren, A.M., and Ormerod, S.J. (1997) Epiphytic and epilithic diatom communities along environmental gradients in the Nepalese Himalaya: implications for the assessment of biodiversity and water quality. *Arch. Hydrobiol.*, **138**, 465–482.

Round, F.E. (1953) An investigation of two benthic algal communities in Malham Tarn, Yorkshire. *J. Ecol.*, **41**, 174–197.

Round, F.E. (1981) *The Ecology of Algae*, Cambridge University Press, Cambridge, 653 pp.

Round, F.E. (1991) Use of diatoms for monitoring rivers, in *The Use of Algae for Monitoring Rivers* (eds B.A. Whitton, E. Rott, and G. Friedrich), Halle, Innsbruck, pp. 25–32.

Round, F.E., and Palmer, J.D. (1966) Persistent, vertical-migration rhythms in benthic micro-flora. II. Field and laboratory studies on diatoms from the banks of the River Avon. *J. Mar. Biol. Assoc. U.K.*, **46**, 191–214.

Roux, A.L. (1975) Température stable et température fluctuante. II. Étude comparative de leurs effets sur la durée d'intermue de Gammaridae femelles. *Verh. Int. Verein. Limnol.*, **19**, 3014–3021.

Rowe, L., Hudson, J., and Berrill, M. (1988) Hatching success of mayfly eggs at low pH. *Can. J. Fish. Aquat. Sci.*, **45**, 1649–1652

Ruffieux, L., Sartori, M., and L'Éplattenier, G. (1996) Palmen body: a reliable structure to estimate the number of instars in Siphlonurus aestivalis (Eaton) (Ephemeroptera: Siphlonuridae). *Int. J. Insect Morphol. Embryol.*, **25**, 341–344.

Ruffieux, L., Sartori, M., and L'Éplattenier, G. (1997) Number of instars in mayfly larvae: a Palmén body method applied to *Siphlonurus aestivalis* (Siphlonuridae), in *Ephemeroptera and Plecoptera Biology–Ecology–Systematics* (eds P. Landolt and M. Sartori), MTL, Fribourg, pp. 389–394.

Rühm, W. (1971) Zur Taxonomie und Morphologie von *Boophthora erythrocephala* de Geer. *Dtsch Entomol. Z. N.F.*, **18**, 149–193.

Rühm, W. (1995) Zur Variabilität meristischer morphologischer Merkmale von *Simulium noelleri* Friedrichs [sic], 1920 (Simuliidae, Diptera). *Entomol. Mitt. Zool. Mus. Hamb.*, **11**, 185–200.

Rühm, W. (1998) Zur Einteilung der Drift der Kriebelmücken (Diptera: Simuliidae). *Entomol. Gen.*, **23**, 223–232.

Rühm, W., and Pegel, M. (1986a) Die Substratbesiedlung durch Kriebelmückenlarven und -puppen (Simuliidae, Diptera). *Arch. Hydrobiol.*, **107**, 75–87.

Rühm, W., and Pegel, M. (1986b) Die Altersstruktur und die Artenzusammensetzung präimaginaler Populationen von Simuliiden auf künstlichen Substraten in Abhängigkeit von der Expositionsdauer. *Arch. Hydrobiol.*, **107**, 261–268.

Rupprecht, R. (1971) Der Kokonbau der Taumelkäferlarve (*Gyrinus substriatus* Stephens). *Mainzer Naturwiss. Arch.*, **10**, 195–202.

Rupprecht, R. (1975) Die Kommunikation von *Sialis* (Megaloptera) durch Vibrationssignale. *J. Insect Physiol.*, **21**, 305–320.

Rupprecht, R. (1981) A new system of communication within Plecoptera and a signal with a new significance. *Biol. Inland Waters*, **2**, 30–35.

Sabater, S., Aremengol, J., Marti, E., Sabater, F., and Guasch, H. (1991) Benthic diatom communities as descriptors of discontinuities in the River Ter, Spain, in *The Use of Algae for Monitoring Rivers* (eds B.A. Whitton, E. Rott, and G. Friedrich), Halle, Innsbruck, pp. 157–164.

Sabater, S., Acuña, V., Giorgi, A., Guerra, E., Muñoz, I., and Romaní, A.M. (2005) Effects of nutrient inputs in a forested Mediterranean stream under moderate light availability. *Arch. Hydrobiol.*, **163**, 479–496.

Sainz-Cantero, C.E., Sanchez-Ortega, A., and Alba-Tercedor, J. (1985) Problematica en la identificacion especifica de *Elmis maugetii* (Latreille, 1798) y *Elmis aenea* (Müller 1806). (Coleoptera: Elmidae). *Acta II Congr. Ibérico Entomol., Bo. Soc. Port. Entomol., Suppl.*, **1**, 269–273.

Salewski, V. (1991) Investigations on ecology and biometry of a population of the brook lamprey (*Lampetra planeri*) in the Odenwald. *Fischökologie*, **4**, 7–22.

Saltveit, S.J. (1979) Field and laboratory studies on the life history of three species of *Amphinemura*. The 6th International Symposium on Plecoptera, Schlitz, August 3–6, 1977. *Gewäss. Abwäss.*, **64**, 30–31.

Saltveit, S.J., and Lillehammer, A. (1984) Studies on egg development in the Fennoscandian *Isoperla* species (Plecoptera). *Ann. Limnol.*, **20**, 91–94.

Sameshima, O., and Sato, H. (1994) Life cycles of *Glossosoma inops* and *Agapetus yasensis* (Trichoptera, Glossosomatidae) at Kii Peninsula, Southern Honshu, Japan. *Aquat. Insects*, **16**, 65–74.

Sander, B.C., and Kalff, J. (1993) Factors controlling bacterial production in marine and freshwater sediments. *Microb. Ecol.*, **26**, 79–99.

Sand-Jensen, K., and Madsen, T.V. (1989) Invertebrates graze on submerged rooted macrophytes in lowland streams. *Oikos*, **55**, 420–423.

Sandrock, F. (1978) Vergleichende Emergenzmessung an zwei Bächen des Schlitzerlandes (Breitenbach und Rohrwiesenbach 1970–1971). Schlitzer produktionsbiologische Studien (24). *Arch. Hydrobiol. Suppl.*, **54**, 328–408.

Sangpradub, N., Giller, P.S., and O'Connor, J.P. (1999) Life history patterns of stream-dwelling caddis. *Arch. Hydrobiol.*, **146**, 471–493.

Sartori, M., Keller, L., Thomas, A.G.B., and Passera, L. (1992) Flight energetics in relation to sexual differences in the mating behaviour of a mayfly, *Siphlonurus aestivalis*. *Oecologia*, **92**, 172–176.

Saunders, D.S. (1976) *Insect Clocks*, Pergamon Press, Oxford, 280 pp.

Savchenko, E.N., Oosterbroek, P., and Starý, J. (1992) Family Limoniidae, in *Catalogue of the Palaearctic Diptera 1* (eds Á. Soós, L. Papp, and P. Oosterbroek), Hungarian Natural History Museum, Budapest, pp. 183–374.

Savolainen, E., Saura, A., and Hantula, J. (1993) Mode of swarming in relation to reproductive isolation in mayflies. *Evolution*, **47**, 1796–1804.

Schäffer, C. (1999) Hydrologische Abflussmodellierung, Sedimenttransport und Gewässerstrukturgüte am Breitenbach. Projektarbeit 3 im Fachgebiet Siedlungswasserwirtschaft, University of Kassel, Kassel, 42 pp.

Schiefele, S., and Schreiner, S. (1991) Use of diatoms for monitoring nutrient enrichment, acidification and impact of salt in rivers in Germany and Austria, in *The Use of Algae for Monitoring Rivers* (eds B.A. Whitton, E. Rott, and G. Friedrich), Backhuys Publishers, Leiden, pp. 103–110.

Schildknecht, H., Krauß, D., Connert, J., Essenbreis, H., and Orfanides, N. (1975) Das Spreitungsalkaloid Stenusin aus dem Kurzflügler *Stenus comma* (Coleoptera: Staphylinidae). *Angew. Chem.*, **87**, 421–422.

Schleiter, I.M., Borchardt, D., Wagner,R., Dapper, T., Schmidt, K.D., Schmidt, H.H., and Werner, H. (1999) Modelling water quality, bioindication and population dynamics in lotic ecosystems using neural networks. *Ecol. Model.*, **120**, 271–286.

Schleiter, I.M., Obach, M., Wagner, R., Werner, H., Schmidt, H.H., and Borchardt, D. (2003) Modelling Ecological Interrelations in Running Water Ecosystems with Artificial Neural Networks, Chapter 7. in *Ecological Informatics. Understanding Ecology by Biologically-Inspired Computation* (ed. F. Recknagel), Springer-Verlag, Berlin, Heidelberg, New York, pp. 109–126,

Schmid, P.E., and Schmid-Araya, J.M. (2002) Trophic relatioships in temporary and permanent freshwater meiofauna, in *Freshwater Meiofauna: Biology and Ecology* (eds S.D. Rundle, A.L. Robertson, and L. Schmid-Araya), Backhuys Publishers, Leiden, pp. 295–319.

Schmidt, G.K. (1970) Numbers of bacteria and algae and their interrelations in some Amazonian waters. *Amazoniana*, **2**, 393–400.

Schmidt, H.H. (1984) Einfluß der Temperatur auf die Entwicklung von *Baetis vernus* Curtis. *Arch. Hydrobiol. Suppl.*, **69**, 364–410.

Schmidt, T. (1991) Faunistische und ökologische Untersuchungen an Fließgewässern im Hessischen Rothaargebirge. Diploma thesis, Philipps-University, Marburg, 175 pp, with Appendix.

Scholz, O., and Marxsen, J. (1996) Sediment phosphatases of the Breitenbach, a first-order Central European stream. *Arch. Hydrobiol.*, **135**, 433–450.

Schönborn, W. (1981) Die Ciliatenproduktion eines Baches. *Limnologica*, **13**, 203–212.

Schönborn, W. (1984) The annual energy transfer from the communities of Ciliata to the population of *Chaetogaster diastrophus* (Gruithuisen) in the River Saale. *Limnologica*, **16**, 15–23.

Schönhals, E. (1954) Die Böden Hessens und ihre Nutzung. *Abh. Hess. L.-Amt Bodenforsch.*, **2**, 1–288.

Schönwiese, C. (2002) Ändert der Mensch das Klima? http://www.energie-fakten.de/pdf/klima.pdf.

Schrempf, H. (2001) Recognition and degradation of chitin by streptomycetes. *Antonie van Leeuwenhoek*, **79**, 285–289.

Schulte, H. (1989) Beiträge zur Biologie und Taxonomie der Gattung *Elmis* Latreille (Insecta, Coleoptera, Elmidae) unter besonderer Berücksichtigung niederbayerischer Vorkommen. *Lauterbornia*, **1**, 23–27.

Schultheis, A.S., Weigt, L.A., and Hendricks, A.C. (2002) Gene flow, dispersal, and nested clade analysis among populations of the stonefly *Peltoperla tarteri* in the southern Appalachians. *Mol. Ecol.*, **11**, 317–327.

Schwank, P. (1979) Two new *Krumbachia* (Turbellaria: Neorhabdocoela) from Highland streams near Schlitz, Hesse. *Int. Rev. Hydrobiol.*, **64**, 801–810.

Schwank, P. (1980) Neue limnische Turbellarien. *Arch Hydrobiol.*, **88**, 463–490.

Schwank, P. (1981a) Turbellarien, Oligochaeten und Archianneliden des Breitenbachs und anderer oberhessischer Mittelgebirgsbäche. I. Lokalgeographische Verbreitung und die Verteilung der Arten in den einzelnen Gewässern in Abhängigkeit vom Substrat. *Arch. Hydrobiol. Suppl.*, **62**, 1–85.

Schwank, P. (1981b) Turbellarien, Oligochaeten und Archianneliden des Breitenbachs und anderer oberhessischer Mittelgebirgsbäche. II. Die Systematik und Autökologie der einzelnen Arten. *Arch. Hydrobiol. Suppl.*, **62**, 86–147.

Schwank, P. (1981c) Turbellarien, Oligochaeten und Archianneliden des Breitenbachs und anderer oberhessischer Mittelgebirgsbäche. III. Die Taxozönosen der Turbellarien und Oligochaeten in Fließgewässern – eine synökologische Gliederung. *Arch. Hydrobiol. Suppl.*, **62**, 191–253.

Schwank, P. (1981d) Turbellarien, Oligochaeten und Archianneliden des Breitenbachs und anderer oberhessischer Mittelgebirgsbäche. IV. Allgemeine Grundlagen der Verbreitung von Turbellarien und Oligochaeten in Fließgewässern. *Arch. Hydrobiol. Suppl.*, **62**, 254–290.

Schwank, P. (1981e) Seltene und wenig bekannte Turbellarien aus Bergbächen und Quellen Mitteleuropas. *Beitr. Naturk. Osthessen*, **17**, 101–131.

Schwank, P. (1983) Breitenbach Grobsubstrat-Kartierung, November 1983. (unpublished manuscript).

Schwank, P. (1985) Differentiation of the coenoses of helminthes and annelida in exposed lotic microhabitats in mountain streams. *Arch. Hydrobiol.*, **103**, 535–543.

Schwarz, P. (1970) Autökologische Untersuchungen zum Lebenszyklus von Setipalpia-Arten (Plecoptera). *Arch. Hydrobiol.*, **67**, 103–140.

Schweder, H. (1985) Experimentelle Untersuchungen zur Ernährungsökologie der Larve von *Ecdyonurus venosus* (Fabr.) (Ephemeroptera: Heptageniidae). Ph.D. thesis, University of Freiburg, Freiburg, 123 pp.

Schwedt, G. (1981) *Fluorimetrische Analyse. Methoden und Anwendungen*. Verlag Chemie, Weinheim, 213 pp.

Schwenk, W., and Schwoerbel, J. (1973) Untersuchungen zur Ernährungsbiologie und Lebensweise der Flußmützenschnecke *Ancylus fluviatilis* (O. F. Müller 1774; Gastropoda Basommatophora). *Arch. Hydrobiol. Suppl.*, **42**, 190–231.

Schwind, R. (1991) Polarization vision in water insects and insects living on a moist substrate. *J. Comp. Physiol. A*, **169**, 531–540.

Schwoerbel, J. (1961) Über die Lebensbedingungen und die Besiedlung des hyporheischen Lebensraumes. *Arch. Hydrobiol. Suppl.*, **25**, 182–214.

Schwoerbel, J. (1967) Das hyporheische Interstitial als Grenzbiotop zwischen oberirdischem und subterranem Ökosystem und seine Bedeutung für die Primärevolution von Kleinsthöhlenbewohnern. *Arch. Hydrobiol. Suppl.*, **33**, 1–62.

Scott, D. (1958) Ecological studies on the Trichoptera of the River Dean, Cheshire. *Arch. Hydrobiol.*, **54**, 340–392.

Sedell, J.R., and Dahm, C.N. (1990) Spatial and temporal scales of dissolved organic carbon in riverine ecosystems, in *Organic Acids in Aquatic Ecosystems* (eds E.M. Perdue and E.T. Gjessing), John Wiley & Sons, Ltd, Chichester, pp. 261–279.

Seibert, P. (1954) Die Wald- und Forstgesellschaften im Graf Görtzischen Forstbezirk Schlitz. *Angew. Pflanzens.*, **9**, 1–63.

Seifert, G. (1999) Tracheensystem, in *Lehrbuch der Entomologie* (eds K. Dettner and W. Peters), Gustav Fischer, Jena, pp. 171–181.

Sephton, D.H., and Hynes, H.B.N. (1982a) The numbers of nymphal instars of several Australian Plecoptera. *Aquat. Insects*, **4**, 153–166.

Sephton, D.H., and Hynes, H.B.N. (1982b) Observations on the first instar nymphs of several Australian stoneflies (Plecoptera). *Aquat. Insects*, **4**, 237–252.

Shcherbakov, D., and Lukashevich, E. (2005) Adult *Ptychoptera* feed on honeydew (Diptera: Nematocera: Ptychopteridae). *Stud. Dipterol.*, **12**, 37–39.

Sheath, R.G., and Burkholder, J.M., (1985) Characteristics of softwater streams in Rhode Island. II. Composition and seasonal dynamics of macroalgal communities. *Hydrobiologia*, **128**, 109–118.

Siebert, Man (1979a) Two New Chironomids (Diptera: Chironomidae) from Germany and Austria. Schlitz Studies on Productivity, No. 37. *Aquat. Insects*, **1**, 165–168.

Siebert, Man (1979b) Description of *Parapsectra wagneri* n. sp. (Diptera: Chironomidae) from Schlitz. Schlitz studies on productivity No. 33. *Aquat. Insects*, **1**, 103–105.

Siebert, Man, (1980) Die Emergenz der Chironomiden im Breitenbach 1969–1973. Schlitzer produktionsbiologische Studien (44). *Arch. Hydrobiol. Suppl.*, **58**, 310–355.

Siebert, Man (1998) Wasserinsekten im Hyporhithral und Epipotamal der Fulda, einst und jetzt. *Lauterbornia*, **33**, 53–83.

Siler, E.R., Wallace, J.B., and Eggert, S.L. (2001) Long-term effects of ressource limitation on stream invertebrate drift. *Can. J. Fish. Aquat. Sci.*, **58**, 1624–1637.

Siltala, A.J. (1907) Trichopterologische Untersuchungen. No. 2. Über die postembryonale Entwicklung der Trichopteren-Larven. *Zool. Jahrb. Suppl.*, **9**, 309–626.

Simon, M., and Azam, F. (1989) Protein content and protein synthesis rates of planktonic marine bacteria. *Mar. Ecol. Prog. Ser.*, **51**, 201–213.

Simon, N.S. (1985) Evaluation and use of a diffusion-controlled sampler for determining chemical and dissolved oxygen gradients at the sediment–water interface. *Hydrobiologia*, **126**, 135–141.

Sinclair, B.J. (1995) Generic revision of the Clinocerinae (Empididae), and description and phylogenetic relations of the Trichopezinae, new status (Diptera: Empidoidea). *Can. Entomol.*, **27**, 665–752.

Sinclair, B.J., and Cumming, J.M. (2006) The morphology, higher-level phylogeny and classification of the Empidoidea (Diptera). *Zootaxa*, **1180**, 1–172.

Sinclair, B.J., and Harkrider, J.R. (2004) The immature stages and rearing observations of the aquatic dance fly *Roederiodes wirthi* Chillcott (Diptera: Empididae: Clinocerinae), with taxonomic notes on the genus. *Stud. Dipterol.*, **11**, 51–61.

Sinsabaugh, R.L., and Findlay, S. (1995) Microbial production, enzyme activity, and carbon turnover in surface sediments of the Hudson River estuary. *Microb. Ecol.*, **30**, 127–141.

Sinsabaugh, R.L., and Linkins, A.E. (1988) Exoenzyme activity associated with lotic epilithon. *Freshwater Biol.*, **20**, 249–261.

Sinsabaugh, R.L., Findlay, S., Franchini, P., and Fisher, D. (1997) Enzymatic analysis of riverine bacterioplankton production. *Limnol. Oceanogr.*, **42**, 29–38.

Smart, K. (1976) A progress report on the building motivation in the caddis larva, *Lepidostoma hirtum*, in *Proc. 1st Int. Symp. Trichoptera, 1974* (ed. H. Malicky), Junk, The Hague, pp. 185–186.

Smilauer, P. (1992) CanoDraw. Microcomputer Power, Ithaca, New York, U.S.A..

Smith, B.J., and Collier, K.J. (2000) Interactions of adult stoneflies (Plecoptera) with riparian zones II. Diet. *Aquat. Insects*, **22**, 285–296.

Smith, D.J., and Underwood, G.J.C. (1998) Exopolymer production by epipelic diatoms. *Limnol. Oceanogr.*, **43**, 1578–1591.

Smock, L.A., and Roeding, C.E. (1986) The trophic basis of production of the macroinvertebrate community of a southeastern U.S.A. blackwater stream. *Hol. Ecol.*, **9**, 165–174.

Sobczak, W.V., Hedin, L.O., and Klug, M.J. (1998) Relationships between bacterial productivity and organic carbon at a soil–stream interface. *Hydrobiologia*, **386**, 45–53.

Solem, J.O., and Solem, T. (1991) Mate location and pre-mating behavior in *Apatania fimbriata* Pictet (Trichoptera Limnephilidae). *Aquat. Insects*, **13**, 1–8.

Sommerhäuser, M. (ed.) (2000) Gewässer ohne Wasser ? Ökologie, Bewertung, Management temporärer Gewässer. Natur- und Umweltschutz-Akademie des Landes Nordrhein-Westfalen Recklinghausen.

Sorokin, J.I. (1965) On the trophic role of chemosynthesis and bacterial biosynthesis in water bodies. *Mem. Ist. Ital. Idrobiol. Suppl.*, **18**, 187–205.

Sorokin, Y.I., and Kadota, H. (eds) (1972) *Techniques for the Assessment of Microbial Production and Decomposition in Fresh Waters*, Blackwell, Oxford, 112 pp.

Southwood, T.R.E. (1977) Habitat, the templet for ecological strategies? *J. Anim. Ecol.*, **46**, 337–365.

Southwood, T.R.E. (1988) Tactics, strategies and templets. *Oikos*, **52**, 3–18.

Sowa, R. (1975) Ecology and biogeography of mayflies (Ephemeroptera) of running waters in the Polish part of the Carpathians. 2. Life cyles. *Acta Hydrobiologica*, **17**, 319–353.

Specht, D.F. (1991) A general regression network; *IEEE Transactions on Neural Networks*; **2** (6), 568–576.

Speir, J.A., and Anderson, N.H. (1974) Use of emergence data for estimating annual production of aquatic insects. *Limnol. Oceanogr.*, **19**, 154–156.

SPSS Inc. (2007) SPSS for Windows, Base system user's guide, Release 15.1.3., Chicago, U.S.A., 740 pp.

Städler, T., Weisner, S., and Streit, B. (1995) Outcrossing rates and correlated matings in a predominantly selfing freshwater snail. *Proc. R. Soc. London B*, **262**, 119–125.

Städler, T., Lowe, M., and Streit, B. (1996) Genetics and mating systems of polyploid freshwater hermaphrodite snails. Molluscan reproduction, *Malacol. Rev. Suppl.*, **6**, 121–127.

Stagliano, D.M., and Whiles, M.R. (2008) Life history and production of the riffle beetle, *Stenelmis crenata* (Say, 1824) (Coleoptera: Elmidae) in a tallgrass prairie stream. *Aquat. Insects*, **30**, 197–204.

Stahls, G., Hippa, H., Rotheray, G., Muona, J., and Gilbert, F. (2003) Phylogeny of Syrphidae (Diptera) inferred from combined analysis of molecular and morphological characters. *Syst. Entomol.*, **28**, 433–450.

Stamer, M. (1987) Reservestoffgehalt palaearktischer Simuliiden (Diptera) bei Ruhehhaltung und Flug. MSc. Thesis, University of Tüningen, Tüningen, 64 pp.

Stanford, J.A., and Gaufin, A.R. (1974) Hyporheic communities of two Montana rivers. *Science*, **185**, 700–702.
Stanley, E.H., and Short, R.A. (1988) Temperature effects on warm-water stream insects: a test of the Thermal Equilibrium Hypothesis. *Oikos*, **51**, 313–320
Starý, J., (1992) Phylogeny and classification of Tipulomorpha with special emphasis on the family Limoniidae. *Acta Zool. Cracov.*, **35**, 11–36.
Statzner, B. (1981) The relation between "hydraulic stress" and microdistribution of benthic macroinvertebrates in a lowland running water system, the Schierenseebrook (North Germany). *Arch. Hydrobiol.*, **91**, 192–218.
Statzner, B. (1987) Characteristics of lotic ecosystems and consequences for future research directions. *Potentials and Limitations of Ecosystem Analyses.* (eds E.D. Schulze and H. Zwölfer). Ecological Studies 61. Springer Verlag, Heidelberg.
Statzner, B., and Higler, B. (1985) Questions and comments on the river continuum concept. *Can. J. Fish. Aquat. Sci.*, **42**, 1038–1044.
Statzner, B., and Higler, B. (1986) Stream hydraulics as a major determinant of benthic invertebrate zonation patterns. *Freshw. Biol.*, **16**, 127–139.
Statzner, B., and Holm, T.F. (1989) Morphological adaptation of shape to flow: microcurrents around lotic macroinvertebrates with known Reynolds numbers at quasi-natural flow conditions. *Oecologia*, **78**, 145–157.
Statzner, B., and Resh, V.H. (1993) Multiple-site and -year analyses of stream insect emergence, a test of ecological theory. *Oecologia*, **96**, 65–79
Statzner, B., Resh, V.H., and Dolédec, S. (eds) (1994) Ecology of the Upper Rhône River: a test of habitat templet theories. *Freshw. Biol.*, **31**, 253–556.
Statzner, B., Gore, J.A., and Resh, V. (1988) Hydraulic stream ecology: observed patterns and potential applications. *J. North Am. Benthol. Soc.*, **7**, 307–360.
Statzner, B., Merigoux, S., and Leichtfried, M. (2005) Mineral grains in caddisfly pupal cases and streambed sediments: resource use and its limitation through conflicting requirements. *Limnol. Oceanogr.*, **50**, 713–721.
Stead, T.K., Schmid-Araya, J.M., and Hildrew, A.G. (2003) All creatures great and small: patterns in the stream benthos across a wide range of metazoan body size. *Freshwater Biol.*, **48**, 532–547.
Stearns, S.C. (1976) Life-history tactics: A review of the ideas. *Q. Rev. Biol.*, **51**, 3–47.
Stearns, S.C. (1981) On measuring fluctuating environments, predictability, constancy, and contingency. *Ecology*, **62**, 185–199.
Stearns, S.C. (1992) *The Evolution of Life Histories*, Oxford University Press, Oxford, 249 pp.
Steffan, A.W. (1958) Die deutschen Arten der Gattungen *Elmis, Esolus, Oulimnius, Riolus, Aptyktophallus* (Coleoptera: Dryopoidea) Genitalmorphologisch-taxonomische Studien an Dryopoidea I. *Beitr. Entomol.*, **8**, 122–178.
Steffan, A.W. (1961) Vergleichend-mikroskopische Genitaluntersuchungen zur Klärung der phylogenetischen Verwandtschaftsverhältnisse der mitteleuropäischen Dryopoidea (Coleoptera). *Zool. Jahrb. Abt. Syst. Geog. Biol. Tiere*, **88**, 255–354.
Steffan, A.W. (1979) 4. Familiengruppe Macrodactylia 42. Familie*: Dryopidae*, in *Die Käfer Mitteleuropas* 6 (eds H. Freude, K.W. Harde, and G.A. Lohse), Goecke and Evers, Krefeld, pp. 265–294.
Steiner, K. (1991) Some notes on feeding of *Protonemura nitida* (Plecoptera), in *Overview and Strategies of Ephemeroptera and Plecoptera* (eds J. Alba-Tercedor and A. Sanchez-Ortega), The Sandhill Crane Press, Inc., Gainesville, pp. 425–438.
Steinman, A.D. (1996) Effects of grazers on freshwater benthic algae, in *Algal Ecology: Freshwater Benthic Ecosystems* (eds R.J. Stevenson, M.L. Bothwell, and R.L. Lowe), Academic Press, San Diego, pp 341–373.
Steinman, A.D., and McIntire, C.D. (1982) Effects of irradiance on the community structure and biomass of algal assemblages in laboratory streams. *Can. J. Fish. Aquat. Sci.*, **44**, 1640–1648.
Steinman, A.D., and McIntire, C.D. (1986) Effects of current velocity and light energy on the structure of periphyton

assemblages in laboratory streams. *J. Phycol.*, **22**, 352–361.

Steinman, A.D., McIntire, C.D., and Lowry, R.R. (1987) Effects of herbivore type and density on chemicalcomposition of algal assemblages in laboratory streams. *J. N. Am. Benthol. Soc.*, **6**, 189–197.

Steinman, A.D., McIntire, C.D., Gregory, S.V., Lamberti, G.A., and Ashkenas, L.R. (1987b) Effects of herbivoretype and density on taxonomic struture and physiognomy of algal assemblages in laboratory streams. *J. N. Am. Benthol. Soc.*, **6**, 175–188.

Steinman, A.D., Mulholland, P.J., and Hill, W.R. (1992) Functional responses associated with growth form in stream algae. *J. N. Am. Benthol. Soc.*, **11**, 229–243.

Steinman, A.D., McIntire, C.D., Gregory, S.V., and Lamberti, G.A. (1989) Effects of irradiance and grazing on lotic algal assemblages. *J. Phycol.*, **25**, 478–485.

Stevens, D.J., Hansell, M.H., Freel, J.A., and Monaghan, P. (1999) Developmental trade-offs in caddis flies: increased investment in larval defense alters adult resource allocation. *Proc. R. Soc. Lond. B*, **266**, 1049–1054.

Stevens, D.J., Hansell, M.H., and Monaghan, P. (2000) Developmental trade-offs and life histories: strategic allocation of resoures in caddis flies. *Proc. R. Soc. Lond. B*, **267**, 1511–1515.

Stevenson, R.J. (1983) Effects of current and conditions simulating autogenically changing microhabitats on benthic diatom immigration. *Ecology*, **64**, 1514–1524.

Stevenson, R.J. (1984) How currents on different sides of substrates in streams affect mechanisms of benthic algal accumulation. *Int. Rev. Ges. Hydrobiol.*, **69**, 241–262.

Stevenson, R.J., and Pan, Y. (1999) Assessing environmental conditions in streams and rivers with diatoms, in *The Diatoms: Applications for the Environmental and Earth Sciences* (eds E.F. Stoermer and J.P. Smol), Cambridge University Press, Cambrige, pp. 11–40.

Stewart, K.W. (1994) Theoretical considerations of mate finding and other adult behaviors of Plecoptera. *Aquat. Insects*, **16**, 95–104.

Stewart, K.W., and Anderson, N.H.. (2010) The life history of *Ostrocerca dimicki* (Frison) in a short-flow, summer-dry Oregon stream. *Illiesia*, **6**, 52–57.

Stewart, K.W., and Harper, P.P. (1996) Plecoptera, in *An Introduction to the Aquatic Insects of North America, 3rd edn, vol 14* (eds R.W. Merritt and K.W. Cummins), Kendull/Hunt, Dubuque, pp. 217–266.

Stewart, K.W., and Stark, B.P. (2002) *Nymphs of North American stonefly genera (Plecoptera)*, 2nd edn, The Caddis Press, Columbus, 510 pp.

Stewart, K.W., and Zeigler, D.D. (1984) The use of larval morphology and drumming in Plecoptera systematics, and further studies of drumming behavior. *Ann. Limnol.*, **20**, 105–114.

Stief, P., and Becker, G. (2005) Structuring of epilithic biofilms by the caddisfly *Tinodes rostocki*: photosynthetic activity and photopigment distribution in and beside larval retreats. *Aquat. Microb. Ecol.*, **38**, 71–79.

Stitz, H., (1931) Planipennia *Biologie der Tiere Deutschlands*, Lief. 33, Teil 35 (ed. P. Schulze), Gebrüder Borntraeger, Berlin, pp. 67–304.

Streit, B. (1975) Experimentelle Untersuchungen zum Stoffhaushalt von *Ancylus fluviatilis* (Gastropoda – Basommatophora). 1. Ingestion, Assimilation, Wachstum und Eiablage. *Arch. Hydrobiol. Suppl.*, **47**, 458–514.

Streit, B. (1976) Energy flow in four different field populations of *Ancylus fluviatilis* (Gastropoda–Basommatophora). *Oecologia*, **22**, 261–273.

Streitfert, A., Grünhage, L., Dörger, G., Hanewald, K., and Wolf, H. (2008) Klimabiomonitoring. http://fss.plone.uni-giessen.de/fss/fbz/fb08/biologie/pflanzenoek/alte_homepage/forschung/Foeinr/UKL/luftmessstation/Infos_HLUG/HLUG/Klimabiomonitoring-Nov08/file/Klimabiomonitoring-Nov08.pdf (accessed 12 December 2010). (Hessisches Landesamt für Umwelt und Geologie ed.)

Studemann, D., and Landolt, P. (1997) Eggs of Ephemerellidae (Ephemeroptera), in *Ephemeroptera and Plecoptera Biology–Ecology–Systematics* (eds P. Landolt and

M. Sartori), MTL – Mauron+Tingeley and Lachat, Fribourg, pp. 362–371.

Studemann, D.P., Landolt, P., Sartori, M., Hefti, D., and Tomka, I. (1992) Ephemeroptera. *Insecta Helvetica, Fauna*, **9**, 176 pp.

Suberkropp, K. (1991) Relationship between growth and sporulation of aquatic hyphomycetes on decomposing leaf litter. *Mycol. Res.*, **95**, 843–850.

Suberkropp, K. (1997) Annual production of leaf decaying fungi in a woodland stream. *Freshwater Biol.*, **38**, 169–178.

Suren, A.M., and Lake, P.S. (1989) Edibility of fresh and decomposing macrophytes to three species of freshwater invertebrate herbivores. *Hydrobiologia*, **178**, 165–178.

Svensson, B.W. (1977) Life cycle, energy fluctuations, and sexual differentiation in *Ephemera danica* (Ephemeroptera), a stream-living mayfly. *Oikos*, **29**, 78–86.

Sweeney, B.W. (1978) Bioenergetic and development response of a mayfly to thermal variation. *Limnol. Oceanogr.*, **23**, 461–477.

Sweeney, B.W. (1984) Factors influencing life-history patterns of aquatic insects, in *The Ecology of Aquatic Insects* (eds V.H. Resh and D.M. Rosenberg), Praeger Publishing, New York, pp. 56–100.

Sweeney, B.W., and Vannote, R.L. (1981) *Ephemerella* mayflies of White Clay Creek: bioenergetic and ecological relationships among six coexisting species. *Ecology*, **62**, 1353–1369.

Sweeney, B.W., and Vannote, R.L. (1984) Influence of food quality and temperature on life history characteristics of the parthenogenetic mayfly, *Cloeon triangulifer*. *Freshwater Biol.*, **14**, 621–630.

Sweeney, B.W., and Vannote, R.L. (1986) Growth and production of a stream stonefly: influences of diet and temperature. *Ecology*, **67**, 1396–1410.

Sweeney, B.W., Jackson, J.K., and Funk, D.H. (1995) Semivoltinism, seasonal emergence, and adult size variation in a tropical stream mayfly (*Euthyplocia hecuba*). *J. N. Am. Benthol. Soc.*, **14**, 131–146.

Sweeney, B.W., Vannote, R.L., and Dodds, P.J. (1986) Effetcs of temperature and food quality on growth and development of a mayfly, *Leptophlebia intermedia*. *Can. J. Fish. Aquat. Sci.*, **43**, 12–18.

Syrnikov, Y.S., Melnitzky, S.I., and Ivanov, V.D. (2005) Effects of feeding on the mating behaviour of Trichoptera, in *Proc. 10th Int. Symp. Trichoptera, 2003* (eds K. Tanida and A. Rossiter), Tokai University Press, pp. 413–419.

Takahashi, K., and Sato, M. (1988) Flight activity of insects sampled with a truck trap. 3. Flight activity of aquatic beetles. *Kontyu*, **56**, 892–895.

Takemon, Y. (1993) Water intake by the adult mayfly *Epeorus ikanonis* (Ephemeroptera: Heptageniidae) and its effect on their longevity. *Ecol. Res.*, **8**, 185–192.

Tank, J.L., Webster, J.R., Benfield, E.F., and Sinsabaugh, R.L. (1998) Effect of leaf litter exclusion on microbial enzyme activity associated with wood biofilms in streams. *J. N. Am. Benthol. Soc.*, **17**, 95–103.

Tavares, A.F., and Williams, D.D. (1990) Life histories, diet, and niche overlap of three sympatric species of Elmidae (Coleoptera) in a temperate stream. *Can. Entomol.*, **112**, 563–577.

Teague, S.A., Knight, A.W., and Teague, B.N. (1985) Stream microhabitat selectivity, resource partitioning, and niche shifts in grazing caddisfly larvae. *Hydrobiologia*, **128**, 3–12.

Teal, J.M. (1957) Community metabolism in a temperate cold spring. *Ecol. Monogr.*, **27**, 283–302.

Teichmann, W. (1982) Lebensläufe und Zeitpläne von Gammariden unter ökologischen Bedingungen. *Arch. Hydrobiol./Suppl.*, **64**, 240–306.

ter Braak, C.J.F. (1988) CANOCO – a FORTRAN program for canonical community ordination by [partial] [detrended] [canonical] correspondence analysis, principal component analysis and redundancy analysis (version 2.1). Agricultural Mathematics Group, Wageningen, 95 pp.

ter Braak, C.J.F. (1990) Update notes: CANOCO version 3.1. Agricultural Mathematics Group, Wageningen, 35 pp.

ter Braak, C.J.F., and Šmilauer, P. (1998) CANOCO Reference Manual and users's guide to CANOCO for Windows. Centre for Biometry, CPRO-DLO, Wageningen, 351 pp.

ter Braak, C.J.F., and Verdonschot, P.F.M. (1995) Canonical correspodence analysis and related multivariate methods in aquatic ecology. *Aquat. Sci.*, **57**, 255–289.

Terry, M.D., and Whiting, M.F. (2003) Phylogeny of Plecoptera: molecular evidence and evolutionary trends. *Entomol. Abh. Mus. Tierk.*, **61**, 130–131.

Terry, M.D., and Whiting, M.F. (2005) Mantophasmatodea and phylogeny of the lower neopterous insects. *Cladistics*, **21**, 240–257.

Teske, A., Alm, E., Regan, J.M., Toze, S., Rittmann, B.E., and Stahl, D.A. (1994) Evolutionary relationships among ammonia- and nitrite-oxidizing bacteria. *J. Bacteriol.*, **176**, 6623–6630.

Tett, P., Gallegos, C., Kelly, M.G., Hornberger, G.M., and Cosby, B.J. (1978) Relationships among substrate, flow, and benthic microalgal pigment density in the Mechums River, Virginia. *Limnol. Oceanogr.*, **23**, 785–797.

Teyrovsky, V. (1939) Studien über den Köcherbau der Trichopterenlarven: Größe versus Gewicht der Teilchen als Baumaterialauslese bestimmender Faktoren bei *Neuronia ruficrus* (SCOP.). *Vestn. Ceske. Zool. Spol. Praze*, **6/7**, 470–475.

Theowald, B. (1978) Tipulidae und Cylindrotomidae, in *Limnofauna Europea, 2nd edn.* (ed. J. Illies), Gustav Fischer Verlag, Stuttgart, pp. 363–366.

Thibault, M. (1971) Le développement des Ephéméroptères d'un ruisseau à truites des Pyrénées Atlantiques, le Lissuraga. *Ann. Limnol.*, **7**, 53–120.

Thienemann, A. (1912) Der Bergbach des Sauerlandes. Faunistisch-biologische Untersuchungen. *Internat. Rev. Biol. Suppl.*, **4**, 1–125.

Thienemann, A. (1925) Die Binnengewässer Mitteleuropas. Die Binnengewässer 1, E. Schweitzerbart'sche Verlagsbuchhandlung Stuttgart, 255 pp.

Thienemann, A. (1936) Alpine Chironomiden. *Arch. Hydrobiol.*, **17**, 221–336.

Thienemann, A. (1955) Die Binnengewässer in Natur und Kultur. Eine Einführung in die theoretische und angewandte Limno-logie, *Verständliche Wissenschaft*, **55**, 156.

Thomas, A.G.B. (1978) Diptères torrenticoles peu connus: V. Les Rhagionidae (genre *Chrysopilus*) du Sud de la France (Brachycera, Orthorrhapha). *Bull. Soc. Hist. Nat. Toulouse*, **114**, 305–331.

Thomas, A.G.B. (1979) Diptères torrenticoles peu connus. VI. Les Dixidae du sud-ouest de la France (Nematocera) (*Dixa puberula* Loew, 1849). Ècologie, microhabitat et intérét pratique pour le dépistage des pollutions par les stations touristiques de montagne. *Bull. Soc. Hist. Nat. Toulouse*, **115**, 242–268.

Thomas, A.G.B. (1997) Rhagionidae and Athericidae, Snipe-flies, in *Aquatic Insects of North Europe – A Taxonomic Handbook. Vol. 2* (ed. A. Nilsson), Apollo Books, Stenstrup, pp. 311–320.

Thomas, E. (1966) Orientierung der Imagines von *Capnia atra* Morton (Plecoptera). *Oikos*, **17**, 278–280.

Thomas, J. (1990) Räumliche und zeitliche Verteilung der β-Glucosidase-Aktivität im Sediment des Breitenbaches (Hessen). 1990, Diploma thesis, University of Gießen, Gießen, 81 pp.

Thompson, F.C., and Rotheray, G.E. (1998) Family Syrphidae, in *Contributions to a Manual of Palaearctic Diptera vol. 3* (eds L. Papp and B. Darvas), Science Herald, Budapest, pp. 81–140.

Thorup, J. (1966) Substrate type and its value as a basis fort the delimination of bottom fauna communities in running waters. *Spec. Publs. Pymatuning Lab. Ecol.*, **4**, 59–74.

Thorup, J. (1970) The influence of a short-termed flood on a springbrook community. *Arch. Hydrobiol.*, **66**, 447–457.

Thurman, E.M. (1985) *Organic Geochemistry of Natural Waters*, in Martinus Nijhoff/Dr W. Junk Publishers, Dordrecht, 497 pp.

Tierno de Figueroa, J.M., and Fochetti, R. (2001) On the adult feeding of several European Stoneflies (Plecoptera). *Entomol. News*, **112**, 130–134.

Tierno de Figueroa, J.M., and Marfil-Daza, C. (2004) Revisión de las caracteristicas de los huevos de los Chloroperlidae (Plecoptera) de Europa, con la descripción del huevo de *Siphonoperla baetica* (Aubert, 1956) [A review of the eggs of European

Chloroperlidae (Plecoptera), with the description of the egg of *Siphonoperla baetica* (Aubert, 1956)]. *Boln SEA*, **35**, 219–224

Tierno de Figueroa, J.M., and Sánchez-Ortega, A. (2000) Imaginal feeding of twelve nemouroidean stonefly species (Insecta, Plecoptera). *Ann. Entomol. Soc. Am.*, **93**, 251–253.

Tierno de Figueroa, J.M., and Sánchez-Ortega, A. (2004) Implications of imaginal-size over the flight period in stoneflies (Plecoptera). *Ann. Soc. Entomol. Fr.*, **40**, 37–40.

Tilzer, M. (1968) Zur Ökologie und Besiedlung des hochalpinen hyporheischen Interstitials im Arlberggebiet (Österreich). *Arch. Hydrobiol.*, **65**, 253–308.

Timm, T. (1988) Die Eibiologie der Kriebelmücken – Potenz, Toleranz und ihre Beziehung zur Habitatbindung (Diptera: Simuliidae). *Arch. Hydrobiol. Suppl.*, **79**, 365–445.

Timm, T. (1993a) Lebensstrategie und Populationsdynamik von *Nevermannia vernum*, einer Kriebelmücke permanenter und periodischer Waldbäche (Diptera, Simuliidae). *Zool. Jahrb. Abt. Syst. Ökol. Geog. Tiere*, **120**, 289–307.

Timm, T. (1993b) Unterschiede in Eibiologie und Habitatbindung zwischen *Prosimulium tomosvaryi* (Prosimuliini) und verschiedenen Simuliini (Diptera, Simuliidae). *Int. Rev. Ges. Hydrobiol.*, **78**, 95–106.

Timm, T. (1994) Reasons for the shift in dominance between *Simulium (N.) vernum* and *Simulium (S.) ornatum* (Diptera: Simuliidae) along the continuum of an unpolluted lowland stream. *Arch. Hydrobiol.*, **131**, 199–210.

Tittizer, T. (1999) 3.7.1. Makrozoobenthos, in *Methoden der Biologischen Gewässeruntersuchung 2* (eds W. von Tümpling and G. Friedrich), Gustav Fischer, Jena, pp. 133–153.

Tobias, W. (1996) Sommernächtliches Schneetreiben am Main. Zum Phänomen des Massenfluges von Eintagsfliegen. *Nat. Mus.*, **126**, 37–54.

Tobias, W., and Tobias, D. (1981) Trichoptera Germanica, Bestimmungstafeln für die deutschen Köcherfliegen – Teil 1: Imagines. *Cour. Forsch.-Inst, Senckenberg*, **49**, 671.

Tolkamp, H.H. (1980) *Organism-substrate relationships in lowland streams.* Centre for Agricultural Publishing and Documentation, Wageningen, 211 pp.

Tomaszewski, C. (1955) Studies on the influence of water microcurrents on *Tinodes waeneri* L. (Trichoptera). *Ecol. Pol.*, **3**, 85–99.

Tomme, P., Warren, R.A.J., and Gilkes, N.R. (1995) Cellulose hydrolysis by bacteria and fungi. *Adv. Microb. Physiol.*, **37**, 1–81.

Towns, D.R. (1981) Life histories of benthic invertebrates in a Kauri forest stream in northern New Zealand. *Aust. J. Mar. Freshwater Res.*, **32**, 191–211.

Townsend, C.R. (1989) The patch dynamics concept of stream community ecology. *J. North Am. Benthol. Soc.*, **8**, 36–50.

Townsend, C.R., and Hildrew, A.G. (1994) Species traits in relation to a habitat templet for river systems. *Freshw. Biol.*, **31**, 265–275.

Townsend, C.R., Dolédec, S., and Scarsbrook, M.R. (1997) Species traits in relation to temporal and spatial heterogeneity in streams: a test of habitat templet theory. *Freshw. Biol.*, **37**, 367–387.

Townsend, C.R., Scarsbrook, M.R., and Dolédec, S. (1997a) Quantifying disturbance in streams: alternative measures of disturbance in relation to macroinvertebrate species traits and species richness. *J. North Am. Benthol. Soc.*, **16**, 531–544.

Townsend, C.R., Scarsbrook, M.R., and Dolédec, S. (1997b) The intermediate disturbance hypothesis, refugia, and biodiversity in streams. *Limnol. Oceanogr.*, **42**, 938–949.

Traunspurger, W. (2000) The biology and ecology of lotic nematodes. *Freshwater Biol.*, **44**, 29–45.

Tuchman, N.C., and Stevenson, R.J. (1991) Effects of selective grazing by snails on benthic algal succession. *J. N. Am. Benthol. Soc.*, **10**, 430–443.

Ullrich, F. (1976) Biologisch-Ökologische Studien an rheophilen Wassermilben (Hydrachnellae, Acari), unter besonderer Berücksichtigung von *Sperchon setiger* (Thor, 1898). PhD. thesis, University of Kiel, Kiel, 241 pp.

Ullrich, F. (1978) Biologisch-ökologische Studien an den Larven rheophiler Wassermilben (Hydrachnellae, Acari), Schlitzer produktionsbiologische Studien (29). *Arch. Hydrobiol. Suppl.*, **54**, 189–255.

Ulrich, H. (1981) *Trichopeza longicornis* (Meigen) (Diptera, Empididae) hat Raubbeine. *Bonn. Zool. Beitr.*, **32**, 237–239.

Vaillant, F. (1955) Recherches sur la faune madicole de France, de Corse et d'Afrique du Nord. Ph.D. thesis Université de Paris, Paris, 252 pp.

Vaillant, F. (1971–1983) Psychodidae-Psychodinae, in *Die Fliegen der paläarktischen Region 9d, Lieferungen 287, 291, 292, 305, 310, 313, 317, 320, 326, 328* (ed. E. Lindner), Schweitzerbart'sche Verlagsbuchhandlung, Stuttgart, 358 pp + 120 tables.

Vaillant, F. (1978) Dolichopodidae, in *Limnofauna Europaea 2nd edition* (ed. J. Illies), Gustav Fischer Verlag, Stuttgart, pp. 470–474.

Vaillant, F. (2002) Insecta: Diptera: Lonchopteridae, in *Süßwasserfauna von Mitteleuropa 21/22, 21/23, Insecta: Diptera: Lonchopteridae und Sciomyzidae* (eds J. Schwoerbel and P. Zwick), Spektrum Akademischer Verlag, Heidelberg, pp. 1–14.

van der Meutter, F., Stoks, R., and de Meester, L. (2006) Lotic dispersal of lentic macroinvertebrates. *Ecography*, **29**, 223–230.

Vannote, R.L., and Sweeney, B.W. (1980) Geographic analysis of thermal equilibria: a conceptual model for evaluation the effects of natural and modified thermal regimes on aquatic insects. *Amer. Nat.*, **115**, 667–695.

Vannote, R.L., Minshal, G.W., Cummins, K.W., Sedell, J.R., and Cushing, C.E. (1980) The river continuum concept. *Can. J. Fish. Aquat. Sci.*, **37**, 130–137.

Vaughn, C.C. (1986) The role of periphyton abundance and quality in the microdistribution of a stream grazer, *Helicopsyche borealis* (Trichoptera: Helicopsychidae). *Freshwater Biol.*, **16**, 485–493.

Verschuren, D. (1987) Modified setae on rheophilous dragonly larva (*Cordulegaster*, Odonata) as tactile receptors of water current during prey detection. *Ann. Soc. R. Zool. Belg.*, **117**, 264.

Vervier, P., Gibert, J., Marmonier, P., and Dole-Olivier, M.J. (1992) A perspective on the permeability of the surface freshwater–groundwater ecotone. *J. N. Am. Benthol. Soc.*, **11**, 93–102.

Waage, J.K. (1984) Sperm competition and the evolution of odonate mating systems, in *Sperm competition and the evolution of animal mating systems* (ed. R.L. Smith), Academic Press, New York, pp. 251–290.

Wagner, F. (2003) Flight behaviour of merolimnic insects from the Leutra River (Thuringia, Germany). *Aquat. Insects*, **25**, 51–62

Wagner, M., Amann, R., Lemmer, H., and Schleifer, K.-H. (1993) Probing activated sludge with proteobacteria-specific oligonucleotides: inadequacy of culture-dependent methods for describing microbial community structure. *Appl. Environ. Microbiol.*, **59**, 1520–1525.

Wagner, R. (1973) Psychodiden aus dem Breitenbach (Diptera, Psychodidae) 1970. *Arch. Hydrobiol.*, **72**, 517–524.

Wagner, R. (1975) Sechs neue Psychodidenarten aus Deutschland und Österreich (Diptera, Psychodidae). *Mitt. Dtsch Entomol. Ges.*, **34**, 1–9.

Wagner, R. (1978) Neue europäische Psychodiden (Diptera: Psychodidae). *Senckenberg. Biol.*, **58**, 157–170.

Wagner, R. (1979) Psychodidenstudien im Schlitzerland. Schlitzer produktionsbiologische Studien (26). *Arch. Hydrobiol. Suppl.*, **57**, 38–88.

Wagner, R. (1980) Die Dipterenemergenz am Breitenbach (1969–1973). *Spixiana*, **3**, 167–177.

Wagner, R. (1983) Aquatische Empididen (Diptera) aus hessischen Mittelgebirgen und angrenzenden Gebieten. *Beitr. Naturk. Osthessen*, **19**, 135–146.

Wagner, R. (1986) Egg development and life cycle of *Chaetopteryx villosa* (Trichoptera). *Hol. Ecol.*, **9**, 294–300.

Wagner, R. (1990a) Influence of temperature, photoperiod and nutrition on growth and consumption of *Chaetopteryx villosa* (Trichoptera). *Hol. Ecol.*, **13**, 247–254.

Wagner, R. (1990b) A laboratory study on the life cycle of *Sericostoma personatum*

(Kirby and Spence), and light dark-dependent food consumption. *Hydrobiologia*, **208**, 201–212.

Wagner, R. (1991) The influence of the diel activity pattern of the larvae of *Sericostoma personatum* (Kirby and Spence) Trichoptera on organic matter distribution in stream-bed sediments – a laboratory study. *Hydrobiologia*, **224**, 65–70.

Wagner, R. (1992) Warum ändern sich Größe und Gewicht adulter Trichopteren entlang eines Fließgewässers? *Mitt. Dtsch Ges. Allg. Angew. Entomol.*, **8**, 320–324.

Wagner, R. (1993) Schutzwürdigkeitsgutachten und Pflegeplan für das Naturschutzgebiet "Breitenbachtal bei Michelsrombach". Im Auftrag des Regierungspräsidiums Kassel, Abteilung für Forsten und Naturschutz (Obere Naturschutzbehörde), 125 pp, maps, tables, (unpublished).

Wagner, R. (2002a) Abundance, size and weight of Trichoptera on the Breitenbach. in *Proc. 10th Int. Symp. Trichoptera* (ed. Mey, W.), *Nova Suppl. Entomol.*, **15**, 607–616.

Wagner, R. (2002b) Zeitliche und räumliche Dynamik der *Wormaldia occipitalis*-Population am Breitenbach (Hessen) [Temporal and spatial dynamics of the *Wormaldia occipitalis*-population in the Breitenbach (Hesse, Germany)]. *Lauterbornia*, **43**, 111–119.

Wagner, R. (2002c) The influence of temperature and food on size and weight of adult *Chaetopteryx villosa* (Fabricius) (Insecta: Trichoptera) along a stream gradient. *Arch. Hydrobiol.*, **154**, 393–411.

Wagner, R. (2002d) Insecta: Diptera: Thaumaleidae, in *Süßwasserfauna von Mitteleuropa, 21/10+11 Chaoboridae und Thaumaleidae* (eds J. Schwoerbel and P. Zwick), Spektrum Akademischer Verlag, Heidelberg, pp. 39–110.

Wagner, R. (2005a) Spatial, temporal and competition effects on size and weight of caddisflies (Insecta, Trichoptera) in emergence traps. *Arch. Hydrobiol.*, **162**, 383–398.

Wagner, R. (2005b) The influence of stream water temperature on size and weight of caddisflies (Insecta, Trichoptera) along the Breitenbach 1983–1991. *Arch. Hydrobiol.*, **163**, 65–79.

Wagner, R., and Gathmann, O. (1996) Long-term studies on aquatic Dance Flies (Diptera, Empididae) 1983–1993: distribution and size patterns along the stream, abundance change between years and the influence of environmental factors on the community. *Arch. Hydrobiol.*, **137**, 385–410.

Wagner, R., and Höchst, M. (2001) Beiträge zum Lebenszyklus von *Ecclisopteryx madida* (Insecta, Trichoptera). *Lauterbornia*, **40**, 109–117.

Wagner, R., and Schmidt, H.H. (2004) Yearly discharge patterns determine species abundance and community diversity: analysis of a 25 year record from the Breitenbach. *Arch. Hydrobiol.*, **161**, 511–540.

Wagner, R., Aurich, M., Reder, E., and Veith, H.J. (1990) Defensive secretions from the larvae of *Apatania fimbriata* (Pictet) (Trichoptera: Limnephilidae). *Chemoecology*, **1**, 96–104.

Wagner, R., Schmidt, H.H., and Marxsen, J. (1993) The hyporheic habitat of the Breitenbach, spatial structure and physicochemical conditions as a basis for benthic life. *Limnologica*, **23**, 285–294.

Wagner, R., Obach, M., Werner, H., and Schmidt, H.H. (2006) Artificial Neural Nets and Abundance Prediction of Aquatic Insects in Small Streams. *Ecol. Inform.*, **1**, 423–430.

Wagner, R., Barták, M., Borkent, A., Courtney, G., Goddeeris, B., Haenni, J.-P., Knutson, L., Pont, A., Rotheray, G.E., Rozkosný, R., Sinclair, B., Woodley, N., Zatwarnicki, T., and Zwick, P. (2008) Global diversity of dipteran families (Insecta Diptera) in freshwater (excluding Simulidae, Culicidae, Chironomidae, Tipulidae and Tabanidae) in (eds Balian, E., Lévêque, C., Segers, H., and Martens, K.), Freshwater Animal Diversity Assessment. *Dev. Hydrobiol.*, **198**, Special Issue of Hydrobiologia, 595, Springer, 489–519.

Wallace, I.D., Wallace, B., and Phillipson, G.N. (2003) *Keys to the Case-Bearing Caddis Larvae of Britain and Ireland*, Scientific Publication 61 (ed Sutcliffe, D.W.), Freshwater Biological Association, Ambleside, 259 pp.

Wallace, J.B., and Anderson, N.H. (1996) Habitat, life history, and behavioral adaptations of aquatic insects, Chapter 5 in *An Introduction to the Aquatic Insects of North America* (eds R.W. Merritt and K.W. Cummins), Kendall/Hunt Publishing Company, Dubuque, pp. 41–73.

Wallace, J.B., and Gurtz, M.E. (1986) Response of *Baetis* mayflies (Ephemeroptera) to catchment logging. *Am. Midl. Nat.*, **115**, 25–41.

Wallis, P.M. (1981) The uptake of dissolved organic matter in groundwater by stream sediments – a case study, in *Perspectives in Running Water Ecology* (eds M.A. Lock and D.D. Williams), Plenum Press, New York, pp. 97–112.

Wallner, G., Amann, R., and Beisker, W. (1993) Optimizing fluorescent in situ hybridization with rRNA-targeted oligonucleotide probes for flow cytometric identification of microorganisms. *Cytometry*, **14**, 136–143.

Wang, G.C.Y., and Wang, Y. (1997) Frequency of formation of chimeric molecules is a consequence of PCR coamplification of 16S rRNA genes from mixed bacterial genomes. *Appl. Environ. Microbiol.*, **63**, 4645–4650.

Wang, Q.G., Garrity, M., Tiedje, J.M., and Cole, J.R. (2007) Naïve Bayesian classifier for rapid assignment of rRNA sequences into the new bacterial taxonomy. *Appl. Environ. Microbiol.*, **73**, 5261–5267.

Wantzen, K.M., and Wagner, R. (2006) Detritus processing by invertebrate shredders: a neotropical-temperate comparison. *J. N. Am. Benthol. Soc.*, **25**, 216–232.

Ward, B.B., Voytek, M.A., and Witzel, K.P. (1997) Phylogenetic diversity of natural populations of ammonia oxidizers investigated by specific PCR amplification. *Microb. Ecol.*, **33**, 87–96.

Ward, J.V. (1985) Thermal characteristics of running waters. *Hydrobiologia*, **125**, 31–46.

Ward, J.V. (1986) Altitudinal zonation in a Rocky Mountain stream. *Arch. Hydrobiol. Suppl.*, **74**, 133–199.

Ward, J.V., and Stanford, J.A. (1983) The intermediate-disturbance hypothesis, an explanation for biotic diversity patterns in lotic ecosystems. *Dynamics of lotic ecosytems* (eds T.D. Fontaine and S.M. Bartell), Ann Arbor Science, Ann Arbor, Michigan, U.S.A., pp.347–356.

Waringer, J., and Graf, W. (1997) *Atlas der Österreichischen Köcherfliegenlarven unter Einschluß der angrenzenden Gebiete*, Facultas-Universitätsverlag, Vienna, 286 pp.

Waringer, J.A. (1989a) Resistance of a cased caddis larva to accidental entry into the drift: the contribution of active and passive elements. *Freshwater Biol.*, **21**, 411–420.

Waringer, J.A. (1989b) Life cycle, horizontal microdistribution and current resistance of *Allogamus auricollis* (Trichoptera: Limnephilidae) in an Austrian mountain brook. *Freshwater Biol.*, **22**, 177–188.

Waringer, J.A. (1993) The drag coefficient of cased caddis larvae from running waters: experimental determination and ecological applications. *Freshwater Biol.*, **29**, 419–427.

Warnecke, F., Amann, R., and Pernthaler, J. (2004) Actinobacterial 16S rRNA genes from freshwater habitats cluster in four distinct lineages. *Environ. Microbiol.*, **6**, 242–253.

Waters, T.F. (1977) Secondary production in inland waters. *Adv. Ecol. Res.*, **10**, 91–164.

Weaver, III, J.S. (1984) The evolution and classification of Trichoptera. Part I: the groundplan of Trichoptera, in *Proc. 4th Int. Symp. Trichoptera, Clemson, South Carolina* (ed. J.M. Morse), Dr. W. Junk Publishers, The Hague, pp. 413–419.

Weaver, III, J.S., and Morse, J.C. (1986) Evolution of feeding and case-making behavior in Trichoptera. *J. N. Am. Benthol. Soc.*, **5**, 150–158.

Webster, D.A., and Webster, P.C. (1943) Influence of water current on case weight in larvae of the caddisfly, *Goera calcarata* Banks. *Can. Entomol.*, **75**, 105–108.

Webster, J.R. and Meyer, J.L. (eds) (1997a) Stream organic matter budgets. *J. N. Am. Benthol. Soc.*, **16**, 3–161.

Webster, J.R., and Meyer, J.L. (1997b) Stream organic matter budgets – introduction. *J. N. Am. Benthol. Soc.*, **16**, 5–13.

Webster, J.R., and Meyer, J.L. (1997c) Organic-matter budgets for streams – a synthesis. *J. N. Am. Benthol. Soc.*, **16**, 141–161.

Weise, W., and Rheinheimer, G. (1978) Scanning electron microscopy and epifluorescence investigation of bacterial colonization of marine sand sediments. *Microb. Ecol.*, **4**, 175–188.

Wellnitz, T., and Poff, N.L. (2006) Herbivory, current velocity and algal regrowth: how does periphyton grow when the grazers have gone? *Freshwater Biol.*, **51**, 2114–2123.

Wendker, S. (1992) Diatom community response to translocation in a small softwater stream. *Nova Hedwigia*, **55**, 397–406.

Werneke, U. (1991) Populationsdynamik und Eiablage von *Baetis rhodani* (Pictet) und *Baetis vernus* Curtis (Insecta: Ephemeroptera). Diploma thesis, University of Tübingen, Tübingen, 98 pp.

Werneke, U. (1994) Untersuchungen zu Emergenz und Eiablage von *Baetis vernus* und *Baetis rhodani* (Insecta: Ephemeroptera) im Breitenbach bei Schlitz/Hessen. *Lauterbornia*, **17**, 19–28.

Werneke, U. (1997) Freilanduntersuchungen zum Einfluß von Nahrungsangebot und Temperatur auf das Wachstum von *Baetis- und Ephemerella-Arten* (Insecta: Ephemeroptera). Ph.D. thesis, University of Marburg, Marburg, 239 pp.

Werneke, U., and Zwick, P. (1992) Mortality of the terrestrial adult and aquatic nymphal life stages of *Baetis vernus* Curtis and *Baetis rhodani* Pictet in the Breitenbach, Germany (Insecta: Ephemeroptera). *Freshwater Biol.*, **28**, 249–255.

Werner, D. (2004) New records of Diptera predators of blackflies (Diptera, Simuliidae). *Entomol. Mon. Mag.*, **140**, 1–25.

Werner, D., and Pont, A.C. (2003) Dipteran predators of Simuliid blackflies: a worldwide review. *Med. Vet. Entomol.*, **17**, 115–132.

Werner, H., and Obach, M. (2001) New neural network types estimating the accuracy of response for ecological modelling. *Ecological Modelling*, **146**, 289–298.

Westlake, D.F. (1963) Comparison of plant productivity. *Biol. Rev.* **38**: 385–425.

Westlake, D.F. (1969) Macrophytes. In: R.A. Vollenweider (ed.) *A manual on methods for measuring primary production in aquatic environments.* IPB handbook No.12. F.A. Davis Co. Philadelphia, PA. pp. 25–32.

Wetzel, R.G. (1975) Primary production, in *River Ecology. Studies in Ecology 2* (ed. B.A. Whitton), University of California Press, Berkeley, 230–247 pp.

Wetzel, R.G. (1983) *Limnology*, 2nd edn, Saunders College Publishing, Philadelphia, 767 pp.

Wetzel, R.G. (1991) Extracellular enzymatic interactions: storage, redistribution, and interspecific communication, in *Microbial Enzymes in Aquatic Environments* (ed. R. Chróst), Brock/Springer Series in Contemporary Bioscience, Springer, New York, pp. 6–28.

Wetzel, R.G. (2001) *Limnology*, 3rd edn, Academic Press, San Diego, 1006 pp.

Wetzel, R.G., Rich, P.H., Miller, M.C., and Allen, H.L. (1972) Metabolism of dissolved and particulate detrital carbon in a temperate hard-water lake. *Mem. Ist. Ital. Idrobiol. Suppl.*, **29**, 185–243

Weyers, H.S., and Suberkropp, K. (1996) Fungal and bacterial production during the breakdown of yellow poplar leaves in 2 streams. *J. N. Am. Benthol. Soc.*, **15**, 408–420.

White, D.S. (1993) Perspectives on defining and delineating hyporheic zones. *J. N. Am. Benthol. Soc.*, **12**, 61–69.

White, T.C.R. (1978) The importance of a relative shortage of food in animal ecology. *Oecologia*, **33**, 71–76.

Whitman, R.L., and Clark, W.J. (1982) Availability of dissolved oxygen in interstitial water of a sandy creek. *Hydrobiologia*, **92**, 651–658.

Whitton, B.A. (1991) Use of phosphatase assays with algae to assess phosphorus status of aquatic environments, in *Bioindicators and Environmental Management* (eds D.W. Jeffrey and B. Madden), Academic Press, London, pp. 295–310.

Wichard, W. (1997) Insekten erobern den aquatischen Lebensraum. *Verh. Westdt. Entomologentag, Düsseldorf*, **1996**, 19–28.

Wichard, W., and Komnick, H. (1974a) Fine structure and function of the rectal chloride epithelia of damselfly larvae. *J. Insect Physiol.*, **20**, 1611–1621.

Wichard, W., and Komnick, H. (1974b) Feinstruktur und Funktion der Analpapillen von aquatischen Käferlarven (Elodidae). *Int. J. Insect Morph. Embryol.*, **3**, 335–341.

Wichard, W., Arens, W., and Eisenbeis, W. (1995) *Atlas zur Biologie der Wasserinsekten*, 1st edn, Gustav Fischer, Stuttgart. 338+3 unpaginated pp.

Wichard, W., Arens, W., and Eisenbeis, W. (2002) *Biological Atlas of Aquatic Insects*, Apollo Books, Stenstrup, 340 pp, (translated from Wichard, Arens, Eisenbeis 1995).

Wiggins, G.B. (1977) *Larvae of North American Caddisfly Genera (Trichoptera)*. University of Toronto Press, Toronto and Buffalo, Canada, 401 pp.

Wiggins, G.B. (1996) *Larvae of the North American caddisfly genera (Trichoptera)*, 2nd edn, University of Toronto Press, Toronto, 457 pp.

Wiggins, G.G. (2001) Construction behavior for new pupal cases by case-making caddis larvae: further comment (Trichoptera: Integripalpia). *Braueria*, **28**, 7–9.

Wigglesworth, V.B. (1972) *Principles of Insect Physiology*, 7th edn, Chapman and Hall, London, 827 pp.

Wilczek, S., Fischer, H., and Pusch, M.T. (2005) Regulation and seasonal dynamics of extracellular enzyme activities in the sediments of a large lowland river. *Microb. Ecol.*, **50**, 253–267.

Wiley, M., and Kohler, S.L. (1984) Behavioral adaptations of aquatic insects, in *The Ecology of Aquatic Insects* (eds V.H. Resh and D.M. Rosenberg), Praeger, New York, pp. 101–133.

Wilhelm, S.W., and Suttle, C.A. (1999) Viruses and nutrient cycles in the sea – viruses play critical roles in the structure and function of aquatic food webs. *Bioscience*, **49**, 781–788.

Williams, D.D. (1989) Towards a biological and chemical definition of the hyporheic zone in two Canadian rivers. *Freshwater Biol.*, **22**, 189–208.

Williams, D.D. (2006) *The Biology of Temporary Waters*. Oxford University Press, Oxford, 348 pp.

Williams, D.D., and Hynes, H.B.N. (1974) The occurrence of benthos deep in the substratum of streams. *Freshwater Biol.*, **4**, 233–256.

Williams, D.D., and Moore, K.A. (1985) The role of epilithon in substrate selection by stream invertebrates. *Arch. Hydrobiol.*, **105**, 105–115.

Williams, D.D., Read, A.T., and Moore, K.A. (1983) The biology and zoogeography of *Helicopsyche borealis* (Trichoptera: Helicopsychidae): a Nearctic representative of a tropical genus. *Can. J. Zool.*, **61**, 2288–2299.

Williams, D.D., Tavares, A.F., and Bryant, E. (1987) Respiratory device or camouflage? – A case for the caddisfly. *Oikos*, **50**, 42–52.

Williams, W.D. (1975) Über das Vorkommen von Wasserasseln in der Fulda, Deutschland. *Gewäss. Abwäss.*, **57/58**, 151–154.

Williamson, N., Brian, P., and Wellington, E.M.H. (2000) Molecular detection of bacterial and streptomycete chitinases in the environment. *Antonie van Leeuwenhook*, **78**, 315–321.

Willoughby, L.G., and Mappin, R.G. (1988) The distribution of *Ephemerella ignita* (Ephemeroptera) in streams: the role of pH and food resources. *Freshwater Biol.*, **19**, 145–155.

Winkler, C. (1992) Einfluss konstanter und fluktuierender Temperaturen und unterschiedlicher Nahrungsqualitäten auf das Wachstum von *Gammarus fossarum* KOCH, 1835 (Crustacea, Amphipoda) im Breitenbach-1992. Diploma thesis, University Bayreuth, 124 pp.

Winterbottom, J.H., Orton, S.E., Hildrew, A.G., and Lancaster, J. (1997) Field experiments on flow refugia in streams. *Freshw. Biol.*, **37**, 569–580.

Winterbourn, M.J. (1978) The macroinvertebrate fauna of a New Zealand forest stream. *N. Z. J. Zool.*, **5**, 157–169.

Witzel, K.P., Krambeck, J., and Overbeck, J. (1981) On the structure of bacterial communities in lakes and rivers – a comparison with numerical taxonomy of isolates. *Verh. Int. Ver. Limnol.*, **21**, 1365–1370.

Witzel, K.P., Overbeck, J., and Moaledj, K. (1982) Microbial communities in Lake Plußsee – An analysis with numerical taxonomy of isolates. *Arch. Hydrobiol.*, **94**, 38–52.

Wlosinski, J.H., and Minshall, G.W. (1983) Predictability of stream ecosytem models of various levels of resolution. *Dynamics of lotic ecosytems* (eds T.D. Fontaine and S.M. Bartell), Ann Arbor Science, Ann Arbor, Michigan, U.S.A., pp.69–86.

Woese, C.R., Weisburg, W.G., Hahn, C.M., Paster, B.J., Zablen, L.B., Lewis, B.J., Macke, T.J., Ludwig, W., and Stackebrandt, E. (1984) The phylogeny of purple bacteria: the beta subdivision. *Syst. Appl. Microbiol.*, **5**, 327–336.

Woese, C.R., Weisburg, W.G., Hahn, C.M., Paster, B.J., Zablen, L.B., Lewis, B.J., Macke, T.J., Ludwig, W., and Stackebrandt, E. (1985) The phylogeny of purple bacteria: the gamma subdivision. *Syst. Appl. Microbiol.*, **6**, 25–33.

Wolf, B. (1991) Nahrungsökologie der Fauna lenitischer Bereiche des Breitenbachs unter besonderer Berücksichtigung der Larven der Faltenmücke *Ptychoptera paludosa* Meigen, 1804 (Diptera). Ph.D. thesis, University of Kassel, Kassel, 123 pp.

Wolf, B., and Zwick, P. (1989) Plurimodal emergence and plurivoltinism of central European populations of *Nemurella pictetii* (Plecoptera: Nemouridae). *Oecologia*, **79**, 431–438.

Wolf, B., and Zwick, P. (2001) Life Cycle, Production and Survival Rates of *Ptychoptera paludosa* (Diptera: Ptychopteridae). *Int. Rev. Hydrobiol.*, **86**, 661–674.

Wolf, B., Zwick, P., and Marxsen, J. (1997) Feeding ecology of the freshwater detritivore *Ptychoptera paludosa* (Diptera, Nematocera). *Freshwater Biol.*, **38**, 375–386.

Wood, P.J., Agnew, M.D., and Petts, G.E. (2000) Flow variations and macroinvertebrate community responses in a small groundwater-dominated stream in South-East England. *Hydrol. Proces.*, **14**, 3133–3147.

Woodward, G., Jones, J.I., and Hildrew, A.G. (2002) Community persistence in Broadstone Stream (U.K.) over three decades. *Freshwater Biol.*, **47**, 1419–1435.

Wool, D. (1979) Genetic and environmental components of morphological variation in gall-forming aphids (Homoptera, Aphididae, Fordinae) in relation to climate. *J. Anim. Ecol.* **46**, 875–889.

Wotton, R.S. (1976) Evidence that blackfly larvae can feed on particles of colloidal size. *Nature*, **261**, 697.

Wuhrmann, K. (1974) Some problems and perspectives in applied limnology. *Mitt. Int. Verein. Limnol.*, **20**, 324–402.

Yoshimura, M. (2003) Relations of intraspecific variations in fecundity, clutch size, and oviposition frequency to the body size in three species of stoneflies, *Sweltsa* sp., *Isoperla aizuana*, and *Stavsolus japonicus*. *Limnology*, **4**, 109–112.

Yoshimura, M., Isobe, Y., and Oishi, T. (2003) The relation among emergence date, days until first effective mating, fecundity and adult longevity in *Isoperla aizuana* and *Sweltsa* sp. (Plecoptera). *Zool. Sci.*, **20**, 471–479.

Zeuthen, E. (1953) Oxygen uptake as related to body size in organisms. *Q. Rev. Biol.*, **28**, 1–12.

Zhang, Z., Schwartz, S., Wagner, L., and Miller, W. (2000) A greedy algorithm for aligning DNA sequences. *J. Comput. Biol.*, **7**, 203–214.

Zhu, M., Fujita, M., and Hashimoto, N. (1994) Application of neural networks to runoff prediction; Stochastic and Statistical Methods in *Hydrology and Environmental Engineering*, 3, 205–216.

Zimmermann, R., and Meyer-Reil, L.A. (1974) A new method for fluorescence staining of bacterial populations on membrane filters. *Kiel. Meeresforsch.*, **30**, 24–27.

Zulka, K.P. (1996) Methodisches Design für die Erfassung und Bewertung von Arthropodenbeständen und Arthropodenlebensräumen am Beispiel der Laufkäferfauna (Coleoptera, Carabidae). *Wiss. Mitt. Niederöst. Landesmus.*, **9**, 341–363.

Zwart, G., Crump, B.C., Agterveld, M.P.K., Hagen, F., and Han, S. (2002) Typical freshwater bacteria: an analysis of available 16S rRNA gene sequences from plankton of lakes and rivers. *Aquat. Microb. Ecol.*, **28**, 141–155.

Zwick, H., and Zwick, P. (1990) Terrestrial mass-oviposition of *Prosimulium*-species

(Diptera: Simuliidae). *Aquat. Insects*, **12**, 33–46.

Zwick, P. (1967) Beitrag zur Kenntnis der Plecopteren-Fauna Schleswig-Holsteins. *Faun.Ökol. Mitt.*, **3**, 108–111.

Zwick, P. (1969) Beitrag zur Plecopterenfauna der Fulda und ihres Einzugsgebietes in der Rhön und dem Vogelsberg. *Beitr. Naturk. Osthessen*, **1**, 65–76.

Zwick, P. (1973) Insecta Plecoptera. Phylogenetisches System und Katalog. *Das Tierreich*, **94**, 465 pp.

Zwick, P. (1977) Plecopteren-Emergenz zweier Lunzer Bäche, 1972–1974 – Schlitzer produktionsbiologische Studien (20). *Arch. Hydrobiol.*, **80**, 458–505.

Zwick, P. (1980) Plecoptera (Steinfliegen), in *Handb. Zool., Berlin*, 4 (2) 2/7, de Gruyter, Berlin, 115 pp.

Zwick, P. (1981) Diapause development of *Protonemura intricata* (Plecoptera: Nemouridae). *Verh. Internat. Verein. Limnol.*, **21**, 1607–1611.

Zwick, P. (1982) *Dictyogenus alpinum* (Plecoptera) im Schreierbach, einem spezialisierten Biotop im Sinne Thienemanns. *Arch. Hydrobiol.*, **95**, 533–548.

Zwick, P. (1986) Notes on the development of *Leuctra digitata* (Plecoptera). Landesentwicklung und Umweltforschung, TU Berlin, 40, 169–171.

Zwick, P. (1990) Emergence, maturation and upstream oviposition flights of Plecoptera from the Breitenbach, with notes on the adult phase as a possible control of stream insect populations. *Hydrobiologia*, **194**, 207–223.

Zwick, P. (1991) Biometric studies of the growth and development of two species of *Leuctra* and of *Nemurella pictetii* (Plecoptera: Leuctridae and Nemouridae), in *Overview and Strategies of Ephemeroptera* (eds J. Alba-Tercedor and A. Sanchez-Ortega), Sandhill Crane Press, Gainesville, pp. 515–526.

Zwick, P. (1992) Fließgewässergefährdung durch Insektizide. *Naturwissenschaften*, **79**, 437–442.

Zwick, P. (1993) Ökologie von Fließgewässern. *Biologie in unserer Zeit*, **23**, 316–321.

Zwick, P. (1996a) Variable egg development of *Dinocras* spp. (Plecoptera, Perlidae) and the stonefly seed-bank-theory. *Freshwater Biol.*, **35**, 81–99.

Zwick, P. (1996b) Capacity of discontinuous egg development and its importance for the geographic distribution of the warm water stenotherm, *Dinocras cephalotes* (Insecta: Plecoptera: Perlidae). *Ann. Limnol.*, **32**, 147–160.

Zwick, P. (1998a) Fließgewässer – Zentren vergessener Biodiversität. *DGL Tagungsbericht 1997 (Frankfurt/M.) Krefeld*, 1, pp. 23–31.

Zwick, P. (1998b) *Micrasema longulum* (Insecta: Trichoptera: Brachycentridae) builds a special pupation chamber. *Ann. Limnol.*, **34**, 437–444.

Zwick, P. (1999a) Egg diapause, egg swelling and mother–child size relationships in Plecoptera (Insecta). *Arch. Hydrobiol., Spec. Iss. Adv. Limnol.*, **54**, 373–386.

Zwick, P. (1999b) Ptychopteridae, in *Checkliste der Dipteren Deutschlands* (eds H. Schumann, R. Bährmann, and A. Stark), Studia Dipterologica, *Suppl.*, 2, Ampyx Verlag, Halle, p. 73.

Zwick, P. (2000) Phylogenetic System and Zoogeography of the Plecoptera. *Annu. Rev. Entomol.*, **45**, 709–746.

Zwick, P. (2003a) Ordnung Plecoptera, Steinfliegen, Frühlingsfliegen, Uferfliegen, in *Kaestner Lehrbuch der Zoologie* (ed. H. Dathe), Spektrum Akademischer Verlag, Heidelberg, Berlin, pp. 144–154.

Zwick, P. (2003b) Shapes and patterns of wingpad development in the Plecoptera, in *Research Update on Ephemeroptera and Plecoptera* (ed. E. Gaino), University of Perugia, Perugia, pp. 477–483.

Zwick, P. (2003c) The stonefly (Insecta: Plecoptera) seed bank theory: new experimental data. *Verh. Int. Ver. Limnol.*, **28**, 1317–1323.

Zwick, P. (2004) Key to the West Palaearctic genera of stoneflies (Plecoptera) in the larval stage. *Limnologica*, **34**, 315–348.

Zwick, P. (2007) *Elodes tricuspis*: description of the larva, and notes on biology (Coleoptera: Scirtidae). *Lauterbornia*, **59**, 85–93.

Zwick, P. (2009) The Plecoptera – who are they? The problematic placement of stoneflies in the phylogenetic system of insects. *Aquat. Insects*, **31** (Suppl. 1), 181–194.

Zwick, P., and Hecht, M. (2008) Life history and instar number of *Elodes minuta*, with notes on additional species (Coleoptera: Scirtidae). *Lauterbornia*, **62**, 79–88.

Zwick, P., and Hohmann, M. (2003) Direct development, no diapause, in *Taeniopteryx nebulosa* (Plecoptera, Taeniopterygidae). *Lauterbornia*, **47**, 141–151.

Zwick, P., and Hohmann, M. (2005) Biological notes on *Isoptena serricornis*, an exceptional stonefly in shifting river sand (Plecoptera: Chloroperlidae). *Lauterbornia*, **55**, 43–64.

Zwick, P., and Teslenko, V.A. (2002) Development and life history of Far Eastern Russian *Pteronarcys* spp. (Plecoptera, Pteronarcyidae). *Arch. Hydrobiol.*, **153**, 503–528.

Zwick, P., and Zwick, H. (2008a) Number of larval instars, early instar structure, and life history of *Scirtes hemisphaericus* (Coleoptera: Scirtidae) in central Germany. *Lauterbornia*, **63**, 87–99.

Zwick, P., and Zwick, H. (2008b) Beitrag zur Morphologie, Systematik und Biologie mitteleuropäischer Scirtidae (Coleoptera). *Entomol. Nachr. Ber.*, **52**, 185–191.

Zwick, P., and Zwick, H. (2010a) Life history and development of *Dictyogenus fontium* (Plecoptera: Perlodidae) in two thermally contrasting streams at Lunz am See, Lower Austria. *Denisia*, **29**, 459–475.

Zwick, P., and Zwick, H. (2010b) Stoneflies and blackflies of the River Fulda, Germany – six decades of study (Plecoptera, Diptera: Simuliidae). With notes on other aquatic insects. *Lauterbornia*, **71**, 113–131.

Index

Central European Stream Ecosystems: The Long Term Study of the Breitenbach, First Edition. Edited by Rüdiger Wagner, Jürgen Marxsen, Peter Zwick, Eileen J. Cox.